Fermentation Microbiology and Biotechnology, Fourth Edition

Fermentation Microbiology and Biotechnology, Fourth Edition

Edited by
Dr. E. M. T. El-Mansi (Lead Editor)
Dr. Jens Nielsen (Senior Editor)
Dr. David Mousdale (Editor) Dr. Tony Allman (Editor)
Dr. Ross Carlson (Editor)

CRC Press
Taylor & Francis Group
Boca Raton London New York

CRC Press is an imprint of the
Taylor & Francis Group, an **informa** business

CRC Press
Taylor & Francis Group
6000 Broken Sound Parkway NW, Suite 300
Boca Raton, FL 33487-2742

© 2019 by Taylor & Francis Group, LLC
CRC Press is an imprint of Taylor & Francis Group, an Informa business

No claim to original U.S. Government works

Printed on acid-free paper

International Standard Book Number-13: 978-1-138-58102-9 (Hardback)

Library of Congress Cataloging-in-Publication Data

Names: El-Mansi, Mansi, editor. | Neilsen, Jens, editor. | Mousdale, David M., editor. | Allman, Tony, editor. | Carlson, Ross, editor.
Title: Fermentation microbiology and biotechnology / editors, E. M. T. El-Mansi, Jens Neilsen, David Mousdale, Tony Allman, and Ross Carlson.
Description: Fourth edition. | Boca Raton : Taylor & Francis, 2018. | Includes bibliographical references and index.
Identifiers: LCCN 2018037045| ISBN 9781138581029 (hardback : alk. paper) | ISBN 9780429506987 (general) | ISBN 9780429015960 (pdf) | ISBN 9780429015953 (epub) | ISBN 9780429015946 (mobi/kindle)
Subjects: | MESH: Industrial Microbiology | Fermentation--physiology | Microbiological Techniques
Classification: LCC TP248.27.M53 | NLM QW 75 | DDC 660/.28449--dc23
LC record available at https://lccn.loc.gov/2018037045

Visit the Taylor & Francis Web site at
http://www.taylorandfrancis.com

and the CRC Press Web site at
http://www.crcpress.com

Dedication

"A man lives not only his personal life as an individual, but also, consciously
or unconsciously, the life of his epoch and his contemporaries."

Thomas Mann

In the history of science, three names stand out prominently and gained added luster with the passage of time for their far-reaching discoveries, which continue to illuminate our passage in life. I refer to (in the order they appear in the photograph from left to wright) Louis Pasteur, Michael Faraday, and Albert Einstein. This fourth edition is dedicated with gratitude to their memories as we continue to enjoy the fruits of their labour.

And, from a personal perspective, to my former mentor
Professor Dr. Mahmoud Ismael Taha, who ignited in me a lifelong passion for biochemistry

Professor Dr. Mahmoud Ismael Taha,
A chemist of exactitude and graceful humility (1924–1981)

Mansi El-Mansi (PhD Biochemistry)
"Lead Editor"
For and on behalf of the editorial team

Contents

SECTION I   Fermentation Microbiology and Biotechnology: Central and Modern Concepts

SECTION II   Fermentation Microbiology and Biotechnology: Control of Carbon Flux to Product Formation in Microbial Cell Factories

SECTION III   Fermentation Microbiology and Biotechnology: Metabolic, Enzymatic, and Genetic Engineering of Microorganisms

SECTION IV   Fermentation Microbiology and Biotechnology: Bioconversion of Renewable Resources to Desirable End Products

SECTION V   Fermentation Microbiology and Biotechnology: Tools, Monitoring, and Control of Fermentation Processes

Preface

"I beseech you to take interest in these sacred domains, so expressively called laboratories. Ask that, there be more and that they be adorned for these are the temples of the future, wealth and wellbeing."

Louis Pasteur

Fermentation has been known and practiced by humankind since prehistoric times, long before the underlying scientific principles were understood. Recently, however, microorganisms, free-living and immobilized, are widely used as biocatalysts in the transformation of many chemical reactions, especially in the production of stereospecific isomers. The high specificity, versatility, and the diverse array of microbial enzymes (proteomes) are currently being exploited for the production of important primary and secondary metabolites including antibiotics, hypercholesterolemia agents, enzyme inhibitors, immunosuppressants, and antitumor therapeutics.

The use of whole-cell biocatalysts for the large-scale production of biomolecules, biopharmaceuticals, fine chemicals, and biofuels has recently been recognized as an urgent need by industrialists as well as academics. The main thrust of recent research in this domain focuses on increasing the efficiency of primary feedstock conversion to desirable end product through improving efflux mechanisms, robustness, and tolerance of whole-cell biocatalysts. In recognition of recent trends and innovations, FIVE new chapters have been commissioned:

1. Control of Metabolite Efflux in Microbial Cell Factories: Current Advances and Future Prospects
2. Increasing Tolerances and Robustness of Industrial Microorganisms: Current Strategies and Future Prospects

3. Engineering Microorganisms for the Production of Pharmaceutical Biomolecules: Current Trends and Future Prospects
4. Recent Advances and Impacts of Microtiter Plate-Based Fermentations in Synthetic Biology and bioprocess Development
5. Biorefineries: Industrial Perspectives and Challenges in Primary and Secondary Metabolism

More exciting advances and discoveries are yet to be unraveled, and the best is yet to come, as we enter a new era in which the exploitations of microorganisms in the conversion of renewable resources to desirable end products and the generation of new therapeutics to combat disease are considered as an urgent need.

The editorial team wishes to stress their readiness to receive your feedback, especially from authors who wish to extend the knowledge base of our book, which is increasingly becoming global with every edition.

As a "lead editor" and after four successful editions, spanning three decades, I feel the time is right to hand over the responsibility to a younger generation. The editorial team of this edition was, therefore, structured to ensure the continuation of this book into its fifth edition and beyond. To this end, I wish to nominate Professor Jens Nielsen, an icon synonymous with excellence in applied biotechnology, and Dr. Ross Carlson, in that order, as the custodian of the fifth edition and beyond.

In future editions, the editorial team will endeavor to keep our readers abreast with recent innovations in this exciting and buoyant field will continue unabatedly.

Mansi El-Mansi
"Lead editor"
For and on behalf of the editorial team

Acknowledgments

This edition (4th) builds on the seminal work presented in earlier editions and owes much to the original and innovative work of our peers and colleagues worldwide. I wish to thank my editorial team and our distinguished authors for their sound contributions and for being very responsive at all times; in particular, I wish to thank Jens Nielsen, whose encouragement and support was instrumental in shaping the identity and personality of this 4th edition.

On behalf of the editorial team and authors, I wish to thank Dr. Chuck Crumly (senior acquisition editor), Jennifer Blaise (senior editorial assistant), and Rachel Cook (Deanta Global) their respective teams at CRC Taylor & Francis for transforming our manuscripts into a high-quality textbook, which I hope meets with your expectations as a reader.

From my own personal perspective (ME-M), I wish to acknowledge my formal mentors David Hopper (UCW, Aberystwyth, United Kingdom) and Harry Holms (Glasgow University, Scotland, United Kingdom) for many stimulating discussions over the years. Like many others at the Department of Biochemistry in the 1980s at Glasgow University, I was touched by the kind support of Martin Smellie, whose leadership and capacity to opening new windows of opportunities for his staff was extraordinary; to him I owe a great deal.

During the course of my employments overseas, I have had the privilege of working with Ravi P. Singh (VC, Sharda University, India) and Chief Michael Ade.Ojo (founder of Elizade University, Ondo State, Nigeria) and V. Aletor (VC, Elizade University) and to whom I owe much respect and admirations for their vision, honesty and genuine desire to build a new generation of entrepreneur graduates that are job creators rather than job seekers.

I also wish to thank my friends Mohamed Atif, Gordon Lang and David Mousdale for being there for me when I needed them the most. It is also befitting to thank my two sons (friends) Adam and Sammy, for their love and the happiness, which they continue to bring to my life in an unparalleled way.

My life has been blessed with some of the most amazing people, some of whom are given above and to whom as well as the readers of this book, I wish to say thank you for being part of my journey.

The editorial team is only too conscious of mistakes and omissions, which may have crept in unnoticed; the credit of producing this book is only partly ours, it is the blame that rests totally with us.

Have a good read.

Mansi El-Mansi
"Lead Editor"
For and on behalf of the editorial team

Editors

Dr. Mansi El-Mansi is a graduate of the University of Assiut, El-Minya, Egypt (BSc, First Hons and MSc Microbiology). He was intrigued and fascinated by the diversity and versatility of microorganisms and soon realized that understanding their physiology demanded a clear understanding of their biochemistry. He made the conscious decision of undertaking his PhD in the field of microbial biochemistry and was fortunate enough to carry it out at UCW, Aberystwyth, United Kingdom, under the supervision of Dr. David J. Hopper, whose meticulous approach to experimental design was a towering influence. During the course of his PhD studies, he became familiar with the work of Stanley Dagley, the father of microbial biochemistry as we know it, and this in turn galvanized his resolve to further his understanding of microorganisms at the molecular level.

Immediately after the completion of his PhD, Dr. El-Mansi joined Dr. Harry Holms at the Department of Biochemistry, University of Glasgow, Scotland, and such a happy and stimulating association continued for the best part of a decade, during which their group was the first to clone and show that the structural gene encoding the bifunctional regulatory enzyme ICDH kinase/phosphatase is indeed a member of the glyoxylate bypass operon. Soon thereafter, Dr. El-Mansi became acquainted with flux control analysis and its immense potential in the fermentation and pharmaceutical industries. His interest in the application of flux control analysis was further stimulated by collaboration with Henrik Kacser, the founder of metabolic control analysis (MCA) theory, at the University of Edinburgh.

During the course of his employment in Edinburgh, Dr. El-Mansi was the first to postulate and unravel the role of HS-CoA in the partition of carbon flux among enzymes of central metabolism during growth of *Escherichia coli* on acetate. He was also the first to provide evidence supporting Dan Kosahland's theory of "ultrasensitivity". His research activities, which span the best part of 30 years, yielded an extensive list of publications of which four are single-author publications in peer-reviewed journals. His most recent achievements include the unraveling of the identity of the "flux-signal" that triggers the reversible inactivation of isocitrate dehydrogenase and, in turn, the interconversion of the topology of central metabolism from acetogenic to gluconeogenic architecture in *E. coli*.

After extensive career in research and teaching in Scotland's premier universities, Dr. El-Mansi felt that the time was right to share his experience with like-minded people in culturally stimulating and enriching environments. He worked as a professor of biotechnology at Sharda University, Greater Noida, India and at Elizade University, Ondo State, Nigeria. He has recently rejoined his former vice Chancellor Professor Valentine Aletor as a professor of biotechnology at the University of Africa Toru-Orua.

Dr. Jens Nielsen has an MSc degree in Chemical Engineering and a PhD degree (1989) in Biochemical Engineering from the Danish Technical University (DTU), Denmark, and after that established his independent research group and was appointed full Professor there in 1998. He was Fulbright visiting professor at MIT in 1995–1996. At DTU he founded and directed Center for Microbial Biotechnology. In 2008 he was recruited as Professor and Director to Chalmers University of Technology, Sweden, where he is currently directing a research group of more than 60 people. At Chalmers he established the Area of Advance Life Science Engineering, a cross-departmental strategic research initiative and was founding chair of the Department of Biology and Biological Engineering, currently employing more than 170 people.

Jens Nielsen has published in excess of 600 papers that have been cited more than 49,000 times (current H-factor 111), coauthored more than 40 books and he is inventor of more than 50 patents. He was identified as a highly cited researcher in 2015, 2016, and 2017 by Thompson Reuter.

Jens Nielsen founded Fluxome A/S that attracted more than M20EUR in venture capital. This company metabolically engineered yeast for production of resveratrol and used this yeast for commercial production of this compound. The company Evolva acquired this process. Jens Nielsen has founded several other biotech companies, including Metabogen AB and Biopetrolia AB, and he has served on the scientific advisory board of a range of different biotech companies in the USA and Europe.

Jens Nielsen has received numerous Danish and international awards including the Villum Kann Rasmussen's Årslegat, Merck Award for Metabolic Engineering, Amgen Award for Biochemical Engineering, Nature Mentor Award, the Gaden Award, the Norblad-Ekstrand gold medal, the Novozymes Prize, the ENI Award, and the Eric and Sheila Samson Prize. He is member of several academies, including the National Academy of Engineering in USA, the Royal Swedish Academy of Science, the Royal Danish Academy of Science and Letters, the Royal Swedish Academy of Engineering Sciences, and the American Academy of Microbiology. He is a founding president of the International Metabolic Engineering Society.

Dr. David Mousdale received his BA in Biochemistry from the University of Oxford (1974), United Kingdom, his PhD degree in analytical plant biochemistry, physiology and phytopathology from the University of Cambridge (1979), United Kingdom, and pursued postdoctoral research in University College Dublin, Ireland, and the University of Glasgow, Scotland, in plant cell culture, hormonal control of plant growth and development, enzymology and the molecular actions of xenobiotics.

Joining the biotech spinout company Bioflux Ltd. (Glasgow) in 1988, he became Technical Director (1991) and Managing Director (1997). The company name was changed in 2002 to beòcarta Ltd. (literally, "life-map"), a combination of words from Celtic and Latin languages to reflect the international nature of the company's work with industrial clients in Europe, the United States, Japan, India, and Korea in microbial and animal cell systems for the commercial production of recombinant proteins, secondary products (antibiotics, antifungals, and enzyme inhibitors), vitamins, amino acids, enzymes, and carboxylic acids.

He is the author of three monographs on biofuels which attempted for the first time to place biofuels in clear economic and sustainability perspectives to complement basic plant and microbial biotechnology. He continues to direct beòcarta's work on the technical and economic improvement of microbial fermentations for established products and novel metabolites, plant biomass systems, bioenergy, and the development of the biorefinery concept.

Since 2008, he has been Independent Expert, Call Vice-Chair and Project Monitor for Marie Skłodowska-Curie FP7 and Horizon 2020 LIFE program research awards (ITN/ETN), Career Reintegration Grants and Individual Fellowships, and Independent Evaluator and Project Monitor for Knowledge-Based Bio-economy. Blue Growth, Algal Biomass, FET-Open Research and Innovation Actions and Bio-Based Industries Joint Undertaking research proposals and grants, European Commission Research Executive Agency, Brussels.

Dr. Tony Allman is a graduate of the University of Liverpool (BSc, PhD), United Kingdom, has been a member of the Institute of Biology and the Society of General Microbiology for more than 25 years. He began his career at Glaxo, where he spent six years carrying out research into the development of subunit bacterial vaccines. During that time, he became acquainted with fermentors and their applications. Subsequently, he acted as a specialist in this area for the U.K. agent of a major European fermentor manufacturer. When Infors U.K. was established in 1987, Tony joined the new company as Product Manager (later Technical Director) and in 2002, he became Fermentation Product Manager for the Swiss parent company, Infors AG. His work involves providing technical support, training, and application expertise in-house and throughout the world.

Dr. Allman is well known among research and industrial communities for having a passion for making fermentation accessible to the wider public. His "extracurricular activities" of devising practical workshops and giving lectures on fermentation technology speak volumes about the active pursuit of this aim.

Dr. Ross Carlson is a professor in the Department of Chemical and Biological Engineering at Montana State University (MSU), Bozeman and is also a member of the MSU Center for Biofilm Engineering and MSU Thermal Biology Institute.

He has an interdisciplinary education with a PhD in chemical engineering, an MS in microbial engineering, and a BS in biochemistry, all from the University of Minnesota, Twin Cities.

Dr. Carlson's research efforts are evenly split between *in silico* analyses and wet bench experiments. His group's research focuses on developing theory and techniques for understanding and engineering metabolisms in monocultures and consortia. The research is applied to medical biofilms, bioprocess platforms for chemical production and microbial communities found in Yellowstone National Park hot springs.

Contributors

César Arturo Aceves-Lara
Université de Toulouse, UPS, INSA, INP, LISBP
Toulouse, France
and
INRA, UMR792, Ingénierie des Systèmes Biologiques et
    des Procédés
Toulouse, France

Tony Allman
Infors UK, Ltd.
Rigate, England, UK

Aristos A. Aristidou
Bioprocess Development
Centennial, Colorado, USA

Namdar Baghaei-Yazdi
Managing Director, Biotech Consultants Limited (BTCL),
    Visiting Fellow, University of Westminster
London, England, UK

Blanca Barquera
Center for Biotechnology and Interdisciplinary Studies,
    Rensselaer Polytechnic Institute
Troy, New York, USA

George N. Bennett
BioSciences Department
Rice University
Houston, Texas, USA

Ross Carlson
Department of Chemical and Biological
    Engineering
Center for Biofilm Engineering
Montana State University
Bozeman, Montana, USA

Corina D. Ceapa
Departamento de Biología Molecular y Biotecnología,
    Instituto de Investigaciones Biomédicas
Universidad Nacional Autónoma de México
Mexico City, Mexico

Surendra K. Chikara
Eurofins Clinical Genetics, India
Doddaaud Bangaluru-560048, Karnataka, India

Yun Chen
Department of Biology and Biological Engineering
Chalmers University of Technology
Gothenburg, Sweden

Boutaib Dahhou
Centre National de la Recherche Scientifique, Laboratoire
    d'Analyse et d'Architecture des Systèmes
Université de Toulouse
Toulouse, France

Arnold L. Demain
Research Institute for Scientists Emeriti
Drew University
Madison, New Jersey, USA

Mansi El-Mansi
Department of Biotechnology, Faculty
    of Science
University of Africa
Toru-Orua (UAT), Bayelsa State, Nigeria

Chris E. French
School of Biological Sciences
University of Edinburgh
Edinburgh, Scotland, UK

Craig J. L. Gershater
Institute of Continuing Education
University of Cambridge
Cambridge, England, UK

Lalitha Devi Gottumukkala
Centre for Biofuels, Biotechnology Division
National Institute for Interdisciplinary Science and
    Technology, CSIR
Trivandrum, India

Brian S. Hartley
Grove Cottage
Cambridge, England, UK

Muhammad Javed
Visiting Fellow, University of Westminster
London, England, UK
CSO, Biotech Consultants Limited (BTCL)
Director, Biorefinings Limited

Toral Joshi
Xcelrislabs, Ltd.
Ahmedabad, India

Douglas B. Kell
School of Chemistry, The Manchester Institute of
    Biotechnology, SYNBIOCHEM
The University of Manchester
Manchester, UK

and
Novo Nordisk Foundation Center for Biosustainability
Technical University of Denmark
DK-2800 Kgs Lyngby, Denmark

Mattheos A. G. Koffas
Center for Biotechnology and Interdisciplinary Studies
Rensselaer Polytechnic Institute
Troy, New York, USA

Zetao Li
Electrical Engineering College
Guizhou University
Guiyang, Guizhou, People's Republic of China

David Mousdale
beòcarta Ltd.
Glasgow, Scotland, UK

Mahmoud M. A. Moustafa
Department of Genetics and Genetic Engineering
Benha University at Moshtohor
Egypt

Murni Halim
Department of Bioprocess Technology, Faculty of Biotechnology and Biomolecular Sciences
Universiti Putra Malaysia
Selangor, Malaysia

Jens Nielsen
Systems and Synthetic Biology
Chalmers University of Technology
Gothenburg, Sweden

Ashok Pandey
Centre for Innovation and Translational Research
CSIR-Indian Institute of Toxicology Research
Lucknow, India

Anil Kumar Patel
Department of Molecular Biosciences and Bioengineering
University of Hawaii at Manoa
Honolulu, HI, USA

Rui Pereira
Department of Biology and Biological Engineering
Chalmers University of Technology
Gothenburg, Sweden

Kuniparambil Rajasree
Centre for Biofuels, Biotechnology Division
National Institute for Interdisciplinary Science and Technology, CSIR
Trivandrum, India

Leonardo Rios-Solis
School of Engineering, Institute for Bioengineering and Centre for Systems and Synthetic Biology (SYNTHSYS)
The University of Edinburgh
Edinburgh, Scotland, UK

Gilles Roux
Centre National de la Recherche Scientifique, Laboratoire d'Analyse et d'Architecture des Systèmes
Université de Toulouse
Toulouse, France

Ka-Yiu San
Department of Bioengineering
Rice University
Houston, Texas, USA

Sergio Sánchez
Departamento de Biología Molecular y Biotecnología, Instituto de Investigaciones Biomédicas
Universidad Nacional Autónoma de México
Mexico City, Mexico

Adilson José da Silva
Center for Biotechnology and Interdisciplinary Studies
Rensselaer Polytechnic Institute
Troy, New York, USA
and
Chemical Engineering Department
Federal University of São Carlos
São Carlos, São Paulo, Brazil

Gregory Stephanopoulos
Department of Chemical Engineering
Massachusetts Institute of Technology
Cambridge, Massachusetts, USA

Beatriz Ruiz-Villafán
Departamento de Biología Molecular y Biotecnología, Instituto de Investigaciones Biomédicas
Universidad Nacional Autónoma de México
Mexico City, Mexico

Reeta Rani Singhania
DBT-IOC Advanced Bio-Energy Research Centre
Indian Oil Corporation, R&D Centre
Faridabad, India

Carlos Ricardo Soccol
Bioprocess Engineering and Biotechnology Division
Federal University of Parana
Curitiba-Parana-Brazil

Ronnie G. Willaert
Department of Structural Biology, Flanders Institute for
    Biotechnology
Vrije Universiteit Brussel
Brussels, Belgium

Patrick Wilk
School of Engineering, Institute for Bioengineering
The University of Edinburgh
Edinburgh, Scotland, UK

Barbara E. Wyslouzil
William G. Lowrie Department of Chemical and
    Biomolecular Engineering
Ohio State University
Columbus, Ohio, USA

Section I

Fermentation Microbiology and Biotechnology

Central and Modern Concepts

1 Fermentation Microbiology and Biotechnology
An Historical Perspective

Mansi El-Mansi

CONTENTS

"I have only two rules, which I regard as principles of conduct. The first is to have no rules. The second is to be independent of the opinion of others."

Albert Einstein

1.1 FERMENTATION: AN ANCIENT TRADITION

Fermentation has been known and practiced by humankind since prehistoric times, long before the underlying scientific principles were understood. That such a useful technology should arise by accident will come as no surprise to those people who live in tropical and subtropical regions, where, as Marjory Stephenson put it, "every sandstorm is followed by a spate of fermentation in the cooking pot" (Stephenson, 1949). For example, the production of bread, beer, vinegar, yogurt, cheese, and wine were well-established technologies in ancient Egypt (Figures 1.1 and 1.2). It is an interesting fact that archaeological studies have revealed that bread and beer, in that order, were the two most abundant components in the diet of ancient Egyptians. Everyone, from the pharaoh to the peasant, drank beer for social as well as ritual reasons. Archaeological evidence has also revealed that ancient Egyptians were fully aware not only of the need to malt the barley or the emmer wheat but also of the need for starter cultures, which at the time may have contained lactic acid bacteria in addition to yeast.

1.2 THE RISE OF FERMENTATION MICROBIOLOGY

With the advent of the science of microbiology, and in particular fermentation microbiology, we can now shed light on these ancient and traditional activities. Consider, for example, the age-old technology of wine making, which relies upon crushing grapes (Figure 1.2) and letting nature take its course (i.e., fermentation). Many microorganisms can grow on grape sugars more readily and efficiently than yeasts, but few can withstand the osmotic pressure arising from the high sugar concentrations. Also, as sugar is fermented, alcohol concentration rises to a level at which only osmotolerant and alcohol-tolerant cells can survive. Hence, inhabitants of ancient civilizations did not need to be skilled microbiologists in order to enjoy the fruits of this popular branch of fermentation microbiology.

In fact, the scientific understanding of fermentation microbiology and, in turn, biotechnology, only began in the 1850s after Louis Pasteur had succeeded in isolating two different forms of amyl alcohol, of which one was optically active (L, or *laevorotatory*) while the other was not. Rather unexpectedly, the optically inactive form resisted all of Pasteur's attempts to resolve it into its two main isomers, the laevorotatory (L) and the *dextrorotatory* (D) forms. It was this observation that led Pasteur into the study of fermentation, in the hope of unraveling the underlying reasons behind his observation, which was

FIGURE 1.1 Bread making as depicted on the wall of an ancient Egyptian tomb dated c. 1400 bc. (Reprinted with the kind permission of the Fitzwilliam Museum, Cambridge, UK.)

contrary to stereochemistry and crystallography understandings at the time.

In 1857, Pasteur published the results of his studies and concluded that fermentation is associated with the life and structural integrity of the yeast cells rather than with their death and decay. He reiterated the view that the yeast cell is a living organism and that the fermentation process is essential for the reproduction and survival of the cell. In his paper, the words *cell* and *ferment* are used interchangeably (i.e., the yeast cell is the ferment). The publication of this classic paper marks the birth of fermentation microbiology and biotechnology as a new scientific discipline. Guided by his critical and unbiased approach to experimental design, Pasteur was able to confidently challenge and reject Liebig's perception that fermentation occurs as a result of contact with decaying matter. He also ignored the well-documented view that fermentation occurs as a result of "contact catalysis," although it is possible that this concept was not suspect in his view. The term "contact catalysis" probably implied that a chain of enzyme-catalyzed reactions brings about fermentation. In 1878, Wilhelm Kühne (1837–1900) was the first to use the term *enzyme*, which is derived from the Greek word ενζυμον

FIGURE 1.2 Grape treading and wine making as depicted on the walls of Nakhte's tomb, Thebes, c. 1400 bc. (Reprinted with the kind permission of AKG, London, UK/Erich Lessing.)

("in leaven") to describe this process. The word *enzyme* was used later to refer to non-living substances such as pepsin and the word *ferment* used to refer to chemical activity produced by living organisms.

Although Pasteur's interpretations were essentially physiological rather than biochemical, they were pragmatically correct. During the course of his further studies, Pasteur was also able to establish not only that alcohol was produced by yeast through fermentation but also that souring was a consequence of contamination with bacteria that were capable of converting alcohol to acetic acid. Souring could be avoided by heat treatment at a certain temperature for a given length of time. Such a treatment eliminated bacterial contaminants without adversely affecting the organoleptic qualities of beer or wine, a process we now know as *pasteurization*.

A second stage in the development of fermentation microbiology and biotechnology began in 1877, when Moritz Traube proposed the theory that fermentation and other chemical reactions are catalyzed by protein-like substances and that, in his view, these substances remain unchanged at the end of the reactions. Furthermore, he described fermentation as a sequence of events in which oxygen is transferred from one part of the sugar molecule to another, culminating in the formation of a highly oxidized product (i.e., CO_2) and a highly reduced product (i.e., alcohol). Considering the limited knowledge of analytical biochemistry in general and enzymology in particular at the time, Traube's remarkable vision was to prove 50 years ahead of its time.

In 1897, Eduard Buchner, two years after Pasteur died, discovered that sucrose could be fermented to alcohol by yeast cell-free extracts and coined the term "zymase" to describe the enzyme that catalyzes this conversion. The term "zymase" is derived from the Greek word "zymosis," which means fermentation. In 1907, he received the Nobel Prize in Chemistry for his biochemical research and his discovery of cell-free fermentation. In the early 1900s, the views of Pasteur were modified and extended to stress the idea that fermentation is a function of a living, but not necessarily multiplying, cell and that fermentation is not a single step but rather a chain of events, each of which is probably catalyzed by a different enzyme.

1.3 DEVELOPMENTS IN METABOLIC AND BIOCHEMICAL ENGINEERING

The outbreak of the First World War provided an impetus and a challenge to produce certain chemicals that, for one reason or another, could not be manufactured by conventional means. For example, there was a need for glycerol, an essential component in the manufacture of ammunition, because no vegetable oils could be imported due to the naval blockade. German biochemists and engineers were able to adapt yeast fermentation, turning sugars into glycerol rather than alcohol. Although this process enabled the Germans to produce in excess of 100 tons of glycerol per month, it was abandoned as soon as the war was over because glycerol could be made very cheaply as a by-product of the soap industry. There was also, of course, a dramatic drop in the level of manufacture of explosives and, in turn, the need for glycerol.

The diversion of carbon flow from alcohol production to glycerol formation was achieved by adding sodium bisulfite, which reacts with acetaldehyde to give an adduct that cannot be converted to alcohol (Figure 1.3). Consequently, NADH accumulates intracellularly, thus perturbing the steady-state redox balance (NAD$^+$:NADH ratio) of the cell. The drop in the intracellular level of NAD$^+$ is accompanied by a sharp drop in the flux through glyceraldehyde-3-phosphate dehydrogenase, which in turn allows the accumulation of the two isomeric forms of triose phosphate (i.e., glyceral-dehyde-3-phosphate and dihydroxyacetone-3-phosphate). Accumulations of the latter together with high intracellular levels of NADH trigger the expression of glycerol-3-phosphate dehydrogenase, which in turn leads to the diversion of carbon flux from ethanol production to glycerol formation, thus restoring the redox balance within the cells by regenerating NAD$^+$ (Figure 1.3). Although this explanation is with the hindsight of modern biochemistry, the process can be viewed as an early example of metabolic engineering.

Following the First World War, research into yeast fermentation was largely influenced by the work of Carl Neuberg and his proposed scheme (biochemical pathway) for the conversion of sugars to alcohol (alcohol fermentation). Although Neuberg's scheme was far from perfect and proved erroneous in many ways, it provided the impetus and framework for many scientists at the Delft Institute, who vigorously pursued research into oxidation/reduction mechanisms and the kinetics of product formation in a wide range of enzyme-catalyzed reactions. Such studies were to prove important in the development of modern biochemistry as well as fermentation biotechnology.

While glycerol fermentation was abandoned immediately after the First World War, the acetone-butanol fermentation process, catalyzed by *Clostridium acetobutylicum*,

flourished. Production lines were modified to accommodate the new approach of "Fill and Spill" (see Box 1.1), which permitted substantial savings in fuels without adversely affecting the output of solvent production during the course of the Second World War. However, as soon as the production of organic solvents as a by-product of the petrochemical industry became economically viable, the acetone-butanol fermentation process was discontinued.

BOX 1.1 "FILL AND SPILL"

This pattern of fermentation is essentially a "batch fermentation" or "fed-batch fermentation" process in which the organism is allowed to grow, and once product formation has reached the maximum level, the fermentation pot is harvested, leaving some 10% of the total volume as an inoculum for the next batch. This process is repeated until the level of contamination becomes unacceptably high.

1.4 DISCOVERY OF ANTIBIOTICS AND GENETIC ENGINEERING

The discoveries of penicillin in the late 1920s and its antibacterial properties in the early 1940s represent a landmark in the development of modern fermentation biotechnology. This discovery, to a country at war, was both sensational and invaluable. However, *Penicillium notatum*, the producing organism, was found to be susceptible to contamination by other organisms, and therefore aseptic conditions were called for. Such a need led to the introduction of so-called stirred-tank bioreactors, which minimize contamination with unwanted

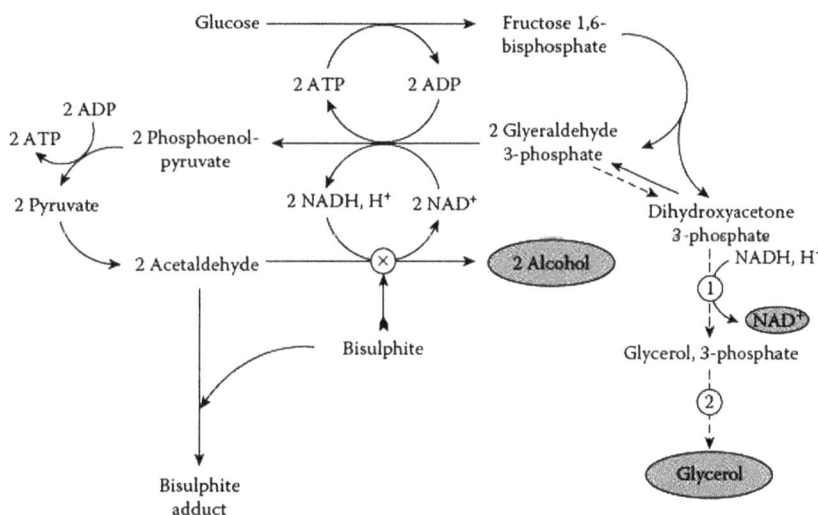

FIGURE 1.3 Diversion of carbon flux from alcohol production to glycerol formation in the yeast *Saccharomyces cerevisiae*. Note that the functional role of bisulfite is to arrest acetaldehyde molecules, thus preventing the regeneration of NAD$^+$ as a consequence of making alcohol dehydrogenase redundant. To redress the redox balance (i.e., the NAD$^+$:NADH ratio), *S. cerevisiae* diverts carbon flow (dashed route) toward the reduction of dihydroxyacetone-3-phosphate to glycerol-3-phosphate, thus regenerating the much-needed NAD$^+$. The glycerol-3-phosphate thus generated is then dephosphorylated to glycerol.

organisms. The demand for penicillin prompted a worldwide screen for alternative penicillin-producing strains, leading to the isolation of *Penicillium chrysogenum*, which produced more penicillin than the original isolate *P. notatum*. *P. chrysogenum* was then subjected to a very intensive program of random mutagenesis and screening. Mutants that showed high levels of penicillin production were selected and subjected to further rounds of mutagenesis, and so on. This approach was successful, as indicated by the massive increase in production from less than 1 g l⁻¹ to slightly more than 20 g l⁻¹ of culture.

Once the antibacterial spectrum of penicillin was determined and found to be far from universal, pharmaceutical companies began the search for other substances with antibacterial activity. These screening programs led to the discovery of many antibacterial agents produced by various members of the actinomycetes. Although the search for new antibiotics is never over, intensive research programs involving the use of genetic and metabolic engineering were initiated with the aim of increasing the productivity and potency of current antibiotics. For example, the use of genetic and metabolic engineering has increased the yield of penicillin substantially.

1.5 THE RISE AND FALL OF SINGLE-CELL PROTEIN

The latter part of the 1960s saw the rise and fall of single-cell protein (SCP) production from petroleum or natural gas. A large market for SCP was forecast, as the population in the third world, the so-called underdeveloped countries, continued to increase despite a considerable shortfall in food supply. However, the development of SCP died in its infancy, largely due to the sharp rise in the price of oil, which made it economically non-viable. Furthermore, improvements in the quality and yields of traditional crops decreased demand for SCP production.

1.6 FERMENTATION BIOTECHNOLOGY AND THE PRODUCTION OF AMINO ACIDS

The next stage in the development of fermentation biotechnology was dominated by success in the use of regulatory control mechanisms for the production of amino acids. The first breakthrough was the discovery of glutamic acid overproduction by *Corynebacterium glutamicum* in the late 1950s and early 1960s, when a number of Japanese researchers discovered that regulatory mutants, isolated by virtue of their ability to resist amino acid analogs, were capable of overproducing amino acids. The exploitation of such a discovery, however, was hampered by the induction of degradative enzymes once the extracellular concentration of the amino acid increased beyond a certain level; for example, accumulation of tryptophan induced the production of tryptophanase, thus initiating the breakdown of the amino acid. This problem was resolved by the use of penicillin, which, with tryptophan as the sole source of carbon in the medium, eliminated the growing cells (i.e., those capable of metabolizing tryptophan, but not those that were quiescent). Following the addition of penicillin, the mutants

that had survived the treatment (3×10^{-4}) were further tested. Enzymic analysis revealed that one mutant was totally devoid of tryptophanase activity. This approach was soon extended to cover the production of other amino acids, particularly those not found in sufficient quantities in plant proteins. The successful use of regulatory mutants stimulated interest in the use of auxotrophic mutants for the production of other chemicals. The rationale is that auxotrophic mutants will negate feedback inhibition mechanisms and in turn allow the accumulation of the desired end product. For example, an arginine-auxotroph was successfully used in the production of ornithine, while a homoserine-auxotroph was used for the production of lysine.

1.7 BIOFUELS AND THE "EVOLUTION" OF BIOREFINERIES

In the last decade, fermentation biotechnology has taken a leap forward in consequence of the rising price of oil and international concern about global warming. Brazil has been a pioneer since the 1970s in developing bioethanol production from its huge sugarcane industry and alongside it an expanding industry for production of ethanol-utilizing cars; as a result, today about 80% of Brazilian vehicles are fueled by > 20% bioethanol–gasoline mixtures.

The US, the largest consumer of gasoline, has expanded its program and increased its bioethanol production from corn by some sixfold over the last decade, thus superseding Brazil's production from sugarcane. However, this is still only a fraction of US gasoline consumption. Ethanol production from corn is much less efficient than from sugarcane, so subsidies are required to market it as a 15% ethanol–gasoline blend, which can be used in conventional car engines without modification. Nevertheless, the rocketing price of oil is close to making unsubsidized corn bioethanol competitive with gasoline. Corn prices are also rising, but the increasing animal feed value of the fermentation residues (distillers' dried grains, or DDGs) almost compensates.

There is therefore a worldwide trend to increase conventional bioethanol production, even if subsidies or tariffs are required and the carbon footprint of the fuel is only marginally better than that of gasoline. The production of bioethanol or the capacity to make it appears to have a buffering capacity against further increases in oil prices. There are many situations in which bioethanol production represents a logical alternative to farming subsidies, such as wheat production in the European Union. There is, however, valid opposition to "first-generation" bioethanol as a sustainable biofuel on grounds of minimal reduction of global warming and competition with food supply. Fortunately, fermentation biotechnology has an answer. We harvest at best 20% of what farmers grow and eat only a fraction of that. Agricultural residues such as corn stover, cereal straws, and palm-oil wastes are lignocellulosic biomass that could dwarf current bioethanol production. Harvested factory residues, such as sugarcane bagasse, corn cobs, wheat bran, palm oilcake, and so on represent an immediate opportunity. Lignocellulosic residues alone could yield sufficient bioethanol to fuel all the cars in the world.

As we will be discussing in Chapter 9, these residues are composed mostly of bundles of long cellulose fibers (40–50% dry weight), waterproofed by a coat of lignin (15–25%), and embedded in a loose matrix of hemicelluloses (25–35%). Most R&D has been directed to cellulose utilization since it can be hydrolyzed to glucose and fermented by yeasts. However, this is a slow and/or energy-intensive process, so cellulosic ethanol is not competitive with corn or wheat bioethanol, let alone cane bioethanol. In contrast, hemicelluloses are easily hydrolyzed to a mixture of C5 and C6 sugars, most of which cannot be fermented by yeasts. Many microorganisms can ferment these sugars but produce lactic acid rather than ethanol. Therefore, much effort has gone into genetic manipulation to divert carbon flux from lactate production to bioethanol formation.

Notable among such microorganisms are thermophilic *Geobacilli* found naturally in compost heaps and/or silage. They can be engineered to produce ethanol from hemicellulosic sugars with yields equivalent to those from yeast fermentations of starch sugars. They have the additional advantages of extremely rapid continuous fermentations at high temperatures in which ethanol vapor can be removed continuously from the broth. Hence, although such fermentations have not yet been commercialized, calculations indicate that the production cost will be well below that of cane bioethanol or gasoline, so the scene is set for the evolution of biorefineries.

By analogy with the emergence of petrochemicals from processing the by-products of oil refineries, new biochemicals will undoubtedly emerge from the processing of biomass in biorefineries. Sugarcane, for example, could yield sugar or ethanol from the juice, ethanol from the hemicellulosic pith, waxes from the external rind, plus heat and electricity from efficient combustion of the lignocellulosic fibers (KTC-Tilby, 2011). Alternatively, the fibers could be used directly for packaging or building board or for paper production after ethanol extraction of the lignin. The lignin extract could make an efficient biodiesel or give rise to a new range of bioaromatics to compete with those currently derived from oil. The residual stillage from ethanol distillation has high animal feed value.

Another advantage of such biorefineries is that they would be self-contained units built close to existing food-processing plants, such as sugar refineries, flour mills, or oil-processing plants. Since rape seed or palm oils are already used for biodiesel production, an intriguing proposal to use the by prod uct glycerol together with hemicellulosic sugars for high-yield bioethanol production would produce biorefineries to rival oil refineries in producing both fuels from a single raw material. The scene is therefore set for the evolution of biorefineries.

1.8 IMPACT OF FUNCTIONAL GENOMICS, PROTEOMICS, METABOLOMICS, AND BIO-INFORMATICS ON THE SCOPE AND FUTURE PROSPECTS OF FERMENTATION MICROBIOLOGY AND BIOTECHNOLOGY

Modern biotechnology, a consequence of innovations in molecular cloning and overexpression in the early 1970s,

started in earnest in the early 1980s after the manufacture of insulin, with the first wave of products hitting the market in the early 1990s. During this early period, the biotechnology companies focused their efforts on specific genes/proteins (natural proteins) that were of well-known therapeutic value and typically produced in very small quantities in normal tissues. Later, monoclonal antibodies became the main products of the biopharmaceutical industry.

> **Functional genomics** is a discipline of biotechnology that attempts to exploit the vast wealth of data produced by genome-sequencing projects. A key feature of functional genomics is their genome-wide approach, which invariably involves the use of a high-throughput approach.

In the mid-1980s, Thomas Roderick coined the term *genomics* to describe the discipline of mapping, sequencing, and analyzing genomic DNA with the view to answering biological, medical, or industrial questions (Jones, 2000). Recent advances in functional genomics and system biology led many scientists to venture from the traditional *in vivo* and *in vitro* research into the *in silico* approach (e.g., Heinemann and Sauer, 2010 and Kotte et al., 2010) to help understand the control of cellular metabolism.

The *in silico* approach employs sophisticated differential equations and computers to store, retrieve, analyze, and compare a given sequence of DNA or protein with those stored in data banks from other organisms. Microbial biotechnologists were quick to realize that the key to successful commercialization of a given sequence relied on the development of an innovative methodology (bio-informatics and system biology) that facilitates the transformation of a given sequence into a diagnostic tool and/or a therapeutic drug, thus bridging the gap between academic research and commercialization.

> **Proteomics** is a new discipline that focuses on the study and exploitation of proteomes. A **proteome** is the complete set of proteins expressed by a given organism under certain conditions.

> **Metabolomics** is a new discipline that focuses on the study and exploitation of metabolomes. A **metabolome** refers to the complete set of primary and secondary metabolites as well as activators, inhibitors, and hormones that are produced by a given organism under certain conditions. It is noteworthy, however, that it is not currently possible to analyze the entire range of metabolites by a single analytical method.

1.9 RECENT INNOVATIONS IN SYSTEMS BIOLOGY

Mapping and identifying regulatory networks that control fluxes among various enzymes of cellular metabolism represents a major challenge to system biologists. Genome-scale metabolic reconstructions together with experimental and modeling data yielded significant insights illustrating how a given metabolic behavior or response. To this end, Heinemann and Sauer (2010), reported that non-transcriptional mechanisms, such as metabolite–protein interactions and reversible phosphorylation, play a significant role in controlling metabolic functions.

1.10   RECENT INNOVATIONS IN MICROBIAL-CELL FACTORIES

The use of whole-cell biocatalysts for the large-scale production of biomolecules, biopharmaceuticals, fine chemicals, and biofuels has recently been recognized as an urgent need by industrialists and academics alike. The impact of industrial systems biology on the synthetic biology of yeast cell factories has recently been assessed by Jens and coworkers (Fletcher et al., 2016) and the findings reported by the aforementioned authors further substantiated the importance of the use of whole-cell biocatalysts in microbial cell factories. The main thrust of recent research in this domain focused on increasing the efficiency of primary feedstock conversion to desirable end product and improving the robustness of whole-cell biocatalysts. Further research revealed that successfully engineered strains were found to be subject to membrane stress, which in turn in adversely affects efflux of the desired products. Damages to cell membranes lead to the loss of cell integrity and in turn failure of the fermentation process. Such damages to cell membranes have recently been observed during the course of producing lipophilic molecules or during fermentation processes in which recombinant proteins are overexpressed in a non-native host.

BOX: LIPOPHILIC MOLECULES

A molecule is described as lipophilic if it dissolves in fat like environments, e.g., oils and hydrocarbons. Such molecules are not water-soluble.

In addition, **p**rotein translocation and in turn excretion has been recognized as an essential feature that needs to be fully understood in order to improve the efficiency of recombinant protein production in native as well as non-native hosts. Translocation of proteins from the cytoplasmic space across the cell membranes is primarily achieved through the activity of specific signal peptide sequences, which drags the protein across the membrane into the extracellular space. Although the excretion of soluble proteins is understandably more problematic, recent studies revealed that the translocation of signal peptide-less protein across the cell membrane of *Escherichia coli* can be induced by subjecting the organism to osmotic and translational stress, thus demonstrating the presence of hitherto unknown and untapped alternative translocation/excretion pathways (Morra et al., 2018). Furthermore, using metabolomic and proteomic analyses of wild type as well as genetic knockouts variants, the aforementioned authors were able to show that the excretion of signal peptide-less proteins is positively controlled by both the large mechano-sensitive channel (MscL) and the alternative ribosome rescue factor A (ArfA).

In recognition of the importance of whole-cell biocatalyst in microbial cell factories, the BBSRC formulated a new initiative, namely, the CBMnet (Crossing Biological Membranes Network in Industrial Biotechnology) to foster collaborative research between academia and industrial concerns. The BBSRC is to be congratulated on such initiative as some projects yielded results that led to the development of new bioprocess at the industrial level; with the scientific principles underlying such developments are being published in the public domain.

The new innovations in functional genomics, proteomics, metabolomics, bio-informatics, systems biology, and more recently in the use of whole-cell biocatalyst in microbial-cell factories will undoubtedly play a major role in transforming our world in an unparalleled way, despite political, ethical, and cultural differences.

In this book, top scientists and industrialists addressed the multidisciplinary nature of fermentation microbiology and biotechnology, and in so doing highlight its many fascinating aspects, thus providing future generations of biotechnologists with the stimulus and the knowledge necessary to innovate and extend the usefulness of microssorganisms as we enter a new era in which the use of whole-cell biocatalyst as a tool for the conversion of renewable resources into desirable end products is recognized as an urgent need.

ACKNOWLEDGMENTS

The author wishes to acknowledge professors Charlie F.A. Bryce, Brian S. Hartley, and Arnold L. Demain for their invaluable comments on this chapter in earlier editions

REFERENCES

Fletcher E, Krivoruchko A, Nielsen J. 2016. Industrial systems biology and its impact on synthetic biology of yeast cell factories. *Biotechnol. Bioeng.* 113(6): 1164–1170.

Heinemann, M, Sauer, U. 2010. Systems biology of microbial metabolism. *Curr. Opin. Microbiol.* 13(3): 337–334.

Jones, P.B.C. 2000. The commercialisation of bioinformatics. *Electron. J. Biotechnol.* 3(2): 33–34.

KTC-Tilby (2011). Sweet sorghum and sugar cane separation technology. www.youtube.com/watch?v=YbQT7Yfmn7s

Kotte O, Zaugg JB, Heinemann M. 2010. Bacterial adaptation through distributed sensing of metabolic fluxes. *Mol. Syst. Biol.* 82(9): 1492–1493.

Morra, R., Del Carratore, F., Muhamadali, H., Horga, L G., Halliwell, S., Goodacre, R., Breitiling, R., Dixon, N. 2018. Translation stress positively regulates MscL-dependent excretion of cytoplasmic proteins. *mBio* 9 (1): 2118–2117.

Stephenson, M. 1949. *Bacterial Metabolism.* London: Longmans Green.

2 Microbiology of Industrial Fermentation
Central and Modern Concepts

Mansi El-Mansi

CONTENTS

"Never regard research as a duty, but as the enviable opportunity and privilege to learn."

Albert Einstein

2.1 INTRODUCTION

Microbial fermentations are currently used for the production of a diverse array of biomolecules including amino acids, fine chemicals, solvents, enzymes, hormones, and antibiotics. Such diversity may be attributed to many factors, including the high surface-to-volume ratio and the ability to utilize a wide spectrum of carbon and nitrogen sources. The high surface-to-volume ratio supports a very high rate of metabolic turnover: for example, protein synthesis in the yeast *Saccharomyces cerevisiae* is several orders of magnitudes faster than that of plants. In addition, microorganisms are capable of utilizing inexpensive renewable resources such as wastes and by-products of the farming and petrochemical industries as a source of carbon and/or nitrogen.

The metabolic route through which glucose is converted to pyruvate, i.e., glycolysis, is universally conserved among all organisms; prokaryotic organisms differ from eukaryotes in their ability to process pyruvate through a diverse array of routes giving rise to a multitude of different end products (Figure 2.1). Such diversity has been fully exploited by fermentation technologists for the production of wine, organic solvents, dairy products, and fine chemicals.

Prokaryotic organisms differ from yeast and fungi as well as other eukaryotes in a number of ways. For example, while DNA is compartmentalized within the nucleus in eukaryotes, it is neatly folded in a nucleoid within the cytoplasm in prokaryotes. Furthermore, unlike eukaryotes where the site of oxidative phosphorylation is associated with mitochondria, the site of oxidative phosphorylation in prokaryotes is associated with cytoplasmic membrane. Moreover, the newly synthesized DNA molecules in prokaryotes need no special assembly to form a chromosome, as the DNA is already attached to the bacterial membrane, a feature that ensure its successful segregations into two daughter cells. Any treatment, e.g., SDS, which compromises the integrity of the DNA's attachment to the cytoplasmic membrane, may lead to failure in DNA segregation.

2.2 BIOSYNTHESIS OF BIOMASS DURING THE COURSE OF FERMENTATION

2.2.1 GROWTH REQUIREMENTS

In addition to carbon, nitrogen, phosphate, potassium, sulfur, irons, and magnesium, the nutrients required for growth also include trace elements (Figure 2.2). The necessity for such a multitude of inputs is paramount as it is required for enzymic activities.

As can be seen from Figure 2.2, the first stage (the fueling reactions) involves the conversion of glucose and inorganic phosphate (Pi) into a whole host of biosynthetic precursors together with reducing powers and ATP.

The second stage (monomerization, biosynthesis of monomers) involves the conversion of biosynthetic precursors monomers (amino acids, nucleotides, sugars, and fatty acids). This stage is carried out through the activities of various enzymes of intermediary metabolism and requires the presence of nitrogen (NH_4) and sulfate (SO_4).

The third stage involves polymerization of monomers into polymers, so that amino acids are converted into protein, while nucleotides are converted to RNA and DNA, and so on. In this stage, following transcription, ribosomes attach to mRNA together with cofactors, enzymes, and complementary tRNA, thus forming a polysome and, in turn, initiating the synthesis of polypeptides. Polysomes are one of the most abundant organelles in growing cells; each polysome contains approximately 20 subunits of ribosomal RNA (Ingraham et al., 1990).

The fourth stage is assembly of polymers into various components of biomass, e.g., nucleoid, inclusion bodies and cell membranes (Figure 2.2).

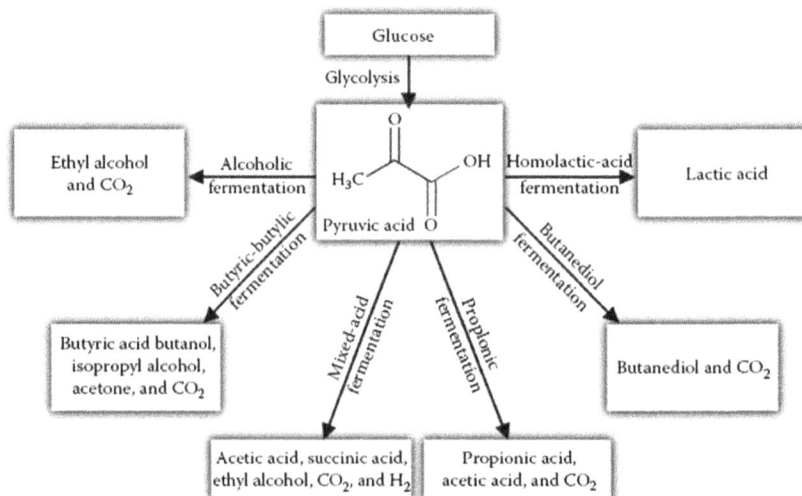

FIGURE 2.1 Diversity of fermentation pathways among microorganisms.

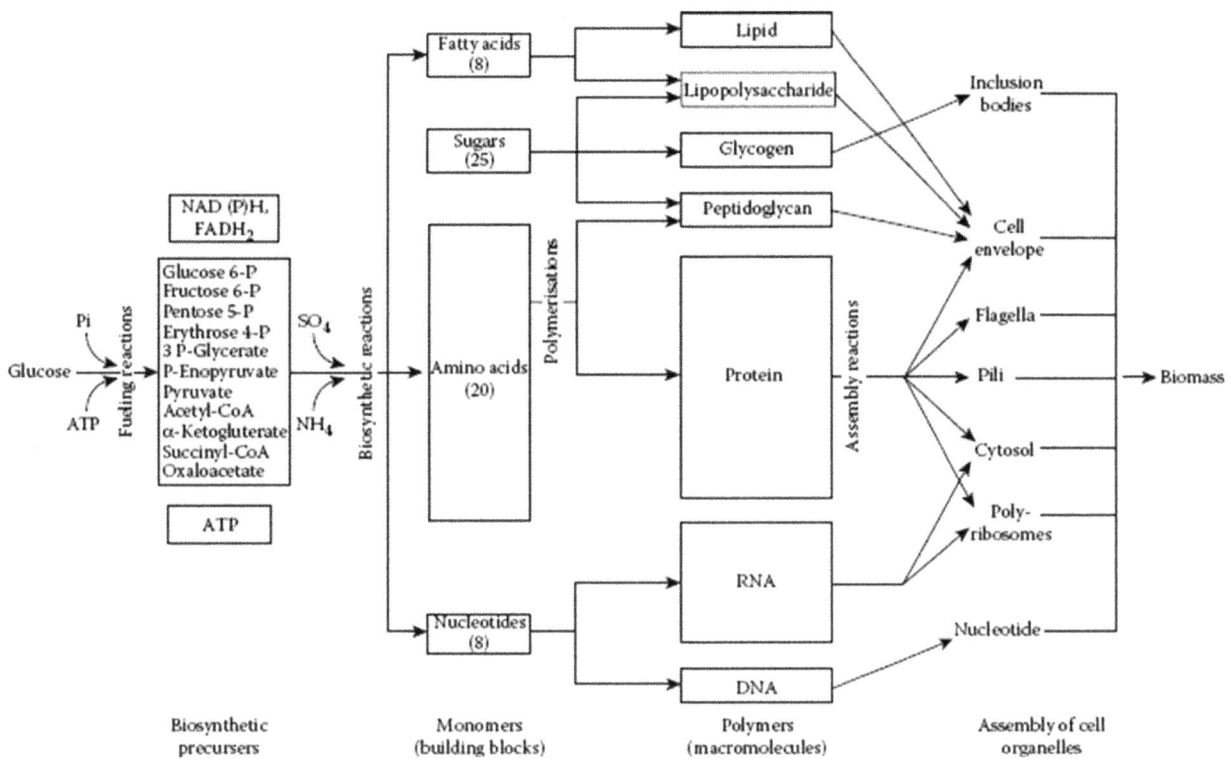

FIGURE 2.2 A diagrammatic representation of the reactions involved in the conversion of glucose and simple salts to biomass of *E. coli* (a slight modification of Ingraham et al., 1990, with permission).

From Figure 2.2, one can clearly see that central metabolism (glycolysis, the pentose phosphate pathway, and the Krebs cycle) is concerned with the generation biosynthetic precursors, ATP, and reducing powers, while intermediary metabolism is concerned with the conversion of biosynthetic precursors into monomers and thence into polymers en route to their assembly into new biomass.

2.2.2 THE METABOLIC PATHWAYS OF CENTRAL METABOLISM

During growth on a glucose minimal medium under aerobic conditions, *Escherichia coli* catabolizes glucose through glycolysis, the pentose phosphate pathway (PPP), and the Krebs cycle (Figure 2.3) to bring about its transformation to biosynthetic precursors; over half of which are phosphorylated, ATP and reducing powers in the shape of NADH, NADPH, and FADH$_2$ as well as esterified CoA derivatives (e.g., succinyl CoA and malonyl CoA).

The central metabolic pathways (Glycolysis, PPP, and the Krebs cycle) fulfill both catabolic (from *cata*, a Greek word for breakdown) and anabolic (from *ana*, a Greek word for buildup) functions and as such may be referred to as *amphibolic pathways*.

As can be seen from Figure 2.3, microorganisms are capable of utilizing a wide range of substrates and afford different entry point for each into central metabolism. As the point of entry into the central metabolic pathways vary from one substrate to another, the makeup of the enzymic machinery necessary for metabolism changes accordingly (Guest and Russell, 1992).

For example, during growth on acetate or fatty acids, *E. coli* expresses uniquely the anaplerotic sequence of glyoxylate bypass (Kornberg, 1966; Holms, 1986; El-Mansi et al., 2006).

2.2.3 IMPACT OF GROWTH RATE AND GROWTH CONDITIONS ON CELL COMPOSITION AND INTRACELLULAR FLUXES

The physiology and bioenergetics as well as cell size and biomass composition including the organism's transcriptome (Nahku et al., 2010) and proteome of a given phenotype of *E. coli* have been found to vary according to growth conditions and growth rate (μ) (Pramanik and Keasling, 1997; Hua et al., 2004; shii et al., 2007). Growth rate (μ) was also found to influence intracellular fluxes in *E. coli* (Zhao and Shimizu, 2003; Kayser et al., 2005) in a manner that is nonlinear (Nanchen et al., 2006). In an effort to unravel the impact of growth rate (μ) on cell composition and intracellular fluxes, Valgepea et al. (2010) simultaneously determined biomass compositions as well as flux and efflux distributions during growth of *E. coli* K-12 MG1655 at different growth rates (μ). Interestingly, the aforementioned authors reported that the shift from one growth rate (μ) to another (faster rate) in an accelerostat was associated with, not only a significant increase in efflux to by-product formation (carbonyl- phosphate, dihydroorotate and orotate), but also with a 36% drop in ATP spilling (generation) and no change in cell composition. Furthermore, the authors suggested that cell composition was balanced by efflux to by-product formation and

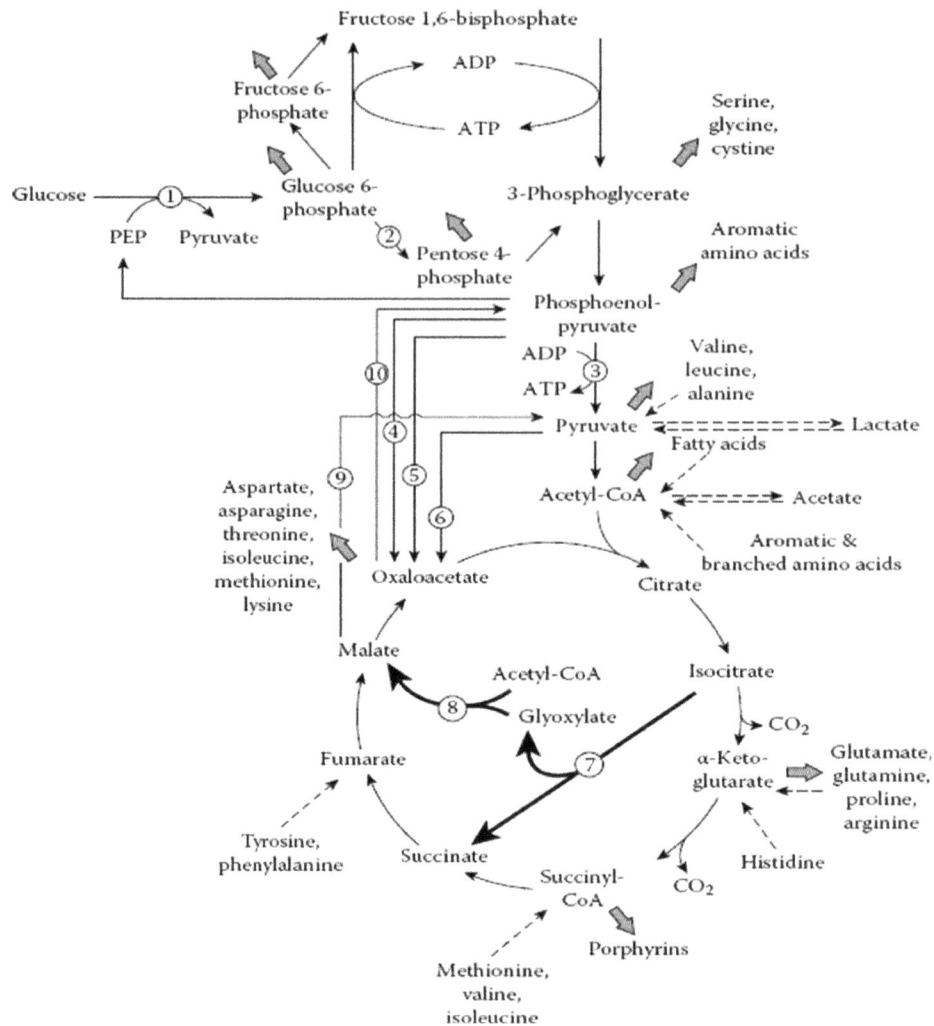

FIGURE 2.3 An overview of the metabolic pathways of central and intermediary metabolism employed by *E. coli* for the conversion of glucose and other substrates to biosynthetic precursors; indicated by heavy-dotted arrows, ATP, and reducing powers. Dashed arrows indicate entry of substrates other than glucose into central metabolism. Key enzymes are as follows: 1, glucose–phosphoenolpyruvate (PEP) phosphotransferase system; 2, the pentose phosphate pathway; 3, pyruvate kinase; 4, PEP carboxylase; 5, PEP carboxy-transphosphorylase; 6, pyruvate carboxylase; 7, isocitrate lyase; 8, malate synthase; 9, malic enzyme; 10, PEP carboxykinase.

that the patterns of carbon efflux are dependent not only on growth rate (μ), but also on the capacity to simultaneously co-utilize glucose and acetate.

BOX: ACCELEROSTAT

Accelerostat is a special type of chemostat in which the transition from growth rate (μ) to another is made smoothly through the use of a special computer program. For more info see Nahku *et al.* (2010) and Kasemets *et al.* (2003).

2.2.4 Transport of Nutrients into the Bacterial Cell

2.2.4.1 During Growth on Rich Media

During growth of *E. coli* on rich media, nutrients are taken up initially through passive diffusion across the cytoplasmic membrane and into the cytosol due to the significant difference in the concentration gradient between the medium and the cytosol. On the other hand, if the extracellular level of nutrients is lower than the intracellular concentration in the cytosol, then the organism must employ active transport mechanisms to ensure continuous uptake of nutrients. For example, when ammonium (NH_4) and phosphate ions (P_i) become scarce, *E. coli* employs AmtB and Pst active transport systems to facilitate their uptake, respectively (Radchenko et al., 2010).

In common with all Gram-negative bacteria, *E. coli* outer and inner cytoplasmic membranes represent the first line of defense against polar and lipophilic molecules. Hydrophilic molecules are taken up through numerous "porin-protein channels" such as LamB glycol-porin channel, which allows *E. coli* not only to scavenge for glucose, but also to permeate disaccharides such as maltose (Nikaido, 2003; Nachen et al., 2008). Interestingly, the relative abundance of OmpC and OmpF porin-channels varies according to growth rate (μ) and growth conditions. Expression of porin genes is under the control of the "two-component signal transduction" EnvZ-OmpR and as such plays a part in sensing signal metabolites (Takatsuka et al., 2010.

2.2.4.2 During Growth on Glucose

Escherichia coli, unlike archaebacteria and eukaryotic organisms, catalyzes the uptake of glucose *via* the PEP-glucose phosphotransferase system (PTS). In this mechanism (Figure 2.4), enzyme 1 (E1) mediates the transfer of phosphate group from PEP to another protein namely histidine carrier protein (HPr) to give phosphohistidine protein (HPr-P); the phosphate group in HPr-P is in turn transferred to glucose specific EIIA-^{glc} to give EIIA-^{glc} P. The phosphate group is then cascaded from EIIA-^{glc} P to another glucose specific enzyme EIIB-^{glc} to give EIIB-^{glc} P. The enzyme EIIC-$^{glc;}$ and a permease together with EIIB-^{glc} P work in concert to bring about the transport of glucose from the periplasm and its conversion to glucose-6P in the cytosol (Tchieu et al., 2001).

At low concentrations of glucose, *E. coli* induces a low-affinity galactose: H$^+$ symporter (GalP), which is capable of transporting glucose from the periplasm into the cytosol directly. The glucose is then phosphorylated to glucose 6-phosphate through the activity of glucokinase at the expense of one mole ATP (Shimizu, 2014). In addition, in carbon-limited chemostat cultures, gluconeogenesis (*maeB*, *sfcA*, and *pck*), the glyoxylate operon (*ace*BAK) and acetyl-CoA synthetase are up regulated. However, it is noteworthy that an acetyl-CoA synthetase-deficient mutant excreted acetate at all dilution rates (Renilla, et al., 2012).

When the extracellular concentration of glucose becomes very low, *E. coli* induces yet another mechanism of transport, the Mgl system. The Mgl system is composed of three proteins, namely, an ATP binding protein (MglA), an integral membrane transporter (MglC), and a glucose periplasmic binding protein (MglB). This Mgl system forms a high-affinity mechanism for the uptake of trace quantities of glucose. Once glucose is taken into the cytosol, it is immediately phosphorylated to glucose 6-phosphate through the activity of glucokinase (Shimizu, 2014).

2.2.5 CONTROL OF GLYCOLYTIC INTERMEDIATES FLOW INTO CENTRAL METABOLISM DURING SUGAR FERMENTATION

Microorganisms are capable of detecting extracellular and intracellular metabolites by virtue of their ability to sense

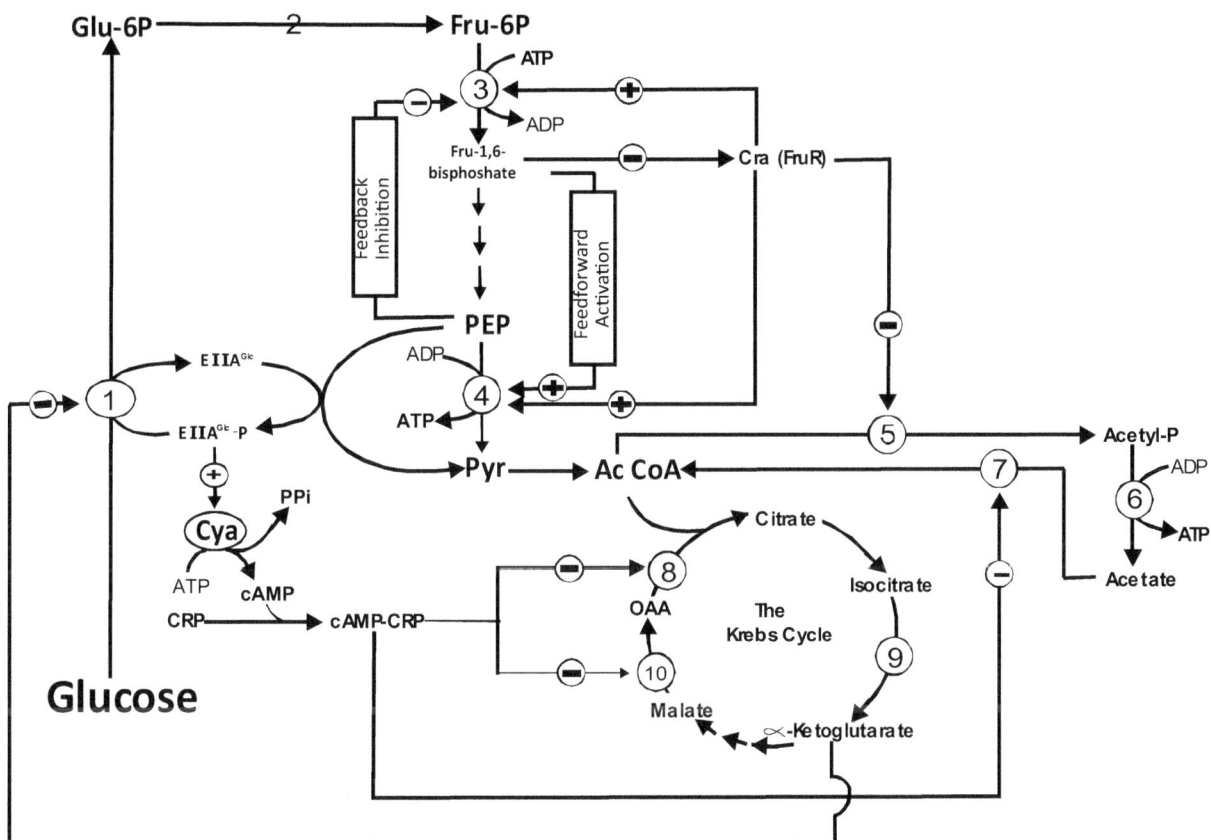

FIGURE 2.4 The role of signal metabolites (fructose 1,6-bisphosphate; PEP; pyruvate and α-ketoglutarate) and transcription factors in the control of flow of glycolytic intermediates into central metabolism during growth of *E. coli* on glucose minimal medium. Enzymes are as follows: 1, PEP-glucose phosphotransferase system (PTS); 2, phosphoglucoisomerase; 3, phosphofructokinase (PFK); 4, pyruvate kinase (PK); 5, phosphotransacetylase; 6, acetate kinase (AK); 7, citrate synthase; 8, isocitrate dehydrogenase (ICDH); 9, isocitrate lyase (ICL); 10, malate synthase (MS).

and transmit signals to sensory control mechanisms, e.g., two-component transduction mechanisms and transcription factors (Yamamoto et al., 2005 Jiang and Ninfa, 2007). Activated transcription factors, in turn, modify (activation or deactivation) the make up of the enzymic machinery of central metabolism in such a way that ensures uninterrupted flow of glycolytic intermediates into central metabolism. It follows any given change in the metabolic environment that leads to a change in the threshold concentrations of signal metabolites will be sensed and transmitted to a specific transcription factor, which in turn activates or deactivates (suppress) the expression of certain genes as illustrated in Figure 2.4. For example, a drop in the intracellular concentration of fructose 1, 6-bisphohate below a certain threshold is sensed and transmitted to *Cra* (*Fru*R) transcription factor, which, in turn, activates the transcription of the structural gene encoding phosphofructokinase, thus facilitating the conversion of fructose 6- phosphate to fructose 1, 6-bisphosphate (Figure 2.4) (Ramseier et al., 1995). In addition, phosphofructokinase (PFK) itself is also subject to feedback inhibition by phosphoenolpyruvate (PEP), which in turn slows down the flow of carbon through the enzyme should the intracellular concentration of PEP increases above a certain threshold (Figure 2.4).

Furthermore, PEP is another signal metabolite that lies at an important node in central metabolism and has been argued to play a significant role in controlling anaplerosis in central metabolism. PEP is converted to pyruvate through the activity of pyruvate kinase (PK; Figure 2.4), which catalyzes an irreversible reaction and is subject to activation at the transcriptional level by *Cra* in a manner similar to that of PFK and is allosterically controlled through feed forward activation by fructose 1,6 bisphosphate (Figure 2.4).

In addition to catalyzing irreversible reactions, PFK and PK are strategically located within the central metabolism to ensure directionality of carbon flow. Needless to say, the aforementioned enzymes are not the only two enzymes that catalyze irreversible reactions in central metabolism, other enzymes exist, e.g., pyruvate dehydrogenase (PDH) and citrate synthase (CS).

2.2.6 THE ROLE OF α-KETOGLUTARATE IN CARBON AND NITROGEN METABOLISM

As can be seen from Figures 2.4 and 2.5, α-ketoglutarate plays a central role in carbon and nitrogen metabolism not only through its ability to provide a carbon skeleton for the glutamate and glutamine family of amino acids, but also for its ability to act as a signal metabolite, which allows the organism to better adapt to changes in the availability of nutrients. It follows that *E. coli* is capable of sensing changes in the intracellular concentration of α-ketoglutarate and responding through the operation of signal transduction systems that monitors the prevailing nutritional status and generates the appropriate metabolic response (Doucette et al., 2011).

Enzymes are as follows:

1, PTS; 2, citrate synthase; 3, isocitrate dehydrogenase; 4, glutamate dehydrogenase; 5, glutamine synthetase;

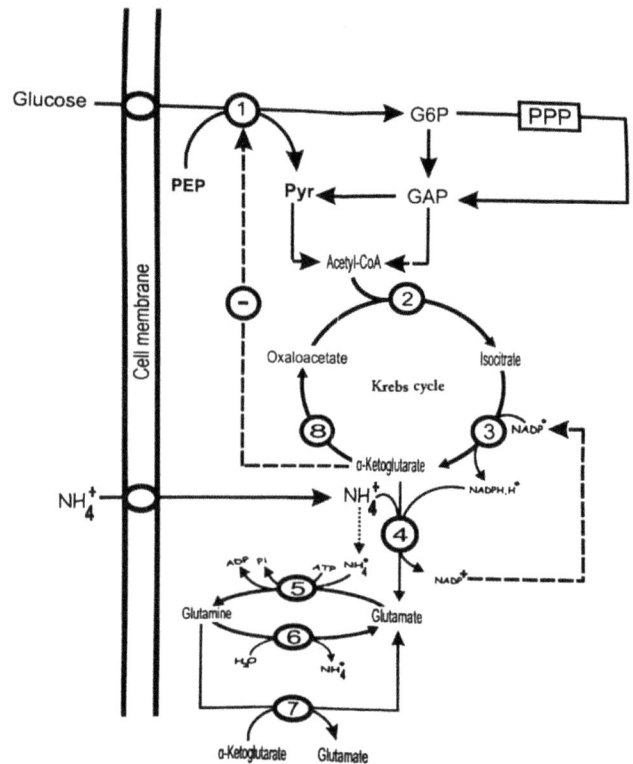

FIGURE 2.5 A schematic representations highlighting the role of α-ketoglutarate as a signal metabolite in central metabolism and the assimilations of carbon and nitrogen in *E. coli*

6, glutaminase; 7, glutamine- α-ketoglutarate aminotransferase; 8, α-ketoglutarate dehydrogenase, succinate dehydrogenase, fumarase, and malate dehydrogenase of the Krebs cycle.

During growth of *E. coli* under nitrogen limitation, the conversion of α-ketoglutarate to glutamate ceases, thus leading to the accumulation of α-ketoglutarate. Should the intracellular concentration of α-ketoglutarate rise above a certain threshold, the organism detects such a rise as a signal and responds by feedback inhibiting the activity of enzyme 1 of the PTS (Figure 2.4), thus restricting the flow of intermediates into the glycolytic pathway until an adequate supply of nitrogen source is made available (Ninfa and Jiang, 2005; Doucette et al., 2011).

As can be seen from Figure 2.4, the directionality of glycolytic intermediates into central metabolism is assured by the operations of irreversible reactions, catalyzed by PFK, PK, PDH, and CS, as well as efficient allosteric control mechanisms (feed forward activation of PK and feedback inhibition of PFK by Fructose 1, 6-bisphosphate and PEP, respectively) that fine control the flow of intermediates at crucial junctions.

2.2.7 ANAPLERTIC REACTIONS IN CENTRAL METABOLISM

During growth on glucose or other acetogenic substrates (i.e., those that support flux to acetate excretion), metabolites of the Krebs cycle, namely, α-ketoglutarate, succinyl CoA, and oxaloacetate are constantly withdrawn for biosynthesis (Figure 2.3), and as such, they must be replenished, otherwise the cycle will grind to a halt. The reactions that fulfill such a function is known as *anaplerotic*; a Greek word for replenishing. The enzymes used to fulfill such a function vary from one organism to another and from one phenotype to another within a given species, i.e., substrate- dependent. For example, growth on glucose, phosphoenolpyruvate carboxylase, and pyruvate carboxylase, as well as phosphoenolpyruvate carboxy-trans-phosphorylase (Figure 2.3), represents the full complement of anaplerotic enzymes that may be used in full or in part depending on the organism under investigation and growth conditions.

During growth on acetate, however, *E. coli* employs the glyoxylate bypass operon enzymes, namely, isocitrate lyase (ICL) and malate synthase (MS), as an anaplerotic sequence (Figure 2.3). Interestingly, however, while the enzymes of the glyoxylate bypass in *E. coli* form an operon (*ace*-BAK) and include the bi-functional regulatory enzyme isocitrate dehydrogenase kinase/phosphatase (ICDH K/P), these enzymes (ICL and MS) are not organized in the same way in other organisms, e.g., *Corynebacterium glutamicum*.

Interestingly, a new pathway, namely the PEP-glyoxylate cycle (Figure 2.6), has recently been found to operate in hungry *E. coli* grown in glucose-limited chemostat (Fischer and Sauer, 2007).

As can be seen from (Figure 2.6), the organism employs the glyoxylate bypass enzymes ICL and MS in conjunction with PEP-carboxykinase to bring about the full oxidation of PEP to CO_2 and ATP. One very interesting aspect of the PEP glyoxylate oxidizing cycle is its ability to redress the redox balance should flux to NADPH generation increases unexpectedly. However, this cycle may not be operational in hungry glycerol grown cultures of *E. coli* because the glycerol phenotype does not employ the acetate switch for the interconversion of central metabolism's topology from acetogenic to gluconeogenic architecture (Mansi El-Mansi, unpublished observations). Confirmation of this, however, awaits further research.

2.3 MICROBIAL GROWTH AND MAINTENANCE REQUIREMENTS DURING FERMENTATION

The rate of product formation in a given industrial process, a significant parameter, is directly proportional to the rate of biomass formation, which is influenced directly or indirectly by a whole host of different environmental factors (e.g., oxygen supply, pH, temperature, and accumulation of inhibitory intermediates). It is, therefore, important that we can describe growth and production in quantitative terms. The study of growth kinetics and growth dynamics involve the formulation and use of differential equations, which has been dealt with extensively in Chapter 3 of this book.

In microbiology, we generally deal with very large numbers, and for convenience we express these numbers as multiples of 10 raised to an appropriate power; for example, 1 million (1,000,000) is written as 1×10^6.

The logarithm of a number is the exponent to which we must raise a base to obtain that number. The exponent is also known as the *index* or the *power*. For example, consider the number 1000 and the base 10; the number 10 is raised to the power 3 to obtain 1000; it follows that the logarithm to the base 10 of 1000 is 3. The base of a logarithm is conventionally

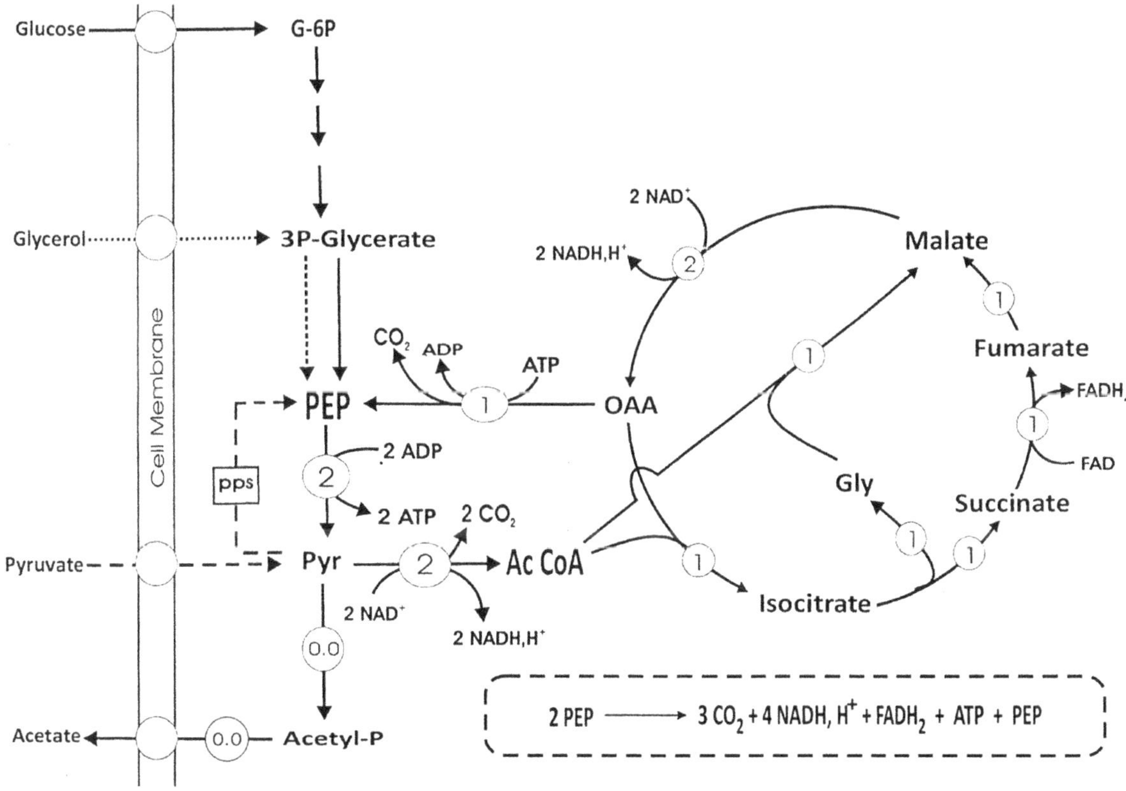

FIGURE 2.6 The PEP-glyoxylate cycle in hungry *E. coli*.

written as a subscript. For example, the logarithm to the base 10 is written as $\log_{10}$ and the logarithm to the base e is written as $\log_e$. If a log is given without a subscript, it denotes a logarithm to the base 10. Logarithms to the base e ($\log_e$) are also called *natural logarithms*, which may also be written as In.

A *semi-logarithmic* (semi-log) graph paper is a graph paper that has an arithmetic scale on one axis and a logarithmic scale on the other. Although it may appear very strange, it is user-friendly and is very useful as it allows you to plot your data directly without having to calculate logs. On the logarithmic scale, there are a number of cycles. Each cycle represents a change in data of an order of magnitude, a factor of 10; any set of factors of 10 can be used but remember that there is no zero on the log scale.

In plotting your data on semi-log paper, the independent variable is plotted on the x-axis (abscissa), while the dependent variable is plotted on the y-axis (ordinate) of the graph. For example, in determining the mean generation time (T) of a given organism under certain conditions, time is the independent variable and is plotted on the x-axis, while the number of cells, which increases exponentially during the course of the logarithmic phase of growth, is plotted on the y-axis. However, it is noteworthy to remember that in determining the minimum inhibitory concentration (MIC) of antibiotics, the drug dosage increases in an exponential fashion, and as such, the log scale here becomes the x-axis.

2.3.1 Mathematical Descriptions of Microbial Growth

Now let us consider the basic equation used to describe microbial growth:

$$\frac{dx}{dt} = \frac{ax}{b} \tag{2.1}$$

Equation 2.1 implies that the rate of biomass (x) formation changes as a function of time (t) and that the rate of change is directly proportional to the concentration of a particular factor (a) such as growth substrate or temperature but is inversely proportional to the concentration of another factor (b) such as inhibitors. In Equation 2.1, both a and b are independent of time t, and the proportionality factor in Equation 2.1 can in effect be ignored. In the early stages of any fermentation process, the increase in biomass is unrestricted, and, as such, the pattern of growth follows an autocatalytic first-order reaction (autocatalytic growth; Figure 2.7) up to a point where either side of Equation 2.1 becomes negative, resulting in autocatalytic death.

> While growth kinetics focuses on the measurement of growth rates during the course of fermentation, growth dynamics relates to the changes in population (biomass) to changes in growth rate and other parameters (e.g., pH and temperature). To unravel such intricate interrelationships, the description of growth kinetics and growth dynamics relies on the use of differential equations.
>
> The term *autocatalytic growth* is generally used to indicate that the rate of increase in biomass formation in a given fermentation is proportional to the original number of cells

present at the beginning of the process, thus reflecting the positive nature of growth.

2.3.2 The Growth Cycle

During batch fermentation, a typical pattern of growth curve, otherwise known as the *growth cycle*, is observed (Figure 2.7). Clearly, a number of different phases of the growth cycle can be differentiated. These are

1. Lag phase
2. Acceleration phase
3. Exponential (logarithmic) phase
4. Deceleration phase
5. Stationary phase
6. Accelerated death phase
7. Exponential death phase
8. Death or survival phase

> The term *growth cycle* is used to describe the overall pattern displayed by microorganisms during growth in batch cultures. It is noteworthy that such a cycle is by no means a fundamental property of the bacterial cell, but rather a consequence of the progressive decrease in food supply or accumulation of inhibitory intermediates in a closed system to which no further additions or removals are made.

The changes in the specific growth rate (μ) as the organism progresses through the growth cycle can also be seen in (Figure 2.7).

We shall now describe the metabolic events and their implications in as far as growth, survival, and productivity are concerned. Naturally, the scenario begins with the first phase of the growth cycle.

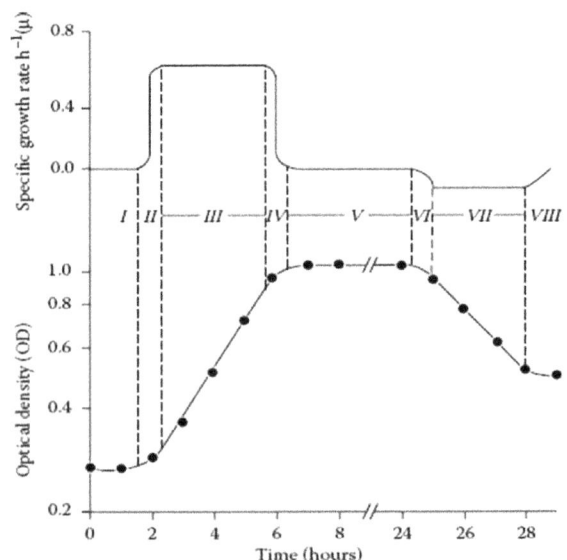

FIGURE 2.7 Typical pattern of growth cycle during the growth of microorganisms in batch cultures; the vertical dotted lines and the roman numerals indicate the changes in specific growth rate (μ) throughout the cycle.

2.3.2.1 The Lag Phase

In this phase, the organism is simply faced with the challenge of adapting to the new environment. Adaptation to other carbon sources, however, may require the induction of a particular set of enzymes that are specifically required to catalyze the transport and hydrolysis of the substrate (e.g., adaptation to lactose or acetate). Irrespective of the mechanisms employed for adaptation, the net outcome at the end of the lag phase is a cell that is biochemically vibrant (i.e., capable of transforming chemicals to biomass).

Entering a lag phase in microbial cell factories is not desirable due to time wasting, cost increase, and loss of revenue, and as such, should be avoided. The question of whether a particular organism has entered a lag phase in a given fermentation process can be determined graphically by simply plotting $\log n$ (biomass) as a function of time, as shown in Figure 2.8.

Note that the transition from the lag phase to the exponential phase involves another phase: the acceleration phase, as illustrated in Figure 2.7. This difficulty can be easily overcome by extrapolating the lag phase sideways and the exponential phase downward as shown, with the point of interception (L) taken as the time at which the lag phase ended. What is also interesting about the graph in Figure 2.8 is that if one continues to extrapolate the exponential phase downward, then the point at which the ordinate is intersected gives the number of cells that were viable and metabolically active at the point of inoculation.

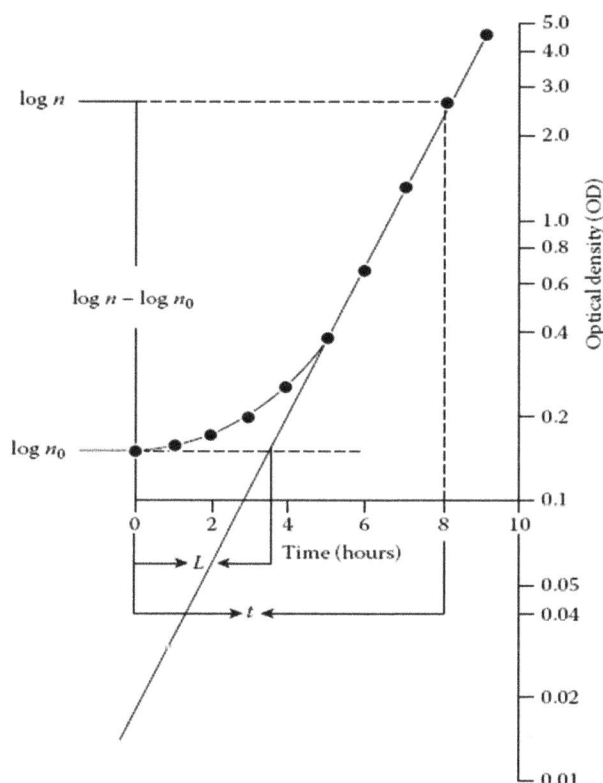

FIGURE 2.8 Graphical determinations of the lag phase and the number of viable cells at the onset of batch fermentation; as the exponential phase is extrapolated downward, it intercepts the extrapolated line of the lag phase and the ordinate, respectively (see text for details).

The question of whether a lag (L) has occurred during the course of the fermentation process and for how long can be easily determined by

$$\frac{\log n - \log n_0}{t - L} = \frac{\log 2}{T} \qquad (2.2)$$

where n is the total number of cells after a given time (t) since the start of fermentation, n_0 is the number of cells at the beginning of fermentation, and T is the organism's mean generation time (doubling time). Equation 2.2 describes the exponential growth, taking into consideration a lag phase in the process. In Section 2.3.2, we shall describe the exponential phase in general and the derivation of Equation 2.2 in particular.

2.3.2.2 Exponential Phase

As soon as *E. coli* adapts to the carbon source happen to be available, it accelerates into the exponential phase (Figure 2.7). Acceleration of growth en route to entry exponential growth is associated with the expression of a nucleotide binding protein, namely, Fis protein. The level of Fis protein increases as a function of growth rate (μ) and peaks during the log phase (Bradley et al., 2007). Once the organism hits the deceleration phase at the end of the exponential phase (Figure 2.7), the organism represses the expression of *Fis* gene in preparation for entering the stationary phase. The mechanism by which *E. coli* controls the expression of Fis involves the binding of a small RNA molecule to Hfq hexamer molecules (Sauer et al., 2012).

In the exponential phase, each cell increases in size, and providing that conditions are favorable, it divides into two, which, in turn, grow and divide; and the cycle continues. During this phase, the cells are capable of transforming the primary carbon source into biosynthetic precursors, reducing power and energy, which is generally trapped in the form of ATP, PEP, and proton gradients. The biosynthetic precursors thus generated are then channeled through various biosynthetic pathways for the biosynthesis of various monomers (amino acids, nucleotides, fatty acids, and sugars) that, in turn, are polymerized to give the required polymers (proteins, nucleic acids, ribonucleic acids, and lipids). Finally, these polymers are assembled in a precise way, and the cell divides to give the new biomass characteristic of each organism. The time span of each cycle (cell division) is known as generation time or doubling time, but because we generally deal with many millions of cells in bacterial cultures, the term *mean generation time* (T) is more widely used to reflect the average generation times of all cells in the culture. Such a rate, providing conditions are favorable, is fairly constant.

> Doubling time or mean generation time (T) is the time required for a given population (N_0) to double in number ($2N_0$).

If a given number of cells (n_0) is inoculated into a suitable medium and the organism was allowed to grow exponentially, then the number of cells after one generation is $2n_0$; at the end of two generations, the number of cells becomes $4n_0$ (or $2^2 n_0$). It follows, therefore, that at the end of a certain number of generations (Z), the total number of cells equals $2^Z n_0$. If the total number of cells, or its log value, at the end of Z generations is known, then

$$n = n_0 2^z \qquad (2.3)$$

$$\log n = \log n_0 + Z \log 2 \qquad (2.4)$$

To determine the number of cell divisions, that is, the number of generations (Z) that have taken place during fermentation, Equation 2.4 can be modified to give

$$Z = \frac{\log n - \log n_0}{\log 2} \qquad (2.5)$$

If T is the mean generation time required for the cells to double in number and t is the time span over which the population has increased exponentially from n_0 to n, then

$$Z = \frac{t}{T} = \frac{\log n - \log n_0}{\log 2} \qquad (2.6)$$

If during the course of a particular fermentation, a lag time has been demonstrated, Equation 2.6 can be modified to take account of this observation. The modified equation is

$$Z = \frac{t - L}{T} = \frac{\log n - \log n_0}{\log 2} \qquad (2.7)$$

Equations 2.6 and 2.7 can be rearranged to give the familiar equations governing the determination of T as follows:

$$\frac{\log n - \log n_0}{t} = \frac{\log 2}{T} \qquad (2.8)$$

$$\frac{\log n - \log n_0}{t - L} = \frac{\log 2}{T} \qquad (2.9)$$

While Equation 2.8 describes the exponential phase of growth in pure terms, Equation 2.9, however, takes into consideration the existence of a lag phase in the process.

The exponential scale (i.e., 2, 4, 8, 16, and so on) demonstrated in Figure 2.8 obeys Equations 2.8 for logarithmic growth. Note that the position of log n_0 on Figure 2.8 lies within the exponential phase, while in Figure 2.7, it lies at the beginning of the lag phase.

Although the use of log to the base 2 has the added advantage of being able to determine the number of generations relatively easily, because an increase of one unit in log 2n corresponds to one generation, the majority of researchers continue to use log to the base 10 ($\log_{10}$). In this case, the slope of the line (Figure 2.9) equals $\mu/2.303$. The relationship between the mean generation time (T) and the specific growth rate (μ) can be described mathematically by

$$\ln 2 = \mu T \qquad (2.10)$$

or

$$\mu = \ln 2 / T \qquad (2.11)$$

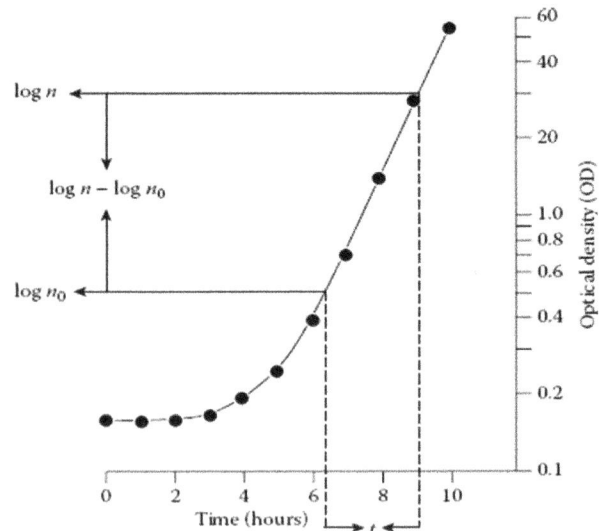

FIGURE 2.9 Graphical determination of the mean generation time (T) during batch fermentation; note the position of log n_0.

Because $\ln 2 = 0.693$, either Equation 2.10 or Equation 2.11 can be rearranged to give

$$T = \frac{0.693}{\mu} \qquad (2.12)$$

During the course of exponential growth, the culture reaches a steady state. As such, the intracellular concentrations of all enzymes, cofactors, and substrates are considered to be constant. During this phase, one can therefore safely assume that all bacterial cells are identical and that the doubling time is constant with no loss in cell numbers due to cell death. The rate of growth of a given population (N) represents, therefore, the rate of growth of each individual cell in the population multiplied by the total number of cells. Such a rate can be described mathematically by this differential equation:

$$\frac{dN}{dT} = \mu N \qquad (2.13)$$

This equation implies that the rate of new biomass formation is directly proportional to the specific growth rate (μ) of the organism under investigation and the number of cells (N). This pattern of growth may be referred to as *autocatalytic*, a term described in this chapter. If Equation 2.13 describes the exponential phase correctly, then a straight line should be obtained when ln N is plotted as a function of time (t). During the exponential phase, it is generally assumed that all cells are identical, and as such the specific growth rate (μ) of individual cells equals that of the whole population. Equation 2.13 can therefore be rearranged to take account of the fact that dN/dt is proportional to the number of cells (N) and that the specific growth rate (μ) is a proportionality factor to give

$$\mu = \frac{1}{N} \frac{dN}{dt} \qquad (2.14)$$

The specific growth rate (μ) is usually expressed in terms of units per hour (h^{-1}).

Although Equations 2.9 and 2.10 describe the exponential phase satisfactorily, in some fermentations, as is the case during growth of *E. coli* on sodium acetate (El-Mansi, unpublished results), the organism fails to maintain a steady state for any length of time and so μ falls progressively with time until the organism reaches the stationary phase. Such a drop in growth rate (μ) can be accounted for by a whole host of different factors, including nutrient limitations, accumulation of inhibitors, and/or the crowding factor (i.e., as the population increases in number, μ decreases).

2.3.2.2.1 Impact of Nutrient Limitations on Growth Rate

To account for the effect of nutrient limitations on growth rate (μ), Monod modified Equation 2.14 so that the effect of substrate concentration (limitations) on μ could be assessed quantitatively. Monod's equation is

$$\mu = \frac{\mu_m S}{K_S + S} \tag{2.15}$$

where S is the concentration of the limiting substrate, μ_m is the maximum specific growth rate, and K_S is the saturation constant (i.e., when $S = K_S$, then $\mu = \mu_m/2$), as illustrated in (Figure 2.10).

During growth in a steady state, Equations 2.16 and 2.17 may be used to describe growth using biomass (X) rather than the number of cells (N) as a measure of growth:

$$\frac{dX}{dt} = \mu X \tag{2.16}$$

$$\frac{dS}{dt} = -\mu \frac{X}{Y} \tag{2.17}$$

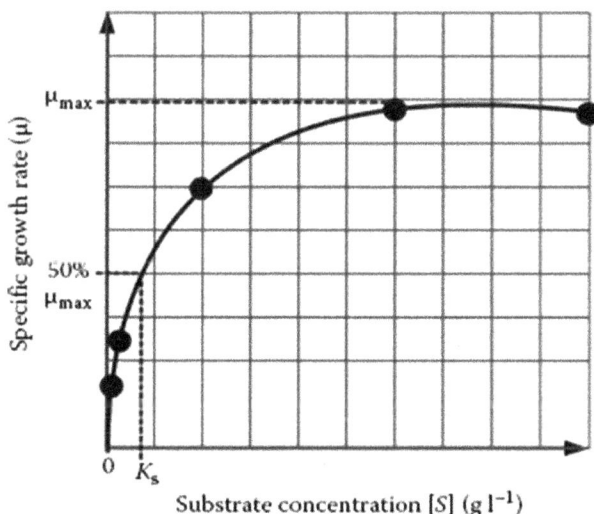

FIGURE 2.10 Graphical determination of specific growth rate (μ) and saturation constant (K_S) during batch fermentations; note that this graph was constructed without taking maintenance energy into consideration. With maintenance in mind, the curve should slide sideways to the right in direct proportion to the fraction of carbon diverted toward maintenance.

where X is the biomass concentration and Y is the growth yield (e.g., biomass generated per gram of substrate utilized). It is noteworthy, however, that during steady-state growth (i.e., in a chemostat or a turbidostat), the terms used to describe the bacterial numbers (N) and biomass (X) in Equations 2.16 and 2.17 are identical. This is not necessarily the case in batch cultures because the size and shape of bacterial cells vary from one stage of growth to another.

While Equation 2.16 addresses the relative change in biomass (the first variable) with respect to time, Equation 2.17 addresses the relative change in substrate concentration (the second variable) as a function of time. Note that following the exhaustion of substrate, $dX/dt = dS/dt = 0$. While Equations 2.16 and 2.17 can surely predict the deceleration and stationary phases of growth, respectively, it should be remembered that nutrient limitation is not the only reason for the deceleration of growth and subsequent entry into the stationary phase. In addition to environmental factors, crowding is reported to have an adverse effect on growth rate (μ), i.e., growth rate diminishes as the size of the population increases.

2.3.2.3 Stationary Phase

As the exponential phase draws to an end, the organism enters the stationary phase and then the death phase. If we assume that the kinetics of death are similar to that of growth, then the specific rate of cell death (λ) can be described mathematically by

$$\frac{dN}{dT} = (\mu - \lambda) N \tag{2.18}$$

Equation 2.18 clearly indicates that if μ is greater than λ, then the organism will grow at a rate equal to μ – λ. If, on the other hand, μ is less than λ, then the population dies at a rate equal to λ – μ. Under conditions where μ equal λ, neither growth nor death is observed, a situation thought to prevail throughout the course of the stationary phase. As the energy supply continues to fall, the equilibrium between λ and μ will finally shift in favor of λ, and consequently cell death begins.

Microorganisms respond differently to nutrient limitations during the stationary phase. For example, while *Bacillus subtilis* and other Gram-positive, spore-forming organisms respond by sporulation, *E. coli* and other Gram-negative, non-spore-forming bacteria cannot respond in the same way, and as such, other mechanisms must have evolved. Although a fraction of the bacterial population dies during this phase of growth, a relatively large number remain viable for a long time despite starvation. The ability of cells to remain viable despite prolonged periods of starvation is advantageous, as most microorganisms in nature are subject to nutrient limitations in one form or another. Our understanding of the molecular mechanisms employed by microorganisms for survival during this phase of growth is rather limited. However, contrary to the notion that microorganisms enter a logarithmic death phase soon after the onset of the stationary phase, some microorganisms such as *S. cerevisiae* and *E. coli* adapt well to starvation, presumably through mutations and induction mechanisms, and do not readily enter a logarithmic phase of death.

2.3.2.4 Death and Survival Phase

Toward the end of the stationary phase, the organism suffers severe a great deal of stress due to; on the one hand, the lack of nutrients, and the accumulation of toxic excreted by-products, on the other. This in turn forces the organism to enter the death phase, during which the majority of cells are either dead or "viable, but non-culturable." The dead cells lyse, thus releasing vital nutrients, which can keep the viable cells as well as the "viable, but non-culturable" cells alive. In *E. coli*, oxidative stress triggers cell lyses of dead cells through a series of reactions initiated by Sigma E and leads to the dismantling of the outer cell membrane (Murata et al., 2012). During starvation, cells endeavor to survive by scavenging the nutrients released from dead cell lysate (Nagamitsu et al., (2013).

On the onset of starvation, *E. coli* induces RpoS; otherwise known as the "master regulator of stationary phase," and this in turn leads to the accumulation of guanosine 3', 5'–bisphosphate (ppGpp), which represses the activity of ribosomal RNA and associated proteins, thus slowing down the rate of protein synthesis of a whole host of different enzymes including those involved in the PTS uptake system, the *fba*B and *pfk*B of glycolysis as well as the nonoxidative part of the pentose phosphate pathway. Such a complex process involves Sigma factors (Shimizu, 2014). The degree to which growth rate (μ) diminishes is a fair reflection of the degree of starvation experiences by the organism. In addition, depending on whether SpoT behaves as a ppGpp synthetase or penta-ppGpp-hydrolase, carbon limitations may also lead to the accumulation of pentaphosphate ppGpp (penta-ppGpp). Stress protein expression induced as a result of activating the Sigma S subunit of RNA polymerase (RpoS) increases the organism's chance of surviving starvation (Kanjee et al., 2012).

The ability of some cells of *S. cerevisiae* to survive prolonged starvation leads to the discovery of a highly conserved family of genes, the snooze genes (SZN). It is interesting to see some researchers using the term *viable but non-culturable* (VBNC) to describe those cells that are viable, but unable to form colonies. However, as colony-forming ability is our only means of assessing whether a particular cell is alive or not, some microbiologists argue the validity of this concept and suggest the term reversibility to indicate as to whether a given cell has the ability to transform from being dormant to being metabolically active. Recent investigations have revealed that resuscitation of stationary phase cells may be aided by the excretion of pheromones as has been demonstrated in *Micrococcus luteus*.

In addition, during the course of starvation, *E. coli* expresses specific genes; the survival genes. A number of *E. coli* strains that cannot survive starvation have been isolated and subsequently designated as survival negative (Sur⁻) strains. Consequently, the genes involved were designated as *sur* genes. While some genes, such as *surA*, are required for survival during "famine" (starvation), others such as *surB* enable the organism to exit the survival mode as the environment changes from "famine" to "feast." Recent studies have also revealed that starvation induces specific transcriptional activator (sigma factor) that is essential for the transcription of *sur* genes by RNA polymerase (Pletnev et al., 2015). The physiological function of the *sur* gene products is to downshift the metabolic demands made on central and intermediary metabolism for maintenance of energy.

As the environment changes from "famine" to "feast" and the nutrients become abundant, the organism expresses the Fis protein, which senses such a change and gives the necessary signal required to initiate growth (Sauer et al., 2012).

2.3.3 Maintenance Energy

Maintenance can be defined as the minimal rate of energy supply required for maintaining the viability of a particular organism without contributing to biosynthesis. The fraction of carbon oxidized in this way is, therefore, expected to end up in the form of carbon dioxide (CO_2). The need for maintenance energy is obvious as the "living state" of any organism, including humans, is remote from equilibrium and as such demands energy expenditure. Moreover, in addition to the carbon processed for maintenance, another fraction of the carbon source may be wasted through excretion (e.g., acetate, α-ketoglutarate, succinate, or lactate). In this context, excretion of metabolites may be seen as an accidental consequence of central metabolism rather than by design to fulfill certain metabolic functions. The fraction of carbon required for maintenance differs from one organism to another and from one substrate to another. For example, maintenance requirements for the lactose phenotype of *E. coli* are greater than those observed for the glucose phenotype as the lac-permease is much more difficult to maintain than the uptake system employed for the transport of glucose, that is, the PTS.

The metabolic interrelationship between maintenance and growth of microorganisms was first discovered in the 1960s by Pirt (1965). In his theoretical treatment of this aspect, he formulated a maintenance coefficient (m) based on the earlier work of Monod and defined the coefficient as the amount of substrate consumed per unit mass of organism per unit time (e.g., g substrate per g biomass per h). If s represents the energy source (substrate), then the rate of substrate consumption for maintenance is governed by

$$-\left(\frac{ds}{dt}\right)_M = mX \tag{2.19}$$

where m is the maintenance coefficient. The yield can therefore be related to maintenance by

$$Y = \frac{Dx}{\left(Ds\right)_G \left(Ds\right)_M} \tag{2.20}$$

where Y is the growth yield, Δx is the amount of biomass generated, $(\Delta s)_G$ is the amount of substrate consumed in biosynthesis, and $(\Delta s)_M$ is the amount of substrate consumed for maintenance of cell viability.

If maintenance energy $(\Delta s)_M$ was determined and found to be zero, then Equation 2.20 reduces to

$$Y_G = \frac{Dx}{(Ds)_G} \qquad (2.21)$$

Equation 2.21 clearly means that growth yield (Y) is a direct function of the amount of substrate utilized. The growth yield $(Y$, gram dry weight per gram substrate) obtained in this case may be referred to as the *true growth yield* to distinguish it from the Y_G where maintenance $(\Delta s)_M$ is involved. However, if a specific fraction of carbon is diverted toward maintenance energy or survival, then one might conclude that the slower the growth rate (μ), the higher the percentage of carbon that is diverted toward maintenance, and the less the growth yield.

On the other hand, during unrestricted growth (i.e., no substrate limitation), the interrelationship between substrate concentration and maintenance can be determined by (Pirt, 1965)

$$ds / dt = \left(ds / dt\right)_M + \left(ds / dt\right)_G \qquad (2.22)$$

In this case, Equation 2.22 implies that the overall rate of substrate utilization equals the rates of substrate utilization for both maintenance and growth. With growth rate expressed in the usual way $(dx/dt = \mu x)$, and assuming that Y equals Y_G, Equations 2.16 and 2.18 can be rearranged to give

$$1 / Y = m / \mu\, 1 / Y_G \qquad (2.23)$$

According to Equation 2.23, if m and Y_G are constants, then the plot of $1/Y$ against $1/\mu$ should give a straight line, the slope of which is maintenance (m) and the point at which the ordinate is intercepted is equal to $1/Y_G$, as illustrated in Figure 2.11. Further analysis of the data shown in Figure 2.11 (El-Mansi, unpublished results) revealed that *E. coli* is capable of shifting down its maintenance requirements during growth on acetate from 5.8 mmol g^{-1} h^{-1} during exponential growth in batch culture or turbidostat to 0.55 mmol g^{-1} h^{-1} during growth in a chemostat at a low growth rate.

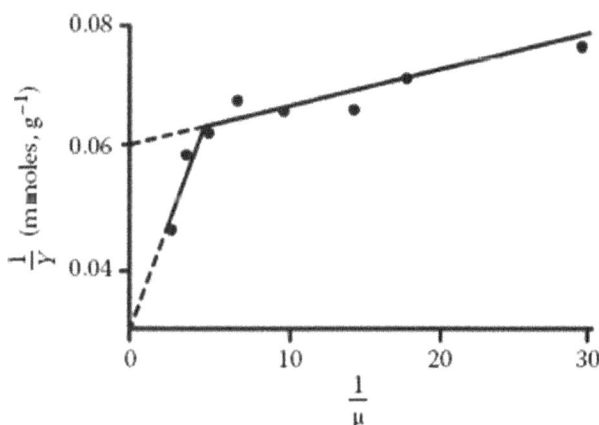

FIGURE 2.11 Graphical determination of maintenance (m) energy.

During the course of fermentation, the minimum rate of energy supply may fall below maintenance requirements due to a shortfall in the supply of phosphorylated intermediates, ATP, and/or reducing powers (NADH, NADPH, $FADH_2$, etc.). Although such a drop in energy supply may not be fatal, it is likely to have an adverse effect on the fitness of the organism; the cells become less capable of taking advantage of favorable changes in the environment, and consequently cells resume growth after a lag period. On the other hand, if the cells were able to grow immediately following inoculation or transfer into another medium (i.e., without lag), then the cells must have been left in optimal conditions. However, if the cells were to enter the lag phase prior to growth, then it is fair to suggest that the organism was left in a suboptimal state, that is, the organism must have suffered a drop in energy supply below its maintenance requirements.

> Ribosome particles, which consist of protein and RNA, are the target to which mRNA and amino acyl-tRNA must bind in preparation for translation (the conversion of mRNA to a polypeptide chain).

It follows, therefore, that the physiological state of the cell at the time when energy supply falls below maintenance is very important and that the deeper the shortfall in energy supply, the longer the lag period. A lag period due to the lack of energy sources is surely different to that observed when the cells are faced with the challenge of changing phenotype, as is the case when glucose phenotype of *E. coli* is forced to change to lactose phenotype. Individual cells in a given population may respond differently when a given drop of energy supply is exerted. The ability of a given population to respond successfully to a "shift-up" or to survive a "shift-down" in nutrients appears to be directly related to the intracellular concentration of ribosomes. For example, in comparison with cells grown in rich medium, a 35-fold drop in the concentration of ribosomes accompanied restricted growth of *Salmonella typhimurium*. In the yeast *S. cerevisiae*, a drop in the cellular concentrations of RNA, DNA, and proteins accompanied the drop in growth rate from $0.4 h^{-1}$ to 0.10 h^{-1} from 12.1%, 0.6%, and 60.1% to 6.3%, 0.4%, and 45%, respectively (Nissen et al., 1997). Furthermore, it has also been demonstrated that the cells of *S. cerevisiae* were capable of "scaling up" or "scaling down" the intracellular level of ribosomes in response to a shift-up or a shift-down in nutrients, respectively.

> *Autogenous regulation*: A regulatory control mechanism that is exerted at the level of translation: the conversion of mRNA to protein, rather than transcription.

The ability to modulate the cellular content of ribosomes is achieved through a mechanism of autogenous (posttranscriptional) regulation. Such a phenomenon is explicable in the light of the proposed hypothesis of "translational couplings" proposed by Nomura et al. (1984). This hypothesis implies that initiation of translation at the first ribosome-binding site is central to exposing the ribosome-binding sites of all other cistrons within a given operon. By the same token, this hypothesis also implies that inhibition or prevention of translation at the first binding site within mRNA means that all other sites within the polycistronic message will remain unexposed and consequently will not be translated.

The argument over the question of whether smaller cells are less able to cope with environmental changes (i.e., nutrient limitations and the drop in energy supply below maintenance) because they contain less ribosomes can now be answered to the satisfaction of everyone as recent research revealed that it is not the number of ribosomes that matter but rather their concentration inside the cell. Furthermore, the need for the cell size to be relatively large during growth under no limitations is not a reflection of high concentration of ribosomes but rather the need to attain a certain mass before replication of the chromosome can be initiated (Donachie, 1968). The essence of Donachie's discovery is that if initiation of replication is triggered in response to a certain volume of cell mass, then the faster the cell divides, the faster the growth rate, and the larger the size of the cell.

In addition to induction and auto-regulatory control mechanisms, successful transition from one phase of growth to another might also involve proteolysis, a mechanism that is particularly significant when the cells have to remove certain proteins at a rate that is significantly faster than their specific growth rate (μ), e.g., regulatory proteins that fluctuate from one phase of growth to another (Grunenfelder et al., 2001) and those that are growth-phase specific and become redundant as the organism enters a new phase (Weichart et al., 2003).

2.3.4 DIAUXIC GROWTH

The ability of microorganisms to display biphasic (diauxic) growth patterns is well documented. For example, *E. coli* displays such a pattern when faced with glucose and lactose (Figure 2.12). Similarly, *Aerobacter aerogenes* displays the same pattern when grown in the presence of glucose and citrate. Both organisms preferentially utilize glucose to either of the competing substrates due to catabolite inhibition and catabolite repression mechanisms elicited by the presence of glucose.

> Catabolite inhibition is a mechanism that, in the presence of glucose, prevents the uptake of lactose, thus allowing preferential utilization of glucose.

> Catabolite repression is a glucose-induced mechanism, which switches off the synthesis of the *lac*-operon enzymes at the level of transcription, thus facilitating preferential utilization of glucose.

2.4 FERMENTATION BALANCES

2.4.1 GROWTH YIELD IN RELATION TO CARBON AND ENERGY CONTENTS OF GROWTH SUBSTRATES

Unlike growth on sugars and polyhydrated alcohols, where a constant yield of 1.1 g dry weight biomass per gram substrate carbon was observed, growth yield on carboxylic acids

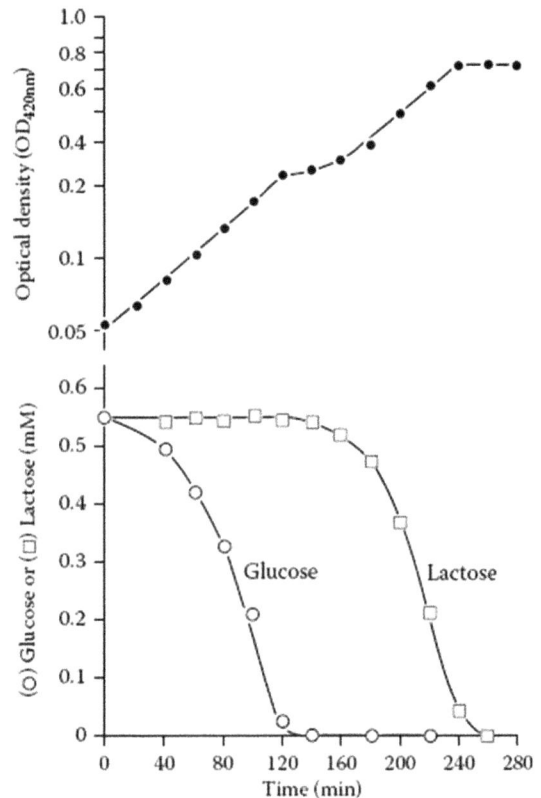

FIGURE 2.12 Diauxic pattern of growth and the pattern of sugar utilization during growth of *E. coli* ML30 on equimolar concentrations (0.55 mM) of glucose and lactose.

gives a much inferior yield. An attempt to resolve this paradox was made (Linton and Stephenson, 1978) by relating the maximum specific growth yield observed on any given carbon source to its carbon and energy content; the latter is defined by the heat combustion (kcal/g substrate carbon). Although the authors recognized the need for using chemostats to achieve steady-state growth and to account for maintenance energy requirements, their analysis of various data in the literature revealed that the growth yield was directly proportional to the heats of combustion up to 11.00 kcal/g substrate carbon and that beyond this level no increase in yield was observed, suggesting the involvement of another mechanism. It is possible, therefore, to argue that during growth on substrates with a heat of combustion value of less than 11.00 kcal/g substrate, the energy generated is insufficient to convert all the available carbon to biomass, and as such growth may be described as energy limited. Further analysis revealed that maximum growth yield is inherently set by the ratio between the biologically available energy and the carbon contents of the substrate. Assuming a bacterial carbon content of 48%, the maximum yield of 1.43g bacterial dry weights per gram substrate carbon, representing a maximum carbon (substrate) to carbon (biomass) conversion of 68%, was observed. This treatment enables the prediction of growth yield with a good degree of accuracy and provides a rational basis for growth limitations; for example, while glycerol- and mannitol-limited growth of *A. aerogenes* may be described as carbon limited, gluconate-limited growth may be viewed as energy limited.

The relative contribution of each substrate to total biomass formation can also be assessed with a good deal of precision providing that growth limitation is a consequence of diminishing carbon supply in a chemostat rather than the accumulation of toxic end products or adverse changes in pH. The relationship between total biomass formation and the concentration of any given substance can therefore be determined from the plot of log n (biomass) as a function of substrate concentration, and providing that all other components are in excess, the organism will enter the stationary phase upon depletion of the substrate or the substance in question. This experiment should obviously be done over a wide range of substrate concentrations, and from the slope obtained, the yield constant or yield coefficient per unit of the substance in question can be determined. If the unit is, say, one mole of a substrate, then the yield coefficient obtained is referred to as the molar growth yield coefficient. This method is used widely as a biological assay for the determination of vitamins, amino acids, purines, and pyridines.

2.4.2 Carbon Balance

Apart from the fraction of carbon used for biomass formation, the remainder is partitioned between products and by-products, including carbon dioxide. The ratio of recovered carbon to that present at the onset of fermentation is referred to as the *carbon balance* or *carbon recovery index*. Such a balance or an index is a measure of efficiency: the higher the index, the higher the carbon recovery, and the more efficient the fermentation process. The carbon balance is generally calculated by working out the number of moles (or mill moles) produced of a given product per 100 moles (or mill moles) of substrate utilized. The number of carbon atoms in each respective molecule can then be multiplied by the value obtained. The resulting values for the products in question can then be totaled and compared with that of the substrate. If the values are equal (i.e., a 1:1 ratio), then a complete recovery of carbon into product formation has been achieved. Although this is theoretically possible, our experience indicates otherwise, as part of this carbon is used for maintenance and assimilation to support growth, however slow. A complete carbon balance can, therefore, be calculated for any given fermentation if the fraction of carbon diverted toward biosynthesis and maintenance is determined. In addition to the carbon balance, some fermentations demand calculation of the redox balance.

2.4.3 Redox Balance

Fermentations of sugars and other primary carbon sources give rise to a whole host of different intermediates; some of which are phosphorylated (energy-rich), while others are not. While biosynthetic intermediates are utilized for the biosynthesis of monomers, other intermediates are produced in excess and this is balanced by their excretion into the medium to redress the redox balance within the cell, as is the case with alcohol excretion or to generate ATP, as is the case with acetate excretion or to regenerate the proton motive force, as is the case with lactate excretion. The stoichiometry of product and by-product formations of any given fermentation process can be ascertained by carefully analyzing the culture filtrates at different stages. From a physiological standpoint and in order for fermentation to go to completion, the redox balance must be maintained.

2.5 EFFICIENCY OF CENTRAL METABOLISM

2.5.1 Impact of Futile Cycling on the Efficiency of Central Metabolism

The efficiency of carbon conversion to biomass and desirable end products is influenced by many different factors (see Chapters 3, 6, and 7 for more details) of which futile cycling is a major contributor. For example, any transient increase in the intracellular concentration of pyruvate in *E. coli* may trigger a futile cycle in the central metabolic pathways at the junction of PEP. Futile cycling at the junction of PEP (Figure 2.13) involves, in addition to pyruvate dehydrogenase and malic enzyme, PEP carboxykinase and pyruvate kinase (ATP-generating reactions) on one hand and PEP-synthetase and PEP-carboxylase (ATP-utilizing reactions) on the other. The operation of this futile cycle may, in addition to wasting ATP, adversely affect the adenylate energy charge within the cell. Transient increase in the intracellular level of pyruvate may also trigger futile cycling in other organisms. For example, *Klebsiella aerogenes* triggers a futile cycle involving two enzymes: the first is the NAD^+-dependent pyruvate reductase, which catalyzes the transformation of pyruvate to D-lactate, and the second is the FAD-dependent D-lactate dehydrogenase, which completes the cycle (Figure 2.14). Flux of carbon through these two enzymes provides a futile cycle in which NADH is oxidized at the expense of reducing FAD^+, thus bypassing the first phosphorylation site in the electron transport chain (Figure 2.13). It follows, therefore, that any changes leading to transient increase in the intracellular level of pyruvate may lead to the operation of this futile cycle and, in turn, energy dissipation. The presence and function of glutamine synthetase and glutamines in ammonia-limited cultures

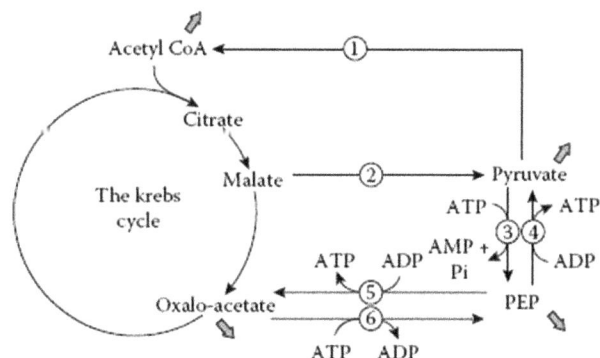

FIGURE 2.13 Futile cycle at the level of phosphoenolpyruvate (PEP) in central metabolism and its role in energy dissipation.

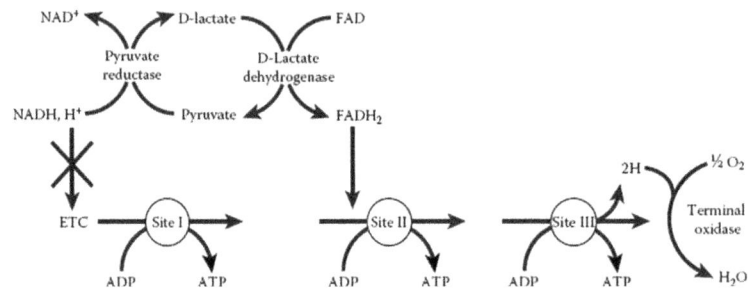

FIGURE 2.14 The role of pyruvate reductase and D-lactate dehydrogenase in bypassing the first phosphorylation site of the electron transport chain (ETC) in the oxidation of NADH, H$^+$.

represent yet another possible futile cycle (Figure 2.15) in which the synthetase generates glutamate while the other affects its hydrolysis with a net loss of one molecule of ATP per turn (Tempest, 1978).

2.5.2 IMPACT OF METABOLITE EXCRETION ON THE EFFICIENCY OF CENTRAL METABOLISM

Excretions of metabolites in general and acetate in particular diminish flux to product formation, which, in turn, adversely affects the efficiency of carbon conversion to desirable end products (Holms, 1986; El-Mansi, 2004). Acetate, particularly in its undissociated form, is a potent uncoupler of oxidative phosphorylation. As such, conditions that promote flux to acetate excretion will, in turn, diminish flux to product formation, as a large fraction of the carbon source has to be diverted toward maintenance requirements (El-Mansi, unpublished observations). While such a role is fully appreciated, it should be remembered that flux to acetate excretion is physiologically significant as it allows faster growth rate and facilitates high cell density growth (El-Mansi, 2004). Flux to acetate excretion is also important as flux through phosphotransacetylase

replenishes central and intermediary metabolism with free-CoA, thus fulfilling an anaplerotic function (El-Mansi, 2005) that is central for the smooth operation of carbon flux through enzymes of central and intermediary metabolism. Interestingly, flux through pyruvate oxidase to acetate excretion, which may appear to be wasteful, looks to be significant as it was found to be equated with high efficacy of growth and energy generation (Guest et al., 2004).

Chemostat: A continuous culture in which growth rate is limited by the rate of nutrients supply.

2.6 CONTINUOUS CULTIVATION OF MICROORGANISMS

The use of continuous cultures in the fermentation industry is, in some cases, preferred to batch cultures for the following reasons:

- It outperforms batch cultures economically by eliminating the inherent downtime that is lost for cleaning, sterilization, and the reestablishing of biomass within the bioreactor.
- Unlike batch cultures where the rate of product formation is at its peak for only a limited period of time, continuous cultures sustain such a period over a much longer span of time.
- The interpretation of data obtained from continuous cultures is much simpler than from batch cultures because continuous cultures afford a steady-state growth in which input and output are in perfect balance, unlike batch cultures where variations in the concentrations of primary carbon and nitrogen sources, product formation, pH, and the redox balance within the cells vary during the course of growth.
- Unlike batch cultures, continuous cultures can be manipulated to create certain environmental conditions to facilitate overexpression of recombinant proteins.

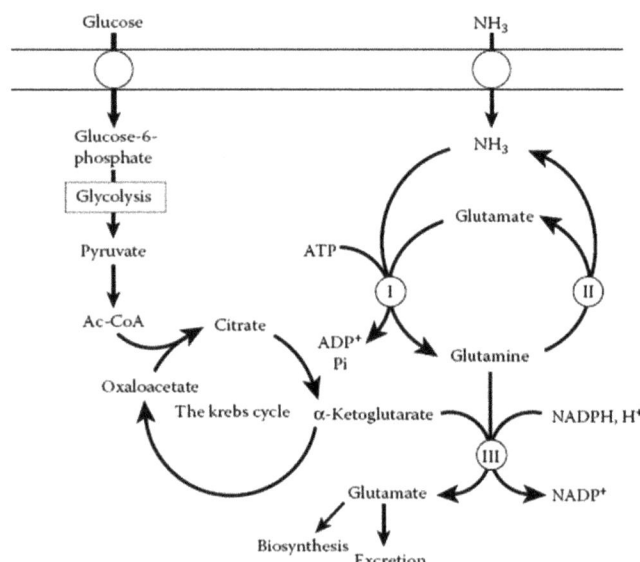

FIGURE 2.15 The role of glutamine synthetase and glutaminase in energy dissipation through the formation of a futile cycle.

Turbidostat: A continuous culture in which the organism grows at its maximum specific growth rate (μ_{max}), thus reaching a steady state of fixed turbidity.

The earliest continuous fermentation process on record is the production of vinegar by *Acetobacter* sp. In this process, a sugar solution is trickled down a bioreactor containing cells of *Acetobacter* sp. adhered to wood shavings, the earliest recorded example of cell immobilization. The acetic acid produced renders the conditions inhospitable for other organisms, thus minimizing the risk of contaminations.

It is noteworthy, however, that the use of continuous cultures in the fermentation industry is much less common than batch processes primarily because quality control regulation stipulates the need for a "batch number." Furthermore, continuous cultivations are much more susceptible to contamination, which is rather risky especially in the production of bioactive pharmaceuticals.

2.6.1 Types of Continuous Cultures

There are three main types of continuous cultures: chemostat, turbidostat, and auxostat; the latter reflects the set parameter (e.g., pH-stat).

In an auxostat, the feeding rate is adjusted to match the rate of cellular metabolism. In that sense, a turbidostat is also an auxostat because the turbidity created as a consequence of bacterial growth is maintained at a set point by adjusting the rate of supply of fresh medium. The most popular auxostat, however, is the pH auxostat (pH-stat), a continuous culture in which pH is maintained at a set point by the addition of NaOH to counterbalance the excreted acetate and other acidic by-products.

Auxostat, nutristat, and *pH-stat* are synonymous names for a continuous culture in which the dilution rate changes to maintain a certain factor constant.

2.6.2 Principles and Theory of Continuous Cultures

Although the calculations of different growth parameters during growth in continuous cultures cultivation of microorganisms are based on the growth kinetics of exponentially growing unicellular organisms, such calculations can be equally applied to multicellular organisms (e.g., filamentous fungi).

A good continuous culture maximizes the number of biochemically active cells in the population while keeping the number of biologically inactive cells to a minimum.

2.6.2.1 Microbial Growth Kinetics in Continuous Cultures

Growth and fermentation kinetics have been dealt with extensively in Chapter 3 and the reader is strongly encouraged to consult it for further details. The theories underlying growth kinetics can be represented by a number of equations, shown here. *Specific growth rate* (μ): This parameter (Figure 2.16) can be calculated using the following equation:

$$\frac{dx}{dt} = \mu \left(h^{-1} \right) \tag{2.24}$$

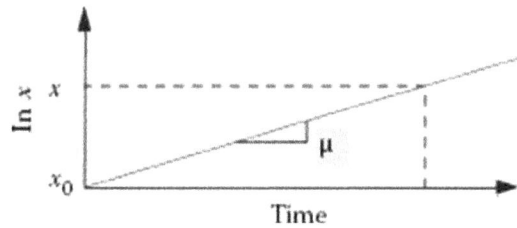

FIGURE 2.16 Graphical determination of specific growth rate (μ).

where dx is the change in biomass (Δ biomass), dt is the duration of time during which the measured increase in biomass has taken place (Δ time), and μ is the specific growth rate h^{-1}.

Doubling time (T_d) *or mean generation time* (T): This parameter can be calculated using the following equation:

$$T_d = \ln 2 / \mu \tag{2.25}$$

where ln 2 is the natural log of 2, which is 0.693, and μ is the specific growth rate.

Dilution rate (D): This parameter can be measured using

$$D = F / V \tag{2.26}$$

where F is the flow rate (ml/min) and V is the volume (ml) of culture in the fermentation vessel.

During growth in continuous culture, the rate of change in the organism's concentration equals the organism's growth rate minus the rate of removal of the organism from the culture vessel. The rate of change in the concentration of biomass can therefore be calculated by

$$dx / dt = \mu - D \tag{2.27}$$

The Monod constant:

The Monod constant (Figure 2.7) can be calculated using Equation 2.28:

$$\mu = \mu_{max} S / S + K_s \tag{2.28}$$

where μ_{max} is the maximum specific growth rate, S is the substrate concentration, and K_S is the substrate concentration at which μ equals 1/2 μ_{max}.

2.6.2.2 Interrelationship Between Growth Rate (μ) and Dilution Rate (D)

During steady-state growth in a chemostat (Equation 2.27), the rate of change in the concentration of both biomass ($dx/dt = 0$) and the limiting substrate ($ds/dt = 0$) equals zero. It follows, therefore, that during growth in a chemostat, where growth is limited by a given nutrient, D equals μ.

During growth in continuous cultures, however, D might exceed (μ), and the point at which D becomes larger than μ (Figure 2.17) is known as the critical dilution rate (D_c).

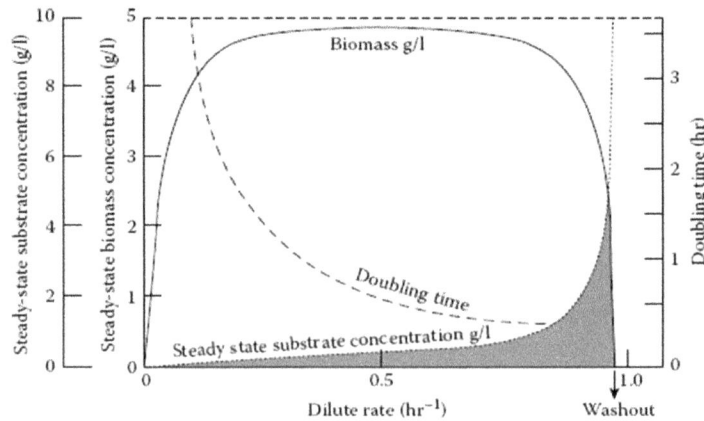

FIGURE 2.17 Impact of dilution rate (D) on doubling time (T), substrate concentration (S), and biomass formation (X) during continuous cultivation of microorganisms.

Whenever $D > \mu$, a wash-out of biomass occurs. It follows that one can determine the maximum specific growth rate, μ_{max}, by increasing the flow rate past the critical dilution rate (D_c) and follow the decrease in biomass concentration over a period of time. The μ_{max} can then be determined using Equation 2.29:

$$\mu_{max} = \left[\ln X - \ln X_0 / t \right] + D \qquad (2.29)$$

This impact of D on biomass formation, the rate of change in biomass formation, the doubling time, and the concentration of substrate illustrated in Figure 2.14 can be described by Equation 2.30:

$$ds / dt = D\left(S_R - S\right) - \mu x / Y \qquad (2.30)$$

where S_R is the concentration of the limiting substrate in the reservoir and Y is the biomass yield on that particular substrate.

Since $dx/dt = 0$ and $(ds/dt = 0)$ during steady-state growth in a chemostat, the steady-state values of biomass (X') and substrate (S') can be calculated using Equation 2.31:

$$X' = Y\left(S_R - K_S D\right) / \left(\mu_{max} - D\right) \qquad (2.31)$$

where K_S is the limiting-substrate constant (the Monod constant).

2.6.2.3 Efficiency and Productivity Of Fermentation Processes

The productivity of a given bioprocess is a significant factor and often plays a central role in deciding which fermentation process (batch, fed batch, or continuous) should be employed. Such a parameter can be assessed by (Pirt, 1975)

$$\text{Productivity index} = \ln X_m / X_0 + 0.693 t_L / t_d \qquad (2.32)$$

where X_0 and X_m are the initial and maximum biomass concentrations, respectively; t_L is the total shutdown time of a batch process; and t_d is the doubling time.

2.7 CURRENT TRENDS IN THE FERMENTATION AND PHARMACEUTICAL INDUSTRY

Although mammalian cell and other eukaryotic cell cultures like the yeast *Pichia* represent an attractive alternative to bacteria for the expression and production of recombinant proteins, especially those requiring posttranslational modification such as glycosylation, recombinant protein expression in bacteria remains the most cost-effective, and as such, from an industrial perspective, it is the most desirable option, particularly since the development of tightly regulated promoters such as the T7 polymerase-specific promoter. In this case, effective control through repression ensures that expression and, in turn, production is turned on only after significant biomass is formed and this is particularly significant if the product is toxic to the cell. Ingenious and effective approaches have recently been developed in bacterial expression systems including modification of ribosome-binding sites to increase yield, and expression at low temperature on its own and in combination with the expression of molecular chaperones, which reduce misfolding and precipitation of proteins, to aid correct folding of recombinant proteins.

2.7.1 "QUIESCENT CELL FACTORY": A NOVEL APPROACH

A nongrowing bacterial culture in which the primary nutrients are directly and quantitatively converted to products represents the ideal scenario for biotechnology. However, this is very difficult to achieve, and an alternative approach is to render transcription and translation processes to be gene specific. Although this approach may seem insurmountable, specific expression of certain proteins in eukaryotic organisms is a well-established phenomenon. Attempts to develop a bacterial system that mimics eukaryotic organisms in that direction led to the discovery of the *E. coli* "Quiescent Cell Protein Expression System," otherwise referred to as the Q-cells (Summer, 2002).

Overexpression of Rcd in the mutant strain bns205, which apart from being defective in *bns* is capable of producing an N-terminal fragment of H-NS, led to a complete cessation of growth within 2–3 hours and the cells entered a quiescent (nongrowing) state (Summer, 2002). Fortuitously, the Q-cells appear to preferentially express the plasmid encoded rather than the chromosomally encoded genes. The reason for this odd but useful feature was unraveled by the same author following DAPI–DNA staining and examination of Q-cells using fluorescence microscopy, which revealed that the bacterial nucleoid was highly condensed; this was a consequence of Rcd/H-NS complex formation, thus preventing its expression in a manner reminiscent of that observed for heterochromatin formation in eukaryotes, which results in global repression of transcription. The Q-cell system has now been shown to function well at the pilot scale and was not unduly sensitive to media composition or culture density. The Q-cell system is currently employed for the production of recombinant proteins. Further developments are currently underway for its adoption in other systems (Summer, 2002).

Nucleoid: A structure within the bacterial cytoplasm containing the genome together with associated proteins. Although the nucleoid does not have a membrane, it may be loosely attached to cellular membrane (El-Mansi et al., 2001).

2.7.2 Applications of Two-Stage Fed-Batch Fermentation in the Production of Biopharmaceuticals: A Robust Approach

The need to develop new methodologies and innovations for the production of biopharmaceuticals is considered an urgent need, so that we can effectively increase their production without compromising their affordability or safety. The prevalence of diabetes is expected to scale new heights; no less than 366 million people are projected to be diabetics by the year 2030. An increase in the production of insulin must therefore be planned to meet the projected rise in demand.

To that end, Gurramkonda et al. (2010) developed a two-stage cultivation process in which a recombinant strain of *Pichia pastoris* (strain X-33) carrying a synthetic insulin-precursor encoding gene fused in frame with the alpha-factor secretory signal peptide of *S. cerevisiae*. The X-33 strain was first grown to high cell density in batch cultures on a low-salt and high-glycerol minimal medium. Following batch culture growth, the expression of the insulin precursor-fusion gene was achieved by growing the organism in a fed-batch culture on methanol (2g l⁻¹); the concentration of methanol was kept constant throughout the production phase. This robust-batch and fed-batch two-stage strategy led to the secretion of nearly 4 g of insulin precursor per liter of culture supernatant. With the aid of immobilized metal ion affinity chromatography (IMAC), a novel approach in protein purification, 95% of the secreted product was recovered. The purified insulin precursor was trypsin digested and further purified, leading to approximately 1.5 g of 99% pure recombinant human insulin

per liter of culture broth. Compared with the highest previously reported value, this approach resulted in an increase in the efficiency of insulin production by a factor of nearly 200%, thus significantly increasing the efficiency of insulin manufacture.

2.8 MICROBIAL FERMENTATIONS AND THE PRODUCTION OF BIOPHARMACEUTICALS

Throughout this chapter, the author outlined the scientific principles underlying microbial fermentation. In the case study that follows, the author illustrates the application of microbial fermentation for the commercial production of useful products.

2.8.1 Production of Insulin: A Case Study

2.8.1.1 History and Background

In addition to maintaining glucose concentration in blood within a certain range, thus avoiding hyperglycemia, insulin also stimulates lipogenesis, activates amino acids uptake, and diminishes lipolysis.

The biologically active insulin is a monomer consisting of two polypeptide chains A and B, 21 and 30 amino acid residues in length, respectively, that are interlinked through two disulfide bridges. However, insulin is first synthesized in the β-cells of the islets of Langerhans in the pancreas, as a single polypeptide that is biologically inactive, otherwise known as *preproinsulin* (Figure 2.18). The preproinsulin is then translocated into the cisternae of the endoplasmic reticulum, where its signal peptide is removed by a specific protease to give proinsulin, which in turn folds into its correct native tertiary structure through the formation of two disulfide bridges (Figure 2.17). Once this is achieved, the proinsulin is packaged into secretory vesicles in the Golgi apparatus. Finally, a specific protease cleaves off the C chain (Figure 2.18). Failure to secret insulin causes diabetes, which if not treated leads to hyperglycemia. The number of diabetics in the world is increasing alarmingly, and as such all possible measures must be taken in order that we satisfy the demands for the provision of insulin. The quantities of exogenous insulin extracted from pig and beef pancreas could never satisfy market needs; a new approach was therefore called for, which was finally realized, after many years of intensive research, through fermentation microbiology and recombinant DNA technology.

2.8.1.2 Cloning and Commercial Production of Insulin

Based on the amino acid sequence of human insulin, Crea et al. (1978) successfully synthesized two oligodeoxyribonucleotides, 77 and 104 base pairs long, encoding the A and B chains, respectively. Each oligos bears single-stranded cohesive termini for *Eco*RI and *Bam*HI to facilitate direct cloning into the pBR322 plasmid vector.

The two oligos encoding the ∝ (A) and β (B) chains were fused into the structural gene encoding β-galactosidase on pBR322 construct, and the resulting recombinant plasmids

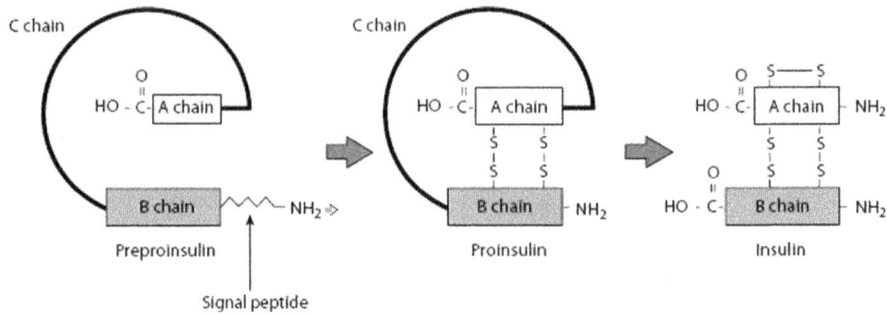

FIGURE 2.18 Posttranslational modifications of preproinsulin to proinsulin and then to biologically active insulin molecule.

were used to transform competent cells of *E. coli* as illustrated in Figure 2.18. Expression of the chimeric plasmid in *E. coli* gave rise to hybrid polypeptides, including the sequence of amino acids corresponding to human insulin A and B chains.

The A and B polypeptide chains were cleaved off from the precursor (the hybrid polypeptide) by the action of cyanogen bromide. After separation and purification of the A and B chains, active human insulin was generated (Figure 2.19) by *in vitro* formation of the correct disulfide bridges between A and B chains.

2.8.2 PROTEIN ENGINEERING OF INSULIN

2.8.2.1 Fast-Acting Insulin Analogs

Fast-acting insulin is therapeutically desirable, and as such, protein engineering of insulin to produce analogs with low association constant is of great commercial interest. Attempts to produce such analogs have been successful; examples include insulin *ispro* and insulin *aspart* (Owens, 2002).

2.8.2.2 Long-Acting Insulin Analogs

The presence of a constant but low (basal) level of insulin in plasma during the fasting state is essential to maintain overall control of glycemia. Engineering insulin to produce long-acting insulin analogs is therefore of therapeutic and commercial interest. Such an objective has now been achieved, and products include insulin glargine, a long-acting insulin, by virtue of introducing two extra arginine residues at the end of the B chain (Arg B31 and Arg B32). After subcutaneous injection of glargine, insulin level rises slowly to a plateau within 6–8 hours and remains essentially unchanged for up to 24 hours (Owens, 2002).

SUMMARY

The diverse array of metabolic networks, together with a very high rate of metabolic turnover, makes microorganisms an ideal tool for the conversion of renewable resources into products of primary and secondary metabolism.

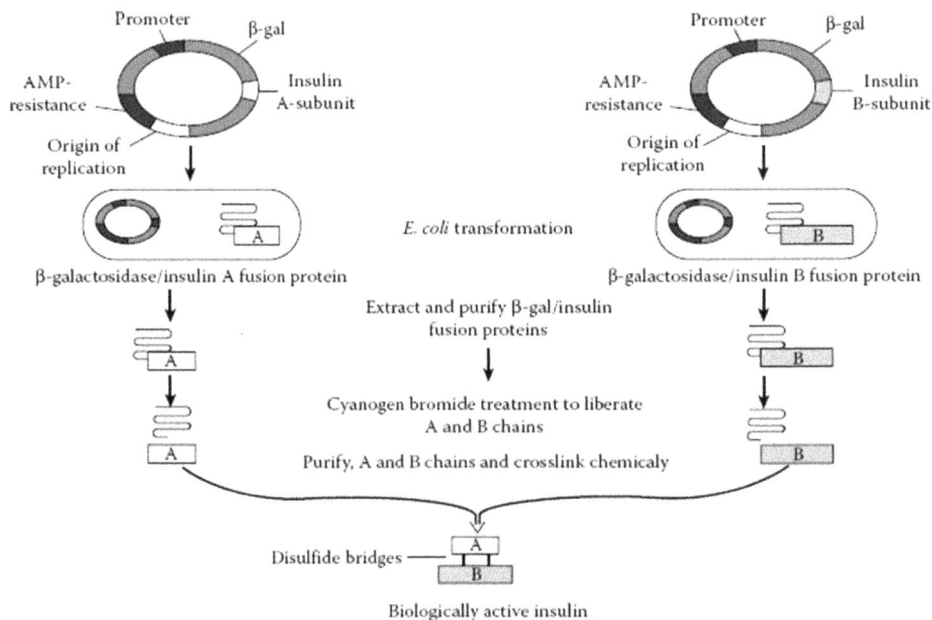

FIGURE 2.19 Applications of fermentation biotechnology and recombinant DNA technology in the commercial production of insulin. As can be seen in the diagram, an oligonucleotide expressing the A-chain and another expressing the B-chain are cloned independently, overexpressed, harvested, purified, and chemically crosslinked to give biologically active insulin.

The conversion of a given carbon source to biomass involves a sophisticated net of biochemical reactions. These reactions can be classified according to their function into fueling reactions, biosynthetic (anabolic) reactions, anaplerotic reactions, polymerization, and assembly reactions.

Microorganisms change in size as the growth rate changes, which is consistent with the view that initiation of replication is dependent on cell mass.

The drop in energy supply below maintenance requirement is not necessarily fatal, but the severity of such a drop is directly proportional to its magnitude and the relative concentration of ribosomes within the cell.

The question of whether a particular organism has lagged following inoculation, together with the number of viable cells at the onset of fermentation, can be ascertained graphically.

The interrelationships among various growth parameters during unrestricted and restricted growth have been described.

Microorganisms adopt different tactics to survive starvation. While Gram-positive organisms resort to sporulation, *S. cerevisiae* and *E. coli* express specific proteins, the primary function of which is scaling down the maintenance requirement on one hand and exiting the stationary phase on the other.

Microorganisms can be grown in batch, fed-batch, or continuous cultures. The latter provides a controlled environment in which a given microorganism can be grown at its maximum specific growth rate in a turbidostat at constant pH in a pH-stat, or under the limitation of certain nutrients in a chemostat. Continuous culture is, therefore, an indispensable tool in the study of growth kinetics and fermentation processes.

The physiology and bioenergetics, as well as cell size and biomass composition including the organism's transcriptome and proteome of a given phenotype of a microorganism, have been found to vary according to growth conditions and growth rate (μ).

Intracellular fluxes among various enzymes of central and intermediary metabolism vary according to growth rate (μ) in a nonlinear manner.

Microorganisms are capable of sensing changes in the nutritional status of their environment through sensor metabolites and specific transcription factors.

Microorganisms ensure unidirectional flow of glycolytic intermediates into central metabolism through the operation of irreversible reaction and a sophisticated net of allosteric control mechanisms.

The use of microorganisms for the production of recombinant proteins and bioactive pharmaceuticals has been enhanced significantly by recent innovations in genetic and protein engineering as evidenced by the production of fast-acting and long-acting recombinant insulin molecules.

Last, but not least, the author, along with many microbiologists and biochemists, enjoyed exercising full control on steady-state growth of microorganisms through the application of the following equation:

OD required = [Inoculum size (ml)/Total volume of culture (ml)] × OD of inoculum × 2 to the power of the number of generations.

Application of the above equation demanded awareness of the mean generation time (T) of the organism on the substrate in question.

REFERENCES

Bradley, M.D., Beach, M.B., de Koning, J.A.P., Pratt, T.S., Osuna, R. 2007. Effects of Fis on *Escherichia coli* gene expression during different growth stages. *Microbiology* 153: 2922–2940.

Crea, R., Kraszewski, A., Hirose, T., Itakura, K. 1978. Chemical synthesis of genes for human insulin. *Proc Natl Acad Sci USA* 75(12): 5765–5769.

Donachie, W.D. 1968. Relationship between cell size and time of initiation of DNA replication. *Nature* 219: 1077–1083.

Doucette, C.D., Schwab, D.J., Wingreen, N.S., Rabinowitz, J.D. (2011). Alpha-ketoglutarate coordinates carbon and nitrogen utilization *via* enzyme I inhibition. *Nat Chem Biol* 7: 894–901.

El-Mansi, E.M.T. 2004. Flux to acetate and lactate excretions in industrial fermentations: Physiological and biochemical implications. *J Ind Microbiol Biotechnol* 31: 295–300.

El-Mansi, E.M.T. 2005. Free-CoA mediated regulation of intermediary and central metabolism: An hypothesis which accounts for the excretion of α-ketoglutarate during aerobic growth on acetate. *Res Microbiol* 156: 874–879.

El-Mansi, M., Cozzone, A.J., Shiloach. J., Eikmanns, B. 2006. Control of carbon flux through enzymes of central and intermediary metabolism during growth of *Escherichia coli* on acetate. *Curr Opin Microbiol* 9: 173–179.

El-Mansi, E.M.T., Anderson, K.J., Inche, C.A., Knowles, L.K., and Platt, D. J. (2001) Isolation and curing of the *Klebsiella pneumoniae* large indigenous plasmid using Sodium Dodecyl Sulphate. *Research in Microbiology* 151: 201–209.

Fischer. E, Sauer, U. 2007. A novel metabolic cycle catalyzes glucose oxidation and anaplerosis in hungry *Escherichia coli*. *J Biol Chem* 278(47): 46446–46451.

Grunenfelder, B., Rummel, G., Vohradsky, J., Roder, D., Langen, H., Jenal, U. 2001. Proteomic analysis of the bacterial cell cycle. *Proc Natl Acad Sci USA* 98: 4681–4686.

Guest, J.R., Russell, G.C. 1992. Complexes and complexities of the citric acid cycle in *Escherichia coli*. *Curr Top Cell Regul* 33: 1–47.

Guest, J.R., Abdel-Hamid, A.M., Auger, G.A., Cunningham, L., Henderson, R.A., Machado, R.S., Attwood, M. M. 2004. Physiological effects of replacing the PDH complex of *E. coli* by genetically engineered variants or by pyruvate oxidase. In F. Gordon and M. S. Patel, eds., *Thiamine: Catalytic mechanisms and role in normal and disease states*, pp. 389–407. New York: Marcel Dekker Inc.

Gurramkonda, C., Polez, S., Skoko, N., Admabn, N., Gabel, T., Chugh, D., Swaminathan, S., Khanna, N., Tisminetzky, S., Rinas., U. 2010. Applications of simple fed-batch to high level secretory production of insulin precursor using *Pichia pastoris*. *Microb Cell Fact* 12: 9–31.

Hua, Q., Yang, C., Oshima, T., Mori, H., Shimizu, K. 2004. Analysis of gene expression in *Escherichia coli* in response to changes of growth-limiting nutrient in chemostat cultures. *Appl Environ Microbiol* 70: 2354–2366.

Holms, W.H. 1986. The central metabolic pathways of *Escherichia coli*: Relationship between flux and control at a branch point, efficiency of conversion to biomass, and excretion of acetate. *Curr Top Cell Regul* 28: 69–105.

Ingraham, J.L., Neidhardt, F.C., Schaechter, M. 1990. *Physiology of the bacterial cell. A molecular approach*. Sunderland, MA: Sinauer Associates, Inc.

Ishii, N., Nakahigashi, K., Baba, T., Robert, M., Soga, T., Kanai, A., Hirasawa, T., Naba, M., Hirai, K., Hoque, A., Ho, P.Y., Kakazu, Y., Sugawara, K., Igarashi, S., Harada, S., Masuda, T., Sugiyama, N., Togashi, T., Hasegawa, M., Takai, Y., Yugi, K., Arakawa, K., Iwata, N., Toya, Y., Nakayama, Y., Nishioka, T., Shimizu, K., Mori, H., Tomita, M. 2007. Multiple high-throughput analyses monitor the response of *E. coli* to perturbations. *Science* 316: 593–597.

Jiang, P., Ninfa, A.J. 2007. *Escherichia coli* PII signal transduction protein controlling Nitrogen assimilation acts as a sensor of adenylate energy charge in vitro. *Biochemistry* 46: 12979–12996.

Kanjee, U., Ogata, K., Houry, W.A. 2012. Direct binding targets of the stringent response alarmone (p)ppGpp. *Mol Micobiol* 85: 1029–1043.

Kasemets, K, Drews, M, Nisamedtinov, I, Paalme, T, Adamberg, K. 2003. Modification of A-stat for the characterization of microorganisms. *J Microbiol Methods*, 55: 187–200.

Kayser, A, Weber, J, Hecht, V, Rinas, U. 2005. Metabolic flux analysis of *Escherichia coli* in glucose-limited continuous culture. I. Growth-rate-dependent metabolic efficiency at steady state. *Microbiology* 151: 693–706.

Kornberg, HL. 1966. The role and control of the glyoxylate cycle in *Escherichia coli*. *Biochem J* 99(1): 1–11.

Linton, J.D., Stephenson, R.J. 1978. A preliminary study of growth yields in relation to the carbon and energy content of various organic growth substrates. *FEMS Microbiol Lett* 3: 95–98.

Murata, M., Noor, R., Nagamitsu, H., Tanaka, S. Yamada, M. 2012. Novel pathway directed by sigma E to cause cell lysis in *Escherichia coli*. *Genes Cells* 17: 234–247.

Nachen, A., Schicker, A., Revelles, O., Sauer, U. 2008. Cyclic AMP-dependent catabolite repression is the dominant control mechanism of metabolic fluxes under glucose limitation in *Escherichia coli*. *J Biotechnol* 190: 2323–2330.

Nagamitsu, H., Murata, M., Kosaka, T., Kawaguchi, J., Mori, H., Yamada, M. 2013. Crucial roles of MicA and RybB as vital factors for σ^E-dependent cell lysis in *Escherichia coli* long-term stationary phase. *J Mol Microbiol Biotechnol* 23: 227–232.

Nanchen, A, Schicker, A, Sauer, U. 2006. Nonlinear dependency of intracellular fluxes on growth rate in miniaturized continuous cultures of *Escherichia coli*. *Appl Environ Microbiol* 72: 1164–1172.

Nahku, R., Valgepea, K., Lahtvee, P.J., Erm, S., Abner, K., Adamberg, K., Vilu, R. 2010. Specific growth rate-dependent transcriptome profiling of *Escherichia coli* K12 MG1655 in accelerostat cultures. *J Biotechnol* 145: 60–65.

Nikaido, H. 2003. Molecular basis of bacterial outer membrane permeability revisited. *Microbiol Mol Biol Rev* 67: 593–656.

Ninfa, J., Jiang, P. (2005). PII signal transduction proteins: Sensors of α-ketoglutarate that regulate nitrogen metabolism. *Curr Opin Microbiol* 8: 168–173.

Nissen, L.N., Schulze, U., Nielsen, J., Villadsen, J. 1997. Flux distribution in anaerobic, glucose-limited continuous cultures of *Saccharomyces cerevisiae*. *Microbiology* 143: 203–218.

Nomura, M., Gourse, R., Baughman, G. 1984. Regulation of synthesis of ribosomes and ribosomal components. *Ann Rev Biochem* 53: 75–89.

Owens, D.R. 2002. New horizons: Alternative routes for insulin therapy. *Nat Rev Drug Discov* 1: 529–540.

Pirt, S.J. 1965. The maintenance energy of bacteria in growing culture. *Proc R Soc London Ser B* 163: 224–231.

Pirt, SJ. 1975. *Principles of microbes and cell cultivation*. Oxford: Blackwell Scientific.

Pletnev, P; Osterman, I; Sergiev, P; Bogdanov, A and Dontsova, O (2015). Survival Guide: Escherichia coli in the stationary phase. ACTA NATURAE 7(4) 22–33.

Pramanik, J., Keasling, J.D. 1997. Stoichiometric model of *Escherichia coli* metabolism: incorporation of growth-rate dependent biomass composition and mechanistic energy requirements. *Biotechnol Bioeng* 56: 398–421.

Radchenko, M.V., Thornton, J., Merrick, M. 2010. Control of AmtB-GlnK complex formation by intracellular levels of ATP, ADP, and 2-oxoglutarate. *J Biol Chem* 285: 31037–31045.

Ramseier, T.M., Bledig, S., Michotey, V., Feghali, R., Saier, M.H. Jr. 1995. The global regulatory protein, FruR, modulates the direction of carbon flow in *Escherichia coli*. *Mol Microbiol* 1995(16): 1157–1169.

Renilla, W., Bernal, V., Fuhrer, T., Castano-Cerezo, S., Iborra, J.L., Sauer, U. 2012. Acetate scavenging activity in *Escherichia coli*: interplay of acetyl–CoA synthetase and the PEP–glyoxylate cycle in chemostat cultures. *Appl Microbiol Biotechnol* 93(5): 2109–2124.

Sauer, E., Schmidt, S., Weichenrieder, O. 2012. Small RNA binding to the lateral surface of Hfq hexamers and structural rearrangements upon mRNA target recognition. *Proc Natl Acad Sci USA* 109: 9396–9401.

Shimizu, K. 2014. Regulation systems of bacteria such as *Escherichia coli* in response to nutrient limitation and environmental stress. *Metabolites* 4: 1–35.

Summer, D. 2002. A quiet revolution in the bacterial cell factory. *Microbiol Today* 29: 76–78.

Takatsuka, Y., Chen, C., Nikaido, H. 2010. Mechanism of recognition of compounds of diverse structures by the multidrug efflux pump AcrB of *Escherichia coli*. *Proc Natl Acad Sci USA* 107: 6559–6656.

Tchieu, J.H., Norris, V., Edwards, J.S., Saier, M.H. 2001. The complete phosphotransferase system in *Escherichia coli*. *J Mol Microbiol Biotcehnol* 3: 329–346.

Tempest, D.W. 1978. The biochemical significance of microbial growth yields: A reassessment. *Trends Biochem Sci* 3: 180–184.

Valgepea, K., Adamberg, K., Nahku, R., Lahtvee, P.J., Arike, L., Vilu, R. 2010. Systems biology approach reveals that overflow metabolism of acetate in *Escherichia coli* is triggered by carbon catabolite repression of acetyl- CoA synthetase. *BMC Syst Biol* 4: 166.

Weichart, D., Querfurth, N., Dreger, M., Hengge-Aronis, R. 2003. Global role for ClpP-containing proteases in stationary phase adaptation of *Escherichia coli*. *J Bacteriol* 185: 115–125.

Yamamoto, K., Hirao, K., Oshima, T., Aiba, H., Utsumi, R., Ishihama, A. 2005. Functional characterization *in vitro* of all two component signal transduction systems from *Escherichia coli*. *J Biol Chem* 280: 1448–1456.

Zhao, J., Shimizu, K. 2003. Metabolic flux analysis of *Escherichia coli* K12 grown on 13C-labeled acetate and glucose using GC-MS and powerful flux calculation method. *J Biotechnol* 101: 101–101.

3 Fermentation Kinetics
Central and Modern Concepts

Jens Nielsen

CONTENTS

"There cannot be a greater mistake than that of looking superciliously upon practical applications of science. The life and soul of science is its practical application."

Lord Kelvin 1883

3.1 INTRODUCTION

Growth of microbial cells is the result of many chemical reactions: fueling reactions that converts nutrients into a set of 12 precursor metabolites, biosynthetic reactions that converts these 12 precursor metabolites into building blocks like amino acids, nucleotides, and fatty acids, and assembly reactions that polymerize building blocks into proteins, DNA, complex lipids, and so on (see Figure 2.3). In preparation for cell division, the cells increase in size (or extend their hyphae, in the case of filamentous microorganisms) as the macromolecules are assembled en route to biomass formation. Biomass formation can be quantified by measuring the increase in dry weight (see Box 3.1), RNA, DNA, and/or proteins. *In situ* measurements of biomass formation during the course of fermentation can also be monitored by

following the increase in turbidity at a given wavelength, as illustrated by Olsson and Nielsen (1997).

Growth of microbial cells is often illustrated with a batchwise growth of a unicellular organism (either a bacterium or yeast). Here the growth occurs in a constant volume of medium with one growth-limiting substrate component that is used by the cells. Cell growth is generally quantified by the so-called specific growth rate μ (h^{-1}), which for such a culture is given by

$$\mu = \frac{1}{x}\frac{dx}{dt} \qquad (3.1)$$

BOX 3.1 STANDARD OPERATING PROCEDURE (SOP) FOR DRY WEIGHT DETERMINATION

Biomass is most frequently determined by dry weight measurements. This can be done either using an oven or a microwave oven, with the latter being the fastest procedure. An important prerequisite for the measurement is that the sample is dried completely, and it is therefore

31

important to apply a consistent procedure. A suggested protocol is as follows:

1. Dry the filter (pore size 0.45 μm for yeast or fungi, 0.20 μm for bacteria) on a glass dish in the microwave oven on 150 W for 10 min. Place a tissue paper between the glass and the filter so that the filter does not stick to the glass.
2. Place the filter in a desiccator and allow to cool for 10–15 min. Weigh the filter.
3. Filter the cell suspension through the filter and wash the cells with demineralized water.
4. Place the filter on the glass dish again and dry in the microwave oven for 15 min at 150 W.
5. Put the filter in a desiccator and allow to cool for 10–15 min. Weigh the filter.
6. If more than 30 mg dry weight is present on the filter, the time in the microwave oven may have to be longer.

where x is the biomass concentration (or cell number). The specific growth rate is related to the *doubling* time t_d(h) of the biomass through

$$t_d = \frac{\ln 2}{\mu} \tag{3.2}$$

The doubling time t_d is equal to the generation time for a cell (i.e., the length of a cell cycle for unicellular organisms), which is frequently used by life scientists to quantify the rate of cell growth.

The design and optimization of a given fermentation process require a quantitative description of the process, which, considering the nature of microbial growth, is generally a complex task. Furthermore, often the product is not the cells themselves, but a compound synthesized by the cells, and depending on the type of product, the kinetics of its formation may vary from one phase of growth to another. Thus, while primary metabolites (see Chapter 4 for more details) are typically formed in conjunction with cellular growth, an inverse relationship between product formation and cell growth is often found in the case of secondary metabolites (see Chapter 5 for more details), and here flux to product formation may be greatest in the stationary phase.

With these differences in mind, it is clear that quantification of product formation kinetics may be a difficult task. However, with the rapid progress in biological sciences, our understanding of cellular function has increased dramatically, and this may form the basis for far more advanced modeling of cellular growth kinetics than seen earlier. Thus, in the literature one may find mathematical models describing events like gene expression, kinetics of individual reactions in central pathways, together with macroscopic models that describe cellular growth and product formation with relatively simple mathematical expressions. These models cannot be compared directly because they serve completely different purposes,

and it is therefore important to consider the aim of the modeling exercise in a discussion of mathematical models.

In this chapter, the applications of kinetic modeling to fermentation and cellular processes will be discussed.

3.2 FRAMEWORK FOR KINETIC MODELS

The net result of the many biochemical reactions within a single cell is the conversion of substrates to biomass and metabolic end products (see Figures 2.3 and 3.1). Clearly, the number of reactions involved in the conversion of, say, glucose into biomass and desirable end products is very large, and it is therefore convenient to adopt the structure proposed by Neidhardt et al. (1990) for describing cellular metabolism, which can be summarized as follows:

Assembly reactions carry out chemical modifications of macromolecules, their transport to prespecified locations in the cell, and, finally, their assembly to form cellular structures such as cell walls, membranes, the nucleus, and so on.

Polymerization reactions represent directed, sequential linkage of activated molecules into long (branched or unbranched) polymeric chains. These reactions lead to the formation of macromolecules from a set of building blocks such as amino acids, nucleotides, and fatty acids.

Biosynthetic reactions produce the building blocks used in the polymerization reactions. They also produce coenzymes and related metabolic factors, including signal molecules. Furthermore, a large number of biosynthetic reactions occur in functional units called biosynthetic pathways, each of which consists of sequential reactions leading to the synthesis of one or more building blocks. Pathways are easily recognized and are often controlled en bloc. In some cases, their reactions are catalyzed by enzymes made from a polycistronic message of messenger RNA (mRNA) transcribed from a set of 12 genes forming an operon. All biosynthetic pathways

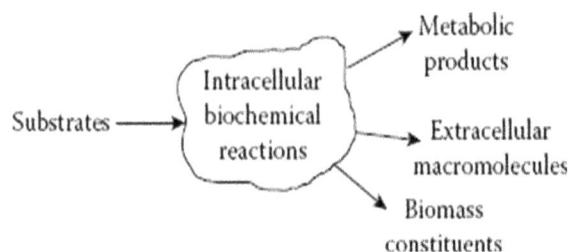

FIGURE 3.1 An overview of the intracellular biochemical reactions in microorganisms; in addition to the formation of biomass constituents, for example, cellular protein, lipids, RNA, DNA, and carbohydrates, substrates are converted into primary metabolites, for example, ethanol, acetate, lactate; secondary metabolites, for example, penicillin; and/or extracellular macromolecules, for example, enzymes, heterologous proteins, polysaccharides.

begin with one of only 12 precursor metabolites from which all building blocks can be synthesized. Some pathways begin directly with such a precursor metabolite, others indirectly by branching from an intermediate or an end product of a related pathway.

Fueling reactions produce the 12 precursor metabolites needed for biosynthesis. Additionally, they generate Gibbs free energy in the form of adenosine-5'-triphosphate (ATP), which is used for biosynthesis, polymerization, and assembling reactions. Finally, the fueling reactions produce the reducing power needed for biosynthesis. The fueling reactions include all biochemical pathways referred to as *catabolic* pathways (degrading and oxidizing substrates).

Thus, the conversion of glucose into cellular protein, for example, proceeds via precursor metabolites formed in the fueling reactions, further via building blocks (in this case amino acids) formed in the biosynthetic reactions, and finally through polymerization of the building blocks (or amino acids). In the fueling reactions, there are many more intermediates than the precursor metabolites, and similarly a large number of intermediates are also involved in the conversion of precursor metabolites into building blocks. The number of cellular metabolites is therefore very large, but still they only account for a small fraction of the total biomass (Table 3.1). The reason for this is the en bloc control of the individual reaction rates in the biosynthetic pathways mentioned above. Furthermore, the high affinity of enzymes for the reactants ensures that each metabolite can be maintained at a very low concentration even at a high flux through the pathway (see Box 3.2).

This control of the individual reactions in long pathways is very important for cell function, but it also means that in a quantitative description of cell growth it is not necessary to consider the kinetics of all the individual reactions, and this obviously leads to a significant reduction in the degree of complexity. Consideration of the kinetics of individual enzymes or reactions is therefore necessary only when the aim of the study is to quantify the relative importance of a particular reaction in a pathway.

3.2.1 STOICHIOMETRY

The first step in a quantitative description of cellular growth is to specify the stoichiometry for those reactions to be considered for analysis. For this purpose, it is important to distinguish between substrates, metabolic products, intracellular metabolites, and biomass constituents (Stephanopoulos et al. 1998):

A **substrate** is a compound present in the sterile medium, which can be further metabolized or directly incorporated into the cell.

A **metabolic product** is a compound produced by the cells and excreted to the extracellular medium.

Biomass constituents are pools of macromolecules that make up the biomass (e.g., RNA, DNA, protein, lipids, and carbohydrates), but also macromolecular products accumulating inside the cell (e.g., a polysaccharide or a nonsecreted heterologous protein).

Intracellular metabolites are all other compounds within the cell (i.e., glycolytic intermediates, precursor metabolites, and building blocks).

Note that this list distinguishes between biomass constituents and intracellular metabolites, because the timescales of their turnover in cellular reactions are very different: intracellular metabolites have a very fast turnover (typically in the range of seconds) compared with that of macromolecules (typically in the range of hours). This means that on the timescale of growth, the intracellular metabolite pools can be assumed to be in pseudo-steady state.

BOX 3.2 CONTROL OF METABOLITE LEVELS IN BIOCHEMICAL PATHWAYS

The level of intracellular metabolites is normally very low. This is due to tight regulation of the enzyme levels and to the high affinity most enzymes have toward the reactants. To illustrate this consider two reactions of a pathway: one forming the metabolite X_i and the other consuming this metabolite:

$$....X_{i-1} \xrightarrow{v_i} X_i \xrightarrow{v_{i+1}} X_{i+1}....$$

Assuming that there is no allosteric regulation of the two enzyme-catalyzed reactions the kinetics can be described with reversible Michaelis-Menten kinetics:

$$v_i = \frac{v_{i,\max}\left(\dfrac{c_{i-1}}{K_{i-1}} - \dfrac{c_i}{K_i}\right)}{1 + \dfrac{c_{i-1}}{K_{i-1}} + \dfrac{c_i}{K_i}}$$

TABLE 3.1
Overall Composition of an Average Cell of *Escherichia Coli*

Macromolecule	% of Total Dry Weight
Protein	55.0
RNA	20.5
rRNA	16.7
tRNA	3.0
mRNA	0.8
DNA	3.1
Lipid	9.1
Lippolysaccharide	3.4
Peptidoglycan	2.5
Glycogen	2.5
Metabolite pool	3.9

Source: Data are taken from Ingraham, J.L., O. Maaloe, and F.C. Neidhardt. 1983. *Growth of the Bacterial Cell.* Sunderland, MA: Sinauer Associates.

where c_i is the metabolite concentration and $v_{i,max}$ expresses the enzyme activity. If the rate of the first reaction increases drastically, for example, due to an increase in the concentration of the metabolite X_{i-1}, the metabolite X_i will accumulate. This will lead to an increase in the second reaction rate and a decrease in the first reaction rate. Consequently, the concentration of metabolite X_i will decrease again. The parameters K_i and K_{i-1} quantify the affinity of the enzyme for the reactant and the product in each reaction, and generally these are in the order of a few μM. Thus, even for low metabolite concentrations (in the order of ten times K_i), the enzyme will be saturated and the reaction rate will be close to $v_{i,max}$, but typically the metabolite concentration is of the order of K_i because hereby the enzyme can respond rapidly to changes in the metabolite level, and metabolite accumulations can be avoided.

With the goal of specifying a general stoichiometry for biochemical reactions, we consider a system where N substrates are converted to M metabolic products and Q biomass constituents. The conversions are carried out in J reactions in which K intracellular metabolites participate as pathway intermediates. The substrates are termed S_i, the metabolic products are termed P_i, the biomass constituents are termed $X_{macro,i}$, and the intracellular metabolites are termed $X_{met,i}$. With these definitions, the general stoichiometry for the jth reaction can be specified as

$$\sum_{i=1}^{N} \alpha_{ji} S_i + \sum_{i=1}^{M} \beta_{ji} P_i + \sum_{i=1}^{Q} \gamma_{ji} X_{macro,i}$$

$$+ \sum_{i=1}^{K} g_{ji} X_{met,i} = 0; \quad j = 1,..,J \tag{3.3}$$

Here, α_{ji} is a stoichiometric coefficient for the ith substrate, β_{ji} is a stoichiometric coefficient for the ith metabolic product, γ_{ji} is a stoichiometric coefficient for the ith macromolecular pool, and g_{ji} is a stoichiometric coefficient for the ith intracellular metabolite. All the stoichiometric coefficients are with sign. Thus, all compounds consumed in the jth reaction have negative stoichiometric coefficients, whereas all compounds that are produced have positive stoichiometric coefficients. Furthermore, compounds that do not participate in the jth reaction have a stoichiometric coefficient of zero.

If there are many cellular reactions (i.e., J is large), it is convenient to write the stoichiometry for all the J cellular reactions in a compact form using matrix notation:

$$\mathbf{AS} + \mathbf{BP} + \mathbf{GX}_{macro} + \mathbf{GX}_{met} = 0 \tag{3.4}$$

where the matrices $\mathbf{A}$, $\mathbf{B}$, $\mathbf{G}$, and $\mathbf{G}$ are stoichiometric matrices containing stoichiometric coefficients in the J reactions for the substrates, metabolic products, biomass constituents,

and pathway intermediates, respectively. In these matrices, rows represent reactions and columns metabolites, that is, the element in the jth row and the ith column of $\mathbf{A}$ specifies the stoichiometric coefficient for the ith substrate in the jth reaction. Formulation of the stoichiometry in matrix form may seem rather complex; however, if the model is simple (i.e., only a few reactions, a few substrates, and a few metabolic products are considered), it is generally more convenient to use the simpler stoichiometric representation in Equation 3.3.

3.2.2 REACTION RATES

The stoichiometry of the individual reaction is the basis of any quantitative analysis. However, of equal importance is specification of the rate of the individual reactions. Normally, the rate of a chemical reaction is given as the forward rate, which, if termed v_i, specifies that a compound that has a stoichiometric coefficient β in the ith reaction is formed with the rate βv_i. Normally, the stoichiometric coefficient for one of the compounds is arbitrarily set to 1, whereby the forward reaction rate becomes equal to the consumption or production of this compound in this particular reaction. For this reason the forward reaction rate is normally specified with the unit moles (or g) h^{-1}, or if the total amount of biomass is taken as reference (so-called specific rates) with the unit moles (or g) (g DW h^{-1}).

For calculation of the overall production or consumption rate, we have to sum the contributions from the different reactions, that is, the total specific consumption rate of the ith substrate equals the sum of substrate consumptions in all the J reactions:

$$r_{s,i} = -\sum_{j=1}^{J} \alpha_{ji} v_j \tag{3.5}$$

The stoichiometric coefficients for substrates are generally negative, that is, the specific formation rate of the ith substrate in the jth reaction given by $\alpha_{ji} v_j$ is negative, but the specific substrate uptake rate is normally considered as positive, and a minus sign is therefore introduced in Equation 3.5. For the specific formation rate of the ith metabolic product, similarly we have

$$r_{p,i} = \sum_{j=1}^{J} \beta_{ji} v_j \tag{3.6}$$

Equations 3.5 and 3.6 specify some very important relations between what can be directly measured: the specific substrate uptake rates and the specific product formation rates, and the rates of the reactions in the metabolic model. If a compound is consumed or formed in only one reaction, it is quite clear that we can get a direct measurement of this reaction rate. For the biomass constituents and the intracellular metabolites, we can specify similar expressions for the net formation rate in all the J reactions:

$$r_{\text{macro},i} = \sum_{j=1}^{J} \gamma_{ji} v_j \qquad (3.7)$$

$$r_{\text{met},i} = \sum_{j=1}^{J} g_{ji} v_j \qquad (3.8)$$

These rates are net specific formation rates, because a compound may be formed in one reaction and consumed in another, and the rates specify the net results of consumption and formation in all the J cellular reactions. Thus, if $r_{\text{met},i}$ is positive there is a net formation of the ith intracellular metabolite, and if it is negative there is a net consumption of this metabolite. Finally, if $r_{\text{met},i}$ is zero, the rates of formation of the ith metabolite exactly balance its consumption.

If the forward reaction rates for the J cellular reactions are collected in the rate vector $\mathbf{v}$, the summations in Equations 3.5 through 3.8 can be formulated in matrix notation as

$$\mathbf{r}_s = -\mathbf{A}^{\mathrm{T}} \mathbf{v} \qquad (3.9)$$

$$\mathbf{r}_p = \mathbf{B}^{\mathrm{T}} \mathbf{v} \qquad (3.10)$$

$$\mathbf{r}_{\text{macro}} = \mathbf{G}^{\mathrm{T}} \mathbf{v} \qquad (3.11)$$

$$\mathbf{r}_{\text{met}} = \mathbf{G}^{\mathrm{T}} \mathbf{v} \qquad (3.12)$$

Here $\mathbf{r}_s$ is a rate vector containing the specific uptake rates of the N substrates, $\mathbf{r}_p$ a vector containing the specific formation rates of the M metabolic products, $\mathbf{r}_{\text{macro}}$ a vector containing the net specific formation rate of the Q biomass constituents, and $\mathbf{r}_{\text{met}}$ a vector containing the net specific formation rate of the K intracellular metabolites. Notice that what appears in the matrix equations are the transposed stoichiometric matrices, which are formed from the stoichiometric matrices by converting columns into rows and vice versa (see Example 3.3a). Equations 3.7 and 3.11 give the net specific formation rate of biomass constituents, and because the intracellular metabolites only represent a small fraction of the total biomass, the specific growth rate μ of the total biomass is given as the sum of formation rates for all the macromolecular constituents:

$$\mu = \sum_{i=1}^{Q} r_{\text{macro},i} = \mathbf{1}_Q^{T} \mathbf{r}_{\text{macro}} = \mathbf{1}_Q^{T} \Gamma^{T} \mathbf{v} \qquad (3.13)$$

where $\mathbf{1}Q$ is a Q-dimensional row vector with all elements being 1. Equation 3.13 is fundamental because it links the information supplied by a detailed metabolic model with the macroscopic (and measurable) parameter μ. It clearly specifies that the formation rate of biomass is represented by a sum of formation of many different biomass constituents (or macromolecular pools), a point that will be discussed further in Section 3.4.3.

3.2.3 Yield Coefficients and Linear Rate Equations

The overall yield (e.g., how much carbon in the glucose ends up in the metabolite of interest) is a very important design parameter in many fermentation processes. This overall yield is normally represented in the form of *yield* coefficients, which can be considered as relative rates (or fluxes) toward the product of interest with a certain compound as reference, often the carbon source or the biomass. These yield coefficients therefore have the units mass per unit mass of the reference (e.g., moles of penicillin formed per mole of glucose consumed or g protein formed per g biomass formed). An often used yield coefficient in the design and operation of aerobic fermentations is the respiratory quotient (RQ), which specifies the moles of carbon dioxide formed per mole of oxygen consumed (see also Example 3.3a). Several different formulations of the yield coefficients can be found in the literature. Here we will use the formulation of Nielsen et al. (2003), where the yield coefficient is stated with a double subscript Y_{ij}, which states that a mass of j is formed or consumed per mass of i formed or consumed. With the ith substrate as the reference compound, the yield coefficients are given by

$$Y_{s_i s_j} = \frac{r_{s,j}}{r_{s,i}} \qquad (3.14)$$

$$Y_{s_i p_j} = \frac{r_{p,j}}{r_{s,i}} \qquad (3.15)$$

$$Y_{s_i x} = \frac{\mu}{r_{s,i}} \qquad (3.16)$$

In the classical description of cellular growth introduced by Monod (1942) (see Section 3.4.2), the yield coefficient Y_{sx} was taken to be constant, and all the cellular reactions were lumped into a single overall growth reaction where substrate is converted to biomass. However, in the late 1950s, it was shown (Herbert, 1959) that the yield of biomass with respect to substrate is not constant. In order to describe this, Herbert introduced the concept of endogenous metabolism and specified substrate consumption for this process in addition to that for biomass synthesis. At the same time, Luedeking and Piret (1959) found that lactic acid bacteria produce lactic acid at nongrowth conditions, which was consistent with an endogenous metabolism of the cells. Their results indicated a linear correlation between the specific lactic acid production rate and the specific growth rate:

$$r_p = a\mu + b \qquad (3.17)$$

In the mid-1960s, Pirt (1965) introduced a similar linear correlation between the specific rate of substrate uptake and the specific growth rate, and suggested the term maintenance, which is currently a widely used concept in endogenous metabolism. The linear correlation of Pirt takes the form of

$$r_s = Y_{xs}^{\text{true}} \mu + m_s \qquad (3.18)$$

where Y_{xs}^{true} is referred to as the true yield coefficient and m_s as the maintenance coefficient. With the introduction of the linear correlations, the yield coefficients can obviously not be constants. Thus, for the biomass yield on the substrate:

$$Y_{sx} = \frac{\mu}{Y_{xs}^{true}\mu + m_s} \qquad (3.19)$$

which shows that Y_{sx} decreases at low specific growth rates where an increasing fraction of the substrate is used to meet the maintenance requirements of the cell. When the specific growth rate becomes large, the yield coefficient approaches the reciprocal of Y_{xs}^{true}. A compilation of true yield and maintenance coefficients for various microbial species is given in Table 3.2.

The empirically derived linear correlations are useful for correlating growth data, especially in steady-state continuous cultures where linear correlations similar to Equation 3.18 were found for most of the important specific rates. The remarkable robustness and general validity of the linear correlations indicate that they have a fundamental basis, and this basis is the continuous supply and consumption of ATP, which are tightly coupled in all cells. Thus, the role of the energy-producing substrate is to provide ATP to drive biosynthesis and polymerization reactions as well as cell maintenance processes according to the linear relationship:

$$r_{ATP} = Y_{xATP}\mu + m_{ATP} \qquad (3.20)$$

which is a formal analog to the linear correlation of Pirt. Equation 3.20 states that ATP produced balances the consumption for growth and for maintenance, and if the ATP yield on the energy-producing substrate is constant (i.e., r_{ATP} is proportional to r_s), it is quite obvious that Equation 3.20 can be used to derive the linear correlation in Equation 3.18, as

illustrated in Example 3.3a. Notice that Y_{xATP} in Equation 3.20 is a true yield coefficient, but it is normally specified without the superscript "true."

The concept of balancing ATP production and consumption can be extended to other cofactors (e.g., NADH and NADPH), and as such, it is possible to derive linear rate equations for three different cases (Nielsen and Villadsen, 1994):

Anaerobic growth where ATP is supplied by substrate-level phosphorylation
Aerobic growth without metabolite formation
Aerobic growth with metabolite formation

For aerobic growth with metabolite formation, the specific substrate uptake rate takes the form of

$$r_s = Y_{xs}^{true}\mu + Y_{ps}^{true}r_p + m_s \qquad (3.21)$$

This linear rate equation can be interpreted as a metabolic model with three reactions:

Conversion of substrate to biomass with a stoichiometric coefficient for the substrate and a forward reaction rate equal to the specific growth rate
Conversion of substrate to the metabolic product with a stoichiometric coefficient for the substrate and a forward reaction rate equal to the specific product formation rate
Metabolism of substrate to meet the maintenance requirements (normally, the substrate is oxidized to carbon dioxide) with the rate m_s

Consequently, the stoichiometry for these three reactions can be specified as

$$-Y_{xs}^{true}S + X = 0; \mu \qquad (3.22)$$

$$-Y_{ps}^{true}S + P = 0; r_p \qquad (3.23)$$

$$-S = 0; m_s \qquad (3.24)$$

With this stoichiometry, the linear rate Equation 3.21 can easily be derived using Equation 3.5, that is, the overall specific substrate consumption rate is the sum of substrate consumption for growth, metabolite formation, and maintenance.

Thus, it is important to distinguish between true yield coefficients (which are rather stoichiometric coefficients) and overall yield coefficients, which can be taken to be stoichiometric coefficients in *one* lumped reaction (often referred to as the black box model; see Section 3.2.4), which represents all the cellular processes:

$$-Y_{xs}S + Y_{xp}P + X = 0; \mu \qquad (3.25)$$

Despite the subscript "true," the true yield coefficients are only parameters for a given cellular system as they simply represent overall stoichiometric coefficients in lumped reactions; for example, reaction (3.22) is the sum of all reactions involved in the conversion of substrate into biomass. If, for

TABLE 3.2
True Yield and Maintenance Coefficients for Different Microbial Species and Growth on Glucose or Glycerol (only for *Aerobacter aerogenes*)

Organism	Substrate	Y_{xs}^{true} [g (g DW)$^{-1}$]	m_s [g (g DW h)$^{-1}$]
Aspergillus nidulans		1.67	0.020
Candida utilis		2.00	0.031
Escherichia coli		2.27	0.057
Klebsiella aerogenes		2.27	0.063
Penicillium chrysogenum		2.17	0.021
Saccharomyces cerevisiae		1.85	0.015
Aerobacter aerogenes	Glycerol	1.79	0.089
Bacillus megatarium		1.67	–
Klebsiella aerogenes		2.13	0.074

Source: Data are taken from Nielsen, J. and J. Villadsen. 1994. *Bioreaction Engineering Principles.* New York: Plenum Press.

example, the environmental conditions change, a different set of metabolic routes may be activated, and this may result in a change in the overall recovery of carbon in each of the three processes mentioned above (i.e., the values of the true yield coefficients change). Even the more fundamental Y_{xATP} cannot be taken to be constant, as illustrated in a detailed analysis of lactic acid bacteria (Benthin et al. 1994).

Example 3.3a: Metabolic Model for Aerobic Growth of *Saccharomyces cerevisiae*

To illustrate the derivation of the linear rate equations for an aerobic process with metabolite formation, we consider a simple metabolic model for the yeast *S. cerevisiae*. For this purpose, we set up a stoichiometric model that summarizes the overall cellular metabolism, and based on assumptions of pseudo-steady state for ATP, NADH, and NADPH, linear rate equations can be derived where the specific uptake rates for glucose and oxygen and the specific carbon dioxide formation rate are given as functions of the specific growth rate. Furthermore, by evaluating the parameters in these linear rate equations, which can be done from a comparison with experimental data, information on key energetic parameters may be extracted.

From an analysis of all the biosynthetic reactions, the overall stoichiometry for synthesis of the constituents of a *S. cerevisiae* cell can be specified (Oura, 1983) as

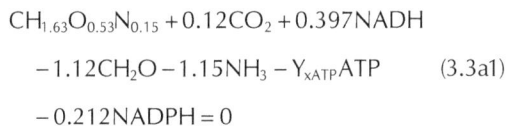

$$CH_{1.63}O_{0.53}N_{0.15} + 0.12CO_2 + 0.397NADH$$
$$- 1.12CH_2O - 1.15NH_3 - Y_{xATP}ATP \quad (3.3a1)$$
$$- 0.212NADPH = 0$$

The stoichiometry (3.3a1) holds for a cell with the composition specified in Table 3.3; the substrate is glucose, and inorganic salts (i.e., ammonia) are the nitrogen source. The stoichiometry is given on a C-mole basis (i.e., glucose is specified as CH_2O), and the elemental composition of the biomass was calculated from the macromolecular composition to be $CH_{1.62}O_{0.53}N_{0.15}$ (see Table 3.3). The ATP and NADPH required for biomass synthesis are supplied by the catabolic pathways, and excess NADH formed in the biosynthetic reactions is, together with NADH formed in the catabolic pathways, reoxidized by transfer of electrons to oxygen via the electron transport chain. Reactions (3.3a2)

TABLE 3.3
Macromolecular Composition of *Saccharomyces Cerevisiae*

Macromolecule	Content [g (g DW)$^{-1}$]
Protein	0.39
Polysaccharides + trehalose	0.39
DNA + RNA	0.11
Phospholipids	0.05
Triacylglycerols	0.02
Sterols	0.01
Ash	0.03

through (3.3a5) specify the overall stoichiometry for the catabolic pathways. Reaction (3.3a2) specifies NADPH formation by the PP pathway, where glucose is completely oxidized to CO_2; reaction (3.3a3) is the overall stoichiometry for the combined EMP pathway and the TCA cycle; reaction (3.3a4) is the fermentative glucose metabolism, where glucose is converted to ethanol (this reaction only runs at high glucose uptake rates); and, finally, reaction (3.3a5) is the overall stoichiometry for the oxidative phosphorylation, where the P/O ratio is the overall (or operational) P/O ratio for the oxidative phosphorylation:

$$CO_2 + 2NADPH - CH_2O = 0 \quad (3.3a2)$$

$$CO_2 + 2NADH + 0.667ATP - CH_2O = 0 \quad (3.3a3)$$

$$CH_3O_{0.5} + 0.5CO_2 + 0.5ATP - 1.5CH_2O = 0 \quad (3.3a4)$$

$$P/O\,ATP - 0.5O_2 - NADH = 0 \quad (3.3a5)$$

Finally, consumption of ATP for maintenance is included simply as a reaction where ATP is used:

$$-ATP = 0 \quad (3.3a6)$$

Note that with the stoichiometry given on a C-mole basis, the stoichiometric coefficients extracted from the biochemistry (e.g., the formation of 2 moles ATP per mole glucose in the EMP pathway) are divided by 6, because glucose contains 6 C moles per mole.

Above, the stoichiometry is written as in Equation 3.3, but we can easily convert it to the more compact matrix notation of Equation 3.4:

$$\begin{pmatrix} -1.120 & 0 \\ -1 & 0 \\ -1 & 0 \\ -1.5 & 0 \\ 0 & -0.5 \\ 0 & 0 \end{pmatrix} \begin{pmatrix} S_{glc} \\ S_{O_2} \end{pmatrix} \begin{pmatrix} 0 & 0.120 \\ 0 & 1 \\ 0 & 1 \\ 1 & 0.5 \\ 0 & 0 \\ 0 & 0 \end{pmatrix} \begin{pmatrix} P_{eth} \\ P_{CO_2} \end{pmatrix}$$
$$+ \begin{pmatrix} 1 \\ 0 \\ 0 \\ 0 \\ 0 \\ 0 \end{pmatrix} X + \begin{pmatrix} Y_{xATP} & 0.397 & -0.212 \\ 0 & 0 & 2 \\ 0.667 & 2 & 0 \\ 0.5 & 0 & 0 \\ P/O & -1 & 0 \\ -1 & 0 & 0 \end{pmatrix} \begin{pmatrix} X_{ATP} \\ X_{NADH} \\ X_{NADPH} \end{pmatrix} = \begin{pmatrix} 0 \\ 0 \\ 0 \\ 0 \\ 0 \\ 0 \end{pmatrix}$$

$$(3.3a7)$$

where X represents the biomass.

We now collect the forward reaction rates for the six reactions in the rate vector v given by

$$v = \begin{pmatrix} \mu \\ v_{PP} \\ v_{EMP} \\ r_{eth} \\ v_{OP} \\ m_{ATP} \end{pmatrix}$$

In analogy with Equation 3.18, we balance the production and consumption of the three cofactors ATP, NADH, and NADPH. This gives the three equations:

$$-Y_{xATP}\mu + 0.667v_{EMP} + 0.5r_{eth}$$
$$+ P/Ov_{OP} - m_{ATP} = 0 \tag{3.3a8}$$

$$0.397\mu + 2v_{EMP} - v_{OP} = 0 \tag{3.3a9}$$

$$-0.212\mu + 2v_{PP} = 0 \tag{3.3a10}$$

Notice that these balances correspond to zero net specific formation rates for the three cofactors, and the three balances can therefore also be derived using Equation 3.12:

$$r_{met} = G^T v$$

$$= \begin{pmatrix} -Y_{xATP} & 0 & 0.667 & 0.5 & P/O & -1 \\ 0.397 & 0 & 2 & 0 & -1 & 0 \\ -0.212 & 2 & 0 & 0 & 0 & 0 \end{pmatrix} \begin{pmatrix} \mu \\ v_{PP} \\ v_{EMP} \\ r_{eth} \\ v_{OP} \\ m_{ATP} \end{pmatrix} \tag{3.3a11}$$

$$= \begin{pmatrix} 0 \\ 0 \\ 0 \end{pmatrix}$$

In addition to the three balances (3.3a8) through (3.3a10), we have the relationships between the reaction rates and the specific substrate uptake rates and the specific product formation rate given by Equations 3.3a5 and 3.3a6, or, using the matrix notation of Equations 3.3a9 and 3.3a10:

$$\begin{pmatrix} r_{glc} \\ r_{O_2} \end{pmatrix} = -\begin{pmatrix} -1.120 & -1 & -1 & -1.5 & 0 & 0 \\ 0 & 0 & 0 & 0 & -0.5 & 0 \end{pmatrix} \begin{pmatrix} \mu \\ v_{PP} \\ v_{EMP} \\ r_{eth} \\ v_{OP} \\ m_{ATP} \end{pmatrix} \tag{3.3a12}$$

$$= \begin{pmatrix} 1.120\mu + v_{PP} + v_{EMP} + 0.5r_{eth} \\ 0.5v_{OP} \end{pmatrix}$$

$$\begin{pmatrix} r_{eth} \\ r_{CO_2} \end{pmatrix} = \begin{pmatrix} 0 & 0 & 0 & 1 & 0 & 0 \\ 0.120 & 1 & 1 & 0.5 & 0 & 0 \end{pmatrix} \begin{pmatrix} \mu \\ v_{PP} \\ v_{EMP} \\ r_{eth} \\ v_{OP} \\ m_{ATP} \end{pmatrix} \tag{3.3a13}$$

$$= \begin{pmatrix} r_{eth} \\ 1.120\mu + v_{PP} + v_{EMP} + 0.5r_{eth} \end{pmatrix}$$

Clearly, the specific ethanol production rate is equal to the rate of reaction (3.3a5) because the stoichiometric coefficient for ethanol in this reaction is 1, and it is the only reaction where ethanol is involved. Using the combined set of Equations 3.3a11 through 3.3a13, the four reaction rates v_{EMP}, v_{PP}, v_{OP}, and m_{ATP} can be eliminated and the linear rate equations 3.3a14 through 3.3a16 can be derived:

$$r_{glc} = (a + 1.226)\mu + (1.5 - b)r_{eth} + c$$
$$= Y_{xs}^{true}\mu + Y_{ps}^{true}r_{eth} + m_s \tag{3.3a14}$$

$$r_{CO_2} = (a + 0.226)\mu + (0.5 - b)r_{eth} + c$$
$$= Y_{xc}^{true}\mu + Y_{pc}^{true}r_{eth} + m_c \tag{3.3a15}$$

$$r_{O_2} = (a + 0.229)\mu + br_{eth} + c$$
$$= Y_{xo}^{true}\mu + Y_{po}^{true}r_{eth} + m_o \tag{3.3a16}$$

The three common parameters a, b, and c are functions of the energetic parameters Y_{xATP}, m_{ATP}, and the P/O ratio according to Equations 3.3a17 through 3.3a19:

$$a = \frac{Y_{xATP} - 0.458P/O}{0.667 + 2P/O} \tag{3.3a17}$$

$$b = \frac{0.5}{0.667 + 2P/O} \tag{3.3a18}$$

$$c = \frac{m_{ATP}}{0.667 + 2P/O} \tag{3.3a19}$$

If there is no ethanol formation, which is the case at low specific glucose uptake rates, Equation 3.3a14 reduces to the linear rate Equation 3.18, but the parameters of the correlation are determined by basic energetic parameters of the cells. It is seen that the parameters in the linear correlations are coupled via the balances for ATP, NADH, and NADPH, and the three true yield coefficients cannot take any value. Furthermore, the maintenance coefficients are the same. This is due to the use of the units C-moles per C-mole biomass per hour for the specific rates. If other units are used for the specific rates, the maintenance coefficients will not take the same values but will remain proportional. This coupling of the parameters shows that there are only three degrees of freedom in the system, and one actually only has to determine two yield coefficients and one maintenance coefficient—the other parameters can be calculated using Equations 3.3a14 through 3.3a16.

The derived linear rate equations are certainly useful for correlating experimental data, but they also allow evaluation of the key energetic parameters Y_{xATP}, m_{ATP}, and the operational P/O ratio. Thus, if the true yield coefficients and the maintenance coefficients of Equations 3.3a14 through 3.3a16 are estimated, the values of a, b, and c can be found, and these three parameters relate the three energetic parameters through Equations 3.3a17 through 3.3a19. Thus, from one of the ethanol yield coefficients, b can be found, and, thereafter, the P/O ratio can be determined. Then m_{ATP} can be found from one of the maintenance coefficients, and finally Y_{xATP} can be found from one of the biomass yield coefficients. In practice, however, it is difficult to extract sufficiently precise values of the true yield coefficients from experimental data to estimate the energetic parameters—especially since the three parameters a, b, and c are closely correlated (especially b and c). However, if either the P/O ratio or Y_{xATP} is known, Equations 3.3a14 through 3.3a16 allow an estimation of the two remaining unknown energetic parameters. Consider the situation where there is no ethanol formation; here the true yield coefficient for biomass is 1.48 C-moles glucose (C-mole biomass)$^{-1}$ and the maintenance coefficient (equal to b) is

0.012 C-moles glucose (C-mole biomass h)$^{-1}$ (both values taken from Table 3.2). Thus, a is equal to 0.254 moles ATP (C-mole^{-1} biomass). If the operational P/O ratio is about 1.5 (which is a reasonable value for *S. cerevisiae*), we find that Y_{xATP} is 1.62 moles ATP (C-mole^{-1} biomass) or about 67 mmoles ATP (g DW^{-1}). Similarly, we find m_{ATP} to be about 2 mmoles (g DW h^{-1}).

In connection with baker's yeast production, it is important to maximize the yield of biomass on glucose:

$$Y_{sx} = \frac{\mu}{(a+1.226)\mu+(1.5-b)r_{eth}+c} \qquad (3.3a20)$$

Clearly, this can best be done if ethanol production is avoided. Thus, the glucose uptake rate is to be controlled below a level where there is respiro-fermentative metabolism. A very good indication of whether there is respiro-fermentative metabolism is the RQ:

$$RQ = \frac{(\alpha+0.226)\mu+(0.5-b)r_{eth}+c}{(a+0.229)\mu-br_{eth}+c} \qquad (3.3a21)$$

If there is no ethanol production, RQ will be close to 1 (independent of the specific growth rate), whereas if there is ethanol production, RQ will be above 1 and will increase with r_{eth} (b is always less than 0.5). From measurements of carbon dioxide and oxygen in the exhaust gas, the RQ can be evaluated. If it is above 1, there is respiro-fermentative metabolism resulting in ethanol formation and hence a low yield of biomass on sugar. This is caused by so-called glucose repression of respiration, where a high sugar concentration causes decreased activity of the respiration resulting in ethanol production due to overflow metabolism. Thus, if RQ is larger than 1 it means that the sugar concentration in the reactor must be reduced, and this can be done by reducing the feed rate to the reactor (typically baker's yeast production is operated as a fed-batch process; see Section 3.4).

3.2.4 BLACK BOX MODEL

In the black box model of cellular growth, all the cellular reactions are lumped into a single reaction. In this overall reaction, the stoichiometric coefficients are identical to the yield coefficients (see also Equation 3.25), and it can therefore be presented as

$$X + \sum_{i=1}^{M} Y_{xp_i} P_i - \sum_{i=1}^{N} Y_{xs_i} S_i = 0 \qquad (3.26)$$

Because the stoichiometric coefficient for biomass is 1, the forward reaction rate is given by the specific growth rate of the biomass, which together with the yield coefficients completely specifies the system. As discussed in Section 3.2.3, the yield coefficients are not constants, and the black box model can therefore not be applied to correlate, for instance, the specific substrate uptake rate with the specific growth rate. However, it is very useful for validation of experimental data because it can form the basis for setting up elemental balances. Thus, in the black box model there are ($M + N + 1$)

parameters: M yield coefficients for the metabolic products, N yield coefficients for the substrates, and the forward reaction rate μ. Because mass is conserved in the overall conversion of substrates to metabolic products and biomass, the ($M + N + 1$) parameters of the black box model are not completely independent but must satisfy several constraints. Thus, the elements flowing into the system must balance the elements flowing out of the system (e.g., the carbon entering the system via the substrates has to be recovered in the metabolic products and biomass). Each element considered in the black box obviously yields one constraint. Thus, a carbon balance gives

$$1 + \sum_{i=1}^{M} f_{p,i} Y_{xp_i} - \sum_{i=1}^{N} f_{s,i} Y_{xs_i} = 0 \qquad (3.27)$$

where $f_{s,i}$ and $f_{p,i}$ represent the carbon content (C-moles mole^{-1}) in the ith substrate and the ith metabolic product, respectively. In the above equation, the elemental composition of biomass is normalized with respect to carbon (i.e., it is represented by the form $CH_aO_bN_c$; see also Example 3.3a). The elemental composition of biomass depends on its macromolecular content and, therefore, on the growth conditions and the specific growth rate (e.g., the nitrogen content is much lower under nitrogen-limited conditions than under carbon-limited conditions; see Table 3.4). However, except for extreme situations, it is reasonable to use the general composition formula $CH_{1.8}O_{0.5}N_{0.2}$ whenever the biomass composition is not exactly known. Often, the elemental composition of substrates and metabolic products is normalized with respect to their carbon content, for example, glucose is specified as CH_2O (see also Example 3.3a). Equation 3.27 is then written on a per C-mole basis as

$$1 + \sum_{i=1}^{M} Y_{xp_i} - \sum_{i=1}^{N} Y_{xs_i} = 0 \qquad (3.28)$$

In Equation 3.28, the yield coefficients have units of C-moles per C-mole biomass. Conversion to this unit from other units is illustrated in Box 3.3. Equation 3.28 is very useful for checking the consistency of experimental data. Thus, if the sum of carbon in the biomass and the metabolic products does not equal the sum of carbon in the substrates, there is an inconsistency in the experimental data.

Similar to Equation 3.27, balances can be written for all other elements participating in the conversion (3.26). Thus, the hydrogen balance will read

$$a_x + \sum_{i=1}^{M} a_{p,i} Y_{xp_i} - \sum_{i=1}^{N} a_{s,i} Y_{xs_i} = 0 \qquad (3.29)$$

where $a_{s,i}$, $a_{p,i}$, and a_x represent the hydrogen content (moles C-mole^{-1} if a C-mole basis is used) in the ith substrate, the ith metabolic product, and the biomass, respectively. Similarly, we have for the oxygen and nitrogen balances

TABLE 3.4

Elemental Composition of Biomass for *Several* Microorganisms

Microorganism	Elemental Composition	Ash Content (w/w %)	Growth Conditions
Candida utilis	$CH_{1.83}O_{0.46}N_{0.19}$	7.0	Glucose limited, $D = 0.05$ h^{-1}
	$CH_{1.87}O_{0.56}N_{0.20}$	7.0	Glucose limited, $D = 0.45$ h^{-1}
	$CH_{1.83}O_{0.54}N_{0.10}$	7.0	Ammonia limited, $D = 0.05$ h^{-1}
	$CH_{1.87}O_{0.56}N_{0.20}$	7.0	Ammonia limited, $D = 0.45$ h^{-1}
Klebsiella aerogenes	$CH_{1.75}O_{0.43}N_{0.22}$	3.6	Glycerol limited, $D = 0.10$ h^{-1}
	$CH_{1.73}O_{0.43}N_{0.24}$	3.6	Glycerol limited, $D = 0.85$ h^{-1}
	$CH_{1.75}O_{0.47}N_{0.17}$	3.6	Ammonia limited, $D = 0.10$ h^{-1}
	$CH_{1.73}O_{0.43}N_{0.24}$	3.6	Ammonia limited, $D = 0.80$ h^{-1}
Saccharomyces cerevisiae	$CH_{1.82}O_{0.58}N_{0.16}$	7.3	Glucose limited, $D = 0.080$ h^{-1}
	$CH_{1.78}O_{0.60}N_{0.19}$	9.7	Glucose limited, $D = 0.255$ h^{-1}
	$CH_{1.94}O_{0.52}N_{0.25}$	5.5	Unlimited growth
Escherichia coli	$CH_{1.77}O_{0.49}N_{0.24}$	5.5	Unlimited growth
	$CH_{1.83}O_{0.50}N_{0.22}$	5.5	Unlimited growth
	$CH_{1.96}O_{0.55}N_{0.25}$	5.5	Unlimited growth
	$CH_{1.93}O_{0.55}N_{0.25}$	5.5	Unlimited growth
Pseudomonas fluorescens	$CH_{1.83}O_{0.55}N_{0.26}$	5.5	Unlimited growth
Aerobacter aerogenes	$CH_{1.64}O_{0.52}N_{0.16}$	7.9	Unlimited growth
Penicillium chrysogenum	$CH_{1.70}O_{0.58}N_{0.15}$		Glucose limited, $D = 0.038$ h^{-1}
	$CH_{1.68}O_{0.53}N_{0.17}$		Glucose limited, $D = 0.098$ h^{-1}
Aspergillus niger	$CH_{1.72}O_{0.55}N_{0.17}$	7.5	Unlimited growth
Average	$CH_{1.81}O_{0.52}N_{0.21}$	6.0	

Source: Compositions for *P. chrysogenum* are taken from Christensen et al. 1995. *J. Biotechnol.* 42:95–107; other data are taken from Roels, J.A. 1983. *Energetics and Kinetics in Biotechnology.* Amsterdam: Elsevier Biomedical Press.

$$b_x + \sum_{i=1}^{M} b_{p,i} Y_{xp_i} - \sum_{i=1}^{N} b_{s,i} Y_{xs_i} = 0 \qquad (3.30)$$

BOX 3.3 CALCULATION OF YIELDS WITH RESPECT TO C-MOLE BASIS

Yield coefficients are typically described as moles (g DW)$^{-1}$ or g (g DW)$^{-1}$. To convert the yield coefficients to a C-mole basis, information on the elemental composition and the ash content of biomass is needed. To illustrate the conversion, we calculate the yield of 0.5 g DW biomass (g glucose)$^{-1}$ on a C-mole basis. First, we convert the g DW biomass to an ash-free basis, that is determine the amount of biomass that is made up of carbon, nitrogen, oxygen, and hydrogen (and, in some cases, also phosphorus and sulfur). With an ash content of 8% we have 0.92 g ash-free biomass (g DW biomass)$^{-1}$, which gives a yield of 0.46 g ash-free biomass (g glucose)$^{-1}$. This yield can now be directly converted to a C-mole basis using the molecular weights in g C-mole^{-1} for ash-free biomass and glucose. With the standard elemental composition for biomass of $CH_{1.8}O_{0.5}N_{0.2}$, we have a molecular weight of 24.6 g ash- free biomass C-mole^{-1},

and therefore find a yield of 0.46/24.6 = 0.0187 C-moles biomass (g glucose)$^{-1}$. Finally, by multiplication with the molecular weight of glucose on a C-mole basis (30 g C-mole^{-1}), a yield of 0.56 C-moles biomass (C-mole glucose)$^{-1}$ is found.

$$c_x + \sum_{i=1}^{M} c_{p,i} Y_{xp_i} - \sum_{i=1}^{N} c_{s,i} Y_{xs_i} = 0 \qquad (3.31)$$

where $b_{s,i}$, $b_{p,i}$, and b_x represent the oxygen content (moles C-mole^{-1}) in the ith substrate, the ith metabolic product, and the biomass, respectively; and $c_{s,i}$, $c_{p,i}$, and c_x represent the nitrogen content (moles C-mole15015^{-1}) in the ith substrate, the ith metabolic product, and the biomass, respectively. Normally, only these four balances are considered; balances for phosphate and sulfate may also be set up, but generally these elements are of minor importance. The four elemental balances (3.28) through (3.31) can be conveniently written by collecting the elemental composition of biomass, substrates, and metabolic products in the columns of a matrix **E**, where the first column contains the elemental composition of biomass, columns 2 through $M+1$ contain the elemental composition of the M metabolic products, and columns $M+2$ to

$M+N+1$ contain the elemental composition of the N substrates. With the introduction of this matrix, the four elemental balances can be expressed as

$$\mathbf{E\,Y} = 0 \qquad (3.32)$$

where $\mathbf{Y}$ is a vector containing the yield coefficients (the substrate yield coefficients are given with a minus sign). With $N+M+1$ variables, $N+M$ yield coefficients and the forward rate of reaction (3.26) and four constraints, the degree of freedom is $F = M+N+1-4$. If exactly F variables are measured, it may be possible to calculate the other rates by using the four algebraic equations given by (3.32), but, in this case, there are no redundancies left to check the consistency of the data. For this reason, it is advisable to strive for more measurements than the degrees of freedom of the system.

Example 3.3b: Elemental Balances in a Simple Black Box Model

Consider the aerobic cultivation of the yeast *S. cerevisiae* on a defined, minimal medium (i.e., glucose is the carbon and energy source and ammonia is the nitrogen source). During aerobic growth, the yeast oxidizes glucose completely to carbon dioxide. However, as mentioned in Example 3.3a, high glucose concentrations cause repression of the respiratory system, resulting in ethanol formation. Thus, at these conditions, both ethanol and carbon dioxide should be considered as metabolic products. Finally, water is formed in the cellular pathways. This is also included as a product in the overall reaction. Thus, the black box model for this system is

$$X + Y_{xe}\,\text{ethanol} + Y_{xc}CO_2 + Y_{xw}H_2O$$
$$- Y_{xs}\,\text{glucose} - Y_{xo}O_2 - Y_{xN}NH_3 \qquad (3.3b1)$$
$$= 0$$

which can be represented with the yield coefficient vector

$$\mathbf{Y} = \left(1\,Y_{xe}\,Y_{xc}\,Y_{xw}\,-Y_{xs}\,-Y_{xo}\,-Y_{xN}\right)T \qquad (3.3b2)$$

We now rewrite the conversion using the elemental composition of the substrates and metabolic products. For biomass, we use the elemental composition of $CH_{1.83}O_{0.56}N_{0.17}$, and therefore we have

$$CH_{1.83}O_{0.56}N_{0.17} + Y_{xe}CH_3O_{0.5} + Y_{xc}CO_2$$
$$+ Y_{xw}H_2O - Y_{xs}CH_2O - Y_{xo}O_2 - Y_{xN}NH_3 \qquad (3.3b3)$$
$$= 0$$

Some may find it difficult to identify $CH_3O_{0.5}$ as ethanol, but the advantage of using the C-mole basis becomes apparent immediately when we look at the carbon balance:

$$1 + Y_{xe} + Y_{xc} - Y_{xs} = 0 \qquad (3.3b4)$$

This simple equation is very useful for checking the consistency of experimental data. Thus, using the classical data of von Meyenburg (1969), we find $Y_{xe} = 0.713$, $Y_{xc} = 1.313$, and $Y_{xs} = 3.636$ at a dilution rate of $D = 0.3$ h^{-1} in a glucose-limited continuous culture. Obviously, the data are not consistent as the carbon balance is not closed. A more elaborate data analysis (Nielsen and Villadsen, 1994) suggests that the missing carbon may be accounted for by ethanol evaporation or stripping due to intensive aeration of the bioreactor.

Similarly, using Equation 3.31, we find that a nitrogen balance gives

$$Y_{xN} = 0.17 \qquad (3.3b5)$$

If the yield coefficients for ammonia uptake and biomass formation do not conform to Equation 3.3b5, an inconsistency is identified in one of these two measurements, or the nitrogen content of the biomass is different from that specified.

We now write all four elemental balances in terms of the matrix equation 3.32:

$$\mathbf{E} = \begin{pmatrix} 1 & 1 & 1 & 0 & 1 & 0 & 0 \\ 1.83 & 3 & 0 & 2 & 2 & 0 & 3 \\ 0.56 & 0.5 & 2 & 1 & 1 & 2 & 0 \\ 0.17 & 0 & 0 & 0 & 0 & 0 & 1 \end{pmatrix} \begin{matrix} \leftarrow \text{carbon} \\ \leftarrow \text{hydrogen} \\ \leftarrow \text{oxygen} \\ \leftarrow \text{nitrogen} \end{matrix} \qquad (3.3b6)$$

where the rows indicate, respectively, the content of carbon, hydrogen, oxygen, and nitrogen and the columns give the elemental composition of biomass, ethanol, carbon dioxide, water, glucose, oxygen, and ammonia, respectively. Using Equation 3.32, we find

$$\begin{pmatrix} 1 & 1 & 1 & 0 & 1 & 0 & 0 \\ 1.83 & 3 & 0 & 2 & 2 & 0 & 3 \\ 0.56 & 0.5 & 2 & 1 & 1 & 2 & 0 \\ 0.17 & 0 & 0 & 0 & 0 & 0 & 1 \end{pmatrix} \begin{pmatrix} 1 \\ Y_{xe} \\ Y_{xc} \\ Y_{xw} \\ -Y_{xs} \\ -Y_{xo} \\ -Y_{xN} \end{pmatrix}$$

$$= \begin{pmatrix} 1 + Y_{xe} + Y_{xc} - Y_{xs} \\ 1.83 + 3Y_{xe} + 2Y_{xw} - 2Y_{xs} - 3Y_{xN} \\ 0.56 + 0.5Y_{xe} + 2Y_{xc} - Y_{xw} - 2Y_{xo} \\ 0.17 - Y_{xN} \end{pmatrix} \qquad (3.3b7)$$

$$= \begin{pmatrix} 0 \\ 0 \\ 0 \\ 0 \end{pmatrix}$$

The first and last rows are identical to the balances derived in Equations 3.3b4 and 3.3b5 for carbon and nitrogen, respectively. The balances for hydrogen and oxygen introduce two additional constraints. However, because the rate of water formation is impossible to measure, one of these equations must be used to calculate this rate (or yield). This leaves only one additional constraint from these two balances.

3.3 MASS BALANCES FOR BIOREACTORS

In Section 3.2, we derived equations that relate the rates of the intracellular reaction with the rates of substrate uptake, metabolic product formation, and biomass formation. These rates are the key elements in the dynamic mass balances for the substrates, the metabolic products, and the biomass, which describes the change in time of the concentration of these state variables in a bioreactor. The bioreactor may be any type of device, ranging from a shake flask to a well-instrumented bioreactor. Figure 3.2 is a general representation of a bioreactor. It has a volume V (unit: L), and it is fed with a stream of fresh, sterile medium with a flow rate F (unit: L h^{-1}). Spent medium is removed with a flow rate of F_{out} (unit: L h^{-1}). The medium in the bioreactor is assumed to be completely (or ideally) mixed, that is, there is no spatial variation in the concentration of the different medium compounds. For small-volume bioreactors (<1 L) (including shake flasks), this can generally be achieved through aeration and some agitation, whereas for laboratory stirred-tank bioreactors (1–10 L) special designs may have to be introduced in order to ensure a homogeneous medium (Sonnleitner and Fiechter, 1988; Nielsen and Villadsen, 1993). The bioreactor may be operated in many different modes of which we will only consider the three most common:

Batch, where $F = F_{out} = 0$ (i.e., the volume is constant)
Continuous, where $F = F_{out} \neq 0$ (i.e., the volume is constant)
Fed-batch (or semibatch), where $F > 0$ and $F_{out} = 0$ (i.e., the volume increases)

These three different modes of reactor operation are discussed separately later, but first we derive general dynamic

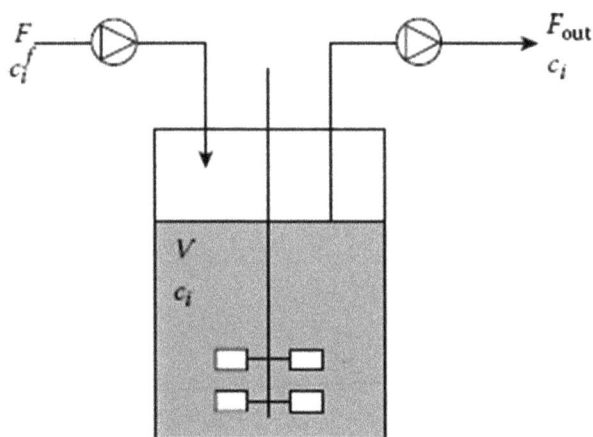

FIGURE 3.2 Bioreactor with addition of fresh, sterile medium and removal of spent medium where c_i^f is the concentration of the ith compound in the feed and c_i is the concentration of the ith compound in the spent medium. The bioreactor is assumed to be very well mixed (or ideal), so that the concentration of each compound in the spent medium becomes identical to its concentration in the bioreactor.

mass balances for the substrates, metabolic products, biomass constituents, and biomass.

3.3.1 DYNAMIC MASS BALANCES

The basis for derivation of the general dynamic mass balances is the mass balance equation

$$\text{Accumulated} = \text{Net formation rate} + \text{In} - \text{Out} \quad (3.33)$$

where the first term on the RHS is given by Equations 3.5 through 3.8 for substrate, metabolic product, biomass constituents, and intracellular metabolites, respectively. The term "In" represents the flow of the compound into the bioreactor, and the term "Out" the flow of the compound out from the bioreactor. In the following, we consider substrates, metabolic products, biomass constituents, intracellular metabolites, and the total biomass separately.

We consider the ith substrate, which is added to the bioreactor via the feed and is consumed by the cells present in the bioreactor. The mass balance for this compound is

$$\frac{d\left(c_{s,i}V\right)}{dt} = -r_{s,i}xV + Fc_{s,i}^f - F_{out}c_{s,i} \quad (3.34)$$

where r_i is the specific consumption rate of the compound (unit: moles (g DW h^{-1})); $c_{s,i}$ is the concentration in the bioreactor, which is assumed to be the same as the concentration in the outlet (unit: moles L^{-1}), $c_{s,i}^f$ (i) is the concentration in the feed (unit: moles L^{-1}); and x is the biomass concentration in the bioreactor (unit: g DW L^{-1}). The first term in Equation 3.34 is the accumulation term, the second term is the consumption (or reaction) term, the third term is accounting for the inlet, and the last term is accounting for the outlet. Rearrangement of this equation gives

$$\frac{dc_{s,i}}{dt} = -r_{s,i}x + \frac{F}{V}c_{s,i}^f - \left(\frac{F_{out}}{V} + \frac{1}{V}\frac{dV}{dt}\right)c_{s,i} \quad (3.35)$$

For a fed-batch reactor,

$$F = \frac{dV}{dt} \quad (3.36)$$

and $F_{out} = 0$. So the term within the parentheses becomes equal to the so-called dilution rate given by

$$D = \frac{F}{V} \quad (3.37)$$

For a continuous and a batch reactor, the volume is constant (i.e., $dV/dt = 0$, and $F = F_{out}$), and so for these bioreactor modes also the term within the parentheses becomes equal to the dilution rate (which, however, is zero for a batch reactor). Equation 3.35 therefore reduces to the mass balance (3.38) for any type of operation:

$$\frac{dc_{s,i}}{dt} = -r_{s,i}x + D\left(c_{s,i}^f - c_{s,i}\right) \tag{3.38}$$

The first term on the right-hand side of Equation 3.38 is the volumetric rate of substrate consumption, which is given as the product of the specific rate of substrate consumption and the biomass concentration. The second term accounts for the addition and removal of substrate from the bioreactor. The term on the left-hand side of Equation 3.38 is the accumulation term, which accounts for the change in time of the substrate, which in a batch reactor (where $D = 0$) equals the volumetric rate of substrate consumption.

Dynamic mass balances for the metabolic products are derived in analogy with those for the substrates and take the form

$$\frac{dc_{p,i}}{dt} = r_{p,i}x + D\left(c_{p,i}^f - c_{p,i}\right) \tag{3.39}$$

where the first term on the right-hand side is the volumetric formation rate of the ith metabolic product. Normally, the metabolic products are not present in the sterile feed to the bioreactor, and $c_{p,i}^f$ is therefore often zero. In these cases, the volumetric rate of product formation in a steady-state continuous reactor is equal to the dilution rate multiplied by the concentration of the metabolic product in the bioreactor (equal to that in the outlet).

With sterile feed, the mass balance for the total biomass is derived directly:

$$\frac{dx}{dt} = (\mu - D)x \tag{3.40}$$

where μ (unit: h^{-1}) is the specific growth rate of the biomass given by Equation 3.13.

For the biomass constituents, we normally use the biomass as the reference (i.e., their concentrations are given with the biomass as the basis). In this case, the mass balance for the ith biomass constituent is derived from (sterile feed is assumed)

$$\frac{d\left(X_{\text{macro},i}xV\right)}{dt} = r_{\text{macro},i}xV - F_{\text{out}}X_{\text{macro},i}x \tag{3.41}$$

where $X_{\text{macro},i}x$ is the concentration of the ith biomass component in the bioreactor (unit: $g\ L^{-1}$) and r_{macro},i is the specific net rate of formation of the ith biomass constituent. Rearrangement of Equation 3.41 gives

$$\frac{dX_{\text{macro},i}}{dt} = r_{\text{macro},i} - \left(\frac{F_{\text{out}}}{V} + \frac{1}{x}\frac{dx}{dt} + \frac{1}{V}\frac{dV}{dt}\right)X_{\text{macro},i} \tag{3.42}$$

Again, we have that for any mode of bioreactor operation:

$$D = \frac{F_{\text{out}}}{V} + \frac{1}{V}\frac{dV}{dt} \tag{3.43}$$

which, together with the mass balance (3.40) for the total biomass concentration, gives the mass balance:

$$\frac{dX_{\text{macro},i}}{dt} = r_{\text{macro},i} - \mu X_{\text{macro},i} \tag{3.44}$$

where $X_{\text{macro},i}$ is the concentration of the ith biomass constituent within the biomass. Different units may be applied for the concentrations of the biomass constituents, but they are normally given as $g\ (g\ DW)^{-1}$, because then the sum of all the concentrations equals 1, that is,

$$\sum_{i=1}^{Q} X_{\text{macro},i} = 1 \tag{3.45}$$

Furthermore, this unit corresponds with the experimentally determined macromolecular composition of cells, where weight fractions are generally used. In Equation 3.44, it is observed that the mass balance for the biomass constituents is completely independent of the mode of operation of the bioreactor (i.e., the dilution rate does not appear in the mass balance). However, there is indirectly a coupling via the last term, which accounts for dilution of the biomass constituents when the biomass expands due to growth. Thus, if there is no net synthesis of a macromolecular pool, but the biomass still grows, the intracellular level decreases.

For intracellular metabolites, it is not convenient to use the same unit for their concentrations as for the biomass constituents. These metabolites are dissolved in the matrix of the cell; therefore, it is more appropriate to use the unit moles per liquid cell volume for the concentrations. The intracellular concentration can then be compared directly with the affinities of enzymes, typically quantified by their Km values, which are normally given with the unit moles per liter. If the concentration is known in one unit, it is, however, easily converted to another unit if the density of the biomass (in the range of 1 g cell per mL cell) and the water content (in the range of 0.67 ml water per ml cell) is known. Even though a different unit is applied, the biomass is still the basis, and the mass balance for the intracellular metabolites therefore takes the same form:

$$\frac{dX_{\text{met},i}}{dt} = r_{\text{met},i} - \mu X_{\text{met},i} \tag{3.46}$$

where X_{met},i is the concentration of the ith intracellular metabolite. It is important to distinguish between concentrations of intracellular metabolites given in moles per liquid reactor volume and in moles per liquid cell volume. If concentrations are given in the former unit, the mass balance will be completely different.

3.3.2 BATCH REACTOR

This is the classical operation of the bioreactor and many life scientists use it because it can be carried out in a relatively simple experimental setup. Batch experiments have the advantage of being easy to perform, and by using shake flasks a large number of parallel experiments can be carried out. The disadvantage is that the experimental data are difficult to interpret because there are dynamic conditions throughout the experiment (i.e., the environmental conditions experienced by the cells vary with time). However, by using well-instrumented bioreactors, at least some variables (e.g., pH and dissolved oxygen tension) may be kept constant.

As mentioned in the previous section, the dilution rate is zero for a batch reactor, and the mass balances for the biomass and the limiting substrate therefore take the form

$$\frac{dx}{dt} = \mu x \quad ; \quad x(t=0) = x_0 \tag{3.47}$$

$$\frac{dc_s}{dt} = -r_s x \quad ; \quad c_s(t=0) = c_{s,0} \tag{3.48}$$

where x_0 indicates the initial biomass concentration, which is obtained immediately after inoculation, and $c_{s,0}$ is the initial substrate concentration. According to the mass balance, the biomass concentration will increase as indicated in Figure 3.3 and the substrate concentration will decrease until its concentration reaches zero and growth stops. Because the substrate concentration is zero at the end of the cultivation, the overall yield of biomass on the substrate can be found from

$$Y_{sx}^{\text{overall}} = \frac{x_{\text{final}} - x_0}{c_{s,0}} \tag{3.49}$$

where x final is the biomass concentration at the end of the cultivation. Normally, $x_0 \ll x_{\text{final}}$, and the overall yield coefficient can therefore be estimated from the final biomass concentration and the initial substrate concentration alone. Notice that the yield coefficient determined from the batch experiment is the overall yield coefficient and not Y_{sx} or $\left(Y_{xs}^{\text{true}}\right)^{-1}$.

The yield coefficient Y_{sx} may well be time dependent as it is the ratio between the specific growth rate and the substrate uptake rate; see Equation 3.16. However, if there is little variation in these rates during the batch culture (e.g., if there is a long exponential growth phase and only a very short declining growth phase), the overall yield coefficient may be very similar to the yield coefficient. The true yield coefficient, on the other hand, is difficult to determine from batch cultivation because it requires information about the maintenance coefficients, which can hardly be determined from a batch experiment. However, in batch cultivation, the specific growth rate is close to its maximum throughout most of the growth phase, and the substrate consumption due to maintenance is therefore negligible. According to Equation 3.17, the true yield coefficient is close to the observed yield coefficient determined from the final biomass concentration.

3.3.3 Chemostat

A typical operation of the continuous bioreactor is the so-called chemostat, where the added medium is designed such that there is a single limiting substrate. This allows for controlled variation in the specific growth rate of the biomass. The advantage of the continuous bioreactor is that a steady state can be achieved, which allows for precise experimental determination of specific rates. Furthermore, by varying the feed flow rate to the bioreactor the environmental conditions can be varied, and valuable information concerning the influence of the environmental conditions on the cellular physiology can be obtained. The continuous bioreactor is attractive for industrial applications because the productivity can be high. However, often the titer (i.e., the product concentration) is lower than can be obtained in the fed-batch reactor, and it is therefore a trade-off between productivity and titer. Furthermore, it is rarely used in industrial processes because it is sensitive to contamination (e.g., via the feed stream) and to the appearance of spontaneously formed mutants that may outcompete the production strain. Other examples of continuous operation besides the chemostat are the pH-stat, where the feed flow is adjusted to maintain the pH constant in the bioreactor, and the turbidostat, where the feed flow is adjusted to maintain the biomass concentration at a constant level.

From the biomass mass balance (3.40), it can easily be seen that in a steady-state continuous reactor, the specific growth rate equals the dilution rate:

$$\mu = D \tag{3.50}$$

Thus, by varying the dilution rate (or the feed-flow rate) in a continuous culture, different specific growth rates can be obtained. This allows detailed physiological studies of the cells when they are grown at a predetermined specific growth rate (corresponding to a certain environment experienced by the cells). At steady state, the substrate mass balance (3.38) gives

$$0 = -r_s x + D\left(c_s^f - c_s\right) \tag{3.51}$$

which, upon combination with Equation (3.50) and the definition of the yield coefficient, directly gives

$$x = Y_{sx}\left(c_s^f - c_s\right) \tag{3.52}$$

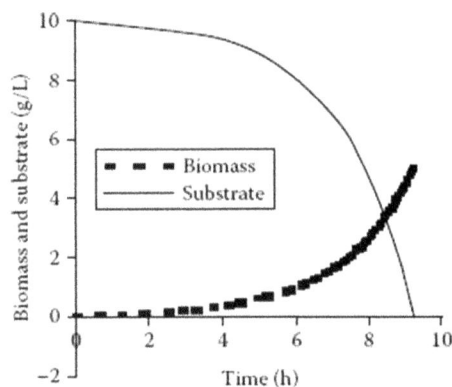

FIGURE 3.3 Batch fermentation described with Monod kinetics. The biomass concentration is found using Equation (c2) and the corresponding substrate concentration is found from Equation (c1); μ_{max} is 0.5 h^{-1}, K_s is 50 mg l^{-1} (a quite high value), and Y_{sx} is 0.5 g g^{-1}. The initial substrate concentration c_s, 0 is 10 g l^{-1}. The substrate concentration decreases from 0.5 g l^{-1} to 0 in less than 5 min, and this is the interesting substrate concentration range for estimation of K_s.

Thus, the yield coefficient can be determined from measurement of the biomass and the substrate concentrations in the bioreactor (the substrate concentration in the feed flow should generally be known as it is determined in the setup of the experiment).

Besides the advantage for obtaining steady-state measurements, the chemostat is well suited to study dynamic conditions because it is possible to perform well-controlled transients. Thus, it is possible to study the cellular response to a sudden increase in the substrate concentration by adding a pulse of the limiting substrate to the reactor or to a sudden change in the dilution rate. These experiments both start and end with a steady state, so the initial and end conditions are well characterized, and this facilitates the interpretation of the cellular response. One type of transient experiment is especially suited to determining an important kinetic parameter, namely, the maximum specific growth rate. By increasing the dilution rate to a value above μ_{max}, the cells will wash out from the bioreactor and the substrate concentration will increase (and eventually reach the same value as in the feed). After adaptation of the cells to the new conditions, they will attain their maximum specific growth rate and the dynamic mass balance for the biomass becomes

$$\frac{dx}{dt} = \left(\mu_{max} - D\right)x \qquad (3.53)$$

$$\frac{x\left(t-t_0\right)}{x\left(t_0\right)} = \exp\left(\left(\mu_{max} - D\right)\left(t - t_0\right)\right) \qquad (3.54)$$

where t_0 is the time at which the cells have become adapted to the new conditions and grow at their maximum specific growth rate. Thus, the maximum specific growth rate is easily determined from a plot of the biomass concentration versus time on a semi-log plot.

3.3.4 FED-BATCH REACTOR

This operation is probably the most common in industrial processes because it allows for control of the environmental conditions, for example, maintaining the glucose concentration at a certain level, as well as enabling the formation of very high titers (up to several hundred grams per liters of some metabolites), which is important for subsequent downstream processing. For a fed-batch reactor, the mass balances for biomass and substrate are given by Equations (3.38) and (3.40). Normally, the feed concentration is very high (i.e., the feed is a very concentrated solution) and the feed flow is low, giving a low dilution rate.

For the fed-batch reactor, the dilution rate is given by

$$D = \frac{1}{V}\frac{dV}{dt} \qquad (3.55)$$

To keep D constant, there needs to be an exponentially increasing feed flow to the bioreactor, which is normally practically impossible as it will result in exponentially increasing rates, for example, growth rate as the biomass concentration is

increasing, and this may lead to oxygen limitations and problems with keeping temperature constant due to limitations in cooling capacity. The feed flow is therefore often adjusted or increased until limitations in the oxygen supply set in, at which point the feed flow is kept constant. This will give a decreasing specific growth rate. However, because the biomass concentration usually increases, the volumetric uptake rate of substrates (including oxygen) may be kept approximately constant. From the above, it is quite clear that there may be many different feeding strategies in a fed-batch process, and optimization of the operation is a complex problem that is difficult to solve empirically. Even when a very good process model is available, calculation of the optimal feeding strategy is a complex optimization problem. In an empirical search for the optimal feeding policy, the two most obvious criteria are

- Keep the concentration of the limiting substrate constant.
- Keep the volumetric growth rate of the biomass (or uptake of a given substrate) constant.

A constant concentration of the limiting substrate is often applied if the substrate inhibits product formation and the chosen concentration is therefore dependent on the degree of inhibition and the desire to maintain a certain growth of the cells. A constant volumetric growth rate (or uptake of a given substrate) is applied if there are limitations in the supply of oxygen or in heat removal.

Fed-batch cultures were used in the production of baker's yeast as early as 1915. The method was introduced by Dansk Gæringsindustri and is therefore sometimes referred to as the Danish method. It was recognized that an excess of malt in the medium would lead to a higher growth rate, resulting in an oxygen demand in excess of what could be met in the fermentors. This resulted in limitations in respiratory metabolism of the yeast, leading to ethanol formation at the expense of biomass production. The yeast was allowed to grow in an initially weak medium to which additional medium was added at a rate less than the maximum rate at which the organism could use it. In modern fed-batch processes for yeast production, the feed of molasses is under strict control, based on the automatic measurement of traces of ethanol in the exhaust gas of the bioreactor. Although such systems may result in low specific growth rates, the biomass yield is generally close to the maximum obtainable, and this is especially important in the production of baker's yeast, where there is much focus on the yield. Apart from the production of baker's yeast, the fed-batch process is used today for the production of secondary metabolites (where penicillin is a prominent example), industrial enzymes, and many other products derived from cultivation processes.

3.4 KINETIC MODELS

Kinetic modeling expresses verbally or mathematically correlations between rates and reactant or product concentrations that, when inserted into the mass balances derived in Section 3.3, permit a prediction of the degree of conversion of substrates and the yield of individual products at other operating conditions. If the rate expressions are correctly set-up, it may be possible to express the course of an entire fermentation experiment based on initial values for the components of the state vector (e.g., concentration of substrates). This leads to simulations, which may finally result in an optimal design of the equipment or an optimal mode of operation for a given system. The basis of kinetic modeling is to express functional relationships between the forward reaction rates **v** of the reactions considered in the model and the concentrations of the substrates, metabolic products, biomass constituents, intracellular metabolites, and/or biomass:

$$v_i = f_i\left(c_s, c_p X_{macro}, X_{met}, x\right) \qquad (3.56)$$

If, during the cultivation, the biomass composition remains constant, then the rates of the internal reactions must necessarily be proportional. This is referred to as balanced growth. In this case the growth process can be described in terms of a single variable that defines the state of the biomass. This variable is quite naturally chosen as the biomass concentration x (g DW 1^{-1}). This is the basis of the so-called unstructured models that have proved adequate during 50 years of practical application to design cultivation processes (especially steady-state or batch cultivations), to install suitable control devices, and to estimate which process conditions are likely to give the best return on the investment in process equipment. However, these unstructured models generally have poor predictive strength and as such are of little value in fundamental studies of cellular function.

3.4.1 Degree of Model Complexity

A typical discussion on the complexity of mathematical modeling of biochemical systems may be initiated by asking the question of whether a mechanistic model or an empirical model should be applied. To illustrate this, consider the fractional saturation y of a protein at a ligand concentration c_l. This may be described by the Hill equation (Hill, 1910):

$$y = \frac{c_l^h}{c_l^h + K} \qquad (3.57)$$

where h and K are empirical parameters. Alternatively, the fractional saturation may be described by the equation of Monod et al. (1965):

$$y = \frac{\left(La\left(1 + \dfrac{ac_l}{K_R}\right)^3 + \left(1 + \dfrac{c_l}{K_R}\right)^3\right)\dfrac{c_l}{K_R}}{L\left(1 + \dfrac{ac_l}{K_R}\right)^4 + \left(1 + \dfrac{c_l}{K_R}\right)^4} \qquad (3.58)$$

where L, a, and K_R are parameters. Both equations address the same experimental problem, but whereas Equation 3.57 is completely empirical with h and K as fitted parameters,

Equation 3.58 is derived from a hypothesis for the mechanism; the parameters therefore have a direct physical interpretation. If the aim of the modeling is to understand the underlying mechanism of the process, Equation 3.57 can obviously not be applied because the kinetic parameters are completely empirical and give no (or little) information about the ligand binding to the protein. In this case, Equation 3.58 should be applied because by estimating the kinetic parameter the investigator is supplied with valuable information about the system and the parameters can be directly interpreted.

If, on the other hand, the aim of the modeling is to simulate the ligand binding to the protein, Equation 3.57 may be as good as Equation 3.58—and this equation may even be preferable because it is simpler in structure, has fewer parameters, and actually often gives a better fit to experimental data than Equation 3.57. Thus, the answer to which model is preferred depends on the aim of the modeling exercise. The same can be said about the unstructured growth models (Section 3.4.2), which are completely empirical but are valuable for extracting key kinetic parameters for growth. Furthermore, they are well suited to simple design problems and for teaching.

If the aim is to simulate dynamic growth conditions, one may turn to simple structured models (Section 4.4.3), for example, the compartment models, which are also useful for an illustration of structured modeling in the classroom. However, if the aim is to analyze a given system in further detail, it is necessary to include far more structure in the model. In this case one often describes only individual processes within the cell, such as a certain pathway or gene transcription from a certain promoter. Similarly, if the aim is to investigate the interaction between different cellular processes (e.g., the influence of a plasmid copy number on chromosomal DNA replication), a single-cell model (Section 3.4.4) has to be applied.

Finally, if the aim is to look into population distributions, which in some cases may have an influence on growth or production kinetics, either a segregated or a morphologically structured model has to be applied (Section 3.5).

3.4.2 Unstructured Models

Even when there are many substrates, one of these substrates is usually limiting (i.e., the rate of biomass production depends exclusively on the concentration of this substrate). At low concentrations, c_s of this substrate μ is proportional to c_s, but for increasing values of c_s an upper value μ_{max} for the specific growth rate is gradually reached. This is the verbal formulation of the Monod (1942) model:

$$\mu = \mu_{max}\frac{c_s}{c_s + K_s} \qquad (3.59)$$

which has been shown to correlate fermentation data for many different microorganisms. In the Monod model, K_s is that value of the limiting substrate concentration at which the specific growth rate is half its maximum value. Roughly speaking, it divides the μ versus c_s plot into a low substrate

TABLE 3.5

Compilation of K_s Values for Sugars

Species	Substrate	K_s (mg l^{-1})
Aerobacter aerogenes	Glucose	8
Escherichia coli	Glucose	4
Klebsiella aerogenes	Glucose	9
	Glycerol	9
Klebsiella oxytoca	Glucose	10
	Arabinose	50
	Fructose	10
Lactococcus cremoris	Glucose	2
	Lactose	10
	Fructose	3

Source: Values are taken from Nielsen, J. and J. Villadsen. 1994. *Bioreaction Engineering Principles.* New York: Plenum Press.

concentration range where the specific growth rate is strongly (almost linearly) dependent on c_s, and a high substrate concentration range where μ is independent of c_s.

When glucose is the limiting substrate, the value of K_s is normally in the micromolar range (corresponding to the mg l^{-1} range), and it is therefore experimentally difficult to determine. Some of the K_s values reported in the literature are compiled in Table 3.5. It should be stressed that the K_s value in the Monod model does not represent the saturation constant for substrate uptake but an overall saturation constant for the entire growth process.

Some of the most characteristic features of microbial growth are represented quite well by the Monod model:

The constant specific growth rate at high substrate concentration

The first-order dependence of the specific growth rate on substrate concentration at low substrate concentrations

In fact, one may argue that the two features that make the Monod model work so well in fitting experimental data are deeply rooted in any naturally occurring conversion process: the size of the machinery that converts substrate must have an upper value, and all chemical reactions will end up as first-order processes when the reactant concentration tends to zero. The satisfactory fit of the Monod model to many experimental data should never be misconstrued to mean that Equation 3.59 is a mechanism of fermentation processes. The Langmuir rate expression of heterogeneous catalysis and the Michaelis-Menten rate expression in enzymatic catalysis are formally identical to Equation 3.59, but the denominator constant has a direct physical interpretation in both cases (the equilibrium constant for dissociation of a catalytic site-reactant complex), whereas K_s in Equation 3.59 is no more than an empirical parameter used to fit the average substrate influence on all cellular reactions pooled into the single reaction by which substrate is converted to biomass.

In the Monod model, it is assumed that the yield of biomass from the limiting substrate is constant; in other words, there is proportionality between the specific growth rate and the specific substrate uptake rate. The Monod model is, however, normally used together with a maintenance consumption of substrate, that is, the specific substrate uptake is described by the linear relation; see Equation 3.18. The Monod model including maintenance is probably the most widely accepted model for microbial growth, and it is well suited for analysis of steady-state data from a chemostat (see Example 3.3c). Often the model is combined with the Luedeking and Piret (1959) model for metabolite production in which the specific rate of product formation is given by Equation 3.17. The Luedeking and Piret model was derived on the basis of an analysis of lactic acid fermentation and is in principle only valid for metabolic products formed as a direct consequence of the growth process (i.e., metabolites of primary metabolism). However, the model may in some cases be applied to other products (e.g., secondary metabolites), but this should not be done automatically.

Example 3.3c: The Monod Model

Despite its simplicity, the Monod model is very useful for extracting key growth parameters, and it generally fits simple batch fermentations with one exponential growth phase and steady-state chemostat cultures (but rarely with the same parameters). We first consider a batch process, where substrate consumption due to maintenance can usually be neglected. In this case, there is an analytical solution to the mass balances for the concentrations of substrate and the limiting substrate (Nielsen and Villadsen, 1994):

$$c_s = c_{s,0} - Y_{xs}\left(x - x_0\right) \tag{3.3c1}$$

$$\mu_{max}t = \left(1 + \frac{K_s}{c_{s,0} + Y_{xs}x_0}\right)\ln\left(\frac{x}{x_0}\right)$$
$$- \frac{K_s}{c_{s,0} + Y_{xs}x_0}\ln\left(1 + \frac{Y_{sx}\left(x_0 - x\right)}{c_{s,0}}\right) \tag{3.3c2}$$

Using this analytical solution, it is in principle possible to estimate the two kinetic parameters in the Monod model, but since K_s generally is very low it is in practice not possible to estimate this parameter from a batch cultivation (see Figure 3.3).

For a steady-state, continuous culture, the mass balance for the biomass, together with the Monod model, gives

$$D = \mu_{max}\frac{c_s}{c_s + K_s} \tag{3.3c3}$$

or

$$c_s = \frac{DK_s}{\mu_{max} - D} \tag{3.3c4}$$

Thus, the concentration of the limiting substrate increases with the dilution rate. When substrate concentration becomes equal to the substrate concentration in the feed, the dilution rate attains its maximum value, which is often called the critical dilution rate:

$$D_{crit} = \mu_{max} \frac{c_s^f}{c_s^f + K_s} \qquad (3.3c5)$$

When the dilution rate becomes equal to or larger than this value, the biomass is washed out of the bioreactor. Equation 3.3b4 clearly shows that the steady-state chemostat is well suited to studying the influence of the substrate concentration on the cellular function (e.g., product formation), because by changing the dilution rate it is possible to change the substrate concentration as the only variable. Furthermore, it is possible to study the influence of different limiting substrates on the cellular physiology (e.g., glucose and ammonia).

Besides quantification of the Monod parameters, the chemostat is well suited to determine the maintenance coefficient. Because the dilution rate equals the specific growth rate, the yield coefficient is given by

$$Y_{sx} = \frac{D}{Y_{xs}^{true}D + m_s} \qquad (3.3c6)$$

or, if we use Equation 3.52,

$$x = \frac{D}{Y_{xs}^{true}D + m_s}\left(c_s^f - c_s\right) \approx \frac{D}{Y_{xs}^{true}D + m_s}c_s^f \qquad (3.3c7)$$

because $c_s^f \gg c_s$ except for dilution rates close to the critical dilution rate. Equation 3.3c7 shows that the biomass concentration decreases at low specific growth rates, where the substrate consumption for maintenance is significant compared with that for growth. At high specific growth rates (high dilution rates), maintenance is negligible and the yield coefficient becomes equal to the true yield coefficient (Figure 3.4). By rearrangement of Equation 3.3c7, a linear relationship between the specific substrate uptake rate and the dilution rate is found, and when using this, the true yield coefficient and the maintenance coefficient can easily be estimated using linear regression.

It is unlikely that the Monod model can be used to fit all kinds of fermentation data. Many authors have tried to improve the Monod model, but generally these empirical models are of little value. However, in some cases, growth is limited either by substrate concentration or by the presence of a metabolic product, which acts as an inhibitor. In order to account for this, the Monod model is often extended with additional terms. Thus, for inhibition by high concentrations of the limiting substrate

$$\mu = \mu_{max} \frac{c_s}{c_s^2 / K_i + c_s + K_s} \qquad (3.60)$$

and for inhibition by a metabolic product

$$\mu = \mu_{max} \frac{c_s}{c_s + K_s} \frac{1}{1 + p / K_i} \qquad (3.61)$$

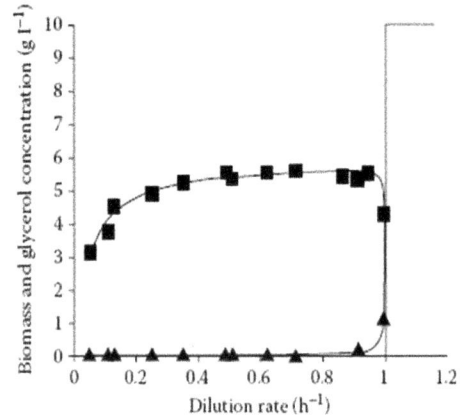

FIGURE 3.4 Growth of *Aerobacter aerogenes* in a chemostat with glycerol as the limiting substrate where the biomass concentration (■) decreases for increasing dilution rate due to the maintenance metabolism, and when the dilution rate approaches the critical value, the biomass concentration decreases rapidly. The glycerol concentration (▲) increases slowly at low dilution rates, but when the dilution rate approaches the critical value it increases rapidly. The lines are model simulations using the Monod model with maintenance, and with the parameter values: $c_s^f = 10$ g L^{-1}; $\mu_{max} = 1.0$ h^{-1}; $K_S = 0.01$ g L^{-1}; $ms = 0.08$ g (g DW h)$^{-1}$; $Y_{xs}^{true} = 1.70$ g$\left(g\ DW\right)^{-1}$.

Equations 3.60 and 3.61 may be a useful way of including product or substrate inhibition in a simple model, and often these expressions are also applied in connection with structured models. Extension of the Monod model with additional terms or factors should, however, be carried out with some restraint because the result may be a model with a large number of parameters but of little value outside the range in which the experiments were made.

3.4.3 Compartment Models

Simple structured models are in one sense improvements to the unstructured models because some basic mechanisms of the cellular behavior are at least qualitatively incorporated. Thus, the structured models may have some predictive strength, that is, they may describe the growth process at different operating conditions with the same set of parameters. But one should bear in mind that "true" mechanisms of the metabolic processes are of course not considered in simple structured models even if the number of parameters is quite large.

In structured models, all the biomass components are lumped into a few key variables (i.e., the vectors $\mathbf{X}_{macro}$ and $\mathbf{X}_{met}$), which are hopefully representative of the state of the cell. The microbial activity thus becomes a function of not only the abiotic variables, which may change with very small time constants, but also the cellular composition, and consequently, the "history" of the cells (i.e., the environmental conditions the cells have experienced in the past).

The biomass can be structured in a number of ways. For example, in simple structured models, only a few cellular components are considered, whereas in highly structured models, up to 20 intracellular components are considered

(Nielsen and Villadsen, 1992). As discussed in Section 3.4.1, the choice of a particular structure depends on the aim of the modeling exercise, but often one starts with a simple structured model onto which more and more structures are added as new experiments are added to the database. Even in highly structured models, many of the cellular components included in the model represent "pools" of different enzymes, metabolites, or other cellular components. The cellular reactions considered in structured models are therefore empirical in nature because they do not represent the conversion between "true" components. Consequently, it is permissible to write the kinetics for the individual reactions in terms of reasonable empirical expressions, with a form judged to fit the experimental data with a small number of parameters. Thus, Monod-type expressions are often used because they summarize some fundamental features of most cellular reactions (i.e., being first order at low substrate concentration and zero order at high substrate concentration). Despite their empirical nature, structured models are normally based on some well-known cell mechanisms, and they therefore have the ability to simulate certain features of experiments quite well.

The first structured models appeared in the late 1960s from the group of Fredrickson and Tsuchiya at the University of Minnesota (Ramkrishna et al. 1966, 1967; Williams, 1967), who also were the first to formulate microbial models within a general mathematical framework similar to that used to describe reaction networks in classical catalytic processes (Tsuchiya et al. 1966; Fredrickson et al. 1967; Fredrickson, 1976). Since this pioneering work, many other simple structured models have been presented (Harder and Roels, 1982; Nielsen and Villadsen, 1992).

In these simple structured models, the biomass is divided into a few compartments. These compartments must be chosen with care, and cell components with similar functions should be placed in the same compartment, for example, all membrane material and otherwise rather inactive components in one compartment, and all active material in another compartment. With the central role of the protein-synthesizing system (PSS) in cellular metabolism, this is often a key component in compartment models. Besides a few enzymes, the PSS consists of ribosomes, which contain approximately 60% ribosomal RNA (rRNA) and 40% ribosomal protein. Because rRNA makes up more than 80% of the total stable RNA in the cell, the level of the ribosomes is easily identified through measurements of the RNA concentration in the biomass. As seen in Figure 3.5, the RNA content of many different microorganisms increases linearly with the specific growth rate. Thus, the level of the PSS is well correlated with the specific growth rate. It is therefore a good representative of the activity of the cell and this is the basis of most simple structured models (see Example 3.3d).

Example 3.3d: Two-Compartment Model

Nielsen et al. (1991a, 1991b) presented a two-compartment model for the lactic acid bacterium *Lactococcus cremoris*. The model is a direct descendant of the model

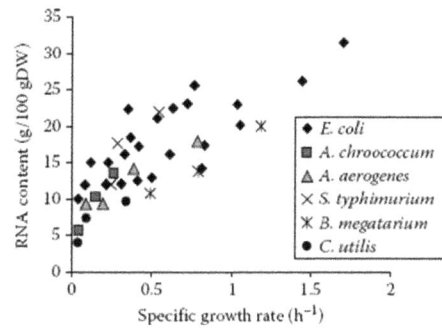

FIGURE 3.5 The level of stable RNA as a function of specific growth rate for different microorganisms. (Data are taken from Nielsen, J. and J. Villadsen. 1994. *Bioreaction Engineering Principles*. New York: Plenum Press)

created by Williams (1967) with similar definitions for the following two compartments:

The active (A) compartment contains the PSS and small building blocks.
The structural and genetic (G) compartment contains the rest of the cell material.

The model considers both glucose and a complex nitrogen source (peptone and yeast extract), but in the following presentation we discuss the model with only one limiting substrate (glucose). The model considers two reactions for which the stoichiometry is

$$\gamma_{11}X_A - s = 0 \tag{3.3d1}$$

$$\gamma_{22}X_G - X_A = 0 \tag{3.3d2}$$

In the first reaction, glucose is converted into small building blocks in the A compartment, and these are further converted into ribosomes. The stoichiometric coefficient γ_{11} can be considered as a yield coefficient because metabolic products (lactic acid, carbon dioxide, etc.) are not included in the stoichiometry. In the second reaction, building blocks present in the A compartment are converted into macromolecular components of the G compartment. In this process, some by-products may be formed and the stoichiometric coefficient γ_{22} is therefore slightly less than 1. The kinetics of the two reactions have the same form:

$$v_i = k_i \frac{c_s}{c_s + K_{s,i}} X_A; \quad i = 1, 2 \tag{3.3d3}$$

From Equation 3.13, the specific growth rate for the biomass is found to be

$$\mu = \begin{pmatrix} 1 & 1 \end{pmatrix} \begin{pmatrix} \gamma_{11} & -1 \\ 0 & \gamma_{22} \end{pmatrix} \begin{pmatrix} v_1 \\ v_2 \end{pmatrix} \tag{3.3d4}$$

$$= \gamma_{11}v_1 - (1 - \gamma_{22})v_2$$

or, with the kinetic expression for v_1 and v_2 inserted,

$$\mu = \left(\gamma_{11}k_1 \frac{c_s}{c_s + K_{s,1}} - (1 - \gamma_{22})k_2 \frac{c_s}{c_s + K_{s,2}} \right) X_A \tag{3.3d5}$$

Thus, the specific growth rate is proportional to the size of the active compartment. The substrate concentration c_s influences the specific growth rate both directly and indirectly by determining the size of the active compartment. The influence of the substrate concentration on the synthesis of the active compartment can be evaluated through the ratio r_1/r_2:

$$\frac{r_1}{r_2} = \frac{k_1}{k_2} \frac{c_s + K_{s,2}}{c_s + K_{s,1}} \tag{3.3d6}$$

If $K_{s,1}$ is larger than $K_{s,2}$, the formation of X_A is favored at high substrate concentration, and it is thus possible to explain the increase in the active compartment with the specific growth rate. Consequently, when the substrate concentration increases rapidly, there are two effects on the specific growth rate:

First a rapid increase in the specific growth rate, which is a result of mobilization of excess capacity in the cellular synthesis machinery.

Thereafter, there is a slow increase in the specific growth rate, which is a result of a slow buildup of the active part of the cell (i.e., additional cellular synthesis machinery has to be formed in order for the cells to grow faster).

This is illustrated in Figure 3.6, which shows the biomass concentration in two independent wash-out experiments. In both cases, the dilution rate was shifted to a value (0.99 h^{-1}) above the critical dilution rate (0.55 h^{-1}), but in one experiment, the dilution rate before the shift was low (0.1 h^{-1}) and in the other experiment it was high (0.5 h^{-1}). The wash-out profile is seen to be very different, with a much faster wash-out when there was a shift from a low dilution rate. When the dilution rate is shifted to 0.99 h^{-1}, the glucose concentration increases rapidly to a value much higher than $K_{s,1}$ and $K_{s,2}$, and this allows growth at the maximum rate. However, when the cells come from a low dilution rate, the size of the active compartment is not sufficiently large to allow rapid growth, and X_A therefore

has to be built up before the maximum specific growth rate is attained. On the other hand, if the cells come from a high dilution rate, X_A is already high and the cells immediately attain their maximum specific growth rate. It is observed that the model is able to correctly describe the two experiments (all parameters were estimated from steady-state experiments), and the model correctly incorporates information about the previous history of the cells.

The model also includes the formation of lactic acid; the kinetics was described using a rate equation similar to Equation 3.3d3. Thus, the lactic acid formation increases when the activity of the cells increases and so it is ensured that there is a close coupling between the formation of this primary metabolite and the growth of the cells.

It is interesting to note that even though the model does not include a specific maintenance reaction, it can actually describe a decrease in the yield coefficient of biomass on glucose at low specific growth rates. The yield coefficient is given by

$$Y_{sx} = \gamma_{21}\left(1 - \left(1 - \gamma_{22}\right)\frac{k_2}{k_1}\frac{c_s + K_{s,1}}{c_s + K_{s,2}}\right) \tag{3.3d7}$$

Because $K_{s,1}$ is larger than $K_{s,2}$, the last term within the parentheses decreases for increasing specific growth rates and the yield coefficient will therefore also increase for increasing substrate concentration.

3.4.4 Single-Cell Models

Single-cell models are in principle an extension of the compartment models, but with the description of many different cellular functions. Furthermore, these models depart from the description of a population and focus on the description of single cells. This allows consideration of characteristic features of the cell and it is therefore possible to study different aspects of cell function:

Cell geometry can be accounted for explicitly, and so it is possible to examine its potential effects on nutrient transport.

Temporal events during the cell cycle can be included in the model, and the effect of these events on the overall cell growth can be studied.

Spatial arrangements of intracellular events can be considered, even though this would lead to significant model complexity.

To set up single-cell models, it is necessary to have a detailed knowledge of the cell, and single-cell models have therefore only been described for well-studied cellular systems such as *Escherichia coli*, *S. cerevisiae*, *Bacillus subtilis*, and human erythrocytes. The most comprehensive single-cell model is the so-called Cornell model set up by Shuler and coworkers (Shuler et al. 1979), which contains 20 intracellular components. This model predicts a number of observations made with *E. coli*, and it has formed the basis for setting up several other models (Nielsen and Villadsen, 1992). Thus, Peretti and Bailey (1987) extended the model to describe

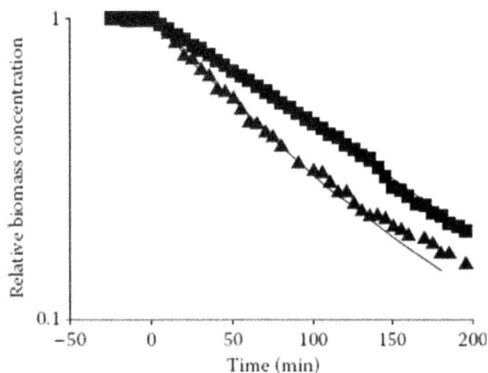

FIGURE 3.6 Measurement of biomass concentration, two transient experiments, of *Lactococcus cremoris* growing in glucose-limited chemostat; the dilution rate was shifted from an initial value of 0.10 h^{-1} (▲) or 0.50 h^{-1} (■) to 0.99 h^{-1}, respectively. The biomass concentration is normalized by the steady-state biomass concentration before the step change, which was made at time zero. The lines are model simulations. (The data are taken from Nielsen, J., K. Nikolajsen, and J. Villadsen. 1991b. *Biotechnol. Bioeng*, 38:11–23.)

plasmid replication and gene expression from a plasmid inserted into a host cell. This allowed study of host-plasmid interactions, especially the effects of copy number, promoter strength, and ribosome binding site strength on the metabolic activity of the host cell and on the plasmid gene expression.

3.4.5 Molecular Mechanistic Models

Despite the level of detail, the single-cell models are normally based on an empirical description of different cellular events (e.g., gene transcription and translation). This is a necessity because the complexity of the model would become very high if all these individual events were to be described with detailed models that include mechanistic information. In many cases, however, it is interesting to study these events separately, and for models where mechanistic information is included, they have to be used. These models are normally set up at the molecular level, and they can therefore be referred to as molecular mechanistic models. Many different models of this type can be found in the literature, but most fall in one of two categories:

> Gene transcription models
> Pathway models

Gene transcription models aim at quantifying gene transcription based on knowledge of the promoter function. The *lac*-promoter of *E. coli* is one of the best studied promoter systems of all, and so this system has been modeled most extensively. Furthermore, this promoter (or its derivatives) is often used in connection with the production of heterologous proteins by this bacterium, because it is an inducible promoter. In a series of papers, Lee and Bailey (1984a, 1984b, 1984c, 1984d) presented an elegant piece of modeling of this system, and through combination with a model for plasmid replication they could investigate, for example, the role of point mutations in the promoter on gene transcription. This promoter system is quite complex, with both activator and repressor proteins, and empirical investigation can therefore be laborious; the detailed mathematical model is a valuable tool to guide the experimental work.

In pathway models the individual enzymatic reactions of a given pathway are described with enzyme kinetic models, and it is therefore possible to simulate the metabolite pool levels and the fluxes through different branches of the pathway. These models have mainly concentrated on glycolysis in *S. cerevisiae* (Galazzo and Bailey, 1990; Rizzi et al. 1997), because much information about enzyme regulation is available for this pathway. However, complete models are also available for other pathways, such as the penicillin biosynthetic pathway (Pissarra et al. 1996). These pathway models are experimentally verified by comparing modeling simulations with measurements of intracellular metabolite pool levels—something that has only been possible with sufficient precision in the last couple of years because it requires rapid quenching of the cellular activity and sensitive measurement techniques.

Pathway models are very useful in studies of metabolic fluxes because they allow quantification of the control of flux by the individual enzymes in the pathway. This can be done by calculation of sensitivity coefficients (or the so-called flux control coefficients; see Box 3.4), which quantify the relative importance of the individual enzymes in the control of flux through the pathway (Stephanopoulos et al. 1998).

BOX 3.4 QUANTIFICATION OF FLUX CONTROL

In a study of flux control in a biochemical pathway, the concept of metabolic control analysis is very useful (Stephanopoulos et al., 1998). Here the flux control of the individual enzymatic reactions on the steady-state flux J through the pathway is quantified by the so-called flux control coefficients (FCC):

$$C_i^J = \frac{v_i}{J}\frac{dJ}{dv_i}$$

where v_i is the rate of the ith reaction. If the enzyme concentration of the ith enzyme is increased, its rate will normally increase and a higher flux through the pathway may be the result. However, it is likely that due to allosteric regulation (or other regulation phenomena) there may be a very small effect of increasing the enzyme concentration. This is exactly what is quantified by the FCC, that is the relative increase in the steady-state flux upon a relative increase in the enzyme activity. Clearly, a step with a high FCC has a large control of the flux through the pathway. If a kinetic model is available for the pathway the flux control coefficients for each step can easily be calculated using model simulations. The FCCs can also be determined experimentally by changing the enzyme concentration (or activity) genetically, by titration with the individual enzymes, or by adding specific enzyme inhibitors (Stephanopoulos et al. 1998).

A general criticism of the application of kinetic models for complete pathways is that despite the level of detail included, they cannot possibly include all possible interactions in the system and therefore only represent one model of the system. The robustness of the model is extremely important, especially if the kinetic model is to be used for predictions, and unfortunately most biochemical models, even very detailed models, are only valid at operating conditions close to those where the parameters have been estimated (i.e., the predictive strength is limited). For analysis of complex systems, it is, however, not necessary that the model gives a quantitatively correct description of all the variables, because even models that give a qualitatively correct description of the most important interactions in the system may be valuable in studies of flux control.

3.5 POPULATION MODELS

Normally, it is assumed that the population of cells is homogeneous (i.e., all cells behave identically). Although this assumption is certainly crude if a small number of cells is considered, it gives a very good picture of certain properties of the cell population because there are billions of cells per ml medium (see Box 3.5).

Furthermore, the kinetics is often linear in the cellular properties (e.g., in the concentration of a certain enzyme), and the overall population kinetics can therefore be described as a function of the average property of the cells (Nielsen et al. 2003). There are, however, situations where cell property distributions influence the overall culture performance, and here it is necessary to consider the cellular property distribution, and this is done in the so-called *segregated* models. In the following, we will discuss two approaches to segregated modeling.

3.5.1 MORPHOLOGICALLY STRUCTURED MODELS

The simplest approach to model distribution in the cellular property is by the so-called morphologically structured models (Nielsen and Villadsen, 1992; Nielsen, 1993). Here the cells are divided into a finite number Q of cell states Z (or morphological forms), and conversion between the different cell states is described by a sequence of empirical metamorphosis reactions. Ideally, these metamorphosis reactions can be described as a set of intracellular reactions, but the mechanisms behind most morphological conversions are largely unknown. Thus, it is not known why filamentous fungi differentiate to cells with a completely different morphology from that of their origin. It is therefore not possible to set up detailed mechanistic models describing these changes in morphology; empirical metamorphosis reactions are therefore introduced. The stoichiometry of the metamorphosis reactions is given by analogy with Equation 3.4:

$$\Delta Z = 0 \qquad (3.62)$$

BOX 3.5 DETERMINISTIC VERSUS STOCHASTIC MODELING

In a description of cellular kinetics macroscopic balances are normally used, that is the rates of the cellular reactions are functions of average concentrations of the intracellular components. However, living cells are extremely small systems with only a few molecules of certain key components and it does not really make sense to talk about "the DNA concentration in the cell" for example, because the number of macromolecules in a cell is always small compared with Avogadro's number. Many cellular processes are therefore stochastic in nature and the deterministic description often applied is in principle not correct. However, the application of a macroscopic (or deterministic) description is convenient and it represents a typical engineering approximation for describing the kinetics in an average cell in a population of cells. This approximation is reasonable for large populations

because the standard deviation from the average "behavior" in a population with elements is related to the standard deviation for an individual cell through

$$\sigma_{pop} = \frac{\sigma}{\sqrt{e}}$$

Thus, with a population of 10^9 cells mL^{-1}, which is a typical cell concentration during a cultivation process, it can be seen that the standard deviation for the population is very small. There are, however, systems where small populations occur, for example at dilution rates close to the maximum in a chemostat, and here one may have to apply a stochastic model.

where Δ is a stoichiometric matrix, Z_q represents both the qth morphological form and the fractional concentration (g qth, morphological form (g DW^{-1})). With the metamorphosis reactions, one morphological form is spontaneously converted to other forms. This is of course an extreme simplification because the conversion between morphological forms is the sum of many small changes in the intracellular composition of the cell. With the stoichiometry in Equation 3.62, it is assumed that the metamorphosis reactions do not involve any change in the total mass and the sum of all stoichiometric coefficients in each reaction is therefore taken to be zero. The forward reaction rates of the metamorphosis reactions are collected in the vector **u**. Each morphological form may convert substrates to biomass components and metabolic products. These reactions may be described by an intracellularly structured model, but in order to reduce the model complexity a simple unstructured model is used for description of the growth and product formation of each cell type (e.g., the specific growth rate of the qth morphological form is described by the Monod model). The specific growth rate of the total biomass is given as a weighted sum of the specific growth rates of the different morphological forms:

$$\mu = \sum_{i=1}^{Q} \mu_i Z_i \qquad (3.63)$$

The rate of formation of each morphological form is determined both by the metamorphosis reactions and by the growth-associated reactions for each form (for derivation of mass balances for the morphological forms, see Nielsen and Villadsen, 1994). The concept of morphologically structured models is well suited to describing the growth and differentiation of filamentous microorganisms (Nielsen, 1993), but it may also be used to describe other microbial systems where a cellular differentiation has an impact on the overall culture performance.

3.5.2 POPULATION BALANCE EQUATIONS

The first example of a heterogeneous description of cellular populations was presented in 1963 by Fredrickson and

Tsuchiya. In their model, single-cell growth kinetics was combined with a set of stochastic functions describing cell division and cell death. The model represents the first application of a completely segregated description of a cell population. In the model, the cell population is described by a number density function $f(X,t)$, where $f(X,t)dX$ is the number of cells with property X being in the interval X to $X + dX$. The dynamic balance for $f(X,t)$ is given by

$$\frac{\partial f(X,t)}{\partial t} + \frac{\partial}{\partial X}\big(f(X,t)v(X,t)\big)$$

$$= 2\int_{X}^{\infty} p(X^*,X,t)f(X^*,t)dX^* \qquad (3.64)$$

$$- b(X,t)f(X,t) - Df(X,t)$$

where v is the net rate of formation of the cell property, X. $b(X,t)$ is the breakage function (i.e., the rate of cell division for cells with property X), and $p(X^*,X,t)$ is the partitioning function (i.e., the probability that a cell with property X is formed upon division of a cell with property X^*). Through the functions p and b, a stochastic element can be introduced into the model, but these functions can also be completely descriptive. The balance equation 3.64 was applied in the original work of Fredrickson and Tsuchiya, but in a later paper, a general framework for segregated population models was presented (Fredrickson et al. 1967). Segregated models represent the complete description of a cell population and they take into account that all cells in a population are not identical. However, complete cellular segregation is rarely applied in cell culture models for two main reasons:

For large populations, the average properties will normally represent the overall population kinetics quite well.

The mathematical complexity of Equation 3.64 is quite substantial, especially if more than a single-cell property is considered (i.e., the number density function becomes multi-dimensional).

If the kinetics for product formation is not zero or first order in a given cell property, application of an average property model will, however, not give the same result as a segregated description. This is the case for production of a heterologous protein in plasmid-containing cells of *E. coli*, where the product formation kinetics is not first order in the plasmid copy number. A segregated model therefore has to be applied to give a good description of the product formation kinetics (Seo and Bailey, 1985). The simplest segregated models are when the cellular property is described by a single variable (e.g., cell age), and in Example 3.3e, the age distribution of an exponentially growing culture is derived from the general balance (Equation 3.64).

Example 3.3e: Age Distribution Model

The simplest segregated population models are those where the cellular property is taken to be described solely by the cell age a. In this case, the rate of increase in the cellular property $v(a,t)$ is equal to 1. Furthermore, if it is assumed that cell division occurs only at a certain cell age $a = t_d$, the two first terms on the right-hand side of Equation 3.64 become equal to zero. At steady state, the balance therefore becomes

$$\frac{d\phi(a)}{da} = -D\phi(a) \qquad (3.3e1)$$

where ϕ is a normalized distribution function:

$$\phi(a) = \frac{f(a)}{n} \qquad (3.3e2)$$

with n being the total cell number, given as the zero moment of the number density function $f(a)$. The solution to Equation 3.3e1 is

$$\phi(a) = \phi(0)e^{-Da} \qquad (3.3e3)$$

Due to the normalization, the 0th moment of $f(a)$ is 1, that is,

$$\int_{o}^{t_d} \phi(a)da = 1 \qquad (3.3e4)$$

which leads to

$$\phi(a) = \frac{D}{1 - e^{-Dt_d}} e^{-Da} \qquad (3.3e5)$$

The cell balance relating to cell division (the so-called renewal equation) is given by

$$\phi(0) = 2\phi(t_d) \qquad (3.3e6)$$

which, together with Equation 3.3e3, directly gives Equation 3.2. Furthermore, when Equation 3.2 is inserted in Equation 3.3e5, we have the simpler expression

$$\phi(a) = 2De^{-Da} \qquad (3.3e7)$$

Thus, the fraction of cells with a given age decreases exponentially with age, and the decrease is determined by the specific growth rate of the culture (equal to the dilution rate at steady state). The average cell age is given as the first moment of $f(a)$:

$$\langle a \rangle = \int_{0}^{t_d} a\phi(a)da = \frac{1 - \ln 2}{D} \qquad (3.3e8)$$

Consequently, the average age of the cells decreases for increasing specific growth rates.

SUMMARY

Understanding microbial growth kinetics is an essential requirement for the design and successful operation of industrial fermentation processes and for obtaining quantitative information about the function of microbial cells.

The primary objective of this chapter was to introduce the reader to the basic principles of the wide-ranging aspects of microbial growth kinetics and dynamics, from the basic principles to the more advanced concept of modeling.

Based on the information given in this chapter, the reader should be able to design fermentation processes for the production of biomass and microbially derived products.

Furthermore, the reader should be able to set up simple mathematical models describing microbial growth as well as evaluate more complex mathematical models.

REFERENCES

Benthin, S., U. Schulze, J. Nielsen, and J. Villadsen. 1994. Growth energetics of *Lactococcus cremoris* FDI during energy, carbon and nitrogen limitation in steady state and transient cultures. *Chem. Eng. Sci.* 49:589–609.

Christensen, L.H., C.M. Henriksen, J. Nielsen, J. Villadsen, and M. Egel-Mitani. 1995. Continuous cultivation of *P. chrysogenum*: Growth on glucose and penicillin production. *J. Biotechnol.* 42:95–107.

Fredrickson, A.G. 1976. Formulation of structured growth models. *Biotechnol. Bioeng.* 18:1481–1486.

Fredrickson, A.G., D. Ramkrishna, and H.M. Tsuchiya. 1967. Statistics and dynamics of procaryotic cell populations. *Math. Biosci.* 1:327–374.

Fredrickson, A.G. and H.M. Tsuchiya. 1963. Continuous propagation of micro-organisms. *AIChE J.* 9:459–468.

Galazzo, J.L. and J.E. Bailey. 1990. Fermentation pathway kinetics and metabolic βux control in suspended and immobilized *Saccharomyces cerevisiae. Enzym. Microb. Technol.* 12:162–172.

Harder, A. and J.A. Roels. 1982. Application of simple structured models in bioengineering. *Adv. Biochem. Eng.* 21:55–107.

Herbert, D. 1959. Some principles of continuous culture. *Recent Prog. Microbiol.* 7:381–396.

Hill, A.V. 1910. The possible effects of the aggregation of the molecules of haemoglobin on its dissociation curves. *J. Physiol. Lond.* 40:4–7.

Ingraham, J.L., O. Maaloe, and F.C. Neidhardt. 1983. *Growth of the Bacterial Cell.* Sunderland, MA: Sinauer Associates.

Lee, S.B. and J.E. Bailey. 1994a. A mathematical model for λdν plasmid replication: Analysis of wild-type plasmid. *Gene* 11:151–165.

Lee, S.B. and J.E. Bailey. 1994b. A mathematical model for λdν plasmid replication: Analysis of copy number mutants. *Gene* 11:166–77.

Lee, S.B. and J.E. Bailey. 1994c. Genetically structured models for *lac* promoter-operator function in the *Escherichia coli* chromosome and in multicopy plasmids: *Lac* operator function. *Biotechnol. Bioeng.* 26:1372–1382.

Lee, S.B. and J.E. Bailey. 1994d. Genetically structured models for *lac* promoter-operator function in the *Escherichia coli* chromosome and in multicopy plasmids: *lac* promoter function. *Biotechnol. Bioeng.* 26:1383–1389.

Luedeking, R. and E.L. Piret. 1959. A kinetic study of the lactic acid fermentation: Batch process at controlled pH. *J. Biochem, Microbiol. Technol. Eng.* 1:393–412.

Meyenburg, K. von. 1969. *Katabolit-Repression Und Der Sprossungszyklus Von* Saccharomyces Cerevisiae. PhD dissertation, ETH, Zurich.

Monod, J. 1942. *Recherches Sur La Croissance Des Cultures Bacteriennes.* Paris: Hermann and Cie.

Monod, J., J. Wyman, and J-P. Changeux. 1965. On the nature of allosteric transitions: A plausible model. *J. Molec. Biol.* 12:88–118.

Neidhardt, F.C., J.L. Ingraham, and M. Schaechter. 1990. *Physiology of the Bacterial Cell: A Molecular Approach.* Sunderland, MA: Sinauer Associates.

Nielsen, J. 1993. A simple morphologically structured model describing the growth of Þlamentous microorganisms. *Biotechnol, Bioeng.* 41:715–27.

Nielsen, J., K. Nikolajsen, and J. Villadsen. 1991a. Structured modeling of a microbial system 1: A theoretical study of the lactic acid fermentation. *Biotechnol. Bioeng.* 38:1–10.

Nielsen, J., K. Nikolajsen, and J. Villadsen. 1991b. Structured modelling of a microbial system 2: Verification of a structured lactic acid fermentation model. *Biotechnol. Bioeng.* 38:11–23.

Nielsen, J. and J. Villadsen. 1992. Modeling of microbial kinetics. *Chem. Eng. Sci.* 47:4225–4270.

Nielsen, J. and J. Villadsen. 1993. Bioreactors: Description and modelling. In Rehm, H.J. and Reed, G., eds., *Biotechnology*, vol. 3, 2nd ed., pp. 77–104. Weinheim: VCR Verlag.

Nielsen, J. and J. Villadsen. 1994. *Bioreaction Engineering Principles.* New York: Plenum Press.

Nielsen, J., J. Villadsen, and G. Lidén. 2003. *Bioreaction Engineering Principles*, 2nd ed. New York: Kluwer Academic/Plenum.

Olsson, L. and J. Nielsen. 1997. On-line and in situ monitoring of biomass in submerged cultivations. *TIBTECH* 15:517–522.

Oura, E. 1983. Biomass from carbohydrates. In Rehm, H-J. and Reed, G., eds., *Biotechnology*, vol. 3, 2nd ed., pp. 3–42. Weinheim: VCR Verlag.

Peretti, S.W. and J.E. Bailey. 1987. Simulations of host-plasmid interactions in *Escherichia coli:* Copy number, promoter strength, and ribosome binding site strength effects on metabolic activity and plasmid gene expression. *Biotechnol. Bioeng.* 29:316–328.

Pirt, S.J. 1965. The maintenance energy of bacteria in growing cultures. *Proc. Roy. Soc. London, Ser. B.* 163:224–231.

Pissarra, P.N., J. Nielsen, and M.J. Bazin. 1996. Pathway kinetics and metabolic control analysis of a high-yielding strain of *Penicillium chrysogenum* during fed-batch cultivations. *Biotechnol. Bioeng.* 51:168–176.

Ramkrishna, D., A.G. Fredrickson, and H.M. Tsuchiya. 1966. Dynamics of microbial propagation: Models considering endogenous metabolism. *J. Gen. Appl. Microbiol.* 12:311–327.

Ramkrishna, D., A.G. Fredrickson, and H.M. Tsuchiya. 1967. Dynamics of microbial propagation: Models considering inhibitors and variable cell composition. *Biotechnol. Bioeng.* 9:129–170.

Rizzi, M., M. Baltes, U. Theobald, and M. Reuss. 1997. *In vivo* analysis of metabolic dynamics in *Saccharomyces cerevisiae*: II. Mathematical model. *Biotechnol. Bioeng.* 55:592–608.

Roels, J.A. 1983. *Energetics and Kinetics in Biotechnology.* Amsterdam: Elsevier Biomedical Press.

Seo, J.H. and J.E. Bailey. 1985. A segregated model for plasmid content and product synthesis in unstable binary fission recombinant organisms. *Biotechnol. Bioeng.* 27:156–165.

Shuler, M.L., S.K. Leung, and C.C. Dick. 1979. A mathematical model for the growth of a single bacterial cell. *Ann. NY Acad. Sci.* 326:35–55.

Sonnleitner, B. and A. Fiechter. 1988. High performance bioreactors: A new generation. *Anal. Chim. Acta* 213:199–205.

Stephanopoulos, G., J. Nielsen, and A. Aristodou. 1998. *Metabolic Engineering.* San Diego, CA: Academic Press.

Tsuchiya, H.M., A.G. Fredrickson, and R. Aris. 1966. Dynamics of microbial cell populations. *Adv. Chem. Eng.* 6:125–206.

Williams, F.M. 1967. Av model of cell growth dynamics. *J. Theor. Biol.* 15:190–207.

Section II

Fermentation Microbiology and Biotechnology

*Control of Carbon Flux to Product Formation
in Microbial Cell Factories*

4 Microbial Synthesis of Primary Metabolites

Current Trends and Future Prospects

Arnold L. Demain and Sergio Sánchez

CONTENTS

"Logic will get you from A to B. Imagination will take you everywhere."

Albert Einstein

4.1 INTRODUCTION

Primary metabolites are microbial products made during the exponential phase of growth whose synthesis is an integral part of the normal growth process. They include intermediates and end products of anabolic pathways leading to the formation of monomers, which are used by the cell as building blocks (e.g., amino acids and nucleotides) for the biosynthesis of polymers (e.g., protein and DNA) and coenzymes (e.g., vitamins). On the other hand, primary metabolites of catabolic pathways (e.g., citric acid, acetic acid, and ethanol) are not biosynthetic precursors but are essential for growth as they are related to energy generation, redox balance, and substrate utilizations. Industrially, the most important primary metabolites are amino acids, nucleotides, vitamins, solvents, and organic acids. These are made by a wide range of bacteria and fungi and have numerous applications in the food, chemical, pharmaceutical, and nutraceutical industries. Many of these metabolites are manufactured by microbial fermentation rather than chemical synthesis because the fermentations are economically competitive and produce biologically active isomers. Several other industrially important chemicals could be manufactured via microbial fermentations (e.g., glycerol and other polyhydroxy alcohols) but are presently synthesized cheaply as petroleum by-products. However, a renewed interest in the microbial production of ethanol, organic acids, and solvents has been triggered as a consequence of successive rises in the price of crude oils.

4.2 CONTROL OF PRIMARY METABOLISM

Microbial metabolism is a conservative process that usually does not expend energy or nutrients to make compounds already available in the environment and does not overproduce components of intermediary metabolism. Coordination of metabolic functions ensures that, at any given moment, only the necessary enzymes, and the correct amounts of each, are made. Once a sufficient quantity of a given metabolite or precursor is made, either the synthesis of the enzymes concerned is turned "off" at the level of transcription (repression) or their activities are turned down (inhibition) through a number of specific regulatory control mechanisms such as feedback inhibition and covalent modifications.

4.2.1 INDUCTION

While the majority of anabolic enzymes are subject to repression, most catabolic enzymes are subject to induction. The latter is a control mechanism by which a substrate (or a compound structurally similar to the substrate) "turns on" the synthesis of the enzymes required for its uptake and initiation of metabolism. Enzymes that are synthesized as a result of genes being turned "on" in response to signal molecules (substrates) are called *inducible enzymes*, with the substance that activates gene transcription being referred to as the *inducer*. Inducible enzymes are produced only in response to the presence of their substrate or substrate analogs, in other words, they are produced only when needed. The inducer molecule renders the repressor molecule unable to bind to the operator region, thus facilitating the binding of RNA polymerase to the promoter region and in turn initiation of transcription and translation into protein. Although most inducers are substrates of catabolic enzymes, products can sometimes function as inducers. For example, fatty acids induce lipase, whereas galacturonic acid induces polygalacturonase. In some cases, however, the synthesis of a particular enzyme may be induced by substrate analogs that are not attacked by the enzyme (e.g., IPTG in the case of β-galactosidase); such analogs are referred to as *gratuitous inducers*.

4.2.2 CATABOLITE INHIBITION AND CATABOLITE REPRESSION

Catabolite inhibition and catabolite repression mechanisms allow microorganisms to preferentially utilize one substrate in preference to another, thus facilitating faster growth through conserving energy and biosynthetic precursors. For example, in the presence of lactose and glucose, *Escherichia coli* preferentially utilizes glucose and prevents the uptake of lactose through catabolite inhibition mechanism. *E. coli* also switches off the transcription of the *lac* operon through catabolite repression mechanism, which is also known as *carbon catabolite repression*. Such a mechanism ensures that the cell produces enzymes to metabolize the assimilated carbon source but represses the synthesis of the enzymes required for the assimilation of the competing substrate until the primary carbon substrate, in this case glucose, is fully exhausted.

4.2.3 NITROGEN SOURCE REGULATION (NSR)

Nitrogen can be assimilated from inorganic or organic sources. Its assimilation from inorganic sources requires reduction to ammonia, followed by incorporation into intracellular metabolites. The appropriate distribution of nitrogen among various pathways usually involves specific or local regulatory mechanisms, such as end product inhibition or end product–mediated transcriptional control. In addition, some global regulators control the expression of genes from several pathways and thereby coordinate metabolism. The ability to assimilate particular inorganic or organic nitrogen sources depends on the particular organism. Organic nitrogen sources are usually the monomeric units of macromolecules (e.g., amino acids or nucleic acids) or compounds derived from them (e.g., agmatine or putrescine). Ammonia usually supports the fastest growth rate and is therefore considered the preferred nitrogen source for *E. coli*. The biochemical basis of this "ammonium preference" is explained by the repression of enzymes acting on the alternative nitrogenous substrates present in the culture medium. NSR is known by many other names such as *nitrogen metabolite repression, nitrogen catabolite repression,* and *ammonia repression*. Enzymes typically under such control are proteases, amidases, and ureases, not to mention those involved in the degradation of amino acids.

4.2.4 PHOSPHORUS SOURCE REGULATION

In natural environments, inorganic phosphorus is commonly a major growth-limiting nutrient. It is hardly surprising therefore to see that microorganisms have evolved specific regulatory control mechanisms to respond to "feast and famine" concentrations of phosphate in their immediate environment. In *E. coli*, the phosphate regulon (pho regulon) is composed of over 30 genes that are transcriptionally activated by phosphorylated PhoB as the level of phosphate in the medium becomes growth limiting. PhoR and PhoB comprise a two-component signal transduction system in which PhoR catalyzes the reversible phosphorylation and activation of PhoB in response to low and high levels of phosphate, respectively. PhoR autophosphorylates and transfers the phosphate to PhoB. The environmental concentration of phosphate is monitored by the periplasmic phosphate-binding protein PstS, which transmits the signal for excess phosphate across the cytoplasmic membrane via PstC, PstA, PstB, and PhoU to PhoR. Phosphorylated PhoB binds to the promoters of 31 genes containing *pho* boxes and interacts with RNA polymerase, allowing the initiation of mRNA synthesis.

In fungi, nucleases and phosphatases are usually repressed when the supply of phosphate is plentiful. Similarly, in *Neurospora*, phosphate represses proteases, isocitrate lyase, aldolase, NADP-specific isocitrate dehydrogenase, and malate dehydrogenase. Phosphate also suppresses the production of riboflavin by *Eremothecium ashbyii*. Phosphate-derepressed mutants can be selected by growth with a phosphate ester (e.g., β-glycerol phosphate) as the sole source of carbon, with phosphate in excess of demands. Of great interest is inorganic polyphosphate (poly P), a linear polymer that carries numerous orthophosphate (P_i) residues linked by high-energy phosphoanhydride bonds. Poly P is found in the cells of all bacteria, archaea, fungi, protozoa, plants, and animals. It is produced by polyphosphate kinase (PPK), which catalyzes the reversible transfer of the terminal phosphate of ATP to form a long-chain polyphosphate. The *E. coli* gene (*ppk*) encoding PPK has been cloned, sequenced, and overexpressed (about 100-fold). The poly P plays a significant role in metabolism, including acting as a substitute for ATP as an energy source, reservoir for inorganic phosphate, chelator of metal ions, and regulator for stress and survival.

4.2.5 SULFUR SOURCE REGULATION

In common with other nutrients, the uptake of sulfur and its subsequent assimilation are controlled at the transcriptional level. However, in this case, the control is achieved through the activity of pleiotropic regulatory proteins. Thus, in *E. coli*, sulfur metabolism is controlled by the CysB transcriptional activator. The cysteine regulon includes most of the genes required for the synthesis of cysteine and genes for uptake of sulfur sources such as l-cystine, sulfate, thiosulfate, and taurine. Transcriptional activation of these genes requires CysB protein, *N*-acety-L-serine as an inducer, and sulfur limitation. CysB's activity is regulated by an efflux pump specific for cysteine and related metabolites. CysB is also an autorepressor (i.e., capable of repressing its own expression at the transcriptional level). Similarly, carbohydrate metabolism and fermentations were found to be adversely affected by CysB. Cysteine inhibits inducer synthesis, resulting in maximal mutations in *cysB* genes. The aforementioned effects are exerted at the transcriptional level and are only partially reversible by exogenous cAMP or sulfur-containing substrates (e.g., cysteine or djenkolate), while growth under sulfur limitation or with poor sulfur sources, such as glutathione, turns "on" (derepresses) the expression of the sulfur regulon.

In *Neurospora crassa*, sulfate uptake is an important point of the regulation of sulfur metabolism as it is subject to sulfur (metabolite) repression in which excess sulfate turns "off" the expression of the sulfate permease-encoding genes. Also, structural genes encoding aryl sulfatase, choline sulfatase, sulfate permeases I and II, a high-affinity methionine permease, and an extracellular protease are turned "on" when sulfur becomes limiting.

4.2.6 FEEDBACK REGULATION

This category of regulation is predominantly used for the regulation of anabolic enzymes involved in the biosynthesis of amino acids, nucleotides, and vitamins. It exerts its function at two levels: enzyme action (feedback inhibition) and enzyme synthesis (feedback repression and attenuation).

In feedback inhibition (see Chapter 2 for further details), the final metabolite of a pathway, when present in sufficient quantities, inhibits the action of the first enzyme of the pathway, thus preventing the synthesis of unwanted intermediates on the one hand and the wasting of energy on the other. Feedback repression involves the turning "off" of enzyme synthesis when the amount of the product has been made in sufficient quantities to satisfy the biosynthetic demands. The end product of the pathway acts as a co-repressor. The apo-repressor specified by the regulator gene is inactive in the absence of its co-repressor and as such is unable to bind to the operator region. However, in the presence of a co-repressor, the inactive apo-repressor is converted to an active repressor that binds to the operator region, thus preventing the binding of RNA polymerase to the promoter region, which in turn brings enzyme biosynthesis to a halt.

4.2.7 ADDITIONAL TYPES OF REGULATION

Other types of regulation include stringent control and regulatory inactivation. The effector of stringent control is the alarmone guanosine 5′-diphosphate 3′-diphosphate (ppGpp). *Regulatory inactivation* refers to the selective inactivation of enzymes by two different mechanisms, namely, modification inactivation and degradative inactivation. In *modification inactivation*, the enzyme remains intact, but its physical state is changed or is covalently modified. Covalent modifications include phosphorylation of a specific serine or threonine residue, nucleotidylation of a specific tyrosine residue, ADP-ribosylation of an arginine residue, methylation of a glutamate or aspartate carboxyl group, acetylation of an ε-amino group of a lysine residue, or tyrosinolation of a protein's terminal carboxyl group. In *degradative inactivation*, at least one peptide bond is broken, which may represent the first step in protein turnover. It is carried out by proteases, which are restricted from nonselective action by confinement in vacuoles or by protease inhibitors. Regulatory inactivation usually occurs after the exponential phase of growth, especially after exhaustion of a source of carbon or nitrogen. This inactivation serves to prevent futile cycles of metabolism, to destroy enzymes no longer needed, and to divert carbon flow at branch points from one branch to another.

4.3   APPROACHES TO STRAIN IMPROVEMENTS

Organisms used today for industrial production of primary metabolites have been developed by programs of intensive random mutagenesis followed by screening and selection of overproducers. Such efforts often start with organisms that have some capacity to make the desired product but that require multiple mutations leading to deregulation in a particular biosynthetic pathway before high productivity can be obtained. Auxotrophic mutants are often very useful (Figures 4.1 and 4.2). The sequential mutations ensure that nutrients are channeled efficiently to the appropriate products without significant deviation to other pathways. These mutations involve not only release of feedback controls but also enhancement of the formation of pathway precursors and intermediates. This approach to strain improvement has been remarkably successful in producing organisms that make industrially significant concentrations of primary metabolites. However, some of the problems with this "brute force" approach include (1) the necessity of screening large numbers of mutants for the rare combination of traits sequentially obtained that lead to overproduction, and (2) the possibility that the vigor of the producing strain may be substantially weakened following several rounds of mutagenesis.

More recent approaches employ recombinant DNA technology to develop strains that are capable of overproducing primary metabolites. This rationale for strain construction relies largely on the same principles of regulation discussed in the previous sections but aims at assembling the appropriate characteristics by means of *in vitro* recombinant DNA techniques. This is particularly valuable in organisms with complex regulatory systems, where deregulation would involve many genetic alterations.

Production of a particular primary metabolite by deregulated organisms may inevitably be limited by the inherent capacity of the particular organism to make the appropriate biosynthetic enzymes, that is, even in the absence of repressive mechanisms, there may not be enough of the enzyme made to obtain high productivity. One way to overcome this is to increase the number of copies of structural genes coding for these enzymes by genetic engineering. Another way often used in combination with this strategy is to increase the frequency of transcription, which is related to the frequency of binding of RNA polymerase to the promoter region. The former can be achieved by cloning the biosynthetic genes *in vitro* into a plasmid that, when introduced into the cell through transformation, will replicate into multiple copies. Increasing the frequency of transcription involves the construction of a recombinant plasmid *in vitro* that contains the structural genes of the biosynthetic enzymes but lacks the regulatory sequences (promoter and operator) normally associated with them. Instead, the structural genes are cloned downstream of an efficient promoter, thus facilitating a higher level of expression. The ideal plasmid for metabolite synthesis would contain a regulatory region with a constitutive phenotype, preferably not subject to nutritional repression. Novel genetic technologies such as *genome-based strain reconstruction* achieve the construction of a superior strain that contains only mutations crucial to hyperproduction, but not other unknown mutations that accumulate by brute-force mutagenesis and screening. This approach has successfully been used to improve lysine production (see Section 5.4.1.2). The directed improvement of product formation or cellular properties via modification of specific biochemical reactions or introduction of new ones with the use of recombinant DNA technology is known as *metabolic engineering*. Analytical methods are combined with molecular biological techniques to quantify fluxes and implement suggested genetic modifications. Different means of analyzing metabolic fluxes are (1) kinetic-based models, (2) control theories, (3) tracer experiment, (4) magnetization transfer, (5) metabolite balancing, (6) enzyme analysis, and (7) genetic analysis. The overall flux through a metabolic

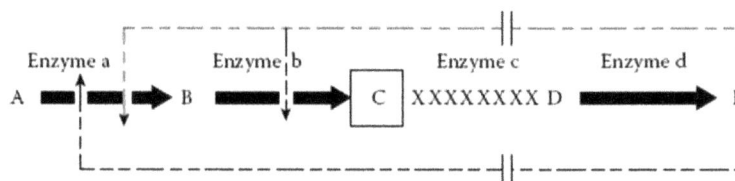

FIGURE 4.1   Overproduction of an intermediate of a linear primary metabolic pathway. Feedback inhibition by the end product inhibits the activity of enzyme a and feedback repression represses the formation of enzymes a and b. By making a genetic block (mutation) at enzyme c, an auxotrophic mutant is made, which cannot grow unless the metabolite E is added to the medium. As long as the amount of E present is not excessive, there will be no feedback effects and the metabolite C will be overproduced.

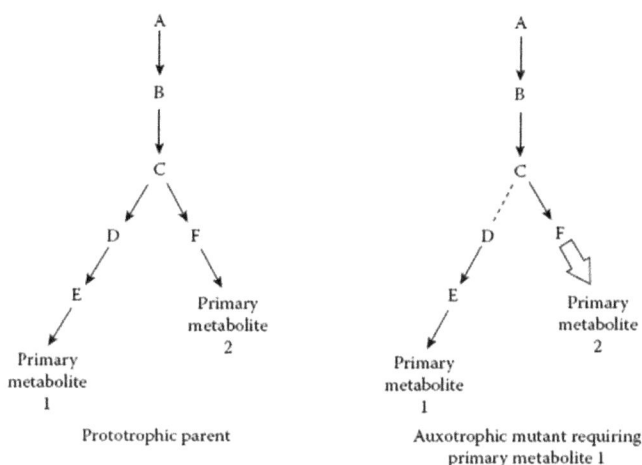

FIGURE 4.2 Use of auxotrophic mutation in a branched pathway. The auxotrophic mutant will require primary metabolite 1 for growth. When the concentration of metabolite 1 is not excessive to biosynthetic demands, primary metabolite 2 will be overproduced.

pathway depends on several steps, not just a single rate-limiting reaction.

A genome-wide transcript expression analysis called *massive parallel signature sequencing* has been successfully used to discover new targets for further improvement of riboflavin production by the fungus *Ashbya gossypii* (see Section 4.4.3.2). The development and combined application of the above technologies will help to develop "inverse metabolic engineering," which in turn will be used to construct certain phenotypes that are ideal for commercial purposes.

Molecular breeding techniques such as DNA shuffling come closer to mimicking natural recombination by allowing *in vitro* homologous recombination. These techniques not only recombine DNA fragments but also introduce point mutations at a very low but controlled rate. Unlike site-directed mutagenesis, this method of pooling and recombining parts of similar genes from different species or strains has yielded remarkable improvements in the catalytic activities of enzymes within a relatively short space of time. *Whole-genome shuffling* is a novel technique for strain improvement combining the advantage of multiparental crossing allowed by DNA shuffling with the recombination of entire genomes.

4.4 PRODUCTION OF PRIMARY METABOLITES

4.4.1 Amino Acids Production

The amino acid market is over US$7 billion and has been growing at 5–10% per year. Production of amino acids amounted to 6.58 million tons in 2014. Produced by fermentation are 2.4 million tons of l-glutamate, 1.5 million tons of L-lysine-HCL, 200,000 tons of l-threonine, 20,000 tons of L-phenylalanine (including that made by chemical synthesis), 3,000 tons of L-glutamine, 1,500 tons of L-arginine, 5,000 tons of L-tryptophan (including enzymatic method), 1,000 tons of L-valine, 1,000 tons of L-leucine (including extraction), 1,000 tons of L-isoleucine (including extraction), 500

tons of L-histidine, 500 tons of L-proline, 400 tons of L-serine, and 200 tons of L-tyrosine; 17,000 tons of L-aspartic acid and 600 tons of L-alanine are made enzymically. Dl-methionine is made chemically at 600,000 tons per year.

Top fermentation titers reported in the literature are as follows: 150 g l^{-1} glutamic acid, 170 g l^{-1} L-lysine-HCL, 124.5 g l^{-1} l-threonine, 51 g l^{-1} L-phenylalanine, 100 g l^{-1} L-arginine, 60 g l^{-1}, 49 g l^{-1} L-glutamine, 58 g L-tryptophan, 105 g l^{-1} L-valine, 34 g l^{-1} L-leucine, 40 g l^{-1} L-isoleucine, 42 g l^{-1} L-histidine, 108 g l^{-1} L-proline, 65 g l^{-1} L-serine, 120 g l^{-1} L-alanine, 131 g l^{-1} l-tyrosine, and 25.5 g l^{-1} l-methionine. Genetic and metabolic engineering have made an impact by use of the following strategies: (1) amplification of the rate-limiting (controlling) enzyme of the pathway, (2) amplification of the first enzyme after a branch point, (3) cloning of a gene encoding an enzyme with more or less feedback regulation, (4) introduction of a gene encoding an enzyme with a functional or energetic advantage as replacement for the normal enzyme, and (5) amplification of the first enzyme leading from the central metabolism to increase carbon flow into the pathway followed by sequential removal of bottlenecks caused by the accumulation of intermediates. Transport mutations are also useful (i.e., mutations decreasing amino acid uptake often allow for improved excretion and lower intracellular feedback control). In cases where excretion is carrier mediated, increase in activity of these carrier enzymes increases production of the amino acid.

Amino acids produced by microbial process are the l-forms. Such stereospecificity makes the process advantageous as compared with synthetic processes. Microbial strains employed for amino acid production are divided into four classes: wild-type strains, auxotrophic mutants, regulatory mutants, and auxotrophic regulatory mutants. Using bacterial mutants, all the essential amino acids except l-methionine can be produced by "direct fermentation" from cheap carbon sources such as carbohydrate materials or acetic acid.

4.4.1.1 Production of L-Glutamic Acid

Monosodium glutamate (MSG) is a potent flavor enhancer, which was first made by fermentation in Japan in the late 1950s. Many organisms belonging to a wide range of taxonomically related genera, including *Micrococcus*, *Corynebacterium*, *Brevibacterium*, and *Microbacterium*, are capable of overproducing glutamate. *Brevibacterium lactofermentum* and *Brevibacterium flavum* are now reclassified as subspecies of *C. glutamicum*. These organisms were shown to possess the Embden–Meyerhof–Parnas glycolytic pathway (EMP), the hexose monophosphate pathway (HMP), the tricarboxylic acid (TCA) cycle, and the glyoxylate bypass (Figure 4.3). The TCA cycle, also widely known as the *Krebs cycle*, requires a continuous replenishment of oxaloacetate in order to replace the intermediates withdrawn for the synthesis of biomass and other amino acids. During growth on glucose and other glycolytic intermediates, the anaplerotic function is fulfilled by phosphoenolpyruvate carboxylase and pyruvate carboxylase.

Normally, glutamic acid overproduction would not be expected to occur due to feedback regulation. Glutamate

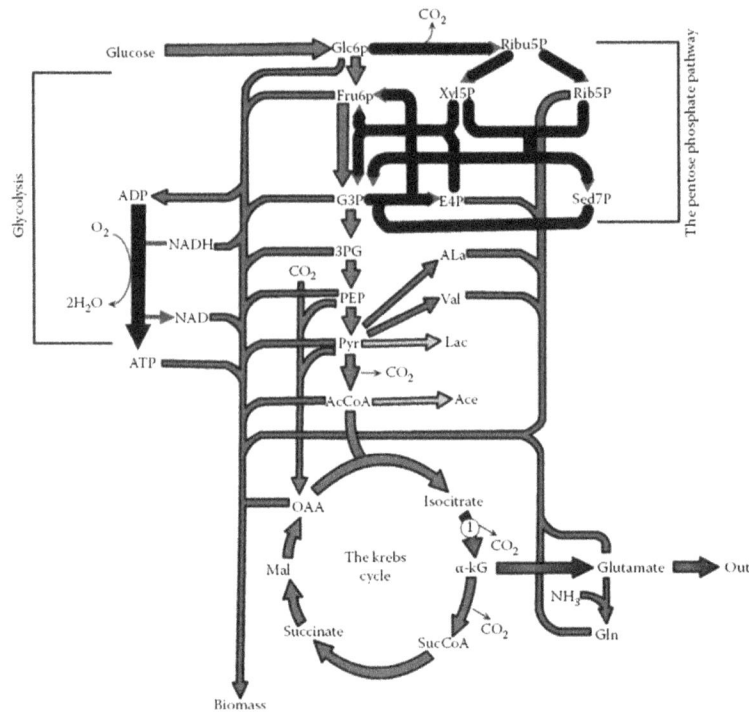

FIGURE 4.3 Microbial biosynthesis of glutamic acid from glucose. (Mansi El-Mansi personal communication; based on Varela et al. 2003. *Appl. Microbiol. Biotechnol.* 60, 547–55, 2003.)

feedback controls include repression of PEP carboxylase, citrate synthase, and NADP-glutamate dehydrogenase; the last-named enzyme is also inhibited by glutamate. However, by decreasing the effectiveness of the barrier to outward passage, glutamate can be pumped out of the cell, thus allowing its biosynthesis to proceed unabated. The excretion of glutamate frees the glutamate pathway from feedback control until a very high level is accumulated; commercial l-glutamate titer is in excess of 150 g l⁻¹.

Glutamate excretion is intentionally influenced by manipulations of growth conditions, biotin limitation was the first means discovered to bring about glutamate overproduction in *C. glutamicum*, and all glutamate overproducers are natural biotin auxotrophs (Figure 4.4). The finding that the addition of penicillin to cells grown in high biotin resulted in excretion of glutamic acid (Figure 4.5) led workers to postulate (1) that growth of the glutamate-overproducing bacterium in the presence of nonlimiting levels of biotin results in a cell membrane permeability barrier that restricts the excretion of glutamate, and (2) that inhibition of cell wall biosynthesis by penicillin alters the permeability properties of the cell membrane and allows glutamate to flow out easily. The commonality in the various manipulations that were found to bring about high-level production of l-glutamic acid (i.e., the limitation of biotin, or addition of penicillin or fatty acid surfactants [e.g., tween 60], to exponentially growing cultures) was recognized. Apparently, all of these manipulations result in a phospholipid-deficient cytoplasmic membrane, which favors active excretion of glutamate from the cell. This view was further substantiated by the discoveries that oleate limitation of an oleate auxotroph and glycerol limitation of a glycerol

auxotroph also bring about glutamate excretion. Furthermore, glutamate-excreting cells were later found to have a very low level of cell lipids, especially phospholipids. In addition, it was found that the various manipulations leading to glutamate overproduction cause increased permeability of the mycolic acid layer of the cell wall. The glutamate-overproducing bacteria are characterized by a special cell envelope containing mycolic acids that surrounds the entire cell as a structured layer and is thought to be involved in the permeation of solutes. The mycolic acids esterified with arabinogalactan and the noncovalently bound mycolic acid derivatives form a second lipid layer of the cell, with the cytoplasmic membrane being the first. Overexpression or inactivation of enzymes that

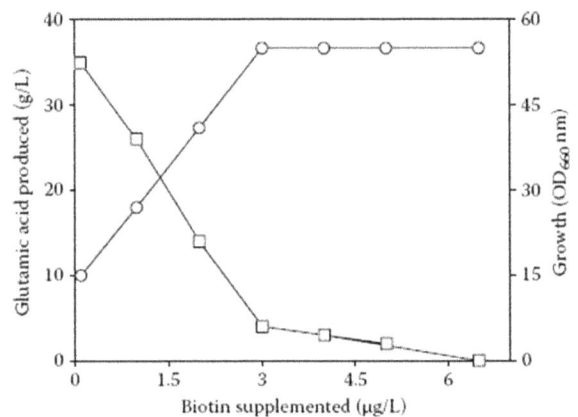

FIGURE 4.4 Effect of biotin concentration on the production of glutamate by *Corynebacterium glutamicum* strains, which are natural biotin auxotrophs.

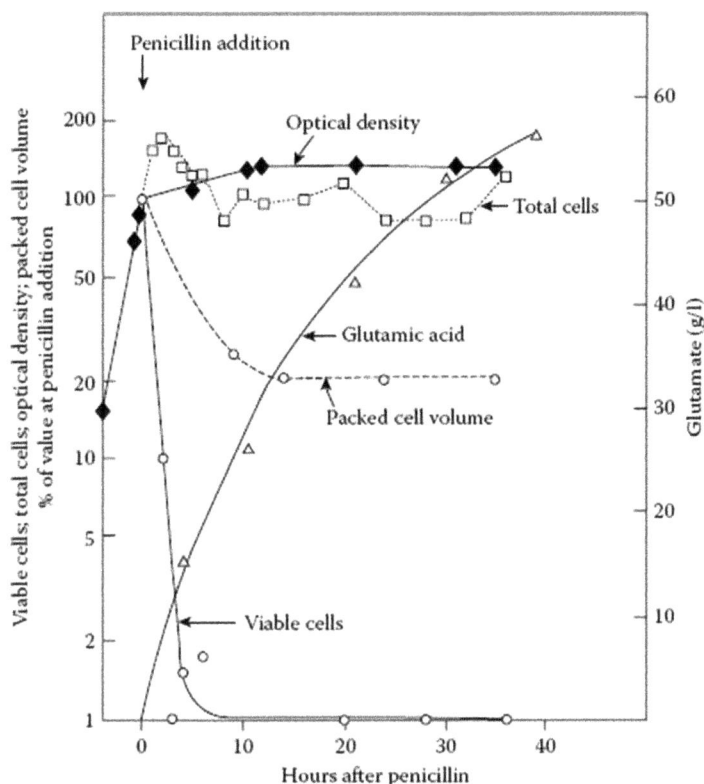

FIGURE 4.5 Effect of penicillin on the overproduction of glutamic acid during the growth of *Corynebacterium glutamicum* in the presence of excess biotin.

are involved in lipid synthesis alters the chemical and physical properties of the cytoplasmic membrane and changes glutamate efflux dramatically.

4.4.1.2 Production of L-Lysine

The bulk of the cereals consumed in the world are deficient in the amino acid L-lysine. This is an essential ingredient for the growth of animals and is an important part of a billion-dollar animal feed industry. Lysine supplementation converts cereals into balanced food or feed for animals including poultry, swine, and other livestock. In addition to animal feed, lysine is used in pharmaceuticals, dietary supplements, and cosmetics.

Lysine is a member of the aspartate family of amino acids (Figure 4.6). It is made in bacteria by a branched pathway that also produces methionine, threonine, and isoleucine. This pathway is controlled very tightly in organisms such as *E. coli*, which contains three aspartate kinases (AKs), each of which is regulated by a different end product. In addition, after each branch point, the initial enzymes are inhibited by their respective end product(s) and no overproduction usually occurs. However, *C. glutamicum*, the organism used for the commercial production of L-*lysine*, contains a single AK that is regulated via concerted feedback inhibition by threonine plus lysine. The relative contribution of carbon flux through the pentose phosphate pathway varies depending on the amino acid being produced; for example, while it contributes only 20% of the total flux in the case of glutamate formation, it contributes 60–70% in the case of lysine production. This is

evidently due to the high level of NADPH required for lysine formation. Use of rDNA technology has shown that the factors that significantly limit the overproduction of lysine are (1) the feedback inhibition of AK by lysine plus threonine, (2) the low level of dihydrodipicolinate synthase (DHPS), (3) the low level of PEP carboxylase, and (4) the low level of aspartase. Much work has been done on auxotrophic and regulatory mutants of the glutamate-overproducing strains for the production of lysine. By genetic removal of homoserine dehydrogenase (HDI), a glutamate-producing wild-type *Corynebacterium* strain was converted into a lysine-overproducing mutant that cannot grow unless methionine and threonine are added to the medium. As long as the threonine supplement is kept low, the intracellular concentration of threonine is limiting, and feedback inhibition of AK is bypassed, leading to excretion of over 70 g l^{-1} of lysine in culture fluids. In some strains, addition of methionine and isoleucine to the medium led to the increase of lysine overproduction. Selection for S-2-aminoethylcysteine (AEC, or thialysine) resistance blocks feedback inhibition of AK. Other antimetabolites useful for the deregulation of AK include a mixture of α-ketobutyrate and aspartate hydroxamate. Leucine auxotrophy can also increase lysine production. L-lysine titers are known to be as high as 170 g l^{-1}.

Excretion of lysine by *C. glutamicum* is by active transport reaching a concentration of several hundred millimolar in the external medium. Lysine, a cation, must be excreted against the membrane potential gradient (outside is positive),

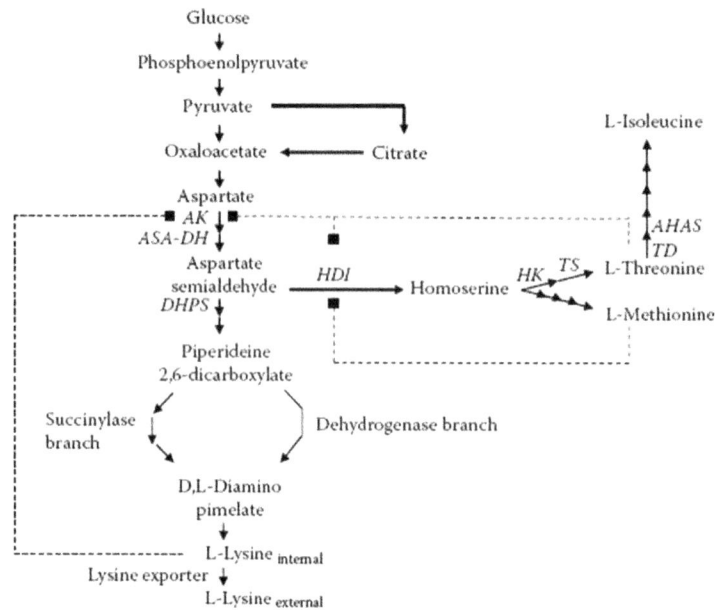

FIGURE 4.6 Biosynthetic pathway to L-lysine, L-threonine, and L-isoleucine productions. Abbreviations for key enzymes are given in the text.

and the excretion is carrier mediated. The system is dependent on electron motive force, not ATP. Genome-based strain reconstruction has been used to improve the lysine production rate of *C. glutamicum* by comparing a high-producing strain (production rate slightly less than 2 g l⁻¹ h⁻¹) and a wild-type strain. Comparison of 16 genes from the production strain, encoding enzymes of the pathway from glucose to lysine, revealed mutations in five of the genes. Introduction of three of these mutations (*hom*, *lysC*, and *pyc* encoding HDI, AK, and pyruvate carboxylase, respectively) into the wild-type created a new strain that produced 80 g l⁻¹ in 27 hours, at a rate of 3 g l⁻¹ h⁻¹, the highest rate ever reported for a lysine fermentation. An additional increase (15%) in L-lysine production was observed by introducing a mutation in the 6-phosphogluconate dehydrogenase gene (*gnd*). Enzymatic analysis revealed that the mutant enzyme was less sensitive than the wild-type enzyme to allosteric inhibition by intracellular metabolites. Isotope-based metabolic flux analysis demonstrated that the *gnd* mutation resulted in an 8% increase in carbon flux through the pentose phosphate pathway during L-lysine production. Several companies, e.g., Ajinomoto Co. Inc., have expanded their facilities for the production of L-lysine. Similarly, strong investments in Brazil, China, and the US saw a significant rise of L-lysine production from the level reported in 2012 (1,950,000 tons).

4.4.1.3 Production of L-Threonine

This amino acid is the second major amino acid used for feeding pigs and poultry. The pathway of threonine biosynthesis is similar in all microorganisms (Figure 4.6). Starting from l-aspartate, the pathway involves five steps catalyzed by five enzymes: AK, aspartate-semialdehyde dehydrogenase (ASA-DH), HDI, homoserine kinase (HK), and threonine synthetase (TS).

Production of L-threonine has been achieved with the use of several microorganisms. In *Serratia marcescens*, construction of a high threonine producer was done by transductional crosses that combined several feedback control mutations into one organism. Three classes of mutants were obtained from the parental strain as the source of genetic material for transduction: (1) one strain in which both the threonine-regulated AK and HD were resistant to feedback inhibition by threonine (it was selected on the basis of β-hydroxynorvaline resistance); (2) a second strain, also selected for β-hydroxynorvaline resistance, in which HDI was resistant to both inhibition and repression and the threonine-regulated AK was constitutively synthesized; and (3) a third strain that was resistant to thialysine, in which the lysine-regulated AK was resistant to feedback inhibition and repression. Since at least one of the three key enzymes in threonine synthesis was still subject to regulation in these strains, each produced only modest amounts of threonine (4.1 to 8.7 g l⁻¹). Recombination of the three mutations by transduction yielded a strain that produced higher levels of threonine (25 g l⁻¹), had AK and HDI activities that were resistant to feedback regulation by threonine and lysine, and was also a methionine bradytroph (leaky auxotroph). Another six regulatory mutations derived by resistance to amino acid analogs were combined into a single strain of *S. marcescens* by transduction. These mutations led to desensitization and derepression of AKs I, II, and III and HDIs I and II. The resulting transductant produced 40 g l⁻¹ of threonine, which was further improved to 63 g l⁻¹ through overexpression of PEP carboxylase. When sucrose was continuously fed to the medium, production levels of 100 g l⁻¹ were obtained at 96 h of incubation.

In *E. coli*, threonine production was increased to 76 g l⁻¹ conventional mutagenesis and by selection and screening techniques. Of major importance were mutations to decrease both regulation of the pathway and degradation of the amino acid. An *E. coli* fed-batch process with methionine and phosphate feeding yielded 98 g l⁻¹ l-threonine at 60 hours. Another *E. coli* strain was developed via mutation and genetic engineering and optimized by inactivation of threonine dehydratase (TD), resulting in a process yielding 100 g l⁻¹ in 36 hours of fermentation.

Threonine excretion by *C. glutamicum* is mainly (>90%) effected by a carrier-mediated export mechanism dependent on membrane potential. Cloning in extra copies of threonine export genes into an *E. coli* strain producing threonine led to increased production. Also increased was resistance to toxic antimetabolites of threonine. Another means of increasing threonine production was reduction in the activity of serine hydroxytransferase, which breaks down threonine to glycine. In *C. glutamicum* ssp. *lactofermentum*, threonine production reached 58 g l⁻¹ when a strain producing both threonine and lysine (isoleucine auxotroph resistant to thialysine, α-amino-β-hydroxyvaleric acid, and S-methylcysteine sulfoxide) was transformed with a recombinant plasmid carrying its own *hom* (encoding HDI), *thrB* (encoding HK), and *thrC* (encoding TS) genes. By using combined feeding strategies, L-threonine levels reached 124.5 g l⁻¹. Global threonine production by 2014 was close to 300,000 tons.

4.4.1.4 Production of L-Isoleucine

Isoleucine is of commercial interest as a food and feed additive and for parenteral nutrition infusions. This branched-chain amino acid is currently produced both by extraction of protein hydrolysates and by fermentation with classically derived mutants of *C. glutamicum*. The biosynthesis of isoleucine by *C. glutamicum* involves 11 reaction steps, of which at least five are controlled with respect to activity or expression (Figure 4.6). L-isoleucine synthesis shares reactions with the lysine and methionine pathways. In addition, threonine is an intermediate in isoleucine formation, and the last four enzymes also carry out reactions involved in valine, leucine, and pantothenate biosynthesis. Therefore, it is not surprising that multiple regulatory steps identified in *C. glutamicum*, as in other bacteria, are required to ensure the balanced synthesis of all these metabolites for cellular demands. In *C. glutamicum*, flux control is exerted by repression of the *homthrB* and *ilvBNC* operons. The activities of AK, HDI, TD, and acetohydroxy acid synthase (AHAS) are controlled by allosteric transitions of the proteins to provide feedback control loops, and HK is inhibited in a competitive manner. Isoleucine increases the Michaelis-Menten Kinetics (Km) of TD from 21 to 78 mM, whereas valine reduces it to 12 mM. The AHAS is 50% feedback inhibited by isoleucine plus valine plus leucine.

Isoleucine processes have been devised in various bacteria such as *S. marcescens*, *C. glutamicum* ssp. *flavum*, and *C. glutamicum*. In *S. marcescens*, resistance to isoleucine hydroxamate and α-aminobutyric acid led to derepressed TD and AHAS and production of 12 g l⁻¹ of isoleucine. Further work involving transductional crosses into a threonine overproducer yielded isoleucine at 25 g l⁻¹. The *C. glutamicum* ssp. *flavum* work employed resistance to α-amino-β-hydroxyvaleric acid,

and the resultant mutant produced 11 g l⁻¹. Mutation to d-ethionine resistance yielded a mutant producing 33.5 g l⁻¹ isoleucine in a fermentation continuously fed with acetic acid. A threonine-overproducing strain of *C. glutamicum* was sequentially mutated to resistance to thiaisoleucine, azaleucine, and α-aminobutyric acid; it produced 10 g l⁻¹ of isoleucine. An improved strain was obtained by cloning multiple copies of *hom* (encoding HDI) and wild-type *ilvA* (encoding TD) into a lysine overproducer, and by increasing HK (encoded by *thrB*), 15 g l⁻¹ isoleucine was produced. Independently, cloning of three copies of the feedback-resistant HDI gene (*hom*) and multiple copies of the deregulated TD gene (ilvA) in a deregulated lysine producer of *C. glutamicum* yielded an isoleucine producer (13 g l⁻¹) with no threonine production and reduced lysine production. Application of a closed-loop control fed-batch strategy raised production to 18 g l⁻¹, which was further amplified using metabolic engineering strategies to 40 g l⁻¹ of isoleucine. The annual production of isoleucine is around 1,000 tons per year.

4.4.1.5 Production of Aromatic Amino Acids

In *C. glutamicum* ssp. *flavum*, 3-deoxy-d-arabino-heptulosonate 7-phosphate synthase (DAHPS) is feedback inhibited concertedly by phenylalanine plus tyrosine and weakly repressed by tyrosine. Other enzymes of the common pathway (Figure 4.7) are not inhibited by phenylalanine, tyrosine, and tryptophan, but the following are repressed: shikimate dehydrogenase (SD), shikimate kinase (SK), and 5-enolpyruvylshikimate-3-phosphate synthase. Elimination of the uptake system for aromatic amino acids in *C. glutamicum* results in increased production of aromatic amino acids in deregulated strains.

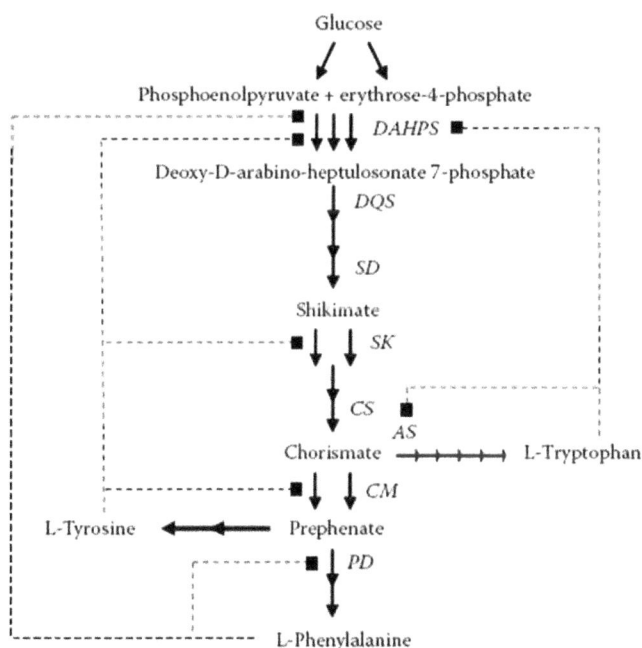

FIGURE 4.7 Biosynthetic pathways for L-tryptophan, L-phenylalanine, and L-tyrosine productions. Abbreviations for key enzymes are given in the text.

A tryptophan process was improved from 8 to 10 g l⁻¹ by mutating the *C. glutamicum* ssp. *flavum* producer to azaserine resistance. Azaserine is an analog of glutamine, the substrate of anthranilate synthase (AS). Such a mutant showed a 2–3-fold increase in the activities of DAHPS, dehydroquinate synthase (DQS), SD, SK, and chorismate synthase (CS). Another mutant, selected for its ability to resist sulfaguanidine, showed additional increases in DAHPS and DQS and tryptophan production. The reason that sulfaguanidine was chosen as the selective agent involves the next limiting step after derepression of DAHPS (i.e., conversion of the intermediate chorismate to anthranilate by AS). Chorismate can also be undesirably converted to *p*-aminobenzoic acid (PABA), and sulfonamides are PABA analogs. A sulfaguanidine-resistant mutant was obtained with *C. glutamicum* ssp. *Flavum*, and production increased from 10 g l⁻¹ tryptophan to 19 g l⁻¹. The sulfaguanidine-resistant mutant was still repressed by tyrosine but showed higher enzyme levels at any particular level of tyrosine. Gene cloning of the tryptophan branch and mutation to resistance to feedback inhibition yielded a *C. glutamicum* strain producing 43 g l⁻¹ L-tryptophan. The genes cloned were those that encoded AS, anthranilate phosphoribosyl transferase, a deregulated DAHPS, and other genes of tryptophan biosynthesis. However, sugar utilization decreased at the late stage of the fermentation, and plasmid stabilization required antibiotic addition. Sugar utilization stopped due to killing by accumulated indole. By cloning in the 3-phosphoglycerate dehydrogenase gene (to increase the production of serine, which combines with indole to form more tryptophan) and by mutating the host cells to deficiency in this enzyme, both problems were solved. The new strain produced 50 g l⁻¹ tryptophan with a productivity of 0.63 g l⁻¹ h⁻¹ and a yield from sucrose of 20%. Further genetic engineering to increase the activity of the pentose phosphate pathway increased production to 58 g l⁻¹. A deregulated strain of *E. coli* in which feedback inhibition and repression controls were removed made 11 g l⁻¹ phenylalanine in a fed-batch culture. Production was increased to 28.5 g l⁻¹ when a plasmid was cloned into *E. coli* containing a feedback inhibition-resistant version of the CM-prephenate dehydratase (PD) gene, a feedback inhibition-resistant DAHPS, and the $O_R P_R$ and $O_L O_L$ operator-promoter system of lambda phage. Further process development of genetically engineered *E. coli* strains brought phenylalanine titers up to 46 g l⁻¹. Independently, genetic engineering based on cloning *aroF* and feedback-resistant *pheA* genes created an *E. coli* strain producing 51 g l⁻¹. A *C. glutamicum* ssp. *lactofermentum* culture, obtained by selection with *m*-fluorophenylalanine, produced 5 g l⁻¹ phenylalanine, 7 g l⁻¹ tyrosine, and 0.3 g l⁻¹ anthranilate and contained desensitized DAHPS and PD. DAHPS in the wild-type was inhibited cumulatively by phenylalanine and tyrosine, whereas PD was inhibited by phenylalanine. Cloning of the gene encoding PD from a desensitized mutant and the gene encoding desensitized DAHPS increased the enzyme activities and yielded a strain producing 18 g l⁻¹ phenylalanine, 1 g l⁻¹ tyrosine, and no anthranilate. Further cloning of a recombinant plasmid expressing desensitized DAHPS increased production to 26 g l⁻¹

phenylalanine. Similarly, *C. glutamicum* strains have been developed, producing up to 28 g l⁻¹ phenylalanine. A genetic engineering approach based on cloning *aroF* and feedback-resistant *pheA* genes created an *E. coli* strain producing 51 g l⁻¹. When SK was cloned into a tyrosine-producing *C. glutamicum* ssp. *lactofermentum* strain, tyrosine production increased from 17 to 22 g l⁻¹. Cloning of desensitized genes encoding DAHPS and CM from a deregulated phenylalanine-producing *C. glutamicum* strain into the deregulated tryptophan producer, *C. glutamicum* KY 10865 (CM-deficient strain, phenylalanine, and tyrosine double auxotroph with a desensitized AS), shifted production from 18 g l⁻¹ tryptophan to 26 g l⁻¹ tyrosine. The use of *E. coli* for tyrosine overproduction was achieved by replacing the *pheLA* genes of a phenylalanine-producing strain with a multigene cassette comprising the *tyrA* gene under the control of the constitutive *trc* promotor and a kanamycin resistance gene. Surprisingly, deletion of the *lacI* repressor led to an increase in *tyrA* expression and a 5-fold increase in tyrosine production to more than 50 g l⁻¹ at a 200:l scale. Using a strain harboring a thermoestable tyrosin phenol lyase in a fed-batch reactor with continuous feeding of phenol, pyruvate, and ammonia, a production level of 131 g l⁻¹ was obtained.

4.4.2 Production Processes for Purines and Pyrimidines, Their Nucleosides and Nucleotides

Commercial interest in nucleotide fermentations is due to the activity of two purine ribonucleoside 5'-monophosphates, namely, guanylic acid (guanosine 5'-monophosphate, or GMP) and inosinic acid (inosine 5'-monophosphate, or IMP) as flavor enhancers. It is quite impressive that a 1:1 mixture of MSG with IMP or GMP gives flavor intensity 30 times stronger than that of MSG alone. Approximately 22,000 tons of GMP and IMP were produced in 2010. The global nucleotide market value in 2014 was US$403.1 million and is expected to reach US$809.3 million by 2022.

The purine residue of IMP is built up on a ribose ring in 10 enzymatic reactions after ribose phosphate pyrophosphokinase (PRPP synthetase) catalyzes the conversion of α-d-ribose-5-phosphate (R5P) and ATP to 5-phosphoribosyl-α-pyrophosphate (PRPP) (Figure 4.8). Adenosine-5'-monophosphate (AMP) and GMP are synthesized from IMP. AMP formation involves participation of two enzymes, adenylosuccinate synthetase and adenylosuccinate. GMP synthesis requires the participation of IMP dehydrogenase and GPM synthetase. PRPP synthetase is feedback inhibited by AMP, GMP, and IMP; adenylosuccinate synthetase is inhibited by AMP; and IMP dehydrogenase is inhibited by xanthosine-5'-monphosphate (XMP) and GMP.

The genes encoding the enzymes of IMP biosynthesis in *Bacillus subtilis* constitute the *pur* operon, whereas the genes encoding the GMP biosynthetic enzymes, *guaA* (GMP synthetase) and *guaB* (IMP dehydrogenase), and the *purA* gene encoding adenylosuccinate synthetase all occur as single units. The *purB* gene encodes an enzyme involved in both IMP and

Phosphoribosyl pyrohosphate (PRPP)

(1) ↓

Phosphoribosylamine

(2) ↓

Glycinamide ribotide

(3) ↓

Formyglycinamide ribotide

(4) ↓

Formylglycinamidine ribotide

(5) ↓

Aminoimidazole ribotide

(6) ↓

Aminoimidazole carboxylic acid ribotide

(7) ↓

Aminoimidazole-N-succino-carboxamide ribotide

(8) ↓

Aminoimidazole carboxamide ribotide (AICAR)

(9) ↓

Formamido-imidazole carboxamide ribotide

(10) ↓

Hypoxanthine ← Inosine ← IMP ←

(11) (12) (15)

Adenylo-   Xanthine → XMP
succinate   Xanthosine ← XMP

(13)   (14)

AMP → Adenosine → Adenine   Guanine ← Guanosine ← GMP

RNA ← ← ←

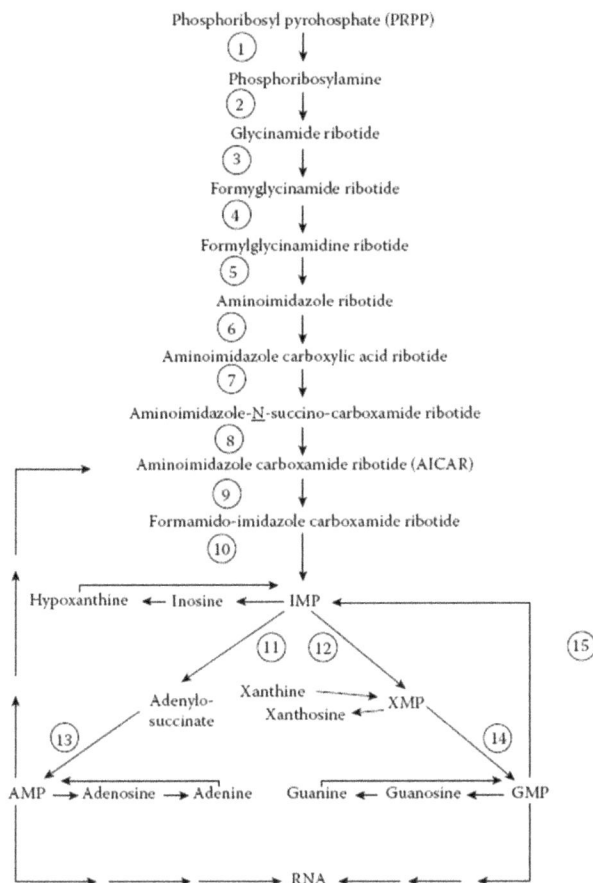

FIGURE 4.8 Biosynthesis of purine nucleotides. Key enzymes are as follows: Enzyme 1 is PRPP amido-transferase, Enzyme 11 is adenylosuccinate synthetase, Enzyme 12 is IMP dehydrogenase, Enzyme 13 is adenylsuccinase, and Enzyme 14 is XMP aminase.

AMP biosynthesis and is located in the *pur* operon. The levels of purine biosynthetic enzymes (except for GMP synthetase) are repressed in cells grown in the presence of purine compounds. Transcription of the *pur* operon is regulated negatively by adenine and guanine compounds including ATP, hypoxanthine, and guanine, which are corepressors. Feedback repression of purine nucleotide biosynthesis in *E. coli* is exerted by binding of the corepressor to the product of the *purR* gene. Hypoxanthine and guanine act cooperatively to change the conformation of *PurR*, thus enhancing its binding to DNA.

Techniques similar to those described above for amino acid fermentations have yielded IMP titers of 30 g l⁻¹. Since only low levels of GMP have been produced by direct fermentation, it is usually made by bioconversion of XMP, reaching GMP titers of about 30 g l⁻¹. Genetic modification of *Corynebacterium ammoniagenes* involving transketolase (an enzyme of the nonoxidative branch of the pentose phosphate pathway) resulted in the accumulation of 39 g l⁻¹ of XMP. This work demonstrates the need for high levels of pentose (ribose) for nucleotide and nucleoside biosynthesis and overproduction.

The key to effective accumulation of purines and their derivatives is the limitation of intracellular AMP and GMP.

This limitation is best achieved by restricting purine supply during growth of the purine auxotrophs. Thus, adenine-requiring mutants lacking adenylosuccinate synthetase accumulate hypoxanthine or inosine that results from the breakdown of intracellularly accumulated IMP. Certain adenine-auxotrophs of *B. subtilis* excrete over 10 g l⁻¹ of inosine. These strains are still subject to GMP repression of enzymes of the common path. To minimize the severity of this regulation, the adenine-auxotrophs are further mutated to eliminate IMP dehydrogenase. These adenine-xanthine double auxotrophs show a 2-fold increase in the specific activity of some common-path enzymes and accumulate inosine up to 15 g l⁻¹ under conditions of limiting adenine and xanthine (or guanosine). Further deregulation is achieved by the selection of mutants resistant to purine analogs. Thus, mutants resistant to azaguanine with requirements for adenine and xanthine produce over 20 g l⁻¹ inosine. Insertional inactivation of the IMP dehydrogenase gene in a *B. subtilis* strain yielded a culture producing inosine at 35 g l⁻¹. Cloning of IMP dehydrogenase has been used to improve guanosine production in *B. subtilis*. The donor strain produced a low level of purine nucleosides, but the distribution was in favor of guanosine (1 g l⁻¹ of inosine vs. 9 g l⁻¹ guanosine). The recipient strain was auxotrophic for adenine, lacked GMP reductase and purine nucleoside phosphorylase, and was resistant to 8-aza-guanine, adenine, and adenosine; it produced 19 g l⁻¹ inosine and 7 g l⁻¹ guanosine. Cloning of IMP dehydrogenase from the first strain into the second strain yielded a recombinant that produced 5 g l⁻¹ inosine and 20 g l⁻¹ guanosine. Other *B. subtilis* mutants produce as much as 23 g l⁻¹ of guanosine. Nucleosides such as inosine and guanosine are then converted to their active nucleotide derivatives chemically, microbiologically, or enzymically.

The *de novo* pyrimidine biosynthetic pathway involves five enzymes and results in uridine-5′-monophosphate (UMP) production. Aspartate transcarbamoylase, the first pathway enzyme committed to pyrimidine biosynthesis, catalyzes the conversion of aspartate and carbamoylphosphate to carbamoylaspartate. The subsequent biosynthetic pathway enzymes are dihydroorotase, dihydroorotate dehydrogenase, orotate phosphoribosyltransferase, and orotidine-5′-monophosphate (OMP) decarboxylase. Uridine triphosphate (UTP) is produced from UMP by the sequential actions of two nucleoside kinases. Cytidine triphosphate (CTP) is formed by the amination of UTP by CTP synthetase. The pyrimidine biosynthetic pathway is usually regulated at the level of gene expression. UMP kinases from *E. coli* and *B. subtilis* are activated by GTP and inhibited by UTP. The selection for antimetabolite resistance has proven to be successful in the development of pyrimidine nucleotide and nucleoside fermentations. Cytidine production by a *B. subtilis* cytidine deaminase-deficient mutant with resistance to fluorocytidine amounted to 10 g l⁻¹. Further mutation to 3-deazauracil resistance increased production to 14 g l⁻¹. By introducing a gene encoding a feedback-resistant carbamyl phosphate synthase, cytidine production was raised to 18 g l⁻¹. Homoserine dehydrogenase (HSD) deficiency in *B. subtilis* increased cytidine production in a deregulated mutant from 9 to 23 g l^{-1}. Increasing the glucose concentration raised production to 30 g l^{-1}. Uridine production by mutants of *B. subtilis* resistant to pyrimidine antimetabolites has reached 55 g l^{-1}.

4.4.3 Production Processes for Vitamins

Vitamins are one of the main products for the international markets of feed, medicine, and health care products. More than half of the vitamins produced commercially are fed to domestic animals. The vitamin market is a multibillion-dollar industry. According to China Marketing Research estimates, in 2016, the global vitamin production was more than 390,000 tons, generating annual revenue in excess of US\$4.0 billion. Microbes produce five vitamins commercially: vitamin B$_{12}$ (cyanocobalamin), ascorbic acid (vitamin C), riboflavin (vitamin B$_2$), pantothenic acid (vitamin B$_5$), and biotin.

4.4.3.1 Production of Vitamin B12

Vitamin B$_{12}$ is produced commercially at about 15 tons per year. Fermentations have to be run under complete or partial anaerobiosis when using species of *Pseudomonas* or *Propionibacterium*. The major industrial organisms are *Pseudomonas denitrificans* and *Propionibacterium shermanii*. Conventional strain improvement has yielded *P. denitrificans* strains producing 150 mg l^{-1}. *Propionibacterium freudenreichii* can produce 206 mg l^{-1}.

Increasing the activity of S-adenosyl-l-methionine uroporphyrinogen III methyltransferase (SUMT) by cloning in a DNA fragment containing this gene (*cobA*) in *P. denitrificans* increased vitamin B$_{12}$ production by 100%. SUMT is at the branch point of the heme and B$_{12}$ biosynthetic pathways. Cloning of the gene *cobI* increased S-adenosyl-l-methionineprecorrin-2-methyltransferase (SP$_2$MT) and B$_{12}$ production by 30%. SP$_2$MT is at the branch point of the siroheme and B$_{12}$ pathways.

4.4.3.2 Production of Riboflavin

Annual production of riboflavin is over 8,000 tons/year by fermentation. Riboflavin overproducers include two yeast-like molds, *E. ashbyii* and *A. gossypii*, which synthesize riboflavin in concentrations greater than 20 g l^{-1}.

With *A. gossypii*, riboflavin production was found to be stimulated 3–4-fold by the addition of the precursors glycine and hypoxanthine. The level of production, which occurs after a growth rate declines, is determined by the activity of the promoter of gene *RIB3*. This gene encodes 3, 4-dihydroxy-2-butanone-4-phosphate (DHBP) synthase, the first enzyme of the pathway. Mutation of *A. gossypii* to resistance to aminomethylphosphonic acid (a glycine antimetabolite) yielded improved producers. Isocitrate lyase (ICL) is important for the use of fatty acids for riboflavin production. Itaconate, an inhibitor of ICL, eliminated the yellow color of *A. gossypii* colonies. A mutant strain, which was yellow on itaconate-containing agar, produced 15% more enzyme and 25-fold more riboflavin.

Processes using recombinant *B. subtilis* strains that produce at least 30 g l^{-1} riboflavin have been developed. In this species, riboflavin formation is regulated by feedback repression, not inhibition. An aporepressor encoded by *ribC*, whose effectors are riboflavin, FMN, and FAD, is responsible for this effect. Mutations of *ribC* lead to riboflavin overproduction. Sequential selection for resistance to 8-azaguanine, decoyinine, methionine sulfoxide, and roseoflavin plus cloning of multiple copies of the *riboflavin biosynthetic rib* operon yielded overproducing mutants. Further improvement was achieved when an extra copy of the *ribA* gene was introduced into the culture. This gene encodes both GTP cyclohydrolase II and 3, 4-dihydroxy-2-butanone 4-phosphate synthase, both of which act to commit precursors GTP and ribulose-5-phosphate to riboflavin biosynthesis.

A *Candida famata* (*Candida flareri*) strain produced 20 g l^{-1} in 200 hours. It was obtained by mutation and selection for resistance to 2-deoxyglucose (DOG), iron, tubercidin (a purine analog), and depleted medium, plus protoplast fusion. The process depends on the addition of glycine and hypoxanthine. Selection for resistance to the adenine antimetabolite 4-aminopyrazolo (3, 4-d) pyrimidine improved production. Threonine showed a 9-fold stimulation in a strain with a cloned threonine aldolase, which converts threonine to glycine.

4.4.3.3 Production of Vitamin C

Vitamin C has a global production of 140,000 tons/year and a market value of US\$820.4 million. It is used for nutrition of humans and animals and as a food antioxidant. The otherwise chemical Reichstein process utilizes one bioconversion reaction, the oxidation of d-sorbitol to l-sorbose by *Gluconobacter oxydans*, as the first step in ascorbic acid production. The biotransformation proceeds at the theoretical maximum (i.e., 200 g l^{-1} of d-sorbitol can be converted to 200 g l^{-1} of l-sorbose) when using a mutant of *G. oxydans* selected for resistance to a high sorbitol concentration. The bioconversion is used rather than a chemical reaction since the latter produces unwanted d-sorbose along with l-sorbose. An excellent fed-batch bioconversion process uses a starting concentration of 100 g l^{-1} of d-sorbitol and achieves production of 280 g l^{-1} of l-sorbose in 16 hours with a productivity of 17.6 g l^{-1} h^{-1}. The Reichstein process converts glucose to 2-keto-l-gulonic acid (2-KLGA) in five steps with a yield of 50%. Then, 2-KLGA is chemically converted to l-ascorbic acid in two more steps.

Fermentation processes are competing with the Reichstein method. A mixed culture of *G. oxydans* (which converts l-sorbose to 2-KLGA) and *Gluconobacter suboxydans* (which converts d-sorbitol to l-sorbose) was able to convert 138 g l^{-1} of d-sorbitol to 112 g l^{-1} of 2-KLGA, with a molecular conversion yield of 75% in two days. A recombinant strain of *G. oxydans* containing genes encoding l-sorbose dehydrogenase and l-sorbosone dehydrogenase from *G. oxydans* was able to produce 2-KLGA effectively from d-sorbitol. Mutation to suppress the l-idonate pathway and improvement of the promoter led to production of 130 g l^{-1} 2-KLGA from 150 g l^{-1} d-sorbitol.

4.4.3.4 Production of Other Vitamins

Recombinant *E. coli*, transformed with genes encoding pantothenic acid (vitamin B_5) biosynthesis and resistant to salicylic and/or other acids, produce 65 g l^{-1} of d-pantothenic acid from glucose using β-alanine as a precursor. Seven thousand tons per year are made chemically and microbiologically. Thiamine (vitamin B_1) is produced synthetically at 4,000 tons per year. Pyridoxine (vitamin B_6) is made chemically at 2,500 tons per year. The vitamin F (polyunsaturated fatty acids) processes of *Mortierella isabellina* or *Mucor circinelloides* yield 5 g l^{-1} of γ-linolenic acid.

4.4.4 Production Processes for Organic Acids

Acetic, citric, gluconic, and lactic acids are the main primary metabolic organic acids with commercial application as chemicals, at least some of which are produced industrially by fermentation. Others include pyruvic, fumaric, succinic, glyceric, shikimic, propionic, and itaconic acids. Production has been improved by classical mutation, by screening and selection techniques, as well as by metabolic engineering. According to Allied Market Research, the global organic acids market was valued at $16.8 billion in 2016 and is expected to reach $29 billion by 2023 with a growth rate of 8.3%.

4.4.4.1 Production of Acetic Acid

The global market for acetic acid is estimated to be around $9.0 billion and is expected to reach nearly $12 billion by 2022. The global production of acetic acid reported in 2016 was nearly 20 million tons. Its production can be achieved chemically by carbonylation of methanol or through microbial fermentations. Vinegar has been produced microbiologically as far back as 4000 bc. Vinegar fermentation is best carried out with species of *Gluconacetobacter* and *Acetobacter*. A solution of ethanol is converted to acetic acid during which 90–98% of the ethanol is attacked, yielding a solution of vinegar containing 12–17% acetic acid. Titers of acetic acid have reached 53 g l^{-1} with genetically engineered *E. coli*, 83 g l^{-1} by a *Clostridium thermoaceticum* mutant, and 97.6 g l^{-1} by an engineered strain of *Acetobacter aceti* ssp. *xylinium*.

4.4.4.2 Production of Citric Acid

Production of citric acid by *Aspergillus niger* and yeasts amounted 1.75 million tons in 2011. The global market of citric acid was $2.6 billion in 2014 and is expected to reach $3.6 billion by 2020. The citric acid produced is mainly consumed by the food and pharmaceutical industries (about 80%). The best strains of *A. niger* make over 230 g l^{-1} of citric acid. Keys to the fermentation are excess carbon source, low levels of pH and dissolved oxygen, and limited concentrations of certain trace metals and phosphate.

Glucose is converted, through the activities of the enzymes of the EMP pathway, to pyruvate, which in turn is converted to acetyl-coenzyme A through the activity of pyruvate dehydrogenase. Acetyl-CoA is then condensed with oxaloacetate, the first committed step in the Krebs cycle to give citric acid through the activity of citrate synthase. Citric acid production by *A. niger* is stimulated by growing the organism in a glucose-rich (10–20%) medium. High concentration of sucrose triggers the accumulation of fructose 2, 6-bisphosphate intracellularly, which in turn activates glycolysis. Another key to successful citric acid fermentation by this fungus is a deficiency of Mn^{2+} ions (cofactor of isocitrate dehydrogenase) as it is necessary for restricting carbon flux through this enzyme, while maintaining an active citrate synthase. Since the equilibrium of the reaction catalyzed by aconitase, which converts citric acid to isocitrate, is markedly in favor of citrate formation, citric acid accumulates. Two isocitrate dehydrogenases, mitochondrial and $NADP^+$ specific, are inhibited by citrate. Since the cofactor of this enzyme is Mg^{2+} or Mn^{2+}, citrate's ability to chelate these bivalent ions restricts the enzymic activity of isocitrate dehydrogenase, which in turn allows citrate to accumulate.

In addition to the Mn^{2+} ion, the metal deficiencies necessary for efficient citric acid production by different strains of *A. niger* include Fe^{2+} and Zn^{2+}. However, it is Mn^{2+} limitation that is paramount for the production of a high titer of citrate. The principal regulatory control site in the reactions from glucose to citrate is phosphofructokinase. This enzyme is inhibited by citrate, an event that would not be favorable for the overproduction of citric acid. However, Mn^{2+} deficiency slows down growth, leading to degradation of intracellular nitrogenous macromolecules and a 5-fold increase in the concentration of NH_4^+ in mycelia. The high ammonium concentration reverses citrate inhibition of phosphofructokinase, thus ensuring the continued conversion of glucose to citrate. Mutants whose phosphofructokinase I is partially desensitized to citrate inhibition are less dependent on low Mn^{2+} for high citric acid production. Citrate inhibition of phosphofructokinase is reversed by fructose 2, 6-diphosphate and AMP. The optimum pH for citric acid production by *A. niger* is 1.7–2.0. At pH values higher than 3.0, oxalic and gluconic acids are produced instead. Low pH inactivates glucose oxidase and prevents gluconate production. Mutants of *A. niger* with greater resistance to low pH are improved citric acid producers. Other selective tools include resistance to high concentrations of citrate and sugars.

Species of the yeast genus *Candida* also excrete large amounts of citric acid and isocitric acid. The key event of the yeast citrate process appears to be a sharp drop in intracellular AMP following nitrogen depletion, inhibiting the AMP-requiring isocitrate dehydrogenase. *Candida guilliermondii* excretes large quantities of citric acid without the undesirable isocitric acid when cultured in the presence of metabolic inhibitors (e.g., sodium fluoroacetate, *n*-hexadecylcitric acid, and *trans*-aconitic acid). These inhibitors block the TCA cycle at the aconitase step. Mutation of *Candida lipolytica* to aconitase deficiency is also effective. The optimum pH for the yeast citrate process is above 5.0. Lower pH values lead to production of polyhydroxy compounds such as erythritol and arabitol. The yeast process utilizes hydrocarbons as substrate and citric acid yields as high as 225 g l^{-1} at a rate of 1.4 g l^{-1} h^{-1}.

High concentrations of citric acid are also produced by *Candida oleophila* from glucose. In chemostats, 200 g l⁻¹ can be made, and more than 230 g l⁻¹ can be made in continuous repeated fed-batch fermentations. This compares with 150 to 180 g l⁻¹ by *A. niger* in industrial-batch or fed-batch fermentation in 6–10 days. The key to the yeast fermentation is nitrogen limitation coupled with an excess of glucose. The citric acid is secreted by a specific energy-dependent transport system induced by intracellular nitrogen limitation. The transport system is selective for citrate over isocitrate. *Yarrowia lipolytica* produces up to 198 g l⁻¹ citric acid in fed-batch fermentations on sunflower oil with a very low production of isocitric acid.

4.4.4.3 Production of Lactic Acid

The global market value for lactic acid was about US$2.08 billion in 2017 and ranked among the high-volume products biologically produced with 2 million tons. *Rhizopus oryzae* is favored for production since it makes stereochemically pure L (+)-lactic acid, whereas lactobacilli produce mixed isomers; furthermore, lactobacilli require yeast extract. However, a mutant strain of *Lactobacillus lactis* has been developed that produces 195 g l⁻¹ of L-Lactic acid from 200 g l⁻¹ of glucose. *R. oryzae* normally converts 60–80% of added glucose to lactate, with the remainder going to ethanol. By increasing lactic dehydrogenase levels via cloning, more lactate and less ethanol are produced. Mutation of wild-type *R. oryzae* led to L (+)-lactic acid production of 131–136 g l⁻¹, a yield from glucose of 86–90% and a productivity of 3.6 g l⁻¹ h⁻¹. This was a 75% improvement over the wild-type strain. The final strain was the result of a six-step mutation sequence. DL-Lactic acid can be made at 129 g l⁻¹ by *Lactobacillus* sp. and D-Lactic acid at 120 g l⁻¹. Using immobilized *R. oryzae* in a fed-batch culture reported a high titer production of 231 g l⁻¹ of lactic acid with a productivity of 1.83 g l⁻¹ h⁻¹. A recombinant *E. coli* strain has been constructed that produces optically active pure D-Lactic acid from glucose at virtually the theoretical maximum yield (e.g., two molecules from one molecule of glucose). The organism was engineered by eliminating genes of competing pathways encoding fumarate reductase, alcohol/aldehyde dehydrogenase, and pyruvate formate lyase, and by a mutation in the acetate kinase gene. Whole-genome shuffling has been used to improve the acid tolerance of a commercial lactic acid-producing *Lactobacillus* sp.

Products in development are nonchlorinated solvent, ethyl lactate, and bioplastic polylactide. Polylactide is made by converting corn starch to dextrose, fermenting dextrose to lactic acid, condensing lactic acid to lactide, and polymerizing lactide.

4.4.4.4 Production of Pyruvic Acid

A recombinant strain of *E. coli* that is a lipoic acid auxotroph and defective in F₁ATPase produces 31 g l⁻¹ of pyruvic acid from 50 g l⁻¹ glucose. The lowering of the energy level in the cell by the F₁ATPase deletion increases the glucose uptake and glycolysis rate, thereby leading to an increase in pyruvate production. An improved fermentation was developed using *Torulopsis glabrata* yielding a pyruvic acid concentration of 77 g l⁻¹, a conversion of 0.80 g g⁻¹ glucose, and a productivity of 0.91 g l⁻¹ h⁻¹ in 85 hours. By directed evolution in a chemostat, a mutant *Saccharomyces cerevisiae* strain was obtained that produces from glucose 135 g l⁻¹ pyruvic acid at a rate of 6 to 7 mmol per gram biomass per hour during exponential growth with a yield of 0.54 g⁻¹.

4.4.4.5 Production of Other Organic Acids

From 120 g l⁻¹ of glucose, *Rhizopus arrhizus* produces 97 g l⁻¹ fumaric acid. The molar yield from glucose is 145% and involves CO_2 fixation from pyruvate to oxaloacetate and the reductive reactions of the TCA cycle. Succinic acid is made chemically at 30,000–50,000 tons per year for commercial use as (1) a surfactant, detergent extender, or foaming agent; (2) an ion chelator in electroplating to prevent metal corrosion and pitting; (3) an acidulant, pH modifier, flavoring agent, or antimicrobial agent for food; and (4) a chemical in the production of pharmaceuticals. The market size is US$400 million per year. Production by fermentation with *Actinobacillus succinogenes* amounts to 106 g l⁻¹. Bioconversion from fumarate yields 85 g l⁻¹ succinate after 24 hours. Using a fed-batch process with *Corynebacterium glutamicum* a production of 146 g l⁻¹ was reported with a productivity of 3.2 g l⁻¹ h⁻¹. Metabolic engineering of *E. coli* yields a shikimic acid overproducer making 84 g l⁻¹ with a 0.33 molar yield from glucose. Production of gluconic acid amounted to 150 g l⁻¹ from 150 g l⁻¹ glucose plus corn steep liquor in 55 hours by *A. niger*. Production level is about 50,000 tons per year. Cloning of fumarase in *S. cerevisiae* remarkably improved the malic acid bioconversion from fumaric acid from 2 g l⁻¹ to 125 g l⁻¹; conversion yield was nearly 90%. Glyceric acid production amounts to 136 g l⁻¹ by *Gluconobacter frateurii* from glycerol, and that of propionic acid is 106 g l⁻¹ from the same carbon source by an acetate-negative mutant of *Propionibacterium acidipropionici*.

4.4.5 Production of Ethanol and Related Compounds

4.4.5.1 Production of Ethanol

Ethanol is a primary metabolite produced by fermentation of sugar or of a polysaccharide that can be depolymerized to a fermentable sugar. *S. cerevisiae* is used for the fermentation of hexoses, whereas the *Kluyveromyces fragilis* or *Candida* species can be used if lactose or a pentose, respectively, is the substrate. Under optimum conditions, approximately 10–12% ethanol by volume is obtained within 5 days. At present, all beverage alcohol is made by fermentation. Industrial ethanol is mainly manufactured by fermentation, but some is still produced from ethylene by the petrochemical industry. Bacteria such as *Clostridia* and *Zymomonas* are being reexamined for ethanol production after years of neglect. *Clostridium thermocellum*, an anaerobic thermophile, can convert waste cellulose and crystalline cellulose directly to ethanol (see Chapter 9 on renewable resources conversion to fine chemicals).

The available cellulosic feedstock in the US could supply 20 billion gallons of ethanol in comparison to the 3 billion gallons currently made from corn. This would be enough to add 10% ethanol to all gasoline used in the US. Other *Clostridia* produce acetate, lactate, acetone, and butanol and will be used to produce these chemicals when the global petroleum supplies begin to become depleted.

Ethyl alcohol is produced in Brazil from cane sugar at over 4.2 billion gallons per year and is used either as a 25% blend or as a pure fuel. Most new cars in Brazil use pure ethanol, whereas the remainder utilizes a blend of 20–25% ethanol in gasoline. In the US, over 3.4 billion gallons of ethanol were made from starchy crops (mainly corn) in 2004. It is chiefly added to gasoline to reduce CO_2 emissions by improving the overall oxidation and performance of gasoline.

Fuel ethanol produced from biomass would provide relief from air pollution caused by the use of gasoline and would not contribute to the greenhouse effect. *E. coli* has been converted into an excellent ethanol producer (43% yield, v/v) by cloning and expressing the alcohol dehydrogenase and pyruvate decarboxylase genes from *Zymomonas mobilis* and *Klebsiella oxytoca*. The recombinant strain was able to convert crystalline cellulose to ethanol in high yield when fungal cellulase was added. Other genetically engineered strains of *E. coli* can produce as much as 60 g l^{-1} of ethanol. An ethanol-resistant mutant of *S. cerevisiae* makes 96 g l^{-1}.

4.4.5.2 Production of Glycerol

Glycerol is widely used in the manufacture of drugs, food, cosmetics, paint, and many other commodities. Over 1.5 million tons of glycerol are produced annually primarily through extraction of materials from the fat and oil industries, or by chemical synthesis from propylene. However, yeast fermentations using *S. cerevisiae* can produce up to 230 g l^{-1}, while osmotolerant *Candida glycerinogenes* can produce 137 g l^{-1} with yields of 63–65% and a productivity of 32 g l^{-1} h^{-1}. Furthermore, *Candida magnoliae* produces 170 g l^{-1} in fedbatch fermentation, and in a similar type of process, *Pichia farinosa* can produce up to 300 g l^{-1}.

4.4.5.3 Production of 1, 3-Propanediol

A strain of *Clostridium butyricum* converts glycerol to 1, 3-propanediol (PDO) at a yield of 0.55 g per gram of glycerol consumed. In a two-stage continuous fermentation, a titer of 41–46 g l^{-1} was achieved with a maximum productivity of 3.4 g l^{-1} h^{-1}. At lower dilution rates, butyrate was produced, and at higher dilution rates, acetate was made. Recent metabolic engineering triumphs have included an *E. coli* culture that grows on glucose and produces PDO at 135 g l^{-1}, with a yield of 51% and a rate of 3.5 g l^{-1} h^{-1}. To do this, eight new genes were introduced to convert dihydroxyacetone phosphate (DHAP) into PDO. These included yeast genes converting dihydroxyacetone to glycerol and *Klebsiella pneumoniae* genes converting glycerol to PDO. Production in the recombinant was improved by modifying 18 *E. coli* genes, including regulatory genes. PDO is the monomer used to chemically synthesize polyurethanes and the polyester fiber Sorono™ by

DuPont. This new bioplastic, polytrimethylene terephthalate (3GT polyester), is made by reacting terephthalic acid with PDO. PDO is also used as a polyglycol-like lubricant and as a solvent. Around 63,000 tons PDO production per year has been reported by DuPont Tate & Lyle Bio Products.

4.4.5.4 Production of Erythritol

The noncariogenic, noncaloric, and diabetic-safe sweetener erythritol is made by fermentation. It has 70–80% the sweetness of sucrose. Osmotic pressure increase was found to raise volumetric and specific production, but to decrease growth. By growing cells first at a low glucose level (i.e., 100 g l^{-1}) and then adding 200 g l^{-1} glucose at 2.5 days, erythritol titer was increased to 45 g l^{-1} as compared to single-stage fermentation with 300 g l^{-1} glucose, which yielded only 24 g l^{-1}. Production of erythritol by a *C. magnoliae* osmophilic mutant yielded a titer of 187 g l^{-1}, a rate of 2.8 g l^{-1} h^{-1}, and 41% conversion from glucose. Other processes have been carried out with *Aureobasidium* sp. (165 g l^{-1} from glucose with a 48% yield), and the osmophile *Trichosporon* sp. (188 g l^{-1} with a productivity of 1.18 g l^{-1} h^{-1} and 47% conversion). Erythritol can also be produced from sucrose by *Torula* sp. at 200 g l^{-1} in 120 h with a yield of 50% and a productivity of 1.67 g l^{-1} h^{-1}. Recent studies have developed a *Pseudozyma tsukubaensis* process yielding 245 g l^{-1} of erythritol from glucose. According to the Mitr Phol Group, global erythritol consumption topped 65,000 tons in 2015, and is forecast to grow at 7–8% annually.

4.4.5.5 Production of Other Compounds

Dihydroxyacetone is used as a cosmetic tanning agent and surfactants. It is produced from glycerol by *Gluconobacter* species with a conversion rate of up to 90%. d-mannitol is used in the food, chemical, and pharmaceutical industries. It is only poorly metabolized by humans, is about half as sweet as sucrose, and is considered a low-calorie sweetener. It is produced mainly by catalytic hydrogenation of glucose-fructose mixtures, but 75% of the product is sorbitol, not mannitol. For this reason, fermentation processes are being considered. Recombinant *E. coli* produces up to 91 g l^{-1} mannitol, and *Leuconostoc* sp. up to 98 g l^{-1}. Mannitol production reached 213 g l^{-1} from 250 g l^{-1} fructose after 110 hours by *C. magnoliae*. Sorbitol, also called d-glucitol, is 60% as sweet as sucrose and has use in the food, pharmaceutical, and other industries. Its worldwide production is 600,000 tons per year, and it is made chemically by catalytic hydrogenation of d-glucose. Toluenized (permeabilized) cells of *Z. mobilis* produce 290 g l^{-1} of sorbitol and 283 g l^{-1} of gluconic acid from a glucose and fructose mixture in 16 hours with yields near 95% for both products. Xylitol is a naturally occurring sweetener with anticariogenic properties used in some diabetes patients. It can be produced chemically by chemical reduction of d-xylose. A mutant of *Candida tropicalis* produces 40 g l^{-1} l from d-xylose in a yield of over 90%. According to Zion Market Research Company, the sorbitol market was valued at US$1.17 billion in 2016 and is expected to grow to US$5.3 billion by 2022.

Early in the nineteenth century, the acetone-butanol fermentation process was a commercial operation but was later replaced by chemical synthesis from petroleum because of economic factors. These included the low concentration of butanol in the broth (1%) and the high cost of butanol recovery. *Clostridium beijerinckii* and *Clostridium acetobutylicum* are the organisms of choice for fermentation. Research on this aspect of fermentation continued over many years, dealing with process engineering, mutation, and metabolic engineering. Butanol-resistant mutants were isolated for their ability to overproduce butanol and acetone. Further research involving biochemical engineering modifications increased the production of acetone, butanol, and ethanol (ABE) to 69 g l^{-1}. A mutant in the presence of added acetate produced 21 g l^{-1} butanol and 10 g l^{-1} of acetone from glucose. Acetate both stimulates production and helps stabilize the culture, which is known to be highly unstable. 2,3-butanediol can be made from glucose at a titer of 130 g l^{-1} by a genetically engineered strain of *K. oxytoca* that no longer produces ethanol and has a lowered level of acetoin formation. The compound has a very high octane rating, making it a potential aviation fuel.

SUMMARY

The microbial production of primary metabolites, through fermentation, contributes significantly to the quality of life that we enjoy today. Microorganisms are capable of converting inexpensive carbon and nitrogen sources into voluble metabolites such as amino acids, nucleotides, organic acids, and vitamins, which can be added to food as a flavor enhancer and/or to increase its nutritional value. Also, many primary metabolites are proving invaluable as biosynthetic precursors for the manufacture of therapeutics.

Overproduction of microbial metabolites is related to developmental phases of microorganisms. Inducers, effectors, inhibitors, and various signal molecules play roles in different types of overproduction. Biosynthesis of enzymes catalyzing metabolic reactions in microbial cells is controlled by well-known positive and negative mechanisms (e.g., induction, nutritional regulation, and feedback regulation).

In the early years of fermentation processes, strain developments depended entirely on classical strain breeding involving intensive rounds of random mutagenesis, followed by an equally strenuous program of screening and selection. However, recent innovations in molecular biology, on one hand, and the development of new tools in functional genomics, transcriptomics, metabolomics, and proteomics, on the other, enabled more rational approaches for strain improvement.

The roles of primary metabolites and, in turn, microbial fermentations stand to grow in stature especially as we enter a new era in which the use of renewable resources is recognized as an urgent need.

REFERENCE

Varela, C., E. Agosin, M. Baez, M. Klapa, and G. Stephanopoulos. 2003. Metabolic flux distribution in *Corynebacterium glutamicum* in response to osmotic stress. *Appl. Microbiol. Biotechnol.* 60:547–555.

FURTHER READING

Abe, H., Y. Fujita, Y. Takaoka, E. Kurita, S. Yano, N. Tanaka, and K.-I. Nakayama. 2009. Ethanol tolerant *Saccharomyces cerevisiae* strains isolated under selective conditions by over-expression of a proofreading DNA polymerase. *J. Biosci. Bioeng.* 108:199–204.

Da Silva, G. P., M. Mack, and J. Contiero. 2009. Glycerol: A promising and abundant carbon source for industrial microbiology. *Biotechnol. Adv.* 27:30–39.

Demain, A. L. 2000. Small bugs, big business: The economic power of the microbe. *Biotechnol. Adv.* 18:499–514.

Demain, A. L., M. Newcomb, and J. H. D. Wu. 2005. Cellulase, clostridia, and ethanol, *Microbiol. Mol. Biol. Rev.* 69:124–154.

Demain, A. L. and J. L. Adrio. 2008. Strain improvement for production of pharmaceuticals and other microbial metabolites by fermentation. *Prog. Drug Res.* 65:253–289.

Eggeling, L., and M. A. Bott. 2015. A giant market and a powerful metabolism: L-lysine provided by *Corynebacterium glutamicum. Appl. Microbiol. Biotechnol.* 9:3387–3394.

Eggeling, L. 2010. Microbial metabolite export. In Flickinger, M.C., ed., *Encyclopedia of industrial biotechnology: Bioprocess, bioseparation, and cell technology*, pp. 1–10. New York: John Wiley & Sons, Inc.

El-Mansi, M. 2004. Flux to acetate and lactate excretions in industrial fermentations: Physiological and biochemical implications, *J. Indust. Microbiol. Biotechnol.* 31:295–300.

Eseji, T., C. Milne, N. D. Price, and H. P. Blaschek. 2010. Achievements and perspectives to overcome the poor solvent resistance in acetone and butanol-producing microorganisms. *Appl. Microbiol. Biotechnol.* 85:1697–1712.

Han, L. and S. Parekh. 2004. Development of improved strains and optimization of fermentation processes. In Barredo, J. L., ed., *Microbial processes and products*, pp. 1–23. Totowa, NJ: Humana.

Hirasawa, T. and H. Shimizu. 2016. Recent advances in amino acid production by microbial cells. *Curr. Opin. Biotechnol.* 42:133–146.

Ikushima, S., T. Fujii, O. Kobayashi, S. Yoshida, and A. Yoshida. 2009. Genetic engineering of *Candida utilis* yeast for efficient production of L-Lactic acid. *Biosci. Biotechnol. Biochem.* 73:1818–1824.

Jaya, M., K.-M. Lee, M. K. Tiwari, J.-S. Kim, P. Gunasekaran, S.-Y. Kim, I.-W. Kim, and J.-K. Lee. 2009. Isolation of a novel high erythritol-producing *Pseudomonas tsukubaensis* and scale-up of erythritol fermentation to industrial level. *Appl. Microbiol. Biotechnol.* 83:225–231.

John, R. P., G. S. Anisha, M. Nampoothirti, and A. Pandey. 2009. Direct lactic acid fermentation: Focus on simultaneous saccharification and lactic acid production. *Biotechnol. Adv.* 27:145–152.

Kinoshita, S. 1987. Thom award address: Amino acid and nucleotide fermentations from their genesis to the current state. *Devel. Indust. Microbiol.* 28(Suppl. 2): 1–12.

Kraemer, R. 2004. Production of amino acids: Physiological and genetic approaches. *Food Biotech.* 18(2): 1–46.

Lee, J.-H., and V. Wendisch. 2017. Production of amino acids – Genetic and metabolic engineering approaches. *Biores. Technol.* 245:1575–1587.

Magnuson, J. K., and L. L. Lasure. 2004. Organic acid production by filamentous fungi. In Lange, J., Lange, L. eds., *Advances in Fungal Biotechnology for Industry, Agriculture, and Medicine*, pp. 307–340. New York: Kluwer Academic/Plenum Publishers.

Ohnishi, J., S. Mitsuhashi, M. Hayashi, S. Ando, H. Yokoi, K. Ochiai, and M. Ikeda. 2002. A novel methodology employing *Corynebacterium glutamicum* genome information to generate a new L-lysine-producing mutant. *Appl. Microbiol. Biotechnol.* 58:217–223.

Sahm, H., L. Eggeling, and A. A. de Graaf. 2000. Pathway analysis and metabolic engineering in *Corynebacterium glutamicum*. *Biol. Chem.* 381:899–910.

Sanchez, S. and A. L. Demain. 2008. Metabolic regulation and overproduction of primary metabolites. *Microb. Biotechnol.* 1:283–319.

Sanchez, S., and A.L. Demain. 2014. Production of amino acids. In Batt, C. A., Tortorello, M. L. eds., *Encyclopedia of food microbiology*, pp. 778–784. New York: Elsevier Ltd, Academic Press.

Tkacz, J. and L. Lange, eds., *Advances in fungal biotechnology for industry, agriculture, and medicine*, pp. 307–340. New York: Kluwer Academic/Plenum.

5 Microbial Synthesis of Secondary Metabolites and Strain Improvement
Current Trends and Future Prospects

Sergio Sánchez, Beatriz Ruiz-Villafán, Corina D. Ceapa, and A.L. Demain

CONTENTS

"Never memorise what you can look up in a book."

Albert Einstein

5.1 INTRODUCTION

Secondary metabolites from microorganisms and plants are typically low-molecular-weight natural products that are generally not essential for survival and growth of the producing organisms but are involved in the interactions of microorganisms and plants with their environment as a result of secondary metabolism regulation. It is generally accepted that secondary metabolites are derived from primary metabolites, often having distinct and versatile physiological functions either induced or regulated by environmental and nutritional factors (first proposed by Albrecht Kossel in 1891; Chapman, 2000). As such, a large number of secondary metabolites of both microbial and plant origin have been found, which have been commercially applied for human health products, industrial biochemicals, agricultural chemicals, food, and nutritional additives. They include antibiotics, antiviral and antitumor agents, cholesterol-regulating drugs, pigments, flavors, fragrances, toxins, effectors of ecological competition and symbiosis, pheromones, enzyme inhibitors, immunomodulating agents, receptor antagonists and agonists, pesticides, and growth promoters of animals and plants (Ruiz et al., 2010). They have unusual chemical structures that are often synthesized and modified in complex multistep metabolic pathways of both biosynthetic steps and post-biosynthetic events. The level of production of these secondary metabolites can be influenced by factors or combination of factors such as the growth phase, the growth rate, nutrients, signal molecules (e.g., hormones and elicitors), feedback control, cultivation conditions (e.g., temperature, pH, light, and dissolved oxygen), and physical microenvironments (e.g., shear stress or mixing) (Schmidt, 2005).

Although a huge variety of secondary metabolites have been identified in microorganisms and plants, especially in the post-genomic era where discovery is facilitated by genome sequencing and bioinformatic search tools and databases (Medema and Fischbach, 2015), the understanding and knowledge of their metabolic pathways and regulation are still very poor. Commonly, most of these products are produced at a very low level in the original producing strains, as it is "secondary" to the living processes of microorganisms and plants. Such a level of production is far lower than the requirement for commercial production, thus creating a challenge for commercial translation of discoveries. This technological challenge is particularly true in the case of production of secondary metabolites using plant cell and tissue culture technologies (Santos et al., 2016). This has resulted in very limited successes in commercial production with only about 200 products having reached the market. In contrast, microbial fermentation technology has been well established with many successful commercial products derived from both primary and secondary metabolism for a variety of applications in the pharmaceutical industry (e.g., antibiotics, and antitumor and antiviral agents), food industry (e.g., amino acids, organic acids, vitamins, and flavoring compounds), agricultural industry (e.g., pesticides, insecticides, and antibacterial agents), chemical and biofuels industry (e.g., ethanol, butanol, methane, and biodiesels), and environmental industry (e.g., bioremediation compounds) (Zhou et al., 2014). However, these commercially successful microbial products still represent only a small fraction of the enormous diversity of secondary metabolites biosynthesized by microorganisms of both terrestrial and aquatic environments.

To develop a successful industrial production process of any useful products by microbial fermentation, mammalian cell culture and/or plant cell culture, it is imperative to develop an elite strain or cell line, and efficient processing strategies that can achieve the following:

1. High and stable product yield and productivity under low-cost process conditions
2. High conversion rate of the cheapest available substrates
3. High product selectivity and specificity, facilitating downstream purification

With these targets in mind, this chapter focuses the key process and strain improvement strategies of both classical and genetic approaches for microbial synthesis of secondary metabolites, drawing examples from the antibiotics industry and new therapeutics of extremely high value to humankind.

5.2 MICROBIAL SYNTHESIS OF SECONDARY METABOLITES AND STRAIN IMPROVEMENT

The major microbial system used for secondary metabolites production and strain improvement is the bacterium *Escherichia coli*, which was the first species to be used to produce a commercial therapeutic protein (recombinant human insulin, approved in 1982). Many simple and unmodified compounds can be manufactured in *E. coli* but more complex compounds requiring posttranslational modifications (such as lasso peptides and lanthipeptides) cannot, unless targeted to the periplasm, and this is not a scalable process. Other *E. coli* production issues include insoluble inclusion

bodies accumulation and the production of endotoxins that can induce septic shock (Santos et al., 2016).

Yeasts are sometimes preferred because their growth patterns are like that of bacteria, but they are eukaryotes and thus support protein folding and modification. Particularly glycosylations that are needed to ensure proper function and activity, by influencing proper charge, solubility, folding, serum half live of the metabolites, *in vivo* activity, correct cellular targeting and immunogenicity, among others, cannot often be fully accomplished in bacterial systems (Martínez et al., 2012). *Saccharomyces cerevisiae* was the first yeast used to express exogenous molecules and it is still used commercially to produce a Hepatitis B virus vaccine, but other yeasts such as *Pichia pastoris* and *Hansenula polymorpha* are now favored for process development due to in-process inducible expression.

Mammalian cells have dominated the biopharmaceutical industry since the 1990s because they can produce high titers (1–10 g/L) of complex metabolites with mammalian glycan structures. They are much more expensive than microbes, but most pharmaceuticals are glycoproteins and the quality of the product is superior when mammalian cells are used. CHO cells are preferred by the industry but others that are widely used include the murine myeloma cells lines NS0 and SP2/0, BHK and HEK-293, and the human retinal line PER-C6. The expenses involved in production, purification, and the risk of contamination constitute major disadvantages of mammalian cells (Santos et al., 2016).

Advances in other microbial platforms are modest but relevant due to their potential for bioprocessing. Cyanobacteria and eukaryotic microalgae are investigated for their ability to synthesize a broad range of molecules from carbon dioxide, their relatively large size with impact on scaling and potentially more appropriate media for end-product recovery. Mixed cultures, on the other hand, are interesting due to their broad spectrum of waste steam usage and for potentially a simpler and cost-effective bioreactor and operation. Pilots to produce bioplastics (polyhydroxyalkanoates, PHA) and carboxylates (volatile fatty acids) are in progress (Cuellar et al., 2015).

Emphasis should also be given to the fact that most bacterial active biomolecules originate from Actinobacteria. Biosynthetic gene clusters (BGCs) and secondary metabolites from Actinobacteria can be expressed in native and various heterologous host production systems. In the native organism, strategies for inducing expression encompass directing metabolic fluxes toward desired products, manipulating regulatory pathways, and engineering cellular translational and transcriptional machineries. The engineering of heterologous hosts includes dynamic metabolic regulation, genome-scale engineering and genome minimization. Additional strategies for host engineering comprise cryptic BGC activation or combinatorial protein expansion or pathway and precursor engineering to generate novel natural product analogs (Zhang et al., 2016).

Although the use of microorganisms in the production of commercially useful products is well established, the general public has yet to fully appreciate the microbial origin of some well-known commodities. For example, citric acid, an organic acid that chemists found very difficult if not impossible to synthesize with the correct stereo-specificity, is now widely produced through fermentation by either *Aspergillus niger* or by the yeast *Yarrowia lipolytica*. It was the expertise gained with such fermentations that allowed the rapid development of the antibiotics industry in the early 1940s, with a consequential leap in the quality of health care.

Another example of this "unrecognized" microbial production is xanthan gum, a product made by *Xanthomonas campestris*, is widely used in the manufacture of ice cream as it prevents foams from collapsing with time. In a slightly modified form, xanthan gum gives paints their unique property of sticking to the brush without dripping but spreading smoothly and evenly from the brush to the surface.

In addition to natural products such as the chemicals mentioned above, recombinant strains of *E. coli*, constructed by genetic engineering, are now used to produce "non-natural products" to the organism, such as human insulin. Biological washing powders contain enzymes drawn from different species, particularly those belonging to the genus *Bacillus*. Microbial products thus have a significant impact on everyday lives in contemporary society.

Significantly, the "new biotechnology" industry is critically dependent on microbial fermentations to produce extremely high-value recombinant therapeutic products (e.g., human insulin and human growth hormone), as well as low-value recombinant products (e.g., prochymosin, which provides an ethically acceptable alternative to rennin [chymosin] in the manufacture of cheese).

In this section, both classical strain improvement strategies and modern recombinant (cloning) strategies will be discussed regarding how to achieve industrially desired properties for the production of microbial secondary metabolites.

5.2.1 ECONOMICS AND SCALE OF MICROBIAL PRODUCT FERMENTATIONS

The type of fermentation used, as well as its size, duration, and nutrient profile, will depend critically on the nature of the microbial product. For "low-value, high-volume" products, such as citric acid and xanthan gum, high-capacity fermentors (often up to 800 m³ in volume) are generally used. However, the duration of the fermentation process and the costs of nutrients and "utilities" (heating, cooling, and air) are the critical factors in the overall profitability of this business.

"Medium-value, medium-volume" products, such as antibiotics, are typically made in fermentors that are considerably smaller (100–200 m³), and again the duration and the utility and nutrient costs are significant factors.

"High-value, low-volume" products, such as recombinant therapeutic proteins, are made in small (approximately 400 L, i.e., 0.4 m³) fermentors for which the cost of the nutrients and utilities is a minor factor in the overall feasibility and profitability.

For all but the high-value products, nutrient costs (especially of the primary carbon source) are critical. Depending on the vagaries of world commodity markets, complicated further by artificially imposed trade tariffs, the availability and, in turn, cost of nutrients can fluctuate at alarming rates. Flexibility in the choice of nutrients is therefore of paramount importance, and strain improvement programs must take this factor into consideration.

In contrast, for the high-value (recombinant) products, the emphasis of the strain improvement program focuses on the following aspects:

- The stability of the strain
- The level of expression
- The overall quality of the product, rather than the cost of the fermentation process *per se*

A fermentation process constitutes the basis for a business that aims to sell the product at an overall profit. This issue positions strain improvement programs at the interface between science and the commercial world and requires a different set of criteria to judge whether a task is worthwhile or not. Strain improvement programs are positioned, and must first be vetted, to establish the cost of the research and development (R&D) that will be necessary to achieve the stated goal, and to set the expected cost against the annual savings likely to ensue should that piece of work deliver the expected gains in productivity. It often comes as a shock to researchers new to this field that projects that are highly innovative, but to which some risk of failure is attached, are not funded because the return on such an investment (set against the risk) is not high enough. For example, a strain improvement program for a "mature product" would be difficult to justify if the annual R&D cost could not be recouped in around three years through the projected increase in productivity. The annual cost of strain improvement programs must, therefore, be no more than three times the projected annual cost savings. Acclimatization to such rigorous reviews of research plans and draconian decision making, constitutes a sharp learning curve for newly recruited research staff.

5.2.2 DIFFERENT PRODUCTS NEED DIFFERENT FERMENTATION PROCESSES

At the "low-value" end of the microbial products business, the margins on profitability are extremely tight. For example, the citric acid titer must be greater than 100 g L⁻¹, with a carbon conversion efficiency (the amount of substrate converted to product on a per gram carbon basis) close to 100% for the process to become economic. Citric acid is produced concurrently with microbial growth, but a fine balance has to be struck between the amount of cells that are made in the fermentation (which consumes some of the carbon source that otherwise could be used to make the product) and the fact that a doubling of the cell mass may result in twice the volumetric rate of citric acid production (as each cell acts as its own "cell factory"), but may hinder the eventual purification of the product. Growth can be arrested by, typically, limiting the amount of nitrogen available to the culture, in which case citric acid is still produced for some time by the cells in the stationary phase. Cheap carbon sources (e.g., unrefined molasses), fast production runs with minimal turn-around time, and a cheap and rapid means of extracting the product are most important, and the strain improvement program of this mature product will be focused on substrate flexibility and rapid production.

Medium-value products, such as antibiotics, are produced by microbes, which, when first isolated from the soil, usually make detectable, but vanishingly small, quantities (a few micrograms) of the bioactive substance. Most commercial fermentation processes will become economically viable only if the strain is eventually capable of making more than 15–20 g L^{-1} of the bioactive product using comparatively cheap carbon sources, such as rapeseed oil, and low-grade sources of protein, such as soya bean or fish meal. Antibiotic production usually occurs following the onset of the stationary phase (i.e., after growth has been arrested), but many antibiotics require nitrogen as well as carbon, and growth is generally limited by the supply of phosphate. A steady supply of nutrients containing carbon and nitrogen, but not phosphate, is fed to the fermentation during the phase of antibiotic production.

Once a new microbial metabolite has been discovered, a campaign of "empirical strain improvement" is undertaken to boost the level of production to a titer at which the process becomes economically viable.

For "high-value" products, such as recombinant therapeutic proteins, the cost of the fermentation broth is not a major issue. To ensure consistency of the final product, expensive well-defined media (either Analar-grade mineral medium or high-specification tryptone hydrolysates) are used along with a high-purity carbon source (usually glucose). Following the onset of the stationary phase, the production of high-value products is triggered, generally in response to an external signal such as temperature shifts or the addition of an exogenous gratuitous inducer. The levels of production may be relatively modest (less than 5 g L^{-1}) to meet commercial targets, but higher levels are always sought. This is because the products are often formulated into injectable medicines, in which there is a fear of raising an immune response if the recombinant protein is not 100% pure or not folded into the correct tertiary structure. A major objective of strain improvement programs, therefore, is to address this important regulatory issue rather than the cost of media and utilities.

5.2.3 Fed-Batch Culture: The Paradigm for Many Efficient Microbial Processes

Simple batch culture is a fairly inefficient way to synthesize a microbial product, as a substantial proportion of the nutrients present in the fermentation is used to make the biomass and the opportunity then to use the biomass as a cell factory to make the product is limited to the time of the growth period. Although very high levels of productivity can often be achieved in continuous culture, this technique has the disadvantages that large volumes of medium (often expensive medium) are required, and the product is made in a dilute stream that must be concentrated before final isolation can

take place. Fed-batch culture is a "halfway" approach. The cells are grown up in batch culture and then the resident biomass, which is no longer growing, is dedicated to product formation by feeding nutrients, except for that chosen to limit growth. This type of culture technique was developed by the antibiotics industry.

To understand the nature of antibiotic fermentations, it is important to consider the life cycle of the producing microorganisms within the natural ecosystem. For instance, Streptomycetes are filamentous bacteria which make up the majority (>60%) of natural antibiotics producers, including streptomycin, tetracyclines, and erythromycin. The life cycles of filamentous eukaryotic organisms, which produce the natural penicillins, are broadly similar. The cycle begins with the spore, which may lie dormant in the soil for many years; when nutrients and water become available, the spore germinates, thus leading to the formation of a mycelium (Figure 5.1). These organisms are said to be "mycelial" in nature, as they colonize soil particles (or agar medium in Petri dishes, if they are in the laboratory) by extending outward in all directions (radial growth) in a fixed branched pattern, often called *vegetative mycelium*. Inevitably, at some point, nutrients or water become in short supply and it is at this point that they differentiate, ultimately to form spores again. Initially, the differentiation process involves the formation of an aerial mycelium in which the cell conglomerate no longer extends out radially but rises up and away from the plane of the radial growth and forms elaborate coiled structures that then septate after a while and the spores are formed again, thus completing the life cycle.

Specific enzymes digest the vegetative mycelium, in part, and the cellular building blocks that are released are used to construct the aerial mycelium. In this way, part of the vegetative mycelium is "sacrificed" to allow the aerial mycelium to be formed in route to sporulation. This process can be viewed as a survival strategy because, when the spores have formed, the organism's DNA is held in an inert state so that the life cycle can begin again when water and nutrients are plentiful. It has been suggested that the reason that these actinobacteria make antibiotics exactly at the same time as the differentiation step is to sterilize the micro-environment around the vegetative mycelia, so that other microbial predators cannot take advantage of the available nutrients released following mycelial lysis (a view that is not held by all scientists in the field).

5.2.4 Nutrient Limitation and the Onset of Secondary Metabolite Formation

Interestingly, antibiotics are made at the time of differentiation between the stage of vegetative mycelia and the aerial mycelial stage (Figures 5.1 and 5.2). This may seem a paradox, as the signal to undertake this differentiation step is lack of nutrients or water, which in turn begs the question of how a microorganism, starved of nutrients, elaborates such a complex structure as the aerial mycelium on one hand and the

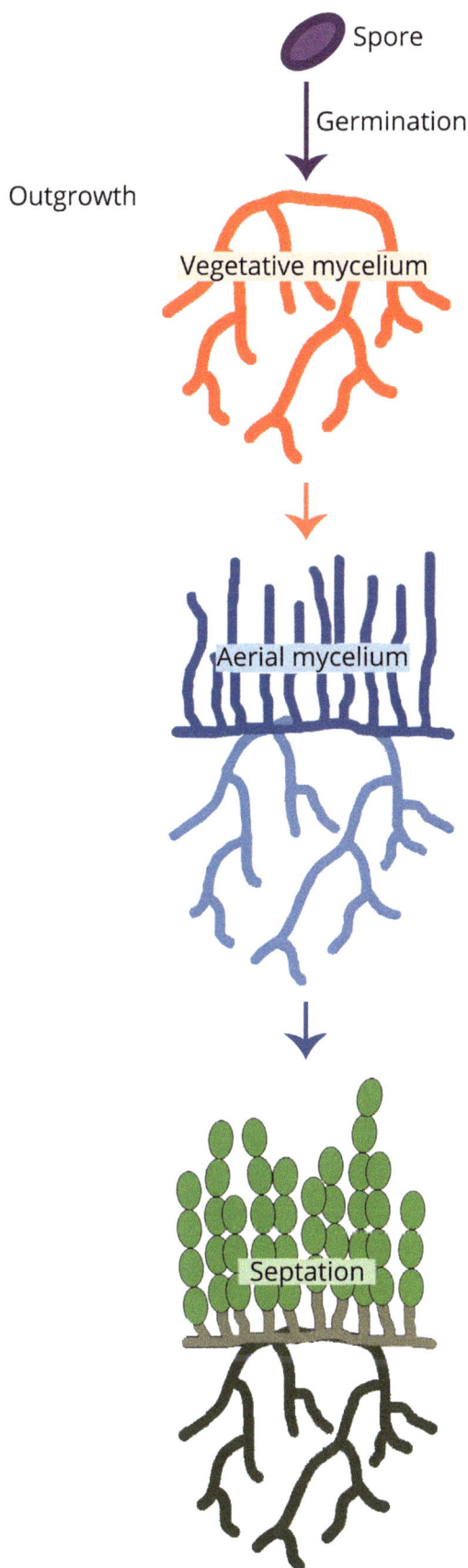

Spore

Germination

Outgrowth

Vegetative mycelium

Aerial mycelium

Septation

FIGURE 5.1 Diagram of the life cycle of *Streptomyces*.

triggering of antibiotic production on the other. The answer to this apparent paradox lies in the ability of the organism to coordinate the production and release of extracellular lytic enzymes (proteolytic, lipolytic, and hydrolytic) and the turning on of the differentiation switch.

5.2.4.1 Role of RelA in Initiating Transcription of Antibiotic Synthesis Genes Under Nitrogen Limitation

Much attention has recently been focused on the role of RelA, the ribosome associated ppGpp synthetase, in initiating secondary metabolism in actinobacteria under nitrogen limitation, but not phosphate. RelA is apparently central to antibiotic production under conditions of nitrogen limitations in *Streptomyces coelicolor* A3 (Chakraburtty and Bibb, 1997; Bentley et al., 2002) and *Streptomyces clavuligerus* (Jin et al., 2004a, 2004b). The question of whether antibiotic production is triggered as a direct consequence of RelA or indirectly as a consequence of slow growth rate through inhibition of rRNA synthesis by ppGpp binding, remains to be established (Bibb, 2005). Evidence supporting the direct participation of ppGpp in activating the transcription of genes involved in the biosynthesis of secondary metabolites (e.g., actII-orf4 genes involved in the biosynthesis of the antibiotic actinorhodin in *S. coelicolor*) has been given (Hesketh et al., 2001). However, the mechanism through which this is achieved has yet to be unraveled (Bibb, 2005).

5.2.4.2 Role of Phosphate in Repressing Transcription of Antibiotic Synthesis Genes

Unlike conditions of nitrogen limitation, the signal that triggers the synthesis of secondary metabolism under phosphate limitation is RelA (ppGpp) independent (Chakraburtty and Bibb, 1997). Under these conditions (i.e., nitrogen limitation), the polyphosphate reserve is hydrolyzed to inorganic phosphate, which in turn represses biosynthetic enzymes, to satisfy growth requirements. The negative role portrayed for polyphosphate kinase (PPK), the enzyme responsible for the biosynthesis of polyphosphate polymer, was ascertained as loss of enzymic activity and accompanied by increased level of antibiotic formation in *S. lividans* (Chouayekh and Virolle, 2002). The following is generally accepted:

- Conditions leading to elevated levels of inorganic phosphate lead to repression of enzymes of secondary metabolism and in turn diminish flux to antibiotic formation.
- Conditions leading to inorganic phosphate deprivation stimulate secondary metabolism and in turn increase flux to antibiotic formation.

However, the question of how elevated levels of inorganic phosphate might interfere with the onset of secondary metabolism and morphological differentiation of actinobacteria awaits further investigation.

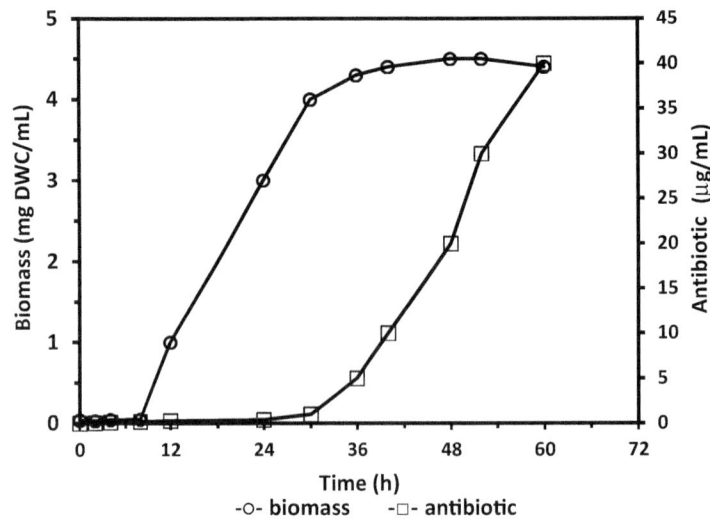

FIGURE 5.2 The time course of a typical *Streptomyces* fermentation for an antibiotic.

5.2.4.3 Role of Quorum Sensing and Extracellular Signals in the Initiation of Secondary Metabolism and Morphological Differentiation in Actinobacteria

It is generally agreed that γ-butyrolactones are produced and, in turn, implicated specifically in the production of secondary metabolites as well as morphological differentiation in several species of *Streptomyces* (Mochizuki et al., 2003), for example, the production of streptomycin by *Streptomyces griseus* and the production of tylosin in *S. fradiae*.

In sharp contrast to γ-butyrolactones-binding proteins, which downregulate the transcription of secondary metabolite gene clusters, another molecule, SpbR, has been shown to stimulate the biosynthesis of pristinamycin (Folcher et al., 2001). Deletion of SpbR not only was accompanied by the abolition of antibiotic formation but also severely impaired growth on agar media, thus lending further support to the positive role of SpbR in antibiotic production and the cellular growth and differentiation in actinobacteria (Bibb, 2005).

Another factor, PI (2, 3-diamino-2, 3-bis (hydroxymethyl)-1, 4-butanediol), has been shown to be able to elicit the biosynthesis of pimaricin in *S. natalensis* (Recio et al., 2004), and its positive role was further substantiated when its addition to P1-defective mutants was accompanied by restoring the organism's ability to produce pimaricin.

5.2.4.4 Positive Activators of Antibiotic Expression

In antibiotic fermentations, temporal expression of antibiotic biosynthesis is regulated tightly as part of the cellular differentiation pathway (Figure 5.2). Production of antibiotics is costly to the cell in terms of carbon and energy. Therefore, it is hardly surprising that tight control of expression has evolved. The genes for antibiotic biosynthesis are invariably clustered together, irrespective of whether the microbe is a prokaryote or eukaryote. In *Streptomyces*, regulation of expression of such gene clusters is controlled by a positive activator—a master gene that, when switched on, makes a protein that

targets itself to the various promoters of the gene cluster and switches them on in concert (Figure 5.3).

In this way, the cell ensures that the full complement of enzymes necessary to make the antibiotic is produced at the same time and at the correct levels.

The great majority of antibiotic pathways are regulated in this way. For example, the *tyl*R system that controls tylosin biosynthesis has been characterized (Stratigopoulos et al., 2004) as a newly discovered regulator for actinorhodin biosynthesis in the model organism *S. coelicolor* (Uguru et al., 2005). Sometimes, the expression of the activator gene is integrated tightly with the developmental pathway, such as the bldA-dependent regulation of ActIIOrf4 in *S. coelicolor* and *bld*G- dependent regulation of *cca*R in *S. clavuligerus* (Bignell et al., 2005), in which CcaR upregulates the biosynthesis of both cephamycin C and clavulanic acid (Perez-Llarena et al., 1997).

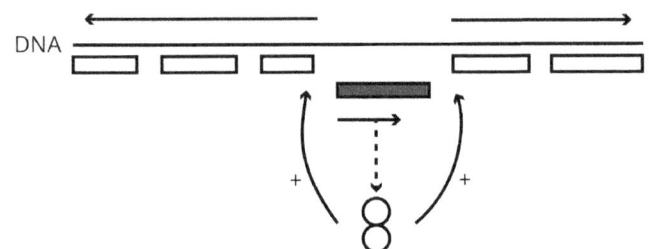

FIGURE 5.3 Schematic diagram to illustrate how antibiotic gene clusters are controlled. The DNA encodes several genes clustered together on the chromosome. The genes (rectangles) are usually transcribed as polycistronic mRNAs (shown by the unbroken straight arrows). Transcription of the production genes (unshaded rectangles) must be dependent on transcription of the positive activator (the master gene, shown by the shaded rectangle and the protein by the two circles) that migrates and binds to the DNA (curved arrows) to allow transcription of the production genes to take place. Without the activator protein, there is no expression of production genes.

Other factors that may act in a pleiotropic manner (e.g., AfsR of *S. coelicolor*) play a key role in integrating multiple signals that are transduced through phosphorylation cascades (Horinouchi, 2003). Another effector is AfsK, which on sensing appropriate signals autophosphorylates itself; autophosphorylated (activated) AfsK phosphorylates cytoplasmic AfsR, which, in turn, through its DNA-binding activity, activates the transcription of *dfsS*, the product of which is apparently capable of stimulating the biosynthesis of several different antibiotics (Bibb, 2005).

5.2.5 TACTICAL ISSUES FOR STRAIN IMPROVEMENT PROGRAMS

In attempting to harness the vast natural potential of the *Streptomyces* and related Actinobacteria for the making of antibiotics on a large scale, the organisms must be grown in large volumes of liquid cultures. Care must be taken so that the essential features of the biology of the antibiotic production process are not lost following the transfer (inoculation) of the organism from a solid medium to liquid medium in the fermentor. One of those features which can be easily lost when growing in a submerged medium is the differentiation of vegetative to aerial mycelium and the subsequent sporulation. However, a greater number of strains have been discovered that are able to differentiate and produce spores in liquid medium (van Dissel et al., 2014).

The physiology of the fermentation process must be adapted to suit the biology of the microbe. As one of the triggers for differentiation in the natural ecosystem is deprivation of nutrients, then the same strategy may also be applied to the fermentation process. All antibiotics are composed of carbon atoms and many contain nitrogen atoms. Therefore, it would be a poor tactic to attempt to trigger the onset of antibiotic production by limiting the cells' supply of either of these elements, because after antibiotic production had commenced, the cells would be starved of one of the most important chemical elements needed to biosynthesize the antibiotic structures. Fortunately, very few antibiotics contain phosphorus in their elemental composition, so the cells are most often limited by the supply of phosphate to trigger antibiotic production. If all other nutrients are supplied in sufficient quantities (but not in vast excess), then antibiotic production will continue for many days (often 8–10 days) at a rate that is linear with time (Figure 5.2). Eventually, this rate starts to tail off. It makes economic sense to terminate the fermentation at this point, recover the maximal amount of product in the shortest possible time, and then prepare the vessel for another round of fermentation.

Fermentations for high-value recombinant therapeutic proteins are best undertaken by fed-batch cultures, which maximize the use of the biomass factory and deliver the product in the most concentrated form. However, the duration of the production phase for these high-value fermentations is considerably shorter than for the antibiotic fermentations, typically around eight hours or less.

In practice, the order in which the experimental strategies for improving strains are used depends on the nature of the desired product and of the fermentation. In the initial stages of developing an antibiotic-producing strain from a "soil isolate," random mutagenesis of a population of the producer microorganism is undertaken and the progeny of the mutagenic treatment are screened for higher levels of the antibiotic. Subsequently, more directed screens are used to further enhance the titer of the antibiotic.

By contrast, improvement of strains making recombinant therapeutic proteins starts with a very directed strategy and finishes with a more random, empirical approach. This is because, in the initial stages of strain improvement, there is a well-defined template of experimental improvements that can be followed. Subsequently, the performance of the strain can be improved further by the empirical approach.

5.2.6 STRAIN IMPROVEMENT: THE RANDOM, EMPIRICAL APPROACH

In this approach a population of microorganisms is subjected to a mutagenic treatment (Figure 5.4), typically with the chemical carcinogen nitrosoguanidine (NTG), but other mutagenic agents such as ultraviolet (UV) light or caffeine may be used (Biot-Pelletier and Martin, 2014). The treatment is tailored so

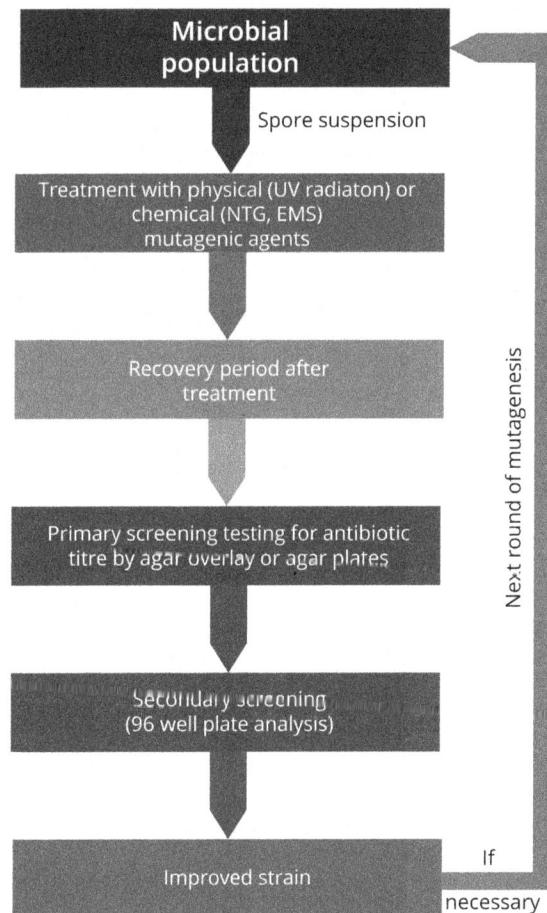

FIGURE 5.4 Random mutagenesis and the empirical approach to strain improvement.

that each cell, on average, has a single mutation induced. The mutagenized population is plated on agar, which will support the growth of the normal (wild-type) strain. Some mutations will be very deleterious to the growth of the organism; cells containing these will not grow up to become colonies. The mutant strains (colonies, each of which is derived from a single cell) that show little or no growth impairment are then tested randomly for their ability to overproduce antibiotics. Classically, this was achieved by overlaying the agar plate containing the surviving colonies after mutagenesis with soft agar containing another microorganism that is sensitive to the antibiotic in question. Colonies producing more biologically active antibiotic will be surrounded with a larger clear zone of inhibition of the culture overlay. However, this technique does not consider the fact that some colonies may be larger (i.e., have a greater diameter) than others. To circumvent this problem, a cylinder of the colony and the agar underneath is cut out (usually with a cork borer) and the "agar plug" placed on a fresh lawn of sensitive bacteria spread on a new agar plate. The cylinders are uniform and contain comparable amounts of cell material. This modification takes account of differences in sizes of different mutagenized colonies. The whole procedure is tedious but must be undertaken in a painstaking manner. In recent times, the pharmaceutical industry has used robots to perform the tasks required in random screening. There has been a tendency to move away from screening for how much antibiotic is made by colonies on a plate, to liquid media-based cultures that reflect the situation in fermentors more accurately. It is not unusual for a screening robot to evaluate the performance of over a million mutants in a year—such "high-throughput screening" is orders of magnitude more efficient than using people for these tasks, leaving the human input focused on the design and efficiency of the overall screening program.

Most of the progeny will make lower levels (Figure 5.5) of the microbial product, and only a few will have larger titers,

often showing 10% (or less) improvement. The reason for this is that most mutations are deleterious to production and only a few mutations will result in slightly elevated levels of production. It is very unusual to isolate mutants that have large (>20%) increases in titer.

These individual isolates are then tested in small-scale fermentations to confirm that they really are improved mutants. Most turn out not to have improved titers and are discarded. Mutants that do show some promise will then be tested intensively at the laboratory and pilot plant levels.

Different companies have different strategies at this stage. Some may carry out extensive laboratory trials at around the 10 L level and scale up the mutants directly to the large production fermentors in one step.

Other companies may adopt a more conservative posture and scale up the volume of fermentation in stages. For a process that is undertaken at a production scale of 100,000 L, the stages in scale-up would typically be from 10 to 500 L, then onward to 5,000 L and finally to the large fermentors. The doctrine is that the operation of these large fermentors is expensive—both in the cost of fermentation broth and in lost opportunity, as that large fermentor might otherwise have been used to make another valuable product. The view of those fermentation technologists who scale up in one step is that time is wasted solving problems associated with the intermediate stages and that a production fermentation that is modeled well at the bench scale should, by definition, predicate immediate translation of the new mutant and its process details to the production fermentor hall.

It is often overlooked that the initial improvement in titer, which is achieved with a new mutant at the production scale, can often be improved still further by process optimization through adjustments to the media and fermentation conditions that better suit the physiology of the new mutant. Process optimization invariably makes an equal, if not greater, contribution to overall increase in titer than the stepwise increase in fermentation performance achieved when the new mutant is adopted initially at the production scale.

5.2.7 Strain Improvement: The Power of Recombination in "Strain Construction"

In the early stages of undertaking a random mutagenesis program for a new product (i.e., starting from a new soil isolate that makes a small amount of product), there are likely to be rapid, concurrent advances in the performances of several strains. These strains, separately, will have individual desirable properties. For example, in an antibiotic program, one strain may give a higher titer, while another may use less carbon source to make a level of antibiotic equivalent to that of the parent (and so be more economical). Yet another strain may show no better performance in terms of titer of product or the amount of nutrients consumed to make it but may carry out the fermentation at a lower viscosity, which will make it easier for the product to be extracted from the fermentation broth and purified. Different "lineages" of the strain are developed very rapidly in these early days in the desire to improve the

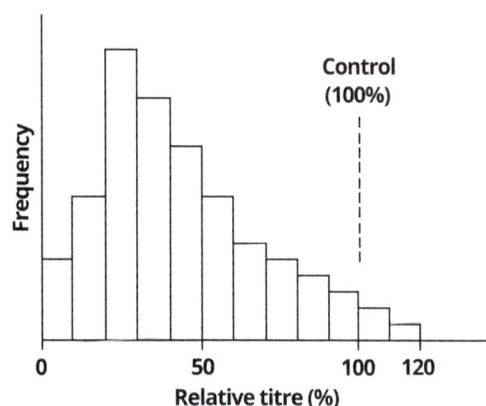

FIGURE 5.5 Antibiotic titers of individual survivors after random mutagenesis. The histogram shows the frequency distribution of titers of survivors after random mutagenesis, for a strain whose lineage has been developed for around two years. Notice that only a few isolates perform better than the control (some of this may be due to error in determining the titer) and that most survivors perform much worse than the control, as the mode is about 25%.

fermentation considerably. However, it soon becomes desirable to construct strains with a combination of several of the desirable properties already present in the individual lineages. To achieve this, desirable traits from each lineage must be "recombined" genetically into a single strain that possesses all of the desirable properties, or "traits."

Most microbial species can exchange genetic information with other members of their species by recombination. Construction of the genetic maps that appear in many textbooks is based on exploitation of these natural recombination processes. For *E. coli* and other enteric Gram-negative bacteria, protocols to undertake genetic recombination are well documented and depend on plasmid-based mobilization of the bacterial chromosome.

However, the producers of many microbial metabolites, including the *Streptomyces*, have a life cycle that involves a sporulation stage. With these species, the most effective method of constructing recombinants that have combinations of several desirable traits is to undertake "spore mating": to mix spores of the parents with the individual traits, germinate them together, allow them to go through an entire life cycle (Figure 5.1), and (from the spores formed at the end of this life cycle) select those that have exchanged some genetic information (i.e., select those that show that they have undertaken some degree of recombination) and then screen them to identify the individuals that have received the desired combination of traits (Figure 5.6).

This screening process is usually conducted empirically, in the same way as screening of strains after random mutagenesis. Unfortunately, a major drawback with this procedure can often arise: as strains with further improved titers are selected and the strain lineage becomes longer, the ability to sporulate is often lost. Although sporulation is associated with the differentiation process, antibiotic production can often become decoupled from sporulation in the advanced high-titer strains. The strains become asporogenous, which makes such a spore-mating strategy impossible. For some strains, the problem can be circumvented by undertaking strain construction by recombination between the vegetative mycelia of different parents. However, not all strains are amenable to this approach. For them, a more involved strategy called *protoplast fusion* (Figure 5.7) must be enacted (Magocha et al., 2018). The most common pitfall with protoplast fusion is the time taken to develop the media conditions and culture techniques that allow the protoplasts to regenerate satisfactorily back into fully competent mycelia (see Figure 5.7). Often, an extensive series of empirical range-finding experiments must be undertaken to optimize the conditions for regeneration, and occasionally it proves impossible to develop a protocol for a particular species.

Recently, the power of recombination has been used to great advantage in combination with high-throughput screening in a "genome-shuffling" strategy that leads to a rapid increase in tylosin titer in the industrially relevant *S. fradiae* (Zhang, et al. 2002). Only two rounds of genome shuffling were needed to achieve results that had previously required at least 20 rounds of classical mutation and screening.

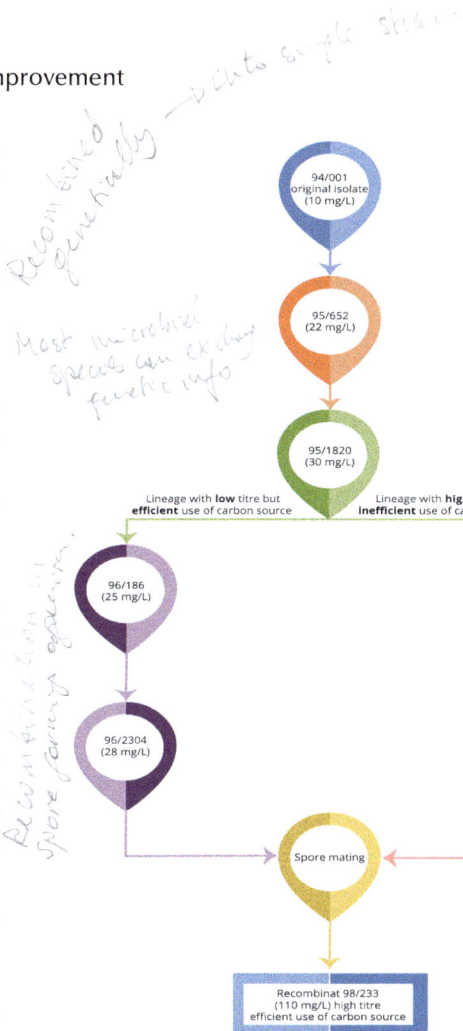

FIGURE 5.6 Strain construction by recombination. In this example, two lineages have been derived from the first soil isolate. One lineage shows a steady improvement in titer with each new strain, but this is accompanied by wasteful utilization of carbon source. The other lineage makes very little antibiotic but does so in a very efficient way. By genetic recombination of these traits, the attributes of both strands of the culture lineage can be combined in a single strain.

5.2.8 DIRECTED SCREENING FOR MUTANTS WITH ALTERED METABOLISM

Although the random mutagenesis and recombination approaches are fruitful in acquiring mutants, which give initial improvements in titers, in due course as the level of microbial product generated increases, it becomes more difficult (in parallel) to isolate further improved strains using the empirical, random approach. For example, in the development of an antibiotic fermentation process, by this time the titer of product will have reached a few grams per liter and considerable background data will have been gathered on the physiology and biochemistry of the fermentation process.

These data can be analyzed to diagnose whether wasteful metabolites are being made during the fermentation process (see Chapters 3, 6, and 7 for further details). For example, many Streptomycete fermentations often consume copious quantities of glucose and excrete pyruvate and α-ketoglutarate, giving low conversion yields of glucose into product. If the

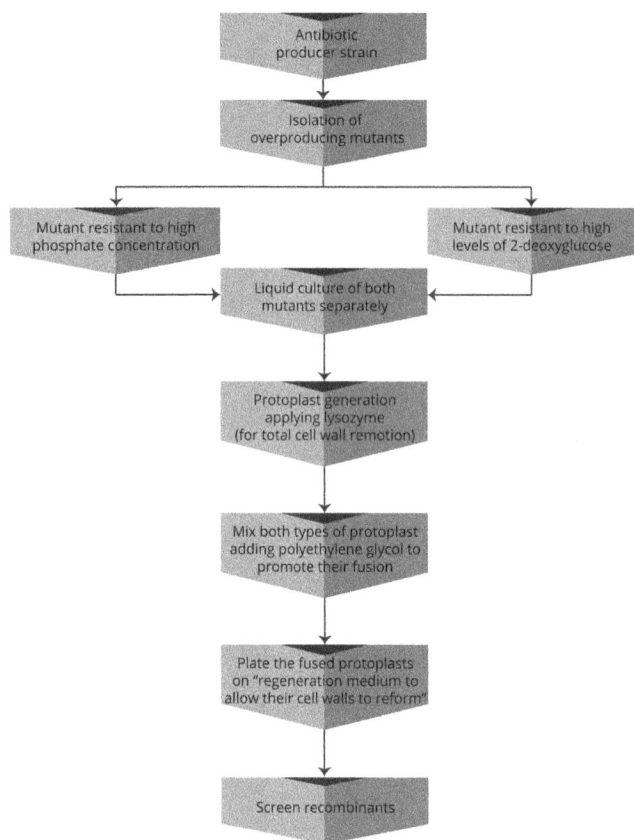

FIGURE 5.7 Schematic diagram of protoplast fusion.

overutilization of glucose can be prevented by careful process control (e.g., limiting the supply of glucose), then the economics of the process will improve substantially. However, the same objective can be achieved by generating mutants that will use the glucose less quickly. Such a strategy is amenable to a "directed screening" approach.

In the case of uncontrolled uptake and wasteful metabolism of glucose, it is known that mutants resistant to 6-deoxyglucose (a toxic analog of glucose) have reduced or impaired glucose uptake and subsequent metabolism. In this strategy, a mutagenic treatment is performed on a population of cells or spores, in the same way as described in Figure 5.3. However, instead of plating all the survivors under nonselective conditions, the survivors are plated directly onto a nutrient agar to which the desired selective pressure can be applied. In this case, the survivors would be plated on media containing 6-deoxyglucose at a level that, under normal conditions, just prevents the growth of the microorganism. The media would also contain a second carbon source (usually glycerol) on which the cultures can normally grow quite vigorously. Out of the millions of survivors of the initial mutagenic treatment, only those which are altered in some aspect of their metabolism of 6-deoxyglucose will survive and grow on such a selective medium. These isolates can then be screened conventionally to identify the individuals among the population that may have the potential in

the fermentor to display a reduced level of glucose uptake and consequently a more balanced metabolism of the sugar, which does not involve the wasteful production of pyruvate and the α-oxo-organic acids.

Many survivors that have become resistant to 6-deoxyglucose will have mutated to confer complete exclusion of the toxic analog from the cell, but in addition, they will no longer be able to utilize glucose at all and will have to be discarded. Strains that are totally incapable of glucose utilization are of no use to the fermentation industry, which is substantially based on cheap, glucose-rich carbon sources.

However, the minor class of analog-resistant survivors, which have impaired but still significant glucose uptake, are the sought-after prizes in this "directed screen." The term *directed* is appropriate because the researcher defines the exact conditions under which such mutants should survive to generate new culture isolates with improved fermentation performance through less wasteful metabolism of glucose.

Directed screening is an extremely important tactic that may be employed to great advantage after the basic details of the fermentation have been worked out and understood. The value of directed screening is the reduction in numbers of mutants to be screened (typically by around 10,000-fold) before an isolate with the desired characteristics is identified. Such a reduction in workload allows time to establish the metabolite production profile of each survivor and to evaluate the data in greater depth. Such is the power of the directed screen that rare spontaneous mutants may be isolated, rather than those generated by mutagenic agents. If 10^6–10^8 cells are plated on a single plate containing a toxic analog, a few spontaneous mutants will always be isolated.

In addition to carbon utilization, the flux of nitrogen (fixation and metabolism of ammonia) and phosphorus (phosphate metabolism) can be altered by specific mutations, for which directed screens can be devised to select for potentially improved mutants. Fixation of ammonia takes place via the enzyme glutamine synthetase in virtually all microbes. The flux through this step is often altered in mutants that are resistant to bialaphos, a toxic compound that specifically inhibits glutamine synthetase. Mutants with altered phosphate metabolism may be obtained by directed screening for resistance to toxic analogs, such as arsenite or dimethyl arsenite.

Fluoroacetate is a classic metabolic inhibitor that poisons the tricarboxylic acid (TCA) cycle by being converted to fluorocitrate, a toxic analog of citric acid. Mutants more resistant (or sometimes those more sensitive) to fluoroacetate often have a TCA cycle with altered properties. As the TCA cycle is a fundamental component of cellular aerobic metabolism (conditions under which most fermentations are conducted), then these mutants can have important properties and be fruitful sources of improved strains.

The building blocks for all microbial products, including the antibiotics, are common metabolic precursors used in other biosynthetic processes of the cell. For example, the three components of penicillin are two amino acids, cysteine and valine, together with adipic acid, which is a precursor of lysine. Thus, penicillin can be viewed biosynthetically as a simple but modified tripeptide (Figure 5.8) of three common amino acids. During microbial growth, cellular metabolism is painstakingly controlled to ensure that supplies of all 20 amino acids needed for growth are available in a balanced fashion.

In a batch-fed culture, such as an antibiotic fermentation, tightly regulated metabolism during the growth phase is followed by the production phase (Figure 5.2), during which the commercial aim is to produce a single product quickly and at high levels—to the exclusion of others. As this microbial product will probably be made from a few key metabolic intermediates (e.g., during production of penicillin, only a supply of the three amino acids will be in high demand), then metabolism must be altered to satisfy this increased demand while minimizing the side reactions of wasteful metabolism. Directed screens can be devised that decouple the usual control strategies of the biosynthetic pathways (such as feedback inhibition and cross-pathway regulation), which normally keep the supply of all precursors just balanced to the needs of the growing cell.

By way of example, consider the supply of adipic acid (part of the lysine pathway) for the biosynthesis of penicillin (Figure 5.8). Normally, the lysine pathway is subject to end-product feedback inhibition. The toxic analog of lysine, δ-(2-amino-ethyl)-l-cysteine, also inhibits the first step of lysine biosynthesis. Mutants resistant to this toxic analog are no longer subject to end-product feedback inhibition of the early part of the biosynthetic pathway. They have an enhanced flux of precursor supply to adipic acid and often produce higher titers of penicillin.

Almost every metabolic pathway that supplies the precursors of microbial products is amenable to this type of directed screening strategy. Thus, the rate of "fuel supply" for biosynthesis of microbial products can be enhanced.

The last example of directed screening relates to the selection of mutants that produce elevated levels of the enzymes responsible for catalysis of the precursor building blocks for the biosynthesis of the backbone structures of antibiotics, such as the tetracyclines and erythromycin. They are polymers of acetyl-CoA and methylmalonyl-CoA, respectively, and both are made by a process that is essentially the same as that for fatty acid biosynthesis. The antibiotic cerulenin targets the enzyme complexes responsible for fatty acid biosynthesis and acts to starve the growing cell of the fatty acids necessary for insertion into the membrane. Mutants resistant to toxic levels of cerulenin have circumvented this problem by making elevated levels of the fatty acid biosynthetic enzymes to "titrate out" the effect of the antibiotic. The close similarity between the enzymes of fatty acid biosynthesis and those that make the backbones of erythromycin and tetracycline allows cerulenin to inhibit the biosynthesis of these antibiotics. Mutants that can still make the antibiotic in the presence of cerulenin have elevated levels of the biosynthetic enzyme complexes. When cerulenin is removed, they retain the high level of biosynthetic capability, which, if supplied with enough of the metabolic precursor "fuel," results in higher titers of the antibiotic. The utility of the directed screening approach is that the survivors of the mutagenic treatment are invariably altered in some aspect of fatty acid or antibiotic biosynthesis.

Individual mutants made by directed screening approaches can be recombined together using genetic recombination, spore mating, or protoplast fusion to combine several desired traits.

5.2.9 RECOMBINANT DNA APPROACHES TO STRAIN IMPROVEMENT FOR LOW- AND MEDIUM-VALUE PRODUCTS

The advent of recombinant DNA techniques, first devised for *E. coli* in the 1970s, has meant that new strategies can be applied to strain improvement for low- and medium-value microbial products. It has taken some time for recombinant techniques to be developed and applied to the commercial strains, such as the filamentous bacteria and filamentous fungi, which are the mainstay of this sector of the fermentation industry.

In addition, regulatory hurdles must be crossed to gain approval to undertake fermentations at the production scale with these "genetically engineered microorganisms." Gaining such approval is time-consuming and costly. Thus, there must be good long-term economic reasons for adopting a recombinant DNA strategy for strain improvement.

By cloning this master gene regulator (see Section 5.2.4) and then expressing it at unnaturally high levels, the cellular complement of the entire biosynthetic machinery for production of an antibiotic may be enhanced. Of course, there must be sufficient fuel (metabolic precursors and energy) to realize the full potential available from this boosted level of biosynthetic machinery. The roles of metabolic flux analysis and process optimization (see Chapters 6 and 7 for more details) in assuring this advantage are very important.

It is also possible to force expression of the master gene during the growth phase and so to produce antibiotics during growth. In some instances, this may be to the advantage of the overall process, but often it is better to mimic the situation in nature: first to focus the design of the process on maximizing the acquisition of biomass, and then to turn that biomass to best advantage by using the cells as a factory (with the maximal level of installed biosynthetic capability) to make the product at a fast rate and to achieve a high titer. Therefore,

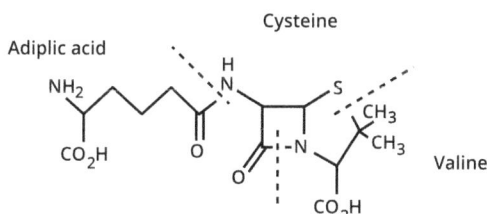

FIGURE 5.8 Diagram of the biosynthetic origin of isopenicillin N. This has three component amino acids, condensed together with peptide bonds. The origins of the amino acids are shown between the dotted lines.

controlled expression of the master gene so that it is switched on decisively at the end of growth is often preferred.

Molecular genetic analysis has also shown that some of the strains from improvement programs, isolated over the years by random mutagenesis and selection, have increased dosages of the biosynthetic genes. Thus, some of the best strains for penicillin production have multiple copies of the critical part of the biosynthetic gene cluster arranged in tandem arrays. This effect can also be achieved by gene-cloning strategies.

The biochemical pathways for antibiotic biosynthesis are long linear series of enzyme reactions. The flux through the entire pathway is governed by the pace of the slowest catalytic step. If the flow of metabolic intermediates through that step can be improved, then there is a good chance that the productivity of the overall process will be improved. Thus, for the antibiotic tylosin, produced by *Streptomyces fradiae*, it was established that the rate of the last step in the pathway, conversion of macrocin to tylosin by macrocin-*O*-methyltransferase (encoded by the gene *tyl*F), limited the overall productivity. As the strain improvement program had developed with time and the production strain lineage was reviewed, it was apparent that mutants with higher titers of tylosin also displayed extremely high levels of macrocin, which was excreted as a shunt metabolite (Figure 5.9) because conversion of macrocin was limiting. Strains that had the methyltransferase gene cloned and expressed at high levels showed improved titers of tylosin.

Often, process analysis shows that some metabolites on the main biosynthetic pathway are being diverted to other shunt products (Figure 5.10). This not only represents a waste of valuable carbon source, but also presents a problem for the ultimate purification of the desired product, as the second metabolite must be purified away.

Genetic manipulation can be used to specifically inactivate the gene for the enzyme that diverts the metabolite to the shunt metabolite. This precise "genetic surgery" enhances the flow through the pathway to the desired product. One last important contribution that genetic manipulation makes is in providing flexibility of utilization of a carbon source. If, for example, there is a plentiful supply of cheap lactose available (from the milk whey industry) but the strain does not use lactose naturally, then lactose utilization genes can be cloned from another species and introduced into it.

At times, the world marketplace has a glut of hydrolyzed sucrose. This cheap carbon source is problematic for fermentation because one of the hydrolysis products (glucose) can inhibit the utilization of the other (fructose). It is possible to overcome this problem by the cloning of the fructose utilization genes, which, in turn, ensures that the two monomeric carbon sources are used concurrently. The overall economics of such a process benefits from being able to use a cheap and plentiful carbon source in an efficient manner, without excess fructose carbon being left in the broth at the end of the fermentation.

5.2.10 Strain Improvement for High-Value Recombinant Products

Strain improvement programs for products derived from cloned genes follow a completely different strategy from those undertaken for their lower value counterparts. The techniques used to clone a gene (or cDNA) usually place it precisely in a vector (most often, a plasmid) in a context that allows continuous high-level expression, or, more advisedly, expression of the gene is controlled and induced at a high level only after growth has ceased—in the same way as antibiotic fermentations are performed (Figure 5.2). A countless number of possibilities arise to tailor the gene within the vector, but this aspect involves only molecular biology and falls outside the remit of this chapter. The biology of the host strain is an equally critical factor in the overall performance of a fermentation process for a recombinant product, and strain improvement programs can have a significant impact on the economics of such a process.

In the early days of the "new biotechnology" industry, considerable difficulty was experienced in "scaling up" recombinant processes from the laboratory to the pilot plant (i.e., from a scale of around 100 ml to 400 L). This was because, at the larger volume, a greater proportion of cells had lost the recombinant plasmid from the cells. Plasmid-deficient cells do not make the recombinant protein, and this reduces the overall productivity. The root cause of the problem is the number of cell generations needed to attain significant growth of the culture at the larger volume, coupled to the inherent instability of engineered plasmids in cells. The growth rate of a cell carrying a plasmid is invariably slower than its counterpart that has shed its plasmid load. Therefore, plasmid-deficient cells will outgrow the plasmid-containing cells in a population, a phenomenon that becomes more significant as the number of

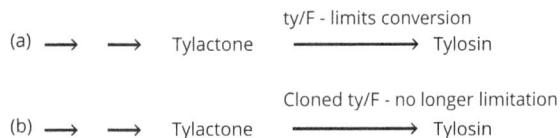

FIGURE 5.9 Schematic diagram highlighting (a) the limitation of tylosin biosynthesis by *Streptomyces fradiae* and (b) the elevation of such limitation using a recombinant strain.

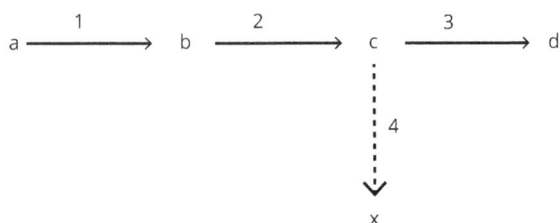

FIGURE 5.10 Schematic diagram highlighting the problem of shunt products and how this can be prevented. The metabolic pathway (from substrate a to product d) is composed of three enzymes (1, 2, 3) with intermediates b and c. Because the carbon flow through steps 1 and 2 is greater than through step 3 (shown by the width of the arrows), the shunt metabolite, X, is formed through the action of enzyme 4. The presence of shunt metabolite, X, may complicate the recovery and purification of the desired product, d. By making a mutant devoid of enzyme 4, this is prevented.

generations in a culture increases: the larger the volume of the fully grown culture, the greater the proportion of plasmid-deficient cells in its population.

Resistance to an antibiotic, encoded by a plasmid-borne gene, is the usual selection strategy for the presence of a plasmid in a recombinant culture. The most commonly used selection is resistance to ampicillin, encoded by β-lactamase. Ampicillin-resistant cells survive because they are protected by β-lactamase, which breaks down the chemical backbone of the antibiotic. Eventually, all the ampicillin in the fermentation broth is broken down and the selective pressure is lost—the longer the fermentation (i.e., the more generations that take place), the more likely it will be that the antibiotic will become inactivated and that plasmid-deficient cells will form.

The frequency at which plasmid-deficient cells are formed, in the absence of an antibiotic, is biased because of their natural tendency to form oligomers within the cell. Consider a theoretical situation in which a cell has a plasmid copy number of four. This may be conceptualized as four separate plasmids, two of which segregate into each daughter cell at division (Figure 5.11). However, because of oligomerization, an equally common situation is that there will be a single tetramer of plasmids within the parent cell (i.e., the copy number will still be four); when the daughter cells are formed, one will receive the plasmid, which consists of four monomers, and the other will not (Figure 5.11b). In the absence of selective pressure (i.e., after all of the antibiotic has been exhausted), both will survive, and the plasmid-deficient daughter will outgrow its plasmid-containing sibling because of the advantage in growth rate.

Engineered plasmids undergo such oligomerization events at alarming rates. However, natural plasmids do not undergo such "segregational instability," because they have a natural tendency to break back down from oligomers to monomers again (Gaimster and Summers, 2015). This ability has been lost during the process of conversion of natural plasmids into genetically engineered vectors. It was discovered that a recombination site, called *cer*, was the missing DNA sequence in the engineered vectors. When *cer* was reintroduced into vectors from the natural plasmids, the segregational stability of the plasmid construct containing the cloned gene to be expressed was improved. However, the *cer* functionality is dependent on the strain; this aspect of the biology of plasmid instability can be addressed by strain improvement programs. The cellular machinery that allows *cer* to operate is rather complex and involves at least three different genetic loci. If any of the three is defective or missing, the result is that plasmids (even those containing a fully competent *cer*) are unstable. In addition to the importance of the right timing to perform the eradication of the oligomers, which clearly must occur before cell division. Strain improvement of such cell lineages is best achieved using a targeted genetic approach, in other words to use the power of genetic recombination to introduce the relevant machinery back into the chromosome of the strain, an extremely focused and targeted task.

Recombinant cultures are prone to infection with bacterial viruses called *bacteriophages*. At a scale of fermentation above a liter, in a production environment, the recombinant cultures are at risk. There are two strain improvement strategies that can be brought into play. First, it is possible to mutate the strain to become resistant to viral infection. This approach works for known viruses, but there is always the risk that infection from a new source will take place. A second, more secure strategy is to introduce a restriction and modification (R/M) system into the cell line. The modification enzyme will alter the host's DNA so that it is no longer a target for the resident restriction system. However, viral DNA that enters the cell will be recognized by the restriction system and degraded, thus preventing the infection. The strategy can be enhanced further by the introduction of a second restriction and modification system as a "backup," should some of the viral DNA escape restriction by the first enzyme system. At present, other bacteriophage resistance systems have been discovered which could potentially be used, such as the abortive infection systems (Abi), argonaute, BREX and DISARM (Ofir et al., 2018).

Hosts for foreign gene expression often recognize the foreign protein as "not natural" and selectively degrade it, thus reducing the overall productivity. The enzyme system that undertakes this function, the Lon protease, works in a similar way in ordinary cells, selectively degrading ordinary cellular proteins that have been made defectively with altered amino acids. By knocking out the *Lon* protease gene, degradation of the foreign protein is prevented, albeit to the detriment of the host cell, which now has a slight growth rate disadvantage. Methods that generate *Lon*-deficient cell lines are well established and extremely targeted (Hunter, 2007).

In each of the examples in this section, there has been no need to use random mutagenesis and screening, as the

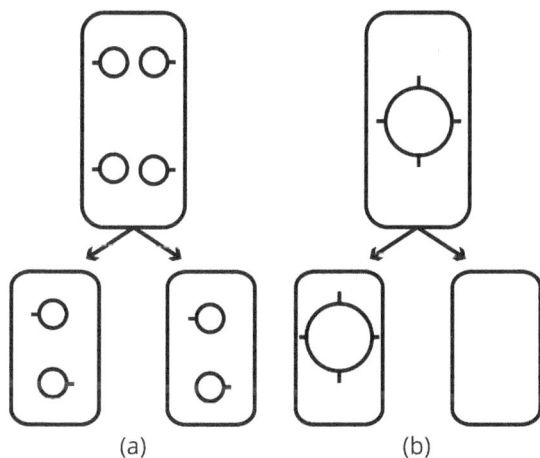

FIGURE 5.11 Plasmid segregation—the complication of plasmid oligomers. In case (a), the plasmids are present as four monomers, and two segregate (on average) into each of the daughter cells. By contrast, in case (b), the plasmids are present as a tetramer. During growth, one daughter cell receives the tetramer, whereas the other does not and, in turn, becomes plasmid deficient.

blueprint to derive the improved strain is direct and straight-forward. After this phase, strain improvement programs use the random approach to fine-tune the genetic content of the production recombinant strain to gain further advantages in productivity. Thus, the order of events is the reverse of that for the lower value counterparts.

SUMMARY

- A wide range of secondary metabolites (e.g., antibiotics and anticancer therapeutics) are currently being produced on a large scale using microorganisms.
- An impressive array of new techniques and methodologies have emerged and proved useful in strain improvement of microorganisms and in increasing metabolic flux to product formation.
- Successful industrial production processes of desirable end products by microbial fermentation rely on the development of a clear and sustainable strategy that ensures the reproducibility of high and stable yield using the cheapest possible feedstock available.
- In the production of new secondary metabolites, it is imperative that we carefully select a strategy that facilitates both production and efficient downstream purification processes.
- The commercial development of new and existing secondary metabolites by microbial fermentation has a very bright future as we enter a new era in which new innovations in techniques and methodology stand to transform our understanding and capabilities.

ACKNOWLEDGMENTS

We thank to Betsabé Linares-Ferrer for skilled assistance in the art work shown in this chapter and M. A. Ortíz-Jiménez for carefully reading the manuscript and help with editorial activities.

REFERENCES

Bentley, S. D., K. F. Chater, A. M. Cerdeño-Tárraga, Challis G. L, Thomson N. R, James K. D, Harris D. E, et al. 2002. Complete genome sequence of the model actinomycetes *Streptomyces coelicolor* A3(2). *Nature* 417:141–147.

Bibb, M. J. 2005. Regulation of secondary metabolism in *Streptomycetes*. *Curr Opin Microbiol* 8:208–215.

Bignell, D. R., K. Tahlan, K. R. Colvin, S. E. Jensen, and B. K. Leskiw. 2005. Expression of *ccaR*, encoding the positive activator of cephamycin C and clavulanic acid production in *Streptomyces clavuligerus* is dependent on *bldG*. *Antimic. Agents. Chemotherapy* 49:1529–1541.

Biot-Pelletier, D., and V. J. J. Martin. 2014. Evolutionary engineering by genome shuffling. *Appl Microbiol Biotechnol* 98:3877–3887.

Chakraburtty, R., and M. J. Bibb. 1997. The ppGpp synthetase gene (relA) of *Streptomyces coelicolor* A3 plays a conditional role in antibiotic production and morphological differentiation. *J Bacteriol* 179:5854–5861.

Chapman, R.F. 2000. Entomology in the twentieth century. *Annual Rev Entomology* 45:261–285.

Chouayekh, H., and M. J. Virolle. 2002. The polyphosphate kinase plays a negative role in the control of antibiotic production in *Streptomyces lividans. Mol Microbiol* 43:919–930.

Cuellar, MC, and L. A. M. van der Wielen. 2015. Recent advances in the microbial production and recovery of apolar molecules. *Curr Opin Biotechnol* 33:39–45.

Folcher, M., H. Gaillard, L. T. Nguyen, K. T. Nguyen, P. Lacroix, N. Bamas-Jacques, M. Rinkel, and C. J. Thompson. 2001. Pleiotropic functions of a *Streptomyces pristinaespiralis* autoregulator receptor in development, antibiotic biosynthesis, and expression of a superoxide dismutase. *J Biol Chem* 276:44297–44306.

Gaimster, H., and D. Summers. 2015. Plasmids in the driving seat: the regulatory RNA Rcd gives plasmid ColE1 control over division and growth of its *E. coli* host. *Plasmid* 78:59–64.

Hesketh, A., J. Sun, and M. J. Bibb. 2001. Induction of ppGpp synthesis in *Streptomyces coelicolor* A3 grown under conditions of nutritional sufficiency elicits actII-ORF4 transcription and actinorhodin bio-synthesis. *Mol Microbiol* 39:136–144.

Horinouchi, S. (2003). AfsR as an integrator of signals that are sensed by multiple serine/threonine kinases in *Streptomyces coelicolor* A3(2). *J Ind Microbiol Biotechnol* 30: 462–467.

Hunter, I. S. 2007. Microbial synthesis of secondary metabolites and strain improvement, in El-Mansi, E.M.T., Bryce, C.F.A., Demain, A.L. and Allman, A.R. (eds.) *Fermentation Microbiology and Biotechnology*, 2nd ed., Boca Raton, New York and London: CRC Press, 131–158.

Jin, W., H. K. Kim, J. Y. Kim, S. G. Kang, S. H. Lee, and K. J. Lee. 2004a. Cephamycin C production is regulated by *relA* and *rsh* genes in *Streptomyces clavuligerus* ATCC27064. *J Biotechnol* 114:81–87.

Jin, W., Y. G. Ryu, S. G. Kang, S. K. Kim, N. Saito, K. Ochi, S. H. Lee, and K. J. Lee. 2004b. Two *relA/spoT* homologous genes are involved in the morphological and physiological differentiation of *Streptomyces clavuligerus. Microbiology* 150:1485–1493.

Magocha, T.A., H. Zabed, M. Yang, J. Yun, H. Zhang, and H. Qi. 2018. Improvement of industrially important microbial strains by genome shuffling: Current status and future prospects. *Bioresour Technol* 257:281–289.

Martínez, J. L., L. Liu, D. Petranovic, and J. Nielsen. 2012. Pharmaceutical protein production by yeast: Towards production of human blood proteins by microbial fermentation. *Curr Opin Biotechnol* 23:965–71.

Medema, M. H., and M. A. Fischbach. 2015. Computational approaches to natural product discovery. *Nat Chem Biol* 11:639–648.

Mochizuki, S., K. Hiratsu, M. Suwa, T. Ishii, F. Sugino, K. Yamada, and H. Kinashi. 2003. The large linear plasmid pSLA2-L of *Streptomyces rochei* has an unusually condensed gene organization for secondary metabolism. *Mol Microbiol* 48:1501–1510.

Ofir, G., S. Melamed, H. Sberro, Z. Mukamel, S. Silverman, G. Yaakov, S. Doron, and R. Sorek. 2018, DISARM is a widespread bacterial defense system with broad anti-phage activities. *Nat Microbiol* 3:90–98.

Perez-Llarena, F. J., P. Liras, A. Rodriguez-Garcia, and J. F. Martin. 1997. A regulatory gene (ccaR) required for cephamycin and clavulanic acid production in *Streptomyces clavuligerus*: amplification results in overproduction of both beta-lactam compounds. *J Bacteriol* 179:2053–2059.

Recio, E., A. Colinas, A. Rumbero, J. F. Aparicio, and J. F. Martin. 2004. PI factor, a novel type quorum sensing inducer elicits pimaricin production in *Streptomyces natalensis. J Biol Chem* 279:41586–41593.

Ruiz, B., A. Chávez, A. Forero, Y. García-Huante, A. Romero, M. Sánchez, D. Rocha, B. Sánchez, R. Rodríguez-Sanoja, S. Sánchez, and E. Langley. 2010. Production of microbial secondary metabolites: Regulation by the carbon source. *Crit Rev Microbiol* 36:146–167.

Santos, R. B., R. Abranches, R. Fischer, M. Sack, and T. Holland. 2016. Putting the spotlight back on plant suspension cultures. *Front Plant Sci* 7:1–12.

Schmidt, F. R. 2005. Optimization and scale up of industrial fermentation processes. *Appl Microbiol Biotechnol* 68:425–435.

Stratigopoulos, G., N. Bate, and E. Cundliffe. 2004. Positive control of tylosin biosynthesis: Pivotal role of TylR. *Molec Microbiol* 54:1326–1334.

Uguru, G. C., K. E. Stephens, J. E. Stead, J. E. Towle, S. Baumberg, and K. J. McDowall. 2005. Transcriptional activation of the pathway-specific regulator of the actinorhodin biosynthetic genes in *Streptomyces coelicolor. Molec Microbiol* 58:131–150.

Van Dissel, D., D. Claessen, and G. P. van Wezel. 2014. Morphogenesis of *Streptomyces* in submerged cultures. *Adv Appl Microbiol* 89:1–45.

Zhang, M. M., Y. Wang, E. L. Ang, and H. Zhao. 2016. Engineering microbial hosts for production of bacterial natural products. *Nat Prod Rep* 33:963–987.

Zhang, Y.-X., K. Perry, V. A. Vinci, K. Powell, W. P. C. Stemmer, and S. B. del Cardayre. 2002. Genome shuffling leads to rapid phenotypic improvement in bacteria. *Nature* 415:644–646.

Zhou, J., G. Du., and J. Chen. 2014. Novel fermentation processes for manufacturing plant natural products. *Curr Opin Biotechnol* 25:17–23.

6 Flux Control Analysis and Stoichiometric Network Modeling
Current Trends and Future Prospects

Mansi El-Mansi, Gregory Stephanopoulos, and Ross Carlson

CONTENTS

"When you can measure what you are speaking about and express it in numbers, you know something about it, and when you cannot measure it, when you cannot express it in numbers, your knowledge is of a meager and unsatisfactory kind."

Lord Kelvin

6.1 INTRODUCTION: TRADITIONAL VERSUS MODERN CONCEPTS

Industrial biotechnologists are generally, but not entirely, of the view that flux through a given pathway is usually limited by one step. Such a step is termed the rate-limiting step or the bottleneck with the enzyme catalyzing such a step being referred to as the "pacemaker." However, the question of how such a step can be identified and quantified in a given pathway remained unanswered, largely because of the lack of an experimental procedure that describes how such a rate-limiting step can be identified and quantified.

The rate-limiting step was defined as the slowest step in a given pathway. Such a definition resulted from the observation made in the mid-1960s that the reaction rate of a sequence of unsaturated enzymes (i.e., where the concentration of substrates is below the K_m value for each of the enzymes involved) depended nonlinearly on the kinetic parameters of all the enzymes involved. However, no theoretical basis was given to validate or substantiate the existence of such a concept.

Clearly, if a rate-limiting step exists in a given pathway, then increasing the activity of the enzyme catalyzing such a step will increase the overall flux through the pathway and, by the same token, varying the activity of any other enzyme will have no effect whatsoever on the overall flux of the pathway in question. Although several studies have been published in support of the concept of the rate-limiting step, by and large such a concept does not appear to be a tenable proposition because it does not adequately explain why flux and, in turn, yield could not be improved after the overexpression of enzymes that are considered to be rate limiting. For example, a 3.5-fold (350%) increase in the activity of phosphofructokinase, the enzyme widely regarded as the "rate-limiting step" in glycolysis, had no appreciable effect on carbon flux through the glycolysis in *Saccharomyces cerevisiae* (Schaaff et al., 1989). Furthermore, the erroneous nature of the concept of the rate-limiting step has also been emphasized in the elegant example given on tryptophan biosynthesis in the yeast *S. cerevisiae* (Cornish-Bowden, 1995; Cornish-Bowden et al., 1995). Biosynthesis of tryptophan in this organism involves the conversion of chorismate to anthranilate through the activity of anthranilate synthase (E1). The latter intermediate is then converted to tryptophan through the activities of four different enzymes: anthranilate phosphoribosyl transferase (E2), phosphoribosyl anthranilate isomerase (E3), indolglycerol phosphate synthase (E4), and tryptophan synthase (E5). In this study, the aforementioned authors revealed that singularly increasing the enzymic activity of any of the five enzymes involved by a factor of up to 50-fold had no significant effect on the flux to tryptophan. However, increasing the concentration of all five enzymes by a factor of 20-fold was accompanied by a significant increase in flux to tryptophan formation.

Several techniques have been devised to assess the relative contribution of each enzyme to the overall flux in a given pathway, and although *in vitro* measurement of the maximal velocity (V_{max}) of a given enzyme is useful, the value obtained does not necessarily reflect the rate of catalysis *in vivo* because of a lack of hard data regarding the intracellular concentrations of substrates and effectors, primarily because of rapid "turnover" of metabolic pools, which, in turn, renders such measurements difficult if not impossible. Therefore, a new quantitative approach was called for, not only to explain the many outstanding observations relating to flux control in industrial fermentations but also to provide a rational basis for the exploitation of the diverse array of metabolic pathways in microorganisms.

Although several approaches have been used as an alternative to the rate-limiting step, it is Kacser's theory of metabolic control analysis (MCA) that has grown in stature since its inception in 1973 and proved, without undermining the intellectual capacity of other approaches, to be the ultimate approach. The controversy over the question of whether a given method can be used successfully to predict or determine the relative contribution of each enzyme to the overall flux in

a given pathway was finally resolved when Kacser and Burns (1973) and Heinrich and Rapaport (1974) independently proposed the theory of MCA. The fundamental difference between the rate-limiting step as a concept and that of the theory of MCA is that whereas the former is a qualitative parameter, the latter examines biological systems in a quantitative way that excludes bias, expectations, or preconceived ideas.

6.2 FLUX CONTROL ANALYSIS: BASIC PRINCIPLES

Kacser's MCA theory facilitates the assessment of not only how perturbation of a particular enzymic activity affects metabolic flux, but also by how much. The response to changes in the concentration of a particular enzyme on flux varies over a wide range. For example, the response could be immediate, with a strong correlation between the increase in flux and the increase in enzymic activity, as is the case for adenylate kinase en route to ATP synthesis. In this case, one might justifiably describe such an enzyme as rate limiting or a pacemaker. However, most enzymes do not enjoy such a high profile and as such an increase in flux may or may not be brought about by increasing the activity of a single enzyme.

Furthermore, the degree of control exerted by a particular enzyme on the overall flux of a given pathway is not purely dependent on the numerical value of its intracellular concentration but rather on whether the enzyme has the capacity for higher throughput, which can only be ascertained from the value of the enzyme's "flux control coefficient," the first pillar of the theory of MCA.

6.2.1 FLUX CONTROL COEFFICIENT

The flux control coefficient is a parameter that describes in quantitative terms the relative contribution of a particular enzyme to flux control in a given pathway. It is not an intrinsic property of the enzyme per se but rather a system property and so is subject to change as the environment changes. It is generally expressed as the fractional change in flux in response to a fractional change in the concentration of the enzyme in question, and its value ranges between 0 and 1.0.

The measurement of the flux control coefficient of a particular enzyme allows an accurate prediction of how flux through a given pathway might fluctuate in response to a specific change in the enzyme's catalytic activity or concentration. Although a change in the concentration can be brought about by cloning and subsequent overexpression of the structural gene encoding the enzyme in question, changes in the catalytic activity of the enzyme without changing its concentration can be brought about through site-directed mutagenesis and protein engineering techniques. For example, consider pyruvate dehydrogenase (PDH) with the view of assessing its effect or influence on flux to acetate excretion in *Escherichia coli* (Figure 6.1). The influence of PDH in that direction can be assessed from the enzyme's flux control coefficient, which can be calculated from the tangent to the curve of a log-log plot of flux (J) as a function of enzymic activity or concentration (E). Assuming that a small increase in the concentration of PDH (dEpdh) was accompanied by a small increase in the

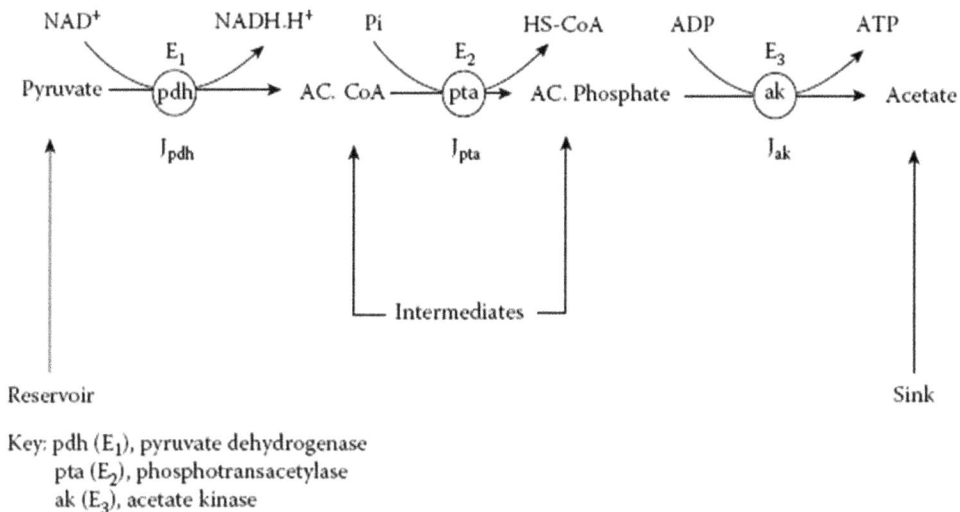

NAD$^+$ NADH.H$^+$ Pi   HS-CoA   ADP   ATP

Pyruvate ——(pdh)——→ AC. CoA ——(pta)→ AC. Phosphate ———(ak)——→ Acetate

J_{pdh} J_{pta} J_{ak}

Intermediates

Reservoir   Sink

Key: pdh (E$_1$), pyruvate dehydrogenase
pta (E$_2$), phosphotransacetylase
ak (E$_3$), acetate kinase

FIGURE 6.1 The enzymes and metabolites en route to acetate excretion.

steady-state flux (J) of the enzyme acetate kinase (dJak), it follows that if we were to change the concentration of PDH very slightly, then the ratio dJak/dEpdh becomes equal to the slope of the tangent to the curve of Jak against Epdh as depicted in Figure 6.2. However, analyzing the data in this way is somewhat imperfect because the numerical value of enzyme concentration and units of enzymic activity will be different from one enzyme to another. This problem could be overcome if we were to relate the fractional changes in flux through acetate kinase to the fractional increase in the concentration of PDH (i.e., dJak/Jak and dEpdh/Epdh), and as such the flux control

coefficient will assume a value between 0 and 1.0, which can then be expressed in terms of a percentage.

However, it is possible for an enzyme to have a flux control coefficient with a negative value, as is the case at branch points where one metabolite has to be partitioned between two enzymes. In such a case, the increase in flux through one branch is generally at the expense of the other, as exemplified in the case study for the partition of carbon flux at the junction of isocitrate (see Section 6.3). At this junction, any increase in the concentration of isocitrate dehydrogenase (ICDH) is concomitant with a decrease in flux through the competing enzyme, namely isocitrate lyase (ICL). It is possible, therefore, to describe ICDH as having a negative flux control coefficient on flux through ICL. Although any increase in the concentration of ICDH is accompanied by a decrease in flux through ICL, the opposite is not true for reasons that will become apparent later on; for further details, see El-Mansi et al. (1994).

6.2.2 SUMMATION THEOREM

The summation theorem, the second pillar of the MCA theory, states that the total sum of flux control coefficients of all enzymes in a given pathway adds up to 1.0. The summation theorem also shows that the flux control coefficient of an enzyme is a system property because any increase in the concentration of a particular enzyme is accompanied by a decrease in its flux control coefficient. According to the summation theorem, such a decrease will have to be balanced by increasing the flux control coefficient of another enzyme—or more than one enzyme—within the same pathway so that the sum of all flux control coefficients remains constant (i.e., 1.0). For example, in a linear pathway consisting of enzymes with usual kinetic properties (i.e., where substrates stimulate and products inhibit reaction rate), the flux control coefficients for every enzyme must be 0 or higher with a total sum of 1.0. If an enzyme were to show a flux control coefficient of 1.0 with all other enzymes showing flux control coefficients of 0, such an enzyme could justifiably be described as rate limiting. The summation theorem also shows that this is not necessarily the

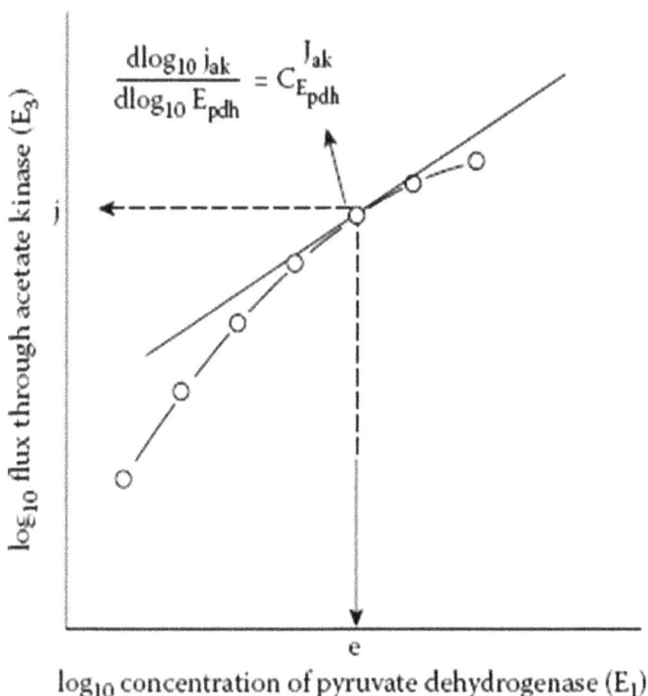

$$\frac{d\log_{10} j_{ak}}{d\log_{10} E_{pdh}} = C^{J_{ak}}_{E_{pdh}}$$

$\log_{10}$ flux through acetate kinase (E$_3$)

$\log_{10}$ concentration of pyruvate dehydrogenase (E$_1$)

FIGURE 6.2 Graphical determination of PDH flux control coefficient with respect to acetate excretion. The graph depicts a typical pattern of variation in flux to acetate; measured as acetate kinase (Jak), in response to changes in the concentration of pyruvate dehydrogenase (Epdh).

case because it is also possible for some or all of the enzymes to have values greater than 0 providing that the total does not exceed 1.0. In practice, we would expect a pathway flux to be influenced mainly by enzymes in that pathway, and to a much lesser extent by closely related pathways, and that distantly connected enzymes would have negligible influence or none at all. In other words, the flux control coefficients of hundreds or even thousands of enzymes that are not directly related or connected to the pathway in question will be zero although flux control is shared among all enzymes.

Another consequence of the highly branched and intricate nature of cellular metabolism is that the central pathways provide biosynthetic precursors and energy for other pathways. So, as biosynthetic precursors are made, some are fed directly into the biosynthetic routes, which in turn diminish flux through the central metabolic pathways. Therefore, it follows that biosynthetic enzymes are likely to have negative flux control coefficients with respect to flux through the central metabolic pathways. According to the summation theorem, if one or more enzymes possess a negative value of flux control coefficient, then it is possible to see some other enzymes displaying a flux control coefficient higher than the numerical value of 1.0. This is because if there are negative flux control coefficients, one or more flux control coefficients would have to be greater than 1.0 so that the total sum adds up to 1.0. This shows that the flux control coefficient is not an intrinsic property of the enzyme itself but rather a property of the whole system.

6.2.3 Elasticity Coefficient

The flux control coefficient of an enzyme is influenced by the enzyme's ability to respond to changes in the concentration of its immediate substrate, as well as its ability to influence the concentrations of other metabolites in the pathway, a linkage that was first demonstrated by Heinrich and Rapaport (1974). The elasticity coefficient, the third pillar of the MCA theory, was therefore introduced to describe how flux is influenced by changes in the concentration of a given metabolite. In other words, elasticity is a parameter that describes, in quantitative terms, the sensitivity and responsiveness of an enzyme to a metabolite.

Unlike the flux control coefficient, elasticity is a property of individual enzymes and not of the pathway. The elasticity of an enzyme to a metabolite is defined by the slope of the curve of enzyme units (reaction rate) plotted as a function of metabolite concentrations, with the measurements taken at the metabolite concentration found *in vivo*. By analogy with the flux control coefficient (Figure 6.2), the value of the elasticity coefficient, which can be calculated from the slope, will depend upon the units used for the measurement of enzymic activities, which may vary from one enzyme to another. This can be avoided, as described earlier for the flux control coefficient, by directly calculating the elasticity coefficient from a log-log plot of catalytic activity versus metabolite concentration to give the fractional change in enzymic activity as a function of the fractional change in the concentration of the substrate. As highlighted in the case study presented in Section 6.3, elasticities have positive values for metabolites that stimulate enzymic activity (substrates, activators) and negative values for those that decrease reaction rate, such as products and inhibitors. Therefore, elasticity is a parameter that describes, in quantitative terms, the sensitivity and responsiveness of an enzyme to a particular metabolite that could be a substrate, a product, or an effector.

6.2.4 Connectivity Theorem

This theorem, the fourth pillar of the MCA theory, addresses the question of how the flux control coefficient of a given enzyme can be related to its kinetic properties. Such an interrelationship is governed by the connectivity theorem, which states that the sum of all connectivity values in a given pathway is zero. The connectivity value for any given enzyme can be calculated by multiplying its flux control coefficient by its elasticity with respect to the metabolite in question. Enzymes not affected by the metabolite in question will naturally have an elasticity of zero and as such will make no contribution toward the final sum obtained. Further analysis of connectivity values has revealed that large elasticities are associated with small flux control coefficients and vice versa. The mathematical equations relating the connectivity theorem to linear pathways, branch points, and cycles have been extensively described and dealt with elsewhere (Fell, 1997).

6.2.5 Response Coefficients

Induction and repression of enzyme synthesis in response to internal or external environmental stimuli are widely distributed in nature and are very effective in "turning on" and "switching off" transcription. Covalent modification through reversible phosphorylation is another mechanism that regulates the activity of existing enzymes by rendering them active or inactive, as is the case for ICDH in *E. coli* during adaptation to acetate (Koshland, 1987; Cozzone, 1988). In addition to degradation of mRNA and proteins, enzymes may also be the subject of allosteric control mechanisms, which change the enzyme's affinity toward its substrate and/or cofactor(s).

The response coefficient, the fifth pillar of the MCA theory, reflects the effectiveness of a particular effector on flux through a given pathway and is dependent on two factors: the flux control coefficient of the target enzyme and the strength of the effector, which is given by its elasticity coefficient. For an effector to have a significant effect on flux, each of the above parameters with respect to the target enzyme clearly has to be of a value higher than zero.

Under circumstances in which a particular effector may activate or inactivate more than one enzyme in a given pathway, the total response will be the sum of the individual responses from each enzyme affected. However, this is only true when the changes in the concentration of the effector are very small because of the nonlinear relationship of the kinetics in metabolic systems (Hofmeyr and Cornish-Bowden, 1991).

Now, let us consider how carbon flux is partitioned at the junction of isocitrate during growth of *E. coli* on acetate.

6.3 CONTROL OF CARBON FLUX AT THE JUNCTION OF ISOCITRATE IN CENTRAL METABOLISM DURING AEROBIC GROWTH OF *E. COLI* ON ACETATE: A CASE STUDY

During aerobic growth on acetate as a sole source of carbon and energy, *E. coli* requires the operation of the anaplerotic sequence of the glyoxylate bypass for the provision of biosynthetic precursors (Kornberg, 1966). Under these conditions, a new junction is created at the level of isocitrate (Figure 6.3), where ICL of the glyoxylate bypass is in direct competition with the Krebs cycle enzyme ICDH. Although ICDH has a much higher affinity for isocitrate, flux through ICL and thence the anaplerotic enzyme malate synthase (MS) is assured by virtue of high intracellular levels of isocitrate and the inactivation of a large fraction (75%) of ICDH (El-Mansi et al., 1985; Cozzone and El-Mansi, 2005). Although the *in vivo* signal that triggers the "acetate switch" and, in turn, the expression of the glyoxylate bypass operon has yet to be unraveled, acetate *per se* can be safely ruled out as a possible signal (El-Mansi, 1998). With the aid of radiolabeled isotopes and NMR spectroscopy, Walsh and Koshland (1984) reported that during steady-state growth of *E. coli* on acetate under aerobic conditions, flux through ICL is 31.0 mmol of isocitrate per minute, which in turn represents 33% of the total carbon processed at this junction.

To assess the relative contribution of each of the above enzymes to the overall distribution of carbon flux among various enzymes of central metabolism, the computer software package Gepasi (Mendes, 1993) was used to calculate the steady-state fluxes and the concentration of various metabolites during growth of *E. coli* on acetate. This computer package also enabled us to formulate the matrices of the elasticity coefficients as well as the control and response coefficients under different steady states. In the next section, we will discuss the data in the light of Kacser's MCA theory as well as the traditional concept of the rate-limiting step.

6.3.1 THE MODEL

To explore the consequences of controlled adjustment of ICL and ICDH enzymic activities on the partition of carbon flux at the junction of isocitrate on one hand and among various enzymes of central metabolism on the other, the complex metabolic pathways of central metabolism used by Walsh and Koshland (1984) was reduced to a skeleton model (Figure 6.4).

It is noteworthy that the model depicted previously for the Krebs cycle and the glyoxylate bypass (Figure 6.4) does not constitute a moiety-conserved cycle (Hofmeyr et al., 1986) because there are reversible sinks to the cycle as well as branching from within the cycles, thus giving rise to metabolite pools that are involved in three different reactions.

According to the summation theorem, flux through ICDH can be affected by other enzymes of central metabolism, with the sum of the flux control coefficients of all enzymes on J_{ICDH} being 1.0. Whereas the flux control coefficient is a measure of the relative change in flux through a particular enzyme in a given pathway in response to a small change in its concentration, the elasticity coefficient of a given enzyme is a measure of the relative change in flux in response to a small change in the concentration of its substrate or cofactor.

It is noteworthy that other metabolic modeling programs are also available and can be obtained free of charge,

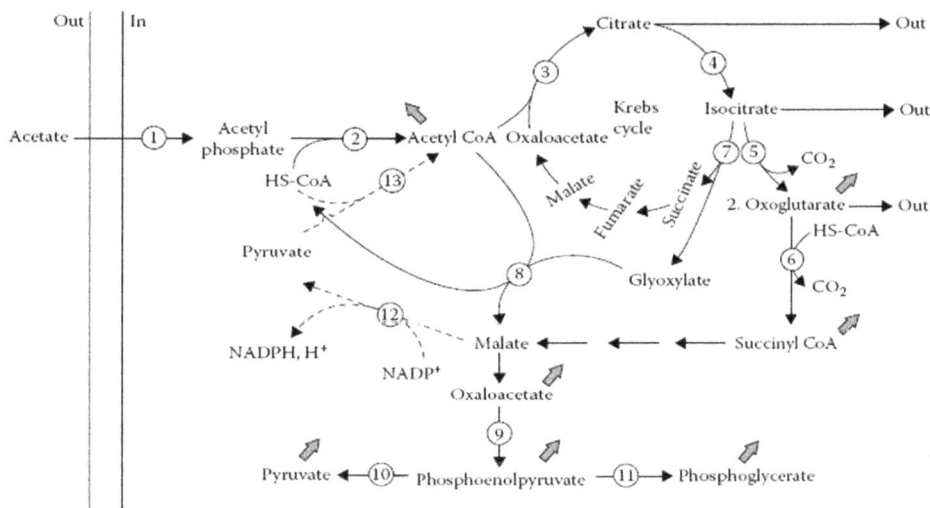

FIGURE 6.3 The metabolic route used by *E. coli* for the assimilation of acetate highlighting the direct competition between the Krebs cycle enzyme ICDH (5) and the glyoxylate bypass enzyme ICL (7). This diagrammatic representation also highlights flux to citrate and isocitrate excretion as well as the biosynthetic role assumed by ICDH under these circumstances for the provision of α-ketoglutarate and succinyl CoA; both are biosynthetic precursors. This metabolic network also accounts for the long-standing observation that expression of malic enzyme (12) and PDH (13) are not essential for growth on acetate. Key enzymes are as follows: 1, acetate kinase; 2, phosphotransacetylase; 3, citrate synthase; 4, aconitase; 5, isocitrate dehydrogenase; 6, α-ketoglutarate dehydrogenase; 7, isocitrate lyase; 8, malate synthase; 9, PEP carboxykinase; 10, pyruvate kinase; 11, enolase; 12, malic enzyme; 13, pyruvate dehydrogenase. (Reproduced from Cozzone and El-Mansi, 2005. With kind permission of Karger AG, Basel.)

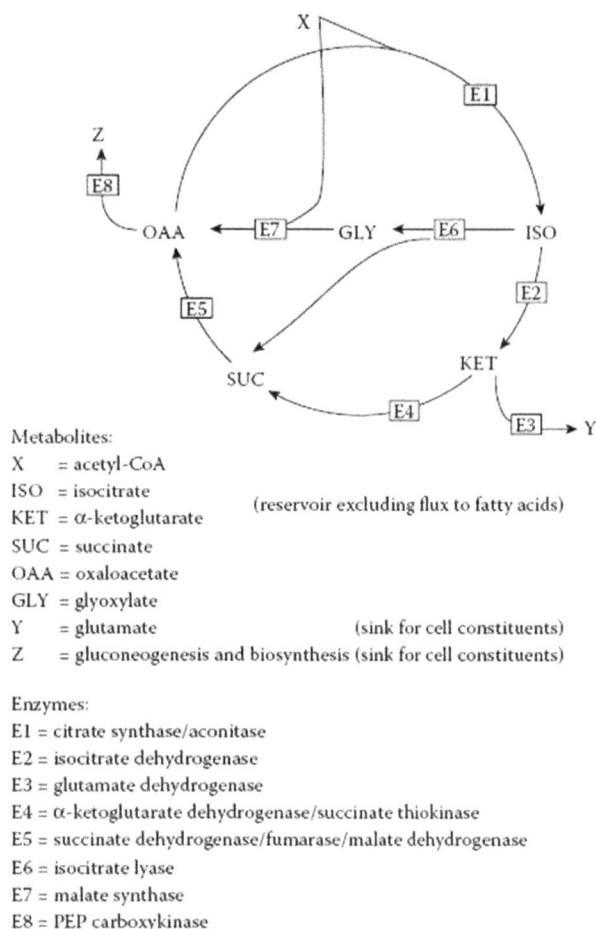

Metabolites:

X   = acetyl-CoA
ISO = isocitrate
KET = α-ketoglutarate   (reservoir excluding flux to fatty acids)
SUC = succinate
OAA = oxaloacetate
GLY = glyoxylate
Y   = glutamate   (sink for cell constituents)
Z   = gluconeogenesis and biosynthesis (sink for cell constituents)

Enzymes:

E1 = citrate synthase/aconitase
E2 = isocitrate dehydrogenase
E3 = glutamate dehydrogenase
E4 = α-ketoglutarate dehydrogenase/succinate thiokinase
E5 = succinate dehydrogenase/fumarase/malate dehydrogenase
E6 = isocitrate lyase
E7 = malate synthase
E8 = PEP carboxykinase

FIGURE 6.4 A skeleton model describing the central metabolic pathways used by *E. coli* during growth on acetate as a sole source of carbon and energy. Key to metabolites and enzymes are given.

6.3.2 MODELING OF CARBON FLUX AT THE JUNCTION OF ISOCITRATE IN CENTRAL METABOLISM

To gain some insight into the consequences of changing the concentration of ICHD and ICL on flux through the enzymes themselves as well as the intracellular concentrations of associated metabolites, Gepasi, version 3.30, a Windows-based application that is user-friendly, was used. Another option for this type of analysis is COPASI: Biochemical System Simulator (copasi.org). The entry of reaction equations and rate laws (kinetics) is straightforward and easily achieved. Several predefined rate laws are available, and it is possible to create user-defined rate laws. The reaction rates and rate laws are first entered followed by the starting concentrations of fixed and variable metabolites.

Once this basic configuration is achieved, the MCA steady-state values can be calculated. Interestingly, the data obtained on the effect of ICL (E6) and ICDH (E2) on the partition of carbon flux at the junction of isocitrate as well as isocitrate concentrations using Gepasi have the added advantage of plotting multifunction analysis as shown in Figures 6.5 and 6.6, respectively. Although the output in the two figures appears to be significantly different with respect to isocitrate concentration and flux through ICDH and ICL, the reaction equation input only differs in the sign used in the denominator; whereas addition (+) was used in Figure 6.5, multiplication (×) was used in Figure 6.6. Although the question of whether which sign is correct is arguable, the *in vivo* analysis of isocitrate concentration is in good agreement with using addition rather than multiplication in the denominator of formula template.

6.4 STRATEGIES FOR MANIPULATING CARBON FLUXES EN ROUTE TO PRODUCT FORMATION IN INTERMEDIARY METABOLISM

6.4.1 VALIDITY OF THE CONCEPT OF THE RATE-LIMITING STEP AS AN APPROACH TO INCREASING FLUX TO PRODUCT FORMATION

Industrialists, driven by a strong faith in genetic engineering, often argue that if one can identify the enzyme catalyzing the

although some are no longer maintained by their respective authors for one reason or another, including MMT (Metabolic Modeling Tool), which unlike other programs uniquely contains a tool that provides a pathway overview and identifies parameter algorithms (Hurlebaus, 2001; www.bioinfo.de/isb/gcb01/poster/hurlebaus.html).

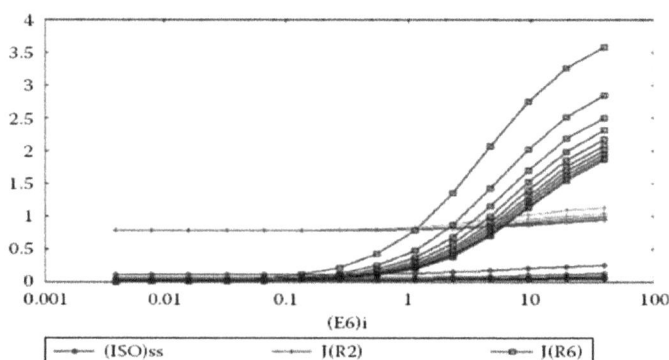

FIGURE 6.5 Effect of ICL (E6) concentrations on flux through the enzyme itself (J6), flux through ICDH (J2), and isocitrate concentration as calculated by Gepasi. Note that the data presented in this graph are composed of a combined scan of E6 (14 values) and E2 (10 values), thus giving rise to 140 separate steady-state data points. The variations of E6 are shown along the abscissa, whereas the variations in E2 are shown as multiple contours of J6, J2, and isocitrate concentrations.

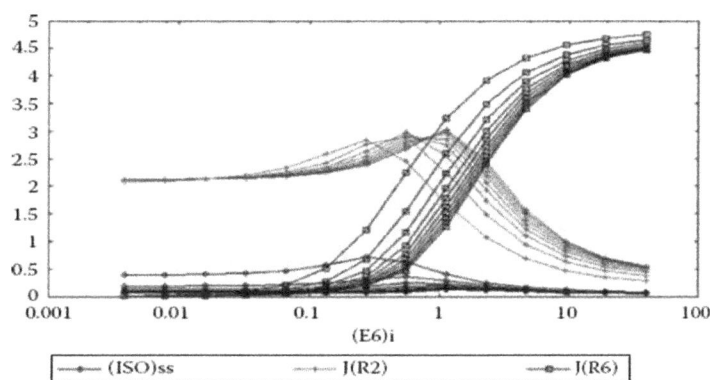

FIGURE 6.6 Effect of ICL (E6) concentrations on flux through the enzyme itself (J6), flux through ICDH (J2), and isocitrate concentration as calculated by Gepasi.

rate-limiting step in a given pathway, then the rest is simple. However, this approach has proved to be, by and large, erroneous, and examples illustrating this notion are given in the introductory section of this chapter. Alternative approaches, which may be beneficiary to the industrialists, include Kacser's theory of MCA. MCA argues that the rate-limiting step is an oversimplification of metabolism and as such should be abandoned; instead, the control of metabolic flux in a given pathway is shared among all of the enzymes involved, albeit to a varying degree of extent as described earlier in this chapter. Although MCA overlooks peculiarities in the metabolic network and ignores regulatory enzymes, "the victim of MCA," as Kacser used to say, it is nevertheless a very powerful tool for examining the general properties of metabolic networks.

6.4.2 Modulation of Carbon Flux En Route to Product Formation

6.4.2.1 The Model

In the model pathway (Figure 6.7), the biosynthesis of two end products, S_{4a} and S_{4b}, from a biosynthetic precursor, X_0, via a branch-point metabolite, S_2, is highlighted. Also note that the precursor metabolite X_0 is external to the system; that is, its concentration is fixed and, unlike all other metabolites within the pathway, does not depend on the properties of the eight enzymes involved in the skeleton model (Figure 6.7) (Cornish-Bowden et al., 1995). Although the numerical values assumed by the aforementioned authors for the eight kinetic equations are arbitrary, the essential features of the

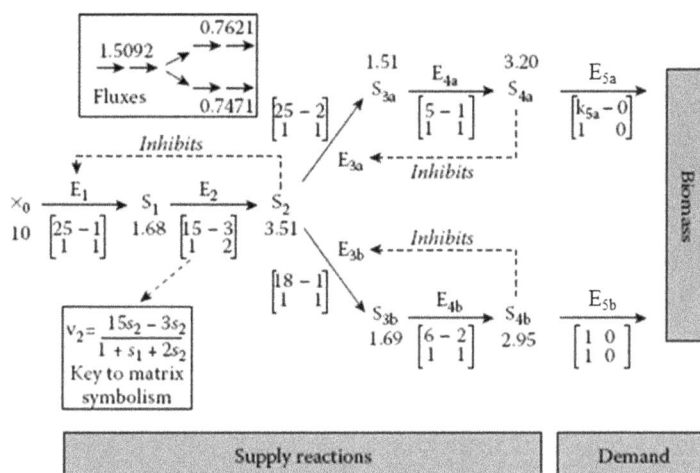

FIGURE 6.7 A model representing a branched biosynthetic pathway that is used for the production of two final end products (i.e., S_{4a} and S_{4b}). Note that the whole pathway can be divided into two main sections (i.e., supply and demand). Although the supply involves enzymes E_1–E_4, the demand section only involves E_5. Also note that each end product S_{4a} and S_{4b} inhibits the enzymic activities of E_{3a} and E_{3b}, respectively, and that the metabolite (S_2) inhibits the enzymic activity of E_1. It was assumed that the reactions obeyed the Michaelis-Menten equation, and that the pool concentration of X_0 was held constant at 10 with the steady-state concentrations of various metabolites; shown underneath each metabolite are calculated values of 1 for K_{5a} as described by Cornish-Bowden, A., Hofmeyr, J.-H.S., and Cardenas, M. L., *Bioorganic Chem.*, 23:439–49, 1995. The steady-state fluxes through the common pathway and the two branches are highlighted by shading (i.e., 1.5092 for the common pathway and 0.7621 and 0.7471 for the fluxes through branches a and b, respectively). This steady state was used as the starting point for examining the effect of different strategies on flux to S_{4a}. (This diagram is a modification of the original model of Cornish-Bowden, A., Hofmeyr, J.-H.S., and Cardenas, M. L., *Bioorganic Chem.*, 23:439–49, 1995, reproduced with the kind permission of Elsevier.)

model are not and correspond to a typical case of sequential feedback inhibition in which each of the two end products (S_{4a} and S_{4b}) inhibits the enzymes (E_{3a} and E_{3b}) catalyzing the first committed step of their formation, respectively, with the branch-point metabolite S_2 inhibiting E_1, the enzyme catalyzing the first step in the whole pathway (Figure 6.7). All simulations were carried out using the latest available version of MetaModel, which was first conceived by Cornish-Bowden and Hofmeyr (1991).

6.4.2.2 Strategies for Manipulating Metabolic Fluxes

Several different strategies (i.e., opposition, oblivion, evasion, suppression, and subversion) have been used in an attempt to increase carbon flux to product formation; in this case, S_{4a} and S_{4b} (Cornish-Bowden et al., 1995). The terms coined for the aforementioned strategies reflect the possible effect of each on cellular regulation (Cornish-Bowden, 1995). Opposition as a strategy is widely used. In opposition, industrialists attempt to increase flux in a given pathway by increasing the relative concentration of the enzyme considered to be rate limiting, often the enzyme that is subject to feedback inhibition (e.g., E_1 and/or E_{3a} in the model system portrayed above) (Figure 6.7). Whereas oblivion strategy utilizes mathematical modeling of certain matrices to determine the degree of change needed to achieve certain increase in flux, evasion utilizes fluxes to calculate the changes needed in enzymic activity to achieve the desired increase in the concentration of end product without adversely affecting the concentrations of intracellular metabolites. Suppression as a strategy makes full use of

the primary regulatory function of feedback inhibition [i.e., to transfer control from the biosynthetic reactions (supply) to polymerization and assembly (demand) (Figure 6.7)] and as such seeks to increase flux to product formation through increasing the demand made on central and intermediary metabolism. Suppression relies on the elimination of feedback loops whereas subversion relies on increasing the demands made on the biosynthetic enzymes.

With the exception of oblivion strategy, all other strategies have been examined *in silico*, and their effect on flux to product formation has been assessed (Table 6.1). Suppression is an "all-or-none" strategy because feedback loops are present or absent and had only a very modest effect on flux. Elimination of the feedback loop to E_{3a} gave a 30% increase in flux, which was accompanied by 17-fold increase in the concentrations of the metabolites in branch a (Figure 6.7). On the other hand, elimination of the feedback loop E_1 yielded a 17% increase in flux with some 50-fold increase in the concentrations of metabolites in the common part of the pathway (Figure 6.7). It is noteworthy to indicate that a large increase in the concentrations of metabolites *in vivo* might adversely affect cell viability and flux to desired end product. In all other strategies, the manipulations were performed with the view of achieving a 5-fold increase in flux through branch a (Figure 6.7). Simulation analysis revealed that opposition as a strategy is largely ineffective because changes in the concentration of one or two enzymes alone was accompanied by a very limited increase in flux to product formation. On the other hand, evasion produced the desired increase in flux (i.e., 5-fold)

TABLE 6.1
Impact of Various Strategies on Metabolites Concentrations and Flux Through Branch a in the Model Shown in Figure 6.7

Strategy	Relative Activity of Each Enzyme								Relative J_{5a}	Relative Metabolite Concentrations			
	E_1	E_2	E_{3a}	E_{4a}	E_{5a}	E_{3b}	E_{4b}	E_{5b}		S_1	S_2	S_{3a}	S_{4a}
Wild-type	1	1	1	1	1	1	1	1	1.00	1.00	1.00	1.00	1.00
Opposition[b]	5	1	1	1	1	1	1	1	1.02	1.50	1.51	1.11	1.11
Opposition[b]	1	1	5	1	1	1	1	1	1.08	1.02	0.99	1.49	1.47
Opposition[b]	5	1	5	1	1	1	1	1	1.10	1.52	1.49	1.66	1.63
Evasion#	3.02	3.02	5	5	5	1	1	1	5.00	1.00	1.00	1.00	1.00
Suppression	*	1	1	1	1	1	1	1	1.17	49.5	47.7	2.64	2.56
Suppression	1	1	*	1	1	1	1	1	1.29	1.06	0.96	17.8	16.9
Suppression	*	1	*	1	1	1	1	1	1.31	47.4	42.7	795	780
Subversion	1	1	1	1	5	1	1	1	4.13	1.74	0.76	3.66	0.53

Source: This table is reproduced from Cornish-Bowden, A., Hofmeyr, J.-H.S., and Cardenas, M. L., *Bioorganic Chem.*, 23:439–49, 1995. With the kind permission of Elsevier.

Note: The factors of 5 and 1 for the enzymes reflect the desired increases in flux. In the common part of the pathway (see Figure 6.10), the factor 3.02 was calculated from the fluxes in the starting steady-state as follows: $(5J_{3a} + J_{3b})/(J_{3a} + J_{3b}) = (5 \times 0.7621 + 0.7471)/(0.7621 + 0.7471) = 3.02$.

[a] In cases marked by asterisk (*), the activities of the enzymes were not changed except for omitting the feedback inhibition terms from the denominators of the rate equations.

[b] The three cases of opposition were also examined with much larger changes in activity, 100-instead of 5-fold in each case. Effects on the flux were only trivially different from those shown, but the effects on concentrations were much larger.

without adversely affecting the intracellular concentrations of metabolites (Table 6.1). Furthermore, although evasion is more effective than subversion as a strategy for increasing fluxes (Table 6.1), it is important. that we recognize that the latter strategy achieved an appreciable increase in flux as a result of manipulating only one enzymic activity rather than all enzymes and that the changes in the intracellular concentrations of metabolites associated with it are modest and as such may be favorable to the organism *in vivo*.

6.5 CONVERSION OF FEEDSTOCK TO BIOMASS AND DESIRABLE END PRODUCTS

The conversion of a given feedstock to desirable end product by microorganisms involves a wide range of metabolic activities depending on the nature of the feedstock and the identity of the desired end product. Unlike products of primary metabolism (see Chapter 4), synthesis of secondary metabolites (see Chapter 5) is usually sequential to cessation of growth, at which point, more feedstock (input) is added intermittently (fed-batch culture) or continuously (continuous cultures, e.g., turbidostat, auxostat, pH-stat) to sustain the metabolic fluxes required for the biosynthesis of desired end product (output) and maintenance of cell viability. During the productive stage, the primary feedstock (input) is generally channeled through specific metabolic routes, and provided that such routes are known, careful measurements of all inputs and outputs in a given fermentation process provide the data from which the fluxes through all of the pathways involved can be calculated (Holms, 1996). A metabolic chart for central and intermediary metabolism can then be constructed that relates input to biosynthesis of desired end product. Flux analysis invariably shows that the efficiency with which a given feedstock is converted to biomass is far from optimal because of excretion of intermediates; usually those upstream of bottlenecks, those that result from the accumulation of intracellular polymers, or both. Analysis of culture filtrates allows for the identification of undesired fluxes, which become targets for metabolic intervention by metabolic engineers (El-Mansi and Holms, 1989; El-Mansi, 2004). Therefore, flux analysis highlights wasteful fluxes, thus enabling the metabolic engineer to develop a rationale for the diversion of wasted fluxes to product formation, a primary target for the discerning biotechnologist.

6.5.1 KACSER'S UNIVERSAL METHOD FOR INCREASING FLUX TO PRODUCT FORMATION WITHOUT PERTURBING STEADY-STATE CONCENTRATIONS OF METABOLITES

In this method, Kacser and Acerenza (1993) provide a sound theoretical framework for increasing *in vivo* flux to product formation by a certain percentage without perturbing steady-state concentrations of metabolites.

6.5.1.1 The Model

In their analysis, Kacser and Acerenza viewed the cell as a "transparent box" (Figure 6.8) connected to external inputs

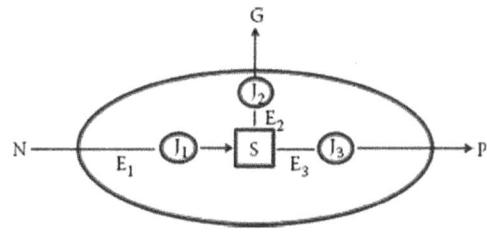

FIGURE 6.8 Kacser's universal model. N, all input; P, desired product; G, flux to biomass and metabolite excretion; S, intracellular concentrations of all metabolites of central and intermediary metabolism.

"N" (carbon, oxygen, nitrogen, phosphate, etc.,) and exports outputs "P," which could be a primary or secondary metabolite. Kacser stressed the following understandings:

- In a nongrowing but metabolically active culture, fluxes through central metabolism are essentially dedicated to satisfy maintenance requirements, and as such fluxes to biosynthesis of macromolecules such as DNA, protein, lipids, etc., are negligible; assumed to be zero.
- However, during the course of exponential growth, fluxes to biosynthesis of macromolecules are large. However, once a steady state is achieved in either case (i.e., nongrowing but metabolically active or exponentially growing cells) a "time-invariant" concentration of all metabolites and fluxes within the cell and out to product formation is realized, thus achieving perfect balance between input and output fluxes.

Kacser used a very simple model (Figure 6.8) to describe his theory. He represented all inputs by "N," desired product by "P," and gathered all other outputs to "G," with "S" representing the intracellular concentrations of all metabolites.

6.5.1.1.1 Assumptions and Scenario

In describing his method, Kacser made the following assumptions:

- That the sequence of reactions involved in the conversion of "N" to "S," "S" to "G," and "S" to "P" are unique to each pathway.
- That the metabolic pathways involved in the conversion of "S" to "G" are complex but will in a steady state reach time invariant with respect to all components, the concentration of which is represented by "S."
- That P and G are excreted into a very large volume of medium, thus rendering their concentrations negligible; that is, zero.

The challenge facing Kacser at the time was how to increase flux to product formation (J_3) without perturbing the steady-state concentration of metabolites or fluxes to metabolite excretions (J_2); Figure 6.8.

To validate his theory, Kacser had to set a target of a certain fold increase in flux through J_3. Therefore, the steady-state fluxes through the scheme shown above (Figure 6.8) are constrained by Equation (6.1), with J^o denoting fluxes in the original state (i.e., wild-type strain):

$$J_1 = J_2 + J_3 \qquad (6.1)$$

Similarly, following genetic modifications using recombinant DNA technology, a new set of fluxes are obtained. Again, these fluxes are constrained by Equation 6.1, with "r" replacing "o" as a superscript.

The increase in flux Δj can be calculated by the following equation:

$$\Delta J_1 = \Delta J_2 + \Delta J_3 \qquad (6.2)$$

However, because no change in the J_2 flux is desired (i.e., $\Delta J_2 = 0$), then

$$\Delta J_1 = \Delta J_3 \qquad (6.3)$$

In his mathematical treatment, Kacser stressed that ΔJ is the absolute change in flux, which will be same for J_1 and J_3. By adding ΔJ to the synthesis of S and its conversion to product, the original value S^0 will remain constant because $\Delta J_2 = 0$. Kacser went on to stress that this is true no matter what the original fluxes J_1 and J_3 are. He also stressed that the ratio of the original fluxes (j_3^0/j_1^0) will be different from (j_3^r/j_1^r) in the recombinant strain. This is the central device that allows for the identification of which enzyme to overexpress.

6.5.1.1.2 Application of the Universal Method: Increasing Flux to Tryptophan Formation in the Yeast S. cerevisiae

In this example, Kacser used the metabolic pathways depicted in Figure 6.9, which are the generally accepted pathway for the production of tryptophan by the yeast *S. cerevisiae* (Niederberger et al., 1992). The aforementioned authors have also experimentally determined fluxes through j_1^0 and j_2^0 as well as j_3^0 and found them to be 3, 2.5, and 0.5 respectively.

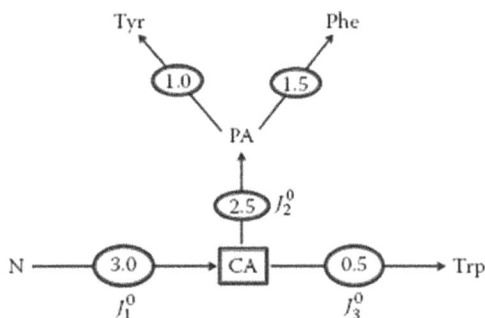

FIGURE 6.9 The metabolic pathways used by the yeast *S. cerevisiae* for the production of tryptophan. N, primary carbon source; CA, chorismate acid; Trp, tryptophan; PA, protocatechuate; Phe, phenylalanine; Tyr, tyrosine.

Using the above information, the universal method can be used to calculate the multiplication factor by which the expression of the enzymes constituting "E1" and "E3" (Figure 6.9) must be increased by to achieve the desired level of increase in flux to product formation by applying the following equations:

$$r_1 = 1 + \Delta J/J_1^0 \qquad (6.4)$$

$$r_3 = 1 + \Delta J/J_3^0 \qquad (6.5)$$

Calculating in the wild-type strain can be achieved as follows:

$$\Delta J = J_1^0 - J_3^0$$
$$\Delta J = 3 - 0.5 = 2.5 \qquad (6.6)$$

The impact of increasing ΔJ from 2.5 (wild type) to 12 in the recombinant strain on flux to product formation can be calculated using the following calculation:

$$r_3 = 1 + \Delta J/J_3 = 1 + 12/0.5 = 1 + 24 = 25\text{-fold}$$

As can be seen from the above calculation, an increase in ΔJ from 2.5 (wild type) to 12 in the recombinant strain was accompanied by a 25-fold increase in flux to product formation. To achieve this increase in flux without perturbing the concentrations of central and intermediary metabolites, a 5-fold increase in the expression of all enzymes involved in the conversion of primary carbon source "N" to "CA" (chorismate) is required as evidenced by the following calculation:

$$r_1 = 1 + \Delta J/J_1 = 1 + 12/3 = 1 + 4 = 5\text{-fold}$$

Now, remember that flux through J_0^r remains constant because $\Delta J_2 = 0$. It therefore follows that a 5-fold increase in the enzymic activities leading to the formation of chorismate (CA) and a 25-fold increase in the activities of the enzymes leading from CA to tryptophan (Figure 6.9) would yield some 25-fold increase in flux to tryptophan without perturbing the intracellular concentration of CA or J_2 fluxes.

However, the question that remains is if we can achieve such a level of overexpression of the entire shikamate pathway (E1) through genetic manipulation? The answer to this is of course we can, and several strategies exist for doing so, including cloning the entire shikamate pathway on a plasmid that has five copies per chromosome and/or cloning downstream of a titratable promoter or indeed using the technique of promoter swapping (McCleary, 2009). Recently, a couple of notable applications of the MCA to the overproduction of primary metabolites (L-phenylalanine or L-methionine) by *E. coli* have been published (Teleki et al., 2017; Weiner et al., 2017). In these studies, the control coefficients of the enzymes involved in the synthesis of the above amino acids were determined and found to vary according to growth conditions and the rate of amino acid synthesis. The aforementioned authors further reported that the control of carbon flux to product formation was to be shared among all enzymes involved in the pathway, which is consistent with the fundamental principles of MCA as described by Kacser and Burns (1973).

6.6 STOICHIOMETRIC ANALYSIS OF METABOLIC NETWORKS

Stoichiometry-based modeling of metabolic pathways is of practical importance because it extracts systemic information such as phenotypes from molecular-level network structure. Stoichiometric analysis requires knowledge of the system's enzymatic reactions and the stoichiometries of their associated substrates and products. Mass conservation is a powerful constraint that restricts how the enzymatic reactions can be integrated to create a functioning steady-state metabolism. Results are specific to metabolic network structure and therefore typically vary from one organism to another. Stoichiometric modeling can utilize various "*omics*," or datasets to build and verify models. Stoichiometry-based methods are in contrast to kinetic modeling methods, which often require large condition-sensitive parameter sets (e.g., v_{max} and K_m for each enzyme). Such parameters are very difficult, if not impossible, to reliably determine *in vivo*, and literature values often vary over an order of magnitude, thus limiting the usefulness of kinetic modeling.

6.6.1 APPLICATIONS OF STOICHIOMETRIC ANALYSIS

To illustrate the usefulness of stoichiometric analysis in biochemical and metabolic engineering studies, let us consider the classical example of phenylalanine production. This example will highlight central concepts of stoichiometric analysis using a relatively simple system that permits manual analysis.

6.6.1.1 Optimization of Phenylalanine Production: A Case Study

In *E. coli*, biosynthesis of phenylalanine and other aromatic amino acids is initiated by the condensation of phosphoenolpyruvate (PEP) and erythrose 4-phosphate (E-4P) to give 3-deoxy-d-arabino-heptulosonate 7-phosphate (DAHP), This condensation reaction is catalyzed through the activities of three different DAHP synthases, the products of *aroF*, *aroG*, and *aroH*, which are subject to feedback inhibition by tyrosine, phenylalanine, and tryptophan, respectively. The metabolic routes to phenylalanine are well established, and assuming that no unknown by-products are being formed, the theoretical yield coefficient for the production of phenylalanine—or any given product—can be calculated (Foberg et al., 1988).

The overall reaction for the formation of phenylalanine can be expressed by

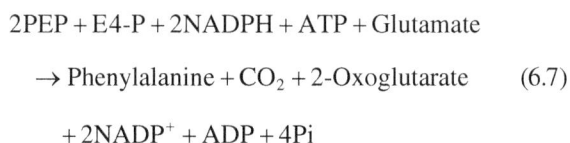

$$2PEP + E4\text{-}P + 2NADPH + ATP + Glutamate$$

$$\rightarrow Phenylalanine + CO_2 + 2\text{-}Oxoglutarate \quad (6.7)$$

$$+ 2NADP^+ + ADP + 4Pi$$

Assuming that E4-P is generated through the activities of transketolase and transaldolase of the pentose phosphate pathway (PPP) and that PEP, which is generated through glycolysis, is required for the uptake of glucose and its activation to glucose 6-phosphate, the overall reaction of phenylalanine production can be described as follows:

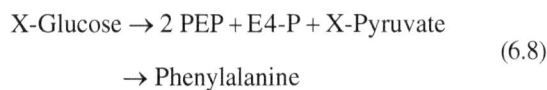

$$X\text{-}Glucose \rightarrow 2\,PEP + E4\text{-}P + X\text{-}Pyruvate$$

$$\rightarrow Phenylalanine \quad (6.8)$$

Taking the number of carbon atoms in each of the above molecules into consideration, the carbon balance of Equation 6.8 can be resolved as follows:

$$X(6) \rightarrow 2(3) + 1(4) + X(3)$$

This, in turn, can be solved as follows:

$$X(6) \rightarrow 6 + 4 + X(3)$$

$$X(6) - X(3) = 6 + 4$$

$$X3 = 10; \text{ it follows}$$

$$X = 10/3 = 3 \text{ and } 1/3rd \text{ moles of glucose.}$$

According to the above stoichiometry, 3.33 moles of glucose (MW 180) are required for the production of 1 mole of phenylalanine (MW 165.2), thus giving a maximum theoretical yield coefficient of 0.275 g of phenylalanine per gram of glucose.

Taking into consideration that the biosynthesis of phenylalanine (Equation 6.7) requires 2 moles of NADPH, which is generated as a consequence of pyruvate oxidation In the Krebs cycle at the level of ICDH, the overall reaction can be rewritten as follows:

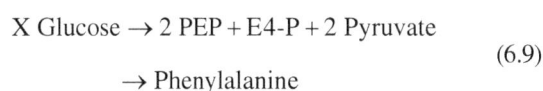

$$X\,Glucose \rightarrow 2\,PEP + E4\text{-}P + 2\,Pyruvate$$

$$\rightarrow Phenylalanine \quad (6.9)$$

Solving Equation 6.9 in the same way as for Equation 6.8, the carbon balance under these circumstances is that 2.67 moles of glucose are required for the production of 1 mole phenylalanine and as such the overall stoichiometry will therefore be as follows:

$$2.67 \text{ moles of glucose}$$

$$\rightarrow 1 \text{ mole of phenylalanine} \quad (6.10)$$

$$+ 7 \text{ moles of } CO_2$$

Further consideration of the stoichiometry balance for the above equation reveals a shortfall of PEP supply by a factor of 0.67 mole. Similarly, on close examination, one can see that 0.66 mole of pyruvate is wasted in the form of CO_2 (Figure 6.10). Diversion of pyruvate to PEP formation rather than CO_2 production therefore becomes a suitable target for metabolic intervention through the expression of PEP synthetase (Figure 6.10). Because this enzyme is subject to catabolite repression by

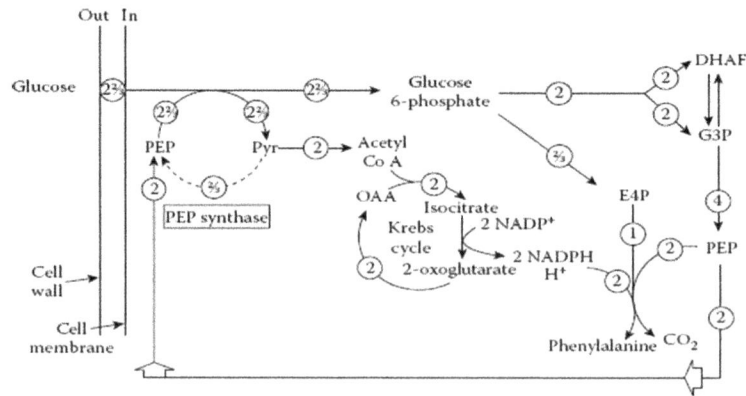

FIGURE 6.10 A diagrammatic representation highlighting the metabolic network used by *E. coli* for the production of phenylalanine from glucose. Note the role of PEP synthase on the recycling of pyruvate to PEP and its consequences on the efficiency of carbon conversion to phenylalanine.

glucose in *E. coli*, the use of recombinant DNA technology to construct a mutant strain that is insensitive to catabolite repression or a recombinant strain that is capable of expressing PEP synthetase constitutively affords a good example in which metabolic and genetic engineering work in concert.

Interestingly, an appreciable increase in flux to DAHP and, in turn, phenylalanine as a result of overexpression of PEP synthetase has been reported in the literature (e.g., Stephanopoulos et al., 1998), thus lending further support to the analysis portrayed above. The effect of overexpression of PEP synthetase on the efficiency of carbon conversion to DAHP and, in turn, phenylalanine production can be assessed quantitatively as illustrated in the stoichiometric diagram shown in Figure 6.10. From Figure 6.10, it can be seen that constitutive expression of PEP synthetase resulted in the formation of a PEP regenerating cycle that, under these circumstances, proved to be advantageous—rather than futile—as evidenced by a 100% increase in flux to DAHP; the biosynthetic precursor for phenylalanine, to give 0.787 g of DAHP per gram of glucose utilized. Therefore, the production of phenylalanine is a classic example in which stoichiometric flux analysis and metabolic and genetic engineering were integrated successfully.

6.6.1.2 *In Silico* Analysis of Stoichiometric Models

The phenylalanine example from the preceding section was evaluated manually using reaction stoichiometry and carbon balances. Biochemical networks are typically highly branched, and fluxes must simultaneously satisfy multiple conservation criteria (e.g., balances of carbon, nitrogen, and reducing equivalents). These systems are often too complex to solve manually. Computational stoichiometric modeling methods are readily scalable and can analyze systems with hundreds or thousands of reactions. These methods can be broadly divided into three major approaches:

1. Metabolic flux analysis (MFA)
2. Constraint-based linear programming including flux balance analysis (FBA)
3. Metabolic pathway analysis including elementary flux mode analysis (EFMA)

All three approaches define a system-specific solution space based on metabolic network structure and conservation relationships; this solution space contains all possible, biologically relevant flux distributions. The three approaches differ in how they select distinct metabolic flux distributions from the solution space; details of these approaches have been reported elsewhere (e.g., Stephanopoulos et al., 1998; Schilling et al., 1999; Schuster et al., 2000; Reed and Palsson, 2003; Trinh et al., 2009).

This section further develops the phenylalanine synthesis case study by constructing a more detailed network model and analyzing it using *in silico* stoichiometric analysis. Phenylalanine synthesis is made possible by a metabolic network that integrates the required carbon, nitrogen, and reducing equivalent fluxes. This network can be defined using a series of conservation relationships linking each reaction's substrates and products. A graphical representation of the current case study network is shown in Figure 6.11a. The conservation equations for each reaction are listed in Figure 6.11b. The presented model balances each reaction for carbon, nitrogen, and reducing equivalents; hydrogen, oxygen, and inorganic phosphate are assumed to be balanced on the basis of implicit cellular maintenance processes. Notable network differences from the case study in Section 6.6.1.1 include consideration of glutamate dehydrogenase in the assimilation of ammonia, the oxidative branch of the pentose phosphate pathway that generates carbon dioxide (CO_2) and NADPH, oxygenic respiration that produces ATP and regenerates NAD^+, anaplerotic reactions, excretion of acetate, and the consumption of ATP for cellular maintenance processes (MainE, reaction R20). To simplify the case study, nonbranching enzyme series have been combined into single model reactions that account for the overall transformation. For instance, reaction R2 accounts for the coupled activities of enzymes *Pgi*, *Pfk*, and *Fba*.

Stoichiometric modeling approaches consider steady-state metabolic scenarios. This treatment necessitates the classification of metabolites as balanced or unbalanced (sometimes referred to as internal and external metabolites, respectively). A balanced metabolite is constrained by the steady-state assumption and is not allowed to accumulate while an unbalanced metabolite serves as a system input or output and is permitted to accumulate. Definitions of balanced and unbalanced metabolites have significant effects on analysis and must be selected carefully. There are seven unbalanced metabolites in this model; that is, system inputs of glucose, ammonium (NH_4) and oxygen (O_2) as well as system outputs of CO_2, phenylalanine, acetate, and maintenance energy.

Conservation equations defining the metabolic reaction network can be written in a matrix format that concisely organizes the reaction information:

$$S \cdot r = C \qquad (6.11)$$

where S is a $m \times n$ matrix containing the substrate and product stoichiometric coefficients for each reaction, r is an n element column vector containing the rate (flux) through each network reaction (moles/L/h), and C is an m element column vector detailing accumulation (time-dependent change in concentration) of each network metabolite (moles/L/h). Reactions

from Figure 6.11b are presented as a stoichiometric matrix S in Table 6.2; note each column corresponds to a reaction (R1–R20) and each row corresponds to a metabolite. Metabolites consumed in a reaction have a negative coefficient, metabolites produced in a reaction have positive coefficient, and metabolites that do not participate in a reaction have a zero coefficient. Unbalanced network metabolites (system inputs and outputs) are highlighted in gray rows.

To prevent accumulation under steady-state functioning, production of balanced metabolites must equal consumption. The system written in Equation 6.11 can be rewritten in terms of just balanced metabolites:

$$S' \cdot r' = C \qquad (6.12)$$

where S' and r' are identical to S and r except the unbalanced metabolites have been removed (gray rows in Table 6.2). The right side of Equation 6.12 is a column vector comprised of zeroes; the balanced metabolite concentrations are not a function of time. All elements of S' are known from the

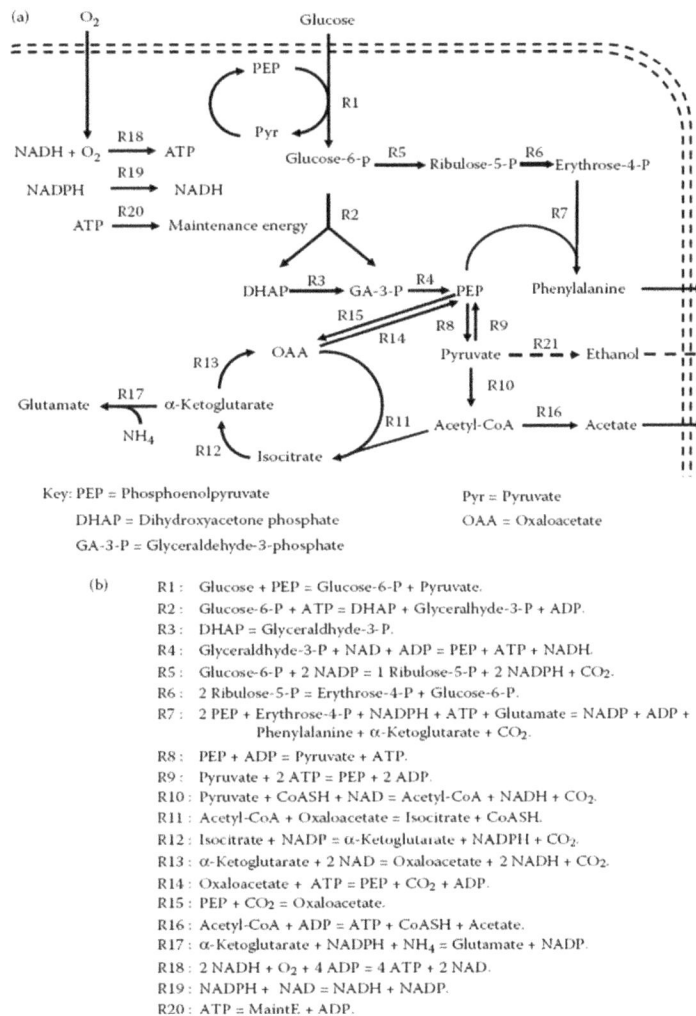

FIGURE 6.11 (a) Graphical representation of case study *E. coli* metabolic network producing phenylalanine from glucose. (b) Stoichiometries for each network reaction. Every reaction is considered irreversible and glucose, O_2, NH_4, CO_2, acetate, phenylalanine, and maintenance energy (MainE) are considered as unbalanced system inputs and outputs.

Key: PEP = Phosphoenolpyruvate
DHAP = Dihydroxyacetone phosphate
GA-3-P = Glyceraldehyde-3-phosphate
Pyr = Pyruvate
OAA = Oxaloacetate

(b)
R1 : Glucose + PEP = Glucose-6-P + Pyruvate.
R2 : Glucose-6-P + ATP = DHAP + Glyceralhyde-3-P + ADP.
R3 : DHAP = Glyceraldhyde-3-P.
R4 : Glyceraldhyde-3-P + NAD + ADP = PEP + ATP + NADH.
R5 : Glucose-6-P + 2 NADP = 1 Ribulose-5-P + 2 NADPH + CO2.
R6 : 2 Ribulose-5-P = Erythrose-4-P + Glucose-6-P.
R7 : 2 PEP + Erythrose-4-P + NADPH + ATP + Glutamate = NADP + ADP + Phenylalanine + α-Ketoglutarate + CO2.
R8 : PEP + ADP = Pyruvate + ATP.
R9 : Pyruvate + 2 ATP = PEP + 2 ADP.
R10 : Pyruvate + CoASH + NAD = Acetyl-CoA + NADH + CO2.
R11 : Acetyl-CoA + Oxaloacetate = Isocitrate + CoASH.
R12 : Isocitrate + NADP = α-Ketoglutarate + NADPH + CO2.
R13 : α-Ketoglutarate + 2 NAD = Oxaloacetate + 2 NADH + CO2.
R14 : Oxaloacetate + ATP = PEP + CO2 + ADP.
R15 : PEP + CO2 = Oxaloacetate.
R16 : Acetyl-CoA + ADP = ATP + CoASH + Acetate.
R17 : α-Ketoglutarate + NADPH + NH4 = Glutamate + NADP.
R18 : 2 NADH + O2 + 4 ADP = 4 ATP + 2 NAD.
R19 : NADPH + NAD = NADH + NADP.
R20 : ATP = MaintE + ADP.

TABLE 6.2
Stoichiometry Matrix (S) for the Reaction Network Employed in the Case Study

	R1	R2	R3	R4	R5	R6	R7	R8	R9	R10	R11	R12	R13	R14	R15	R16	R17	R18	R19	R20
PEP	-1	0	0	1	0	0	-2	-1	0	0	0	0	0	1	-1	0	0	0	0	0
Glucose-6-P	1	-1	0	0	-1	1	0	1	0	0	0	0	0	0	0	1	0	0	0	0
ATP	0	-1	0	1	0	0	-1	1	-2	0	0	0	0	-1	0	1	0	4	0	-1
ADP	0	1	0	-1	0	0	1	-1	2	0	0	0	0	1	0	-1	0	-4	0	1
Pyruvate	1	0	0	0	0	0	0	1	-1	-1	0	0	0	0	0	0	0	0	0	0
Glyceraldehyde-3-P	0	1	1	-1	0	0	0	0	-1	0	0	0	0	0	0	0	0	0	0	0
DHAP	0	1	-1	0	0	0	0	0	0	0	0	0	0	0	0	0	0	0	0	0
NADH	0	0	0	1	0	0	0	0	0	1	0	0	2	0	0	0	0	-2	1	0
NAD	0	0	0	-1	0	0	0	0	0	-1	0	0	-2	0	0	0	0	2	-1	0
NADPH	0	0	0	0	2	0	-2	0	0	0	-1	1	0	0	0	-1	-1	0	-1	0
NADP	0	0	0	0	-2	0	2	0	0	0	1	-1	0	0	0	1	1	0	1	0
Acetyl-CoA	0	0	0	0	0	0	0	0	0	1	-1	0	0	0	0	-1	0	0	0	0
CoASH	0	0	0	0	0	0	0	0	0	-1	1	0	0	0	0	1	0	0	0	0
Isocitrate	0	0	0	0	0	0	0	0	0	0	1	-1	0	0	0	0	0	0	0	0
Oxaloacetate	0	0	0	0	0	0	0	0	0	-1	-1	0	1	0	0	1	0	0	0	0
α-Ketoglutarate	0	0	0	0	0	0	1	0	0	0	0	1	-1	-1	1	0	-1	0	0	0
Erythrose-4-P	0	0	0	0	0	1	-1	0	0	0	0	0	0	0	0	0	0	0	0	0
Ribulose-5-P	0	0	0	0	1	-2	0	0	0	0	0	0	0	0	0	0	0	0	0	0
Glutamate	0	0	0	0	0	0	-1	0	0	0	0	0	0	0	0	0	1	0	0	0
Glucose	-1	0	0	0	0	0	0	0	0	0	0	0	0	0	0	0	0	0	0	0
O₂	0	0	0	0	0	0	0	0	0	0	0	0	0	0	0	0	0	-1	0	0
NH4	0	0	0	0	0	0	0	0	0	0	0	0	0	0	0	0	-1	0	0	0
CO₂	0	0	0	0	1	0	1	0	0	1	0	1	0	1	0	0	0	0	0	0
Phenylalanine	0	0	0	0	0	0	1	0	0	0	0	0	0	0	0	0	0	0	0	0
Acetate	0	0	0	0	0	0	0	0	0	0	0	0	0	0	0	1	0	0	0	0
Maintenance ATP	0	0	0	0	0	0	0	0	0	0	0	0	0	0	0	0	0	0	0	1

Note: The network reactions (R1–R20) are listed as columns and network metabolites are listed as rows. Unbalanced network metabolites are highlighted in gray rows. G3P, glyceraldehyde 3-phosphate; CoASH, coenzyme A; DHAP, dihydroxyacetone phosphate.

reaction stoichiometries. The elements of the rate vector $\mathbf{r}'$ are unknown, and identifying their magnitudes is the typical aim of Equation 6.12. As presented, $\mathbf{r}'$ has 20 unknown rates whereas $\mathbf{S}'$ is comprised of 19 balanced metabolite conservation relationships. The rank of $\mathbf{S}'$ is 15, indicating that there are only 15 linearly independent relationships that can be used to identify the 20 unknown elements of $\mathbf{r}'$. Most metabolic networks, because of their high connectivity, have more unknowns (elements in vector $\mathbf{r}'$) than linearly independent relationships (rank of $\mathbf{S}'$). This class of linear algebra problem is termed "underdetermined" and more than one solution exists. A solution would represent a metabolic flux distribution (denoted in vector $\mathbf{r}'$) that balances system inputs with outputs. There are two widely used approaches for identifying solutions to this problem: constraint-based linear programming such as FBA (Reed and Palsson, 2003; Maarleveld et al., 2013) and metabolic pathway analysis such as EFMA (Schuster et al., 2000; von Kamp et al., 2017); EFMA will be presented here.

EFMA identifies all possible, biologically significant solutions to Equation 6.12 based on a minimal set of nondecomposable network reaction fluxes. These reaction combinations are called elementary flux modes (EFMs) and can be viewed as mathematically defined biochemical pathways. Each EFM is genetically independent because each uses a unique combination of genes/reactions (Schuster et al., 2000). EFMs represent the simplest balanced flux unit for a biochemical network operating at steady state and therefore are ideal starting points for understanding metabolic organization and phenotypes. EFMs are fundamental biochemical building blocks that can define any steady state. EFMA can be performed using publicly available software (Terzer and Stelling, 2008; von Kamp et al., 2017). The work presented here used Metatool version 5.0 (http://pinguin.biologie.uni-jena.de/bioinformatik/networks/index.html). Additional EFMA material including sample models and worked examples can be found at: www.chbe.montana.edu/biochemenglab/EFMAworkshop.html.

The model described here possesses 13 unique EFMs that balance the system inputs with outputs. The 13 EFMs (EFM1-EFM13) are listed as rows in Table 6.3a. Each column entry is a relative flux through the corresponding network reaction. As mentioned earlier, each EFM is genetically independent and uses a unique combination of reactions. There are seven distinct EFMs that produce phenylalanine (reaction R7). The efficiency of each EFM to produce phenylalanine can be determined from yields. Because each column in Table 6.3a represents a relative reaction flux, calculating yields only requires calculating flux ratios between phenylalanine production and either glucose consumption or O_2 consumption. The phenylalanine glucose yield ranges from 0.25 to 0.5 moles phenylalanine synthesized per mole glucose consumed and from 0.08 to 0.5 moles phenylalanine synthesized per mole O_2 consumed. Six EFMs do not produce phenylalanine but instead represent other metabolic functioning. Table 6.3b lists the overall transformation for each EFM written in terms of the unbalanced system metabolites (inputs and outputs).

All EFMs satisfy the network steady-state relationship in Equation 6.12 as do any nonnegative linear combination of EFMs. For instance, 2*EFM1 + 3*EFM5 + 1*EFM8 is a solution to Equation 6.12. This flux distribution would produce phenylalanine with a glucose yield of 0.25 moles phenylalanine per mole of glucose.

6.6.1.3 *In Silico* Design of Recombinant Systems

EFMA identifies a set of indivisible biochemical pathways that define a network's steady-state metabolic capabilities. The output explicitly lists which reactions are necessary and unnecessary for each EFM. This information provides metabolic engineering targets for increasing or forcing flux toward a desired product. For instance, using the phenylalanine case study (Table 6.3) it is possible to identify a single gene deletion (*pyk*, reaction R8) that would force the network to produce phenylalanine. Reaction R8 is used by all EFMs that do not make phenylalanine, although it would also prevent functioning of EFM3, which makes phenylalanine. The deletion would alter network structure so that only phenylalanine-producing fluxes would be possible under steady-state conditions. It should be noted that EFM1 and EFM2 would require expression of PEP synthetase (reaction R9); as mentioned in Section 6.6.1.1, this enzyme is typically associated with gluconeogenesis and activity is repressed by glucose.

EFMA is a useful tool for analyzing potential fluxes through recombinant metabolic pathways (Liao et al., 1996; Carlson et al., 2002, 2005; Wlaschin et al., 2006). EFMA identifies how recombinant fluxes can integrate into a host's native metabolic structure; the approach can also identify gene deletion strategies for diverting *in vivo* fluxes through the recombinant pathway; for example, the case study network can be modified to consider recombinant expression of the pyruvate decarboxylase and alcohol dehydrogenase of *Zymomonas mobilis*. This experimental *E. coli* system has been built and used to produce significant amounts of ethanol (Ingram et al., 1987; Ohta et al., 1991). To model expression of this pathway, the following reaction can be added to the model listed in Figure 6.11b:

$$R21: Pyr + NADH = Ethanol + CO_2 + NAD. \quad (6.13)$$

This reaction represents the combined activity of both *Z. mobilis* enzymes. Ethanol is treated as an unbalanced system output. Analyzing this modified model results in 18 unique EFMs. Five EFMs produce ethanol (Table 6.4) and the remaining 13 are identical to the results from the original simulation (Table 6.3). The output explicitly lists which reactions are required and not required for ethanol production. For instance, removing genes associated with R9 and R10 (*ppsA* and *aceE*, respectively) would force the network to produce ethanol under steady-state conditions.

The *E. coli* central metabolism is a highly robust system capable of many unique metabolic fluxes, and multiple gene deletions are often required to construct a desired strain. Models containing millions of EFMs and examples of strain

TABLE 6.3

(a) EFMA Results for Case Study Network. (b) Overall Transformation of Each EFM Written in Terms of Unbalanced System Inputs and Outputs

(a)

Reaction	R1	R2	R3	R4	R5	R6	R7	R8	R9	R10	R11	R12	R13	R14	R15	R16	R17	R18	R19	R20	Yp/g	Yp/O$_2$
EFM1	2	1	1	2	2	1	1	0	2	0	0	0	0	4	4	0	1	2	2	0	0.50	0.50
EFM2	2	1	1	2	2	1	1	0	2	0	0	0	0	0	0	0	1	2	2	4	0.50	0.50
EFM3	2	1	1	2	2	1	1	4	6	0	0	0	0	0	0	0	1	2	2	0	0.50	0.50
EFM4	1	1	1	2	0	0	0	1	0	2	0	0	0	12	12	2	0	2	0	0	0.00	0.00
EFM5	1	1	1	2	0	0	0	1	0	2	2	2	2	22	22	0	0	5	2	0	0.00	0.00
EFM6	1	1	1	2	0	0	0	1	0	2	0	0	0	0	0	2	0	2	0	12	0.00	0.00
EFM7	1	1	1	2	0	0	0	1	0	2	2	2	2	0	0	0	0	5	2	22	0.00	0.00
EFM8	1	1	1	2	0	0	0	13	12	2	0	0	0	0	0	2	0	2	0	0	0.00	0.00
EFM9	1	1	1	2	0	0	0	23	22	2	2	2	2	0	0	0	0	5	2	0	0.00	0.00
EFM10	4	3	3	6	2	1	1	0	0	4	0	0	0	30	30	4	1	6	2	0	0.25	0.17
EFM11	4	3	3	6	2	1	1	0	0	4	4	4	4	50	50	0	1	12	6	0	0.25	0.08
EFM12	4	3	3	6	2	1	1	0	0	4	0	0	0	0	0	4	1	6	2	30	0.25	0.17
EFM13	4	3	3	6	2	1	1	0	0	4	4	4	4	0	0	0	1	12	6	50	0.25	0.08

(b)

Overall Transformation

EFM	
EFM1	2 Glucose + 2 O$_2$ + NH$_4$ = 3 CO$_2$ + Phenylalanine
EFM2	2 Glucose + 2 O$_2$ + NH$_4$ = 4 MaintE + 3 CO$_2$ + Phenylalanine
EFM3	2 Glucose + 2 O$_2$ + NH$_4$ = 3 CO$_2$ + Phenylalanine
EFM4	Glucose + 2 O$_2$ = 2 CO$_2$ + 2 Acetate
EFM5	Glucose + 5 O$_2$ = 6 CO$_2$
EFM6	Glucose + 2 O$_2$ = 12 MaintE + 2 CO$_2$ + 2 Acetate
EFM7	Glucose + 5 O$_2$ = 22 MaintE + 6 CO$_2$
EFM8	Glucose + 2 O$_2$ = 2 CO$_2$ + 2 Acetate
EFM9	Glucose + 5 O$_2$ = 6 CO$_2$
EFM10	4 Glucose + 6 O$_2$ + NH$_4$ = 7 CO$_2$ + Phenylalanine + 4 Acetate
EFM11	4 Glucose + 12 O$_2$ + NH$_4$ = 15 CO$_2$ + Phenylalanine
EFM12	4 Glucose + 6 O$_2$ + NH$_4$ = 30 MaintE + 7 CO$_2$ + Phenylalanine + 4 Acetate
EFM13	4 Glucose + 12 O$_2$ + NH$_4$ = 50 MaintE + 15 CO$_2$ + Phenylalanine

Note: The 13 EFMs are listed as rows; network reactions are listed as columns with each entry representing the relative flux through the reaction. Yp/g, phenylalanine yield on glucose (mol/mol); Yp/O$_2$, is the phenylalanine yield on O$_2$ (mol/mol). MainE, maintenance energy.

TABLE 6.4

(a) EFMA Results for Recombinant, Ethanologenic Network. (b) Overall Transformation of each EFM Written in Terms of Unbalanced System Inputs and Outputs

(a)

Reaction	R1	R2	R3	R4	R5	R6	R7	R8	R9	R10	R11	R12	R13	R14	R15	R16	R17	R18	R19	R20	R21
EFM14	1	1	1	2	0	0	0	1	0	0	0	0	0	2	2	0	0	0	0	0	2
EFM15	1	1	1	2	0	0	0	1	0	0	0	0	0	0	0	0	0	0	0	2	2
EFM16	1	1	1	2	0	0	0	3	2	0	0	0	0	0	0	0	0	0	0	0	2
EFM17	4	3	3	6	2	1	1	0	0	0	0	0	0	10	10	0	1	2	2	0	4
EFM18	4	3	3	6	2	1	1	0	0	0	0	0	0	0	0	0	1	2	2	10	4

(b)

EFM14	Glucose = 2 CO_2 + 2 Ethanol
EFM15	Glucose = 2 MaintE + 2 CO_2 + 2 Ethanol
EFM16	Glucose = 2 CO_2 + 2 Ethanol
EFM17	4 Glucose + 2 O_2 + NH_4 = 7 CO_2 + Phenylalanine + 4 Ethanol
EFM18	4 Glucose + 2 O_2 + NH_4 = 10 MaintE + 7 CO_2 + Phenylalanine + 4 Ethanol

Note: The five EFMs that produce ethanol are listed as columns; network reactions are listed as rows; with each entry representing the relative flux through the reaction. The five listed EFMs are in addition to the 13 EFMs identified for the base model. MainE, maintenance energy.

TABLE 6.5
E. coli Biomass Macromolecular Composition Listed as Mass Fraction Dry Weight (Gray Columns) and Stoichiometric Coefficients for Biomass Generating Reactions as a Function of Culture Doubling Time

| Doubling Time (min) | Mass Fraction | | | | glc-6-p | rib-5-p | ery-4-p | pep | pyr | ace-coa | akg | Oxalo | NADH | NH4 | CO$_2$ | ATP | C-mole Biomass |
	Protein	DNA	RNA	Other													
30	52	2	20	26	4	13	5	32	38	41	14	24	178	139	-2	547	565
40	55	3	18	25	4	13	5	34	41	42	16	26	192	148	-3	584	599
50	58	3	17	22	4	14	6	39	50	43	19	31	218	171	-4	678	686
60	61	3	16	20	4	16	7	45	58	44	22	36	249	198	-6	786	777
80	66.5	4.5	15	14	4	22	11	64	87	49	32	53	347	291	-10	1153	1091
100	70.5	5.5	14	10	4	32	17	91	129	55	47	78	487	424	-15	1674	1554
200	78	6	10	6	4	46	31	156	237	72	86	139	856	731	-35	2921	2652

Note: The "other" category includes lipopolysaccharides, peptidoglycan, polysaccharides, and lipids. C-mole biomass refers to the mass of cells produced by the reaction stoichiometry. See Carlson and Srienc (2004a) for References. glc-6-p, glucose-6-P; rib-5-p, ribulose-5-P; ery-4-p, erythrose-4-P; pep, phosphoenolpyruvate; pyr, pyruvate; Ace-coa, acetyl-CoA; Akg, alpha-ketoglutarate (oxoglutarate); Oxalo, oxaloacetate.

construction through targeted gene deletions have been reported (Trinh et al., 2006; Wlaschin et al., 2006; Carlson, 2007).

6.6.1.4 Modeling Cellular Growth

Bioprocess synthesis of desired chemicals is often linked to cellular growth. Growth requires a combination of maintenance energy and biomass macromolecule synthesis. The case study considers maintenance energy (MainE, reaction R20) but not biomass synthesis. Metabolic network models are often used to analyze metabolite fluxes associated with cellular growth. There are two widely utilized strategies for predicting fluxes associated with growth. The first approach is based on the observations of Neidhardt and colleagues who deduced that all major biomass macromolecules (protein, RNA, DNA, lipids, polysaccharides) are derived from a few biosynthetic precursors drawn from central metabolism (Neidhardt et al., 1990). Using this information, it is possible to consider biomass synthesis by accounting for the metabolic draw of these biosynthetic precursors as a function of specific growth rate (Pramanik and Keasling, 1997; Carlson and Srienc, 2004a). Table 6.5 lists the *E. coli* growth-rate-dependent macromolecular composition and appropriate growth-rate-dependent stoichiometries for synthesizing the corresponding biomass using a dozen intermediates drawn from central metabolism. The ATP consumption requirements listed in Table 6.5 are solely for biosynthetic purposes, including polymerization of monomers into polymers (amino acids to proteins, nucleotides into DNA, etc.); maintenance energy requirements in the form of ATP are not included. An integrated analysis of biomass synthesis and maintenance-energy-associated fluxes can be found in Carlson and Srienc (2004b). For convenience, each overall biomass synthesis reaction is normalized to four glucose-6-phosphate molecules and therefore represents a different physical mass of cells; the cellular mass is listed as C-moles

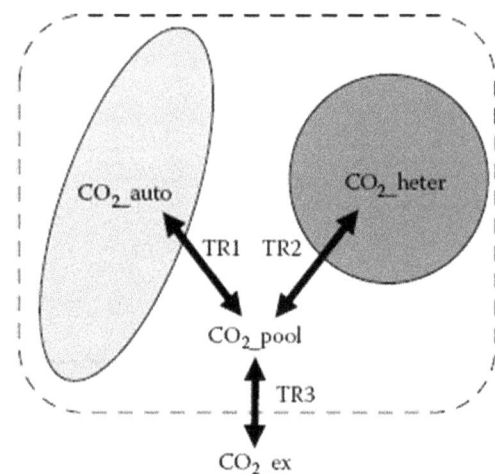

FIGURE 6.12 Diagram illustrating the partitioning of CO$_2$ between four distinct physical compartments indicated by suffixes: "auto" refers to an autotrophic bacterium, "heter" refers to a heterotroph bacterium, "pool" refers to a balanced extracellular compartment, and "ex" refers to an unbalanced external compartment.

(carbon moles) in Table 6.5 and is further illustrated in Section 6.6.1.5.

The second approach for modeling biomass synthesis considers the explicit metabolic reactions and metabolites necessary to construct each biomass constituent, including macromolecules, vitamins, cofactors, and other minor constituents. This approach, depending on the level of detail considered, can require hundreds or thousands of reactions and metabolites (Magnusdottir et al., 2017; Feist et al., 2007). The detailed treatment of biomass is an excellent means of analyzing microbial growth medium requirements and anabolic pathways, but it requires significant a priori information, including accurate gene annotation and detailed biomass composition measurements. Subject to the focus of the research, it is often reasonable to use the simpler biomass treatment. Although the central metabolism is highly robust, most biosynthetic pathways are linear series of enzyme-catalyzed reactions without alternative routes. In addition, constituents such as vitamins and cofactors represent a minor fraction of total cellular biomass, typically within the error of experimental biomass measurements.

6.6.1.5 *In Silico* Analysis of Compartmentalized Systems: From Eukaryotes to Microbial Consortia

Eukaryotes such as yeast and fungi are widely used by industrial microbiologists for the production of a wide range of chemicals, including primary and secondary metabolites. In addition, in the natural habitat, microbial populations exist as a well balanced and interacting consortium and there is growing interest in exploiting microbial consortia as a platform in bioprocess technology (Kleerebezem and van Loosdrecht,

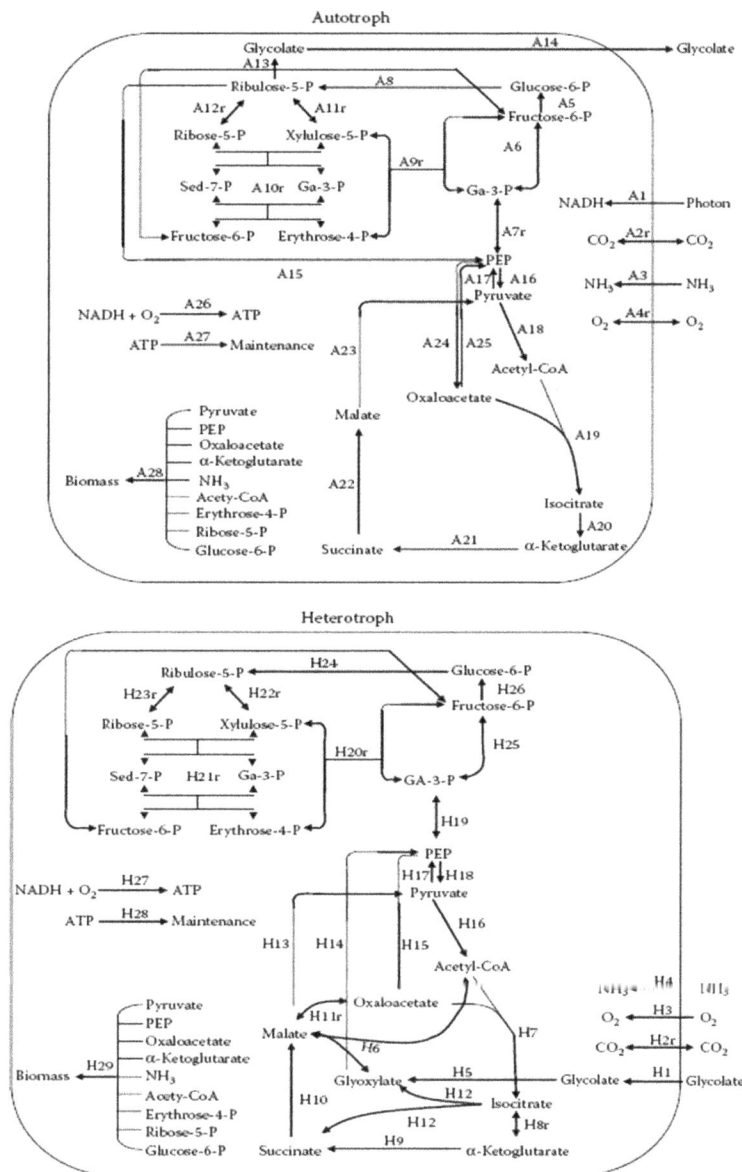

FIGURE 6.13 Graphical illustrations of photoautotroph and heterotroph metabolic networks from an interacting microbial consortium. The photoautotroph secretes glycolate, which is the sole reduced carbon and energy source for the heterotroph. Reaction designations are defined in Table 6.6.

TABLE 6.6
Reaction Network Model for Interacting Microbial Consortium

A1: 8 hv_ex + 3 ADP_auto + 2 NAD_auto = 3 ATP_auto + 2 NADH_auto + O_2_auto.

A2r: CO_2_auto = CO_2_pool.

A3: NH_3_pool = NH_3_auto.

A4r: O_2_auto = O_2_pool.

A5: fru6p_auto = glc6p_auto.

A6r: ATP_auto + fru6p_auto = 2 ga3p_auto + ADP_auto.

A7r: ga3p_auto + ADP_auto + NAD_auto = ATP_auto + NADH_auto + PEP_auto.

A8: glc6p_auto + 2 NAD_auto = CO_2_auto + 2 NADH_auto + rbl5p_auto.

A9r: ery4p_auto + xll5p_auto = fru6p_auto + ga3p_auto.

A10r: rbo5p_auto + xll5p_auto = ery4p_auto + fru6p_auto.

A11r: rbl5p_auto = xll5p_auto.

A12r: rbl5p_auto = rbo5p_auto.

A13: ATP_auto + rbl5p_auto + O_2_auto = PEP_auto + glyc_auto + ADP_auto.

A14: glyc_auto = glyc_pool.

A15: ATP_auto + CO_2_auto + rbl5p_auto = 2 PEP_auto + ADP_auto.

A16: PEP_auto + ADP_auto = ATP_auto + pyr_auto.

A17: 2 ATP_auto + pyr_auto = PEP_auto + 2 ADP_auto.

A18: pyr_auto + CoASH_auto + NAD_auto = ac_CoA_auto + CO_2_auto + NADH_auto.

A19: ac_CoA_auto + oaa_auto = icit_auto + CoASH_auto.

A20: icit_auto + NAD_auto = akg_auto + CO_2_auto + NADH_auto.

A21: akg_auto + ADP_auto + NAD_auto = ATP_auto + CO_2_auto + NADH_auto + succ_auto.

A22: O_2_auto + 2 succ_auto + 2.5 ADP_auto = 2.5 ATP_auto + 2 mal_auto.

A23: mal_auto + NAD_auto = CO_2_auto + NADH_auto + pyr_auto.

A24: CO_2_auto + PEP_auto = oaa_auto.

A25: ATP_auto + oaa_auto = CO_2_auto + PEP_auto + ADP_auto.

A26: 2 NADH_auto + O_2_auto + 5 ADP_auto = 5 ATP_auto + 2 NAD_auto.

A27: ATP_auto = MainE_auto_ex + ADP_auto.

A28: 4 glc6p_auto + 46 rbo5p_auto + 31 ery4p_auto + 156 PEP_auto + 237 pyr_auto + 72 ac_CoA_auto + 86 akg_auto + 139 oaa_auto + 731 NH3_auto + 856 NADH_auto + 2921 ATP_auto = bm_auto_ex + 35 CO_2_auto + 72 CoASH_auto + 2921 ADP_auto + 856 NAD_auto.

T1r: CO_2_pool = CO_2_ex.

T2: glyc_pool = glyc_ex.

T3r: O_2_pool = O_2_ex.

T4: NH_3_ex = NH_3_pool.

H1: glyc_pool = glyc_heter.

H2r: CO_2_heter = CO_2_pool.

H3: O_2_pool = O_2_heter.

H4: NH_3_pool = NH_3_heter.

H5: glyc_heter + NAD_heter = glyox_heter + NADH_heter.

H6r: ac_CoA_heter + glyox_heter = mal_heter + CoASH_heter.

H7: ac_CoA_heter + oaa_heter = icit_heter + CoASH_heter.

H8r: icit_heter + NAD_heter = akg_heter + CO_2_heter + NADH_heter.

H9: akg_heter + ADP_heter + NAD_heter = ATP_heter + CO_2_heter + NADH_heter + succ_heter.

H10: succ_heter + 0.5 O_2_heter + 1.25 ADP_heter = 1.25 ATP_heter + mal_heter.

H11r: mal_heter + NAD_heter = NADH_heter + oaa_heter.

H12: icit_heter = glyox_heter + succ_heter.

H13: mal_heter + NAD_heter = CO_2_heter + NADH_heter + pyr_heter.

H14: ATP_heter + 2 glyox_heter + NADH_heter = CO_2_heter + PEP_heter + NAD_heter + ADP_heter.

H15: CO_2_heter + PEP_heter = oaa_heter.

H16: pyr_heter + CoASH_heter + NAD_heter = ac_CoA_heter + CO_2_heter + NADH_heter.

H17: 2 ATP_heter + pyr_heter = PEP_heter + 2 ADP_heter.

H18: PEP_heter + ADP_heter = ATP_heter + pyr_heter.

H19r: ATP_heter + NADH_heter + PEP_heter = ga3p_heter + ADP_heter + NAD_heter.

H20r: ery4p_heter + xll5p_heter = fru6p_heter + ga3p_heter.

(Continued)

TABLE 6.6 (CONTINUED)

Reaction Network Model for Interacting Microbial Consortium

H21r: rbo5p_heter + xll5p_heter = ery4p_heter + fru6p_heter.

H22r: rbl5p_heter = xll5p_heter.

H23r: rbl5p_heter = rbo5p_heter.

H24: glc6p_heter + 2 NAD_heter = CO_2_heter + 2 NADH_heter + rbl5p_heter.

H25r: ATP_heter + fru6p_heter = 2 ga3p_heter + ADP_heter.

H26: fru6p_heter = glc6p_heter.

H27: 2 NADH_heter + O_2_heter + 5 ADP_heter = 5 ATP_heter + 2 NAD_heter.

H28: ATP_heter = MainE_heter_ex + ADP_heter.

H29: 4 glc6p_heter + 46 rbo5p_heter + 31 ery4p_heter + 156 PEP_heter + 237 pyr_heter + 72 ac_CoA_heter + 86 akg_heter + 139 oaa_heter + 731 NH3_ heter + 856 NADH_heter + 2921 ATP_heter = bm_heter_ex + 35 CO_2_ heter + 72 CoASH_heter + 2921 ADP_heter + 856 NAD_heter.

Note: Reactions labeled with "A" are associated with the photoautotroph whereas reactions labeled with "H" are associated with the heterotroph. reactions labeled with a "T" are transport reactions moving metabolites between the balanced "pool" and unbalanced "external" extracellular spaces. Reactions with a lower-case "r" were considered reversible for metatool simulations. Unbalanced system input and outputs were hv_ex (photon), CO_2_ex, NH3_ ex, bm_auto_ex (photoautotroph biomass), bm_heter_ex (heterotroph biomass), O_2_ex, MainE_auto_ex (photoautotroph maintenance energy), MainE_ heter_ex (heterotroph maintenance energy), and glyc_ex (glycolate). All other metabolites were treated as balanced and constrained by the no-accumulation assumption.

2007; Bernstein and Carlson, 2012; Succurro and Ebenhoeh, 2018; Song et al., 2018). EFMA can be used to consider metabolite transformation and fluxes in systems with complex physical architectures, including eukaryotes with multiple subcellular compartments and interactions in microbial consortia. Modeling these systems requires the assignment of reactions and metabolites to distinct physical compartments. For instance, Figure 6.12 illustrates an interacting microbial consortium with respect to CO_2 exchange between the organisms on one hand and the environment on the other. The extracellular environment is treated in two manners, including a balanced "pool" compartment constrained by steady-state, no-accumulation constraints and as an external, unbalanced compartment that permits CO_2 to enter or leave the system. Transport reactions (TR1-TR3) permit transfer of CO_2 from one compartment to another. It should be noted that model reactions do not necessarily have to be enzyme-catalyzed reactions. CO_2, a small uncharged molecule, readily diffuses across cellular membranes.

Stoichiometric modeling can analyze fundamental metabolic interactions in microbial populations (Stoylar et al., 2007; Taffs et al., 2009). An example model based on interactions between a *photoautotroph* and a *heterotroph* is shown in Figure 6.13 with explicit reactions listed in Table 6.6. The consortium model is comprised of four distinct compartments analogous to Figure 6.12. Metabolites associated with each compartment have a suffix designating their location. Metabolites and therefore reactions associated with the photoautotroph have the suffix "auto" whereas metabolites associated with the heterotrophs have the suffix "heter," metabolites in a balanced extracellular compartment are labeled with the suffix "pool," and unbalanced, external metabolites that serve as system inputs and outputs have the suffix "ex." Metabolites such as glycolate or oxygen can be transferred from the photoautotroph to the heterotroph by first transiting the balanced

pool compartment, or they can leave the system passing from the pool compartment to the external compartment.

The photoautotroph uses energy from sunlight to extract electrons from water, releasing O_2 as a by-product in a biological process known as *oxygenic photosynthesis* (reaction A1). The electrons can be used to fix CO_2 into organic carbon compounds (reaction A15). Some of this organic carbon (e.g., glycolate) can be secreted, which feeds associated heterotrophic organisms (reactions A14 and H1, respectively). The consortium model system considers growth of both microbes. The biomass-synthesizing reactions (reactions A28, H29) use the 200-min doubling time biomass composition from Table 6.5. Although these organisms are not *E. coli*, basic macromolecular composition for many bacteria is similar and biosynthetic pathways are highly conserved. For modeling purposes only, photosynthetically available *photons* are considered (hv_ex). These photons would be collected by the appropriate photoantennae. The presented ecological template of primary productivity (photoautotrophy) and associated heterotrophy is globally widespread and is relevant to many proposed CO_2 remediation strategies and to proposed cyanobacterial bioprocess platforms (Burja et al., 2001; Beck et al., 2017).

EFMA of the consortium model identifies 720 unique EFMs with 560 EFMs synthesizing photoautotroph or heterotroph biomass. Every EFM utilizes energy absorbed from the photoautotroph's photosynthetic apparatus to fuel the consortium. All heterotrophic growth is based on oxidation of glycolate secreted by the photoautotroph. Plotting the results from EFMA quickly visualizes trends in system behavior. For instance, Figure 6.14 examines photoefficiency (C-mole biomass/photons absorbed) of the 560 biomass-synthesizing EFMs as a function of O_2 and CO_2 flux at the RuBisCO enzyme (reactions A15 and A13, respectively). RuBisCO (ribulose-1,5-bisphosphate carboxylase oxygenase) is a major CO_2-fixing enzyme that transfers

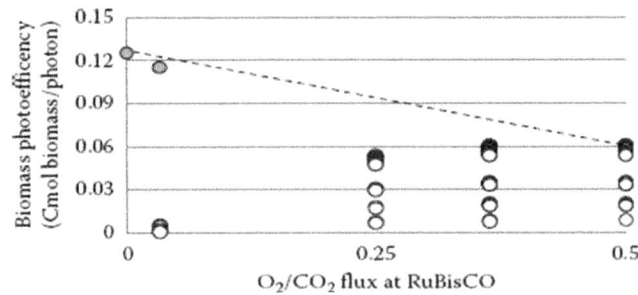

FIGURE 6.14 Consortium biomass photoefficiency as a function of O_2 competition at the RuBisCO active site. Each circle represents one distinct EFM, which produces biomass (gray circle, photoautotroph biomass; white circle, heterotroph biomass). Overall consortium functioning can be described by nonnegative linear combinations of distinct EFMs. The dotted line represents strategies defined by combinations of the highest yielding photoautotroph biomass strategy and the highest yielding heterotroph strategy. Community functioning, as defined by position along this line, is determined by the O_2 competition at the RuBisCO active site (ratio of O_2 to CO_2 fluxes).

reducing equivalents produced during photosynthetic water-splitting to CO_2, fixing an additional carbon mole (C-mole). O_2, a by-product of photosynthetic water splitting, can compete with CO_2 for the RuBisCO active site. When O_2 binds, the oxidized organic metabolite glycolate is formed instead of an additional fixed C-mole. Glycolate is typically secreted by the photoautotroph as an undesirable by-product. Glycolate can be used as a reduced carbon and energy source by heterotrophs.

Figure 6.14 demonstrates that increased RuBisCO O_2 competition reduces photoautotroph biomass synthesis efficiency by diverting reduced carbon toward glycolate biosynthesis, thus favoring heterotrophic growth. Overall consortium functioning can be modeled as linear combinations of distinct EFMs. The line on Figure 6.14 demonstrates the most efficient photoautotroph biomass synthesis (left end point) and the most efficient heterotroph biomass synthesis (right end point). The figure also shows the effect of varying levels of O_2 competition at the RuBisCO active site on biomass formation in the consortium. A more detailed account of a similar ecological scenario can be found in Taffs et al. (2009).

The metabolite partitioning approach presented for an interacting microbial consortium has already been used to model metabolic interactions between eukaryotic organelles and the cytosol (Fell and Small, 1986; Carlson et al., 2002; Forster et al., 2003; Mardinoglu and Nielsen, 2015). Enzymes and metabolites must be assigned to different physical subcellular compartments. The use of convenient suffixes such as "ex," "cyt," or "mit" would facilitate editing of modeling files and analysis of results.

SUMMARY

It is proposed that qualitative terms used to describe an enzyme as "rate-limiting," "bottleneck," or "pacemaker" should be abandoned and replaced by the term "rate-controlling" and further qualified by the flux control coefficient to describe, in quantitative terms, the capacity of the enzyme in question with respect to flux control within a given pathway.

The flux control coefficient of a particular enzyme is a system property (i.e., its value is not entirely independent of other enzymes in the pathway). The interrelationship between the flux control coefficients in a given pathway is governed by the summation theorem, which dictates that the total sum of all flux control coefficients adds up to 1.

In a steady state, the influence of a particular metabolite on flux through a given enzyme on the one hand and the whole pathway on the other can be determined from the enzyme's elasticity coefficient, a quantitative term that is directly related to the kinetic properties of the enzyme involved.

The metabolic interrelationship between the flux control coefficient and the elasticity coefficient is described by the connectivity theorem, which takes into account the kinetic properties of each of the enzymes involved.

The action of external effectors on metabolic flux can be assessed by measuring the response coefficient, which, to a large extent, is dependent on the flux control coefficient and the elasticity coefficient of the enzyme with respect to the effector. For an effector to be able to influence the flux through a certain enzyme, the values of the aforementioned coefficients must be relatively high.

Activating or increasing the catalytic activity of a single enzyme is not usually accompanied by a significant increase in flux (productivity), even with enzymes possessing a relatively large flux control coefficient. This is simply because the flux control shifts to other enzymes as the target enzyme is activated. This, in turn, implies that amplification of single enzymic activity is not a viable option for increasing productivity. However, this limitation does not apply to the reduction of catalytic activity because reduction or inactivation is generally accompanied by a considerable drop in flux.

A universal method has been developed to calculate the exact degree of overexpression required for increasing flux to product formation by a certain factor without perturbing the steady-state fluxes in other enzymes.

Increasing the concentration of an effector that activates all enzymes in the pathway will be accompanied by an appreciable increase in flux and, in turn, yield. Furthermore, an effector that stimulates the activity of more than one enzyme in a given pathway may lead to increase in flux (productivity), particularly if those enzymes share a relatively high flux control coefficient.

The case study presented in this chapter on the partition of carbon flux between ICDH and ICL revealed that during growth on acetate as sole source of carbon and energy, ICDH is not rate-controlling and that flux through isocitrate ICL is essential not only to replenish central metabolism with biosynthetic precursors but also to sustain a high intracellular level of isocitrate. Furthermore, above a certain threshold concentration of ICL, the Krebs cycle and the glyoxylate bypass can work in concert without the need for the inactivation of ICDH.

Among the five different strategies reported for the manipulation of carbon flux *in silico*, evasion and subversion are the most suitable strategies for increasing fluxes to product formation without adversely affecting the intracellular concentrations of metabolites; although evasion has the added advantage of being applicable to all metabolic pathways because it does not make any assumption about the regulatory control systems used *in vivo*.

However, subversion affords a practical strategy because it only involves the manipulation of one enzymic activity and as such may prove more of an attractive proposition for industrialists. Subverting feedback inhibition by the desired end product can simply be achieved by generating a mutant that leaks the end product into the medium. Such a system is highly desirable because the recovery and downstream processing become more effective and less expensive.

Flux models, in association with the universal method of Kacser, allow for the development of a rationale for strain improvement.

Stoichiometric modeling approaches do not require extensive kinetic parameters and can be broadly subdivided into three main categories: MFA, FBA, and EFMA.

EFMA dissects complex and highly sophisticated metabolic networks into their simplest, indivisible flux units known as EFMs. EFMs can be viewed as mathematically defined biochemical pathways and represent a practical starting point for analyzing metabolic network structure and function.

EFMA is a convenient tool for metabolic engineering. EFMA identifies which network reactions are and are not necessary for desired cellular function highlighting targets for gene deletions and overexpression.

EFMA can also be used to analyze expression of recombinant metabolic pathways and for identifying genetic manipulations to divert flux through the pathway.

The conversion of 12 biosynthetic precursors, drawn from central metabolism, into biomass can be used to model cellular growth. By varying the ratios of these intermediates, it is possible to consider different growth-rate-dependent biomass macromolecular compositions.

Flux through systems with complex physical architectures including microbial communities and eukaryotes can be modeled by assigning metabolites and reactions to distinct compartments.

ACKNOWLEDGMENT

The authors wish to thank Gordon Lang, for skilled assistance in the compilation of flux matrices and for stimulating discussions over the years.

REFERENCES

Beck, A.E., H.C. Bernstein, and R.P. Carlson. 2017. Stoichiometric network analysis of cyanobacterial acclimation to photosynthesis-associated stresses identifies heterotrophic niches. *Processes* 5(2):32.

Bernstein, H., and R.P. Carlson. 2012. Microbial consortia engineering for cellular factories: *In vitro* to *in silico*. *Comp Struc Biotechnol J* 3(4):e201210017.

Burja, A.M., B. Banaigs, E. Abou-Mansour, J.G. Burgess, and P.C. Wright. 2001. Marine cyanobacteria: A prolific source of natural products. *Tetrahedron* 57:9347–9377.

Carlson, R.P. 2007. Metabolic systems cost-benefit analysis for interpreting network structure and regulation. *Bioinformatics* 23:1258–1264.

Carlson, R., D.A. Fell, and F. Srienc. 2002. Metabolic pathway analysis of a recombinant yeast for rational strain development. *Biotechnol Bioeng* 79:121–134.

Carlson, R., and F. Srienc. 2004a. Fundamental *Escherichia coli* biochemical pathways for biomass and energy production: Identification of reactions. *Biotechnol Bioeng* 85:1–19.

Carlson, R., and F. Srienc. 2004b. Fundamental *Escherichia coli* biochemical pathways for biomass and energy production: Creation of overall flux states. *Biotechnol Bioeng* 86:149–162.

Carlson, R., A.P. Wlaschin, and F. Srienc. 2005. Kinetic studies and biochemical pathway analysis of anaerobic poly-(R)-3-hydroxybutyric acid synthesis in *Escherichia coli*. *Appl Environ Microbiol* 71:713–720.

Cornish-Bowden, A. 1995. In H.-J. Rehm and G. Read, eds., *Biotechnology: A comprehensive treatise*, 2nd ed., Vol. 9, pp. 121–136. Weinheim, Germany: Springer-Verlag.

Cornish-Bowden, A., J.-H.S. Hofmeyr, and M.L. Cardenas. 1995. Strategies for manipulating metabolic fluxes in biotechnology. *Bioorganic Chem* 23:439–449.

Cornish-Bowden, A., and J.-H.S. Hofmeyr. 1991. MetaModel: A program for modelling and control analysis of metabolic pathways on the IBM PC and compatibles. *Comput. Applic. Biosci.* 7:89–93.

Cozzone, A.J. 1988. Protein phosphorylation in prokaryotes. *Ann Rev Microbiol* 42:97–125.

Cozzone, A.J., and E.M.T. El-Mansi. 2005. Control of isocitrate dehydrogenase catalytic activity by protein phosphorylation in *Escherichia coli*. *J Mol Microbiol Biotechnol* 9:132–146.

El-Mansi, E.M.T. 1998. Control of metabolic interconversion of isocitrate dehydrogenase between the catalytically active and inactive forms in *Escherichia coli*. *FEMS Microbiol. Lett* 166:333–339.

El-Mansi, EMT (2004) Flux to acetate and lactate excretions in industrial fermentations: Physiological and biochemical implication. *Journal of Industrial Microbiology and Biotechnology* 31(7):295–300.

El-Mansi, E.M.T., and W.H. Holms. 1989. Control of carbon flux to acetate excretion during growth of *Escherichia coli* in batch and continuous cultures. *J Gen Microbiol* 135:2875–2883.

El-Mansi, E.M.T., G.C. Dawson, and C.F.A. Bryce. 1994. Steady-state modelling of metabolic flux between the tricarboxylic acid cycle and the glyoxylate bypass in *Escherichia coli*. *Comp Appl Biosci* 10:295–299.

El-Mansi, E.M.T., H.G. Nimmo, and W.H. Holms. 1985. The role of isocitrate in control of the phosphorylation of isocitrate dehydrogenase in *Escherichia coli*. *FEBS Lett* 183:251–255.

Feist, A.M., C.S. Henry, J.L. Reed, M. Krummenacker, A.R. Joyce, P.D. Karp, L.J. Broadbelt, V. Hatzimanikatis, and B.O. Palsson. 2007. A genome-scale metabolic reconstruction of *Escherichia coli* K-12 MG1655 that accounts for 1260 ORFs and thermodynamic information. *Mol Sys Biol* 3:e121.

Fell, D. 1997. *Understanding the Control of Metabolism*. London: Portland Press.

Fell, D.A., and J.R. Small. 1986. Fat synthesis in adipose tissue: An examination of stoichiometric constraints. *Biochem J* 238:781–786.

Foberg, C., T. Eliaeson, and L. Haggstrom. 1988. Correlation of theoretical and experimental yields of phenylalanine from non-growing cells of a recombinant *Escherichia coli* strain. *J Biotechnol* 7:319–331.

Forster, J., I. Famili, P. Fu, B.O. Palsson, and J. Nielsen. 2003. Genome-scale reconstruction of the *Saccharomyces cerevisiae* metabolic network. *Genome Res* 13:244–253.

Heinrich, R., and T. Rapaport. 1974. A linear steady-state treatment of enzymatic chains. *Eur J Biochem* 42:89–95.

Hofmeyr, J.H.S., and A. Cornish-Bowden. 1991. Quantitative assessment of regulation in metabolic systems. *Eur J Biochem* 200:223–236.

Hofmeyr, J.-H.S., H. Kacser, and K.J. Merwe. 1986. Metabolic control analysis of moiety-conserved cycles. *Eur J Biochem* 155:631–41.

Holms, W.H. 1996. Flux analysis and control of the central metabolic pathways in *Escherichia coli*. *FEMS Microbiol Rev* 19:85–116.

Hurlebaus, J. 2001. A pathway modeling tool for metabolic engineering. PhD thesis, University of Bonn, 2001.

Ingram, L.O., T. Conway, D.P. Clark, G.W. Sewell, and J.F. Preston. 1987. Genetic engineering of ethanol production in *Escherichia coli*. *Appl Environ Microbiol* 53:2420–2425.

Kacser, H., and L. Acerenza. 1993. A universal method for achieving increases in metabolite production. *Eur J Biochem* 216:361–367.

Kacser, H., and J. Burns. 1973. The control of flux. *Symp Soc Exp Biol* 27:65–104. (Reprinted in *Biochem Soc Trans* 23: 341–66, 1995.)

Kleerebezem, R., and M.C.M. van Loosdrecht. 2007. Mixed culture biotechnology for bioenergy production. *Curr Opin Biotechnol* 18:207–212.

Kornberg, H.L. 1966. The role and control of the glyoxylate cycle in *Escherichia coli*. *Biochem J* 99:1–11.

Koshland, D.E., Jr. 1987. Switches, thresholds and ultrasensitivity. *Trends Biochem Sci* 12: 225–229.

Liao, J.C., S.Y. Hou, and Y.P. Chao. 1996. Pathway analysis, engineering and physiological considerations for redirecting central metabolism. *Biotechnol Bioeng* 52:129–140.

Maarleveld, T.R., R.A. Khandelwal, B.G. Olivier, B. Teusink, and F.J. Bruggeman. 2013. Basic concepts and principles of stoichiometric modeling of metabolic networks. *Biotech J* 8(9):997–1008.

Magnusdottir, S. et al. 2017. Generation of genome-scale metabolic reconstructions for 773 member of the human gut microbiota. *Nat Biotech* 35: 81–89.

Mardinoglu, A., and J. Nielsen. 2015. New paradigms for metabolic modeling of human cells. *Curr Opin Biotechnol* 34:91–97.

McCleary, W.R., 2009. Applications of promoter swapping techniques to control of expression of chromosomal genes. *Appl Microbiol Biotechnol* 84:641–648.

Mendes, P. 1993. GEPASI: A software package for modelling the dynamics, steady states and control of biochemical and other systems. *Comput Appl Biosci* 9:563–571.

Niederberger, P., R. Prasad, G. Miozzari, and H. Kacser. 1992. A strategy for increasing an in vivo flux by genetic manipulations: The tryptophan system of yeast. *Biochem. J.* 287:473–480.

Neidhardt, F. C., J. Ingraham, and M. Schaechter. 1990. *Physiology of the Bacterial Cell: A Molecular Approach*. Sunderland, MA: Sinauer Associates.

Ohta, K., D.S. Beall, J.P. Mejia, K.T. Shanmugan, and L.O. Ingram. 1991. Genetic improvements of *E. coli* for ethanol production: Chromosomal integration of *Zymomonas mobilis* gene encoding pyruvate decarboxylase and alcohol dehydrogenase II. *Appl Environ Microbiol* 57:893–900.

Pramanik, J., and J.D. Keasling. 1997. Effect of *Escherichia coli* biomass composition on central metabolic fluxes predicted by a stoichiometric model. *Biotechnol Bioeng* 60:230–238.

Reed, J, and B.O. Palsson. 2003. Thirteen years of building constraint-based *in silico* models of *Escherichia coli*. *J Bacteriol* 185:2692–2699.

Schaaf, I., J. Heinisch, and F.K. Zimmermann. 1989. Overproduction of glycolytic enzymes in yeast. *Yeast* 5(4) 285–290.

Schilling, C.H., S. Schuster, B.O. Palsson, and R. Heinrich. 1999. Metabolic pathway analysis: Basic concepts and scientific applications in the post-genomic era. *Biotechnol Prog* 15:296–303.

Schuster, S., D.A. Fell, and T. Dandekar. 2000. A general definition of metabolic pathways useful for systematic organization and analysis of complex metabolic networks. *Nat Biotechnol* 18:326–332.

Song, H.S., W.C. Nelson, J.Y. Lee, R.C. Taylor, C.S. Henry, A.S. Beliaev, D. Ramkrishna, and H.C. Bernstein. 2018. Metabolic network modeling for computer-aided design of microbial interactions. In H.N. Chang, ed., *Emerging Areas in Bioengineering*, 793–801 doi:10.1002/9783527803293.ch45.

Stephanopoulos G.N., A.A. Aristidou, and J. Nielsen. 1998. *Metabolic Engineering. Principles and Methodologies*, New York: Academic Press.

Stoylar, S., S. Van Dien, K.L. Hillesland, N. Pinel, T.J. Leigh, and D.A. Stahl. 2007. Metabolic modeling of a mutualistic microbial community. *Mol Syst Biol* 3:e92.

Succurro, A., and O. Ebenhoeh. 2018. Review and perspective on mathematical modeling of microbial ecosystems. *Biochem Soc Trans* 46(2):403–412. BST20170265.

Taffs, R., J.E. Aston, K. Brileya, Z. Jay, C.G. Klatt, S. McGlynn, N. Mallette, S. Montross, R. Gerlach, W.P. Inskeep, D.M. Ward, and R.P. Carlson. 2009. *In silico* approaches to study mass and energy flows in microbial consortia: A syntrophic case study. *BMC Sys Biol* 3:e114.

Teleki, A., M. Rahnert, O. Bungart, B. Gann, I., Ochrombel, and R. Takors. 2017. Robust identification of metabolic control for microbial L-methionineproduction following an easy-to-use puristic approach. *Metab Eng* 41:159–172.

Terzer, M., and J. Stelling. 2008. Large-scale computation of elementary flux modes with bit pattern trees. *Bioinformatics* 24:2229–2235.

Trinh, C.T., R. Carlson, A. Wlaschin, and F. Srienc. 2006. Design, construction and performance of the most efficient biomass producing *E. coli* bacterium. *Metab Eng* 8:628–638.

Trinh, C.T., A. Wlaschin, and F. Srienc. 2009. Elementary mode analysis: A useful metabolic pathway analysis tool for characterizing cellular metabolism. *Appl Microbiol Biotechnol* 81:813–826.

Von Kamp, A., S. Thiele, O. Hadicke, and S. Klamt. 2017. Use of CellNetAnalyzer in biotechnology and metabolic engineering. *J Biotech* 261:221–228.

Walsh, K., and D. E. Koshland. 1984. Determination of flux through the branch point of two metabolic cycles: The tricarboxylic acid cycle and the glyoxylate shunt. *J Biol Chem* 259:9646–9654.

Weiner, M., J. Trondle, C. Albermann, G.A. Sprenger, and D. Weuster-Botz. 2017. Metabolic control analysis of l-phenylalanine production from glycerol with engineered *E. coli* using data from short-term steady-state perturbation experiments. *Biochem Engin J* 126:86–100.

Wlaschin, A.P., C.T. Trinh, R. Carlson, and F. Srienc. 2006. The fraction contributions of elementary modes to the metabolism of *Escherichia coli* and their estimation from reaction entropies. *Metab Eng* 8:338–352.

7 Control of Metabolite Efflux in Microbial Cell Factories

Current Advances and Future Prospects

Douglas B. Kell

CONTENTS

"It has been said that we always overestimate what we can do in two years and underestimate what we can do in twenty."

P. Ball and L. Garwin, Science at the atomic scale*

7.1 INTRODUCTION

In a typical fermentation, substrates are provided externally to the cells catalyzing the fermentation and converted to products; initially (at least in most cases) product formation occurs within the cells. However, for a relative density (specific gravity) of 1, cells occupy 1 mL.g^{-1} wet weight so for a fermentation that achieves even 100 mg wet cell wt.mL^{-1} only approximately one tenth of the total volume is intracellular (Figure 7.1). Commonly, intracellular concentrations during the development of processes for industrial bioengineering can soon become toxic [1, 2]. Since thermodynamics dictates that it is standard free energies and concentrations that control the eventual outcome, we need to recognize (Fig 1) that outwith a solid substrate fermentation in which the biomass is the product (as in [3]) the overall titer of product will be much enhanced if its internal concentrations can be decreased by secretion into the larger extracellular space. In addition, it is much easier to purify products if cells are not present [4]. Hence the desirability for product efflux, that is the focus of this chapter. Although much the same can be said of protein

* Ball, P. & Garwin, L. (1992). Science at the atomic scale. *Nature* 355, 761–766.

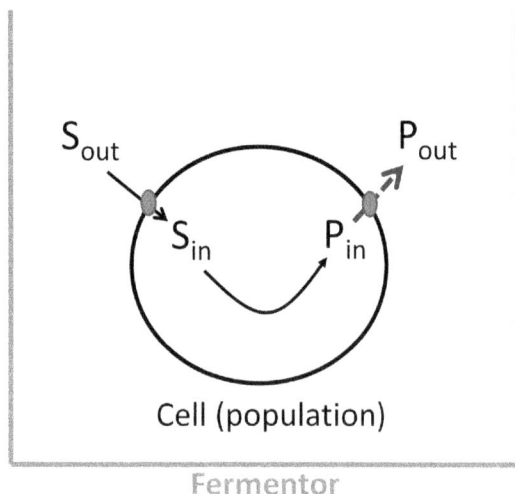

FIGURE 7.1 The ability to efflux an intracellular product is of great significance in biotechnology, as the extracellular space is normally much greater than the intracellular space in a typical fermentation.

production, we focus here solely on small molecules. An overview of the chapter is given as a mind map [5] in Figure 7.2.

7.2 MEMBRANE TRANSPORTERS

Although this is textbook material, we first rehearse briefly the different types of membrane transporter. A first distinction is whether they are equilibrative (i.e., permitting "facilitated diffusion") or whether their activities are coupled to an external free energy source such as ATP (hydrolysis) or electrochemical gradients; the latter kinds of transporter may then be

concentrative in terms of changing the transmembrane ratio of their substrate concentrations (properly, activities) away from 1. Those lacking secondary coupling are referred to as uniporters, symporters cotransport co-substrates, while antiporters act to exchange substrates in opposite directions in a coupled manner. "Group transfer" reactions involve the direct coupling of a chemical motif to the substrate, as in the PEP-dependent glucose transferases [6, 7] whose external substrate is glucose but whose internal product is glucose-6-phosphate.

We also note here the use and misuse of the term "passive" to describe transport activities; this term has two common but orthogonal meanings. The first is thermodynamic, and means "equilibrative," while the second is mechanistic and is then taken to mean "transport through a bilayer." When the mechanism involves a transporter, it is properly known as "facilitated diffusion." Since these two uses of "passive" are often conflated, and consequently cause much unnecessary confusion, we recommend that the term 'passive' is simply dropped in the context of transporters [8].

In a similar vein, it is common to refer to "influx" and "efflux" transporters on the basis of the direction of substrate flux observed in their most typical operating conditions. Clearly, however, any reaction is in principle thermodynamically reversible (even if free energy changes are large and negative). We note in particular therefore that while a particular transporter might "normally" be an "influx" transporter if its substrate is provided externally, there is no reason of principle, especially if it is equilibrative, why it would not become an "efflux" transporter if large amounts of the same substrate are made intracellularly in a biotechnological process. Consequently, while the focus is on 'efflux' transporters,

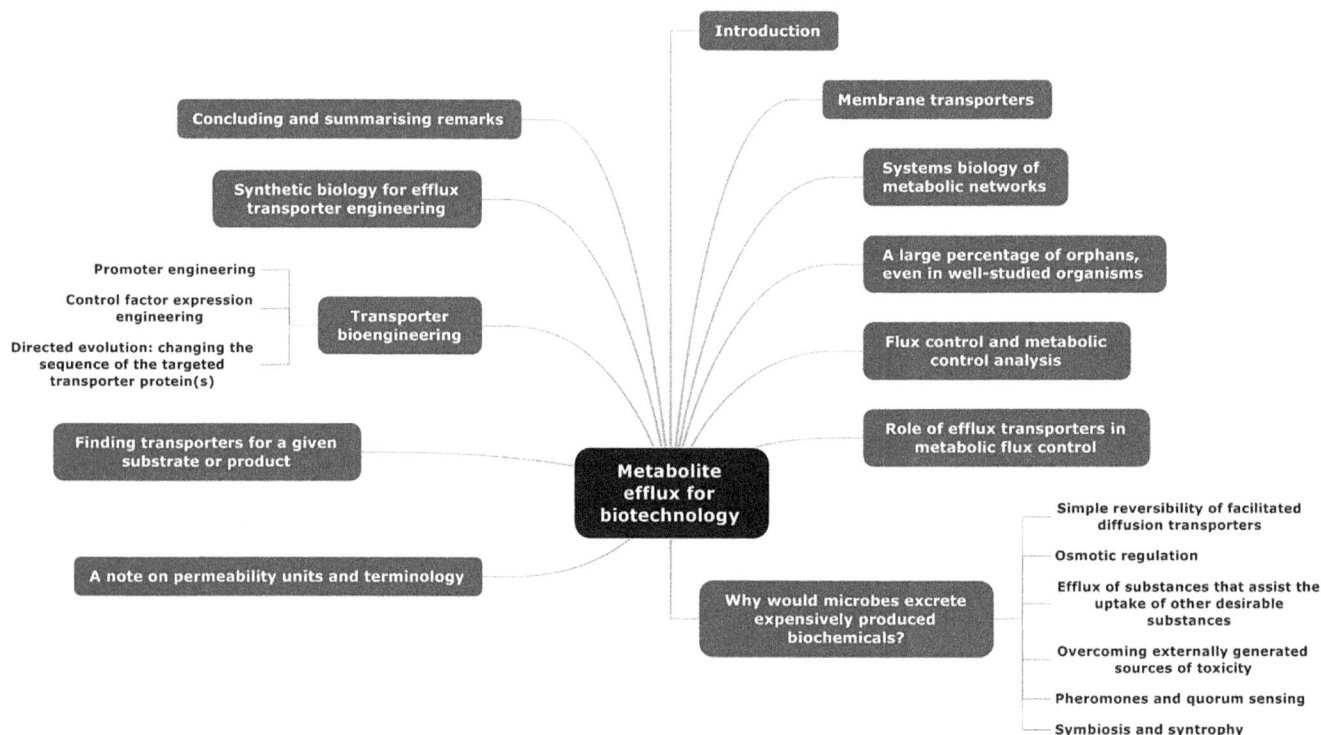

FIGURE 7.2 A "mind map" [5] of the layout of this paper.

we shall have plenty to say about the more widely studied "influx" transporters as well.

In general, despite their prevalence (transporters account for one eighth of *E. coli* genes [9], and see later), transporters are comparatively little studied [10] and the substrates of many (and in some organisms most) of them are unidentified [11, 12]. Although our focus here is on microorganisms (and see [13]), the situation is also acute for mammalian transporters, as these underpin the distributions of xenobiotic pharmaceutical drugs within and between cells [8, 14–20].

One possible reason for the comparative lack of study is the widespread (but erroneous) textbook belief that many substances can actually cross the phospholipid layer of biological membranes at non-negligible rates *without* the use of protein transporters. This belief is based in part on the fact that such "transport" *can* indeed be observed in liposomes or across "black" (bilayer) lipid membranes [21, 22] that are simply leaky to such molecules; however, this is not the case in real biomembranes, with a typical protein: phospholipid mass ratio of 3:1. Winter and colleagues [23] provide a particularly nice example, showing that at least 99.5% of the mammalian cell uptake of a drug known as sepantronium bromide occurs *only* via the SLC35F2 transporter, meaning that maximally 0.5% can be transported by *any* other means, including through the bilayer.

Although most commentators recognize that charged substances require transporters (including for so-called lipophilic cations [24]), we now know that membrane transporters exist (and are required) even for all kinds of small *uncharged* molecules that were once widely believed to be capable of crossing biomembranes without transporters; (Table 7.1) provides a set of examples. Notably, the discovery of water transporters (aquaporins) attracted a Nobel Prize (see [25] and [26]).

Since the role of biomembranes in general is to keep things inside and outside of the compartments that they surround, it is easy to imagine that transporters have an important role in the *selective* influx and efflux of molecules. Consequently, they attract our focus within the context of metabolic network or systems biology. Indeed, having the metabolic network is a *sine qua non* for developing such studies, so we briefly rehearse the systems biology approach to metabolic biotechnology.

7.3 SYSTEMS BIOLOGY OF METABOLIC NETWORKS

To understand metabolic networks, it is first necessary to reconstruct them [43]. This is done in four main steps [44, 45]. The first two are qualitative: they define the network in terms of the actors that are involved (enzymes, metabolites and effectors), and the means by which they are connected (mathematically as a 'graph'). The result of this is the familiar biochemical wallchart or the kinds of qualitative representation visible in an online resource such as KEGG [46]. The third step defines the kinetic rate equations governing each step, while the fourth step parametrizes them. Because of the extreme stoichiometric constraints [43], i.e., the requirement that in the absence of alchemy all atoms are conserved, a surprisingly large amount can be done with the qualitative network alone, and we shall tend to focus on this in the present chapter. Obviously, transporter steps are to be seen simply as "extra" enzymatic steps, as in (Figure 7.1). Reconstructions have become quite advanced by now, with strategies including "jamborees" [47–49], helped by the availability of digital tools (e.g., [50]), including for the reconstruction of non-natural pathways for the biosynthesis of "novel" products [51, 52]. In very few cases are all the kinetic rate equations known with any precision (see e.g., [53] for yeast glycolysis), but when they are one may resort to ODE-type modeling, using tools such as Copasi [54, 55]. Another strategy is to use surrogate kinetic rate equations for each step, lin-log being both common and effective [56]. "Convenience" kinetics [57] are also widely used for this.

7.4 A LARGE PERCENTAGE OF ORPHANS, EVEN IN WELL-STUDIED ORGANISMS

All of this said, however, in many cases we still lack the "parts list" [58]; in other words, we do not *know* the substrates for substantial fractions of ("uptake" *or* "efflux") membrane transporters, and even when we know some they may not be the most active or "natural" or "real" substrates. The systematic genome sequencing of microbes finally allowed at least a start to be made on the parts list and increased it considerably. Thus in *E. coli*, sequencing caused the number of proposed genes to be increased from ~1700 to over 4000 [59]. This 2006 survey of *E. coli* [59] listed some 4453 genes, of which 591 were transporters (13.3%) (see also [9, 12]); 337 were "known" and 254 (43% of the total) were "predicted" (i.e. orphan) transporters. In 2009, Hu and colleagues [60] found that about one-third of *E. coli* genes were functionally orphans. Ecogene (www.ecogene.org/) listed 326 transporters, of which 113 were "y-genes" (unassigned) (see also [61]). A Table (https://ecocyc.org/group?id=biocyc17-4655-3682299 327) at EcoCyc [62] gives 163 "putative" transporters, including 124 y-genes.

TABLE 7.1

Some small, neutral molecules for which transporters are known

Molecular Class	Selected Reference(s)
Alkanes	[27, 28]
Ammonia (NH_3)	[29]
Carbon dioxide (CO_2)	[30, 31]
Ethanolamine	[32]
Fatty acids	[33, 34]
Glycerol	[35]
Hydrogen peroxide (H_2O_2)	[36]
Hydroxyurea	[37]
Nitric oxide (NO)	[38]
(Di)oxygen (O_2)	[39]
Urea	[40, 41]
Water	[35, 42]

In 1998, Paulsen *et al.* [63] summarized this for *S. cerevisiae*, concluding that "among the 258 yeast transporters … a total of 139 {35%} lack genetic or biochemical names and thus lack either a demonstrated transport function or a recognizable physiological function." The latest version of transportDB [64] www.membranetransport.org/transportDB2/overview.html, the most comprehensive listing, gives 341 transporters for *S. cerevisiae* S288c, and shows that improvements in identification have been less than substantial. In general, the efflux transporters seem to have evolved to remove natural toxins from the yeast's environment [65, 66], are highly promiscuous [67, 68], and many are known as pleiotropic drug transporters (PDRs) [69].

The general conclusion of this is that at least 30% of transporters (and in many cases a great deal more) are genuine orphans in terms of actual experimental assessment; *deorphanization is thus the key first step* for improving our understanding of them, even in supposedly well-characterized organisms [10, 70].

7.5 FLUX CONTROL AND METABOLIC CONTROL ANALYSIS

El-Mansi and colleagues give an excellent description in this volume of the metabolic control analysis (/theory/formalism) of Kacser, Burns, Heinrich, and Rapoport [71, 72]. Other reviews can be found in [73–76] and an online tutorial at http://dbkgroup.org/metabolic-control-analysis/. Although not at all new, it still provides an excellent formalism for describing the extent (their flux control coefficients, FCC) to which individual reactions, including transporter reactions, control metabolic fluxes in biochemical networks. We note that those that do have reasonably high FCCs catalyze reactions that in the steady state are held far from their thermodynamic equilibrium, meaning (by definition) that these steps exert regulatory control. Increasing the rate of such steps will by definition increase fluxes. Both influx and efflux transporters tend to have this property (for an early example, see [77]).

7.6 ROLE OF EFFLUX TRANSPORTERS IN METABOLIC FLUX CONTROL

Since we reviewed the role of transporters in biotechnology not too long ago [13], with many historical examples, we start with just one main generic example here, followed by a couple not previously discussed by us in any detail.

7.6.1 AMINO ACIDS

The classic example of the role of efflux transporters in improving the yield of an important fermentation product (more than 2M tonnes p.a. [78]) is represented by the glutamate fermentation carried out in particular by *Corynebacterium glutamicum* [78–80]. Following the initial discovery of the microbial/fermentative production of glutamate [81], various empirical findings in the 1960s and 1970s

[82, 83] showed that a variety of treatments, involving biotin limitation, or the addition of weak surfactants such as "acetylated corn oil" or Tween, or the use of certain auxotrophs, would enhance the efflux of glutamate in producer strains. Soon enough, however, it was recognized that this was not due to a general membrane-leakiness, because it was very selective for glutamate (and was even against a glutamate concentration gradient!), but that it was due to a change in membrane tension that activated a mechanosensitive glutamate efflux pump encoded by a gene called NCgl1221 (a homologue of the *E. coli yggB* gene, now known as *mscS*, the mechanosensitive channel of small conductance) [84–89]. Similar efflux pumps are now known to be involved in the export of product during a variety of other amino acid fermentations [79, 90], such as those for lysine [91–95], isoleucine [96, 97], serine [98], threonine [99–102], cysteine [103], methionine [104], alanine [105], and others [103, 106, 107]. Thus, the general role of product exporters in enhancing fermentation yields (as per Figure 7.1) is clear [2, 103, 108]. We return to this later.

7.6.2 FATTY ACIDS

In contrast to amino acids, it used to be widely believed that both short- and long-chain fatty acids could simply diffuse through (the bilayer portion) of cellular membranes. Since that is not the case (see above) we devote a short section to longer-chain fatty acid uptake and efflux.

In mammalian systems, fatty acids are taken up by members of the SLC27 family [109], also known as fatty acid transport proteins (FATPs), of which there are six types with different expression profiles). All FATP members have a highly conserved, 311-amino acid signature sequence known as the FATP sequence.

Many microbes will use fatty acids as sole source of carbon, and thus microbes also contain fatty acid transporters (e.g., [110]), the main one in Gram-negative organisms being known as FadL [33, 34, 111, 112]. Yeasts also contain fatty acid transporters; these are mainly homologues of mammalian FAT1 [113–115]. It is of particular interest that the Fat1p homologue in the oleaginous yeast *Yarrowia lipolytica* is involved in the export of fatty acids [116–119].

7.6.3 RIBOFLAVIN

Riboflavin is a vitamin and nutritional supplement that is nowadays produced biosynthetically [120, 121]. It is of interest to us here because a main producer, *Ashbyi gossypii*, contains a very active (influx and) efflux transporter for it [122]. It is also of interest, and a general point worth nothing for fungi, that disruption of its vacuolar uptake system (by disrupting the vacuolar ATPase) ensures that all of the biosynthesized riboflavin is indeed excreted [123]. The other major industrial producer is *B. subtilis* [120], where an efflux transporter from *Streptomyces davawensis* has been used to improve productivity [124]. Of course, since a vitamin is by definition not synthesized therein, all human cells require [125] and express

[126] the relevant uptake transporters, in this case of the SLC52 family [127].

7.7 WHY WOULD MICROBES EXCRETE EXPENSIVELY PRODUCED BIOCHEMICAL?

In many cases, possibly the majority, efflux transporters are involved in the removal of potentially cytotoxic drugs or xenobiotics that happened to have been taken up by influx transporters, especially in mammalian cells where they represent a major problem, e.g., for cancer chemotherapy (e.g., [128–130]). In microbes, where they have also presumably evolved to remove environmental toxins that had been taken up [131], they can play a leading role in microbial resistance to anti-infectives [132–139]. In each of these cases, it is easy to understand why natural evolution would select for such activities.

By contrast, for substances that are actively biosynthesized by the host, it is rather less obvious why they might evolve an efflux transporter for them instead of simply lowering the rate of synthesis to a level that is adequate to satisfy the requirements of the host. Needless to say, evolution has in fact *selected* for this active efflux, and it is of interest to seek to understand its basis (if only to replicate it in the selection schemes of the biotechnologist). A number of examples can be given.

7.7.1 SIMPLE REVERSIBILITY OF FACILITATED DIFFUSION TRANSPORTERS

Although there can be an apparent kinetic irreversibility that follows from the Haldane relation [140], *any* equilibrative transporter might serve as a selective efflux transporter when the intracellular concentration of its substrates begin to exceed those of the extracellular ones. This appears to be the case, for instance, in at least one of the riboflavin examples given above [122]. It probably becomes true for any substance that cells can both take up and biosynthesize if the uptake transporter is not completely turned off in the latter case. Possibly this counts as neutral evolution [141], but by definition if such activities remain then natural selection was involved. To this end, it provides a rationale for adding the relevant equilibrative "*influx*" transporters to the cells of interest, even when it is efflux that is desired.

7.7.2 OSMOTIC REGULATION

Osmotic stress can modulate the activity of membrane transporters (e.g., [77]), and this is the basis for the mechanosensitive efflux pumps, e.g., for glutamate (see above). Thus, as in plants [142], the sudden advent of a raindrop or other source of water [143] under previously dry or drier conditions to which the cells had acclimatized can create a huge osmotic stress for a cell [144]; only cells that can respond with an almost instantaneous secretion of internal osmolytes can easily survive this, and one can imagine (as is the case, see above) that these kinds of channels work precisely because they are "mechanically" sensitive to the osmotic stress generated across the membrane.

7.7.3 EFFLUX OF SUBSTANCES THAT ASSIST THE UPTAKE OF OTHER DESIRABLE SUBSTANCES

Fermentative production of citric acid by the fungus *Aspergillus niger* is a long-standing and very large-scale process [145]. It too involves active export of the product from the producer strain using a proton-symporting transporter [146]. Here it would appear that in nature both plants and fungi have evolved citrate efflux to serve as a chelator for metal ions that are necessary for the growth of the host [147–149]. Depending on the pH, an equilibrative transporter may be sufficient [150]. Siderophores are of course another well-known class of compounds excreted by plants and microbes to assist their uptake of iron [151–154]. Dicarboxylate efflux transporters are also known [155], potentially (and experimentally) serving similar purposes [148, 156].

7.7.4 OVERCOMING EXTERNALLY GENERATED SOURCES OF TOXICITY

Microbes are necessarily exposed to many cytotoxic stresses, including both physical stresses (such as UV radiation) and chemicals. This can account for excretion of substances that compete with uptake transporters for uptake of the toxins (just as anti-metabolites can be used to select high producers, see [157]). It is likely that redox-active colored compounds such as riboflavin are indeed produced for these purposes; the same is true for astazanthin production by *Phaffia rhodozyma* [158, 159].

7.7.5 INTERCELLULAR SIGNALING: PHEROMONES AND QUORUM SENSING

"A pheromone is a chemical excreted by an organism into the environment that acts to elicit a specific response from other organisms of the same species. The importance of pheromones in the life cycle of various species of mammals, insects and fungi is well known. In the past decade, it has become apparent that pheromones influence the behavior and development of prokaryotes" [160]. Pheromones are thus well recognized as molecules that are secreted by organisms in order to elicit physiological responses from their genetic relatives. It is probable that secreted "secondary" metabolites (so named because their distribution between clades is rather restricted [161]) are made by microbes precisely for these purposes [162–165]. This said, some are clearly intended to send signals (including cytotoxic molecules) to *other* organisms, and many examples (e.g., [166]) are known in which the presence of organism 1 induces organism 2 to synthesize molecules that are normally cryptic. In enclosed environments, a molecule that is secreted and promotes both its own synthesis and secretion in another cell of the same species can achieve a steady-state concentration of the pheromone that depends on the concentration of cells. Consequently, and for historical reasons, such (pheromonal) behavior has become known, in prokaryote biology, as "quorum sensing" (see e.g., [167–169]). It provides another general example of secretory processes involving small molecules that have selective advantage for the host and has applications in biotechnology [170, 171].

7.7.6 Symbiosis and Syntrophy

As well as *signaling* molecules, many natural microbial and other ecosystems involve the exchange of *nutrients* between organisms, often of different types, necessarily involving secretion or metabolite efflux. Thus, lichens are symbioses between algae and fungi [172, 173], and stable consortia can develop even in novel ecosystems by learning to assist each other [174]. Since this is not a review of microbial ecology, our purpose is only to recognize the role of effluxers in complex ecosystems [175]; in some cases, the dependence of at least one organism is absolute, for thermodynamic reasons involving the removal by a second organism of a product secreted by the first; this is referred to as syntrophy [176–178].

In microbes that reproduce by binary fission, selection is usually for growth *rate* [179], and if circumstances can be arranged to select for it, improved efflux may lead to improved growth rate that can be selected for in a turbidostat (e.g., [180, 181]). In conclusion, as ever in biology, evolutionary considerations can give useful insights into circumstances in which the secretion of a biosynthesized metabolite is, perhaps surprisingly, beneficial for the host.

7.8 A NOTE ON PERMEABILITY UNITS AND TERMINOLOGY

Biochemists (and modelers) are used to expressing fluxes in terms of units such as nmol.(min.mg protein)$^{-1}$. If the volume of the compartment of interest is known this is sometimes rendered as nM.min^{-1} or its dimensional equivalent. However, the convention in transporter studies, especially those in epithelial cells such as Caco-2 cells, is to express such rates in terms of so-called permeability coefficients (P$_{app}$), with the dimensionality (c)m.s^{-1}. (Many pharmaceutical drugs have values in Caco-2 cells in the range 1–10 cm.s^{-1} when expressed in these units [182].) Transport efficiency is defined [183, 184] as the initial rate of uptake (μmol.min^{-1}) (or μmol.(min.mg^{-1})) divided by substrate concentration (nmol.cm^{-3}). The former thus has the dimensionality of the usual measure of enzyme kinetic power (k$_{cat}$/K$_m$) [185] and the SI dimensions of m^3.t^{-1}, and when this is divided by an estimate of the membrane area in (c)m^2 we obtain the (apparent) permeability P$_{app}$ with the dimensions length.time^{-1} (see also [186]).

7.9 FINDING TRANSPORTERS FOR A GIVEN SUBSTRATE OR PRODUCT

In line with our "four-stage" systems biology strategy, the first step is to establish qualitatively which transporters might be responsible for the transport of particular substrates (in shorthand: '**have substrate, seek transporter(s)**'). A different but related question. albeit with a similar endpoint, ('**have**

transporter, seek substrate(s)'), enquires as to which substrates are used by a given transporter. We shall consider both. One strategy (see e.g., [103]) for the former is purely *in silico* and uses sequence similarity to suggest transporter substrates. However, this is not particularly reliable as most transporters have undergone divergent evolution [187] such that moderately similar sequences or motifs can have considerably different substrates. (Note that efflux systems do seem to be more strongly conserved than are influx systems [188] and are often highly promiscuous [68, 134].)

A more reliable method for both questions seeks to assess a co-variation of uptake with the loss or gain of a transporter activity. This is particularly convenient if there can be a (growth) selection step, since if the transporter is not essential for cell growth, and sufficient substrate can be added that it is toxic, cells that survive the presence of normally cytotoxic concentrations of the substrate may do so because they lack the relevant uptake transporter (e.g., [23, 157]) or have increased activity of an efflux transporter (e.g., [132, 133, 135, 136, 189–192]) or a factor controlling its expression (see below). This then allows their identification, nowadays typically by sequencing survivors directly.

In the former (uptake) case, we [157] exploited the availability [193] of a knockout collection of barcoded, haploid cell lines in baker's yeast, and assessed those exhibiting resistance to substrates at concentration capable of decreasing the wild-type growth rate by 90%. These could easily be observed in 18/26 cases, the others being considered to have multiple uptake transporters such that deleting them individually was without major effect (or any such effects were subsumed by pleiotropy [194, 195]). The parallel analyses of the barcoded mutants were mirrored by a robotic analysis of individual mutant strains.

Superti-Furga and colleagues [23] used a similar but more elegant strategy, exploiting the availability of a (nearly entirely) haploid *mammalian* cell line KBM7 [196] into whose genome a retrovirus could insert randomly, thereby knocking out a particular gene. They challenged the cells with a cytotoxic concentration of the candidate anticancer drug sepantronium bromide (YM155), noting that on 122/122 occasions the gene that was knocked out to enable cell survival was an uptake transporter known as SLC35F2. These knockouts were some 500 times more resistant than was the wild-type, and the sensitivity of different (diploid) cell lines correlated strongly and negatively (P=0.0007) with SLC35F2 expression levels (R=−0.77) as judged by transcriptome analysis.

These strategies work well when there is a screening (or better selection) step available for cells that survive a specific treatment, or can at least be differentiated from the rest, e.g., by flow cytometry (see e.g., [159, 197] and below). A more general strategy (Figure 7.3) is to expose different cell lines, in which the transporter of interest is differentially expressed (including naturally [126]), to a large cocktail of substrates, and see which substrates are differentially taken up. These may then be tested individually. A classic example is that of Gründemann and colleagues [184, 198, 199], who were interested in the function of a transporter previously named

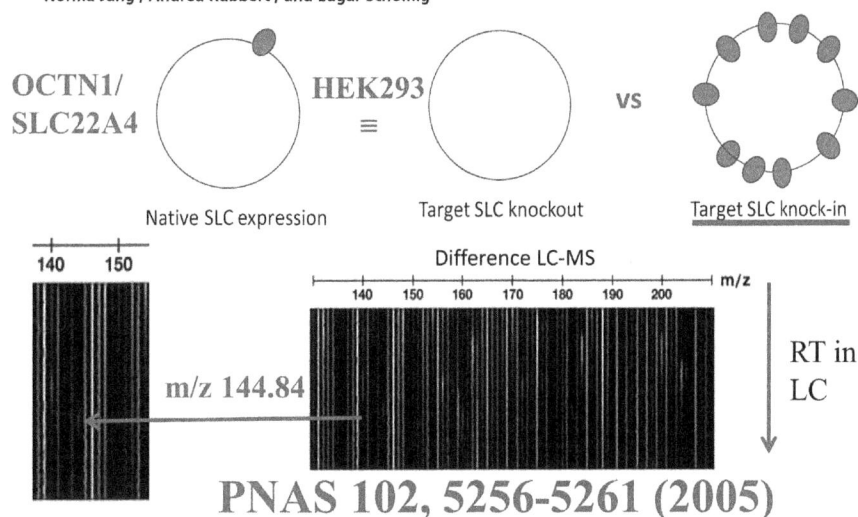

FIGURE 7.3 An untargeted metabolomics strategy for determining the substrates of "orphan" transporters whose "true" substrates are considered not known. Based on [184]

OCTN1 (organic cation transporter N1), now referred to as SLC22A4 [200]. It was at the time alleged to be a transporter of the cations carnitine and (the unnatural substrate) tetraethyl ammonium. However, Gründemann and colleagues [184] recognized that the rates of uptake of these substrates were in fact rather miserable (see also [182]). They used what would now (see e.g., [201–205]) be referred to as "untargeted metabolomics" to determine the differential uptake of substances from pooled serum when it was incubated with HEK293 cells either lacking measurable amounts, or containing cloned-up levels, of SLC22A4. A specific mass of m/z 144.84 was detected as being particularly taken up in the transporter-containing cells, and this was identified as the dipeptide proline betaine (aka stachydrine), a characteristic constituent of citrus fruits and their juices [206–208]. Gründemann and colleagues [184] then performed some elementary cheminformatics (as in [209]) to look for other molecules that were chemically (structurally) similar to stachydrine, recognizing that among them was ergothioneine (also known as 2-mercaptohistidine trimethylbetaine; IUPAC name (2S)-3-(2-Thioxo-2,3-dihydro-1H-imidazol-4-yl)-2-(trimethylammonio)propanoate) (Figure 7.4). When tested individually, ergothioneine turned out to be by far the best substrate, being taken up ~100x more efficiently than was carnitine, and being accumulated ca 180-fold (its uptake was Na^+-coupled, presumably with a stoichiometry of at least $1-2Na^+$ per ergothioneine). Although not our subject here, the biology of ergothioneine (an antioxidant) is very interesting (see e.g., [126, 199, 210–212]), with $SLC22A4^{-/-}$ cells being very sensitive to oxidative stress, but the key point is that this analysis shows the likelihood that many transporters were actually selected on the basis of their ability to take up dietary, bioactive compounds that are effectively xenobiotics [213].

Cloning large libraries of chromosome fragments [94, 214], and seeing their effects on export or productivity, or assessing genes with increased transcription in response to increased external metabolite concentrations [108, 215] provide for other methods (see e.g., [103]).

Competition assays provide yet another means of assessing which transporters take up which drugs if an uptake assay is available for at least one, known substrate. Traditionally, radio-isotopically labeled drugs would be used, but nowadays fluorescent assays (e.g., [216–220]) or mass spectrometric assays [221] are more common.

Finally, a major trend for the detection of the production or uptake of a substance, especially in microbes, is the development of screening or selection based on various types of biosensors (e.g., [12, 197, 222–236]). These powerful techniques, that give optical readouts amenable to cell sorting techniques [159], are now becoming much more common.

7.10 THE IMPORTANCE OF QSARs (QUANTITATIVE STRUCTURE-ACTIVITY RELATIONSHIPS) FOR DRUG TRANSPORTERS

Ultimately, we would like some kind of a mathematical model that could predict the rate of uptake of *any* small molecule by a particular transporter molecule (i.e., its turnover number k_{cat}), with the total maximum uptake rate per cell or organelle at saturating external concentration (V_{max}) being given by the

(A) **Ergothioneine**

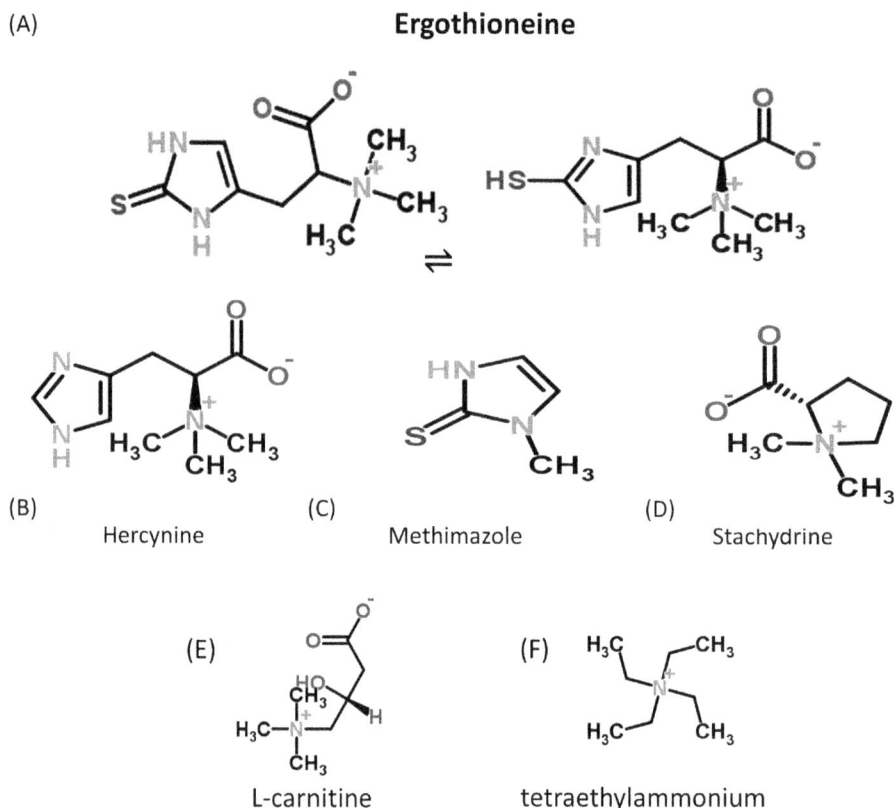

(B) Hercynine
(C) Methimazole
(D) Stachydrine

(E) L-carnitine
(F) tetraethylammonium

FIGURE 7.4 The structures of L-carnitine, ergothioneine, and some related molecules.

product of its expression level and the k_{cat}. We note, of course, that transporter kinetics are particularly complex, with strong interactions between internal and external concentrations of the various substrates and inhibitors. However, the usual practice is to perform experiments under initial rate conditions, and to assume (or determine the) Michaelis-Menten kinetics. Given at least one molecule that is a substrate, it is possible (as with the example of Gründemann and colleagues [184] mentioned above) to assess structurally related molecules, on the basis (the "molecular similarity" principle, e.g., [237]) that structurally similar molecular will tend to have similar effects.

This is not a review of cheminformatics (see e.g., [238, 239]), but, in brief, a similarity comparison is typically done as follows. The structure of the interrogating molecule, provided in the form of a SMILES [240] or InChI [241–243] string, is encoded therefrom as a string of 1 s and 0 s. Modern software suites such as RDKit [244] (www.rdkit.org/), CDK [245], or KNIME (which includes them both) [246–248] allow one to do this automatically. The same is done for the molecule(s) with which the interrogating molecule is to be compared. Although more complex comparisons are occasionally used (e.g., [249–251]), strings are typically compared on the basis of the number of bits they have in common relative to the total, a true metric (between 0 and 1) known as the Jaccard or Tanimoto similarity. Although this is a continuous function, Tanimoto similarities above 0.8 are commonly found (i) to be resistant to the precise encoding used, and (ii) indeed to have broadly similar effects in most cases (review

at [213]). An alternative encoding uses calculated molecular "descriptors" or parametrized properties [252] such as polarity, number of hydrogen bond donors, and so on (CDK includes 22, for instance [253]). Of course, fingerprints can also be combined with descriptors. Note, however, that "pure" structural similarity analyses do not take any pharmacological activities into account, and such methods are referred to as "unsupervised" learning methods (Figure 7.5), an important subset of which includes clustering methods (see e.g., [253–256]). QSAR analysis counts as a "supervised" method (Figure 7.5), in which a set of candidate drugs, that may be

Unsupervised and Supervised Learning methods

Unsupervised
- No objective function
- Typically used in assessing similarities
- Related to clustering methods
- Not susceptible to overtraining, though many algorithms are not deterministic

Supervised
- Objective (output) function used
- Trained with paired inputs and outputs
- Can (if done properly) generalise and extrapolate
- Can be susceptible to overtraining

FIGURE 7.5 Unsupervised and supervised learning methods in cheminformatics

chosen "actively" on the basis of previous knowledge, are assessed for their potency in an assay of interest, e.g., the determination of the k_{cat} of a transporter for which they may be a substrate or the K_i if they are inhibitors. This "potency" (or whatever it is that we are trying to predict) is known as the output or objective function. The result of a series of such measurements using various drugs is a set of paired values of structure (encoded as a bitstring of 1 s and 0 s as described above) and the value of the objective function. Any number of the modern methods of machine learning may then be used to construct a nonlinear mapping of these inputs onto the output; in [248] we compared three, viz a multivariate linear statistical method known as partial least squares [257], a potent general method known as random forests [258], and an evolutionary algorithm known as genetic programming (see e.g., [259–262]). Overtraining is avoided by testing the model on a subset of molecules not used in the training; if successful, the model is said to generalize, and can with reasonable confidence (but see [263–265]) be used to make predictions *in silico* on any unseen molecules (that can, of course, themselves be tested experimentally). Possibly the most extensive uptake transporter studies (e.g., [266, 267]) have been carried out on SLC15 [268] members (previously known as PEPT1 and PEPT2), and responsible e.g., for the intestinal uptake of penicillins [269] and cepaholsporins [270, 271].

As mentioned, vitamin transporters, by definition, are required by essentially all mammalian cells (and are expressed therein; see e.g., data in [126]). Another example here is provided by Ray and colleagues [272, 273], who noted that a variety of molecules, including Janus kinase (JAK) inhibitors such as fedratinib, were transported by and/or were inhibitors of two of the three main thiamine transporters (SLC19 family [274]), thereby inducing Wernicke's encephalopathy (due to thiamine deficiency). A kind of QSAR model (in this case a pharmacophore model) suggested that 2,4-diaminopyrimidine–containing compounds can in fact adopt a conformation matching several key features of thiamine. This led to the discovery that the antibiotic trimethoprim also potently inhibits thiamine uptake [272]. Metformin is also both a substrate and an inhibitor of one of the transporters [275].

In general, provided we know the QSAR of relevant transporters and their expression profiles, we can expect to be able to predict their utility in the cellular uptake or efflux of *any* substrates of interest whose structure is known and can thus be encoded in a similar way to that used in the construction of the model.

7.11 TRANSPORTER BIOENGINEERING

Armed with knowledge of an (uptake or) efflux transporter we wish to incorporate or improve in our host organism, three main non-empirical strategies for transporter engineering can be identified:

• Varied (usually increased) expression by promoter engineering

• Varied (usually increased) expression by other control factor (e.g., transcription factor) engineering
• Varied (usually increased) activity per molecule via directed evolution or synthetic biology, this also potentially including varying substrate specificities

We consider each in turn.

7.11.1 PROMOTER ENGINEERING

This is nowadays very well known and does not demand much space. Well-established inducible promoter systems include various forms of tet [276, 277], while in *E. coli*, IPTG is still widely used at small scales. Varying the promoter sequence can allow an almost continuous tuning of expression levels [278–282], even in difficult high-G+C-containing actinobacteria [283, 284]. However, the overall limit to protein expression levels in an organism does place limits on the extent to which this strategy alone is possible.

7.11.2 CONTROL FACTOR EXPRESSION ENGINEERING

As well as promoter sequences, a great many other features can determine the steady-state expression level of a target protein of interest (e.g., [285–288]). These include mRNA stability [289, 290], codon usage [291–294], riboswitches [12, 295], ribosome binding site potency [296, 297] and even the RNA polymerase itself [298] (including photoswitchable variants [299]).

By definition, transcription factors can have major effects on the expression of multiple genes and hence pathway fluxes, a particularly clear example being the role of the myb transcription factor driving phenylpropanoid and anthocyanin biosynthesis in plants [300, 301]. Transcription factor engineering rather lags behind in prokaryotes (see e.g., [197, 302, 303]), but an example of present interest is the use of marA to improve solvent tolerance (geraniol) in *E. coli* [304], as this acts, at least in part, by increasing the expression level of the enormous (770kDa) [305] and otherwise somewhat intractable (but see [306]) tolC/acrAB efflux transporter. Mutations in *marR* [307] and σ^{70} (*rpoD*) [308] can have similar effects. Of course, one cannot also fail to mention the variants of CRISPR/Cas9 gene editing technology that allow almost unlimited tinkering (e.g., in yeast [309, 310]).

While microbes even of the same genotype are purposely highly heterogeneous physiologically [311, 312], even in the same media, the biotechnologist might wish to turn on and off pathways in "all" cells at once; "quorum-sensing" methods are one means to seek to do this [169, 313].

Finally, we would also recognize the potential benefits of engineering the production or uptake (e.g., [314–318]) of compatible solutes [319–322].

7.11.3 DIRECTED EVOLUTION: CHANGING THE SEQUENCE OF THE TARGET TRANSPORTER PROTEIN(S)

Ever since the recognition of mutation itself, it has been obvious that changes in the primary sequence of a protein are more or less tightly coupled to changes in its properties [323, 324], whether they be k_{cat} [325], thermotolerance [326], substrate specificity [327], solvent tolerance [328], or any other properties.

Examples involving transporter engineering are legion [13]. A very striking one includes the cloning and engineering of an NTP transporter to allow *E coli* to take up non-natural nucleotide triphosphates, to encode "a form of life that can stably store genetic information using a six-letter, three-base-pair alphabet" [329]. As to efflux transporters, the engineering toward influx of substances normally pumped out, for instance via uncoupled variants of the LmrP "efflux" transporter in lactobacilli [330–333] is notable. Other examples have changed the specificity [334] and promiscuity [335] of transporters.

With regard to efflux transporters more generally, one sixth of the transporters of *E. coli* are efflux transporters [336]), including as many as 37 "multidrug-resistant" (MDR) transporters [337], most commonly from the major facilitator superfamily [338, 339]. Any of these are candidates for manipulation to cause the efflux of biotechnological products. Known ones include those for dipeptides [340], antibiotics [132, 133, 135, 136], and solvents [133, 341, 342]. Broadly similar statements are true in *S. cerevisiae*, where a variety of efflux pumps help remove intracellular toxicants of all kinds [343–348]. Given the importance of horizontal gene transfer in natural evolution (e.g., [74, 349–351]), it is not surprising that even interkingdom transfer of efflux genes can be effective, e.g., the use of yeast genes in causing xenobiotic tolerance in plants [352].

Table 7.2 gives some examples in which efflux has been selected by exposing cells to potentially inhibitory concentrations of the substances to be effluxed.

Consequently, when the product of interest is not close in structure to a known substrate (product) of any native transporter, it may be worth starting with a transporter for which it is; various resources allow one to seek these (e.g., [367–373]). In particular, modern methods of synthetic biology are far more powerful than are the classical methods of directed evolution, and clearly represent the future.

TABLE 7.2
Some substances for which efflux pumps have been selected via tolerance to the substances in question

Class of Substance	Selected Reference(s)
Reviews	[353]
Alkanes	[189, 354–358] (see also [28])
Arenes	[359–361]
Short-chain alcohols	[190, 306]
Short-chain fatty acids	[362, 363]
Long-chain fatty acids	[110, 364]
Terpenoids	[189, 365, 366]

7.12 SYNTHETIC BIOLOGY FOR EFFLUX TRANSPORTER ENGINEERING

Whichever of the above general strategies are chosen (promoters, other control elements, the target protein itself), improving them always involves changing the host's DNA sequence. In the past, and partly because of the enormous number of *possible* sequences [374–376] this was done rather empirically, using methods such as error-prone PCR (ePCR) [377–379] to introduce mutations. Although showing the utility of the general directed evolution strategy, this had three highly undesirable consequences: (i) there was no control over which mutations were made, (ii) the search could only be *local*, as high mutation rates necessarily introduced stop codons [379, 380], and (iii) the reliance on selection of local "winners" as starting points for the next generation *inevitably* meant that search was soon trapped in local minima from which it was impossible to escape (as was evident from many published studies showing a lack of further improvement after three or so generations, despite quite poor k_{cat} values) [324].

The *conceptual* solution, well known to those studying evolutionary and related algorithms for purposes of optimization (e.g., [324, 376, 381–385]), is that one has to combine exploitation (local search) with exploration (wider forays), and that consequently it can be helpful to know where one is in the search space (i.e. the genotype [262, 386]).

The practical experimental solution to this is to make the DNA in a statistically deterministic manner, which means synthetically [387], as part of a synthetic biology (synbio) pipeline (Figure 7.6) [324, 388]. Thus, we have described methods for the controlled generation and assembly of DNA/protein sequences [288, 294, 389–391] designed to navigate these very large search spaces "intelligently" [324]. While specificity is largely (but not at all completely) based on residues at or near the active site, we note in particular that raising k_{cat} requires contributions from residues that are farther away from the active site (e.g., [324, 325, 392, 393]). The opportunities afforded by "deep mutational scanning" [394–399], the coupled deep sequencing, and the "deep" learning [400, 401] of structure(sequence)-activity relationships [324, 402] are enormous.

A particularly nice example of the application of synbio to transporter engineering comes from Sommer and colleagues [12], who deployed and evolved riboswitches that could detect either thiamine or xanthine, thereby enabling a selection for microbes presented with metagenomes that had acquired transporters for those substrates therefrom. Presumably, a related strategy based on transcriptional events that were turned off rather than on by a riboswitch [403] could equally be applied to efflux transporters. Related strategies serve to highlight other important genes in the network of interest [404].

SUMMARY

In this comparatively short review (albeit we have tried to give many citations), the aim has been to rehearse the value

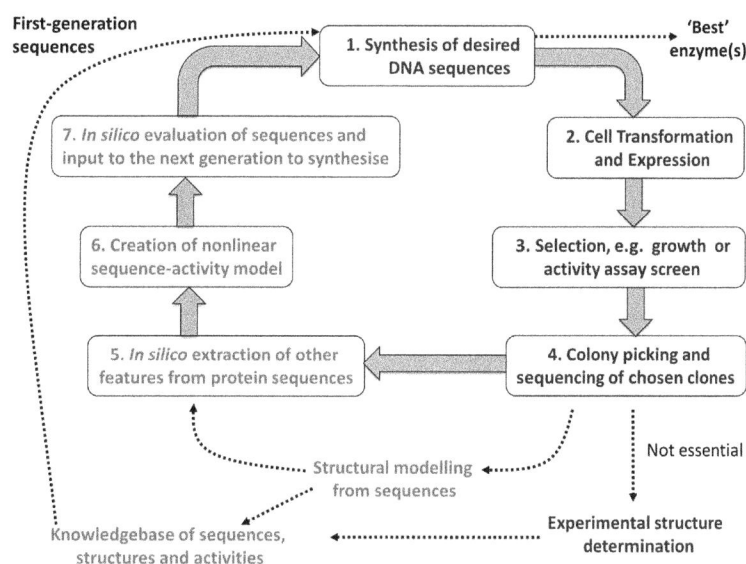

FIGURE 7.6 A generic scheme for modern synthetic biology; parts in red are computational, in **blue** experimental. Based on a Figure in [324] via a CC-BY licence

of efflux transporter engineering to biotechnology. The key points are as follows:

- Essentially nothing "floats freely" through any phospholipid bilayer that may be present in the plasma or other membrane of producer cells.
- Consequently, there is typically a kinetic restriction or barrier to effluxing product by producers of high intracellular concentrations thereof, unless a suitable transporter is, or can be arranged, to be present.
- The first step in a systems biology strategy is to make a model of the organism of interest, and discover which transporters might have the desired activity, and whether native activities can be increased or if it is necessary to add exogenous genes.
- Even equilibrative transporters (often labeled as "influx" transporters) can be useful for these purposes if their kinetics are sufficiently great.
- There can be sound evolutionary (natural selection) reasons why a cell might naturally choose to efflux expensively synthesized product; the biotechnologist is wise to make use of these where they exist
- In favorable cases, it may be possible to pump out the product of interest using an efflux transporter that is coupled to cellular sources of free energy.
- The methods of synthetic biology offer almost unlimited opportunities for efflux transporter engineering, and thereby for learning the sequence-structure-activity relationships of transporters and their substrates of interest

ACKNOWLEDGMENTS

I thank the BBSRC (grants BB/P009042/1 and BB/M017702/1) for financial support, and Irina Borodina, Behrooz Darbani, Zachary King, and Bernhard Palsson for useful discussions.

REFERENCES

1. Fletcher E, Krivoruchko A, Nielsen J. Industrial systems biology and its impact on synthetic biology of yeast cell factories. *Biotechnol Bioeng* 2016;113(6):1164–1170.
2. Jezierska S, Van Bogaert INA. Crossing boundaries: The importance of cellular membranes in industrial biotechnology. *J Ind Microbiol Biotechnol* 2017; 44 (4–5):721–733.
3. Peñaloza W, Davey CL, Hedger JN, Kell DB. Physiological studies on the solid-state quinoa tempe fermentation, using on-line measurements of fungal biomass production. *J Sci Food Agric* 1992;59:227–235.
4. Krämer R. Production of amino acids: Physiological and genetic approaches. *Food Biotechnol* 2004;18(2):171–216.
5. Buzan T. *How to Mind Map*. London: Thorsons; 2002.
6. Erni B. Group translocation of glucose and other carbohydrates by the bacterial phosphotransferase system. *Int Rev Cytol* 1992;137:127–148.
7. Gabor E, Gohler AK, Kosfeld A, Staab A, Kremling A, et al. The phosphoenolpyruvate-dependent glucose-phosphotransferase system from Escherichia coli K-12 as the center of a network regulating carbohydrate flux in the cell. *Eur J Cell Biol* 2011;90(9):711–720.
8. Kell DB, Oliver SG. How drugs get into cells: Tested and testable predictions to help discriminate between transporter-mediated uptake and lipoidal bilayer diffusion. *Front Pharmacol* 2014;5:231.
9. Ren Q, Paulsen IT. Large-scale comparative genomic analyses of cytoplasmic membrane transport systems in prokaryotes. *J Mol Microbiol Biotechnol* 2007;12 (3–4):165–179.
10. César-Razquin A, Snijder B, Frappier-Brinton T, Isserlin R, Gyimesi G, et al. A call for systematic research on solute carriers. *Cell* 2015;162(3):478–487.
11. Arai M, Okumura K, Satake M, Shimizu T. Proteome-wide functional classification and identification of prokaryotic transmembrane proteins by transmembrane topology similarity comparison. *Protein Sci* 2004;13(8):2170–2183.

12. Genee HJ, Bali AP, Petersen SD, Siedler S, Bonde MT, et al. Functional mining of transporters using synthetic selections. *Nat Chem Biol* 2016;12(12):1015–1022.

13. Kell DB, Swainston N, Pir P, Oliver SG. Membrane transporter engineering in industrial biotechnology and whole-cell biocatalysis. *Trends Biotechnol* 2015;33:237–246.

14. Dobson PD, Kell DB. Carrier-mediated cellular uptake of pharmaceutical drugs: An exception or the rule? *Nat Rev Drug Disc* 2008;7:205–220.

15. Dobson P, Lanthaler K, Oliver SG, Kell DB. Implications of the dominant role of cellular transporters in drug uptake. *Curr Top Med Chem* 2009;9:163–184.

16. Kell DB, Dobson PD, Bilsland E, Oliver SG. The promiscuous binding of pharmaceutical drugs and their transporter-mediated uptake into cells: What we (need to) know and how we can do so. *Drug Disc Today* 2013;18(5/6):218–239.

17. Kell DB. Finding novel pharmaceuticals in the systems biology era using multiple effective drug targets, phenotypic screening, and knowledge of transporters: where drug discovery went wrong and how to fix it. *FEBS J* 2013;280:5957–5980.

18. Kell DB. What would be the observable consequences if phospholipid bilayer diffusion of drugs into cells is negligible? *Trends Pharmacol Sci* 2015;36(1):15–21.

19. Kell DB. The transporter-mediated cellular uptake of pharmaceutical drugs is based on their metabolite-likeness and not on their bulk biophysical properties: Towards a systems pharmacology. *Perspect Sci* 2015;6:66–83.

20. Kell DB. How drugs pass through biological cell membranes – a paradigm shift in our understanding? *Beilstein Magazine* 2016;2(5): www.beilstein-institut.de/download/628/609_kell.pdf.

21. Jain MK. *The Bimolecular Lipid Membrane.* New York: Van Nostrand Reinhold; 1972.

22. Tien HT, Ottova-Leitmannova A (editors). *Planar Lipid Bilayers (Blms) And Their Applications.* New York:Elsevier;2003.

23. Winter GE, Radic B, Mayor-Ruiz C, Blomen VA, Trefzer C, et al. The solute carrier SLC35F2 enables YM155-mediated DNA damage toxicity. *Nat Chem Biol* 2014;10:768–773.

24. Barts PWJA, Hoeberichts JA, Klaassen A, Borst-Pauwels GWFH. Uptake of the lipophilic cation dibenzyldimethylammonium into *Saccharomyces cerevisiae.* Interaction with the thiamine transport system. *Biochim Biophys Acta* 1980;597(1):125–136.

25. Agre P. Aquaporin water channels (Nobel lecture). *Angew Chem Int Ed Engl* 2004;43(33):4278–4290.

26. Benga G. The first discovered water channel protein, later called aquaporin 1: Molecular characteristics, functions and medical implications. *Mol Aspects Med* 2012;33(5–6):518–534.

27. Call TP, Akhtar MK, Baganz F, Grant C. Modulating the import of medium-chain alkanes in *E. coli* through tuned expression of FadL. *J Biol Eng* 2016;10:5.

28. Grant C, Deszcz D, Wei YC, Martinez-Torres RJ, Morris P et al. Identification and use of an alkane transporter plug-in for applications in biocatalysis and whole-cell biosensing of alkanes. *Sci Rep* 2014;4:5844.

29. Wang J, Fulford T, Shao Q, Javelle A, Yang H, et al. Ammonium transport proteins with changes in one of the conserved pore histidines have different performance in ammonia and methylamine conduction. *PLoS One* 2013;8(5):e62745.

30. Kaldenhoff R, Kai L, Uehlein N. Aquaporins and membrane diffusion of CO_2 in living organisms. *Biochim Biophys Acta* 2014;1840:1592–1595.

31. Kai L, Kaldenhoff R. A refined model of water and CO_2 membrane diffusion: Effects and contribution of sterols and proteins. *Sci Rep* 2014;4:6665.

32. Penrod JT, Mace CC, Roth JR. A pH-sensitive function and phenotype: Evidence that EutH facilitates diffusion of uncharged ethanolamine in *Salmonella enterica. J Bacteriol*, Research Support, US Government, P.H.S. 2004;186(20):6885–6890.

33. Black PN, DiRusso CC. Transmembrane movement of exogenous long-chain fatty acids: Proteins, enzymes, and vectorial esterification. *Microbiol Mol Biol Rev* 2003;67(3):454–472.

34. van den Berg B. The FadL family: Unusual transporters for unusual substrates. *Curr Opin Struct Biol*, Review 2005;15(4):401–407.

35. Ishibashi K, Kondo S, Hara S, Morishita Y. The evolutionary aspects of aquaporin family. *Am J Physiol Regul Integr Comp Physiol* 2011;300(3):R566–576.

36. Bienert GP, Chaumont F. Aquaporin-facilitated transmembrane diffusion of hydrogen peroxide. *Biochim Biophys Acta* 2014;1840(5):1596–1604.

37. Walker AL, Franke RM, Sparreboom A, Ware RE. Transcellular movement of hydroxyurea is mediated by specific solute carrier transporters. *Exp Hematol* 2011;39(4):446–456.

38. Herrera M, Garvin JL. Aquaporins as gas channels. *Pflugers Arch* 2011;462(4):623–630.

39. Wang Y, Tajkhorshid E. Nitric oxide conduction by the brain aquaporin AQP4. *Proteins* 2010;78(3):661–670.

40. Shayakul C, Clémençon B, Hediger MA. The urea transporter family (SLC14): Physiological, pathological and structural aspects. *Mol Aspects Med* 2013;34(2–3):313–322.

41. Strugatsky D, McNulty R, Munson K, Chen CK, Soltis SM et al. Structure of the proton-gated urea channel from the gastric pathogen *Helicobacter pylori. Nature*, Research Support, N.I.H., Extramural Research Support, Non-US Government Research Support, US Government, Non-P.H.S. 2013;493(7431):255–258.

42. Day RE, Kitchen P, Owen D, Bland C, Marshall L et al. Human aquaporins: Regulators of transcellular water flow. *Biochim Biophys Acta* 2013;1840:1492–1506.

43. Palsson BØ. *Systems Biology: Properties of Reconstructed Networks.* Cambridge: Cambridge University Press; 2006.

44. Kell DB, Knowles JD. The role of modeling in systems biology. In: Szallasi Z, Stelling J, Periwal V (editors). *System modeling in cellular biology: From concepts to nuts and bolts.* Cambridge: MIT Press; 2006. pp. 3–18.

45. Kell DB. Metabolomics, modelling and machine learning in systems biology: Towards an understanding of the languages of cells. The 2005 Theodor Bücher lecture. *FEBS J* 2006;273:873–894.

46. Kanehisa M, Goto S, Sato Y, Kawashima M, Furumichi M, et al. Data, information, knowledge and principle: back to metabolism in KEGG. *Nucleic Acids Res* 2014;42(1):D199–205.

47. Herrgård MJ, Swainston N, Dobson P, Dunn WB, Arga KY, et al. A consensus yeast metabolic network obtained from a community approach to systems biology. *Nat Biotechnol* 2008;26(10):1155–1160.

48. Thiele I, Palsson BØ. Reconstruction annotation jamborees: A community approach to systems biology. *Mol Syst Biol* 2010;6:361.

49. Thiele I, Swainston N, Fleming RMT, Hoppe A, Sahoo S, et al. A community-driven global reconstruction of human metabolism. *Nat Biotechnol* 2013;31(5):419–425.

50. Swainston N, Smallbone K, Mendes P, Kell DB, Paton NW. The SuBliMinaL Toolbox: Automating steps in the reconstruction of metabolic networks. *Integrative Bioinf* 2011;8(2):186.

51. Carbonell P, Parutto P, Baudier C, Junot C, Faulon JL. Retropath: Automated Pipeline for Embedded Metabolic Circuits. *Acs Synth Biol* 2014;3(8):565–577.

52. Carbonell P, Parutto P, Herisson J, Pandit SB, Faulon JL. XTMS: Pathway design in an eXTended metabolic space. *Nucleic Acids Res* 2014;42:W389–W394.

53. Smallbone K, Messiha HL, Carroll KM, Winder CL, Malys N, et al. A model of yeast glycolysis based on a consistent kinetic characterization of all its enzymes. *FEBS Lett* 2013;587:2832–2841.

54. Bergmann FT, Hoops S, Klahn B, Kummer U, Mendes P, et al. COPASI and its applications in biotechnology. *J Biotechnol* 2017;261:215–220.

55. Hoops S, Sahle S, Gauges R, Lee C, Pahle J, et al. COPASI: A complex pathway simulator. *Bioinformatics* 2006;22(24):3067–3074.

56. Smallbone K, Simeonidis E, Broomhead DS, Kell DB. Something from nothing: bridging the gap between constraint-based and kinetic modelling. *FEBS J* 2007;274:5576–5585.

57. Liebermeister W, Klipp E. Bringing metabolic networks to life: Convenience rate law and thermodynamic constraints. *Theor Biol Med Model* 2006;3:41.

58. Hanson AD, Pribat A, Waller JC, de Crécy-Lagard V. 'Unknown' proteins and 'orphan' enzymes: the missing half of the engineering parts list--and how to find it. *Biochem J*, Research Support, N.I.H., Extramural Research Support, Non-US Government Research Support, US Government, Non-P.H.S. Review 2010;425(1):1–11.

59. Riley M, Abe T, Arnaud MB, Berlyn MK, Blattner FR et al. *Escherichia coli* K-12: A cooperatively developed annotation snapshot--2005. *Nucleic Acids Res* 2006;34(1):1–9.

60. Hu P, Janga SC, Babu M, Diaz-Mejia JJ, Butland G et al. Global functional atlas of *Escherichia coli* encompassing previously uncharacterized proteins. *PLoS Biol* 2009;7(4):e96.

61. Ito M, Baba T, Mori H, Mori H. Functional analysis of 1440 *Escherichia coli* genes using the combination of knock-out library and phenotype microarrays. *Metab Eng* 2005;7(4):318–327.

62. Keseler IM, Mackie A, Santos-Zavaleta A, Billington R, Bonavides-Martínez C et al. The EcoCyc database: Reflecting new knowledge about *Escherichia coli* K-12. *Nucleic Acids Res* 2017;45(D1):D543–D550.

63. Paulsen IT, Sliwinski MK, Nelissen B, Goffeau A, Saier MH, Jr. Unified inventory of established and putative transporters encoded within the complete genome of *Saccharomyces cerevisiae*. *FEBS Lett* 1998;430 (1–2):116–125.

64. Elbourne LDH, Tetu SG, Hassan KA, Paulsen IT. TransportDB 2.0: A database for exploring membrane transporters in sequenced genomes from all domains of life. *Nucleic Acids Res* 2017;45(D1):D320–D324.

65. Del Sorbo G, Schoonbeek HJ, De Waard MA. Fungal transporters involved in efflux of natural toxic compounds and fungicides. *Fungal Genetics and Biology* 2000;30(1):1–15.

66. Perlin MH, Andrews J, Toh SS. Essential letters in the fungal alphabet: Abc and mfs transporters and their roles in survival and pathogenicity. *Advances in Genetics* 2014;85:201–253.

67. Prasad R, Banerjee A, Khandelwal NK, Dhamgaye S. The ABCs of *Candida albicans* Multidrug Transporter Cdr1. *Eukaryot Cell* 2015;14(12):1154–1164.

68. Wong K, Ma J, Rothnie A, Biggin PC, Kerr ID. Towards understanding promiscuity in multidrug efflux pumps. *Trends Biochem Sci* 2014;39(1):8–16.

69. Lamping E, Baret PV, Holmes AR, Monk BC, Goffeau A, et al. Fungal PDR transporters: Phylogeny, topology, motifs and function. *Fungal Genetics and Biology* 2010;47(2):127–142.

70. Diallinas G. Dissection of transporter function: From genetics to structure. *Trends Genet* 2016;32(9):576–590.

71. Kacser H, Burns JA. The control of flux. In: Davies DD (editor). *Rate Control of Biological Processes Symposium Of The Society For Experimental Biology Vol 27*. Cambridge: Cambridge University Press; 1973. pp. 65–104.

72. Heinrich R, Rapoport TA. A linear steady-state treatment of enzymatic chains. General properties, control and effector strength. *Eur J Biochem* 1974;42:89–95.

73. Kell DB, Westerhoff HV. Metabolic control theory: its role in microbiology and biotechnology. *FEMS Microbiol Rev* 1986;39:305–320.

74. Kell DB, van Dam K, Westerhoff HV. Control analysis of microbial growth and productivity. *Symp Soc Gen Microbiol* 1989;44:61–93.

75. Heinrich R, Schuster S. *The Regulation of Cellular Systems*. New York: Chapman & Hall; 1996.

76. Fell DA. *Understanding the Control Of Metabolism*. London: Portland Press; 1996.

77. Walter RP, Morris JG, Kell DB. The roles of osmotic stress and water activity in the inhibition of the growth, glycolysis and glucose phosphotransferase system of *Clostridium pasteurianum*. *J Gen Microbiol* 1987;133:259–266.

78. Sano C. History of glutamate production. *Am J Clin Nutr*, Historical Article Review 2009;90(3):728S–732S.

79. Eggeling L, Sahm H. New ubiquitous translocators: Amino acid export by *Corynebacterium glutamicum* and *Escherichia coli*. *Arch Microbiol* 2003;180(3):155–160.

80. Tryfona T, Bustard MT. Mechanistic understanding of the fermentative L-glutamic acid overproduction by *Corynebacterium glutamicum* through combined metabolic flux profiling and transmembrane transport characteristics. *J Chem Technol Biotechnol* 2004;79(12):1321–1330.

81. Kinoshita S, Udaka S, Shimamoto M. Studies on amino acid fermentation, part I. Production of L-glutamic acid by various microorganisms. *J Gen Appl Microbiol* 1957;3:193–205.

82. Hirasawa T, Kim J, Shirai T, Furusawa C, Shimizu H. Molecular Mechanisms and Metabolic Engineering of Glutamate Overproduction in *Corynebacterium glutamicum*. *Subcellular Biochem* 2012;64:261–281.

83. Vertès AA, Inui M, Yukawa H. The biotechnological potential of *Corynebacterium glutamicum*, from Umami to Chemurgy. *Microbiol Monogr* 2013;23:1–49.

84. Nakamura J, Hirano S, Ito H, Wachi M. Mutations of the *Corynebacterium glutamicum* NCgl1221 gene, encoding a mechanosensitive channel homolog, induce L-glutamic acid production. *Appl Environ Microbiol* 2007;73(14): 4491–4498.

85. Hashimoto K, Nakamura K, Kuroda T, Yabe I, Nakamatsu T, et al. The protein encoded by NCgl1221 in *Corynebacterium glutamicum* functions as a mechanosensitive channel. *Biosci Biotechnol Biochem* 2010;74(12):2546–2549.

86. Hashimoto K, Murata J, Konishi T, Yabe I, Nakamatsu T, et al. Glutamate is excreted across the cytoplasmic membrane through the NCgl1221 channel of *Corynebacterium glutamicum* by passive diffusion. *Biosci Biotechnol Biochem* 2012;76(7):1422–1424.

87. Nakayama Y, Yoshimura K, Iida H. A gain-of-function mutation in gating of *Corynebacterium glutamicum* NCgl1221 causes constitutive glutamate secretion. *Appl Environ Microbiol* 2012;78(15):5432–5434.

88. Yamashita C, Hashimoto K, Kumagai K, Maeda T, Takada A, et al. L-Glutamate secretion by the N-terminal domain of the *Corynebacterium glutamicum* NCgl1221 mechanosensitive channel. *Biosci Biotechnol Biochem* 2013;77(5): 1008–1013.

89. Yao W, Deng X, Liu M, Zheng P, Sun Z, et al. Expression and localization of the *Corynebacterium glutamicum* NCgl1221 protein encoding an L-glutamic acid exporter. *Microbiol Res* 2009;164(6):680–687.

90. Mitsuhashi S. Current topics in the biotechnological production of essential amino acids, functional amino acids, and dipeptides. *Curr Opin Biotechnol* 2014;26:38–44.

91. Bröer S, Krämer R. Lysine excretion by *Corynebacterium glutamicum*. 1. Identification of a specific secretion carrier system. *Eur J Biochem*, Research Support, Non-US Government 1991;202(1):131–135.

92. Bröer S, Krämer R. Lysine excretion by *Corynebacterium glutamicum*. 2. Energetics and mechanism of the transport system. *Eur J Biochem*, Research Support, Non-US Government 1991;202(1):137–143.

93. Kelle R, Laufer B, Brunzema C, Weuster-Botz D, Krämer R, et al. Reaction engineering analysis of L-lysine transport by *Corynebacterium glutamicum*. *Biotechnol Bioeng* 1996;51(1):40–50.

94. Vrljic M, Sahm H, Eggeling L. A new type of transporter with a new type of cellular function: L-lysine export from *Corynebacterium glutamicum*. *Mol Microbiol* 1996;22(5):815–826.

95. Bellmann A, Vrljić M, Pátek M, Sahm H, Krämer R, et al. Expression control and specificity of the basic amino acid exporter LysE of *Corynebacterium glutamicum*. *Microbiology*, Research Support, Non-US Government 2001;147(Pt 7):1765–1774.

96. Hermann T, Krämer R. Mechanism and Regulation of isoleucine excretion in *Corynebacterium glutamicum*. *Appl Environ Microbiol* 1996;62(9):3238–3244.

97. Xie X, Xu L, Shi J, Xu Q, Chen N. Effect of transport proteins on L-isoleucine production with the L-isoleucine-producing strain *Corynebacterium glutamicum* YILW. *J Ind Microbiol Biotechnol* 2012;39(10):1549–1556.

98. Mundhada H, Seoane JM, Schneider K, Koza A, Christensen HB, et al. Increased production of L-serine in *Escherichia coli* through adaptive laboratory evolution. *Metab Eng* 2017;39:141–150.

99. Lee KH, Park JH, Kim TY, Kim HU, Lee SY. Systems metabolic engineering of *Escherichia coli* for L-threonine production. *Mol Syst Biol* 2007;3:149.

100. Diesveld R, Tietze N, Fürst O, Reth A, Bathe B, et al. Activity of exporters of *Escherichia coli* in *Corynebacterium glutamicum*, and their use to increase L-threonine production. *J Mol Microbiol Biotechnol* 2009;16(3–4):198–207.

101. Dong X, Quinn PJ, Wang X. Metabolic engineering of Escherichia coli and Corynebacterium glutamicum for the production of L-threonine. *Biotechnol Adv* 2011;29(1):11–23.

102. Dong X, Quinn PJ, Wang X. Microbial metabolic engineering for L-threonine production. *Subcell Biochem* 2012;64:283–302.

103. Eggeling L. Exporters for production of amino acids and other small molecules. *Adv Biochem Eng Biotechnol* 2017;159:199–225.

104. Trötschel C, Deutenberg D, Bathe B, Burkovski A, Krämer R. Characterization of methionine export in *Corynebacterium glutamicum*. *J Bacteriol* 2005;187(11):3786–3794.

105. Hori H, Yoneyama H, Tobe R, Ando T, Isogai E, et al. Inducible L-alanine exporter encoded by the novel gene ygaW (alaE) in *Escherichia coli*. *Appl Environ Microbiol* 2011;77(12):4027–4034.

106. Marin K, Krämer R. Amino acid transport systems in biotechnologically relevant bacteria. *Microbiol Monogr* 2007;5:289–325.

107. Van Dyk TK. Bacterial efflux transport in biotechnology. *Adv Appl Microbiol* 2008;63:231–247.

108. Kind S, Kreye S, Wittmann C. Metabolic engineering of cellular transport for overproduction of the platform chemical 1,5-diaminopentane in *Corynebacterium glutamicum*. *Metab Eng* 2011;13(5):617–627.

109. Anderson CM, Stahl A. SLC27 fatty acid transport proteins. *Mol Aspects Med* 2013;34(2–3):516–528.

110. Villalba MS, Alvarez HM. Identification of a novel ATP-binding cassette transporter involved in long-chain fatty acid import and its role in triacylglycerol accumulation in *Rhodococcus jostii* RHA1. *Microbiology* 2014;160(Pt 7):1523–1532.

111. Dirusso CC, Black PN. Bacterial long chain fatty acid transport: Gateway to a fatty acid-responsive signaling system. *J Biol Chem* 2004;279(48):49563–49566.

112. van den Berg B, Black PN, Clemons WM, Jr., Rapoport TA. Crystal structure of the long-chain fatty acid transporter FadL. *Science* 2004;304(5676):1506–1509.

113. Færgeman NJ, DiRusso CC, Elberger A, Knudsen J, Black PN. Disruption of the *Saccharomyces cerevisiae* homologue to the murine fatty acid transport protein impairs uptake and growth on long-chain fatty acids. *J Biol Chem* 1997;272(13):8531–8538.

114. Zou Z, DiRusso CC, Ctrnacta V, Black PN. Fatty acid transport in *Saccharomyces cerevisiae*. Directed mutagenesis of FAT1 distinguishes the biochemical activities associated with Fat1p. *J Biol Chem* 2002;277(34):31062–31071.

115. Lorenz S, Guenther M, Grumaz C, Rupp S, Zibek S, et al. Genome sequence of the Basidiomycetous fungus *Pseudozyma aphidis* DSM70725, an efficient producer of biosurfactant Mannosylerythritol Lipids. *Genome Announcements* 2014;2(1).

116. Dulermo R, Gamboa-Melendez H, Dulermo T, Thevenieau F, Nicaud JM. The fatty acid transport protein Fat1p is involved in the export of fatty acids from lipid bodies in Yarrowia lipolytica. *FEMS Yeast Res* 2014;14(6):883–896.

117. Dulermo R, Gamboa-Melendez H, Ledesma-Amaro R, Thevenieau F, Nicaud JM. Unraveling fatty acid transport and activation mechanisms in Yarrowia lipolytica. *Biochim Biophys Acta* 2015;1851(9):1202–1217.

118. Ledesma-Amaro R, Kerkhoven EJ, Revuelta JL, Nielsen J. Genome scale metabolic modeling of the riboflavin overproducer *Ashbya gossypii*. *Biotechnol Bioeng* 2014;111(6):1191–1199.

119. Ledesma-Amaro R, Dulermo R, Niehus X, Nicaud JM. Combining metabolic engineering and process optimization to improve production and secretion of fatty acids. *Metab Eng* 2016;38:38–46.

120. Schwechheimer SK, Park EY, Revuelta JL, Becker J, Wittmann C. Biotechnology of riboflavin. *Appl Microbiol Biotechnol* 2016;100(5):2107–2119.

121. Revuelta JL, Ledesma-Amaro R, Lozano-Martínez P, Díaz-Fernández D, Buey RM, et al. Bioproduction of riboflavin: A bright yellow history. *J Ind Microbiol Biotechnol* 2017; 44(4–5): 659–665.

122. Förster C, Revuelta JL, Krämer R. Carrier-mediated transport of riboflavin in *Ashbya gossypii*. *Appl Microbiol Biotechnol* 2001;55(1):85–89.

123. Förster C, Santos MA, Ruffert S, Krämer R, Revuelta JL. Physiological consequence of disruption of the VMA1 gene in the riboflavin overproducer *Ashbya gossypii*. *J Biol Chem* 1999;274(14):9442–9448.

124. Hemberger S, Pedrolli DB, Stolz J, Vogl C, Lehmann M et al. RibM from *Streptomyces davawensis* is a riboflavin/roseoflavin transporter and may be useful for the optimization of riboflavin production strains. *BMC Biotechnol* 2011;11:119.

125. Udhayabanu T, Subramanian VS, Teafatiller T, Gowda VK, Raghavan VS, et al. SLC52A2 [p.P141T] and SLC52A3 [p.N21S] causing Brown-Vialetto-Van Laere Syndrome in an Indian patient: First genetically proven case with mutations in two riboflavin transporters. *Clin Chim Acta* 2016;462:210–214.

126. O'Hagan S, Wright Muelas M, Day PJ, Lundberg E, Kell DB. GeneGini: Assessment via the gini coefficient of reference "housekeeping" genes and diverse human transporter expression profiles *Cell Syst* 2018;6(2):230–244.

127. Yonezawa A, Inui K. Novel riboflavin transporter family RFVT/SLC52: identification, nomenclature, functional characterization and genetic diseases of RFVT/SLC52. *Mol Aspects Med* 2013;34(2–3):693–701.

128. Montanari F, Ecker GF. Prediction of drug-ABC transporter interaction—recent advances and future challenges. *Adv Drug Deliv Rev* 2015;86:17–26.

129. Zinzi L, Contino M, Cantore M, Capparelli E, Leopoldo M, et al. ABC transporters in CSCs membranes as a novel target for treating tumor relapse. *Front Pharmacol* 2014;5:163.

130. Bugde P, Biswas R, Merien F, Lu J, Liu DX, et al. The therapeutic potential of targeting ABC transporters to combat multidrug resistance. *Expert Opin Ther Targets* 2017;21(5):511–530.

131. Deparis Q, Claes A, Foulquié-Moreno MR, Thevelein JM. Engineering tolerance to industrially relevant stress factors in yeast cell factories. *FEMS Yeast Res* 2017;17(4).

132. Blair JM, Richmond GE, Piddock LJV. Multidrug efflux pumps in Gram-negative bacteria and their role in antibiotic resistance. *Future Microbiol* 2014;9(10):1165–1177.

133. Fernandes P, Ferreira BS, Cabral JM. Solvent tolerance in bacteria: role of efflux pumps and cross-resistance with antibiotics. *Int J Antimicrob Agents* 2003;22(3):211–216.

134. Lewinson O, Adler J, Sigal N, Bibi E. Promiscuity in multidrug recognition and transport: the bacterial MFS Mdr transporters. *Mol Microbiol*, Research Support, Non-US Government Review 2006;61(2):277–284.

135. Li XZ, Plésiat P, Nikaido H. The challenge of efflux-mediated antibiotic resistance in Gram-negative bacteria. *Clin Microbiol Rev* 2015;28(2):337–418.

136. Maira-Litrán T, Allison DG, Gilbert P. An evaluation of the potential of the multiple antibiotic resistance operon (*mar*) and the multidrug efflux pump acrAB to moderate resistance towards ciprofloxacin in *Escherichia coli* biofilms. *J Antimicrob Chemother* 2000;45(6):789–795.

137. Pence MA, McElvania TeKippe E, Burnham CA. Diagnostic assays for identification of microorganisms and antimicrobial resistance determinants directly from positive blood culture broth. *Clinics In Laboratory Medicine* 2013;33(3):651–684.

138. Rahman T, Yarnall B, Doyle DA. Efflux drug transporters at the forefront of antimicrobial resistance. *Eur Biophys J* 2017;46(7):647–653.

139. Saidijam M, Benedetti G, Ren Q, Xu Z, Hoyle CJ, et al. Microbial drug efflux proteins of the major facilitator superfamily. *Curr Drug Targets* 2006;7(7):793–811.

140. *Fundamentals Of Enzyme Kinetics*, 2nd ed., (1995).

141. Kimura M. *The Neutral Theory of Molecular Evolution*. Cambridge: Cambridge University Press; 1983.

142. Kell A, Glaser RW. On the mechanical and dynamic properties of plant cell membranes: Their role in growth, direct gene transfer and protoplast fusion. *J Theor Biol* 1993;160(1):41–62.

143. Ouyang Y, Li X. Recent research progress on soil microbial responses to drying–rewetting cycles. *Acta Ecol Sin* 2013;33:1–6.

144. Morbach S, Krämer R. Body shaping under water stress: Osmosensing and osmoregulation of solute transport in bacteria. *ChemBioChem* 2002;3(5):384–397.

145. Soccol CR, Vandenberghe LPS, Rodrigues C, Pandey A. New perspectives for citric acid production and application. *Food Technol Biotechnol* 2006;44(2):141–149.

146. García J, Torres N. Mathematical modelling and assessment of the pH homeostasis mechanisms in *Aspergillus niger* while in citric acid producing conditions. *J Theoret Biol* 2011;282(1):23–35.

147. Franz A, Burgstaller W, Schinner F. Leaching with *Penicillium simplicissimum*: Influence of metals and buffers on proton extrusion and citric acid production. *Appl Environ Microbiol* 1991;57(3):769–774.

148. Jones DL. Organic acids in the rhizosphere - a critical review. *Plant and Soil* 1998;205(1):25–44.

149. Durrett TP, Gassmann W, Rogers EE. The FRD3-mediated efflux of citrate into the root vasculature is necessary for efficient iron translocation. *Plant Physiol* 2007;144(1):197–205.

150. Burgstaller W. Thermodynamic boundary conditions suggest that a passive transport step suffices for citrate excretion in Aspergillus and Penicillium. *Microbiology* 2006;152(Pt 3):887–893.

151. Kell DB. Iron behaving badly: Inappropriate iron chelation as a major contributor to the aetiology of vascular and other progressive inflammatory and degenerative diseases. *BMC Med Genom* 2009;2:2.

152. De Serrano LO. Biotechnology of siderophores in high-impact scientific fields. *Biomolecular concepts* 2017;8(3–4):169–178.

153. Saha M, Sarkar S, Sarkar B, Sharma BK, Bhattacharjee S et al. Microbial siderophores and their potential applications: A review. *Environmental science and pollution research international* 2016;23(5):3984–3999.

154. Wilson BR, Bogdan AR, Miyazawa M, Hashimoto K, Tsuji Y. Siderophores in iron metabolism: From mechanism to therapy potential. *Trends Mol Med* 2016;22(12):1077–1090.

155. Chen J, Zhu X, Tan Z, Xu H, Tang J, et al. Activating C_4-dicarboxylate transporters DcuB and DcuC for improving succinate production. *Appl Microbiol Biotechnol* 2014;98(5):2197–2205.

156. Li XZ, Zhang WM, Wu MK, Xin FX, Dong WL, et al. Performance and mechanism analysis of succinate production under different transporters in *Escherichia coli*. *Biotechnol Bioproc Eng* 2017;22(5):529–538.

157. Lanthaler K, Bilsland E, Dobson P, Moss HJ, Pir P, et al. Genome-wide assessment of the carriers involved in the cellular uptake of drugs: A model system in yeast. *BMC Biol* 2011;9:70.

158. Schroeder WA, Johnson EA. Carotenoids protect *Phaffia rhodozyma* against singlet oxygen damage. *J Ind Microbiol* 1995;14(6):502–507.

159. Davey HM, Kell DB. Flow cytometry and cell sorting of heterogeneous microbial populations: The importance of single-cell analysis. *Microbiol Rev* 1996;60:641–696.

160. Stephens K. Pheromones among the prokaryotes. *CRC Crit Rev Microbiol* 1986;13:309–334.

161. Bu'lock JD. Intermediary metabolism and antibiotic synthesis. *Adv Microbial Physiol* 1961;3:293–333.

162. Kell DB, Kaprelyants AS, Grafen A. On pheromones, social behaviour and the functions of secondary metabolism in bacteria. *Trends Ecol Evolution* 1995;10:126–129.

163. Goh EB, Yim G, Tsui W, McClure J, Surette MG, et al. Transcriptional modulation of bacterial gene expression by subinhibitory concentrations of antibiotics. *Proc Natl Acad Sci U S A* 2002;99(26):17025–17030.

164. Raaijmakers JM, Mazzola M. Diversity and natural functions of antibiotics produced by beneficial and plant pathogenic bacteria. *Annu Rev Phytopathol* 2012;50:403–424.

165. Yim G, Wang HH, Davies J. The truth about antibiotics. *Int J Med Microbiol* 2006; 296 (2–3): 163–170.

166. Peiris D, Dunn WB, Brown M, Kell DB, Roy I et al. Metabolite profiles of interacting mycelial fronts differ for pairings of the wood decay basidiomycete fungus, *Stereum hirsutum* with its competitors *Coprinus micaceus* and *Coprinus disseminatus*. *Metabolomics* 2008;4(1):52–62.

167. Fuqua WC, Winans SC, Greenberg EP. Quorum sensing in bacteria - the *luxR-luxI* family of cell density-responsive transcriptional regulators. *J Bacteriol* 1994;176:269–275.

168. Dandekar AA, Chugani S, Greenberg EP. Bacterial quorum sensing and metabolic incentives to cooperate. *Science*, Research Support, N.I.H., Extramural Research Support, Non-US Government 2012;338(6104):264–266.

169. Whiteley M, Diggle SP, Greenberg EP. Progress in and promise of bacterial quorum sensing research. *Nature* 2017;551(7680):313–320.

170. Choudhary S, Schmidt-Dannert C. Applications of quorum sensing in biotechnology. *Appl Microbiol Biotechnol* 2010;86(5):1267–1279.

171. Mangwani N, Dash HR, Chauhan A, Das S. Bacterial quorum sensing: Functional features and potential applications in biotechnology. *J Mol Microbiol Biotechnol* 2012;22(4):215–227.

172. Crittenden PD, Porter N. Lichen-forming fungi - potential sources of novel metabolites. *Trends Biotechnol* 1991;9:409–414.

173. Lutzoni F, Pagel M, Reeb V. Major fungal lineages are derived from lichen symbiotic ancestors. *Nature* 2001;411(6840): 937–940.

174. Senior E, Bull AT, Slater JH. Enzyme evolution in a microbial community growing on the herbicide Dalapon. *Nature* 1976;263(5577):476–479.

175. Xiao Y, Angulo MT, Friedman J, Waldor MK, Weiss ST, et al. Mapping the ecological networks of microbial communities. *Nat Commun* 2017;8(1):2042.

176. McInerney MJ, Sieber JR, Gunsalus RP. Syntrophy in anaerobic global carbon cycles. *Curr Opin Biotechnol* 2009;20(6): 623–632.

177. Sieber JR, McInerney MJ, Gunsalus RP. Genomic insights into syntrophy: the paradigm for anaerobic metabolic cooperation. *Annu Rev Microbiol* 2012;66:429–452.

178. Rotaru AE, Woodard TL, Nevin KP, Lovley DR. Link between capacity for current production and syntrophic growth in *Geobacter* species. *Front Microbiol* 2015; 6:744.

179. Westerhoff HV, Hellingwerf KJ, van Dam K. Thermodynamic efficiency of microbial growth is low but optimal for maximal growth rate. *Proc Natl Acad Sci USA* 1983;80(1): 305–309.

180. Markx GH, Davey CL, Kell DB. The permittistat: a novel type of turbidostat. *J Gen Microbiol* 1991;137:735–743.

181. Davey HM, Davey CL, Woodward AM, Edmonds AN, Lee AW et al. Oscillatory, stochastic and chaotic growth rate fluctuations in permittistatically-controlled yeast cultures. *Biosystems* 1996;39:43–61.

182. O'Hagan S, Kell DB. The apparent permeabilities of Caco-2 cells to marketed drugs: Magnitude, and independence from both biophysical properties and endogenite similarities *PeerJ* 2015;3:e1405.

183. Gründemann D, Liebich G, Kiefer N, Köster S, Schömig E. Selective substrates for non-neuronal monoamine transporters. *Mol Pharmacol* 1999;56(1):1–10.

184. Gründemann D, Harlfinger S, Golz S, Geerts A, Lazar A, et al. Discovery of the ergothioneine transporter. *Proc Natl Acad Sci* 2005;102(14):5256–5261.

185. Fersht A. *Structure And Mechanism In Protein Science: A Guide To Enzyme Catalysis And Protein Folding*. San Francisco: W.H. Freeman; 1999.

186. Mendes P, Oliver SG, Kell DB. Fitting transporter activities to cellular drug concentrations and fluxes: Why the bumblebee can fly. *Trends Pharmacol Sci* 2015;36:710–723.

187. Höglund PJ, Nordström KJV, Schiöth HB, Fredriksson R. The solute carrier families have a remarkably long evolutionary history with the majority of the human families present before divergence of Bilaterian species. *Mol Biol Evol*, Research Support, Non-US Government 2011;28(4):1531–1541.

188. Hassan KA, Fagerlund A, Elbourne LDH, Vörös A, Kroeger JK et al. The putative drug efflux systems of the *Bacillus cereus* group. *PLoS One* 2017;12(5):e0176188.

189. Foo JL, Leong SSJ. Directed evolution of an *E. coli* inner membrane transporter for improved efflux of biofuel molecules. *Biotechnol Biofuels* 2013;6:81.

190. Foo JL, Jensen HM, Dahl RH, George K, Keasling JD et al. Improving microbial biogasoline production in *Escherichia coli* using tolerance engineering. *MBio* 2014;5(6):01932–01914.

191. Ramos JL, Sol Cuenca M, Molina-Santiago C, Segura A, Duque E et al. Mechanisms of solvent resistance mediated by interplay of cellular factors in *Pseudomonas putida*. *FEMS Microbiol Rev* 2015;39(4):555–566.

192. Mohamed ET, Wang S, Lennen RM, Herrgard MJ, Simmons BA et al. Generation of a platform strain for ionic liquid tolerance using adaptive laboratory evolution. *Microb Cell Fact* 2017;16(1):204.

193. Giaever G, Chu AM, Ni L, Connelly C, Riles L et al. Functional profiling of the *Saccharomyces cerevisiae* genome. *Nature* 2002;418(6896):387–391.

194. Featherstone DE, Broadie K. Wrestling with pleiotropy: genomic and topological analysis of the yeast gene expression network. *Bioessays* 2002;24(3):267–274.

195. Eraly SA. Striking differences between knockout and wild-type mice in global gene expression variability. *PLoS One* 2014;9(5):e97734.

196. Bürckstümmer T, Banning C, Hainzl P, Schobesberger R, Kerzendorfer C, et al. A reversible gene trap collection empowers haploid genetics in human cells. *Nat Methods* 2013;10(10):965–971.

197. Skjoedt ML, Snoek T, Kildegaard KR, Arsovska D, Eichenberger M, et al. Engineering prokaryotic transcriptional activators as metabolite biosensors in yeast. *Nat Chem Biol* 2016;12(11):951–958.

198. Gründemann D. The ergothioneine transporter controls and indicates ergothioneine activity--a review. *Prev Med* 2012;54 Suppl:S71–74.

199. Kerley RN, McCarthy C, Kell DB, Kenny LC. The potential therapeutic effects of ergothioneine in pre-eclampsia. *Free Radic Biol Med* 2018;117:145–157.

200. Koepsell H. The SLC22 family with transporters of organic cations, anions and zwitterions. *Mol Aspects Med* 2013;34(2–3):413–435.

201. Dunn WB, Broadhurst D, Begley P, Zelena E, Francis-McIntyre S et al. Procedures for large-scale metabolic profiling of serum and plasma using gas chromatography and liquid chromatography coupled to mass spectrometry. *Nat Protoc* 2011;6(7):1060–1083.

202. Cho K, Mahieu NG, Johnson SL, Patti GJ. After the feature presentation: Technologies bridging untargeted metabolomics and biology. *Curr Opin Biotechnol* 2014;28:143–148.

203. Dunn WB, Erban A, Weber RJM, Creek DJ, Brown M et al. Mass Appeal: Metabolite identification in mass spectrometry-focused untargeted metabolomics.. *Metabolomics* 2013;9:S44–S66.

204. Martin JC, Maillot M, Mazerolles G, Verdu A, Lyan B, et al. Can we trust untargeted metabolomics? Results of the metaboring initiative, a large-scale, multi-instrument inter-laboratory study. *Metabolomics* 2015;11(4):807–821.

205. Di Guida R, Engel J, Allwood JW, Weber RJ, Jones MR, et al. Non-targeted UHPLC-MS metabolomic data processing methods: a comparative investigation of normalisation, missing value imputation, transformation and scaling. *Metabolomics* 2016;12:93.

206. Lloyd AJ, Beckmann M, Favé G, Mathers JC, Draper J. Proline betaine and its biotransformation products in fasting urine samples are potential biomarkers of habitual citrus fruit consumption. *Br J Nutr* 2011;106(6):812–824.

207. Heinzmann SS, Brown IJ, Chan Q, Bictash M, Dumas ME, et al. Metabolic profiling strategy for discovery of nutritional biomarkers: Proline betaine as a marker of citrus consumption. *Am J Clin Nutr* 2010;92(2):436–443.

208. Lang R, Lang T, Bader M, Beusch A, Schlagbauer V, et al. High-throughput quantitation of proline betaine in foods and suitability as a valid biomarker for citrus consumption. *J Agric Food Chem* 2017;65(8):1613–1619.

209. O'Hagan S, Swainston N, Handl J, Kell DB. A 'rule of 0.5' for the metabolite-likeness of approved pharmaceutical drugs. *Metabolomics* 2015;11(2):323–339.

210. Cheah IK, Halliwell B. Ergothioneine; antioxidant potential, physiological function and role in disease. *Biochim Biophys Acta* 2012;1822(5):784–793.

211. Cheah IK, Tang RMY, Yew TSZ, Lim KHC, Halliwell B. Administration of pure ergothioneine to healthy human subjects: Uptake, metabolism, and effects on biomarkers of oxidative damage and inflammation. *Antioxid Redox Signal* 2017;26(5):193–206.

212. Halliwell B, Cheah IK, Drum CL. Ergothioneine, an adaptive antioxidant for the protection of injured tissues? A hypothesis. *Biochem Biophys Res Commun* 2016;470(2):245–250.

213. O'Hagan S, Kell DB. Consensus rank orderings of molecular fingerprints illustrate the 'most genuine' similarities between marketed drugs and small endogenous human metabolites, but highlight exogenous natural products as the most important 'natural' drug transporter substrates. *ADMET & DMPK* 2017;5(2):85–125.

214. Daßler T, Maier T, Winterhalter C, Böck A. Identification of a major facilitator protein from *Escherichia coli* involved in efflux of metabolites of the cysteine pathway. *Mol Microbiol* 2000;36(5):1101–1112.

215. Fukui K, Koseki C, Yamamoto Y, Nakamura J, Sasahara A, et al. Identification of succinate exporter in *Corynebacterium glutamicum* and its physiological roles under anaerobic conditions. *J Biotechnol* 2011;154(1):25–34.

216. Holle SK, Ciarimboli G, Edemir B, Neugebauer U, Pavenstädt H, et al. Properties and regulation of organic cation transport in freshly isolated mouse proximal tubules analyzed with a fluorescence reader-based method. *Pflugers Arch* 2011;462(2):359–369.

217. Kido Y, Matsson P, Giacomini KM. Profiling of a prescription drug library for potential renal drug-drug interactions mediated by the organic cation transporter 2. *J Med Chem* 2011;54(13):4548–4558.

218. Chang HC, Yang SF, Huang CC, Lin TS, Liang PH, et al. Development of a novel non-radioactive cell-based method for the screening of SGLT1 and SGLT2 inhibitors using 1-NBDG. *Mol Biosyst* 2013;9(8):2010–2020.

219. Tegos GP, Evangelisti AM, Strouse JJ, Ursu O, Bologa C, et al. A high throughput flow cytometric assay platform targeting transporter inhibition. *Drug Disc Today Technol* 2014;12:e95-e103.

220. Ugwu MC, Pelis R, Esimone CO, Agu RU. Fluorescent organic cations for human OCT2 transporters screening: Uptake in CHO cells stably expressing hOCT2. *ADMET & DMPK* 2017;5(2):135–145.

221. Liang Y, Li S, Chen L. The physiological role of drug transporters. *Protein Cell* 2015;6(5):334–350.

222. Binder S, Schendzielorz G, Stäbler N, Krumbach K, Hoffmann K, et al. A high-throughput approach to identify genomic variants of bacterial metabolite producers at the single-cell level. *Genome Biol* 2012;13(5):R40.

223. D'Ambrosio V, Jensen MK. Lighting up yeast cell factories by transcription factor-based biosensors. *FEMS Yeast Res* 2017;17(7).

224. Liu D, Evans T, Zhang F. Applications and advances of metabolite biosensors for metabolic engineering. *Metab Eng* 2015;31:35–43.

225. Leavitt JM, Wagner JM, Tu CC, Tong A, Liu Y, et al. Biosensor-enabled directed evolution to improve muconic acid production in *Saccharomyces cerevisiae*. *Biotechnol J1600687* 2017;12(10):1600687.

226. Libis V, Delépine B, Faulon JL. Sensing new chemicals with bacterial transcription factors. *Curr Opin Microbiol* 2016;33:105–112.

227. Mahr R, Gätgens C, Gätgens J, Polen T, Kalinowski J, et al. Biosensor-driven adaptive laboratory evolution of L-valine production in *Corynebacterium glutamicum*. *Metab Eng* 2015;32:184–194.

228. Mannan AA, Liu D, Zhang F, Oyarzún DA. Fundamental design principles for transcription-factor-based metabolite biosensors. *Acs Synth Biol* 2017;6(10):1851–1859.

229. Mustafi N, Grunberger A, Kohlheyer D, Bott M, Frunzke J. The development and application of a single-cell biosensor for the detection of L-methionine and branched-chain amino acids. *Metab Eng* 2012;14(4):449–457.

230. Qian S, Cirino PC. Using metabolite-responsive gene regulators to improve microbial biosynthesis. *Current Opinion in Chemical Engineering* 2016;14:93–102.

231. Rogers JK, Taylor ND, Church GM. Biosensor-based engineering of biosynthetic pathways. *Curr Opin Biotechnol* 2016;42:84–91.

232. Schallmey M, Frunzke J, Eggeling L, Marienhagen J. Looking for the pick of the bunch: High-throughput screening of producing microorganisms with biosensors. *Curr Opin Biotechnol* 2014;26:148–154.

233. Siedler S, Stahlhut SG, Malla S, Maury J, Neves AR. Novel biosensors based on flavonoid-responsive transcriptional regulators introduced into Escherichia coli. *Metab Eng*, Research Support, Non-US Government 2014;21:2–8.

234. Siedler S, Khatri NK, Zsohár A, Kjærbølling I, Vogt M, et al. Development of a bacterial biosensor for rapid screening of yeast p-Coumaric acid production. *Acs Synth Biol* 2017;6(10):1860–1869.

235. Younger AKD, Dalvie NC, Rottinghaus AG, Leonard JN. Engineering modular biosensors to confer metabolite-responsive regulation of transcription. *Acs Synth Biol* 2017;6(2):311–325.

236. Zhang J, Jensen MK, Keasling JD. Development of biosensors and their application in metabolic engineering. *Curr Opin Chem Biol* 2015;28:1–8.

237. Bender A, Glen RC. Molecular similarity: A key technique in molecular informatics. *Org Biomol Chem* 2004;2(22):3204–3218.

238. Gasteiger J (editor) *Handbook of Chemoinformatics: From Data to Knowledge.* Weinheim: Wiley/VCH; 2003.

239. Gasteiger J. *Basic Chemoinformatics: A Textbook.* Weinheim: Wiley/VCH; 2003.

240. Weininger D. SMILES, A chemical language and information system .1. Introduction to methodology and encoding rules. *J Chem Inf Comput Sci* 1988;28(1):31–36.

241. Bachrach SM. InChI: A user's perspective. *J Cheminform*, Editorial 2012;4(1):34.

242. Heller S, McNaught A, Stein S, Tchekhovskoi D, Pletnev I. InChI - the worldwide chemical structure identifier standard. *J Cheminform* 2013;5(1):7.

243. Williams AJ. InChI: connecting and navigating chemistry. *J Cheminform* 2012;4(1):33.

244. Riniker S, Landrum GA. Open-source platform to benchmark fingerprints for ligand-based virtual screening. *J Cheminform* 2013;5(1):26.

245. Beisken S, Meinl T, Wiswedel B, de Figueiredo LF, Berthold M, et al. KNIME-CDK: Workflow-driven cheminformatics. *BMC bioinformatics* 2013;14:257.

246. Berthold MR, Cebron N, Dill F, Gabriel TR, Kötter T et al. KNIME: The Konstanz Information Miner. In: Preisach C, Burkhardt H, Schmidt-Thieme L, Decker R (editors). *Data Analysis, Machine Learning and Applications*. Berlin: Springer; 2008. pp. 319–326.

247. Mazanetz MP, Marmon RJ, Reisser CBT, Morao I. Drug discovery applications for KNIME: An open source data mining platform. *Curr Top Med Chem*, Review 2012;12(18):1965–1979.

248. O'Hagan S, Kell DB. The KNIME workflow environment and its applications in Genetic Programming and machine learning. *Genetic Progr Evol Mach* 2015;16:387–391.

249. Arif SM, Holliday JD, Willett P. Comparison of chemical similarity measures using different numbers of query structures. *J Inf Sci* 2013;39(1):7–14.

250. Willett P. The calculation of molecular structural similarity: Principles and practice. *Mol Inform* 2014;33(6–7):403–413.

251. O'Hagan S, Kell DB. MetMaxStruct: A Tversky-similarity-based strategy for analysing the (sub)structural similarities of drugs and endogenous metabolites. *Front Pharmacol* 2016;7:266.

252. Todeschini R, Consonni V. *Molecular Descriptors For Cheminformatics*. Weinheim: Wiley-VCH Verlag GmbH; 2009.

253. O'Hagan S, Kell DB. Analysing and navigating natural products space for generating small, diverse, but representative chemical libraries. *Biotechnol J* 2018;13(1):1700503.

254. Handl J, Knowles J, Kell DB. Computational cluster validation in post-genomic data analysis. *Bioinformatics* 2005;21:3201–3212.

255. Xu R, Wunsch D, 2nd. Survey of clustering algorithms. *IEEE Trans Neural Netw*, Comparative Study Research Support, Non-US Government Research Support, US Government, Non-P.H.S. Review 2005;16(3):645–678.

256. Hastie T, Tibshirani R, Friedman J. *The Elements of Statistical Learning: Data Mining, Inference And Prediction, 2nd edition*. Berlin: Springer-Verlag; 2009.

257. Wold S, Sjöström M, Eriksson L. PLS-regression: A basic tool of chemometrics. *Chemometr Intell Lab* 2001;58(2):109–130.

258. Breiman L. Random forests. *Machine Learning* 2001;45(1):5–32.

259. Koza JR. *Genetic Programming: On the Programming of Computers by Means of Natural Selection*. Cambridge, MA: MIT Press; 1992.

260. Langdon WB. *Genetic Programming and Data Structures: Genetic Programming + Data Structures = Automatic Programming!* Boston: Kluwer; 1998.

261. Langdon WB, Poli R. *Foundations of Genetic Programming*. Berlin: Springer-Verlag; 2002.

262. Kell DB. Genotype:phenotype mapping: Genes as computer programs. *Trends Genet* 2002;18(11):555–559.

263. Golbraikh A, Tropsha A. Beware of q²! *J Mol Graph Model* 2002;20(4):269–276.

264. Golbraikh A, Muratov E, Fourches D, Tropsha A. Data set modelability by qsar. *J Chem Inf Model* 2014;54(1):1–4.

265. Fujita T, Winkler DA. Understanding the roles of the "Two QSARs.". *J Chem Inf Model* 2016;56(2):269–274.

266. Bailey PD, Boyd CA, Collier ID, George JP, Kellett GL, et al. Affinity prediction for substrates of the peptide transporter PepT1. *Chem Commun (Camb)* 2006(3):323–325.

267. Foley DW, Rajamanickam J, Bailey PD, Meredith D. Bioavailability through PepT1: The role of computer modelling in intelligent drug design. *Curr Comput Aided Drug Des* 2010;6(1):68–78.

268. Smith DE, Clémençon B, Hediger MA. Proton-coupled oligopeptide transporter family SLC15: Physiological, pharmacological and pathological implications. *Mol Aspects Med* 2013;34(2–3):323–336.

269. Bretschneider B, Brandsch M, Neubert R. Intestinal transport of beta-lactam antibiotics: Analysis of the affinity at the H⁺/peptide symporter (PEPT1), the uptake into Caco-2 cell monolayers and the transepithelial flux. *Pharm Res* 1999;16(1):55–61.

270. Mitsuoka K, Kato Y, Kubo Y, Tsuji A. Functional expression of stereoselective metabolism of cephalexin by exogenous transfection of oligopeptide transporter PEPT1. *Drug Metab Dispos* 2007;35(3):356–362.

271. Hironaka T, Itokawa S, Ogawara K, Higaki K, Kimura T. Quantitative evaluation of PEPT1 contribution to oral absorption of cephalexin in rats. *Pharm Res* 2009;26(1):40–50.

272. Giacomini MM, Hao J, Liang X, Chandrasekhar J, Twelves J et al. Interaction of 2,4-diaminopyrimidine-containing drugs including fedratinib and trimethoprim with thiamine transporters. *Drug Metab Dispos* 2017;45(1):76–85.

273. Lepist EI, Ray AS. Beyond drug-drug interactions: Effects of transporter inhibition on endobiotics, nutrients and toxins. *Expert Opin Drug Metab Toxicol* 2017;13(10):1075–1087.

274. Zhao R, Goldman ID. Folate and thiamine transporters mediated by facilitative carriers (SLC19A1–3 and SLC46A1) and folate receptors. *Mol Aspects Med* 2013;34(2–3):373–385.

275. Liang X, Chien HC, Yee SW, Giacomini MM, Chen EC et al. Metformin Is a Substrate and Inhibitor of the Human Thiamine Transporter, THTR-2 (SLC19A3). *Mol Pharm* 2015;12(12):4301–4310.

276. Loew R, Heinz N, Hampf M, Bujard H, Gossen M. Improved Tet-responsive promoters with minimized background expression. *BMC Biotechnol* 2010;10:81.

277. Urlinger S, Baron U, Thellmann M, Hasan MT, Bujard H et al. Exploring the sequence space for tetracycline-dependent transcriptional activators: Novel mutations yield expanded range and sensitivity. *Proc Natl Acad Sci U S A* 2000;97(14):7963–7968.

278. Jensen PR, Hammer K. Artificial promoters for metabolic optimization. *Biotechnol Bioeng* 1998;58(2–3):191–195.

279. Hammer K, Mijakovic I, Jensen PR. Synthetic promoter libraries--tuning of gene expression. *Trends Biotechnol* 2006;24(2):53–55.

280. Braatsch S, Helmark S, Kranz H, Koebmann B, Jensen PR. *Escherichia coli* strains with promoter libraries constructed by Red/ET recombination pave the way for transcriptional fine-tuning. *Biotechniques* 2008;45(3):335–337.

281. Wagner S, Klepsch MM, Schlegel S, Appel A, Draheim R, et al. Tuning *Escherichia coli* for membrane protein overexpression. *Proc Natl Acad Sci U S A* 2008;105(38):14371–14376.

282. Rohlhill J, Sandoval NR, Papoutsakis ET. Sort-seq approach to engineering a formaldehyde-inducible promoter for dynamically regulated *Escherichia coli* growth on methanol. *Acs Synth Biol* 2017.

283. Rytter JV, Helmark S, Chen J, Lezyk MJ, Solem C, et al. Synthetic promoter libraries for *Corynebacterium glutamicum*. *Appl Microbiol Biotechnol* 2014;98(6):2617–2623.

284. Shen J, Chen J, Jensen PR, Solem C. A novel genetic tool for metabolic optimization of *Corynebacterium glutamicum*: Efficient and repetitive chromosomal integration of synthetic promoter-driven expression libraries. *Appl Microbiol Biotechnol* 2017;101(11):4737–4746.

285. Hockney RC. Recent developments in heterologous protein production in *Escherichia coli*. *Trends Biotechnol* 1994;12(11):456–463.

286. Brawand D, Soumillon M, Necsulea A, Julien P, Csárdi G, et al. The evolution of gene expression levels in mammalian organs. *Nature* 2011;478(7369):343–348.

287. GTEx Consortium, Battle A, Brown CD, Engelhardt BE, Montgomery SB. Genetic effects on gene expression across human tissues. *Nature* 2017;550(7675):204–213.

288. Swainston, N., Dunstan, M., Jervis, A. J., Robinson, C. J., Carbonell, P., Williams, A. R., Breitling, R., Faulon, J.-L., Scrutton, N. S. & Kell, D. B. (2018). PartsGenie: an integrated tool for optimising and sharing synthetic biology parts. *Bioinformatics* 34:2327–2329.

289. Kudla G, Murray AW, Tollervey D, Plotkin JB. Coding-sequence determinants of gene expression in *Escherichia coli*. *Science* 2009;324(5924):255–258.

290. Jones DL, Brewster RC, Phillips R. Promoter architecture dictates cell-to-cell variability in gene expression. *Science* 2014;346(6216):1533–1536.

291. Angov E. Codon usage: Nature's roadmap to expression and folding of proteins. *Biotechnol J* 2011;6(6):650–659.

292. Angov E, Legler PM, Mease RM. Adjustment of codon usage frequencies by codon harmonization improves protein expression and folding. *Methods Mol Biol* 2011;705:1–13.

293. Plotkin JB, Kudla G. Synonymous but not the same: The causes and consequences of codon bias. *Nat Rev Genet* 2011;12(1):32–42.

294. Swainston N, Currin A, Day PJ, Kell DB. GeneGenie: Optimised oligomer design for directed evolution. *Nucleic Acids Res* 2014;12:W395–W400.

295. Winkler WC, Nahvi A, Roth A, Collins JA, Breaker RR. Control of gene expression by a natural metabolite-responsive ribozyme. *Nature* 2004;428(6980):281–286.

296. Salis HM, Mirsky EA, Voigt CA. Automated design of synthetic ribosome binding sites to control protein expression. *Nat Biotechnol* 2009;27(10):946–950.

297. Salis HM. The ribosome binding site calculator. *Methods Enzymol* 2011;498:19–42.

298. Alper H, Stephanopoulos G. Global transcription machinery engineering: A new approach for improving cellular phenotype. *Metab Eng* 2007;9(3):258–267.

299. Han T, Chen Q, Liu H. Engineered photoactivatable genetic switches based on the bacterium phage T7 RNA polymerase. *Acs Synth Biol* 2017;6(2):357–366.

300. Butelli E, Titta L, Giorgio M, Mock HP, Matros A, et al. Enrichment of tomato fruit with health-promoting anthocyanins by expression of select transcription factors. *Nat Biotechnol* 2008;26(11):1301–1308.

301. Zhang Y, Butelli E, Alseekh S, Tohge T, Rallapalli G, et al. Multi-level engineering facilitates the production of phenylpropanoid compounds in tomato. *Nat Commun* 2015;6: 8635.

302. Lange C, Mustafi N, Frunzke J, Kennerknecht N, Wessel M, et al. Lrp of *Corynebacterium glutamicum* controls expression of the *brnFE* operon encoding the export system for L-methionine and branched-chain amino acids. *J Biotechnol* 2012;158(4):231–241.

303. Fang X, Sastry A, Mih N, Kim D, Tan J, et al. Global transcriptional regulatory network for *Escherichia coli* robustly connects gene expression to transcription factor activities. *Proc Natl Acad Sci U S A* 2017;114(38):10286–10291.

304. Shah AA, Wang C, Chung YR, Kim JY, Choi ES, et al. Enhancement of geraniol resistance of *Escherichia coli* by MarA overexpression. *J Biosci Bioeng* 2013;115(3):253–258.

305. Du D, Wang Z, James NR, Voss JE, Klimont E, et al. Structure of the AcrAB-TolC multidrug efflux pump. *Nature* 2014;509(7501):512–515.

306. Fisher MA, Boyarskiy S, Yamada MR, Kong N, Bauer S, et al. Enhancing tolerance to short-chain alcohols by engineering the *Escherichia coli* AcrB efflux pump to secrete the nonnative substrate n-butanol. *Acs Synth Biol* 2014;3(1):30–40.

307. Pourahmad Jaktaji R, Ebadi R, Karimi M. Study of organic solvent tolerance and increased antibiotic resistance properties in *E. coli* gyrA mutants. *Iran J Pharm Res* 2012;11(2):595–600.

308. Zhang F, Qian X, Si H, Xu G, Han R, et al. Significantly improved solvent tolerance of *Escherichia coli* by global transcription machinery engineering. *Microb Cell Fact* 2015;14:175.

309. Ronda C, Maury J, Jakociunas T, Jacobsen SA, Germann SM, et al. CrEdit: CRISPR mediated multi-loci gene integration in *Saccharomyces cerevisiae*. *Microb Cell Fact* 2015;14:97.

310. Stovicek V, Holkenbrink C, Borodina I. CRISPR/Cas system for yeast genome engineering: Advances and applications. *FEMS Yeast Res* 2017.

311. Kell DB, Ryder HM, Kaprelyants AS, Westerhoff HV. Quantifying heterogeneity: Flow cytometry of bacterial cultures. *Antonie van Leeuwenhoek* 1991;60:145–158.

312. Kell DB, Potgieter M, Pretorius E. Individuality, phenotypic differentiation, dormancy and 'persistence' in culturable bacterial systems: Commonalities shared by environmental, laboratory, and clinical microbiology. *F1000Res* 2015;4:179.

313. Williams TC, Averesch NJH, Winter G, Plan MR, Vickers CE, et al. Quorum-sensing linked RNA interference for dynamic metabolic pathway control in *Saccharomyces cerevisiae*. *Metab Eng* 2015;29:124–134.

314. Csonka L. Physiological and genetic responses of bacteria to osmotic stress. *Microbiol Rev* 1989;53(1):121–147.

315. Farwick M, Siewe RM, Krämer R. Glycine betaine uptake after hyperosmotic shift in *Corynebacterium glutamicum*. *J Bacteriol* 1995;177(16):4690–4695.

316. Kempf B, Bremer E. Uptake and synthesis of compatible solutes as microbial stress responses to high-osmolality environments. *Arch Microbiol* 1998;170(5):319–330.

317. Weinand M, Krämer R, Morbach S. Characterization of compatible solute transporter multiplicity in *Corynebacterium glutamicum*. *Appl Microbiol Biotechnol*, Research Support, Non-US Government 2007;76(3):701–708.

318. Ochrombel I, Becker M, Krämer R, Marin K. Osmotic stress response in *C. glutamicum*: impact of channel- and transporter-mediated potassium accumulation. *Arch Microbiol* 2011;193(11):787–796.

319. Brown AD. Compatible solutes and extreme water stress in eukaryotic micro-organisms. *Adv Microb Physiol* 1978;17:181–242.

320. Hohmann S, Krantz M, Nordlander B. Yeast osmoregulation. *Methods in enzymology*, Research Support, Non-US Government Review 2007;428:29–45.

321. Fahnert B. Folding-promoting agents in recombinant protein production. *Methods Mol Biol* 2004;267:53–74.

322. Fahnert B. Using folding promoting agents in recombinant protein production: a review. *Methods Mol Biol* 2012;824:3–36.

323. Voigt CA, Kauffman S, Wang ZG. Rational evolutionary design: The theory of *in vitro* protein evolution. *Adv Prot Chem* 2001;55:79–160.

324. Currin A, Swainston N, Day PJ, Kell DB. Synthetic biology for the directed evolution of protein biocatalysts: Navigating sequence space intelligently. *Chem Soc Rev* 2015;44(5):1172–1239.

325. Jiménez-Osés G, Osuna S, Gao X, Sawaya MR, Gilson L, et al. The role of distant mutations and allosteric regulation on LovD active site dynamics. *Nat Chem Biol* 2014;10:431–436.

326. Rocklin GJ, Chidyausiku TM, Goreshnik I, Ford A, Houliston S, et al. Global analysis of protein folding using massively parallel design, synthesis, and testing. *Science* 2017;357(6347):168–175.

327. Herter S, Medina F, Wagschal S, Benhaïm C, Leipold F, et al. Mapping the substrate scope of monoamine oxidase (MAO-N) as a synthetic tool for the enantioselective synthesis of chiral amines. *Bioorg Med Chem* 2017.

328. Savile CK, Janey JM, Mundorff EC, Moore JC, Tam S, et al. Biocatalytic asymmetric synthesis of chiral amines from ketones applied to sitagliptin manufacture. *Science* 2010;329(5989):305–309.

329. Zhang Y, Lamb BM, Feldman AW, Zhou AX, Lavergne T, et al. A semisynthetic organism engineered for the stable expansion of the genetic alphabet. *Proc Natl Acad Sci U S A* 2017;114(6):1317–1322.

330. Mazurkiewicz P, Poelarends GJ, Driessen AJM, Konings WN. Facilitated drug influx by an energy-uncoupled secondary multidrug transporter. *J Biol Chem* 2004;279(1):103–108.

331. Mazurkiewicz P, Driessen AJM, Konings WN. Energetics of wild-type and mutant multidrug resistance secondary transporter LmrP of *Lactococcus lactis*. *Biochim Biophys Acta* 2004;1658(3):252–261.

332. Mazurkiewicz P, Driessen AJM, Konings WN. What do proton motive force driven multidrug resistance transporters have in common? *Curr Issues Mol Biol*, Review 2005;7(1):7–21.

333. Schaedler TA, van Veen HW. A flexible cation binding site in the multidrug major facilitator superfamily transporter LmrP is associated with variable proton coupling. *FASEB J*, Research Support, Non-US Government 2010;24(10):3653–3661.

334. Madej MG, Dang S, Yan N, Kaback HR. Evolutionary mix-and-match with MFS transporters. *Proc Natl Acad Sci* 2013;110(15):5870–5874.

335. Khersonsky O, Tawfik DS. Enzyme promiscuity: A mechanistic and evolutionary perspective. *Annu Rev Biochem* 2010;79:471–505.

336. Daley DO, Rapp M, Granseth E, Melen K, Drew D et al. Global topology analysis of the *Escherichia coli* inner membrane proteome. *Science*, Research Support, Non-US Government 2005;308(5726):1321–1323.

337. Nishino K, Yamaguchi A. Analysis of a complete library of putative drug transporter genes in *Escherichia coli*. *J Bacteriol* 2001;183(20):5803–5812.

338. Holdsworth SR, Law CJ. Functional and biochemical haracterization of the *Escherichia coli* major facilitator superfamily multidrug transporter MdtM. *Biochimie* 2012;94(6):1334–1346.

339. Reddy VS, Shlykov MA, Castillo R, Sun EI, Saier MH, Jr. The major facilitator superfamily (MFS) revisited. *FEBS J* 2012;279(11):2022–2035.

340. Hayashi M, Tabata K, Yagasaki M, Yonetani Y. Effect of multidrug-efflux transporter genes on dipeptide resistance and overproduction in *Escherichia coli*. *FEMS Microbiol Lett* 2010;304(1):12–19.

341. Segura A, Molina L, Fillet S, Krell T, Bernal P et al. Solvent tolerance in Gram-negative bacteria. *Curr Opin Biotechnol* 2012;23(3):415–421.

342. Udaondo Z, Duque E, Fernandez M, Molina L, de la Torre J, et al. Analysis of solvent tolerance in Pseudomonas putida DOT-T1E based on its genome sequence and a collection of mutants. *FEBS Lett* 2012;586(18):2932–2938.

343. Goffeau A, Park J, Paulsen IT, Jonniaux JL, Dinh T, et al. Multidrug-resistant transport proteins in yeast: complete inventory and phylogenetic characterization of yeast open reading frames with the major facilitator superfamily. *Yeast* 1997;13(1):43–54.

344. Balakrishnan R, Park J, Karra K, Hitz BC, Binkley G et al. YeastMine--an integrated data warehouse for Saccharomyces cerevisiae data as a multipurpose tool-kit. *Database (Oxford)* 2012;2012:bar062.

345. Cherry JM, Hong EL, Amundsen C, Balakrishnan R, Binkley G et al. *Saccharomyces* Genome Database: the genomics resource of budding yeast. *Nucleic Acids Res* 2012;40(Database issue):D700–705.

346. Rogers B, Decottignies A, Kolaczkowski M, Carvajal E, Balzi E et al. The pleiotropic drug ABC transporters from *Saccharomyces cerevisiae*. *J Mol Microbiol Biotechnol* 2001;3(2):207–214.

347. Dos Santos SC, Teixeira MC, Dias PJ, Sá-Correia I. MFS transporters required for multidrug/multixenobiotic (MD/MX) resistance in the model yeast: understanding their physiological function through post-genomic approaches. *Front Physiol*, Review 2014;5:180.

348. dos Santos SC, Sa-Correia I. Yeast toxicogenomics: Lessons from a eukaryotic cell model and cell factory. *Curr Opin Biotechnol* 2015;33:183–191.

349. Elena SF, Whittam TS, Winkworth CL, Riley MA, Lenski RE. Genomic divergence of *Escherichia coli* strains: evidence for horizontal transfer and variation in mutation rates. *International Microbiology: The Official Journal of the Spanish Society For Microbiology* 2005;8(4):271–278.

350. Gluck-Thaler E, Slot JC. Dimensions of horizontal gene transfer in eukaryotic microbial pathogens. *PLoS Pathog* 2015;11(10):e1005156.

351. Koonin EV, Makarova KS, Aravind L. Horizontal gene transfer in prokaryotes: Quantification and classification. *Annu Rev Microbiol* 2001;55:709–742.

352. Remy E, Nino-Gonzalez M, Godinho CP, Cabrito TR, Teixeira MC, et al. Heterologous expression of the yeast Tpo1p or Pdr5p membrane transporters in *Arabidopsis* confers plant xenobiotic tolerance. *Sci Rep* 2017;7(1):4529.

353. Dunlop MJ, Dossani ZY, Szmidt HL, Chu HC, Lee TS et al. Engineering microbial biofuel tolerance and export using efflux pumps. *Mol Syst Biol* 2011;7:487.

354. Doshi R, Nguyen T, Chang G. Transporter-mediated biofuel secretion. *Proc Natl Acad Sci*, Research Support, US Government, Non-P.H.S. 2013;110(19):7642–7647.

355. Ling H, Chen B, Kang A, Lee JM, Chang MW. Transcriptome response to alkane biofuels in *Saccharomyces cerevisiae*: Identification of efflux pumps involved in alkane tolerance. *Biotechnol Biofuels* 2013;6(1):95.

356. Chen B, Ling H, Chang MW. Transporter engineering for improved tolerance against alkane biofuels in *Saccharomyces cerevisiae*. *Biotechnol Biofuels* 2013;6(1):21.

357. Nishida N, Ozato N, Matsui K, Kuroda K, Ueda M. ABC transporters and cell wall proteins involved in organic solvent tolerance in *Saccharomyces cerevisiae*. *J Biotechnol* 2013;165(2):145–152.

358. Lee JO, Cho KS, Kim OB. Overproduction of AcrR increases organic solvent tolerance mediated by modulation of SoxS regulon in *Escherichia coli*. *Appl Microbiol Biotechnol* 2014;98(20):8763–8773.

359. Sun X, Zahir Z, Lynch KH, Dennis JJ. An antirepressor, SrpR, is involved in transcriptional regulation of the SrpABC solvent tolerance efflux pump of *Pseudomonas putida* S12. *J Bacteriol* 2011;193(11):2717–2725.

360. Fillet S, Daniels C, Pini C, Krell T, Duque E, et al. Transcriptional control of the main aromatic hydrocarbon efflux pump in *Pseudomonas*. *Environmental Microbiology Reports* 2012;4(2):158–167.

361. Mingardon F, Clement C, Hirano K, Nhan M, Luning EG, et al. Improving olefin tolerance and production in *E. coli* using native and evolved AcrB. *Biotechnol Bioeng* 2014;112, 879–888.

362. Moschen I, Bröer A, Galić S, Lang F, Bröer S. Significance of short chain fatty acid transport by members of the monocarboxylate transporter family (MCT). *Neurochem Res* 2012;37(11):2562–2568.

363. Sá-Pessoa J, Paiva S, Ribas D, Silva IJ, Viegas SC, et al. SATP (YaaH), a succinate-acetate transporter protein in *Escherichia coli*. *Biochem J*, Research Support, Non-US Government 2013;454(3):585–595.

364. Lin MH, Khnykin D. Fatty acid transporters in skin development, function and disease. *Biochim Biophys Acta*, Research Support, NIH, Extramural Review 2014;1841(3):362–368.

365. Mukhopadhyay A. Tolerance engineering in bacteria for the production of advanced biofuels and chemicals. *Trends Microbiol* 2015;23(8):498–508.

366. Turner WJ, Dunlop MJ. Trade-offs in improving biofuel tolerance using combinations of efflux pumps. *Acs Synth Biol* 2015;4(10):1056–1063.

367. Rahman SA, Cuesta SM, Furnham N, Holliday GL, Thornton JM. EC-BLAST: A tool to automatically search and compare enzyme reactions. *Nat Meth*, Research Support, Non-US Governmentt 2014;11(2):171–174.

368. Lam SD, Dawson NL, Das S, Sillitoe I, Ashford P, et al. Gene3D: Expanding the utility of domain assignments. *Nucleic Acids Res* 2016;44(D1):D404–409.

369. Dawson NL, Lewis TE, Das S, Lees JG, Lee D, et al. CATH: An expanded resource to predict protein function through structure and sequence. *Nucleic Acids Res* 2017;45(D1):D289–D295.

370. Carbonell P, Wong J, Swainston N, Takano E, Turner N, et al. Selenzyme: Enzyme selection tool for pathway design. *Bioinformatics* 2018:in press.

371. Koch M, Duigou T, Carbonell P, Faulon JL. Molecular structures enumeration and virtual screening in the chemical space with RetroPath2.0. *J Cheminform* 2017;9(1):64.

372. Swainston N, Batista-Navarro R, Carbonell P, Dobson PD, Vinaixa M, et al. Biochem4j: Integrated and extensible biochemical knowledge through graph databases. *PLoSOne* 2017;12(7):e0179130.

373. Delépine B, Duigou T, Carbonell P, Faulon JL. RetroPath2.0: A retrosynthesis workflow for metabolic engineers. *Metab Eng* 2018;45:158–170.

374. Moore JC, Jin HM, Kuchner O, Arnold FH. Strategies for the *in vitro* evolution of protein function: Enzyme evolution by random recombination of improved sequences. *J Mol Biol* 1997;272(3):336–347.

375. Kell DB. Scientific discovery as a combinatorial optimisation problem: how best to navigate the landscape of possible experiments? *Bioessays* 2012;34(3):236–244.

376. Kell DB, Lurie-Luke E. The virtue of innovation: innovation through the lenses of biological evolution. *J R Soc Interface* 2015;12(2):20141183.

377. Copp JN, Hanson-Manful P, Ackerley DF, Patrick WM. Error-prone PCR and effective generation of gene variant libraries for directed evolution. *Methods Mol Biol* 2014;1179:3–22.

378. McCullum EO, Williams BA, Zhang J, Chaput JC. Random mutagenesis by error-prone PCR. *Methods Mol Biol* 2010;634:103–109.

379. Pritchard L, Corne DW, Kell DB, Rowland JJ, Winson MK. A general model of error-prone PCR. *J Theoret Biol* 2004;234(4):497–509.

380. Oates MJ, Corne DW, Kell DB. The bimodal feature at large population sizes and high selection pressure: implications for directed evolution. In: Tan KC, Lim MH, Yao X, Wang L (editors). *Recent advances in simulated evolution and learning.* Singapore: World Scientific; 2003. pp. 215–240.

381. Sacks J, Welch W, Mitchell T, Wynn H. Design and analysis of computer experiments (with discussion). *Statist Sci* 1989;4:409–435.

382. Bäck T, Fogel DB, Michalewicz Z (editors). *Handbook of evolutionary computation.* Oxford: IOPPublishing/Oxford University Press; 1997.

383. Corne D, Dorigo M, Glover F (editors). *New Ideas in Optimization.* London: McGraw Hill; 1999.

384. Goldberg DE. *The Design of Innovation: Lessons From And For Competent Genetic Algorithms.* Boston: Kluwer; 2002.

385. Knowles J. Closed-loop evolutionary multiobjective optimization. *IEEE Computational Intelligence Magazine* 2009;4(3):77–91.

386. O'Hagan S, Knowles J, Kell DB. Exploiting genomic knowledge in optimising molecular breeding programmes: Algorithms from evolutionary computing. *PLoS ONE* 2012;7(11):e48862.

387. Heinemann M, Panke S. Synthetic biology—putting engineering into biology. *Bioinformatics* 2006;22(22):2790–2799.

388. Carbonell P, Currin A, Dunstan M, Fellows D, Jervis A, et al. SYNBIOCHEM - a SynBio foundry for the biosynthesis and sustainable production of fine and speciality chemicals. *Biochem Soc Trans* 2016;44(3):675–677.

389. Currin A, Swainston N, Day PJ, Kell DB. SpeedyGenes: A novel approach for the efficient production of error-corrected, synthetic gene libraries. *Protein Eng Design Sel* 2014;27:273–280.

390. Currin A, Swainston N, Day PJ, Kell DB. SpeedyGenes: Exploiting an improved gene synthesis method for the efficient production of error-corrected, synthetic protein libraries for directed evolution. *Meth Mol Biol* 2017;1472:63–78.

391. Swainston N, Currin A, Green L, Breitling R, Day PJ, et al. CodonGenie: Optimised ambiguous codon design tools. *Peer J Comp Sci* 2017;3:e120.

392. Romero-Rivera A, Garcia-Borràs M, Osuna S. Computational tools for the evaluation of laboratory-engineered biocatalysts. *Chem Comm* 2017;53(2):284–297.

393. Welch GR, Somogyi B, Damjanovich S. The role of protein fluctuations in enzyme action: A review. *Prog Biophys Mol Biol* 1982;39(2):109–146.

394. Araya CL, Fowler DM. Deep mutational scanning: Assessing protein function on a massive scale. *Trends Biotechnol* 2011;29(9):435–442.

395. Shin H, Cho BK. Rational protein engineering guided by deep mutational scanning. *Int J Mol Sci* 2015;16(9):23094–23110.

396. Rubin AF, Gelman H, Lucas N, Bajjalieh SM, Papenfuss AT, et al. A statistical framework for analyzing deep mutational scanning data. *Genome Biol* 2017;18(1):150.

397. Starita LM, Fields S. Deep Mutational Scanning: Library construction, functional selection, and high-throughput sequencing. *Cold Spring Harb Protoc* 2015;2015(8):777–780.

398. Kitzman JO, Starita LM, Lo RS, Fields S, Shendure J. Massively parallel single-amino-acid mutagenesis. *Nat Methods* 2015;12(3):203–206.

399. Matuszewski S, Hildebrandt ME, Ghenu AH, Jensen JD, Bank C. A statistical guide to the design of deep mutational scanning experiments. *Genetics* 2016;204(1):77–87.

400. LeCun Y, Bengio Y, Hinton G. Deep learning. *Nature* 2015;521(7553):436–444.

401. Schmidhuber J. Deep learning in neural networks: An overview. *Neural networks: the official journal of the International Neural Network Society* 2015;61:85–117.

402. Knight CG, Platt M, Rowe W, Wedge DC, Khan F, et al. Array-based evolution of DNA aptamers allows modelling of an explicit sequence-fitness landscape. *Nucleic Acids Res* 2009;37(1):e6.

403. Rugbjerg P, Genee HJ, Jensen K, Sarup-Lytzen K, Sommer MOA. Molecular buffers permit sensitivity tuning and inversion of riboswitch signals. *Acs Synth Biol* 2016;5(7):632–638.

404. Cardinale S, Tueros FG, Sommer MOA. Genetic-metabolic coupling for targeted metabolic engineering. *Cell Rep* 2017;20(5):1029–1037.

Section III

Fermentation Microbiology and Biotechnology

Metabolic, Enzymatic, and Genetic Engineering of Microorganisms

8 Increasing Tolerance and Robustness of Industrial Microorganisms
Current Strategies and Future Prospects

Rui Pereira and Yun Chen

CONTENTS

"Actually, the answer to the question of which type of science to fund is quite simple: Since all science is problem driven, it should be judged by the quality of the problems posed, and the quality of the solutions provided."

Brenner, S. 1998. The impact of society on Science. *Science* 282:1411–1412.

8.1 INTRODUCTION

The heart of industrial biotechnology is fermentation conducted by living catalysts—microorganisms. Obtaining efficient cell factories is therefore an imperative need for a successful industrial production process of any useful products by microbial fermentation. However, the tricky thing is that we deal with living catalysts—the cells that can change characteristics as they adapt to stressful situations that arise during industrial production. This is one of the biggest challenges that needs to be overcome to attain a high yield and productivity for developing industrial microorganisms. This challenge results from the low inherent tolerance of microbial cells to various harsh industrial conditions, for instance,

low pH, high temperature, fermentation inhibitors that may be carried over from lignocellulosic biomass hydrolysis, and/or may be accumulated as imbalanced pathway intermediates and produced as the desired end products. These stressful conditions and/or the presence of these toxic compounds can adversely affect growth rates and reduce cell viability, which results in lower productivity and ultimately diminishes cost competitiveness. Hence, this chapter focuses on strategies of both (semi)rational and evolutionary engineering approaches for increasing tolerance and robustness of industrial microorganisms, drawing examples mostly from *Escherichia coli* and *Saccharomyces cerevisiae*.

8.2 RANDOM MUTAGENESIS AND SELECTION

Random mutagenesis followed by selection has traditionally been, and still is, used in the biotechnological industry for strain development, such as for improvement of yield and increasing tolerance to products. This method usually involves construction of a cell library by introducing random mutations, followed by selection of mutants with improved characteristics, for example, faster growth with increasing concentrations of product or analogs.

Most frequently used mutagens are ultraviolet radiation, ethyl methane sulfonate (EMS), and N-methyl-N-nitroso-N'-nitroguanidine (NTG), as they are technically simple and widely applicable to almost any organism. One of the challenges in random mutagenesis screening is selecting an optimum dose of mutagen to increase the genetic diversity in the surviving populations. In some cases, phenotypic changes can be ranked to help determine the optimum dosage, but for complex or hard-to-detect phenotypes determining the optimum dosage can be difficult. Often, dosages slightly above the optimum are used to increase the proportion of productive mutants in the surviving population. However, neutral or potentially harmful mutations can be also introduced into the selected mutants at such dosages, which makes it more difficult to identify the governing mutations related to the desired phenotypes.

As one example to increase tolerance toward L-arginine, *Corynebacterium glutamicum* production strains were treated with NTG and ultraviolet for mutagenesis, and then were spread on agar plates containing increasing concentrations of

two arginine analogs, arginine hydroxamate and canavanine (Figure 8.1), in sequence. Through this random mutagenesis experiment, a mutant was obtained showing resistance against 10 g/L of arginine hydroxamate and 30 g/L of canavanine. In this case, the tolerant phenotype strain also gives better performance for production of final product L-arginine, resulted in the production of 34.2 g/L of L-arginine, which is more than 2-fold higher than that of the parental strain under the same conditions (Park et al., 2014). The mutant performance can be explained by removal of the negative-feedback regulation in the arginine biosynthesis induced by the selection in the L-arginine analogs.

As shown in Figure 8.2, the repressors ArgR and FarR, can be bind to the arg operon when the intracellular concentration of L-arginine rises above a critical threshold, and thus represses the expression of the *argC*, *argJ*, *argB*, *argD*, and *argF* genes. By selecting mutant strains with a higher capability of resistance toward arginine analogs, there will be at least a partial relief of this negative regulation and thereby allow better growth under higher concentrations of these analogs. By comparative genome sequencing analysis of both parental and mutant strains, non-synonymous single-nucleotide polymorphisms were found in *argR*, encoding the arginine repressor, and *argF*, encoding the ornithine carbamoyl transferase.

This is however not always the case as enhanced product- or analog-tolerance at high concentrations does not necessarily correlate with an ability to synthesize product at increased specific productivity and yield.

8.3 RATIONAL AND SEMI-RATIONAL ENGINEERING

Although random mutagenesis has been shown as a powerful approach for generation of genetic diversity of any

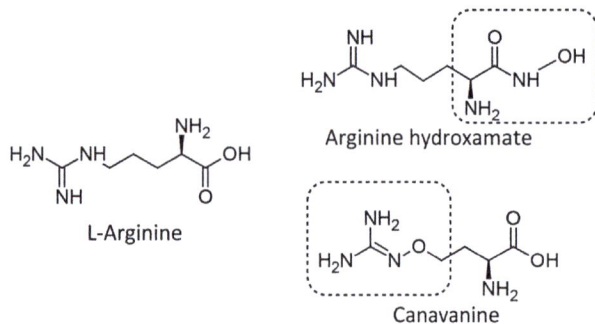

FIGURE 8.1 Structure of L-arginine and its analogs: arginine hydroxamate and canavanine. The differences between L-arginine and its analogs are marked with dashed line boxes.

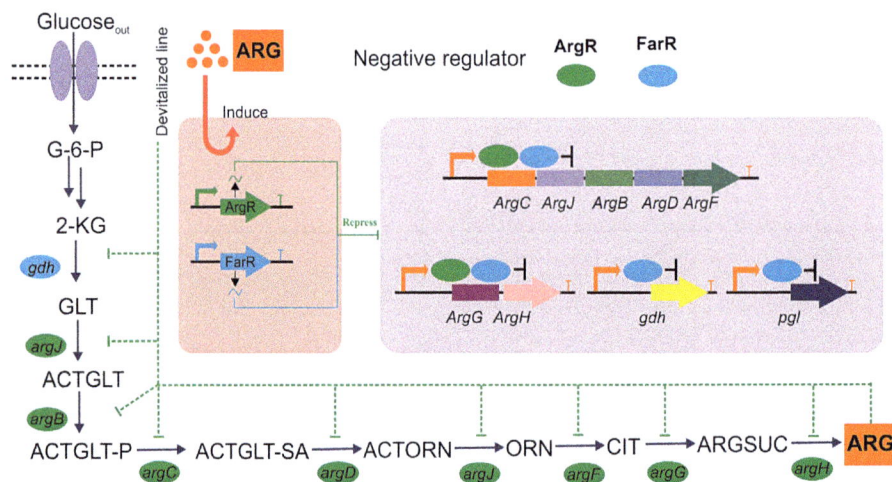

FIGURE 8.2 Schematic illustration of feedback regulation of L-arginine on its biosynthesis in *Corynebacterium glutamicum*. ArgR and FarR are repressors, which can bind to the arg operon upon increases in intracellular L-arginine, resulting in repression of the *argC*, *argJ*, *argB*, *argD*, and *argF* genes, as well as *argG* and *argF* genes. G-6-P, glucose -6-phosphate, α-KG, α-ketoglutarate, GLT, L-glutamate, ACTGLT, acetylglutamate, ACTGLT-P, acetylglutamyl phosphate, ACTGLT-SA, acetylglutamate semialdehyde, ACTORN, acetylornithine, ORN, L-ornithine, CIT, L-citrulline, ARGSUC, arginosuccinate, ARG, L-arginine. Dotted lines indicate the feedback repression by L-arginine. Adapted from Park et al. (2014).

microorganisms, the direct selection of productive mutants can be very time-consuming and labor-intensive. Moreover, the resulting phenotypes of the mutagenized strains are sometimes difficult to transfer into naive strains. In some cases, an alternatively strategy, like rational or semi-rational engineering, can be an effective approach to improve industrial microorganism tolerance and robustness. These approaches are especially useful in circumstances where the mechanisms behind the toxicity are at least partially known. In this section, some examples of rational engineering are discussed to illustrate the usefulness of this strategy.

8.3.1 Engineering Regulatory Circuits

Microorganisms have evolved and optimized their cellular systems through fine-tuning regulation of a multitude of pathways using many different transcriptional regulatory circuits to cope with different environmental changes and/or stresses. Therefore, engineering transcriptional regulatory circuits can be a feasible strategy for increasing tolerance phenotypes of industrial microorganisms.

Examples of engineering transcriptional regulatory circuits have been demonstrated in the yeast *S. cerevisiae* to increase the tolerance against high concentrations of ethanol and glucose (Alper et al., 2006). These strain properties are important for the ethanol industry, where very high gravity fermentations are usually employed, giving rise to high sugar concentrations (which leads to high osmotic stress) at the beginning of the processes and high ethanol concentrations at the end of batch. Achieving such complex traits does not seem feasible with a monogenic modification approach, but instead may require multigenic perturbations, which may well be achieved by engineering transcriptional regulatory circuits. One example of such regulatory circuit is the effect of the TATA-binding protein (TBP) on SAGA-dependent and TATA-containing genes, which comprise more than 15% of all genes in yeast (Ansari et al., 2012; Paul et al., 2015). These TBP-regulated genes are involved in many different metabolic processes; thus, mutations in the TATA-binding protein would change the preference of RNA polymerases for these genes, resulting in globally altered transcriptomic profiles that can be screened for improved tolerance phenotypes.

In yeast, about 75 components have been classified as gene-specific activators, coactivators, or general transcription factors of the RNA polymerase II system (Hahn, 2004). Among these components is the TATA-binding protein, encoded by *SPT15*, and 14 other associated factors that collectively thought to be the main DNA binding proteins regulating promoter specificity in yeast (Chasman et al., 1993; Kim et al., 1993; Hahn, 2004). Two mutant libraries were constructed from either *SPT15*, which encodes the TATA-binding protein, or *TAF25*, which encodes one of the TATA-binding protein-associated factors, via error-prone polymerase chain reaction (PCR) mutations. Libraries containing a plasmid-born, mutated version of either *SPT15* or *TAF25* were transformed into wild-type yeast and screened by subculturing

cells with increased concentrations of both ethanol and glucose. Thereafter, single isolates were subcultured on plates to isolate plasmids containing mutant genes. To confirm that the improved phenotypes were indeed conferred by the mutant factor, the isolated plasmids containing mutated genes were retransformed into a fresh background strain and evaluated for their phenotypes in the presence of elevated glucose and ethanol levels. Interestingly, the best mutant *spt*15-300 harboring just three amino-acid mutations (Phe177Ser, Tyr195His, and Lys218Arg) had a significant growth improvement, up to 13-fold, in the presence of 6% (v/v) ethanol and 100 g/L of glucose, compared with that of the control strain at the same concentrations. Unfortunately, the strain used in this study is a standard laboratory yeast strain. It would be interesting to explore this method in industrial or isolated yeast exhibiting naturally higher starting ethanol tolerances.

Similarly, random mutagenesis of the principal σ-factor has been proved to work in an industrial *Lactobacillus plantarum* strain for enhanced lactic and inorganic acid tolerance (Klein-Marcuschamer and Stephanopoulos, 2008), as well as in *E. coli* for increased tolerance to ethanol and sodium dodecyl sulphate (Alper and Stephanopoulos, 2007).

In addition to transcription factor engineering, other methods focused on alteration of global regulators exist, such as manipulation of global signaling compounds for improved tolerance phenotypes, as was reported for butanol tolerance in *E. coli* by mutagenesis of cyclic AMP receptors (CRP) (Zhang et al., 2012). CRP is a well-known global regulator that controls more than 400 genes in *E. coli*. Random mutagenesis was similarly applied to this regulator via error-prone-PCR or DNA shuffling to create CRP mutant libraries. The libraries were screened to select for butanol tolerance improved phenotypes. The growth rate of the best mutant was found twice fast as the wild-type in 1.2% (v/v) butanol, and its resistance to other alcoholic solvents such as isobutanol increased as well.

8.3.2 Engineering Efflux Pumps

Compared with global rewiring metabolic networks, pumping out the toxic product using exporters would be a more direct strategy for the cell to elicit the tolerance phenotype. In addition, alleviation of toxicity via efflux pumps would also decrease intracellular accumulation of final products and thus may function as a pull force for further increasing productivity and yield.

Large amounts of membrane proteins have been described for a variety of compounds and can be classified into several families according to their functions, such as the major facilitator superfamily, the small drug resistance superfamily, the ATP-binding cassette superfamily, the resistance-nodulation cell division superfamily, and the multidrug and toxic compound extrusion superfamily (Mukhopadhyay, 2015). Almost all known solvent-resistant exporters in Gram-negative bacteria fall into the hydrophobe/amphiphile efflux (HAE1) family of resistance-nodulation-division pumps (Tseng et al., 1999). A comparative search of HAE1 pumps from all sequenced bacterial genomes was performed, resulting into

43 candidate efflux pumps based on their sequence specificities. The library of 43 pumps was thus heterologously expressed in *E. coli*, and tested for resistance of seven representative biofuels, such as n-butanol, geraniol, α-pinene, and limonene. It was found that the native *E. coli* pump AcrAB and a previously uncharacterized pump from *Alcanivorax borkumensis* showed higher survival in the presence of limonene. Interestingly, strains expressing the *A. borkumensis* pump also produced significantly more limonene than those without it (Dunlop et al., 2011).

Another example of efflux pump engineering is the manipulation of ionic membrane gradients of *S. cerevisiae* for ethanol tolerance (Lam et al., 2014). In the production of fuel ethanol, the toxicity of ethanol and other alcohols to the yeast *S. cerevisiae* is a primary factor that limits titer and productivity in industrial production. The elevation of extracellular potassium and pH was found to increase resistance to higher alcohols and ethanol fermentation in commercial (PE-2 from Brazil and Ethanol Red from the US, both are used in bioethanol industry) and laboratory yeast strains under industrial-like conditions. Based on this, the plasma membrane potassium (Trk1) and proton (Pma1) pumps were overexpressed in a laboratory strain and increased ethanol titer by 27% over the wildtype. These improvements in output mirrored the enhancements observed in net cell viability, affirming the coupled nature of production and tolerance. Interestingly, the increased ethanol production in the laboratory strain, by strengthening the ion pump activities, surpassed the performance of the two industrial production strains. Notably, these genetic modifications of *TRK1* and *PMA1* were not reported in industrial production strains.

Though the use of efflux pumps is a feasible strategy to alleviate toxicity through upregulation of exporters, their overexpression may lead to cell membrane stress and thus inhibit cell growth, as shown in Figure 8.3. Therefore, expression of exporters should be carefully tuned to minimize membrane burden and cellular toxicity. For this purpose, an example of negative-feedback loop was constructed to dynamically control the expression of butanol exporter AcrBv2 under the stress-responsive promoter PgntK in *E. coli*, which showed increased butanol tolerance (Boyarskiy et al., 2016). This design also improved butanol production by 40% in comparison to a control strain. This approach shows a great promise for pump protein expression because the cell would be able to sense the membrane environment and respond accordingly to express the efflux pumps at an optimal level, without requiring an inducer supplementation (Figure 8.3).

8.3.3 Engineering the Plasma Membrane

Microbial membranes are permeable to small neutral molecules for growth and removal of waste products but create a large free energy barrier to diffusion of large and highly polar or charged molecules (Sandoval and Papoutsakis, 2016). Due to its gatekeeper function, the ability to maintain plasma membrane integrity and fluidity is crucial for energy generation and other cellular functions. Damage to the cell membrane has long been implicated in the toxicity of various chemicals, for example, modifying the fatty acid profile of the phospholipid bilayer under stress conditions. Therefore, membrane engineering is also a feasible strategy to increase stress tolerance by maintaining membrane stability and integrity.

Generally, saturated fatty acids increase membrane stability by packing the acyl chains tightly while *cis*-unsaturated fatty acids increase membrane fluidity by breaking up this tightly packed configuration (Figure 8.4). To restore normal membrane saturation levels and increase tolerance to overproduction of free fatty acids (FFAs) in *E. coli*, a *Geobacillus* thioesterase that primarily hydrolyzes unsaturated medium chain-length FFAs was expressed in the production strain. The strategy successfully reduced unsaturated fatty acid content in the membrane and improved viability in a FFA producing strain (Lennen and Pfleger, 2013).

Unlike *cis*-unsaturated fatty acids, *trans*-unsaturated fatty acids are more linear in form and consequently better at maintaining membrane rigidity, as illustrated in Figure 8.4. In some bacteria, the periplasmic *cis-trans* isomerase (Cti) can convert unsaturated fatty acids from the *cis* to the *trans* conformation,

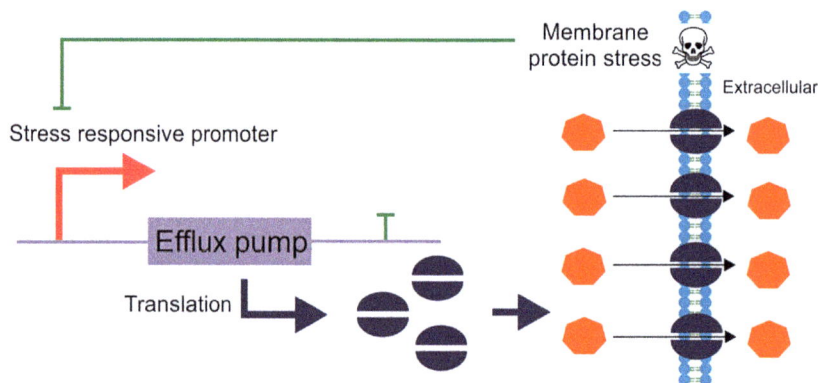

FIGURE 8.3 Illustration of dynamic regulation of efflux pump protein expression for increasing tolerance and robustness of microbial cells. Membrane stress-responsive promoters are used to control the expression of efflux pump genes, allowing for creation of a negative-feedback loop that autonomously controls pump protein expression to minimize cellular toxicity. One example can be found in Boyarskiy et al. (2016).

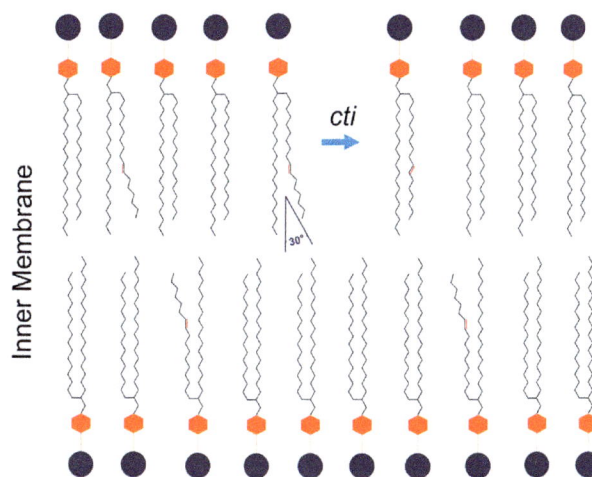

FIGURE 8.4 Simplified illustration of bacterial cytoplasmic membrane consisting phospholipids with varying degrees of *cis*-unsaturated fatty acids. Due to a kink with an angle of 30° at double bond position, *cis*-unsaturated fatty acid in the membrane will form a disorderly membrane and increase its fluidity; *cis*-unsaturated fatty acid can be converted into the straight-chain *trans*-unsaturated fatty acid by, e.g., the periplasmic *cis-trans* isomerase (Cti) in some bacteria. Enabling *cis-trans* transition via expression of the Cti enzyme has been shown to increase *E. coli* robustness in adverse conditions typically faced in a bioproduction environment. More information can be found in Tan et al. (2016).

providing a rapid mechanism to increase phospholipid density. In one example, a *Pseudomonas* Cti was expressed in *E. coli* and proved to enhance tolerance and production of octanoic acid. Moreover, the Cti expression strain has showed improved tolerance to a range of chemical stressors such as alcohols, organic acids, aromatic compounds, and a variety of industrial-like stress conditions including low pH, high temperature, and osmotic pressure (Tan et al., 2016). Given the presence of a 30° angle at the double bond in *cis*-unsaturated fatty acid, the straight-chain *trans*-unsaturated fatty acid will pack more tightly than the *cis* (Figure 8.4). Therefore, the increase in straight-chain *trans*-unsaturated fatty acids causes increased membrane rigidity and increased stress tolerance.

8.4 ADAPTIVE LABORATORY EVOLUTION

Rational engineering is an effective approach to improve microbial tolerance to toxic compounds. However, applications of rational strategies in tolerance engineering may be limited unless there is clear understanding of toxicity and tolerance mechanisms. Alternatively, evolutionary engineering, also known as adaptive laboratory evolution, is a complementary strategy to improve certain features of common industrial strains, such as inhibitor tolerance, substrate utilization or growth temperature range, without requiring knowledge of any underlying genetic mechanisms, as long as the desired trait can be coupled with growth. At its core, adaptive evolution simply involves the extended propagation of a microbial strain, typically for hundreds of generations, under the influence of the desired selective pressure. Mutants with enhanced growth rates, either due to increased tolerance or faster consumption of limiting nutrients, will occasionally arise and become enriched in the population.

The basic concept of directed evolution is also evident in classical, empirical strain development by random

mutagenesis followed by direct selection on plates. However, adaptive evolution can be a more powerful approach, since it allows for a large population cells to adapt to the selective pressure over many generations.

8.4.1 ADAPTIVE LABORATORY EVOLUTION METHODS

There two major methodologies to perform adaptive laboratory evolution and both are based on the same principle: prolonged growth under selective conditions. The first protocol consists in cultivating the target microorganism until a certain cell density is achieved and then transferring part of the cell suspension to a vessel containing fresh culture medium. This process of sequential serial passages can be repeated until the microorganism reaches the desired increase in fitness. One of the main advantages of this protocol is that it can be performed with routine laboratory equipment. The length of the experiment is usually measured in number of culture transfers or number of generations. A typical experiment will last 1–2 months and consist of 200–500 generations, but this will depend on how complicated the phenotype will be to evolve. Notably, the process of almost daily culture transfers for fast growing microorganism can present some logistical challenges. Because the perfect timing for the culture transfers might be difficult to predict, the cultures may need to be frequently monitored by checking the cell density and assess their status. If multiple cultures are being evolved in parallel, this process can become time-consuming and require an operator to be physically present at the right moment to transfer the cultures to fresh medium. This process can be made less dependent on the operator by incorporating online cell density measuring devices and automatizing the culture transfer mechanism—see Oud et al. (2013) for an example of an automated setup using bioreactors or Sandberg et al. (2016) for an automated culture transfer system.

A second option for performing adaptive laboratory evolution consists in using continuous cultivation in bioreactors. Using pumps to control the inflow and outflow of medium, it is possible to maintain the cells in steady growing conditions. The main advantages of this setup are the lower maintenance required from an operator and the possibility of controlling the pH and oxygen concentration in the medium. The operator just needs to make sure that fresh medium reservoir is replenished routinely and that medium waste containers are emptied. Despite the advantages when compared to sequential cultivations, the use of continuous culture for adaptive evolution requires specialized equipment that not all laboratories are equipped with.

8.4.1.1 Creation of Genetic Diversity

Any evolutionary process is ultimately dependent on genetic variations that occur either via native mutation or by environmental or genetic manipulations. Native spontaneous mutations in microbial populations occur at a rate close to 0.003-point mutations per genome and round of replication (Drake, 1991). The low frequency of genetic diversity could potentially lower the chance to acquire a desired phenotype. However, due to their small cell size, microbial population size can be large in number, e.g., exceeding 10^{11} cells per liter, so that an efficient evolution process can still be achieved.

In order to set up a successful adaptive laboratory evolution experiment, it is important to understand what types of mutations can arise during the process and the chance of complex phenotypes to appear. Taking *S. cerevisiae* as an example, it has been estimated that the mutation rate of this microorganism is around 5.0×10^{-10} mut/bp/division (Lang and Murray, 2008); the precise number depends on the genome location. This information can be used to compute a rough estimation of how many mutants are expected to appear per cell division (Figure 8.5).

The mutation frequencies shown in Figure 8.5 can be translated into probabilities of obtaining certain phenotypes. For example, the effective size for loss of function mutations in the *URA3* gene has been determined to be 125 base pairs (Lang and Murray, 2008). This number can be used to estimate the number of loss of function mutants that should arise in one cultivation by multiplying it by the number of mutations per base pair ($1.8 \times 125 = 224$). Furthermore, one can also calculate the probability that the mutant of interest to be transferred to the next cultivation. Assuming that 1/100th of the culture is transferred into a new cultivation flask, then two

loss of function mutants should be transferred. The magnitude of number of loss of function mutants suggest that this type of event can occur frequently in an evolution experiment and be selected for during the process. Phenotypes that require more complex mutations will be more difficult to obtain when relying on the natural mutation frequency of microorganisms.

For most evolution experiments, the basal rate of mutation present in microorganisms is sufficient to generate enough mutants for selective pressure to act upon. However, if the desired phenotypes require more complex mutation combinations, it is possible (and desirable) to accelerate the mutation process by using other strategies. These methods include but not limited to:

- Random mutagenesis via chemicals or radiation treatment, which has been frequently used in classical, empirical strain development, proven to be an effective method to induce a great variety of molecular alterations given an optimum dose of mutagen.
- Mutator strains, realized by, for example, engineering strains defective in DNA repair proteins or expressing error-prone polymerases. These strains show frequencies of spontaneous mutagenesis that are orders of magnitude higher than usual. However, such high mutation frequency can lead to a higher chance of accumulation of deleterious mutations that may reduce overall fitness. Therefore, mutators may not necessarily accelerate the pace of adaptive evolution.
- Transposon mutagenesis, a wide range of transposable elements have been found or engineered to catalyze their own movements to a location within a chromosome or within extrachromosomal elements. These transposable elements can be used for parallel gene inactivation and random gene overexpression. However, most available transposons display certain degree of target preference and their capability for multiple random insertion mutagenesis within one strain is usually limited.
- Mating is a very powerful approach for breeding independently improved variants, for example, by creating a diploid cell from two haploids. The offspring from this chimeric diploid cell can then be screened to select improved combinations of both haploid variants. This method has been traditionally used in industrial strain development. For

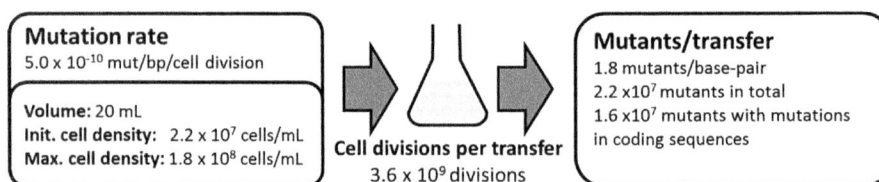

FIGURE 8.5 Illustration of the number of mutants expected to arise in a representative cultivation of *S. cerevisiae*. The numbers shown are approximations based on the mutation rates determined by Lang and Murray (2008).

example, desired traits such as increased growth rates or enhanced sporulation capability have been introduced into high-yielding production strains via mating.

It should be borne in mind that with increased mutagenesis rates, the difficulty in the identification of the genetic basis that confers a desired trait will also increase.

8.4.1.2   Selection of Desirable Traits

Selection is at the heart of adaptive laboratory evolution and can be a decisive factor in the success of evolution experiments. One of the most important factors to take into account is the level of selection imposed during the evolution process: if the selective pressure is too low, the mutants displaying the desired phenotype will not become enriched in the population; if the selective pressure is too high, growth can be inhibited and slow down the evolution process. Consequently, it is important to fine-tune the level of the selection pressure applied during the evolution process to allow high enough growth rates, while leaving a good margin of fitness improvement to permit mutants with the desired phenotype to outcompete the wild-type strain.

In adaptive laboratory evolution experiments where the desired phenotype is directly associated with growth (e.g., tolerance to chemicals, physical stress, co-consumption of two substrates), selection can be easily applied and fine-tuned by varying the level of the toxic compound or manipulating the concentration of the limiting substrates (see the following sections for details on different types of evolution experiments). Even if the desired phenotype is not directly associated with growth, it is still possible to engineer the coupling between growth and other phenotypical traits. For example, biosensors based on transcription factors can be used to couple the expression of an essential gene (e.g., antibiotic resistance) to the presence of a certain metabolite. Using this strategy, by increasing the concentration of antibiotic in the medium it is possible to select for mutant strains that have increased expression of the antibiotic detoxifying enzyme as a result of higher levels of the metabolite of interest. Successful implementation of such a strategy has been used to improve the flux to aromatic amino acids in *S. cerevisiae* by expressing a geneticin resistance gene under the control of a promoter based on the upstream sequence of the *ARO9* gene (Leavitt et al., 2017). This strategy allowed the selection of strains with improved availability of aromatic amino acids by culturing the strains containing the mentioned genetic construct in increasing amounts of geneticin. Furthermore, selected evolved strains from this screen were tested as production strains for muconic acid, which shares precursors with the aromatic acid pathway. An improvement of 3-fold was observed in the production of muconic acid for the evolved strain tested.

A common drawback of cultivating microbes for an extended period of time in stable conditions is the appearance of condition-dependent phenotypes. For example, nonconstitutive phenotypes can appear during evolution, i.e., before displaying the phenotype of interest the cells require a period of adaptation to the condition. To minimize the appearance of these phenotypes it is possible design experiments with alternating levels of selective pressure. For example, to promote constitutive tolerance to a certain stress factor, the microorganism would be cultivated in selective conditions and nonselective conditions alternatingly. In this way, the mutants that can display the tolerance phenotype faster have a fitness advantage in comparison to the ones that need an adjustment period. *S. cerevisiae* has been evolved for constitutive acetic acid tolerance using a strategy based on alternating cultivation in nonselective conditions and selective conditions containing toxic amounts of acetic acid (González-Ramos et al., 2016). Unlike other evolution experiments that resulted in nonconstitutive tolerance to acetic acid, the mutants obtained using alternating conditions displayed their tolerance even after being cultivated in the absence of acetic acid. The analysis of the evolved strains using whole-genome sequencing revealed causal mutations in four genes *(ASG1, ADH3, SKS1, and GIS4)*.

Although most adaptive laboratory evolution protocols rely on the selective pressure being applied during the cultivation procedure, it is also possible to perform a manual selection of improved individuals after the growth phase. For example, fluorescence-activated cell sorting can be used to select individuals that exhibit a fluorescent phenotype coupled to a desirable trait. Previously, *C. glutamicum* has been modified genetically to include a biosensor based on a transcription factor responsive to L-valine coupled with the expression of a fluorescent protein (Mahr et al., 2015). An increase of up to 100% for L-valine production was obtained by subjecting this modified strain to several cycles of cultivation followed by fluorescence-activated cell sorting to enrich for cells with increased L-valine production.

8.4.2   Nutrient Adaption

Substrate costs are one of the major contributors to determine if a bioprocess will be viable industrially. Therefore, the development of strains that can assimilate cheaper substrates and have no special growth requirements is a hot research topic. Adaptive laboratory evolution can be used to find mutant strains with improved fitness in desirable medium compositions for industrial purposes. Furthermore, detailed analysis of the adaptations present in these strains can elucidate the physiological mechanisms responsible for the improved fitness, which can help generating rational hypotheses for further improvements.

The most challenging aspect of nutrient adaption experiments is making a microorganism assimilate a new substrate. Besides glucose, which is the preferred carbon source for many organisms, lignocellulosic biomass hydrolysates also contain considerable amounts of xylose. *S. cerevisiae* is the best choice for the production of ethanol for biofuel applications, but it cannot naturally ferment xylose. In order to produce as much ethanol as possible from the mixture of sugars in biomass hydrolysates, *S. cerevisiae* needs to be engineered for this purpose. For example, after introducing the genes necessary for xylose consumption into *S. cerevisiae*, the resulting strain was evolved for 1,000 hours (7 transfers) in medium containing only xylose as the carbon source (Peng et al., 2012). The evolved strain showed an improved growth rate (0.06 h⁻¹) and a good yield of ethanol on xylose (0.39 g/g). Another evolution experiment involving xylose utilization in *S. cerevisiae* also resulted in mutants with improved characteristics (Shen et al., 2012). Transcriptomic characterization of the evolved strains revealed the upregulation of genes involved in glycolysis and glutamate synthesis and the downregulation of genes from glycogen and trehalose synthesis pathways.

Improvement of the growth characteristics in suboptimal carbon sources is also relevant to many industrial bioprocesses. *E. coli* can grow on glycerol as sole carbon source, but the growth rate is modest. Adaptive evolution experiments using sequential serial passages have shown that the growth-rate of this microorganism in medium with glycerol could improve from 0.23 h⁻¹ to 0.55 h⁻¹ at 30°C and from 0.46 h⁻¹ to 0.65 h⁻¹ at 37°C (Ibarra et al., 2002). Similar experiments in *E. coli* using continuous culture evolution for 217 generations resulted in mutants with increased growth-rate and higher biomass yield on glycerol (Sonderegger and Sauer, 2003).

Another desirable feature for industrial microorganisms is the co-consumption of multiple substrates simultaneously. Using a mixture of transporter engineering and adaptive evolution, *S. cerevisiae* has been made capable of fermenting glucose and xylose simultaneously (Nijland et al., 2014). The strain improvement started with a strain that could metabolize xylose proficiently on its own, but when glucose is present the transport of xylose into the cell is reduced due to the endogenous transporter's preference for glucose. In this study, the glucose transporters were deleted in this strain and a single xylose/glucose transporter (Hxt36) was left intact. Subsequently, evolution was performed in medium containing xylose and increasing amounts of glucose, which selected for mutations in the *HXT36* gene that improve its xylose transporting activity without being inhibited by glucose. Rational engineering of the mutated amino-acid residues allowed the design of a mutant strain of *S. cerevisiae* capable of fermenting glucose and xylose simultaneously.

8.4.3 Environmental Stress

The optimal physical environment (e.g., temperature, pH, oxygen availability) for microbial growth depends on the microorganism being used and it is the subject of considerable optimization in an industrial setting. In addition to the impact on the microbial growth and production of target compounds, the physical environment can have an influence on the cost competitiveness of a bioprocess.

Consequently, shifting the optimum environment for a certain microorganism to overlap with the industrial requirements for a more cost-competitive bioprocess is of paramount importance. Adaptive laboratory evolution is an attractive methodology to create strains with improved fitness in the required conditions. For example, ethanol production using *S. cerevisiae* benefits from a higher than optimal process temperature.

Besides reducing the cooling costs, the production of ethanol at higher temperatures can prevent contaminations and allow the simultaneous saccharification and fermentation process to be run simultaneously. Adaptive laboratory evolution has been used to generate heat tolerant strains of *S. cerevisiae* and help to elucidate the mechanisms that can make yeast tolerate higher than optimal temperatures (Caspeta et al., 2014). The characterization of the thermotolerant strains at the genetic level using whole-genome sequencing revealed mutations in the C-5 sterol desaturase, which caused the sterols present in the membrane to change from ergosterol to fecosterol.

Another form of optimizing the environment for industrial bioprocesses concerns the composition of the culture medium. For example, the lack of essential nutrients in medium or the presence of inhibitors can have a severe impact on the growth of the microorganism and result in lower product formation. Adaptive laboratory evolution can be used to create microorganisms that are less dependent on the addition of extra nutrients, thus decreasing environmental stress and saving on media costs. For example, *S. cerevisiae* has been engineered to grow independent of biotin supplementation by evolving it using the continuous cultivation methodology (Bracher et al., 2017). The genome sequence of the evolved strains revealed that in order to grow independently of the biotin supplementation, the evolved strain of *S. cerevisiae* required the amplification of the *BIO1* gene (encoding a putative pimeloyl-CoA synthetase) and inactivating mutations in *TPO1* (polyamine transporter) and *PDR12* (ABC transporter involved in acid tolerant).

Another cause of environmental stress is the presence of growth inhibitors in the culture media. These inhibitors are also problematic because they can severely hamper growth and result in lower productivity and product formation. The production of lignocellulosic ethanol is one example where the presence of inhibitors in biomass hydrolysates would benefit from using adaptive laboratory evolution to find strains with improved tolerance to the complex medium composition used. By evolving *S. cerevisiae* in hydrolysates of lignocellulosic biomass, it was possible to obtain strains with improved tolerance to acetic acid, 5-hydroxymethylfurfural and furfural (Almario et al., 2013). The best strain obtained could grow 56% faster on lignocellulosic hydrolysates and transcriptional characterization showed changes in the expression of genes such as *PDR1* (transcription factor that regulates the pleiotropic drug response), *ALD4* (mitochondrial aldehyde dehydrogenase), and *YRR1* (which activates genes involved in multidrug resistance).

8.4.4 Product Stress

Microbial strains with increased production of chemicals are usually engineered for that purpose using a combination of rational and traditional approaches. An important aspect of production strains is the capacity to tolerate the presence of the product in the medium at least until the maximum titer is achieved. Organic acids are examples of products that induce toxicity when the pH of the medium is lower than the acid pKa. In such conditions the undissociated acid can cross the cellular membrane into the cytosol and there, because the intracellular pH is close to neutral, the acid dissociates and contributes to the acidification of the intracellular compartment. Although this type of product toxicity can be minimized by maintaining the medium pH above the acid pKa, this can be costly at the industrial level and not optimal for the producing microorganism. Therefore, adaptive laboratory evolution can help elucidate mechanisms that allow production strains to achieve high acid titers at low pH. Tolerance to lactic acid and 3-hydroxypropionic acid are just a couple examples of how adaptive laboratory evolution helps finding genetic tolerance targets.

Lactic acid tolerant mutants of *S. cerevisiae* have been obtained by repeated cultivations on increasing selective amounts of lactic acid (Fletcher et al., 2017). Whole-genome sequencing of the tolerant strains revealed common mutations on the gene encoding a high affinity Ca^{2+}/Mn^{2+} P-type ATPase (*PMR1*) and the transcription factor required for septum destruction after cytokinesis (*ACE2*). Reproducing the loss of function mutation of the *PMR1* gene in the wild-type strain did not lead to an increase tolerance to lactic acid and inactivation of *ACE2* only partially improved the growth rate in these conditions. The inactivation of *ACE2* was found to lead to the formation of cell clumps, which seem to protect the cells from the toxic environment and improve their growth rate under stressing conditions.

Evolution of tolerance to 3-hydroxypropionic acid was obtained using repeated cultivations in medium containing this acid at low pH (3.5) (Kildegaard et al., 2014). The tolerant mutants obtained after roughly 200 generations showed frequent mutations in S-(hydroxymethyl) glutathione dehydrogenase (*SFA1*) and CoA-transferase (*ACH1*). The analysis of the mutated genes revealed that the main cause of 3-hydroxypropionic acid toxicity was mediated by its aldehyde (hydroxypropionic aldehyde). Furthermore, the mutations in *SFA1* were necessary for the S-(hydroxymethyl) glutathione dehydrogenase to be capable of detoxifying this toxic compound.

Ideally, product stress is tested before building production strains for non-native products, since most industrial microorganisms have not evolved in the presence of many chemicals of industrial importance. Often the mechanism of toxicity of a given product is poorly understood in a given organism, which makes rational engineering difficult. In this case, the use of adaptive laboratory evolution can become an essential step in the construction of production strains. Butanol, for example, can be toxic to *S. cerevisiae* and adaptive laboratory evolution has been used to mitigate this toxicity (Ghiaci et al., 2013). The evolved mutant, which can tolerate 3% of 2-butanol, was characterized using proteomic analysis and the results showed that among many changes there was an increase in the glycerol 3-phosphatase isoform (Gpp2). Further validation of this target for increasing the tolerance to 2-butanol was performed by overexpressing *GPP2* on the reference strain. The decrease in product stress observed in the example above can be used to increase product titer and productivity.

8.4.5 Fitness Trade-Off

When analyzing the fitness landscape of microorganisms, it is evident that evolution has selected for the conservation of traits that allow the extraction of resources from the environment and protection against common stress factors encountered during an organism's life cycle. After the phenotype of an organism is reprogrammed using adaptive laboratory evolution, the fitness landscape of the host will change to better tolerate the applied selective pressure. However, during this process, it is common for other traits to be compromised or even lost. This trade-off is difficult to predict but, in some cases, it is unavoidable that some phenotypical traits must be lost to accommodate the new one.

One example of fitness trade-off happened when *S. cerevisiae* was evolved in carbon limiting conditions but lost some of its fitness when cultivated under carbon abundance (Wenger et al., 2011). Sequencing analysis of the evolved strains revealed amplification of high-affinity glucose transporters (Hxt6/7), which helped the mutant strains compete for limiting resources (glucose). However, most of the strains evolved for competing in glucose limiting conditions showed decreased fitness when cultivated in glucose sufficient conditions.

Another case of a trade-off involving substrate uptake was reported by Sonderegger and Sauer (2003). In this study, *S. cerevisiae* was evolved to grow anaerobically on xylose for 460 generations. Interestingly, one of the evolved strains, which could grow on mixtures of xylose and glucose, lost part of its fitness when grown on glucose alone.

Tolerance phenotypes are also vulnerable to trade-offs as shown by evolved strains of *S. cerevisiae* tolerant to high temperature (Caspeta and Nielsen, 2015). When the thermotolerant strains were cultivated in the presence of hydrogen peroxide and acetic acid, their growth rate was significantly decreased in comparison with the wild-type strain. This shows that adaptive laboratory evolution experiments should be designed carefully to avoid generating strains with one improved trait at the cost of losing other important industrial traits.

8.5 PERSPECTIVES

With prior knowledge on toxicity and tolerance mechanisms available, rational approaches, such as engineering regulatory circuits, harnessing efflux pumps, and manipulating plasma membranes can be applied to improve tolerance phenotypes. However, these methods are mostly based on conventional gene (over)expression or inactivation. With rapid development in synthetic biology, more advanced strategies are expected to be developed and adopted in engineering tolerance phenotypes in the future. One emerging advanced strategy is the use of a dynamic regulation system to control the accumulation of toxic intermediates. An example of this was reported using farnesyl pyrophosphate (FPP)-responsive promoters in an amorphadiene-producing *E. coli* strain to drive the expression of genes responsible for the synthesis and consumption of toxic intermediate FPP (Dahl et al., 2013). Another promising strategy to be explored is spatial engineering and compartmentalization, which can be used to reduce the toxicity of metabolic intermediates as well. Synthetic scaffolds have been developed with the assistance of, e.g., protein (Dueber et al., 2009), and RNA (Delebecque et al., 2011), for assembling biosynthetic pathways. Notably, DNA scaffolds have attracted more attention due to highly predictable and stable structures. One example of this is using DNA scaffolds for assembling both 1,2-propanediol and mevalonate pathways in *E. coli*, which could reduce toxic intermediates and increase product titers (Conrado et al., 2012). Further, developing protein shells has emerged as a new tool for compartmentalizing multistep biosynthetic pathways (Chen and Silver, 2012; Lee et al., 2012), which could also be used to reduce toxic intermediates.

Traditionally, when the molecular basis for a certain desired phenotype is lacking, simple empirical methods based on random mutagenesis and direct selection on agar plates or microtiter plates has been effective for obtaining complex phenotypes. The biggest challenge with this method is the identification of genetic basis that benefit the desired phenotype due to multiple, random genetic changes at the genome level. However, recent technological advances that enable cost-effective mass sequencing and functional genomics are likely to change this situation. One can compare global responses of phenotypic variations at different omics levels, such as genomics, transcriptomics, proteomics, metabolomics and fluxomics. With omics data and systems biology tools to integrate and analyze this information, one could infer the genes most likely to be involved in a particular phenotype.

As discussed in Section 8.4, adaptive evolution offers an interesting alternative to empirical methods, and is likely to gain more relevance for industrial strain development in the future. One reason is that the desired phenotypes can be obtained with accumulation of beneficial mutations under selective pressure by adaptive evolution. At present, systems biology approaches are being successfully applied to study the evolved strains, allowing scientists to uncover the causative genetic targets. These targets can be experimentally verified by subsequent genetic manipulation in naive strains, and thus endow useful phenotypes on other industrial production strains.

SUMMARY

To meet commercial demands, microbial cells often need to be engineered to achieve certain metrics such as titer, yield, and productivity. Equally, these metabolically rewired cell factories must be able to withstand the harsh conditions encountered in industrial processes and thus maintain high yield and high productivity.

Based on prior understanding of toxicity and tolerance mechanisms, rational approaches, such as engineering regulatory circuits, harnessing efflux pumps, and manipulating plasma membranes, can be used to improve tolerance phenotypes.

Evolutionary engineering is a powerful approach to improve tolerance phenotypes, without requiring knowledge of underlying genetic mechanisms.

Combined with systems biology approaches, the genetic basis enabling improved tolerance phenotypes in the evolved strains can be elucidated and thus endowed on other microbial hosts.

ACKNOWLEDGMENTS

The authors wish to thank Tyler Doughty and Xiaowei Li for critical reading of the manuscript.

REFERENCES

Almario, M.P., L.H. Reyes, and K.C. Kao. 2013. Evolutionary engineering of *Saccharomyces cerevisiae* for enhanced tolerance to hydrolysates of lignocellulosic biomass. *Biotechnol Bioeng* 110:2616–2623.

Alper, H., J. Moxley, E. Nevoigt, G.R. Fink, and G. Stephanopoulos. 2006. Engineering yeast transcription machinery for improved ethanol tolerance and production. *Science* 314: 1565–1568.

Alper, H., and G. Stephanopoulos. 2007. Global transcription machinery engineering: a new approach for improving cellular phenotype. *Metab Eng* 9:258–267.

Ansari, S.A., M. Ganapathi, J.J. Benschop, F.C.P. Holstege, J.T. Wade, and R.H. Morse. 2012. Distinct role of mediator tail module in regulation of SAGA-dependent, TATA-containing genes in yeast. *EMBO J* 31:44–57.

Boyarskiy, S., S. Davis Lopez, N. Kong, and D. Tullman-Ercek. 2016. Transcriptional feedback regulation of efflux protein expression for increased tolerance to and production of n-butanol. *Metab Eng* 33:130–137.

Bracher, J.M., E. de Hulster, C.C. Koster, M. van den Broek, J.G. Daran, A.J.A. van Maris, and J.T. Pronk. 2017. Laboratory evolution of a biotin-requiring *Saccharomyces cerevisiae* strain for full biotin prototrophy and identification of causal mutations. *Appl Environ Microbiol* 83:e00892–17.

Caspeta, L., Y. Chen, P. Ghiaci, A. Feizi, S. Buskov, B.M. Hallstrom, D. Petranovic, and J. Nielsen. 2014. Altered sterol composition renders yeast thermotolerant. *Science* 346:75–78.

Caspeta, L., and J. Nielsen. 2015. Thermotolerant yeast strains adapted by laboratory evolution show trade-off at ancestral temperatures and preadaptation to other stresses. *mBio* 6(4):e00431.

Chasman, D.I., K.M. Flaherty, P.A. Sharp, and R.D. Kornberg. 1993. Crystal structure of yeast TATA-binding protein and model for interaction with DNA. *Proc Natl Acad Sci USA* 90:8174–8178.

Chen, A.H., and P.A. Silver. 2012. Designing biological compartmentalization. *Trend Cell Biol* 22:662–670.

Conrado, R.J., G.C. Wu, J.T. Boock, H. Xu, S.Y. Chen, T. Lebar, J. Turnek, N. Tomšič, M. Avbelj, R. Gaber, T. Koprivnjak, J. Mori, V. Glavnik, I. Vovk, M. Beninča, V. Hodnik, G. Anderluh, J.E. Dueber, R. Jerala, and M.P. Delisa. 2012. DNA-guided assembly of biosynthetic pathways promotes improved catalytic efficiency. *Nucleic Acid Res* 40:1879–1889.

Dahl, R.H., F. Zhang, J. Alonso-Gutierrez, E. Baidoo, T.S. Batth, A.M. Redding-Johanson, C.J. Petzold, A. Mukhopadhyay, T.S. Lee, P.D. Adams, and J.D. Keasling. 2013. Engineering dynamic pathway regulation using stress-response promoters. *Nat Biotechnol* 31:1039–1046.

Delebecque, C.J., A.B. Lindner, P.A. Silver, and F.A. Aldaye. 2011. Organization of intracellular reactions with rationally designed RNA assemblies. *Science* 333:470–474.

Drake, J.W. 1991. A constant rate of spontaneous mutation in DNA-based microbes. *Proc Natl Acad Sci USA* 88:7160–7164.

Dueber, J.E., G.C. Wu, G.R. Malmirchegini, T.S. Moon, C.J. Petzold, A.V. Ullal, K.L. Prather, and J.D. Keasling. 2009. Synthetic protein scaffolds provide modular control over metabolic flux. *Nat Biotechnol* 27:753–759.

Dunlop, M.J., Z.Y. Dossani, H.L. Szmidt, H.C. Chu, T.S. Lee, J.D. Keasling, M.Z. Hadi, and A. Mukhopadhyay. 2011. Engineering microbial biofuel tolerance and export using efflux pumps. *Mol Syst Biol* 7:487.

Fletcher, E., A. Feizi, M.M.M. Bisschops, B.M. Hallstrom, S. Khoomrung, V. Siewers, and J. Nielsen. 2017. Evolutionary engineering reveals divergent paths when yeast is adapted to different acidic environments. *Metab Eng* 39:19–28.

Ghiaci, P., J. Norbeck, and C. Larsson. 2013. Physiological adaptations of *Saccharomyces cerevisiae* evolved for improved butanol tolerance. *Biotechnol Biofuels* 6:101.

González-Ramos, D., A.R. Gorter de Vries, S.S. Grijseels, M.C. van Berkum, S. Swinnen, M. van den Broek, E. Nevoigt, J.-M.G. Daran, J.T. Pronk, and A.J.A. van Maris. 2016. A new laboratory evolution approach to select for constitutive acetic acid tolerance in *Saccharomyces cerevisiae* and identification of causal mutations. *Biotechnol Biofuel* 9(1):173.

Hahn, S. 2004. Structure and mechanism of the RNA polymerase II transcription machinery. *Nature Structural Mol Biol* 11:394–403.

Ibarra, R.U., J.S. Edwards, and B.O. Palsson. 2002. *Escherichia coli* K-12 undergoes adaptive evolution to achieve *in silico* predicted optimal growth. *Nature* 420:186–189.

Kildegaard, K.R., B.M. Hallstrom, T.H. Blicher, N. Sonnenschein, N.B. Jensen, S. Sherstyk, S.J. Harrison, J. Maury, M.J. Herrgard, A.S. Juncker, J. Forster, J. Nielsen, and I. Borodina. 2014. Evolution reveals a glutathione-dependent mechanism of 3-hydroxypropionic acid tolerance. *Metab Eng* 26: 57–66.

Kim, J.L., D.B. Nikolov, and S.K. Burley. 1993. Co-crystal structure of TBP recognizing the minor groove of a TATA element. *Nature* 365:520–527.

Klein-Marcuschamer, D., and G. Stephanopoulos. 2008. Assessing the potential of mutational strategies to elicit new phenotypes in industrial strains. *Proc Natl Acad Sci USA* 105:2319–2324.

Lam, F.H., A. Ghaderi, G.R. Fink, and G. Stephanopoulos. 2014. Engineering alcohol tolerance in yeast. *Science* 346:71–75.

Lang, G.I., and A.W. Murray 2008. Estimating the per-base-pair mutation rate in the yeast *Saccharomyces cerevisiae*. *Genetics* 178:67–82.

Leavitt, J.M., J.M. Wagner, C.C. Tu, A. Tong, Y. Liu, and H.S. Alper. 2017. Biosensor-enabled directed evolution to improve muconic acid production in *Saccharomyces cerevisiae*. *Biotechnol J* 12.1600687.

Lee, H., W.C. DeLoache, and J.E. Dueber. 2012. Spatial organization of enzymes for metabolic engineering. *Metab Eng* 14:242–251.

Lennen, R.M., and B.F. Pfleger. 2013. Modulating membrane composition alters free fatty acid tolerance in *Escherichia coli*. *PloS One* 8(1):e54031.

Mahr, R., C. Gätgens, J. Gätgens, T. Polen, J. Kalinowski, and J. Frunzke. 2015. Biosensor-driven adaptive laboratory evolution of l-valine production in *Corynebacterium glutamicum*. *Metab Eng* 32:184–194.

Mukhopadhyay, A. 2015. Tolerance engineering in bacteria for the production of advanced biofuels and chemicals. *Trend Microbiol* 23:498–508.

Nijland, J., H. Shin, R. de Jong, P. de Waal, P. Klaassen, and A. Driessen. 2014. Engineering of an endogenous hexose transporter into a specific D-xylose transporter facilitates glucose-xylose co-consumption in *Saccharomyces cerevisiae*. *Biotechnol Biofuel* 7(1):168.

Oud, B., V. Guadalupe-Medina, J.F. Nijkamp, D. de Ridder, J.T. Pronk, A.J. van Maris, and J.M. Daran. 2013. Genome duplication and mutations in ACE2 cause multicellular, fast-sedimenting phenotypes in evolved *Saccharomyces cerevisiae*. *Proc Natl Acad Sci USA* 110:E4223–E4231.

Park, S.H., H.U. Kim, T.Y. Kim, J.S. Park, S.-S. Kim, and S.Y. Lee. 2014. Metabolic engineering of *Corynebacterium glutamicum* for L-arginine production. *Nat Commun* 5:art. no. 4618.

Paul, E., Z.I. Zhu, D. Landsman, and R.H. Morse. 2015. Genome-wide association of mediator and RNA polymerase II in wild-type and mediator mutant yeast. *Mol Cell Biol* 35:331–342.

Peng, B., Y. Shen, X. Li, X. Chen, J. Hou, and X. Bao. 2012. Improvement of xylose fermentation in respiratory-deficient xylose-fermenting *Saccharomyces cerevisiae*. *Metab Eng* 14:9–18.

Sandberg, T.E., C.P. Long, J.E. Gonzalez, A.M. Feist, M.R. Antoniewicz, and B.O. Palsson. 2016. Evolution of *E. coli* on [U-13C]glucose reveals a negligible isotopic influence on metabolism and physiology. *PLoS One* 11(3):e0151130.

Sandoval, N.R., and E.T. Papoutsakis. 2016. Engineering membrane and cell-wall programs for tolerance to toxic chemicals: Beyond solo genes. *Curr Opin Microbiol* 33:56–66.

Shen, Y., X. Chen, B. Peng, L. Chen, J. Hou, and X. Bao. 2012. An efficient xylose-fermenting recombinant *Saccharomyces cerevisiae* strain obtained through adaptive evolution and its global transcription profile. *Appl Microbiol Biotechnol* 96:1079–1091.

Sonderegger, M., and U. Sauer. 2003. Evolutionary engineering of *Saccharomyces cerevisiae* for anaerobic growth on xylose. *Appl Environ Microbiol* 69:1990–1998.

Tan, Z., J.M. Yoon, D.R. Nielsen, J.V. Shanks, and L.R. Jarboe. 2016. Membrane engineering via *trans* unsaturated fatty acids production improves *Escherichia coli* robustness and production of biorenewables. *Metab Eng* 35:105–113.

Tseng, T.-T., K.S. Gratwick, J. Kollman, D. Park, D.H. Nies, A. Goffeau, and M.H. Saier Jr. 1999. The RND permease superfamily: an ancient, ubiquitous and diverse family that includes human disease and development proteins. *J Mol Microbiol Biotechnol* 1:107–125.

Wenger, J.W., J. Piotrowski, S. Nagarajan, K. Chiotti, G. Sherlock, and F. Rosenzweig. 2011. Hunger artists: Yeast adapted to carbon limitation show trade-offs under carbon sufficiency. *PLoS Genetics* 7(8): e1002202.

Zhang, H., H. Chong, C.B. Ching, H. Song, and R. Jiang. 2012. Engineering global transcription factor cyclic AMP receptor protein of *Escherichia coli* for improved 1-butanol tolerance. *Appl Microbiol Biotechnol* 94:1107–1117.

9 Engineering Microorganisms for the Production of Pharmaceutical Biomolecules
Current Trends and Future Prospects

Adilson José da Silva, Blanca Barquera, and Mattheos A. G. Koffas

CONTENTS

"You never fail until you stop trying."

Albert Einstein

Recent developments in the fields of metabolic engineering, systems biology, and synthetic biology have enabled the engineering of microbial cells to produce natural and nonnatural biomolecules with special biological activities. Many of these compounds are natural products isolated originally in tiny amounts from plants, animal tissues, or microorganisms. Due to difficulties extracting and purifying these products in significant amounts from their native sources, scientists have been pursuing the production of these compounds in recombinant microbial hosts. This chapter brings a brief overview of three classes of biomolecules that have been produced in engineered microbial cells and present potential pharmaceutical applications.

9.1 FLAVONOIDS

Flavonoids are secondary metabolites widespread in the plant kingdom. These colorful natural products are low-molecular-weight molecules composed of a simple 15 carbon backbone. According to the position of linkage of B-ring to the C-ring (benzopyrano moiety) they can be classified as flavonoids, isoflavonoids, and neoflavonoids. Some examples of different flavonoids that have already been produced by microbial hosts can be seen in Figure 9.1.

This class of compounds is known to possess preventive and therapeutic action against cardiovascular and neurodegenerative diseases caused by free radicals among other factors. Their medicinal properties come from their ability to scavenge free radicals such as reactive oxygen species (ROS) and reactive nitrogen species (RNS) and inhibit damage and propagation through metal chelation (Heim et al., 2002).

Moreover, it was found that flavonoids also play roles in the prevention and onset of diabetes and cancers. The antioxidant, anti-inflammatory and antibacterial effects of these compounds are also being explored for pharmaceutical purposes (Wang, 2000; Ding et al., 2013; Xiao and Shao, 2013; Xiao and Tundis, 2013).

Flavonoids have been the target of metabolic engineers for 10 to 15 years. To date, these compounds have been successfully produced and modified in microbial hosts, especially *Escherichia coli* and *Saccharomyces cerevisiae*. Although various approaches have been used to produce flavonoids in these hosts, the volumetric production achieved is still too low to be considered for industrial production. However, significant increments in the production have been achieved along the way and currently researchers make use of modern metabolic engineering and synthetic biology tools to improve the capabilities of microbial cell factories and expand the array of flavonoids being produced.

The flavonoids biosynthetic pathway in plants is well studied and was found to have a common core unit, chalcone, from the phenylpropanoid pathway. By action of many enzymes like oxidoreductases, isomerases, hydroxylases, glycosyltransferases, methyltransferases, acyltransferases, and so on, this core unit is converted into more than 8,000 known flavonoids (Veitch and Grayer, 2011; Iwashina, 2015) and approximately the 2,000 known isoflavonoids (Veitch, 2007).

L-tyrosine or L-phenylalanine are the precursors of the phenylpropanoid pathway (Figure 9.2). Both amino acids are initially converted into their respective carboxylic acids, *p*-coumaric acid and cinnamic acid, by the enzymes tyrosine ammonia lyase (TAL) or phenylalanine ammonia lyase (PAL), respectively. Cinnamic acid is then converted to *p*-coumaric acid by the enzyme cinnamic acid 4-hydroxylase (C4H). The next step converts *p*-coumaric acid into activated

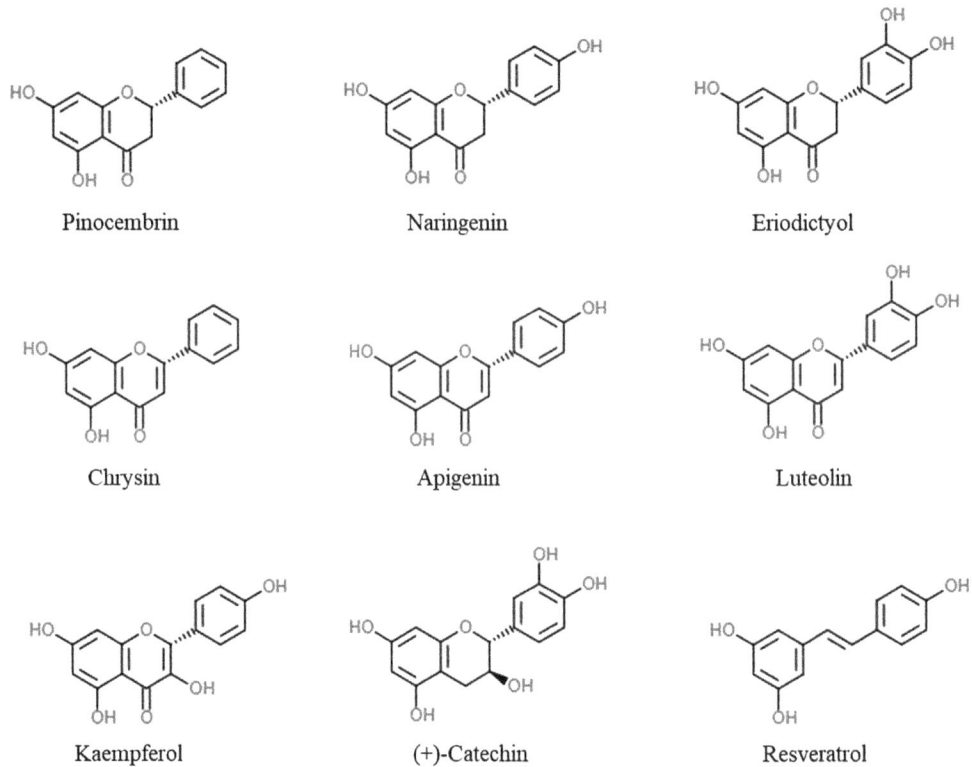

FIGURE 9.1 Examples of some heterogeneous flavonoids produced in microorganisms.

p-coumaroyl-CoA at the expense of one molecule of ATP and one molecule of coenzyme-A (CoA). This activated molecule combines with three molecules of malonyl-CoA, catalyzed by chalcone synthase (CHS), to form naringenin chalcone which is the C6-C3-C6 backbone of all flavonoids. After a ring-closing step by chalcone isomerase (CHI), naringenin chalcone is converted into naringenin which can be further transformed into different classes of flavonoids, isoflavonoids, aurones, and other compounds. Because naringenin is the major intermediate in flavonoids biosynthesis, it has been the focus of much research (Wu et al., 2014).

Malonyl-CoA is an important precursor for flavonoids biosynthesis and other compounds of interest like polyketides and fatty acids. Malonyl-CoA is mainly produced from acetyl-CoA by acetyl-CoA carboxylase (ACC) complex. One of the first attempts to increase the cytosolic pool of malonyl-CoA was done by overexpression of ACC complex genes (Leonard et al., 2007; Wattanachaisaereekul et al., 2008). A complementary approach consisted of deleting competing reactions that convert acetyl-CoA to other products like ethanol or acetate, via acetaldehyde desidrogenase (adhE) and acetate kinase (acka), respectively (Zha et al., 2009). Additionally, inhibition of fatty acid biosynthesis enzymes (FabB and FabF) by addition of cerulenin has increased malonyl-CoA availability for flavonoids production (Subrahmanyam et al., 1998). Recently, novel strategies to enhance malonyl-CoA have been proposed using genome-scale metabolic models (Lim et al., 2011; Xu et al., 2011; Bhan et al., 2013).

De novo pathway engineering for flavonoids production in microorganisms has been explored. Horinouchi's group

published the production of pinocembrin and naringenin from phenylalanine and tyrosine, respectively using *E. coli* as the host (Hwang et al., 2003). This was accomplished by the assembly of a three-gene pathway composed of *PAL* from *Rhodotorula rubra*, *CCL* from *S. coelicolor* A3(2), and *CHS* from *Glycyrrhiza echinata*. In the following year, *S. cerevisiae* was first employed to express plant phenylpropanoid pathway genes from *P. trichocarpa* x *P. deltoides* by Ro and Douglas (2004). Later on, our group showed the production of pinocembrin, naringenin, and eriodictyol in *S. cerevisiae* cultures fed with their respective precursors cinnamic acid, *p*-coumaric acid and caffeic acid (Yan et al., 2005b). Ralston and colleagues (2005) also used yeast cells to produce flavonoid compounds. Taken together along with additional studies (Watts et al., 2004; Jiang et al., 2005; Miyahisa et al., 2005), these pioneering works settled the first platforms for microbial production of flavonoids and related compounds.

Other flavonoid subclasses were also produced in microbial hosts. For example, Trantas and colleagues (2009) reported various combinations of 11 genes from 4 different plant species that were assembled in *S. cerevisiae* for production of stilbenes, flavanones, isoflavones, and flavonols. Koffas group showed the production of flavone chrysin, apigenin and luteolin in *S. cerevisiae* by expressing flavone synthases from parsley or snapdragon in the previously constructed flavanane producing strain (Leonard et al., 2005; Fowler and Koffas, 2009). The production of unstable anthocyanins (Yan et al., 2005a) or rare catechins for the first time (Chemler et al., 2007) also represent other pioneering works reporting *de novo* biosynthesis of flavonoids from simple precursors.

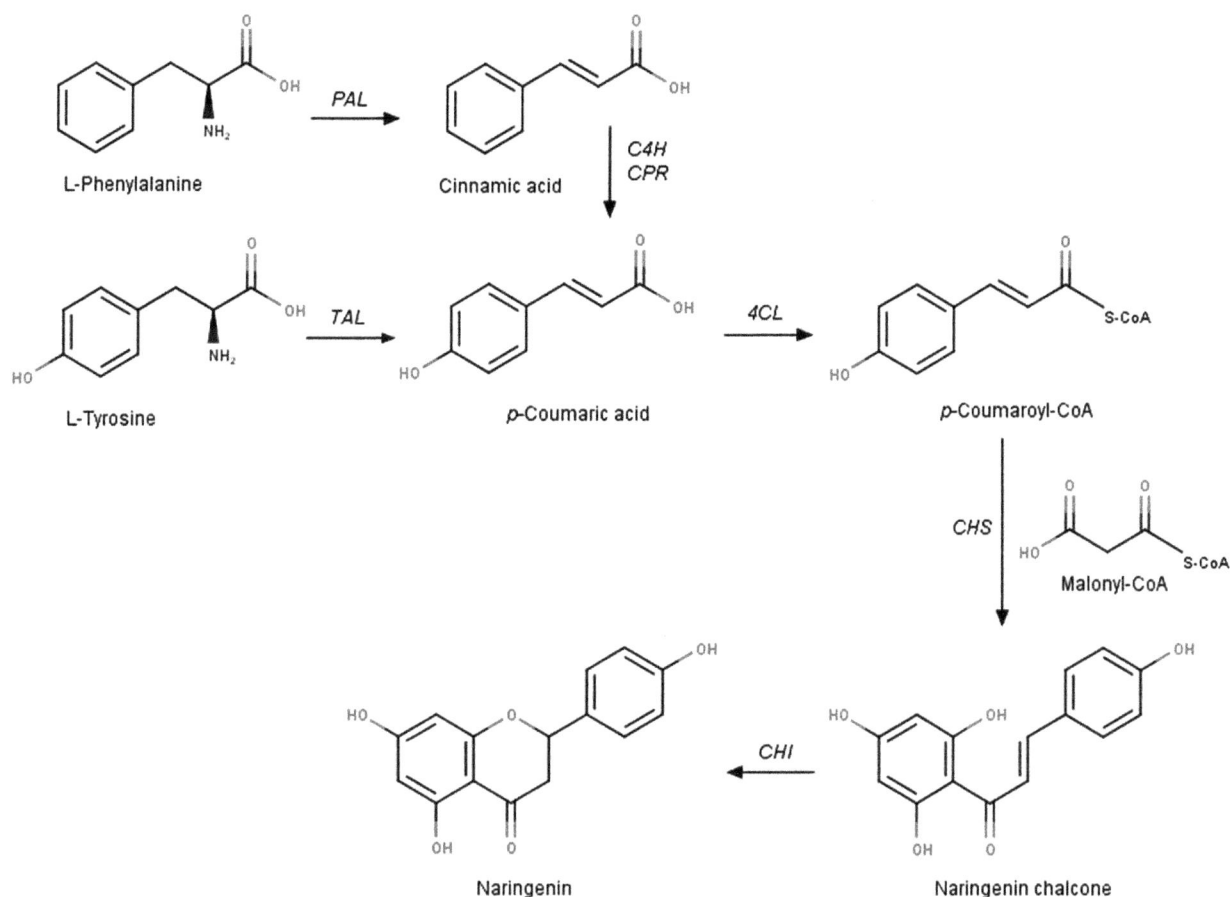

FIGURE 9.2 Biosynthetic pathway of naringenin, the common precursor of flavonoids. Abbreviations of enzymes: PAL: phenyl ammonia lyase; TAL: tyrosine ammonia lyase; C4H: cinnamic acid 4-hydroxylase; CPR: cytochrome P450 reductase; 4CL: p-coumaroyl-CoA ligase; CHS: chalcone synthase; CHI: chalcone isomerase.

Added to that, there is the production of resveratrol and some of its methylated derivatives using an artificial biosynthetic pathway reported by Hong and his group (Kang et al., 2014). Another good example of *de novo* biosynthesis of flavonoids in *S. cerevisiae* was reported by Rodriguez and colleagues (2017), where six flavonoids were produced including reso-kaempferol and fisetin, which were produced in yeast for the first time. Some of these examples are listed in Table 9.1. An extensive list of different flavonoids produced by microbial hosts can be found elsewhere (Pandey et al., 2016).

With the use of modern synthetic biology approaches, improved strains for flavonoid production were designed. Guided by simulations of genome-scale metabolic models for *E. coli*, improved strains with enhanced carbon flux toward malonyl-CoA were built (Fowler et al., 2009; Xu et al., 2011). These studies employed different algorithms which suggested deletion of different subsets of genes along with overexpression of malonyl-CoA biosynthetic pathway related enzymes. Using the CiED algorithm, Fowler and colleagues engineered an *E. coli* strain deficient in *sdhA*, *adhE*, *brnQ* and *citE* and overexpression of ACC complex, the biotin ligase *birA* from *P. luminescens*, as well as the *coaA* and *acs* genes. Xu and colleagues constructed an *E. coli* strain based on the *OptForce* computational tool that included knocking out of *fumC* and

sucC and overexpression of *accABCD*, *pgk*, *aceEF*, and *lpdA* genes. Both strains showed significant increase of intracellular malonyl-CoA production that reflected in high titers of naringenin (270 mg/L and 472 mg/L, respectively) after supplementation of *p*-coumaric acid (Xu et al., 2011). Fowler and colleagues also showed the production of 150 mg/L of eriodictyol when caffeic acid was fed as precursor (Fowler et al., 2009). Moreover, using the same tool, these authors also designed an *E. coli* strain deficient in *pgi*, *pldA*, and *ppc* genes that showed improved accumulation of NADPH and was used to produce leucoanthocyanidin and (+)-catechin (Fowler et al., 2009). Using CRISPRi/dCas9, the downregulation of *fumB*, *fumC*, *acnA*, and *fadR* were also confirmed to improve flavanone production as previously predicted by OptForce (Cress et al., 2015).

After integrating genomic and transcriptomic analysis, both scutellarin and apigenin-7-O-glucuronide were produced in *S. cerevisiae* engineered cells (Liu et al., 2018). These are two major components of the flavonoid extract from *Erigeron breviscapus*, which is used in China to treat cardio and cerebrovascular diseases. By high-throughput functional screening, the authors mapped the breviscapine biosynthetic pathway in *E. breviscapus* and cloned the corresponding genes in *S. cerevisiae* to produce the desired flavonoids from

TABLE 9.1

Examples of Flavonoids Produced in Engineered *E. coli* and *S. cerevisiae* Cells

Product	Precursor	Overexpressed Genes	Host	Titer (mg/L)	References
Naringenin	Tyrosine	*R. rubra* (*PAL*), *S. coelicolor* A3(2) (*CCL*), *G.*	*E. coli*	57	Miyahisa et
Pinocembrin	Phenylalanine	*echinata* (*CHS*), *P. lobata* (*CHI*), *C.*		58	al., 2005
		glutamicum (*accBC* and *dtsR1*)			
Cyanidin 3-*O*-glucoside	Catechin	ANS-3-GT fusion protein,	*E. coli*	104	
Pelargonidin 3-*O*-glucoside	Afzelechin	*E. coli* (*pyrE, pyrF, cmk, ndk, pgm, galU*)		113	
Pelargonidin 3-*O*-glucoside	Naringenin	*M. domestica* (*F3H*), *A. andraeanum* (*DFR*), *P.*		0.98	Yan et al.,
Cyanidin 3-*O*-glucoside	Eriodictyol	*hybrida* (*ANS*), *A. thaliana* (*3GT*),		2.07	2005a
		Desmodium uncinatum (*LAR*)			
Pelargonidin	Naringenin	Same as for Pelargonidin 3-*O*-glucoside (from		0.10	
3-*O*-6″-*O*-malonylglucoside	Eriodictyol	naringenin) plus *Dahlia variabilis* (*3MaT*)		0.12	
Cyanidin					
3-*O*-6″-*O*-malonylglucoside					
Pelargonidin 3-*O*-glucoside	Afzelechin	Fusion of *P. hybrida* (*ANS*) and *A. thaliana*		78.9	
Cyanidin 3-*O*-glucoside	(+)-Catechin	(*3GT*), *E. coli* (*Pgm, galU*)		70.7	
Resveratrol	*p*-Coumaric acid	*A. thaliana* (*4CL1*), *A. hypogaea* (*STS*)	*E. coli*	104.5	Watts et al.,
Piceatannol	Caffeic acid			13.3	2006
Apigenin	L-Tyrosine	*R. rubra* (*PAL*), *S. coelicolor* A3(2) (*CCL*), *G.*	*E. coli*	13	Miyahisa et
Chrysin	Phenylalanine	*echinata* (*CHS*), *P. lobata* (*CHI*), *P. crispum*		9.4	al., 2006
		(*FS1*) *C. glutamicum* (*accBC* and *dtsR1*)			
Kaempferol	L-Tyrosine	Same as for Apigenin/Chrysin plus *Citrus*		15.1	
Galangin	Phenylalanine	species (*F3H, FLS*)		1.1	
(+)-Catechin	Eriodictyol	*M. domestica* (*F3H*), *P. crispum* (*4CL-2*), *D.*	*E. coli*	8.8	Chemler
(+)-Afzelechin	Naringenin	*uncinatum* (*LAR*), *A. andraeanum* (*DFR1*)		0.7	et al., 2007
Pinocembrin	Cinnamic acid	*R. trifolii* (*matBC*), *P. crispum* (*4CL2*), *P.*	*E. coli*	710	Leonard et
Naringenin	*p*-Coumaric acid	*hybrida* (*CHS*), *M. sativa* (*CHI*)		186	al., 2008
Eriodictyol	Caffeic acid	(Cerulenin added)		54	
Chrysin	Cinnamic acid	*R. trifolii* (*matBC*), *P. crispum* (*4CL, FS1*), *P.*		5	
Apigenin	*p*-Coumaric acid	*hybrida* (*CHS*), *M. sativa* (*CHI*)		110	
Luteolin	Caffeic acid	(Cerulenin added)		4	
Resveratrol	Phenylalanine	*P. trichocarpa* x *P. deltoides* (*PAL, CPR*), *G.*	*S. cerevisiae*	0.29	Trantas
	p-Coumaric acid	*max* (*C4H, 4CL*), *V. vinifera* (*RS*)		0.31	et al., 2009
Naringenin	Phenylalanine	*P. trichocarpa* x *P. deltoides* (*PAL, CPR*), *G.*		8.9	
	p-Coumaric acid	*max* (*C4H, 4CL, CHS, CHI*)		15.6	
Genistein	Phenylalanine	Same as for naringenin plus *G. max* (*IFS*)		0.1	
	p-Coumaric acid			0.14	
	Naringenin			7.7	
Kaempferol	Phenylalanine	Same as for naringenin plus *G. max* (*F3H*), *S.*		1.3	
	p-Coumaric acid	*tuberosum* (*FLS*)		0.9	
	Naringenin			4.6	
Quercetin	Phenylalanine	Same as for naringenin plus *G. max* (*F3H,*		–	
	p-Coumaric acid	*F3'H*), *S. tuberosum* (*FLS*)		0.26	
	Naringenin			0.38	
Naringenin	*p*-Coumaric acid	*P. crispum* (*4CL2*), *P. hybrida* (*CHS*), *M.*	*E. coli*	270	Fowler
Eriodictyol		*sativa* (*CHI*)		150	et al., 2009
		E. coli (*coaA, acs*), *P. luminescens* (*ACC, BirA*)			
7-*O*-methyl aromadendrin	*p*-Coumaric acid	*P. crispum* (*4Cl-2*), *P. hybrida* (*CHS*), *M.*	*E. coli*	2.7	Malla et al.,
		sativa (*CHI*), *A. thaliana* (*F3H*), *S. avermitilis*			2012
		(*OMT*), *N. farcinica* (*AccBC, birA, acs*)			
3,5-dihydroxy 4'-methoxy stilbene	Glucose	*S. espanaensis* (codon optimized *TAL*), *S.*	*E. coli*	2.5	Kang et al.,
3,4'dimethoxy -5-hydroxy stilbene		*bicolor* (*OMT1, OMT3*), *A. hypogaea* (*STS*) *S.*		0.17	2014
3,5,4'-trimethoxy stilbene		*coelicolor* A3(2) (*CCL*)		0.05	

(Continued)

TABLE 9.1 (CONTINUED)

Examples of Flavonoids Produced in Engineered *E. coli* and *S. cerevisiae* Cells

Product	Precursor	Overexpressed Genes	Host	Titer (mg/L)	References
Naringenin	Glucose	*P. crispum (4CL), P. hybrida (CHS), M. sativa (CHI)*	*S. cerevisiae*	1.6	Rodriguez et al., 2017
Kaempferol		Same as for naringenin plus *A. mongholicus (F3H), A. thaliana (FLS)*		26.6	
Quercetin		Same as for kaempferol plus *P. hybrida (FMO)* or *F. ananassa (FMO), C. roseus (CPR)*		20.4	
Liquiritigenin		Same as for naringenin plus *A. mongholicus (CHR)*		5.3	
Resokaempferol		Same as for liquiritigenin plus *A. mongholicus (F3H), A. thaliana (FLS)*		0.5	
Fisetin		Same as for resokaempferol plus *P. hybrida (FMO)* or *F. ananassa (FMO), C. roseus (CPR)*		2.3	

glucose. After optimization, titers of 108 mg/L for scutellarin and 185 mg/L for apigenin-7-O-glucuronide were reached (Liu et al., 2018).

Other researchers have made use of modular metabolic engineering to enhance flavonoid production in microbial hosts. By this approach, the biosynthetic pathway is designed as a combination of a few modules, where each of them is composed of a subset of all the metabolic reactions. The production of the final product can be enhanced by optimizing each module individually or by combining them in an optimized way. Wu and colleagues reported the use of modular metabolic engineering to produce different flavonoids. In one example, the resveratrol biosynthetic pathway was divided into three modules: the first one contained *RgTAL* and *Pc4CL* for coumaroyl-CoA production, the second module was composed of *VvSTS* for resveratrol synthesis and the last module was composed of *matB* and *matC* to improve the malonyl-CoA production. Using two different promoters and four different backbone plasmids, 18 different combinations of the 3 modules were evaluated for resveratrol biosynthesis and its production reached 35 mg/L for the best combination (Wu et al., 2013). Other researchers have explored co-culture systems, where two different engineered strains produce complementary parts of the biosynthetic pathway of interest. Koffas and his group reported the first co-culture system for microbial flavonoid biosynthesis, where the production of flavon-3-ols by two engineered *E. coli* strains resulted in a 970-fold improvement in titer over previously published monoculture production (Jones et al. 2016). Another example of co-culture system was developed for *de novo* resveratrol production where one strain was designed to produce and excrete the precursor *p*-coumaric acid into the medium while the second strain was harboring the corresponding genes to convert this precursor to the final product (Camacho-Zaragoza et al., 2016). A further step was reported by Jones and colleagues where four different modules were assembled in four separated strains for the *de novo* microbial production of the

anthocyanidin-3-*O*-glucoside callistephin for the first time. In this polyculture system, *E. coli* cells were collectively expressing 15 exogenous or modified pathway enzymes from different plant and microbial sources. By dividing this pathway into four modules and coordinating the expression of each module by a different strain in a consortium-based system, the authors were able to express a lengthy pathway while managing the associated metabolic burden (Jones et al., 2017).

Post-modification of flavonoids has been shown to improve the pharmacokinetic properties and biological activities of these molecules, and these reactions are currently being carried out by microbial cells. Instead of carrying out simple biotransformation reactions with actinomycetes and eukaryotic cells, the use of engineered hosts for this particular purpose can minimize side products formation and improve regioselectivity of desired modifications. Expressing these functional post-modification enzymes into hosts harboring flavonoid pathways has proved to be an effective way to obtain these compounds (Kim et al., 2015; Luo et al., 2015; Zhang et al., 2016).

Glycosylation, which is the transfer of a sugar unit from an activated sugar nucleotide diphosphate to an acceptor molecule, is the most explored post-modification reaction for flavonoids. Plants as well as bacterial glycosyl transferases (GTs) can be used for glycosylation of flavonoids, and the initial approach to producing such compounds used to be the expression of GTs from *Arapdopsis* or other sources in *E. coli* cells (Kim et al., 2006c, 2010; He et al., 2008). Using this approach, diverse flavonoid O-glucosides derivatives were reported, as for example, quercetin glucosides, kaempferol 3-*O*-glucoside, luteolin glucosides, genistein 7-*O*-glucosid, and so on (Lim et al., 2004; Kim et al., 2006c; He et al., 2008). Another common strategy consists on engineering nucleotide sugar biosynthetic pathway to direct the glycosylation of supplemented flavonoids in a desired manner. By this approach, for instance, quercetin 3-*O*-xyloside was produced by an *E. coli* strain with deletion of glucose-6-phosphate utilizing pathway genes and overexpression of UDP-xylose biosynthetic pathway genes

(Pandey et al., 2013). Strains where the utilization of glucose-6-phosphate was blocked were also used to produce kaempferol 3-*O*-glucoside from naringenin (Malla et al., 2013) or different flavonols with four novel conjugated aminodeoxy sugars after further engineering (Pandey et al., 2015).

Methylation, which is another common post-modification of many flavonoids, has been carried out mainly by the expression of functional O-methyltransferases (OMTs) from different sources in *E. coli*, *S. cerevisiae*, and *Streptomyces*. Ahn and his group have shown the production of O-methylated flavonols, flavones, flavanones, and isoflavonoids by expressing OMTs derived from rice, poplar and *G. max* in their engineered flavonoid producer *E. coli* strains (Kim et al., 2005, 2006a, 2006c). Production of methylated flavonoids was also reported by expression of OMTs from *Streptomyces* species (Kim et al., 2006b, Koirala et al., 2014). A recent review brings a wide overview of flavonoids O and C methylation (Koirala et al., 2016).

9.2 POLYSACCHARIDES

Glycosaminoglycans (GAGs) are polysaccharides widely used for pharmaceutical purposes. These molecules are linear hetero-polysaccharides containing amino-sugars such as glucosamine and galactosamine, and acidic sugars such as glucuronic acid and iduronic acid. According to their structural features, different GAGs present the ability to interact with different subsets of cellular proteins, triggering diverse biological activities. Two relevant examples among the major commercial GAGs are heparin and chondroitin (Figure 9.3).

Heparin is a GAG that figures among the most used anticoagulants in current medicine. This molecule was also described to have other relevant biological activities like anti-infectious,

anti-inflammatory, anti-atherosclerotic, anti- and pro-proliferative, and anti-metastatic properties (Onishi, 2016).

The structure of heparin is mostly composed of a trisulfated (TriS) repeating disaccharide unit composed of 2-*O*-sulfo-α-L-iduronic acid (IdoA2 S) 1,4-linked to 6-*O*-sulfo-*N*-sulfo-α-D-glucosamine (GlcNS6S) (Figure 9.3). Other varied minor disaccharide units complete the heparin structure making it a complex GAG. Other GAG family members include heparin sulfate (HS), chondroitin sulfate (CS), keratan sulfate (KS), and hyaluronan (HA).

Heparin is found as a mixture of polysaccharide chains composed of 16 to 160 saccharide units, approximately. The average length of heparin chain is around 30 to 40 residues, giving heparin a molecular weight of approximately 20 kDa (Edens et al., 1992). But low (4–6 kDa) and ultra-low (2–3 kDa) molecular weight heparins are also produced by chemical or enzymatic depolymerization of heparin. These LMW heparins are composed of 6–10 disaccharide residues and are more suitable for certain therapeutic uses due to their better bioavailability and pharmacodynamics properties (Xu et al., 2014). Ultra-low-molecular-weight (ULMW) heparins can be prepared by further depolymerization of heparin until the formation of small polymers composed of 3–5 disaccharide residues (Lima et al., 2013). Some of these compounds can also be produced by chemical synthesis, as the pharmaceutical product Arixtra® (fondaparinux) (Petitou et al., 1986).

The anticoagulant activity of heparin is associated to a pentasaccharide sequence containing a central 3-*O*-sulfo group (GlcN3S) residue that can bind to antithrombin III (AT). This interaction causes a conformational change in AT that enhances its inhibitory activity over factor IIa (thrombin), factor Xa and other coagulation cascade serine proteases (Linhardt, 2003).

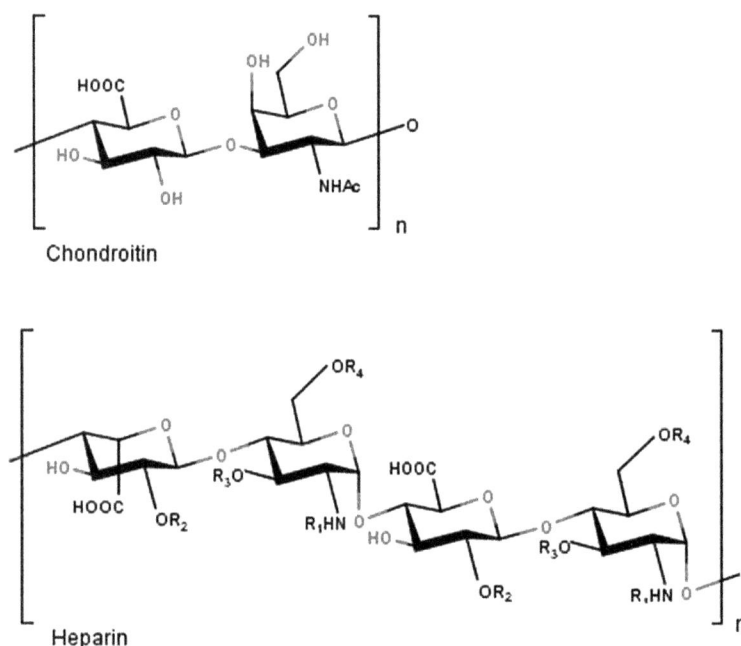

FIGURE 9.3 Chondroitin and heparin repeating units. Heparin: (80%IdoA; 20%GlcA; average 2.5 SO_3H/disaccharide) $R_1 = SO_3H/Ac/H$, $R_2 = SO_3H/H$, $R_3 = H/SO_3H$, $R_4 = SO_3H/H$.

Currently, heparin is produced by extraction and purification from porcine intestine, with an average yield of 300 mg of heparin per animal (Bhaskar et al., 2012). The initial extraction step consists of using proteases to solubilize the intestines, from where heparin can be precipitated by ammonium salts or captured by anion exchange resins. Raw heparin is then prepared by repeated steps of precipitation using alcohol and this partially purified product is further processed under good manufacturing practices (cGMP) to generate pharmaceutical heparin (Liu et al., 2009).

New production routes for heparin have been pursued in the last few years by the scientific community as an effort to replace the use of animal tissues and to attend the increasing demand for this life-saving drug. Among the new heparin producing technologies, the chemoenzymatic synthesis appears as a promising platform. This process is composed of the preparation of the polysaccharide backbone followed by chemical or enzymatic modifications on the attached groups. These modifications include removal of N-acetyl groups and addition of N-sulfo groups to the resulting GlcN residues, epimerization of most of the GlcA residues to IdoA residues and introduction of O-sulfo groups to the 2-position of IdoA residues and 6- and 3- positions of GlcN residues (Vaidyanathan et al., 2017).

Due to the highly complex structure of heparin, its production in microbial cells is currently not feasible. However, these organisms have been explored to produce heparosan, which is a polysaccharide capsular component of bacterial cells that can be used as the polysaccharide backbone for heparin synthesis (Suflita et al., 2015). Also, chondroitin, which is the backbone polysaccharide precursor of CS and has been used as anti-inflammatory drug for the treatment of osteoarthritis and rheumatism (McAlindon et al., 2000), is another product of bacterial cells that has been produced by fermentation (He et al., 2015; Jin et al., 2016).

E. coli capsular polysaccharide (CPS) from strains K4 and K5 are naturally composed of fructosylated chondroitin and heparosan, respectively, which in turn can be used as backbones for production of chondroitin and heparin.

The heparosan CPS of E. coli K5 is a high molecular weight polysaccharide (75–150 kDa) composed of repeating units of GlcA 1,4 linked to GlcNAc. The reduction of heparosan molecular weight generates the backbone for heparin biosynthesis, and this depolymerization process can be carried out by chemical, enzymatic or photochemical treatment (Higashi et al., 2011; Masuko et al., 2011). The enzymatic process employs the K5 lyase which promotes the release of the polysaccharide and its shortening by β-elimination mechanism.

The heparosan backbone is then modified by chemical or enzymatic reactions to be converted to heparin. Generally, heparosan is treated with a strong base for N-deacetylation and Et_3N-SO_3 to N-sulfonation forming an N-sulfo-N-acetyl heparosan (Wang et al., 2011). Different hydrolysis conditions result in varying amounts of NA (repeating units with multiple N-acetyl groups) and NS (repeating units with multiple N-sulfo groups). Alternatively, enzymatic treatment

with N-deacetylase/N-sulfotransferase enzymes NDST-1 or NDST-2 generates N-sulfo-N-acetyl heparosan with different domains arising according to the different specificities of the NDST isoforms (Aikawa et al., 2001).

Production of heparosan by E. coli K5 fermentation has reached 15 g/L of product at a volumetric productivity of 0.4 g/(L.h) by growing the cells in fed-batch mode employing a glucose exponential feeding regime (Wang et al., 2010). Similar results were obtained by the same group in a 15 L bioreactor culture using a pH-stat fed-batch strategy (Wang et al., 2011).

Recombinant E. coli strains have also been constructed to avoid using the pathogenic K5 strain for heparosan production. Synthesis and exportation of capsular E. coli K5 heparosan are codified by a cluster of genes composed of three regions. Region 2 contains KfiA, KfiB, KfiC, and KfiD genes, which are related to the biosynthesis of heparosan polysaccharide (Barreteau et al., 2012). Zhang and colleagues (2012) cloned different combinations of E. coli K5 Kfi genes into BL21, a widely used and safe E. coli strain. The authors found that heparosan could be produced by expressing only KfiA together with KfiC, but the co-expression of KfiB and KfiD favored heparosan production. The recombinant strain sABCD containing all four Kfi genes showed the highest production among the six strains constructed, reaching a titer of 1.88 g/L in a DO-stat fed-batch process. The obtained heparosan was shown to be identical to E. coli K5 heparosan by NMR analysis but the recombinant product showed different average molecular weights for each strain. The absence of K5 lyase in the E. coli BL21 genome may explain the higher molecular weights observed for heparosan produced by three out of four recombinant strains (Zhang et al., 2012).

E. coli K-12 was also used for heparosan production. The genes kfiABCD from E. coli K5 were cloned in pairs (kfiAB and kfiCD) in low copy and compatible plasmids (Barreteau et al., 2012). After running a fed-batch culture using minimal medium and glycerol as carbon source, a titer of 1 g/L of heparosan was attained from the E. coli K-12 strain harboring both plasmids. The product accumulated in the cytoplasm of host cells and was composed of a high molecular weight heparosan (105 kDa). Because this average molecular weight is higher than what is observed for E. coli K5 (50–80 kDa) as well as for a recombinant E. coli K-12 expressing the complete three gene clusters for heparosan biosynthesis, it was hypothesized that the exportation system impacts the actual polysaccharide length (Barreteau et al., 2012). Cloning of an E. coli K5 naturally occurring lyase gene (elmA) in a third compatible plasmid led to production of various heparosan derived oligosaccharides in vivo using an E. coli K-12 strain harboring the three plasmids (Barreteau et al., 2012).

Chondroitin production from CPS of E. coli K4 represents a promising alternative to animal-sourced chondroitin and derivatives. E. coli K4 CPS is formed by a fructosylated chondroitin but an unfructosylated product can be obtained by knocking-out kfoE gene (He et al., 2015). The E. coli K4 CPS biosynthetic pathway is composed of three genes kfoA, kfoF, and kfoC, coding for a UDP-glucose-4-epimerase, a

UDP-glucose dehydrogenase, and chondroitin polymerase, respectively. The three-gene pathway was transferred to *E. coli* Star BL21(DE3) by cloning the respective genes into the ePathBrick vectors. This expression system supports the modular assembly of pathway components under the control of the T7 promoter (Xu et al., 2012). Using the isocaudamer pairs of restriction sites present in the vector, the three genes for CPS biosynthesis were assembled in operon, pseudo-operon, and monocistronic configurations. The pseudo-operon form, where each gene is preceded by a promoter, but all genes share the same terminator for all the mRNA transcripts, showed best results, and the cloned genes were then assembled into eight different combinatorial orders. Maximum production was observed for the genes arranged in the order *kfoC*, *kfoA*, *kfoF* where 213 mg/L of chondroitin were produced in shake flasks and 2.4 g/L in fed-batch culture (He et al., 2015). The chondroitin production can be increased by engineering transcriptional regulators such as *SlyA* and the anti-termination transcriptional factor *rfaH*. Wu and colleagues observed an increase of 1.5- to 1.8-fold in CPS production after homologously overexpressing the transcriptional regulator SlyA in *E. coli* K4 (Wu et al., 2013). An increase of 40% to 140% on CPS production by *E. coli* K4 was observed after homologous expression of *rfaH*, which is involved in an antitermination process in capsule expression and may have an impact on the intracellular concentration of UDP-sugar precursors (Cimini et al., 2013).

More recently, food-grade *Bacillus subtilis* was used as a recombinant host for both heparosan and chondroitin production (Jin et al., 2016). For heparosan biosynthesis, the genes *kfiA* and *kfiC* from *E. coli* K5 were cloned and integrated into the chromosome of *B. subtilis* 168 strain. A similar strategy was followed for chondroitin production: *kfoA* and *kfoC* chondroitin pathway genes of *E. coli* K4 were cloned and integrated into the chromosome of the same *B. subtilis* strain. Overexpression of *B. subtilis* tuaD gene that encodes

for the UDP-glucose dehydrogenase increased yields of both products by favoring the production of UDP-glucuronic acid, which is a common precursor for heparosan and chondroitin. Heparosan producing cells reached titers of 2.65 g/L in shake flasks and 5.82 g/L in fed-batch culture. Titers of 2.54 g/L and 5.82 g/L of chondroitin were attained for flask and bioreactor cultures, respectively (Jin et al., 2016). A new study from the same group showed the production of chondroitin sulfate A and C. The authors reported the optimization of chondroitin production in *Bacillus subtilis* reaching 7.15 g/L of product from sucrose. The produced chondroitin was then enzymatically converted to chondroitin sulfate A and C at high conversion rates by the action of purified aryl sulfotransferase IV (ASST IV), chondroitin 4-sulfotransferase (C4ST), and chondroitin 6-sulfotransferase (C6ST) (Zhou et al., 2018).

9.3 ISOPRENOIDS

Isoprenoids, also known as terpenes, represent a class of compounds composed of two or more units of a five-carbon building block called isoprene (2-methyl-1,3-butadiene), which is a branched-chain unsaturated hydrocarbon. This diverse class of natural products has been largely used for different applications, including pharmaceutical purposes. In this context, one important isoprenoid molecule that is currently produced by microbial cells is artemisinin (Figure 2). This compound is an anti-malarial drug used worldwide for the treatment of malaria, being recommended by the World Health Organization (WHO). Artemisinin was originally produced by extracts from the plant *Artemisia annua*, but the fragility of this supply chain has caused great price and availability fluctuations, suggesting the need for a new production method for this important bioproduct. A semi-synthetic process was developed to attend this demand, where engineered *S. cerevisiae* cells produce artemisinic acid which is chemically converted to artemisinin (Figure 9.4).

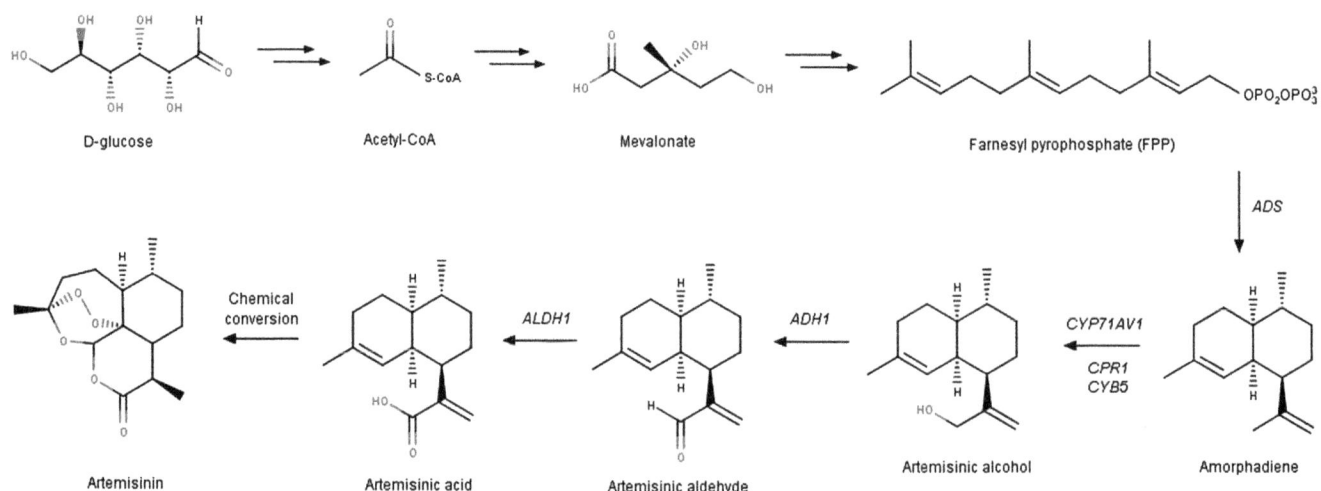

FIGURE 9.4 Semi-synthetic route for artemisinin production. Artemisinic acid is produced by engineered *S. cerevisiae* cells expressing *A. annua* enzymes. After extraction from the fermentation broth, artemisinic acid is chemically converted to artemisinin. Enzymes abbreviations: ADS: amorphadiene synthase; CYP71AV1: cytochrome P450 that converts amorphadiene to artemisinic alcohol; CPR1: cytochrome P450 reductase; CYB5: cytochrome b5; ADH1: artemisinic alcohol dehydrogenase; ALDH1: artemisinic aldehyde dehydrogenase.

The natural biosynthetic pathway of artemisin in *A. annua* starts with farnesyl diphosphate (FPP), which is the universal sesquiterpene precursor. By the action of amorphadiene synthase (ADS), FPP is converted to the alkane amorphadiene. The latter compound is then oxidized to dihydroartemisinic acid from which the final product artemisinin is obtained after a spontaneous photochemical rearrangement (Paddon and Keasling, 2014).

The process development involved an extensive knowledge of the artemisinin pathway in *A. annua*, and one of the first important discoveries was about the role of a cytochrome P450 enzyme (designated CYP71AV1) in the conversion of amorphadiene to artemisinic acid (Teoh et al., 2006). Because P450 is well expressed in yeast cells, *S. cerevisiae* was the host of choice for artemisinin production.

S. cerevisiae naturally produces the precursor FPP from the mevalonate pathway for ergosterol production. So, the first step was to increase FPP availability, which was accomplished by overexpressing integrated copies of all the mevalonate pathway genes under the control of the strong promoters *GAL1* or *GAL10*. Using a high-copy plasmid, the enzyme ADS from *A. annua*, which is responsible for the conversion of FPP to amorphadiene, was cloned into yeast cells under control of *GAL1* promoter. At this stage, after several optimization studies, amorphadiene production was increased from 100 mg/L to over 40 g/L (Westfall et al., 2012). This high titer was seen as crucial to make the process economically competitive.

The next step involving the oxidation of amorphadiene to artemisinic acid required further investigation about the *A. annua* enzymes involved in artemisinin biosynthesis. The first important discovery came when researchers found out that the reaction carried out by CYP71AV1 required the presence of a cytochrome P450 reductase (CPR1) to provide the two electrons for the monooxygenase reaction. By co-expressing these two enzymes there was artemisinin formation, but also a highly toxic level of reactive oxygen species (ROS) that had a great impact on cell's viability. This toxicity problem was solved after scientists discovered another enzyme involved in amorphadiene oxidation. This enzyme was cytochrome B5 (CYB5), which has an important role in the electron transfer reaction.

At this point, although there was a high production of amorphadiene, titers of artemisinin were still very low. The reason behind this low conversion was discovered when the oxidation process of amorphadiene to artemisinic acid was further investigated. This oxidation reaction was composed of three steps, with amorphadiene being first converted to artemisinic alcohol which was then converted to artemisinic aldehyde and finally to artemisinic acid. After discovery and expression of another two dehydrogenases (*ADH1* and *ALDH1*) from *A. annua* that were found to be responsible for the last two oxidation reactions (Bertea et al., 2005), titers of 25 g/L of arteminisic acid were reached by the engineered yeast cells (Tsuruta et al., 2009). With the development of a chemical reaction to convert artemisinic acid to the final product artemisinin, a cost-competitive process was established to provide artemisinin.

Taxol is another important isoprenoid which had its precursor biosynthesis developed in microbial hosts. Taxol (paclitaxel) is an anticancer drug approved by FDA that was isolated from the Pacific yew tree *Taxus brevifolia*. Because Taxol production by plant extraction or cultured plant cells is an expensive process, scientists have been trying to produce this compound heterologously at yields that can make Taxol production from microbial cells economically feasible.

Taxol biosynthetic pathway is very complex and starts from the universal isoprenoid substrates isopentenyl diphosphate (IPP) and dimethylallyl diphosphate (DMAPP) followed by polyisoprene formation of a geranylgeranyl diphosphate (GGPP) (Figure 9.5). The downstream reactions to form Taxol are composed of 19 steps from GGPP, involving many complex reactions like oxygenations and ring formation steps (Croteau et al., 2006).

Stephanopoulos and his group studied the production of taxadiene, the first committed Taxol intermediate, in engineered *E. coli* cells. The taxadiene pathway was divided into two modules, where the first module comprised the native reactions of the upstream methylerythritol-phosphate (MEP) pathway that generates IPP, and the second module was composed of the heterologous downstream pathway that leads to taxadiene. By a multivariate search that also minimized the accumulation of indole, which is an inhibitory intermediate, the authors reported a 15,000-fold increase in production reaching approximately 1 g/L of taxadiene. Moreover, they also engineered a P450-mediated 5α-oxidation of taxadiene to taxadien-5α-ol, which is the next step in Taxol biosynthesis (Ajikumar et al., 2010). The same group also evaluated the taxadiene production in K- and B- derived *E. coli* strains. Using different promoters (T7, Trc, and T5) and cellular backgrounds, they found a 2.5-fold higher production of taxadiene in a K-derivative *E. coli*. Temperature did not play a significant role, but 22°C was described as the optimal production temperature. Major differences between the studied strains were found in pyruvate metabolism and the higher sensitivity of K-derived cells to indole (Boghigian et al., 2012).

Meng and colleagues reported an *in silico* analysis comparing the maximum theoretical IPP yields and the thermodynamic properties of two different metabolic pathways (DXP and MVA) using different hosts and carbon sources. The MVA pathway is native to yeasts like *S. cerevisiae* and plants while the DXP is a bacterial pathway. Computational results showed that the DXP pathway can give better yields of IPP and combining genetic manipulation of the DXP pathway and chromosomal engineering, the authors reported a titer of 876 mg/L taxadiene production in *E. coli* (Meng et al., 2011). However, further studies are still necessary to increase the titers of taxadiene production in microbial cells.

SUMMARY

The production of complex molecules like flavonoids, polysaccharides, and isoprenoids by microbial cells illustrates the enormous potential of metabolic engineering to provide alternative routes to plant and animal-tissue derived

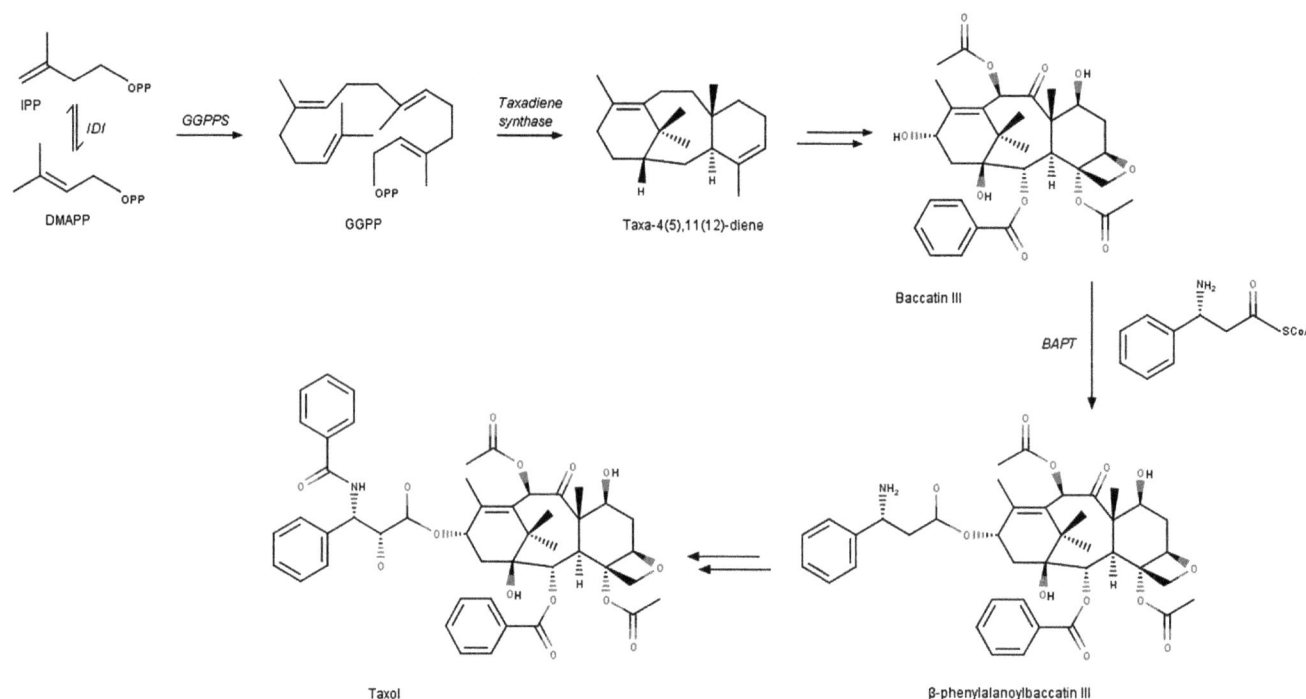

FIGURE 9.5 Simplified representation of Taxol biosynthetic pathway in *Taxus brevifolia*. Enzymes abbreviations: GGPPS: geranylgeranyl diphosphate synthase; BAPT: bacattin III:3-amino-3-phenylpropanoyltransferase.

compounds. Despite the intrinsic metabolic limitations imposed by the heterologous host cells, the use of synthetic biology tools, high-throughput technologies, and combinatorial approaches, among other advances have made possible the production of increasingly complex molecules by engineered microorganisms.

Semi-synthetic routes, where key precursors are produced by fermentation and further converted to final products by chemical or enzymatic reactions, can be successfully used to replace difficult and costly conventional chemical synthesis. This strategy, demonstrated by heparin, chondroitin, and artemisinin examples, may permit the production of extremely complex pharmaceutical molecules by economically feasible and more sustainable processes.

The production of flavonoids and Taxol precursors remind us that reaching high titers still represents a major challenge when developing new fermentation processes. However, cases of success as illustrated by the artemisinin production encourage us and highlight important aspects that can have a major impact on final yields. In this aspect, a comprehensive understanding of the metabolic pathway can significantly increase the chances of success: a good knowledge about the required enzymatic reactions to convert the precursors into the final products can dictate the best host choice, the demand of cofactors and redox balancing, and the need to eliminate competing pathways and bottlenecks, as well as to circumvent regulatory constraints. Fortunately, there is an increasing set of tools and new approaches to deal with each of these demands as, for example, bioinformatics tools to model and simulate metabolic networks or to mine genomic information

to search for alternative enzymes for each desired metabolic reaction. Cheaper DNA synthesis and sequencing, faster and simpler cloning techniques, CRISPR-based technologies, and synthetic biology tools that allow fine-tune of metabolic pathways also play an important role in making it possible to rationally construct efficient cellular factories. Taken together, all these advances in the metabolic engineering field hold a great potential to pave the way for using engineered microorganisms for production of molecules with pharmaceutical properties in the near future.

REFERENCES

Aikawa J, Grobe K, Tsujimoto M, Esko JD. Multiple Isozymes of Heparan Sulfate/Heparin GlcNAcN-Deacetylase/GlcN N-Sulfotransferase structure and activity of the fourth member, NDST4. *J Biol Chem.* 2001;276(8):5876–5882.

Ajikumar PK, Xiao WH, Tyo KEJ, Wang Y, Simeon F, Leonard E, Mucha O, Phon TH, PfeiferB, Stephanopoulos G. Isoprenoid pathway optimization for taxol precursor overproduction in *Escherichia coli. Science.* 2010;330:70–74.

Barreteau H, Richard E, Drouillard S, Samain E, Priem B. Production of intracellular heparosan and derived oligosaccharides by lyase expression in metabolically engineered *E. coli* K-12. *Carbohydr Res.* 2012;360:19–24.

Bertea CM, Freije JR, van der Woulde H, Perk L, Marquez V, De Kraker JW, Posthumus MA, et al. Identification of intermediates and enzymes involved in the early steps of artemisinin biosynthesis in *Artemisia annua. Planta Med.* 2005;71:40–47.

Bhan N, Xu P, Khiladi O, Koffas MA. Redirecting carbon flux into malonyl-CoA to improve resveratrol titers: Proof of concept for genetic interventions predicted by optForce computational framework. *Chem Eng Sci.* 2013;103:109–114.

Bhaskar U, Sterner E, Hickey AM, Onishi A, Zhang F, Dordick JS, Linhardt RJ. Engineering of routes to heparin and related polysaccharides. *Appl Microbiol Biotechnol*. 2012;93(1):1–16.

Boghigian BA, Salas D, Ajikumar PK, Stephanopoulos G, Pfeifer BA. Analysis of heterologous taxadiene production in K- and B-derived *Escherichia coli*. *Appl Microbiol Biotechnol*. 2012;93:1651–1661.

Camacho-Zaragoza JM, Hernández-Chávez G, Moreno-Avitia F, Ramírez-Iñiguez R, Martínez A, Bolívar F, Gosset G. Engineering of a microbial coculture of *Escherichia coli* strains for the biosynthesis of resveratrol. *Microb Cell Fact*. 2016;15:163.

Chemler JA, Lock LT, Koffas MA, Tzanakakis ES. Standardized biosynthesis of flavan-3-ols with effects on pancreatic beta-cell insulin secretion. *Appl Microbiol Biotechnol*. 2007; 77:797–807.

Cimini D, Rosa M, Carlino E, Ruggiero A, Schiraldi C. Homologous overexpression of *rfaH* in *E. coli* K4 improves the production of chondroitin-like capsular polysaccharide. *Microb Cell Fact*. 2013;12:46.

Cress BF, Toparlak ÖD, Guleria S, Lebovich M, Stieglitz JT, Englaender JA, Jones JA, et al. CRISPathBrick: Modular combinatorial assembly of type II-A CRISPR arrays for dCas9-mediated multiplex transcriptional repression in E. coli. *ACS Synth Biol*. 2015;4:987–1000.

Croteau R, Ketchum RE, Long RM, Kaspera R, Wildung MR. Taxol biosynthesis and molecular genetics. *Phytochem Rev Proc Phytochem Soc Eur*. 2006;5:75–97.

Ding X, Ouyang M, Liu X, Wang RZ. Acetylcholinesterase inhibitory activities of flavonoids from the leaves of *Ginkgo biloba* against brown plant hopper. *J Chem*. 2013; Article ID 645086, 1–4. DOI:10.1155/2013/645086.

Edens RE, al-Hakim A, Weiler JM, Rethwisch DG, Fareed J, Linhardt RJ. Gradient polyacrylamide gel electrophoresis for determination of molecular weights of heparin preparations and low-molecular-weight heparin derivatives. *J Pharm Sci*. 1992;81(8):823–827.

Fowler ZL, Gikandi WW, Koffas MA. Increased malonyl coenzyme biosynthesis by tuning the *Escherichia coli* metabolic network and its application to flavanone production. *Appl Environ Microbiol*. 2009;75:5831–5839.

Fowler ZL, Koffas MA. Biosynthesis and biotechnological production of flavanones: Current state and perspectives. *Appl Microbiol Biotechnol*. 2009;83:799–808.

He W, Fu L, Li G, Andrew Jones J, Linhardt RJ, Koffas M. Production of chondroitin in metabolically engineered *E. coli*. *Metab Eng*. 2015;27:92–100.

He XZ, Li WS, Blount JW, Dixon RA. Regioselective synthesis of plant (iso)flavone glycosides in *Escherichia coli*. *Appl Microbiol Biotechnol*. 2008;80:253–60.

Heim KE, Tagliaferro AR, Bobilya DJ. Flavonoid antioxidants: Chemistry, metabolism and structure–activity relationships. *J Nutr Biochem*. 2002;13:572–584.

Higashi K, Ly M, Wang Z, Masuko S, Bhaskar U, Sterner E, Zhang F, et al. Controlled photochemical depolymerization of K5 heparosan, a bioengineered heparin precursor. *Carbohydr Polym*. 2011;86(3):1365–1370.

Hwang EI, Kaneko M, Ohnishi Y, Horinouchi S. Production of plant-specific flavanones by *Escherichia coli* containing an artificial gene cluster. *Appl Environ Microbiol*. 2003;69:2699–706.

Iwashina T. Contribution of flower colors of flavonoids including anthocyanins: A review. *Nat Prod Commun*. 2015;10:529–544.

Jiang H, Wood KV, Morgan JA. Metabolic engineering of the phenylpropanoid pathway in *Saccharomyces cerevisiae*. *Appl Environ Microbiol*. 2005;71:2962–2969.

Jin P, Zhang L, Yuan P, Kang Z, Du G, Chen J. Efficient biosynthesis of polysaccharides chondroitin and heparosan by metabolically engineered Bacillus subtilis. *Carbohydr Polym*. 2016;140:424–432.

Jones JA, Vernacchio VR, Collins SM, Shirke AN, Xiu Y, Englaender JA, Cress BF, McCutcheon CC, Linhardt RJ, Gross RA, Koffas MAG. Complete biosynthesis of anthocyanins using *E. coli* polycultures. *MBio*. 2017;8:e00621–17.

Jones JA, Vernacchio VR, Sinkoe AL, Collins SM, Ibrahim MH, Lachance DM, Hahn J, et al. Experimental and computational optimization of an Escherichia coli co-culture for the efficient production of flavonoids. *Metab Eng*. 2016;35:55–63

Kang SY, Lee JK, Choi O, Kim CY, Jang JH, Hwang BY, Hong YS. Biosynthesis of methylated resveratrol analogs through the construction of an artificial biosynthetic pathway in *E. coli*. *BMC Biotechnol*. 2014;14:67.

Kim BG, Jung BR, Lee Y, Hur HG, Lim Y, Ahn JH. Regiospecific flavonoid 7-*O*-methylation with *Streptomyces avermitilis* *O*-methyltransferase expressed in *Escherichia coli*. *J Agric Food Chem*. 2006b;54:823–828.

Kim BG, Kim H, Hur HG, Lim Y, Ahn JH. Regioselectivity of 7-*O*-methyltransferase of poplar to flavones. *J Biotechnol*. 2006a;126:241–247.

Kim BG, Sung SH, Jung NR, Chong Y, Ahn JH. Biological synthesis of isorhamnetin 3-*O*-glucoside using engineered glucosyltransferase. *J Mol Catal B: Enzym*. 2010;63:194–199.

Kim DH, Kim BG, Lee Y, Ryu JY, Lim Y, Hur HG, Ahn JH. Regiospecific methylation of naringenin to ponciretin by soybean *O*-methyltransferase expressed in *Escherichia coli*. *J Biotechnol*. 2005;119:155–162.

Kim E, Moore BS, Yoon YJ. Reinvigorating natural product combinatorial biosynthesis with synthetic biology. *Nat Chem Biol*. 2015; 11:649–59.

Kim JH, Shin KH, Ko JH, Ahn JH. Glucosylation of flavonols by *Escherichia coli* expressing glucosyltransferase from rice (*Oryza sativa*). *J Biosci Bioeng*. 2006c;102:135–137.

Koirala N, Pandey RP, Parajuli P, Jung HJ, Sohng JK. Methylation and subsequent glycosylation of 7,8-dihydroxyflavone. *J Biotechnol*. 2014;184:128–137.

Koirala N, Thuan NH, Ghimire GP, Thang DV, Sohng JK. Methylation of flavonoids: Chemical structures, bioactivities, progress and perspectives for biotechnological production. *Enz Microb Technol*. 2016;86:103–116.

Leonard E, Lim KH, Saw PN, Koffas MA. Engineering central metabolic pathways for high-level flavonoid production in *Escherichia coli*. *Appl Environ Microbiol*. 2007;73:3877–3886.

Leonard E, Yan Y, Chemler J, Matern U, Martens S, Koffas MA. Characterization of dihydroflavonol 4-reductases for recombinant plant pigment biosynthesis applications. *Biocatal Biotransfor*. 2008;26:243–251.

Leonard E, Yan Y, Lim KH, Koffas MA. Investigation of two distinct flavone synthases for plant-specific flavone biosynthesis in *Saccharomyces cerevisiae*. *Appl Environ Microbiol*. 2005;71:8241–8248.

Lim CG, Fowler ZL, Hueller T, Schaffer S, Koffas MA. High-yield resveratrol production in engineered *Escherichia coli*. *Appl Environ Microbiol*. 2011;77:3451–3460.

Lim EK, Ashford DA, Hou B, Jackson RG, Bowles DJ. *Arabidopsis* glycosyltransferases as biocatalysts in fermentation for regioselective synthesis of diverse quercetin glucosides. *Biotechnol Bioeng*. 2004;87:623–631.

Lima MA, Viskov C, Herman F, Gray AL, de Farias EH, Cavalheiro RP, Sassaki GL, et al. Ultra-low-molecular-weight heparins: precise structural features impacting specific anticoagulant activities. *Thromb Haemost*. 2013;109(3):471–478.

Linhardt RJ. Claude S. Hudson award address in carbohydrate chemistry. Heparin: structure and activity. *J Med Chem.* 2003;46(13):2551–2564.

Liu H, Zhang Z, Linhardt RJ. Lessons learned from the contamination of heparin. *Nat Prod Rep.* 2009;26(3):313–321.

Liu X, Cheng J, Zhang G, Ding W, Duan L, Yang J, Kui L, et al. Engineering yeast for the production of breviscapine by genomic analysis and synthetic biology approaches. *Nat Commun.* 2018;9:448.

Luo Y, Li BZ, Liu D, Zhang L, Chen Y, Jia B, Zeng BX, Zhao H, Yuan YJ. Engineered biosynthesis of natural products in heterologous hosts. *Chem Soc Rev.* 2015;44:5265–5290.

Malla S, Koffas MA, Kazlauskas RJ, Kim BG. Production of 7-*O*-methyl aromadendrin, a medicinally valuable flavonoid, in *Escherichia coli. Appl Environ Microbiol.* 2012;78: 684–694.

Malla S, Pandey RP, Kim BG, Sohng JK. Regiospecific modifications of naringenin for astragalin production in *Escherichia coli. Biotechnol Bioeng.* 2013;110:2525–2535.

Masuko S, Higashi K, Wang Z, Bhaskar U, Hickey AM, Zhang F, Toida T, et al. Ozonolysis of the double bond of the unsaturated uronate residue in low-molecular-weight heparin and K5 heparosan. *Carbohydr Res.* 2011;346(13):1962–1966.

McAlindon TE, LaValley MP, Gulin JP, Felson DT. Glucosamine and chondroitin for treatment of osteoarthritis: A systematic quality assessment and meta-analysis. *J Am Med Assoc.* 2000; 283:1469–1475.

Meng H, Wang Y, Hua Q, Zhang S, Wang X. *In silico* analysis and experimental improvement of taxadiene heterologous biosynthesis in *Escherichia coli. Biotechnol Bioprocess Eng.* 2011;16:205–215.

Miyahisa I, Funa N, Ohnishi Y, Martens S, Moriguchi T, Horinouchi S. Combinatorial biosynthesis of flavones and flavonols in *Escherichia coli. Appl Microbiol Biotechnol.* 2006;71: 53–58.

Miyahisa I, Kaneko M, Funa N, Kawasaki H, Kojima H, Ohnishi Y, Horinouchi S. et al. Efficient production of (2*S*)-flavanones by *Escherichia coli* containing an artificial biosynthetic gene cluster. *Appl Microbiol Biotechnol.* 2005;68:498–504.

Onishi A. Heparin and anticoagulation. *Front Biosci.* 2016;21: 1372–1392.

Paddon CJ, Keasling JD. Semi-synthetic artemisinin: A model for the use of synthetic biology in pharmaceutical development. *Nat Rev Microbiol.* 2014;12:355–367.

Pandey RP, Malla S, Simkhada D, Kim BG, Sohng JK. Production of 3-*O*-xylosyl quercetin in *Escherichia coli. Appl Microbiol Biotechnol.* 2013;97:1889–1901.

Pandey RP, Parajuli P, Chu LL, Darsandhari S, Sohng JK. Biosynthesis of amino deoxy-sugar-conjugated flavonol glycosides by engineered *Escherichia coli. Biochem Eng J.* 2015;101:191–199.

Pandey RP, Parajuli P, Koffas MAG, Sohng JK. Microbial production of natural and non-natural flavonoids: Pathway engineering, directed evolution and systems/synthetic biology. *Biotechnol Adv.* 2016; 34:634–662.

Petitou M, Duchaussoy P, Lederman I, Choay J, Sinaÿ P, Jacquinet JC, Torri G, et al. Synthesis of heparin fragments. A chemical synthesis of the pentasaccharide O-(2-deoxy-2-sulfamido-6-O-sulfo-alpha-D-glucopyranosyl)-(1-4)-O-(beta-D-glucopyranosyluronic acid)-(1-4)-O-(2-deoxy-2-sulfamido-3,6-di-O-sulfo-alpha-D-glu copyranosyl)-(1-4)-O-(2-O-sulfo-alpha-L-idopyranosyluronic acid)-(1-4)-2-deoxy-2-sulfamido-6-O-sulfo-D-glucopyranose decasodium salt, a heparin fragment having high affinity for antithrombin III. *Carbohydr Res.* 1986;147(2):221–236.

Ralston L, Subramanian S, Matsuno M, Yu O. Partial reconstruction of flavonoid and isoflavonoid biosynthesis in yeast using soybean type I and type II chalcone isomerases. *Plant Physiol.* 2005;137:1375–1388.

Ro DK, Douglasm CJ. Reconstitution of the entry point of plant phenylpropanoid metabolism in yeast (*Saccharomyces cerevisiae*): Implications for control of metabolic flux into the phenylpropanoid pathway. *J Biol Chem.* 2004;279:2600–2607.

Rodriguez A, Strucko T, Stahlhut ST, Kristensen M, Svenssen DK, Forster J, Nielsen J, Borodina I. Metabolic engineering of yeast for fermentative production of flavonoids. *Biores Technol.* 2017;245:1645–1654.

Subrahmanyam S, Cronan JE Jr. Overproduction of a functional fatty acid biosynthetic enzyme blocks fatty acid synthesis in *Escherichia coli. J Bacteriol.* 1998; 180:4596–4602.

Suflita M, Fu L, He W, Koffas M, Linhardt RJ. Heparin and related polysaccharides: Synthesis using recombinant enzymes and metabolic engineering. *Appl Microbiol Biotechnol.* 2015;99(18):7465–7479.

Teoh KH, Polichuk DR, Reed DW, Nowak G, Covello PS. *Artemisia annua L.* (Asteraceae) trichome-specific cDNAs reveal CYP71AV1, a cytochrome P450 with a key role in the biosynthesis of the antimalarial sesquiterpene lactone artemisinin. *FEBS Lett.* 2006;580:1411–1416.

Trantas E, Panopoulos N, Ververidis F. Metabolic engineering of the complete pathway leading to heterologous biosynthesis of various flavonoids and stilbenoids in *Saccharomyces cerevisiae. Metab Eng.* 2009;11:355–366.

Tsuruta H, Paddon CJ, Eng D, Lenihan JR, Horning T, Anthony LC, Regentin R, Keasling JD, Renninger NS, Newman JD. High-level production of amorpha-4,11-diene, a precursor of the antimalarial agent artemisinin, in *Escherichia coli. PLoS One.* 2009;4.

Vaidyanathan D, Williams A, Dordick JS, Koffas MAG, Linhardt RJ. Engineered heparins as new anticoagulant drugs. *Bioeng Translat Med* 2017;2:17–30.

Veitch NC, Grayer RJ. Flavonoids and their glycosides, including anthocyanins. *Nat Prod Rep.* 2011;28:1626–1695.

Veitch NC. Isoflavonoids of the Leguminosae. *Nat Prod Rep.* 2007;24:417–464.

Wang HK. The therapeutic potential of flavonoids. *Expert Opin Investig Drugs.* 2000; 9:2103–2119.

Wang Z, Li J, Cheong S, Bhaskar U, Akihiro O, Zhang F, Dordick JS, et al. Response surface optimization of the heparosan N-deacetylation in producing bioengineered heparin. *J Biotechnol.* 2011;156(3):188–196.

Wang Z, Ly M, Zhang F, Zhong W, Suen A, Hickey AM, Dordick JS. et al. E. coli K5 fermentation and the preparation of heparosan, a bioengineered heparin precursor. *Biotechnol Bioeng.* 2010;107(6):964–973.

Wattanachaisaereekul S, Lantz AE, Nielsen ML, Nielsen J. Production of the polyketide 6-MSA in yeast engineered for increased malonyl-CoA supply. *Metab Eng.* 2008;10:246–254.

Watts KT, Lee PC, Schmidt-Dannert C. Biosynthesis of plant-specific stilbene polyketides in metabolically engineered *Escherichia coli. BMC Biotechnol.* 2006;6:22.

Watts KT, Lee PC, Schmidt-Dannert C. Exploring recombinant flavonoid biosynthesis in metabolically engineered *Escherichia coli. Chembiochem.* 2004;5:500–507.

Westfall PJ, Pitera DJ, Lenihan JR, Eng D, Woolard FX, Regentin R, Horning T, et al. Production of amorphadiene in yeast, and its conversion to dihydroartemisinic acid, precursor to the antimalarial agent artemisinin. *Proc Natl Acad Sci USA.* 2012;109:111–118.

Wu J, Du G, Zhou J, Chen J. Systems metabolic engineering of microorganisms to achieve large-scale production of flavonoid scaffolds. *J Biotechnol.* 2014;188C:72–80.

Wu J, Liu P, Fan Y, Bao H, Du G, Zhou J, Chen J. Multivariate modular metabolic engineering of *Escherichia coli* to produce resveratrol from L-tyrosine. *J Biotechnol.* 2013;167: 404–411.

Wu Q, Yang A, Zou W, Duan Z, Liu J, Chen J, Liu L. Transcriptional engineering of Escherichia coli K4 for fructosylated chondroitin production. *Biotechnol Prog.* 2013;29(5):1140–1149.

Xiao JB, Shao R. Natural products for treatment of Alzheimer's disease and related dis-eases: Understanding their mechanism of action. *Curr Neuropharmacol.* 2013;11:337.

Xiao JB, Tundis R. Natural products for Alzheimer's disease therapy: Basic and application. *J Pharm Pharmacol.* 2013;65:1679–1680.

Xu P, Ranganathan S, Fowler ZL, Maranas CD, Koffas MA. Genome-scale metabolic network modeling results in minimal interventions that cooperatively force carbon flux towards malonyl-CoA. *Metab Eng.* 2011;13:578–587.

Xu P, Vansiri A, Bhan N, Koffas MAG. ePathBrick: A synthetic biology platform for engineering metabolic pathways in E. coli. *ACS Synth Biol.* 2012;1(7):256–266.

Xu Y, Cai C, Chandarajoti K, Hsieh PH, Li L, Pham TQ, Sparkenbaugh EM, et al. Homogeneous low-molecular-weight heparins with reversible anticoagulant activity. *Nat Chem Biol.* 2014;10(4):248–250.

Yan Y, Chemler J, Huang L, Martens S, Koffas MA. Metabolic engineering of anthocyanin biosynthesis in *Escherichia coli. Appl Environ Microbiol.* 2005a;71:3617–3623.

Yan Y, Kohli A, Koffas MA. Biosynthesis of natural flavanones in *Saccharomyces cerevisiae. Appl Environ Microbiol.* 2005b;71:5610–5613.

Zha W, Rubin-Pitel SB, Shao Z, Zhao H. Improving cellular malonyl-CoA level in Escherichia coli via metabolic engineering. *Metab Eng.* 2009;11:192–198.

Zhang C, Liu L, Teng L, Chen J, Liu J, Li J, Du G. Metabolic engineering of *Escherichia coli* BL21 for biosynthesis of heparosan, a bioengineered heparin precursor. *Metab Eng.*ss 2012;14(5):521–527.

Zhang S, Wang S, Zhan J. Engineered biosynthesis of medicinally important plant natural products in microorganisms. *Curr Top Med Chem.* 2016;16:1740–1754.

Zhou Z, Li Q, Huang H, Wang H, Wang Y, Du G, Chen J, Kang Z. A microbial–enzymatic strategy for producing chondroitin sulfate glycosaminoglycans. *Biotechnol Bioeng.* 2018;115: 1561–1570.

10 Enzyme and Cofactor Engineering
Current Trends and Future Prospects in the Pharmaceutical and Fermentation Industries

George N. Bennett and Ka-Yiu San

CONTENTS

"Logic will get you from A to B. Imagination will take you everywhere."

Albert Einstein

10.1 INTRODUCTION

The effect of molecular genetics and recombinant DNA technology on industrial biotechnology is well documented. In this chapter, we will address some general themes and use a few examples to illustrate the concepts involved in enzyme and cofactor engineering.

In considering a biological process for metabolic and enzyme engineering purposes, two main points must be addressed to identify the right enzyme for metabolic intervention. In addition to the flux control coefficient (see Chapter 7 for more details), there are two other main approaches to identifying target enzymes. Whereas the first makes use of the diversity in the properties of enzymes, the other relies on the enzymes' unique structural properties. By making deliberate structural alterations through rational design or random mutagenesis, a desired enzyme variant can be obtained. In some cases, a combination of both processes, that of identifying a natural enzyme from an organism that performs that reaction well, and then modifying it to optimize it for the industrial process.

10.2 TYPES OF MAJOR INDUSTRIAL ENZYMES AND DESIRED MODIFICATIONS

10.2.1 DESIRED TARGETS FOR ENZYME ENGINEERING

In identifying targets for enzyme engineering, several factors need to be considered. These include:

- The projected long-term market value of the product(s) and the inherent economic advantages that follow.

- The applicability of the engineered enzyme to other industrial processes.
- The availability and validity of suitable strategies for engineering, screening, and detection of the desired variant.

One general area in which enzymes with engineered properties may become more important is the production of optically active amines as precursors for bioactive pharmaceuticals. Particular additional classes include esterases, racemases, and redox enzymes (oxidoreductases) including peroxidases and cytochrome P450. The use of enzymatic procedures that are stereospecific can enhance the overall synthetic process. Specific processes in which the chemical reaction yields a mixture of isomers that are hard to separate are attractive candidates for enzyme engineering.

10.2.1.1 Hydrolytic Enzymes

Most enzymes used in large quantity are hydrolytic enzymes. Proteases are widely used as additives to remove protein-rich stains or blood stains from laundry. The *Bacillus* alkaline serine protease subtilisin is the most common enzyme used for this application. Desirable properties for such enzymes include stability and activity at high temperature and high pH associated with typical laundry conditions. Stability in the face of other harsh substances and chelating agents is also a desirable property. Amylases represent another class of hydrolytic enzymes that are widely used in starch liquefaction and the removal of starchy foods from clothing in laundries. Alpha-amylases are derived from *Bacillus* species whereas the major beta-amylases are derived from *Bacillus* and plants. The ability to act on crude substrates under harsh conditions or extreme environments is also a desirable property.

The production of high-fructose corn syrup represents a classic example in which glucose isomerase is used to convert glucose into fructose, thus considerably enhancing the level of sweetness. Under industrial conditions, the glucose isomerase reaction requires high temperature and low pH to avoid side reactions. Various enzymes from *Bacillus coagulans*, *Streptomyces rubiginusus*, and other bacterial species have been used for this purpose. With the increased interest in degrading plant biomass to soluble sugars that can be used as a feedstock for biofuels and chemical production, a good deal of attention has been diverted toward enhancing the properties and the efficacy of cellulases (Blumer-Schuette et al., 2008; Doi, 2008; Dowe, 2009; Maki et al., 2009) and xylanases (Khandeparker and Numan, 2008), which breakdown the cellulose and hemicelluloses in plant cell wall into sugars. Several companies produce a blend of enzymes for lignocllulose degradation and other specialized sources of biomass, e.g., seaweed (Sharma and Horn, 2016).

Cheese-making is yet another example in which proteases play a central role not only in the formation of the curd, which is catalyzed by the enzyme chymosin (Rennin), but also for cheese ripening and maturation, that give the particular cheese its special flavor characteristics. Although chymosin used to be made from the gastric juice of the fourth stomach chamber of weaning calves or fungi, it is now recombinant chymosin that enjoys center stage, especially at the industrial level.

Lipases are another group of hydrolytic enzymes used in various ways, including removing grease stains and clogs; hydrolysis of fats in the food industry is a major concern. Lipases can also be used under conditions of low water content in an organic solvent to modify oils and fats for food use by transesterification of the acyl chains to generate a product with a desirable pattern (e.g., level of saturation or chain length).

Another widely used hydrolytic enzyme is penicillin acylase. In the production of various β-lactam antibiotics, penicillinase or cephalosporin acylase has been successful in the production of useful derivatives (Sio and Quax, 2004). It is also noteworthy that the hydration of certain molecules in a specific fashion is desirable industrially. For example, nitrile hydratase (Chen et al., 2009) from *Pseudomonas chlororaphis* B23 has been used to convert nitrile into acrylamide on a large scale; such a useful reaction has environmental and economic advantages.

10.2.1.2 Specialty Enzymes for Pharmaceuticals

In the pharmaceutical industry, much interest has been focused on the formation of compounds with a specific chiral structure for the synthesis of stereospecific biopharmaceuticals. The ability to interconvert one stereoisomer into another is of special interest because often only one of the two forms has the desired pharmaceutical properties and it is very difficult to separate from the other isomer. Different enantiomers may also have different metabolic and pharmacokinetic properties and side effects. Therefore, efforts to selectively form the functional chiral product have intensified. The use of enzymatic steps can also improve yields and reduce the use of hazardous chemicals to provide a more "green chemistry" process (Tao and Xu, 2009). In addition to the specific synthetic pharmaceuticals, the nutritional area of natural amino acids and vitamins involves many processes that require chiral production systems. Therefore, enzymes that can catalyze stereospecific reactions on complex molecules, particularly those such as sterols, alkaloids, etc., are used in many synthetic processes. Another use of enzymes exhibiting stereospecific catalytic activity is in the separation or isomerization of a racemic mixture after achiral chemical synthesis; for example, the separation of (R,S)·-Naproxen 2,2,2-trifluoroethyl thioester to produce the desired (S)-Naproxen, an anti-inflammatory drug (Ng and Tsai, 2005). The general aspects of formation of chiral products and specificity have been discussed (Turner, 2003; Jaeger and Eggert, 2004).

Enantiomers: Enantiomers of a chemical compound are equivalent in composition and bonding pattern of the atoms but differ in the relative positioning of a group on one of the atoms. The two forms are mirror images like right and left hands. This difference allows each to interact differently with other molecules.

A large family of proteins that has received much attention in recent times is the cytochrome P450 family. Redox-active enzymes that are central in detoxifications of radicals and aromatics in the body and as such testing lead compounds against the activities of these enzymes has become an important parameter in drug discovery programs (Purnapatre et al., 2008).

Sterols: Sterols are a wide group of lipid molecules with the same basic four-fused ring system of steroids but encompass a wider class, including plant and fungal sterols as well as the animal sterols and steroids.

10.3 ALTERATION OF PHYSICAL PROPERTIES OF ENZYMES FOR PROCESS APPLICATIONS

Thermostability of enzymes is a prerequisite for successful applications in most industrial processes. Recent studies revealed that thermostability is due to the dense hydrophobic core of amino acids along with an increased density of intramolecular hydrogen bonds and ionic interactions (Matsui and Harata, 2007). Examples of widely used thermostable enzymes include glucoisomerases (Asboth and Naray-Szabo, 2000; Hartley et al., 2000) and glucoamylases (Sauer et al., 2000). The engineering of thermostable subtilisin using molecular biology tools has been described (Zhao and Arnold, 1999). Similarly, psychrophilic (cold-loving) enzymes are useful, especially if the enzyme is to be inactivated after the completion of the reaction. Successful adaptations to cold temperatures are linked to increased flexibility around the active site of the enzymes (Feller, 2003; Georlette et al., 2004). Examples of enzymes that are active at cold temperature have been cited (Georlette et al., 2004; Siddiqui and Cavicchioli, 2006; Joseph et al., 2008), and protein engineering of cold-adapted subtilisin has been described (Taguchi et al., 2000). The cold-adapted enzymes are useful in the food industry where many processes are conducted under refrigerated conditions.

Compatibility of enzymic activity with conditions of high salt or organic solvents is also a challenge, and the isolation and engineering of enzymes that operate effectively under these conditions has been described (Luetz et al., 2008; Gupta and Khare, 2009).

10.4 TOOLS AND METHODOLOGIES OF ENZYME ENGINEERING

10.4.1 SITE-SPECIFIC MUTAGENESIS

Site-specific mutagenesis has been developed and practiced by many researchers. In this method, a primer incorporating the desired nucleotide change is constructed and, in turn, annealed to a circular single-stranded template vector containing the gene to be mutated; the primer is then extended by the action of DNA polymerase in the presence of deoxynucleoside triphosphates. After extension of the primer by DNA polymerase, the double-stranded molecule is introduced into a host cell (e.g., *Escherichia coli*). The strand made by extending the primer is preferentially replicated, and the template strand is preferentially degraded by host enzymes. The template strand is previously isolated from a cell that incorporates deoxyuridine instead of thymidine making it susceptible to degradation in the transformed host cell. This method yields a high proportion of the progeny vector with the desired mutation (Kunkel et al., 1991). This method has been adapted in various ways for use with other plasmids (Jung et al., 1992).

Recently, new methods based on PCR incorporation of designed primer sequences carrying the desired mutation and standard kits have become commercially available. In these methods, a specific codon change is introduced into the sequence of the designed primers, and this, in turn, ensures the desired change in the DNA sequence of the cloned gene. Replication of the modified strand can be enhanced through selective digestion of the parent strand by enzymes, or through using a special set of primers that facilitate selection on selective medium. The commercial kits allow for rapid site-specific mutagenesis and generate a relatively high proportion of recombinant colonies bearing the desired mutation (Figure 10.1). These methods, coupled with the greater availability of low-cost synthesized oligonucleotides and sequencing services, have made site-specific mutagenesis a routine technique in many laboratories.

10.4.2 CASSETTE MUTAGENESIS

To simultaneously introduce several nearby mutations, the technique of cassette mutagenesis has been devised (Wells et al., 1985; Kegler-Ebo et al., 1996). This method allows for a short segment of the gene to be cut out and replaced by a specifically designed oligonucleotide segment carrying different nucleotides at a given position. After substitution and replication of the target codon, several mutants carrying different amino acids can be isolated. In this way, several different amino acid residues at a particular position in the protein could be introduced and individually tested for their effect on enzyme properties. The use of segments with more extensively changed nucleotide sequences could allow for the formation of protein variants with several amino acid changes within the section replaced, giving a localized patch on the protein where there are many alterations. This approach is especially useful near active sites to modify the specificity of the enzyme or at areas of the protein's surface to alter its interaction with other proteins.

As the technology for oligonucleotide synthesis has greatly improved, the opportunity of synthesizing entire genes, or even genomes, has become possible. With this approach, the gene encoding the desired protein could be designed in such a way that it contains several unique restriction sites for replacement of different segments of the protein by cassette mutagenesis, which is illustrated in Figure 10.2.

Schematic of site-specific mutagenesis

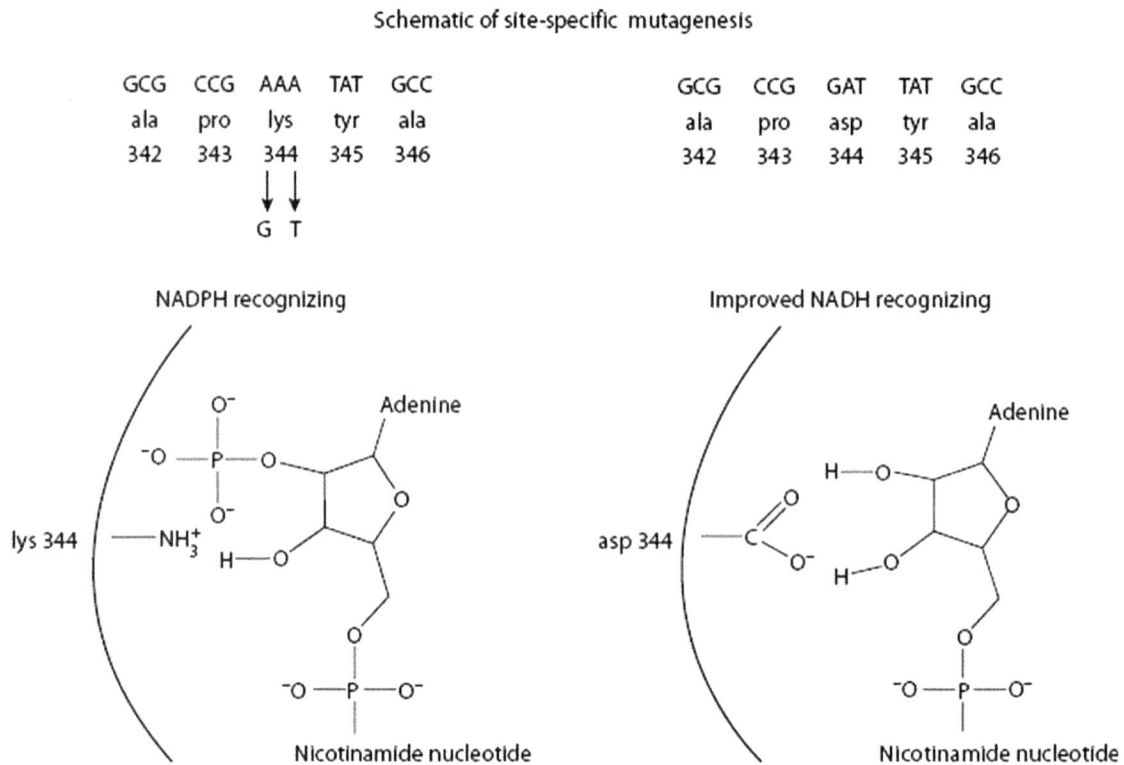

FIGURE 10.1 In the example, the cofactor specificity of ICDH has been changed from NADPH to NADH by (Hurley et al., 1996). The lys-344 is central to NADPH binding because lysine is a positively charged residue and as such can interact with the negatively charged phosphate group of NADPH. When the codon encoding lys-344 was changed to GAT (the codon specifying aspartic acid, which is a negatively charged amino acid residue) the enzyme could no longer bind NADPH. Under these conditions, another kind of interaction occurs with the hydroxyl groups of the nonphosphorylated ribose to form hydrogen bonds with the carboxylic group of aspartic acid. Such an interaction would favor binding of NADH rather than NADPH at the active site. As with most enzymes there are several amino acids that contribute to specificity at the active site. More details are given in Chen et al. (1995), Hurley et al. (1996), and Yaoi et al. (1996), and other alterations affecting substrate specificity are described in a crystallographic study (Doyle et al., 2001).

10.4.3 GENE SYNTHESIS (SYNTHETIC BIOLOGY)

As synthesis of oligonucleotides has become more affordable and as the technology for their assembly has become more accessible, it is possible to create and engineer new genes for specific purposes. Such advent has led to the creation of a new branch in science otherwise termed "synthetic biology" (Keasling, 2008; Xiong et al., 2008; Tian et al., 2009). For example, if a change in the kinetic property or specificity of a given protein is desired, several genes from various organisms can be synthesized and placed into the desired expression system for *in vivo* testing of performance. The choice of genes for synthesis and testing would be made based on data from the sequence databases from organisms likely to have a robust activity for that reaction and the *in vitro* activity and characteristics of enzymes provided by databases such as the Brenda enzyme database (www.brenda-enzymes.org/). During the synthesis of the gene, several variants at specific positions can be introduced and the mixture can be cloned, and individual variants examined for performance. The use of this synthetic biology approach has been reviewed, and wider implications of designer circuits for appropriate expression within a host are discussed (Keasling, 2008; Carothers et al., 2009; Landrain et al., 2009). This approach is used by companies

where many gene variants expressing the desired enzyme can be synthesized and tested using automated equipment.

10.4.4 TERTIARY STRUCTURE AND SPECIFIC MUTATIONS

If one wishes to enhance a particular property of an enzyme through site-specific mutagenesis, the three-dimensional 3D structures of the enzyme before and after binding to substrate analogue must be examined to gain an understanding of the positioning of the substrate in the active site and to identify the amino acid residues needed for catalysis (Benkovic and Hammes-Schiffer, 2003). Computational approaches to analyzing mechanistic aspects of enzyme action have been developed and are widely used in protein engineering (van der Kamp and Mulholland, 2008). An example of a change in substrate specificity of the protease subtilisin illustrates this approach. When the binding cavity of subtilisin is reduced in size by substituting glycine 127 (a small amino acid) with a larger amino acid, the enzyme exhibits a distinct preference for smaller substrates (Takagi et al., 1996). In general, expanding the size of the substrate binding site can lead to a broadening of the specificity of the enzyme because a wider range of substrates may be able to fit within the active site. However, the affinity of the enzyme for the substrate may

Cassette mutagenesis

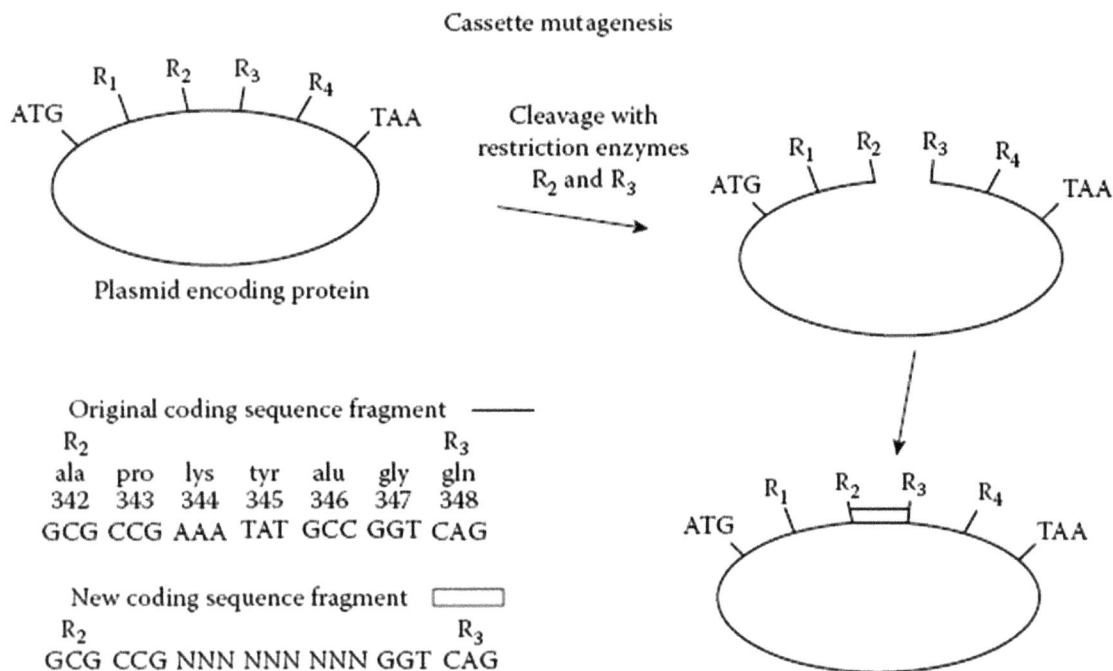

FIGURE 10.2 In the initial plasmid, the ATG indicates the codon encoding the first amino acid of the protein (N-terminus), whereas the TAA indicates the stop codon, which terminates synthesis at the C-terminus of the protein. The sites R_1, R_2, R_3, and R_4 denote the positions of cleavage of the plasmid at a unique place by the restriction enzymes R_1, R_2, R_3, and R_4, respectively. Cleavage by R_2 and R_3 then removes a defined segment, and this segment can be replaced by a synthetic segment having a mixture of nucleotides (N for a mixture of A, T, G, C) at a particular position within the segment. In the example shown, the nine positions specifying amino acids 344, 345, and 346 are mixed and could then encode any amino acid at these positions. Replacement of the R_2-R_3 segment in the plasmid would generate various plasmid molecules with different possibilities for codons at positions 344, 345, and 346 in this protein. After transformation and replication of individual molecules of the plasmid and plating to generate individual colonies, the protein from each colony could be isolated and examined and the nucleotide sequence of the specific R_2–R_3 segment determined.

decrease (i.e., requiring a higher concentration of the substrate for effective catalysis) (Zhang et al., 2008).

Another important specificity is cofactor requirement. The pools of NADH and NADPH in the cell influence the redox potential within the cell on one hand and product formation on the other. Generally speaking, NADH is the preferred cofactor for *in vitro* oxidation/reduction reactions because of its greater stability. The conversion of the NADPH-specific 2,5-diketo-d-gluconic acid reductase A, an enzyme used in the synthesis of vitamin C, into an NADH-specific enzyme has been reported (Banta et al., 2002b). Similarly, the NADPH-dependent carbonyl reductase that is used for the synthesis of optically active alcohols has also been modified to become NADH-specific (Morikawa et al., 2005). However, in some cases, it may be preferable to have NADPH specificity to avoid interference with other reducing enzymes, and examples of the conversion of NADH specificity to that of NADPH have been reported for xylitol dehydrogenase (Watanabe et al., 2005) and lactate dehydrogenase (Holmberg et al., 1999).

10.4.5 DIRECTED EVOLUTION AND DNA SHUFFLING

The ability to make and screen many random mutations within a localized region is referred to as "directed evolution." This approach can be used without knowledge of the three-dimensional structure of the enzyme and can be applied

to a wider range of sequences and structures to generate the desired protein sequence with appropriate functional properties. Successful application of directed evolution, which makes use of error-prone PCR and DNA shuffling techniques, has recently been reviewed (Labrou, 2010). However, it must be born in mind that such an approach demands an efficient screening and selection program (Jestin and Kaminski, 2004; Boersma et al., 2007).

The effect of PCR and DNA shuffling techniques on biomolecular engineering has been reported, and reviews that specifically focus on directed evolution of industrially important organisms have been published (Cherry and Fidantsef, 2003; Hult and Berglund, 2003; Arnold, 2017).

As the use of this approach continues to grow, many enzymes are likely to be modified in one way or another to improve their performance in industrial fermentations.

Error-prone PCR (Figure 10.3) takes advantage of the misincorporation of nucleotides into DNA during polymerization to yield a newly synthesized strand with random changes. The level of misincorporation can be manipulated by the use of different polymerases, inclusion of different metal ions, or addition of allosteric substrates that affect the structure of the polymerase or its ability to discriminate base pairing (Cirino and Arnold, 2002; Cirino et al., 2003). Theoretical models relating the error rate to the overall distribution of base changes in the final population of molecules have been reported (Bessler et al., 2003).

Error prone PCR

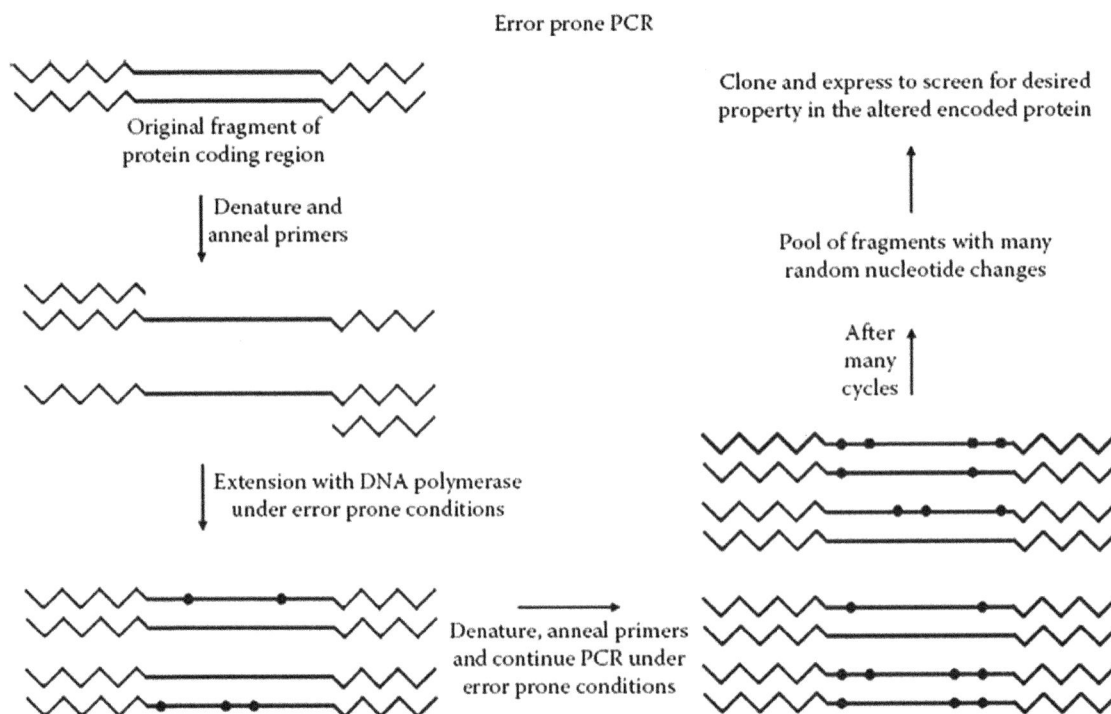

FIGURE 10.3 In error-prone PCR, the template is annealed with primers that allow for extension (by DNA polymerase) and incorporation of the dNTPs, deoxynucleotide triphosphates, in the incubation mixture. The condition of the extension and the enzyme used will define how frequently a mistake or misincorporation (e.g., A for G) is made. Denaturation of the strand at high temperature followed by reannealing of the primers can then allow the strands with mistakes to be copied, and the original strands and further errors by misincorporation are introduced, giving a product that may have many variations within the product molecules. Conditions can be used that will give an appropriate level of altered molecules (e.g., 1 mutation per 100 or 1000 residues on average). Because the molecules can be cloned and replicated independently within a plasmid, individual colonies with separate variants can again be isolated and screened for enzyme properties (e.g., high-temperature activity) and the specific DNA sequence of interesting variants can be analyzed.

DNA shuffling technique utilizes fragments from closely related genes in a PCR misincorporation process to generate a high diversity of full-length genes (Figure 10.4). This technique allows for a wider expanse of diverse sequences to be created than the original error-prone PCR technique because of the presence of a greater variety of genes. However, because many variants are generated, developing a suitable protocol for screening and selection is a significant consideration. The applicability of this technique to pharmaceutical and vaccine production (Patten et al., 1997; Locher et al., 2005) and improvement of industrial enzymes (Powell et al., 2001) has been reviewed.

Where possible, it is useful to couple molecular evolution with an effective screening strategy for optimized enzyme activity. In the case of growth coupling, the adapted strain may display significant improvement in performance. The changes in genotypes that are associated with enhanced performance can be unraveled by genome sequencing of high-performing variants. This technique has successfully been applied to kanamycin (Yanai et al., 2006) and lysine (Ohnishi et al., 2002) production. This approach has also been applied to improving the ability of *E. coli* to grow on glycerol as a sole source of carbon and energy (Fong et al., 2005; Herring et al., 2006).

Although details of screening systems are not addressed here, the usual practice is to make use of a substrate or a coupled reaction that yields a chromogenic compound that serves as an indicator of the desired genotype. The conditions of assay and detection can be modified to include the desired physical parameters such as pH, temperature, or the presence of organic solvents or other inhibitors. The use of high-throughput detection systems in combination with a multiwell plate format allows for identification of modified proteins with enhanced activity. In some cases, the use of fluorescence-activated cell sorting (FACS) or microdroplet screening processes can allow cells to be screened rapidly for improved variants (Kwon et al., 2004; Davids et al., 2013).

The development of the direct use of oligonucleotides in mutagenesis of the chromosome of bacteria (Grogan and Stengel, 2008; Weiss, 2008) or yeast (Storici et al., 2001) has generated a new way for targeted mutagenesis in certain organisms. This approach can either be used to alter a specific nucleotide in the chromosome in a manner similar to that of site-specific mutagenesis or a multiple of nucleotides. In either case, a screening process follows to identify colonies with desired phenotypes. This combinatorial approach has been used to generate *E. coli* strains able to form a high level of lycopene (Wang et al., 2009). And many uses of similar methods have been used in other organisms in the newly developing field of "genome editing" (Sauer et al., 2016).

DNA shuffling

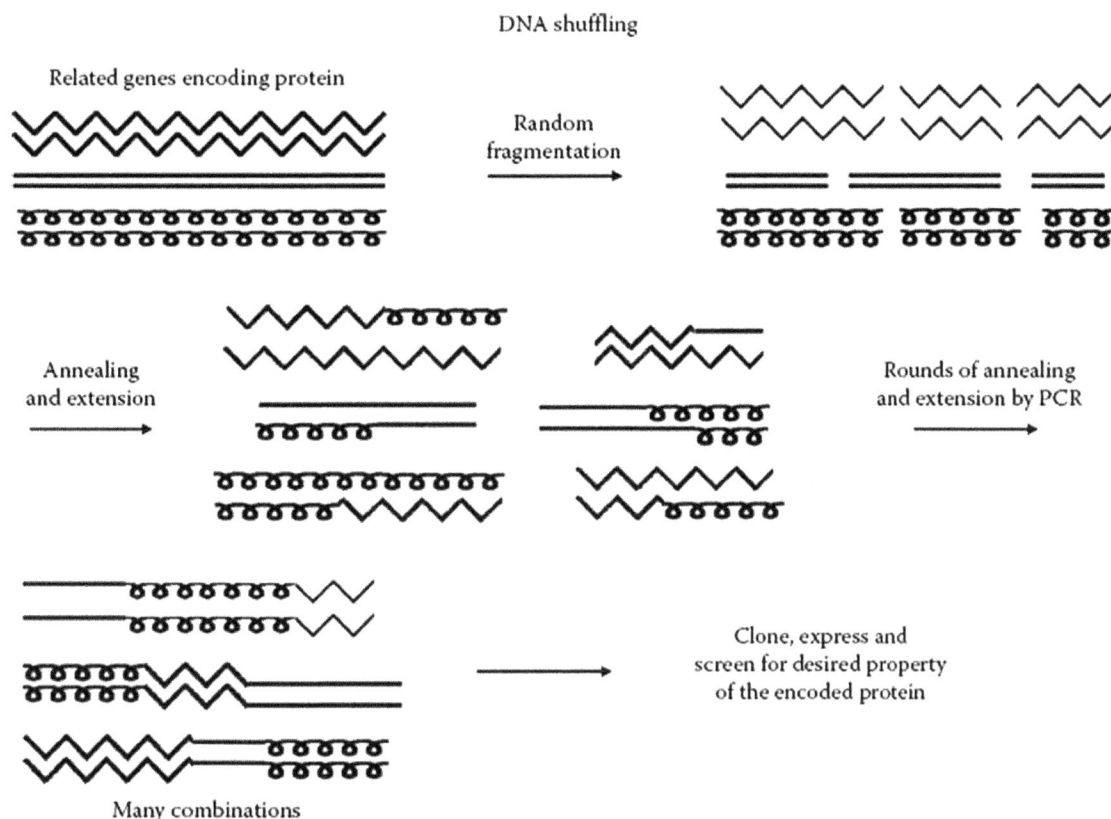

FIGURE 10.4 In the DNA shuffling procedure, several gene fragments encoding similar proteins are used. The segments must have a high enough sequence similarity to anneal well in heterologous combinations so that the polymerase can extend the various duplexes formed. The repeated denaturation and reannealing give a chance for different sequences to be represented. The combination of various sections and the individual base-pair mutations introduced during the repeated process generates a great variety of altered sequences. These DNA fragments can be cloned, the individual variant genes encoding protein are expressed in individual colonies, and screening schemes or selections impressed on individual colonies are used to identify variants with desired properties. The DNA encoding the variant gene can be isolated and sequenced to identify the changes producing the effect.

10.5 MODIFICATION OF PHARMACEUTICAL PROPERTIES OF PROTEIN AGENTS

In the case of therapeutic proteins, protein engineering is used to produce a form with reduced immunogenicity, improved stability of therapeutic action, and increased half-life in circulation (Shanafelt, 2005). Such modifications can also alter the solubility, cofactor specificity, and susceptibility to degradation by proteases. Modifications such as pegylation (i.e., addition of polyethylene glycol) or glycosylation (i.e., addition of carbohydrate moiety) to a protein often improves its pharmacological properties such as absorption, circulation lifetime, and reduced clearance rate (Sola and Griebenow, 2010) and may be especially useful in conjunction with delivery through microcapsules or bioimplants (Pai et al., 2009; Zilberman et al., 2010). Site-specific modifications that allow pegylation or glycosylation to occur on the protein is one important avenue that can be explored to enhance the pharmaceutical properties of proteins (Graddis et al., 2002). The effect of pegylation on the properties of drugs has recently been reviewed (Veronese and Mero, 2008; Bailon and Won, 2009). The addition of the soluble polymer prevents clearance from the bloodstream by the kidney, reduces enzymatic degradation, and reduces immunogenic problems. Specific proteins where this has been advantageous include interferon alpha-2a. The enhanced pharmacokinetic properties of the modified protein can allow for more convenient and appropriate dosing schedules for patients. Recent examples of reviews in this area are cited here (Dicker and Strasser, 2015; Wang et al., 2015; Griswold and Bailey-Kellogg, 2016; Lopes et al., 2017). The fusion to an albumin binding domain allows for longer life of the administered protein in serum in animal trials (Jacobs et al., 2015).

Pegylation: Pegylation refers to the attachment of polyethyleneglycol (PEG) to a protein or small molecule, typically to improve its lifetime for pharmaceutical purposes. PEG can be specifically attached to a protein through binding with amino or sulfhydryl groups on the surface of the protein.

Glycosylation: Glycosylation is an abundant form of post-translational modification that covalently attaches an oligosaccharide chain containing 4–15 sugars through the activity of oligosaccharyl transferase. It plays a significant role in *cell-cell recognition* and affords a "*steric*" protection against digestion by proteases. In this mechanism a carbohydrate moiety is attached to a protein, typically through linkage to exposed asparagine (N-link) or serine and threonine (O-link) residues on the protein.

Certain drugs can be better agents if they are modified to enhance uptake or pharmacokinetic properties. For example, the acylation of one of the two hydroxyl groups of the drug lobucavir improved its antiviral potency.

Acylation: The binding of an acyl group to a protein or other molecule, typically forming an amide link with the side-chain amino group of a surface lysine residue or with a hydroxyl group of a small molecule.

10.6 MODIFICATION OF ENZYMES FOR *IN VIVO* BIOSYNTHETIC PROCESSES

The application of enzyme engineering to biomolecules, whole-cell biocatalysts, or immobilized enzymes in the manufacture of pharmaceuticals has been an area of expansion in recent years; examples include β-lactam acylases, which have been modified and in turn used to carry out reactions beyond their usual hydrolytic specificities on penicillin G and cephalosporin C.

Goals in this area include enhancing performance of immobilized enzymes and creating suitable specialized reaction conditions that are compatible with downstream processing. Because the chiral nature of products is central in many cases, the resolution of racemic mixtures into its component parts is of great importance. Selective hydrolysis of acyl groups has been used to facilitate the separation of isomers. In this case, the racemic mixture is typically treated with an enzyme that can only hydrolyze one stereoisomer, and after the reaction is completed, the acylated form is easily separated from the unacylated form using biochemical techniques.

Another group of antibiotics is the polyketides, and there has been considerable work on altering the complex enzymes involved in the biosynthesis of these compounds (e.g., Khosla et al., 2007; Khosla, 2009; Khosla et al., 2009; Zhan, 2009). In these articles, the modularity of the biosynthetic pathways together with the formation of large multifunctional proteins and multiprotein complexes were discussed in great depth. The modification of the individual functional domains and switching of these modules among the gene clusters specifying the biosynthesis of different antibiotics have given new approaches to industry involving advanced genetic techniques for the synthesis of novel polyketide-based pharmaceuticals.

Steroids: Steroids are lipid molecules with four interconnected ring systems—three fused six-member rings and one five-member ring—that are commonly found in membranes and hormones.

Sterols are an important group of compounds with many uses as pharmaceuticals. Yeast has been engineered to make pregnenolone and progesterone (Duport et al., 1998), hydrocortisone (Szczebara et al., 2003; Brocard-Masson and Dumas, 2006), and triterpenes (Kirby et al., 2008), and metabolic flux through the pathways involved has been analyzed (Maczek et al., 2006) and subsequently manipulated (Paradise et al., 2008).

Cytochrome P450's reactions on **steroids** as a substrate is of medical importance, and the broad potential for engineering yeast for sterol modifications has been reviewed (Veen and Lang, 2004). The value of engineering yeast for biosynthesis of terpenes and sterol compounds have been recently reviewed and yeast's ability to form the precursor molecule, isopentenyl pyrophosphate in high amount and the ability to express many specific cytochrome P450s for desired oxidation reactions are key factors in the production of these pharmaceuticals and pharmaceutical intermediates by these organisms (Wriessnegger and Pichler, 2013; Paramasivan and Mutturi, 2017).

Plants have also been engineered for improved production of phytosterols and related compounds (Seo et al., 2005; Liao et al., 2016). Furthermore, synthesis and applications of such biotransformation have been reviewed (Malaviya and Gomes, 2008). Several other organisms are used to make specific chiral modifications of the basic sterol structure and are used for chemical conversions or assays (MacLachlan et al., 2000). Protein engineering of specific amino acids at key positions in particular enzymes has altered the activities of steroid dehydrogenases involved in steroid hormone metabolism; in particular, the role of oxygenated sterols has been discussed with regard to various medical and biotechnological applications (Brown and Jessup, 2009; Lordan et al., 2009; Pasqualini, 2009; Pollegioni et al., 2009).

Carotenoids represent a group of compounds containing several conjugated double bonds and are of interest for their color, nutritional, and pharmaceutical values. These compounds are made by various microorganisms and the genes encoding the enzymes of the biosynthetic pathways are known. Recent work has involved the expression of biosynthetic enzymes in different hosts and in unique combinations. The possibility of evolving novel pathways for the formation of a new family of carotenoids is being explored using a combinatorial approach (Tanaka and Ohmiya, 2008; Chemier et al., 2009; Harada and Misawa, 2009; Liu et al., 2009; Muntendam et al., 2009; Nishihara and Nakatsuka, 2010).

The biosynthetic pathways involved in the formation of carotenoids and flavonoids require reducing power in the form of NADPH or often reduce CoA conjugates of the desired compound by reactions with reduced cofactors. While the cytochrome P450s are used in oxidation reactions of sterols and other compounds with O2, the enzymes require recycling by a reductase, commonly coupled with NADPH. Cofactor engineering is therefore directly relevant to pathway engineering.

10.7 WHOLE-CELL BIOCATALYSTS

Whole-cell biocatalysts that incorporate genes producing specific enzymes for catalysis of an entire biosynthetic pathway have become more widely used. In designing and implementing such metabolic engineering schemes, a two-stage strategy involving the removal and/or deregulation of competing metabolic pathways as well as the introduction of novel pathways into the organism of choice is often used. In the first stage, gene "knockouts" are constructed to eliminate activity of the competing pathway using recombination methods routinely used in *E. coli* or yeast genetics (Datsenko and Wanner, 2000). In the second stage, a novel pathway is introduced, or alternatively tandem copies of the gene of interest (Tyo et al., 2009), cloned downstream of a powerful promoter, are used to transform the organism under investigation (Yuan et al., 2006; McCleary, 2009). Expression of genes brought in from other organisms can be done in the same fashion as those from the native host, but additional problems may be encountered related to the codon preference of the host versus the heterologous donor, mRNA stability and structure, protein stability or aggregation, and location or interaction with other proteins or cell constituents. In these complex engineering endeavors, it is important to not only account for the flow of carbon through the network but also to consider the flow and stoichiometry of oxidation/reduction equivalents to redox balance. Strains constructed using this approach are currently used for the production of biofuels, specialty chemicals, and biopharmaceuticals. A number of large molecules have now been made through metabolic engineering that require more steps and more complex or difficult reactions to be catalyzed. Examples of such large pharmaceutical compounds include the recent work on producing the anti-cancer compound Taxol, in metabolically engineered microbes. Recent reviews have summarized the complexity of engineering these long pathways, and detail considerations of adequately expressing the genes in a foreign host organism and efforts of protein engineering to stabilize and retain activity in the enzyme complexes required for the multistep synthesis of such compounds (O'Connor, 2015; Li et al., 2015; Zhang et al., 2016; Dziggel et al., 2017). Another prominent example is the effort to synthesize opioids such as thebaine, hydrocodone and noscapine (Galanie et al., 2015; Li and Smolke, 2016).

These efforts to produce complex pharmaceutical molecules in an industrially well-defined organism require a combination of discovering new genes and enzymes in the natural producer that are essential for the pathway, enhancing the activity of known and new enzymes by protein engineering, and optimization of the host and culture conditions to form the precursor molecules in the required quantity.

One aspect of enhancing a new or artificial pathway is forming protein complexes among the enzymes of the pathway so intermediates can be more efficiently channeled to give the final product without loss or disruption of other critical cell metabolism. This can be done by using the natural protein complexes from the native organism that produces the compound, or in the case of a novel compound by artificially creating protein complexes by attaching the catalytically active protein to a protein domain that allows specific association with the protein catalyzing the next reaction of the pathway. Such channeling complexes have been made through attachments to RNA scaffolds (Myhrvold and Silver, 2017), packaging within metabolosome or microcompartments (Huber et al., 2017) or fusing the enzymatically active protein domain to a protein domain module that can facilitate the desired co-localization (Howorka, 2011; Gao et al., 2014).

This localization approach to optimization of pathway flux to product is especially important if dealing with unstable intermediate compounds in the pathway or pathway intermediates that may build up to toxic levels if the series of reactions in the pathway are unbalanced.

There are also applications for carrying out an individual redox process in which regeneration of the cofactor is an important consideration, and these have become useful industrial processes. Some of these features have been reviewed and discussed (Carballeira et al., 2009). Such processes use an active cofactor-regenerating system to ensure the regeneration of the desired cofactor. In this system, a specific precursor molecule, for example an alcohol or acid that can be readily oxidized by a cellular enzyme in the engineered cells yields a reduced cofactor, NADH or other, to ensure the continuation of the regenerating cycle without supporting growth.

Research on in vitro mimics of whole-cell biocatalysts has advanced and use purified enzymes or sets of specialized cell extracts to catalyze a series of reactions. Such in vitro systems have been applied to several step processes and have also given attention to including all recycling reactions for energy and cofactors needed for the pathway. This area has been gaining interest and several publications have shown the ability to produce a chemical efficiently. Some examples of publications on this direction for biocatalysis have been recently published (Fessner, 2015; Schmidt-Dannert and Lopez-Gallego, 2016; Karim et al., 2017; Opgenorth et al., 2017).

At the other extreme of defined systems for biocatalytic processes are efforts to better understand and utilize mixed cultures that are common in for example, traditional fermented foods, the waste processing industry, degradative microbial consortia and host associated microbiomes to achieve improved performance and more efficient and stable processing or synthesis of desired products. Engineering the microbiome in a pharmaceutical or health sense has opened new avenues for pharmaceutical development through manipulation of effects of the microbiome and its naturally released metabolites on host metabolism and organ function. The engineering of microbiomes of humans, domesticated animals, and plants has been brought to more widespread attention due to observations of the effects of microbiome distortions on host processes and health. The theoretical framework for modeling and design of multi-organism systems or microbial community metabolism is providing a fundamental basis for developments in this emerging area.

Some areas where microbial consortia are particularly desired are those where pathways converge and the product of one organism becomes a valuable feed for another organism and thus affects the overall energetics of the microbial community metabolism. Manipulation of gene and enzyme controls may allow optimization of specific desired products, or the use of novel or low value feedstocks for production. The use of multiple organisms may be needed where a combination of several favorable attributes may be not readily engineered into one organism. Some reviews on the analysis and impact of the microbial community on human health include

Heinken and Thiele (2015), Rooks and Garrett (2016), and Widder et al. (2016).

10.8 COFACTOR ENGINEERING

The term "cofactor engineering" was coined in the early 1990s in a general article (Duine, 1991) that drew attention to the essential role of cofactors in different reactions. It had long been known that the level of biotin could greatly affect the production of glutamate by *Corynebacterium glutamicum* (Kimura, 2003). A related experience was noted in the early production of citric acid with respect to iron, which is normally a constituent of certain enzymes or hemes.

Specific oxidation reactions and their bioprocess considerations have been reviewed (Buhler and Schmid, 2004), and the limitation of cofactor regeneration has been considered (Schroer et al., 2009). The cofactor most studied is the reductant used in many reactions (i.e., NADH or NADPH). Although reductions are very important, and the chiral nature of the product is often crucial, there is less industrial usage of specific oxido-reduction pathways in whole cell biocatalysts because of the lack of cofactor regenerating systems *in vivo*. However, *in vitro*, the cofactor can be regenerated by another enzyme that can form the reduced cofactor by oxidation of a readily available substrate that will not interfere with the desired reaction. An enzyme often used in the *in vitro* systems to form NADH is formate dehydrogenase. This enzyme can use formate to produce the reduced form of the cofactor NADH and carbon dioxide (CO_2). The CO_2 is easily removed and does not generally interfere with redox reactions (Wandrey, 2004). This and other cofactor regenerating systems have been reviewed (Wichmann and Vasic-Racki, 2005; Weckbecker et al., 2010). For NADPH regeneration, glucose dehydrogenase has been used, and with the use of permeabilized cells efficient coupled processes have been developed (Zhang et al., 2009). In some cases, a large amount of an alcohol can be supplied and a rather nonspecific alcohol dehydrogenase can perform the desired reduction and the regeneration of cofactor if the enzyme parameters and appropriate concentrations are used (Hollrigl et al., 2008; Schroer et al., 2009). Because NADH is more stable and generally available in higher quantities, it is preferred to NADPH. Thus, for *in vitro* processes, it is often considered useful to convert an NADPH-dependent enzyme to an NADH-utilizing enzyme (see above example in Figure 10.1). This is also true for some *in vivo* processes in which reaction rates are dependent on the availability of NADPH on one hand and/or the NADH-NADPH ratio. A specific example of this is the formation of xylitol as a by-product in the fermentation of xylose to ethanol by engineered yeasts, where the xylose is first converted to xylulose so it can be metabolized by the yeast. However, the activity of the NAD$^+$-dependent xylitol dehydrogenase forms xylitol and lowers the yield of ethanol. This imbalance has been redressed by altering the specificity of the xylitol reductase (Petschacher and Nidetzky, 2008), and this, in turn, allows for better utilization of mixed substrates and yields a low level of xylitol formation (Krahulec et al., 2010)

Xylose: A five-carbon sugar, the d form of which commonly found in hemicellulose is shown. Xylose exists in two different forms: linear or cyclic.

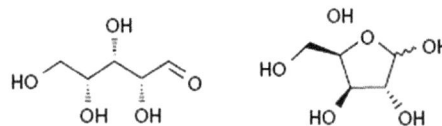

Xylulose: Xylulose is a five-carbon sugar but is a ketose rather than aldose, as is the case with xylose.

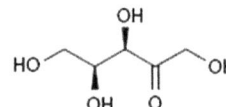

Xylitol: All of xylitol's oxygen is in the hydroxyl form.

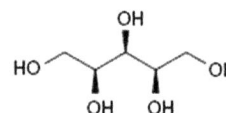

Another large-scale commercial metabolite that is affected by availability of NADPH is lysine. This amino acid is produced in high demand since many foods for humans and animals are low in this amino acid. Two routes have been explored to provide a higher level of NADPH for the biosynthesis of lysine. One is to increase the flux through the pentose pathway, which is a major route for NADPH generation relative to the typical NADH provided from glycolysis. This factor has been discussed for *pgi* (phosphoglucose isomerase) mutations in *Corynebacterium glutamicum*, a major industrial producer of lysine (Marx et al., 2003), employment of a feedback-resistant 6-phosphogluconate dehydrogenase gene that increased carbon flux through the pentose phosphate pathway, (Ohnishi et al., 2005). The overexpression of fructose 1,6-bisphosphatase (FBPase) redirects carbon flux from glycolysis toward the pentose phosphate pathway and increased lysine yield (Becker et al., 2005) and overexpression of the *zwf* gene, encoding G6P dehydrogenase, increased flux and when combined with other pentose pathway focused strategies lead to a major increase in lysine yield.

The other route to increased NADPH availability uses a different glycolytic pathway step to replace the usual NADH-producing step of glycolysis, the NAD-dependent glyceraldehyde 3-phosphate dehydrogenase (GapA) with a nonphosphorylating NADP-dependent glyceraldehyde 3-phosphate dehydrogenase (GapN) (Takeno et al., 2010; Takeno et al., 2016), and a natural or modified NADPH-dependent ATP forming enzyme has also been used in some applications (Wang et al., 2013; Bommareddy et al., 2014; Xu et al., 2014). This strategy has been employed with or without an NAD kinase to stimulate production of products highly dependent on NADPH-requiring reactions in their pathway (Lee et al., 2013).

Other redox cofactors in the membrane, ubiquinone and menaquinone, have been studied for their specificity in different oxidative or reducing reactions and the levels of these membrane-bound cofactors have been manipulated leading to different patterns of metabolites. The effect on the electron transport chain and operation of the cell in a more anaerobic or aerobic mode has been exploited for metabolic engineering purposes (Wu et al., 2015).

Studies of the actions and specificity of protein electron carriers (ferredoxins and flavodoxins, etc.) have examined their specificity in different types of cells and reactions and how modification of levels of these electron carriers in specific cells may be used to modify metabolism for engineering purposes (Atkinson et al., 2016).

10.8.1 NADH VERSUS NADPH SPECIFICITY OF ENZYMES

Isocitrate dehydrogenase (ICDH), an NADPH-dependent enzyme from *E. coli*, has been engineered, on the basis of data derived from its three-dimensional structure, by the introduction of seven mutations that change the specificity from favoring NADPH by 7000-fold to an enzyme that prefers NADH by 200-fold (Hurley et al., 1996). In general, a major difference between the NADP$^+$ and NAD$^+$ utilizing enzymes is the presence of a carboxylate side chain that chelates the diol group at the ribose near the adenine of NAD$^+$, whereas in enzymes that utilize NADP$^+$, there is frequently a highly positively charged arginine side chain oriented toward the adenine that interacts with the negatively charged ribose 3'-phosphomonoester (Carugo and Argos, 1997). See the example of mutagenesis illustrated in Figure 10.1 and the associated literature (Chen et al., 1995; Hurley et al., 1996; Yaoi et al., 1996).

In the important group of cytochrome P450 enzymes, the NADPH specificity of an enzyme was altered so the enzyme could use NADH for recycling (Dohr et al., 2001). *Corynebacterium's* 2,5-diketo-d-gluconic acid (2,5-DKG) reductase, an enzyme used in the synthesis of vitamin C, has been mutated at four different sites giving rise to the F22Y/K232G/R238H/A272G quadruple mutant that is capable of utilizing NADH and NADPH (Banta et al., 2002c). The consequences of such alterations on vitamin C production have been modeled and were claimed to be economically beneficial through cost savings (Banta et al., 2002a).

Although *in vitro* regeneration of NADH has been addressed for some time (Wichmann and Vasic-Racki, 2005), alteration of *in vivo* metabolism by use of cofactor manipulation is a recent development (Schroer et al., 2009; Weckbecker et al., 2010). An illustration of the importance of redox cofactor balance was illustrated in the engineering of an escape valve for excess NADPH accumulation in phosphoglucose-isomerase-deficient yeast. This mutant yeast cannot grow on glucose in absence of an NADPH oxidizing system. Accumulation of NADPH was prevented through cofactor engineering by altering the organism's own butanediol dehydrogenase specificity from NADH to NADPH (Ehsani et al., 2009).

Enzymatic reactions involved in the formation of flavor components vanillin, gamma-decalactone, carboxylic acids, C6 aldehydes and alcohols ("green notes"), esters, and 2-phenylethanol, including the need for cofactors and efficient recycling, have been discussed (Schrader et al., 2004).

10.8.2 MANIPULATION OF NADH AND NADPH *IN VIVO*

Expression of NADH oxidase *in vivo* was used to change the pattern of metabolic products formed by *Lactococcus lactis* (Lopez de Felipe et al., 1998). In such a strain, preference was for the production of more oxidized final products because of restricted supply in NADH, which is essential for the reduction of oxidized intermediates. Several cofactor engineering alterations have been reported in the literature (San et al., 2002), including a system for using formate dehydrogenase (FDH) to recycle NADH *in vivo* as well as the manipulation of CoA levels in the formation of products derived from acetyl-CoA. The *in vivo* use of the *Candida boidinii* FDH system (Allen and Holbrook, 1995; Sakai et al., 1997) was further investigated. The coupling of FDH to l-amino acid dehydrogenases could produce optically active amino acids from α-keto acids in the presence of ammonium formate and a whole-cell biocatalyst (Galkin et al., 1997). Enhancement of ethanol production during growth on glucose was found in cells expressing *C. boidinii* FDH grown under anaerobic conditions. In fact, some ethanol could be formed under aerobic conditions when formate was added to an FDH-expressing strain (Berrios-Rivera et al., 2002a, 2002b). The increased availability of NADH in an organism could provide potential for reducing other molecules present in the cell or added to the media (Figure 10.5). The coupling of FDH with a transhydrogenase to increase the intracellular level of NADPH for chiral alcohol production has been reported (Weckbecker and Hummel, 2004). Engineering of the enzyme by mutagenesis to aid stabilization of the enzyme to oxidation and other improved characteristics of FDH have been reported (Slusarczyk et al., 2000; Tishkov and Popov, 2006) and the temperature stability of this enzyme from *Pseudomonas sp. 101* has been improved (Rojkova et al., 1999). Several FDHs from other species have been recently cloned, characterized, and altered (Karaguler et al., 2007). A comparison between FDH and glucose dehydrogenase (GDH) revealed that the latter enzyme was more effective in the regeneration of NADH required for the reduction of 4-chloroacetoacetate to form (S) 4 chloro 3 hydroxybutanoate, an important pharmaceutical intermediate (Yamamoto et al., 2004). The use of GDH to supply reduced cofactor for reactions *in vivo* such as the production of chiral alcohols has also been reported (Kataoka et al., 2003).

Mannitol can be produced by the enzymatic reduction of fructose (Liu et al., 2005), with NADH being generated through the coupling of mannitol dehydrogenase with FDH (Kaup et al., 2003, 2004). The extension of the system to the formation of other chiral compounds such as conversion of

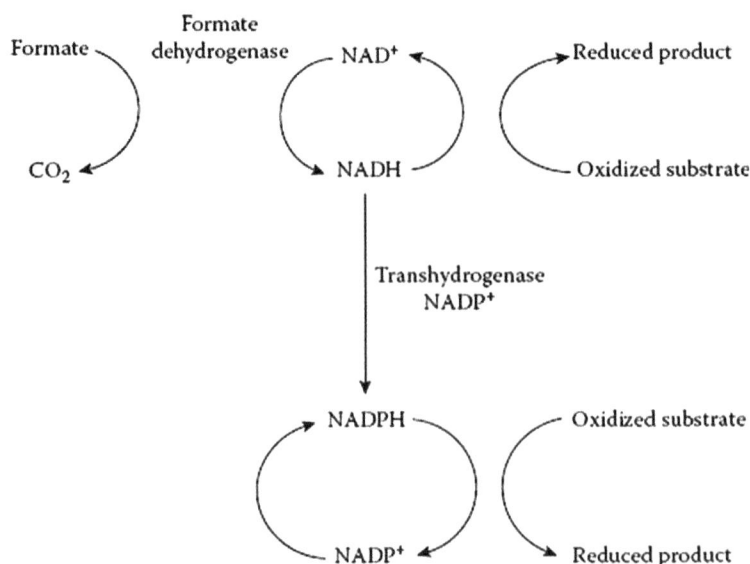

FIGURE 10.5 Coupling of *in vivo* activity of FDH to other redox reactions. Through internally produced or externally added formate, the reducing power of formate is used to produce NADH by the action of FDH. The NADH formed can be used by another enzyme in the cell, either an existing enzyme such as alcohol dehydrogenase or a heterologous enzyme from another organism capable of reducing a desired substrate. Various oxidoreductases can be produced in a host organism by introduction of appropriate genes by recombinant DNA techniques. The reduced product formed can be derived from a substrate normally produced by the growing cell from metabolism of a carbon source or it could be an added substrate molecule that is not otherwise metabolized by the organism. For reduced products, which require the participation of NADPH-dependent enzymes, an NADPH regenerating system can be constructed with the aid of transhydrogenase.

methyl acetoacetate to chiral hydroxy acid derivatives [methyl (R)-3-hydroxy butanoate] was developed by using the *fdh* gene from *Mycobacterium vaccae* N10. The gene encoding the NAD⁺-dependent FDH was co-expressed in *E. coli* with an alcohol dehydrogenase from *Lactobacillus brevis* that his able to catalyze a highly regioselective and enantioselective reduction (Ernst et al., 2005).

Cofactor regeneration has also been used to increase the intracellular levels of NADPH and NADH, through the expression of a soluble transhydrogenase, for the production of hydro-morphone (Boonstra et al., 2000). In the utilization of xylose by *Saccharomyces cerevisiae* for ethanol production, an imbalance of NADPH was observed. Such imbalance can be redressed through changing the cofactor specificity of the enzyme as illustrated earlier (Petschacher and Nidetzky, 2008).

In the ammonia pathway, conversion of the NADPH-dependent glutamate dehydrogenase to an NADH-dependent (GDH2), and its subsequent expression, was accompanied by increase in flux to ethanol formation, while reducing flux to xylitol formation. This approach seems promising for improving batch culture performance. However, overexpression of the GS-GOGAT complex modestly improved ethanol yield only under special continuous culture conditions. An improvement of ethanol production was found when GDP1, which codes for a fungal NADP⁺-dependent d-glyceraldehyde-3-phosphate dehydrogenase (NADP-GAPDH), was used to reduce the cofactor imbalance (Verho et al., 2003).

In some examples of the practical application of redox cofactors in pharmaceutical production, the conversion of

ketoisophorone (2,6,6-trimethyl-2-cyclohexen-1,4-dione) to (6R)-levodione (2,2,6-trimethylcyclohexane-1,4-dione) was stimulated by the overexpression of old yellow enzyme (Oye) from *Candida macedoniensis* in *E. coli*. This enzyme is a well-characterized oxidoreductase. It was reported that the use of *E. coli* BL21 (DE3) cells co-expressing *oye* and *gdh* as a whole-cell catalyst was advantageous for the practical synthesis of (6R)-levodione (Kataoka et al., 2004).

10.8.3 CoA Compounds and S-Adenosyl Methionine

10.8.3.1 Acetyl-CoA Levels

The key intermediate acetyl-CoA is involved in the formation of many products of interest from isoprene compounds, carotenoids, flavanones, and polyketides to fatty acids and biofuel-related molecules. Manipulation of the acetyl-CoA levels for improved formation of esters derived from condensation of an alcohol with the acetyl-CoA has been reported (Vadali et al., 2004a, 2004b, 2004c). The high levels of CoA were produced by overexpression of a pantothenate kinase, and increased acetyl-CoA was achieved by overexpression of pyruvate dehydrogenase. In the experiments described, the acetyl-CoA was condensed with an alcohol by an acetyl-CoA-alcohol transferase to form an ester. This reaction could serve as a useful reporter of available acetyl-CoA without disrupting other areas of metabolism. The recycling and balance of CoA-utilizing pathways can serve as a possible means to influence the flux to other sinks for this intermediate and affect the overall pattern of metabolites (Figure 10.6).

Formation and uses of acetyl-CoA

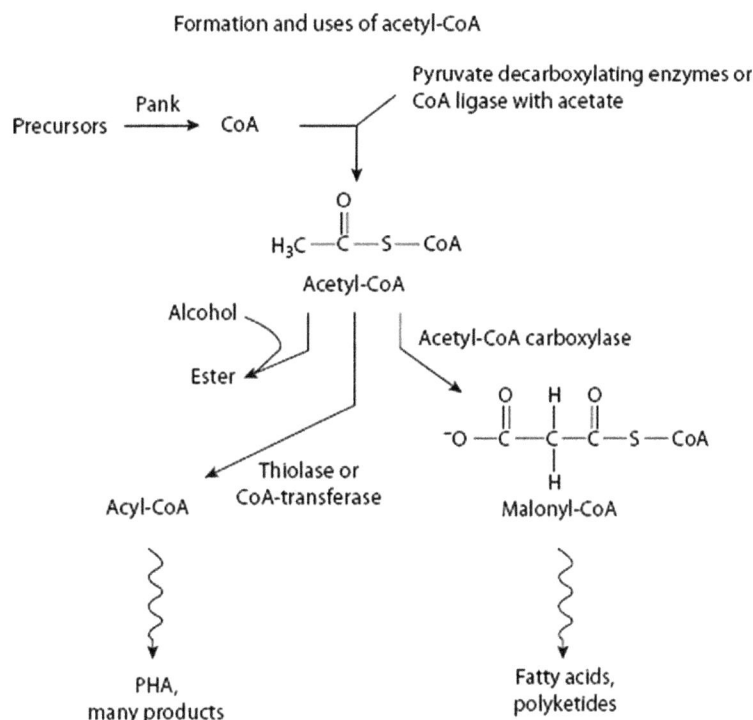

FIGURE 10.6 The CoA level can be raised by overexpression of a key enzyme in the biosynthetic pathway of CoA from pantothenate. The CoA can then be more effectively converted to acetyl-CoA or other acyl-CoA by pyruvate dehydrogenases or CoA ligases; the acetyl-CoA is then more available for conversion to esters.

Although feedback systems generally control these CoA levels, they can be manipulated to some extent, which can contribute to enhanced production of acetyl-CoA-derived compounds. Alteration of the acetyl-CoA levels could have wide application because many pathways for production of larger molecules rely on the sequential addition of acyl-CoA groups. Because the mevalonate pathway compounds synthesize the intermediate isopentanyl pyrophosphate and many valuable compounds including terpenes and sterols are derived for this precursor, this pathway has been effectively introduced into *E. coli* (Yoon et al., 2009). Efforts to engineer the acetyl-CoA carboxylase-catalyzed carboxylation of acetyl-CoA to malonyl-CoA have been studied (Zha et al., 2009), and the engineering of this component on the production of flavanones showed a strong effect on the levels of flavanones obtained in engineered *E. coli* (Fowler et al., 2009).

One area where the level of acyl-CoA compounds is important is in the production of poly-hydroxybutyrate and its various derivatives. In this process, hydroxyacyl-CoA compounds are polymerized into a large polymer. Different physical properties within the polymer are obtained depending on the composition of monomers incorporated. Therefore, copolymerization of hydroxy-butyryl-CoA with other acyl-CoA compounds has been investigated extensively (Steinbuchel and Schlegel, 1991; Liu and Chen, 2007; Chung et al., 2009; Elbahloul and Steinbuchel, 2009; Jian et al., 2010). To obtain high levels of the nonstandard acyl-CoA substrates, the mechanisms of uptake and activation of the corresponding acids, including the use of the *prpE* gene, were targeted (Valentin et al., 2000;

Aldor and Keasling, 2001) Expression of a combination of a butyrate kinase and phosphotransbutyrylase (Liu et al., 2003) and development of a means to take advantage of the connection in formation of acyl-CoA compounds between the pathways of fatty acid degradation and those acyl-CoA compounds needed for desired product formation have also been investigated (Liu and Chen, 2007; Chung et al., 2009; Jian et al., 2010). For example, the use of CoA transferases to bring other acids into the polymerization process has also been reported with 4-hydroxybutyrate-containing polymers (Song et al., 2005). The use of glycerol as a feedstock for the formation of poly(3-hydroxypropionate) in engineered *E. coli* (Andreessen et al., 2010) and the engineering/evolution of polyhydroxyalkanoate synthase have generated enzymes with higher activity and altered specificity (Amara et al., 2002) and even allowed for the formation of polylactate in *E. coli* (Jung et al., 2010; Matsumoto and Taguchi, 2010; Yang et al., 2010). The introduction of acyltransferases that can transfer the long-chain fatty acid moiety from acyl-CoA compounds to alcohols has allowed for the formation of long-chain esters relevant to biofuels, e.g. "microdiesel" (Stoveken and Steinbuchel, 2008; Elbahloul and Steinbuchel, 2010).

A large group of compounds built from acyl-CoA precursors are the polyketides, a group of bioactive structures [e.g., avermectin (Ikeda et al., 2001)]. In these reactions, malonyl-CoA is a major precursor. Malonyl-CoA is formed from acetyl-CoA by addition of a carboxyl group and is a precursor for fatty acid synthesis. In the case of polyketide synthesis, the group is transferred to an acyl carrier protein for incorporation into the chain as the longer-chain polyketide is synthesized (Khosla, 2009). The acyl carrier proteins require activation by addition of a 4'-phosphopantotheinyl group from CoA by specific enzymes, PPTases (Mofid et al., 2004). The analysis and overexpression of acetyl-CoA carboxylase from various sources have led to increased levels of malonyl-CoA (Davis et al., 2000; Gande et al., 2004). Manipulation of the acyl-CoA pools or use of selective acyl-transferases can lead to alterations of the polyketide product (Liou and Khosla, 2003). Providing enzymes such as CoA ligases for forming methylmalonyl-CoA or propionyl-CoA has increased the pool of the corresponding acyl-CoAs and improved synthesis of the erythromycin precursor, 6-deoxyerythronolide B (Murli et al., 2003). Furthermore, the use of CoA ligases with broad specificities has also been exploited for the precursor pools (Pohl et al., 2001; Arora et al., 2005).

10.8.3.2 S-Adenosyl Methionine Levels

The modification by addition of a methyl group to generate methoxy side chains is a widely used event in many secondary metabolites. This and the volatility of biologically important methylate compounds generated a good deal of interest in exploiting methylation reactions using *S*-adenosyl methionine (SAM) as the methyl donor. In a biotechnological context, methyl transfer strategy has been used to generate methyl halides in engineered strains of *E. coli* and yeast (Bayer et al., 2009). The engineering of methionine adenosyltransferase by DNA shuffling of *E. coli*, *S. cerevisiae*, and *Streptomyces spectabilis* MAT genes resulted in a more active enzyme and allowed for a high level SAM productivity during the course of large-scale fermentation (Hu et al., 2009).

SUMMARY

Recent innovations in functional genomics, proteomics, transcriptomics, metabolomics, and bioinformatics have revolutionized the scope of enzyme engineering and its role in the production of biopharmaceuticals. Many enzymes have been targeted individually and in combinations with the view of improving industrial processes. In addition to random mutagenesis coupled with high-throughput screening, other techniques including site-directed mutagenesis, molecular breeding, and synthetic biology are currently being exploited to improve the physical and pharmokinetic properties of key industrial enzymes.

The ability to manipulate the intracellular levels of cofactors to favor the formation of desired products further extends and expands the repertoire of tools available for enzyme and metabolic engineering.

The coupling of these experimental methods with better understanding of global cellular metabolism via advanced analytical techniques and mathematical treatments bodes well for further developments in the applications of enzymes and microbial biocatalysts in industrial and pharmaceutical bioprocesses.

REFERENCES

Aldor, I., and J.D. Keasling. 2001. Metabolic engineering of poly(3-hydroxybutyrate-co-3-hydroxyvalerate) composition in recombinant *Salmonella enterica* serovar typhimurium. *Biotechnol Bioeng* 76:108–114.

Allen, S.J., and J.J. Holbrook. 1995. Isolation, sequence and overexpression of the gene encoding NAD-dependent formate dehydrogenase from the methylotrophic yeast *Candida methylica*. *Gene* 162:99–104.

Amara, A.A., A. Steinbuchel, and B.H. Rehm. 2002. In vivo evolution of the Aeromonas punctata polyhydroxy-alkanoate (PHA) synthase: Isolation and characterization of modified PHA synthases with enhanced activity. *Appl Microbiol Biotechnol* 59:477–482.

Andreessen, B., A.B. Lange, H. Robenek, and A. Steinbuchel. 2010. Conversion of glycerol to poly(3-hydroxy-propionate) in recombinant *Escherichia coli*. *Appl Environ Microbiol* 76:622–626.

Arnold, F.H. 2017. Bringing new chemistry to life. *Angew Chem Int Ed Engl* Oct 24. doi:10.1002/anie.201708408.

Arora, P., A. Vats, P. Saxena, D. Mohanty, and R.S. Gokhale. 2005. Promiscuous fatty acyl CoA ligases produce acyl-CoA and acyl-SNAC precursors for polyketide biosynthesis. *JAm Chem Soc* 127:9388–9389.

Asboth, B., and G. Naray-Szabo. 2000. Mechanism of action of D-xylose isomerase. *Curr Protein Pept Sci* 1:237–254.

Atkinson, J.T., I. Campbell, G.N. Bennett, and J.J. Silberg. 2016. Cellular assays for ferredoxins: A strategy for understanding electron flow through protein carriers that link metabolic pathways. *Biochemistry* 55:7047–7064.

Bailon, P., and C.Y. Won. 2009. PEG-modified biopharmaceuticals. *Expert Opin Drug Deliv* 6:1–16.

Banta, S., M. Boston, A. Jarnagin, and S. Anderson. 2002a. Mathematical modeling of in vitro enzymatic production of 2-keto-L-gulonic acid using NAD(H) or NADP(H) as cofactors. *Metab Eng* 4:273–284.

Banta, S., B.A. Swanson, S. Wu, A. Jarnagin, and S. Anderson. 2002b. Alteration of the specificity of the cofactor-binding pocket of Corynebacterium 2,5-diketo-D-gluconic acid reductase A. *Protein Eng* 15:131–140.

Banta, S., B.A. Swanson, S. Wu, A. Jarnagin, and S. Anderson. 2002c. Optimizing an artificial metabolic pathway: Engineering the cofactor specificity of Corynebacterium 2,5-diketo-D-gluconic acid reductase for use in vitamin C biosynthesis. *Biochemistry* 41:6226–6236.

Bayer, T.S., D.M. Widmaier, K. Temme, E.A. Mirsky, D.V. Santi, and C.A. Voigt. 2009. Synthesis of methyl halides from biomass using engineered microbes. *J Am Chem Soc* 131:6508–6515.

Becker, J., C. Klopprogge, O. Zelder, E. Heinzle, and C. Wittmann. 2005. Amplified expression of fructose 1,6-bisphosphatase in *Corynebacterium glutamicum* increases in vivo flux through the pentose phosphate pathway and lysine production on different carbon sources. *Appl Environ Micro* 71:8587–8596.

Benkovic, S.J., and S. Hammes-Schiffer. 2003. A perspective on enzyme catalysis. *Science* 301:1196–1202.

Berrios-Rivera, S.J., G.N. Bennett, and K.Y. San. 2002a. Metabolic engineering of *Escherichia coli*: Increase of NADH availability by overexpressing an NAD(+)-dependent formate dehydrogenase. *Metab Eng* 4:217–229.

Berrios-Rivera, S.J., G.N. Bennett, and K.Y. San. 2002b. The effect of increasing NADH availability on the redistribution of metabolic fluxes in *Escherichia coli* chemostat cultures. *Metab Eng* 4:230–237.

Bessler, C., J. Schmitt, K.H. Maurer, and R.D. Schmid. 2003. Directed evolution of a bacterial alpha-amylase: Toward enhanced pH-performance and higher specific activity. *Protein Sci* 12:2141–2149.

Blumer-Schuette S.E., I. Kataeva, J. Westpheling, M.W. Adams, and R.M. Kelly. 2008. Extremely thermophilic microorganisms for biomass conversion: Status and prospects. *Curr Opin Biotechnol* 19:210–217.

Boersma, Y.L., M.J. Droge, and W.J. Quax. 2007. Selection strategies for improved biocatalysts. *FEBS J* 274:2181–2195.

Bommareddy, R.R., Z. Chen, S. Rappert, and A.P. Zeng. 2014. A *de novo* NADPH generation pathway for improving lysine production of *Corynebacterium glutamicum* by rational design of the coenzyme specificity of glyceraldehyde 3-phosphate dehydrogenase. *Metab Eng* 25:30–37.

Boonstra, B., D.A. Rathbone, C.E. French, E.H. Walker, and N.C. Bruce. 2000. Cofactor regeneration by a soluble pyridine nucleotide transhydrogenase for biological production of hydromorphone. *Appl Environ Microbiol* 66:5161–5166.

Brocard-Masson, C., and B. Dumas. 2006. The fascinating world of steroids: *S. cerevisiae* as a model organism for the study of hydrocortisone biosynthesis. *Biotechnol Genet Eng Rev* 22:213–252.

Brown, A.J., and W. Jessup. 2009. Oxysterols: Sources, cellular storage and metabolism, and new insights into their roles in cholesterol homeostasis. *Mol Aspects Med* 30:111–122.

Buhler, B., and A. Schmid. 2004. Process implementation aspects for biocatalytic hydrocarbon oxyfunctionalization. *J Biotechnol* 113(1–3):183–210.

Carballeira J. D., M.A. Quezada, P. Hoyos, Y. Simeo, M.J. Hernaiz, A.R. Alcantara, and J.V. Sinisterra. 2009. Microbial cells as catalysts for stereoselective red-ox reactions. *Biotechnol Adv* 27:686–714.

Carothers, J.M., J.A. Goler, and J.D. Keasling. 2009. Chemical synthesis using synthetic biology. *Curr Opin Biotechnol* 20:498–503.

Carugo, O., and P. Argos. 1997. NADP-dependent enzymes. I: Conserved stereochemistry of cofactor binding. *Proteins* 28:10–28.

Chemier, J.A., Z.L. Fowler, M.A. Koffas, and E. Leonard. 2009. Trends in microbial synthesis of natural products and biofuels. *Adv Enzymol Relat Areas Mol Biol* 76:151–217.

Chen, J., R.C. Zheng, Y.G. Zheng, and Y.C. Shen. 2009. Microbial transformation of nitriles to high-value acids or amides. *Adv Biochem Eng Biotechnol* 113:33–77.

Chen, R., A. Greer, A.M. Dean. 1995. A highly active decarboxylating dehydrogenase with rationally inverted coenzyme specificity. *Proc Natl Acad Sci U S A* 92:11666–11670.

Cherry, J.R., and A.L. Fidantsef. 2003. Directed evolution of industrial enzymes: An update. *Curr Opin Biotechnol* 14:438–443.

Chung, A., Q. Liu, S.P. Ouyang, Q. Wu, and G.Q. Chen. 2009. Microbial production of 3-hydroxydodecanoic acid by pha operon and fadBA knockout mutant of *Pseudomonas putida* KT2442 harboring tesB gene. *Appl Microbiol Biotechnol* 83:513–519.

Cirino, P.C., and F.H. Arnold. 2002. Protein engineering of oxygenases for biocatalysis. *Curr Opin Chem Biol* 6:130–135.

Cirino, P.C., K.M. Mayer, and D. Umeno. 2003. Generating mutant libraries using error-prone PCR. *Methods Mol Biol* 231:3–9.

Datsenko, K.A., and B.L. Wanner. 2000. One-step inactivation of chromosomal genes in *Escherichia coli* K-12 using PCR products. *Proc Natl Acad Sci USA* 97:6640–6645.

Davids, T., M. Schmidt, D. Böttcher, and U.T. Bornscheuer. 2013. Strategies for the discovery and engineering of enzymes for biocatalysis. *Curr Opin Chem Biol* 17:215–220.

Davis, M.S., J. Solbiati, and J.E. Cronan, Jr. 2000. Overproduction of acetyl-CoA carboxylase activity increases the rate of fatty acid biosynthesis in *Escherichia coli*. *J Biol Chem* 275:28593–28598.

Dicker, M., and R. Strasser. 2015. Using glyco-engineering to produce therapeutic proteins. *Expert Opin Biol Ther* 15:1501–1516.

Dohr, O., M.J. Paine, T. Friedberg, G.C. Roberts, and C.R. Wolf. 2001. Engineering of a functional human NADH-dependent cytochrome P450 system. *Proc Natl Acad Sci USA* 98:81–86.

Doi, R.H. 2008. Cellulases of mesophilic microorganisms: Cellulosome and noncellulosome producers. *Ann N Y Acad Sci* 1125:267–279.

Dowe, N. 2009. Assessing cellulase performance on pretreated lignocellulosic biomass using saccharification and fermentation-based protocols. *Methods Mol Biol* 581:233–245.

Doyle, S.A., P.T. Beernink, and D.E. Koshland, Jr. 2001. Structural basis for a change in substrate specificity: Crystal structure of S113E isocitrate dehydrogenase in a complex with isopropyl-malate, Mg^{2+}, and NADP. *Biochemistry* 40:4234–4241.

Duine, J.A. 1991. Cofactor engineering. *Trends Biotechnol* 9:343–346.

Duport, C., R. Spagnoli, E. Degryse, and D. Pompon. 1998. Self-sufficient biosynthesis of pregnenolone and progesterone in engineered yeast. *Nat Biotechnol* 16:186–189.

Dziggel, C., H. Schäfer, and M. Wink. 2017. Tools of pathway reconstruction and production of economically relevant plant secondary metabolites in recombinant microorganisms. *Biotechnol J* 12:doi:10.1002/biot.201600145.

Ehsani, M., M.R. Fernandez, J.A. Biosca, and S. Dequin. 2009. Reversal of coenzyme specificity of 2,3-butanediol dehydrogenase from *Saccharomyces cerevisae* and in vivo functional analysis. *Biotechnol Bioeng* 104:381–389.

Elbahloul, Y., and A. Steinbuchel. 2009. Large-scale production of poly(3-hydroxyoctanoic acid) by *Pseudomonas putida* GPo1 and a simplified downstream process. *Appl Environ Microbiol* 75:643–651.

Elbahloul, Y., and A. Steinbuchel. 2010. Pilot-scale production of fatty acid ethyl esters by an engineered *Escherichia coli* strain harboring the p(Microdiesel) plasmid. *Appl Environ Microbiol* 76:4560–4565.

Ernst, M., B. Kaup, M. Muller, S. Bringer-Meyer, and H. Sahm. 2005. Enantioselective reduction of carbonyl compounds by whole-cell biotransformation, combining a formate dehydrogenase and a (R)-specific alcohol dehydrogenase. *Appl Microbiol Biotechnol* 66:629–634.

Feller, G. 2003. Molecular adaptations to cold in psychrophilic enzymes. *Cell Mol Life Sci* 60:648–662.

Fessner, W. D. 2015. Systems biocatalysis: Development and engineering of cell-free "artificial metabolisms" for preparative multi-enzymatic synthesis. *N Biotechnol* 25:658–664.

Fong, S.S., A.R. Joyce, and B.O. Palsson. 2005. Parallel adaptive evolution cultures of *Escherichia coli* lead to convergent growth phenotypes with different gene expression states. *Genome Res* 15:1365–1372.

Fowler, Z.L., W.W. Gikandi, and M.A. Koffas. 2009. Increased malonyl coenzyme A biosynthesis by tuning the *Escherichia coli* metabolic network and its application to flavanone production. *Appl Environ Microbiol* 75:5831–5839.

Galanie, S., K. Thodey, I.J. Trenchard, M. Filsinger Interrante, and C.D. Smolke. 2015. Complete biosynthesis of opioids in yeast. *Science* 349:1095–1100.

Galkin, A., L. Kulakova, T. Yoshimura, K. Soda, and N. Esaki. 1997. Synthesis of optically active amino acids from alpha-keto acids with *Escherichia coli* cells expressing heterologous genes. *Appl Environ Microbiol* 63:4651–4656.

Gande, R., K.J. Gibson, A.K. Brown, K. Krumbach, L.G. Dover, H. Sahm, S. Shioyama, T. Oikawa, G.S. Besra, and L. Eggeling. 2004. Acyl-CoA carboxylases (accD2 and accD3), together with a unique polyketide synthase (Cg-pks), are key to mycolic acid biosynthesis in Corynebacterianeae such as *Corynebacterium glutamicum* and *Mycobacterium tuberculosis*. *J Biol Chem* 279:44847–44857.

Gao, X., S. Yang, C. Zhao, Y. Ren, and D. Wei. 2014. Artificial multienzyme supramolecular device: Highly ordered self-assembly of oligomeric enzymes in vitro and in vivo. *Angew Chem Int Ed Engl* 53:14027–14030.

Georlette, D., V. Blaise, T. Collins, S. D'Amico, E. Gratia, A. Hoyoux, J.C. Marx, G. Sonan, G. Feller, and C. Gerday. 2004. Some like it cold: Biocatalysis at low temperatures. *FEMS Microbiol Rev* 28:25–42.

Graddis, T.J., R.L. Remmele, Jr. and J.T. McGrew. 2002. Designing proteins that work using recombinant technologies. *Curr Pharm Biotechnol* 3:285–297.

Griswold, D.W., and C. Bailey-Kellogg. 2016. Design and engineering of deimmunized biotherapeutics. *Curr Opin Struct Biol* 39:79–88.

Grogan, D.W., and K.R. Stengel. 2008. Recombination of synthetic oligonucleotides with prokaryotic chromosomes: Substrate requirements of the *Escherichia coli*/lambda red and *Sulfolobus acidocaldarius* recombination systems. *Mol Microbiol* 69:1255–1265.

Gupta, A., and S.K. Khare. 2009. Enzymes from solvent-tolerant microbes: Useful biocatalysts for non-aqueous enzymology. *Crit Rev Biotechnol* 29:44–54.

Harada, H., and N. Misawa. 2009. Novel approaches and achievements in biosynthesis of functional isoprenoids in *Escherichia coli*. *Appl Microbiol Biotechnol* 84:1021–1031.

Hartley, B.S., N. Hanlon, R.J. Jackson, and M. Rangarajan. 2000. Glucose isomerase: Insights into protein engineering for increased thermostability. *Biochim Biophys Acta* 1543:294–335.

Heinken, A., and I. Thiele. 2015. Systematic prediction of health-relevant human-microbial co-metabolism through a computational framework. *Gut Microbes* 6:120–130.

Herring, C.D., A. Raghunathan, C. Honisch, T. Patel, M.K. Applebee, A.R. Joyce, T.J. Albert, F.R. Blattner, D. van den Boom, C.R. Cantor, and B.O. Palsson. 2006. Comparative genome sequencing of *Escherichia coli* allows observation of bacterial evolution on a laboratory timescale. *Nat Genet* 38:1406–1412.

Hollrigl, V., F. Hollmann, A.C. Kleeb, K. Buehler, and A. Schmid. 2008. TADH, the thermostable alcohol dehydrogenase from Thermus sp. ATN1: A versatile new biocatalyst for organic synthesis. *Appl Microbiol Biotechnol* 81:263–273.

Holmberg, N., U. Ryde, and L. Bulow. 1999. Redesign of the coenzyme specificity in L-lactate dehydrogenase from bacillus stearothermophilus using site-directed mutagenesis and media engineering. *Protein Eng* 12:851–856.

Howorka, S. 2011. Rationally engineering natural protein assemblies in nanobiotechnology. *Curr Opin Biotechnol* 22:485–491.

Hu, H., J. Qian, J. Chu, Y. Wang, Y. Zhuang, and S. Zhang. 2009. DNA shuffling of methionine adenosyl-transferase gene leads to improved S-adenosyl-L-methionine production in *Pichia pastoris*. *J Biotechnol* 141:97–103.

Huber, I., D. Palmer, K.N. Ludwig, I.R. Brown, M.J. Warren, and J. Frunzke. 2017. Construction of recombinant PDU **metabolosome** shells for small molecule production in corynebacterium glutamicum. *ACS Syn Biol* 6:2145–2156.

Hult, K., and P. Berglund 2003. Engineered enzymes for improved organic synthesis. *Curr Opin Biotechnol* 14:395–400.

Hurley, J.H., R. Chen, and A.M. Dean. 1996. Determinants of cofactor specificity in isocitrate dehydrogenase: Structure of an engineered $NADP^+ \rightarrow NAD^+$ specificity-reversal mutant. *Biochemistry* 35:5670–5678.

Ikeda, H., T. Nonomiya, and S. Omura. 2001. Organization of biosynthetic gene cluster for avermectin in *Streptomyces avermitilis*: Analysis of enzymatic domains in four polyketide synthases. *J Ind Microbiol Biotechnol* 27:170–176.

Jacobs, S.A., A.C. Gibbs, M. Conk, F. Yi, D. Maguire, C. Kane, and K.T. O'Neil. 2015. Fusion to a ahighly stable consensus albumin binding domain allows for tunable pharmacokinetics. *Protein Eng Des Sel* 28:385–393.

Jaeger, K.E., and T. Eggert. 2004. Enantioselective biocatalysis optimized by directed evolution. *Curr Opin Biotechnol* 15:305–313.

Jestin, J.L., and P.A. Kaminski. 2004. Directed enzyme evolution and selections for catalysis based on product formation. *J Biotechnol* 113:85–103.

Jian, J., Z.J. Li, H.M. Ye, M.Q. Yuan, and G.Q. Chen. 2010. Metabolic engineering for microbial production of polyhydroxyalkanoates consisting of high 3-hydroxyhexanoate content by recombinant *Aeromonas hydrophila*. *Bioresour Technol* 101:6096–6102.

Joseph, B., P.W. Ramteke, and G. Thomas. 2008. Cold active microbial lipases: Some hot issues and recent developments. *Biotechnol Adv* 26:457–470.

Jung, R., M.P. Scott, L.O. Oliveira, and N.C. Nielsen. 1992. A simple and efficient method for the oligodeoxy-ribonucleotide-directed mutagenesis of double-stranded plasmid DNA. *Gene* 121:17–24.

Jung, Y.K., T.Y. Kim, S.J. Park, and S.Y. Lee. 2010. Metabolic engineering of *Escherichia coli* for the production of polylactic acid and its copolymers. *Biotechnol Bioeng* 105:161–171.

Karaguler, N.G., R.B. Sessions, B. Binay, E.B. Ordu, and A.R. Clarke. 2007. Protein engineering applications of industrially exploitable enzymes: *Geobacillus stearothermophilus* LDH and *Candida methylica* FDH. *Biochem Soc Trans* 35:1610–1615.

Karim, A.S., J.T. Heggestad, S.A. Crowe, and M.C. Jewett. 2017. Controlling cell-free metabolism through physiochemical perturbations. *Metab Eng* 45:86–94.

Kataoka, M., K. Kita, M. Wada, Y. Yasohara, J. Hasegawa, and S. Shimizu. 2003. Novel bioreduction system for the production of chiral alcohols. *Appl Microbiol Biotechnol* 62:437–445.

Kataoka, M., A. Kotaka, R. Thiwthong, M. Wada, S. Nakamori, and S. Shimizu. 2004. Cloning and overexpression of the old yellow enzyme gene of *Candida macedoniensis*, and its application to the production of a chiral compound. *J Biotechnol* 114:1–9.

Kaup, B., S. Bringer-Meyer, and H. Sahm. 2003. Metabolic engineering of *Escherichia coli*: Construction of an efficient biocatalyst for D-mannitol formation in a whole-cell biotransformation. *Commun Agric Appl Biol Sci* 68:235–240.

Kaup, B., S. Bringer-Meyer, and H. Sahm. 2004. Metabolic engineering of *Escherichia coli*: Construction of an efficient biocatalyst for D-mannitol formation in a whole-cell biotransformation. *Appl Microbiol Biotechnol* 64:333–339.

Keasling, J.D. 2008. Synthetic biology for synthetic chemistry. *ACS Chem Biol* 3:64–76.

Kegler-Ebo, D.M., G.W. Polack, and D. DiMaio. 1996. Use of codon cassette mutagenesis for saturation mutagenesis. *Methods Mol Biol* 57:297–310.

Khandeparker, R., and M.T. Numan. 2008. Bifunctional xylanases and their potential use in biotechnology. *J Ind Microbiol Biotechnol* 35:635–644.

Khosla, C. 2009. Structures and mechanisms of polyketide synthases. *J Org Chem* 74:6416–6420.

Khosla, C., S. Kapur, and D.E. Cane. 2009. Revisiting the modularity of modular polyketide synthases. *Curr Opin Chem Biol* 13:135–143.

Khosla, C., Y. Tang, A.Y. Chen, N.A. Schnarr, and D.E. Cane. 2007. Structure and mechanism of the 6-deoxyerythronolide B synthase. *Annu Rev Biochem* 76:195–221.

Kimura, E. 2003. Metabolic engineering of glutamate production. *Adv Biochem Eng Biotechnol* 79:37–57.

Kirby, J., D.W. Romanini, E.M. Paradise, and J.D. Keasling. 2008. Engineering triterpene production in *Saccharomyces cerevisiae*-beta-amyrin synthase from Artemisia annua. *FEBS J* 275:1852–1859.

Krahulec, S., B. Petschacher, M. Wallner, K. Longus, M. Klimacek, and B. Nidetzky. 2010. Fermentation of mixed glucose-xylose substrates by engineered strains of *Saccharomyces cerevisiae*: Role of the coenzyme specificity of xylose reductase, and effect of glucose on xylose utilization. *Microb Cell Fact* 9:16.

Kunkel, T.A., K. Bebenek, and J. McClary. 1991. Efficient site-directed mutagenesis using uracil-containing DNA. *Methods Enzymol* 204:125–139.

Kwon, S.J., R. Petri, A.L. DeBoer, and C. Schmidt-Dannert. 2004. A high-throughput screen for porphyrin metal chelatases: Application to the directed evolution of ferrochelatases for metalloporphyrin biosynthesis. *Chembiochem* 5:1069–1074.

Labrou, N.E. 2010. Random mutagenesis methods for in vitro directed enzyme evolution. *Curr Protein Pept Sci* 11:91–100.

Landrain, T. E., J. Carrera, B. Kirov, G. Rodrigo, and A. Jaramillo. 2009. Modular model-based design for heterologous bioproduction in bacteria. *Curr Opin Biotechnol* 20:272–279.

Lee, W.H., M.D. Kim, Y.S. Jin, and J.H. Seo. 2013. Engineering of NADPH regenerators in *Escherichia coli* for enhanced biotransformation. *Appl Microbiol Biotechnol* 97:2761–2772.

Li, Y., G. Zhang, and B.A. Pfeifer. 2015. Current and emerging options for taxol production. *Adv Biochem Eng Biotechnol* 148:405–425.

Li, Y., and C.D. Smolke. 2016. Engineering biosynthesis of the anti-cancer alkaloid noscapine in yeast. *Nat Commun* 7:12137.

Liao, P., A. Hemmerlin, T.J. Bach, and M.L. Chye. 2016. The potential of the mevalonate pathway for enhanced isoprenoid production. *Biotechnol Adv* 34:697–713.

Liou, G.F., and C. Khosla. 2003. Building-block selectivity of polyketide synthases. *Curr Opin Chem Biol* 7:279–284.

Liu, G.N., Y.H. Zhu, and J.G. Jiang. 2009. The metabolomics of carotenoids in engineered cell factory. *Appl Microbiol Biotechnol* 83:989–999.

Liu, S., B. Saha, and M. Cotta. 2005. Cloning, expression, purification, and analysis of mannitol dehydrogenase gene *mtlK* from *Lactobacillus brevis*. *Appl Biochem Biotechnol* 121–124:391–401.

Liu, S.J., T. Lutke-Eversloh, and A. Steinbuchel. 2003. Biosynthesis of poly (3-mercaptopropionate) and poly (3-mercaptoprop ionate-co-3-hydroxybutyrate) with recombinant *Escherichia coli*. *Sheng Wu Gong Cheng Xue Bao* 19:195–199.

Liu, W., and G.Q. Chen. 2007. Production and characterization of medium-chain-length polyhydroxyalkanoate with high 3-hydroxytetradecanoate monomer content by *fadB* and *fadA* knockout mutant of *Pseudomonas putida* KT2442. *Appl Microbiol Biotechnol* 76:1153–1159.

Locher, C.P., M. Paidhungat, R.G. Whalen, and J. Punnonen. 2005. DNA shuffling and screening strategies for improving vaccine efficacy. *DNA Cell Biol* 24:256–263.

Lopes, A.M., L. Oliveira-Nascimento, A. Ribeiro, C.A. Tairum Jr., C.A. Breyer, M.A. Oliveira, G. Monteiro, C.M. Souza-Motta, P.O. Magalhães, J.G. Avendaño, A.M. Cavaco-Paulo, P.G. Mazzola, C.O. Rangel-Yagui, L.D. Sette, A. Converti, and A. Pessoa. 2017. Therapeutic l-asparaginase:upstream, downstream and beyond. *Crit Rev Biotechnol* 37:82–99.

Lopez de Felipe, F., M. Kleerebezem, W.M. de Vos, and J. Hugenholtz. 1998. Cofactor engineering: A novel approach to metabolic engineering in *Lactococcus lactis* by controlled expression of NADH oxidase. *J Bacteriol* 180:3804–3808.

Lordan, S., J.J. Mackrill, and N.M. O'Brien. 2009. Oxysterols and mechanisms of apoptotic signaling: Implications in the pathology of degenerative diseases. *J Nutr Biochem* 20:321–336.

Luetz, S., L. Giver, and J. Lalonde. 2008. Engineered enzymes for chemical production. *Biotechnol Bioeng* 101:647–653.

MacLachlan, J., A.T. Wotherspoon, R.O. Ansell, and C.J. Brooks. 2000. Cholesterol oxidase: Sources, physical properties and analytical applications. *J Steroid Biochem Mol Biol* 72:169–195.

Maczek, J., S. Junne, P. Nowak, and P. Goetz. 2006. Metabolic flux analysis of the sterol pathway in the yeast *Saccharomyces cerevisiae*. *Bioprocess Biosyst Eng* 29:241–252.

Maki, M., K.T. Leung, and W. Qin. 2009. The prospects of cellulase-producing bacteria for the bioconversion of lignocellulosic biomass. *Int J Biol Sci* 5:500–516.

Malaviya, A., and J. Gomes. 2008. Androstenedione production by biotransformation of phytosterols. *Bioresour Technol* 99:6725–6737.

Marx, A., S. Hans, B. Möckel, B. Bathe, A.A. de Graaf, A.C. McCormack, C. Stapleton, K. Burke, M. O'Donohue, and L.K. Dunican. 2003. Metabolic phenotype of phosphoglucose isomerase mutants of *Corynebacterium glutamicum*. *J Biotechnol* 104:185–197.

Matsui, I., and K. Harata. 2007. Implication for buried polar contacts and ion pairs in hyperthermostable enzymes. *FEBS J* 274:4012–4022.

Matsumoto, K., and S. Taguchi. 2010. Enzymatic and whole-cell synthesis of lactate-containing polyesters: Toward the complete biological production of polylactate. *Appl Microbiol Biotechnol* 85:921–932.

McCleary, W.R. 2009. Application of promoter swapping techniques to control expression of chromosomal genes. *Appl Microbiol Biotechnol* 84:641–648.

Mofid, M.R., R. Finking, L.O. Essen, and M.A. Marahiel. 2004. Structure-based mutational analysis of the 4'-phosphopantetheinyl transferases Sfp from *Bacillus subtilis*: Carrier protein recognition and reaction mechanism. *Biochemistry* 43:4128–4136.

Morikawa, S., T. Nakai, Y. Yasohara, H. Nanba, N. Kizaki, and J. Hasegawa. 2005. Highly active mutants of carbonyl reductase S1 with inverted coenzyme specificity and production of optically active alcohols. *Biosci Biotechnol Biochem* 69:544–552.

Muntendam, R., E. Melillo, A. Ryden, and O. Kayser. 2009. Perspectives and limits of engineering the isoprenoid metabolism in heterologous hosts. *Appl Microbiol Biotechnol* 84:1003–1019.

Murli, S., J. Kennedy, L.C. Dayem, J.R. Carney, and J.T. Kealey. 2003. Metabolic engineering of *Escherichia coli* for improved 6-deoxyerythronolide B production. *J Ind Microbiol Biotechnol* 30:500–509.

Myhrvold, C., and P. Silver. 2015. Using synthetic RNAs as scaffolds and regulators. *Nat Struct Mol Biol* 22:8–10.

Ng, I.S., and S.W. Tsai. 2005. Hydrolytic resolution of (R,S)-naproxen 2,2,2-trifluoroethyl thioester by *Carica papaya* lipase in water-saturated organic solvents. *Biotechnol Bioeng* 89:88–95.

Nishihara, M., and T. Nakatsuka. 2010. Genetic engineering of novel flower colors in floricultural plants: Recent advances via transgenic approaches. *Methods Mol Biol* 589:325–347.

O'Conner, S.E. 2015. Engineering secondary metabolism. *Ann Rev Genet* 49:71–94.

Ohnishi, J., S. Mitsuhashi, M. Hayashi, S. Ando, H. Yokoi, K. Ochiai, and M. Ikeda. 2002. A novel methodology employing *Corynebacterium glutamicum* genome information to generate a new L-lysine-producing mutant. *Appl Microbiol Biotechnol* 58:217–223.

Ohnishi, J., R. Katahira, S. Mitsuhashi, S. Kakita and M. Ikeda. 2005. A novel *gnd* mutation leading to increased L-lysine production in *Corynebacterium glutamicum*. *FEMS Microbiol Lett* 242:265–274.

Opgenorth, P.H., T.P. Korman, L. Iancu and J.U. Bowie. 2017. A molecular rheostat maintains ATP levels to drive a synthetic biochemistry system. *Nat Chem Biol* 13:938–942.

Pai, S.S., R.D. Tilton, and T.M. Przybycien. 2009. Poly(ethylene glycol)-modified proteins: Implications for poly(lactide-co-glycolide)-based microsphere delivery. *AAPS J* 11:88–98.

Paradise, E.M., J. Kirby, R. Chan, and J.D. Keasling. 2008. Redirection of flux through the FPP branchpoint in *Saccharomyces cerevisiae* by down-regulating squalene synthase. *Biotechnol Bioeng* 100:371–378.

Paramasivan, K., and S. Mutturi. 2017. Progress in terpene synthesis strategies through engineering of *Saccharomyces cerevisiae*. *Crit Rev Biotechnol* 37:974–989.

Pasqualini, J.R. 2009. Breast cancer and steroid metabolizing enzymes: The role of progestogens. *Maturitas* 65(Suppl 1):S17–S21.

Patten, P.A., R.J. Howard, and W.P. Stemmer. 1997. Applications of DNA shuffling to pharmaceuticals and vaccines. *Curr Opin Biotechnol* 8:724–733.

Petschacher, B., and B. Nidetzky. 2008. Altering the coenzyme preference of xylose reductase to favor utilization of NADH enhances ethanol yield from xylose in a metabolically engineered strain of *Saccharomyces cerevisiae*. *Microb Cell Fact* 7:9.

Pohl, N.L., M. Hans, H.Y. Lee, Y.S. Kim, D.E. Cane, and C. Khosla. 2001. Remarkably broad substrate tolerance of malonyl-CoA synthetase, an enzyme capable of intracellular synthesis of polyketide precursors. *J Am Chem Soc* 123:5822–5823.

Pollegioni, L., L. Piubelli, and G. Molla. 2009. Cholesterol oxidase: Biotechnological applications. *FEBS J* 276:6857–6870.

Powell, K.A., S.W. Ramer, S.B. Del Cardayre, W.P. Stemmer, M.B. Tobin, P.F. Longchamp, and G.W. Huisman. 2001. Directed evolution and biocatalysis. *Angew Chem Int Ed Engl* 40:3948–3959.

Purnapatre, K., S.K. Khattar, and K.S. Saini. 2008. Cytochrome P450s in the development of target-based anticancer drugs. *Cancer Lett* 259:1–15.

Rojkova, A.M., A.G. Galkin, L.B. Kulakova, A.E. Serov, P.A. Savitsky, V.V. Fedorchuk, and V.I. Tishkov. 1999. Bacterial formate dehydrogenase. Increasing the enzyme thermal stability by hydrophobization of alpha-helices. *FEBS Lett* 445:183–188.

Rooks, M.G., and W.S. Garrett. 2016. Gut microbiota, metabolites and host immunity. *Nat Rev Immunol* 16:341–352.

Sakai, Y., A.P. Murdanoto, T. Konishi, A. Iwamatsu, N. Kato. 1997. Regulation of the formate dehydrogenase gene, FDH1, in the methylotrophic yeast *Candida boidinii* and growth characteristics of an FDH1-disrupted strain on methanol, methylamine, and choline. *J Bacteriol* 179:4480–4485.

San, K.Y., G.N. Bennett, S.J. Berrios-Rivera, R.V. Vadali, Y.T. Yang, E. Horton, F.B. Rudolph, B. Sariyar, and K. Blackwood. 2002. Metabolic engineering through cofactor manipulation and its effects on metabolic flux redistribution in *Escherichia coli*. *Metab Eng* 4:182–192.

Sauer, J., B.W. Sigurskjold, U. Christensen, T.P. Frandsen, E. Mirgorodskaya, M. Harrison, P. Roepstorff, and B. Svensson. 2000. Glucoamylase: Structure/function relationships, and protein engineering. *Biochim Biophys Acta* 154:275–293.

Schmidt-Dannert, C., and F. Lopes-Gallego. 2016. A roadmap for biocatalysis – functional and spatial orchestration of enzyme cascades. *Microb Biotechnol* 9:601–609.

Sauer, N.J., Narváez-Vásquez, J., Mozoruk, J., Miller, R.B., Warburg, Z.J., Woodward, M.J., Mihiret, Y.A., Lincoln, T.A., Segami, R.E., Sanders, S.L., Walker, K.A., Beetham, P.R., Schöpke, C.R., Gocal, G.F. 2016. Oligonucleotide-mediated genome editing provides precision and function to engineered nucleases and antibiotics in plants. *Plant Physiol*. 170(4):1917–1928.

Schrader, J., M.M. Etschmann, D. Sell, J.M. Hilmer, and J. Rabenhorst. 2004. Applied biocatalysis for the synthesis of natural flavour compounds—Current industrial processes and future prospects. *Biotechnol Lett* 26:463–472.

Schroer, K., B. Zelic, M. Oldiges, and S. Lutz. 2009. Metabolomics for biotransformations: Intracellular redox cofactor analysis and enzyme kinetics offer insight into whole cell processes. *Biotechnol Bioeng* 104:251–260.

Seo, J.W., J.H. Jeong, C.G. Shin, S.C. Lo, S.S. Han, K.W. Yu, E. Harada, J.Y. Han, and Y.E. Choi. 2005. Overexpression of squalene synthase in *Eleutherococcus senticosus* increases phytosterol and triterpene accumulation. *Phytochemistry* 66:869–877.

Shanafelt, A.B. 2005. Medicinally useful proteins—Enhancing the probability of technical success in the clinic. *Expert Opin Biol Ther* 5:149–151.

Sharma, S., and S.J. Horn. 2016. Enzymatic saccharification of brown seaweed for production of fermentable sugars. *Bioresour Technol* 213:155–161.

Siddiqui, K.S., and R. Cavicchioli. 2006. Cold-adapted enzymes. *Annu Rev Biochem* 75:403–433.

Sio, C.F., and W.J. Quax. 2004. Improved beta-lactam acylases and their use as industrial biocatalysts. *Curr Opin Biotechnol* 15:349–355.

Slusarczyk, H., S. Felber, M.R. Kula, and M. Pohl. 2000. Stabilization of NAD-dependent formate dehydrogenase from *Candida boidinii* by site-directed mutagenesis of cysteine residues. *Eur J Biochem* 267:1280–1289.

Sola, R.J., and K. Griebenow. 2010. Glycosylation of therapeutic proteins: An effective strategy to optimize efficacy. *BioDrugs* 24:9–21.

Song, S.S., H. Ma, Z.X. Gao, Z.H. Jia, and X. Zhang. 2005. Construction of recombinant *Escherichia coli* strains producing poly (4-hydroxybutyric acid) homopolyester from glucose. *Wei Sheng Wu Xue Bao* 45:382–386.

Steinbuchel, A., and H.G. Schlegel. 1991. Physiology and molecular genetics of poly(beta-hydroxy-alkanoic acid) synthesis in *Alcaligenes eutrophus*. *Mol Microbiol* 5:535–542.

Storici, F., L.K. Lewis, and M.A. Resnick. 2001. In vivo site-directed mutagenesis using oligonucleotides. *Nat Biotechnol* 19:773–776.

Stoveken, T., and A. Steinbuchel. 2008. Bacterial acyltransferases as an alternative for lipase-catalyzed acylation for the production of oleochemicals and fuels. *Angew Chem Int Ed Engl* 47:3688–3694.

Szczebara, F.M., C. Chandelier, C. Villeret, A. Masurel, S. Bourot, C. Duport, S. Blanchard, A. Groisillier, E. Testet, P. Costaglioli, G. Cauet, E. Degryse, D. Balbuena, J. Winter, T.

Achstetter, R. Spagnoli, D. Pompon, and B. Dumas. 2003. Total biosynthesis of hydrocortisone from a simple carbon source in yeast. *Nat Biotechnol* 21:143–149.

Taguchi, S., S. Komada, and H. Momose. 2000. The complete amino acid substitutions at position 131 that are positively involved in cold adaptation of subtilisin BPN'. *Appl Environ Microbiol* 66:1410–1415.

Takagi, H., T. Maeda, I. Ohtsu, Y.C. Tsai, and S. Nakamori. 1996. Restriction of substrate specificity of subtilisin E by introduction of a side chain into a conserved glycine residue. *FEBS Lett* 395:127–132.

Tanaka, Y., and A. Ohmiya. 2008. Seeing is believing: Engineering anthocyanin and carotenoid biosynthetic pathways. *Curr Opin Biotechnol* 19:190–197.

Takeno, S., R. Murata, R. Kobayashi, S. Mitsuhashi and M. Ikeda. 2010. Engineering of *Corynebacterium glutamicum* with an NADPH-generating glycolytic pathway for L-lysine production. *Appl Environ Microbiol* 76:7154–7160.

Takeno, S., K. Hori, S. Ohtani, A. Mimura, S. Mitsuhashi and M. Ikeda. 2016. L-Lysine production independent of the oxidative pentose phosphate pathway by *Corynebacterium glutamicum* with the *Streptococcus mutans gapN* gene. *Metab Eng* 37:1–10.

Tao, J., and J.H. Xu. 2009. Biocatalysis in development of green pharmaceutical processes. *Curr Opin Chem Biol* 13:43–50.

Tian, J., K. Ma, and I. Saaem. 2009. Advancing high-throughput gene synthesis technology. *Mol Biosyst* 5:714–722.

Tishkov, V.I. and V.O. Popov. 2006. Protein engineering of formate dehydrogenase. *Biomol Eng* 23:89–110.

Turner, N.J. 2003. Controlling chirality. *Curr Opin Biotechnol* 14:401–406.

Tyo, K.E., P.K. Ajikumar, and G. Stephanopoulos. 2009. Stabilized gene duplication enables long-term selection-free heterologous pathway expression. *Nat Biotechnol* 27:760–765.

Vadali, R.V., G.N. Bennett, and K.Y. San. 2004a. Applicability of CoA/acetyl-CoA manipulation system to enhance isoamyl acetate production in *Escherichia coli*. *Metab Eng* 6:294–299.

Vadali, R.V., G.N. Bennett, and K.Y. San. 2004b. Cofactor engineering of intracellular CoA/acetyl-CoA and its effect on metabolic flux redistribution in *Escherichia coli*. *Metab Eng* 6:133–139.

Vadali, R.V., G.N. Bennett, and K.Y. San. 2004c. Enhanced isoamyl acetate production upon manipulation of the acetyl-CoA node in *Escherichia coli*. *Biotechnol Prog* 20:692–697.

Valentin, H.E., T.A. Mitsky, D.A. Mahadeo, M. Tran, and K.J. Gruys. 2000. Application of a propionyl coenzyme A synthetase for poly(3-hydroxypropionate-co-3-hydroxybutyrate) accumulation in recombinant *Escherichia coli*. *Appl Environ Microbiol* 66:5253–5258.

van der Kamp, M.W., and A.J. Mulholland. 2008. Computational enzymology: Insight into biological catalysts from modelling. *Nat Prod Rep* 25:1001–1014.

Veen, M., and C. Lang. 2004. Production of lipid compounds in the yeast *Saccharomyces cerevisiae*. *Appl Microbiol Biotechnol* 63(6):635–46.

Verho, R., J. Londesborough, M. Penttila, and P. Richard. 2003. Engineering redox cofactor regeneration for improved pentose fermentation in *Saccharomyces cerevisiae*. *Appl Environ Microbiol* 69:5892–5897.

Veronese, F.M., and A. Mero. 2008. The impact of PEGylation on biological therapies. *BioDrugs* 22:315–329.

Wandrey, C. 2004. Biochemical reaction engineering for redox reactions. *Chem Rec* 4:254–265.

Wang, H.H., F.J. Isaacs, P.A. Carr, Z.Z. Sun, G. Xu, C.R. Forest. and G.M. Church. 2009. Programming cells by multiplex genome engineering and accelerated evolution. *Nature* 460:894–898.

Wang, Q., M. Stuczynski, Y. Gao and M.J. Betenbaugh. 2015. Strategies for engineering protein N-glycosylation pathways in mammalian cells. *Methods Mol Biol* 1321:287–305.

Wang, Y., K.Y. San and G.N. Bennett. 2013. Improvement of NADPH bioavailability in *Escherichia coli* by replacing NAD(+)-dependent glyceraldehyde-3-phosphate dehydrogenase GapA with NADP(+)-dependent GapB from *Bacillus subtilis* and addition of NAD kinase. *J Ind Microbiol Biotechnol* 40:1449–1460.

Watanabe, S., T. Kodaki, and K. Makino. 2005. Complete reversal of coenzyme specificity of xylitol dehydrogenase and increase of thermostability by the introduction of structural zinc. *J Biol Chem* 280:10340–10349.

Weckbecker, A., H. Groger, and W. Hummel. 2010. Regeneration of nicotinamide coenzymes: Principles and applications for the synthesis of chiral compounds. *Adv Biochem Eng Biotechnol* 120:195–242.

Weckbecker, A., and W. Hummel. 2004. Improved synthesis of chiral alcohols with *Escherichia coli* cells coexpressing pyridine nucleotide transhydrogenase, NADP+-dependent alcohol dehydrogenase and NAD+-dependent formate dehydrogenase. *Biotechnol Lett* 26:1739–1744.

Weiss, B. 2008 Removal of deoxyinosine from the *Escherichia coli* chromosome as studied by oligonucleotide transformation. *DNA Repair (Amst)* 7:205–212.

Wells, J.A., M. Vasser, and D.B. Powers. 1985. Cassette mutagenesis: An efficient method for generation of multiple mutations at defined sites. *Gene* 34:315–323.

Wichmann, R., and D. Vasic-Racki. 2005. Cofactor regeneration at the lab scale. *Adv Biochem Eng Biotechnol* 92:225–260.

Widder, S., R.J. Allen, T. Pfeiffer, T.P. Curtis, C. Wiuf, W.T. Sloan, O.X. Cordero, S.P. Brown, B. Momeni, W. Shou, H. Kettle, H.J. Flint, A.F. Haas, B. Laroche, J.U. Kreft, P.B. Rainey, S. Freilich, S. Schuster, K. Milferstedt, J.R. van der Meer, T. Großkopf, J. Huisman, A. Free, C. Picioreanu, C. Quince, I. Klapper, S. Labarthe, B.F. Smets, H. Wang, Isaac Newton Fellows, and O.S. Soyer. 2016. Challenges in microbial ecology: Building predictive understanding of community function and dynamics. *ISME J* 10:2557–2568.

Wriessnegger, T., and Pichler. 2013. Yeast metabolic engineering-targeting sterol metabolism and terpenoid formation. *Prog Lipid Res* 52:277–293.

Wu, H, L. Tuli, G.N. Bennett, and K.Y. San. 2015. Metabolic transistor strategy for controlling electron transfer chain activity in *Escherichia coli*. *Metab Eng* 28:159–168.

Xiong, A.S., R.H. Peng, J. Zhuang, F. Gao, Y. Li, Z.M. Cheng, and Q.H. Yao. 2008. Chemical gene synthesis: Strategies, softwares, error corrections, and applications. *FEMS Microbiol Rev* 32:522–540.

Xu, J., M. Han, T Zhang, Y. Guo, and W. Zhang. 2014. Metabolic engineering *Corynebacterium glutamicum* for the L-lysine production by increasing the flux into L-lysine biosynthetic pathway. *Amino Acids* 46:2165–2175.

Yamamoto, H., K. Mitsuhashi, N. Kimoto, A. Matsuyama, N. Esaki, and Y. Kobayashi. 2004. A novel NADH-dependent carbonyl reductase from *Kluyveromyces aestuarii* and comparison of NADH-regeneration system for the synthesis of ethyl (S)-4-chloro-3-hydroxybutanoate. *Biosci Biotechnol Biochem* 68:638–649.

Yanai, K., T. Murakami, and M. Bibb. 2006. Amplification of the entire kanamycin biosynthetic gene cluster during empirical strain improvement of *Streptomyces kanamyceticus*. *Proc Natl Acad Sci USA* 103:9661–9666.

Yang, T.H., T.W. Kim, H.O. Kang, S.H. Lee, E.J. Lee, S.C. Lim, S.O. Oh, A.J. Song, S.J. Park, and S.Y. Lee. 2010. Biosynthesis of polylactic acid and its copolymers using evolved propionate CoA transferase and PHA synthase. *Biotechnol Bioeng* 105:150–160.

Yaoi, T., K. Miyazaki, T. Oshima, Y. Komukai, and M. Go. 1996. Conversion of the coenzyme specificity of isocitrate dehydrogenase by module replacement. *J Biochem (Tokyo)* 119:1014–1018.

Yoon, S.H., S.H. Lee, A. Das, H.K. Ryu, H.J. Jang, J.Y. Kim, D.K. Oh, J.D. Keasling, and S.W. Kim. 2009. Combinatorial expression of bacterial whole mevalonate pathway for the production of beta-carotene in *E. coli. J Biotechnol* 140:218–226.

Yuan, L.Z., P.E. Rouviere, R.A. Larossa, and W. Suh. 2006. Chromosomal promoter replacement of the isoprenoid pathway for enhancing carotenoid production in *E. coli. Metab Eng* 8:79–90.

Zha, W., S.B. Rubin-Pitel, Z. Shao, and H. Zhao. 2009. Improving cellular malonyl-CoA level in *Escherichia coli* via metabolic engineering. *Metab Eng* 11:192–198.

Zhan, J. 2009. Biosynthesis of bacterial aromatic polyketides. *Curr Top Med Chem* 9:1958–1610.

Zhang, K., M.R. Sawaya, D.S. Eisenberg, and J.C. Liao. 2008. Expanding metabolism for biosynthesis of nonnatural alcohols. *Proc Natl Acad Sci USA* 105:20653–20658.

Zhang, S., S. Wang and J. Zhan. 2016. Engineered biosynthesis of medicinally important plant natural products in microorganisms. *Curr Top Med Chem* 16:1740–1754.

Zhang, W., K. O'Connor, D.I. Wang, and Z. Li. 2009. Bioreduction with efficient recycling of NADPH by coupled permeabilized microorganisms. *Appl Environ Microbiol* 75:687–694.

Zhao, H., and F.H. Arnold. 1999. Directed evolution converts subtilisin E into a functional equivalent of thermitase. *Protein Eng* 12:47–53.

Zilberman, M., A. Kraitzer, O. Grinberg, and J.J. Elsner. 2010. Drug-eluting medical implants. *Handb Exp Pharmacol* 197:299–341.

Section IV

Fermentation Microbiology and Biotechnology

Bioconversion of Renewable Resources to Desirable End Products

11 Conversion of Renewable Resources to Biofuels and Fine Chemicals
Current Trends and Future Prospects

Muhammad Javed, Namdar Baghaei-Yazdi, Aristos Aristidou, and Brian S. Hartley

CONTENTS

"There's enough alcohol in one year's yield of an acre of potatoes to drive the machinery necessary to cultivate the fields for one hundred years."

Henry Ford

11.1 INTRODUCTION

Fossil fuel has been considered as one of the most important resources, which is directly or indirectly used to meet our fuel and fine chemical needs. But its use has also posed environmental challenges such as global warming. Furthermore, a rapid increase in human population and a rise in living standards have resulted in the fast depletion of world's reserve resources. It is, therefore, essential to look for alternative sustainable resources and technologies to meet our fuel and chemical needs. Recent advances in fermentation microbiology, industrial biotechnology, and synthetic biology are poised to revolutionize the way we produce our energy and the chemicals needs by substituting fossilized organic matter, such as petroleum with renewable organic biomass. These new technologies allow various industries to use renewable resources, such as biomass to produce fuels and chemicals which otherwise would be obtained from the fossil fuel. A large number of chemicals can be derived from the biomass. In this chapter, we concentrate more specifically on platform chemicals, and especially on ethanol. These chemicals are used in industry to produce other fuels and polymers such

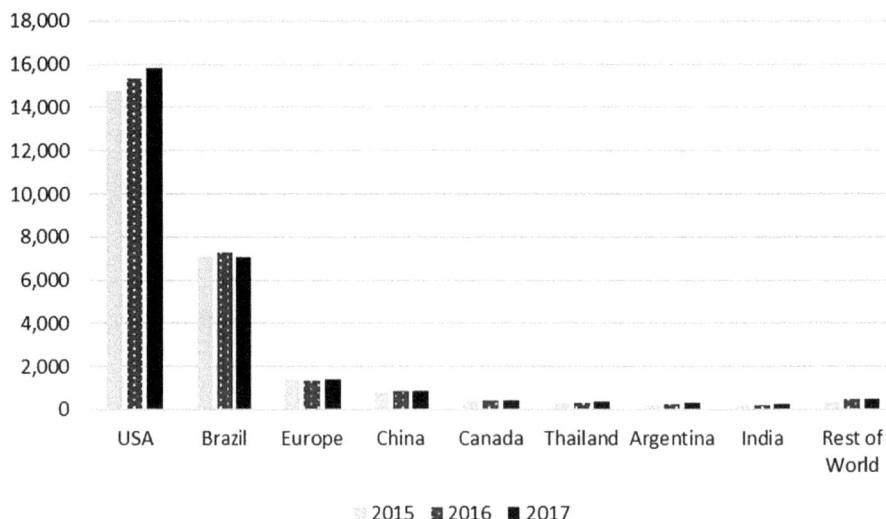

FIGURE 11.1 Examples of lignocellulosic biomass and relative macromolecular composition. (Renewable Fuels Association (RFA) analysis of public and private data sources (www.ethanolrfa.org/resources/industry/statistics/#1454098996479)).

as polylactic acid (PLA); or fine or bulk chemicals such as organic acids, polyols, or esters. The abundance of available biomass makes it the only foreseeable sustainable source of organic fuels. Success in this area will have major implications on human society by reducing the world's dependence on oil, thus minimizing carbon dioxide (CO_2) emissions and the climatic changes associated with it.

Bioethanol production (Figure 11.1) currently relies predominantly on the utilization of glucose derived from cornstarch in the US or sucrose from sugarcane (in Brazil). However, the relatively high value and increasing demand of these feedstocks for food production is a major manufacturing cost that renders the current bioethanol production uncompetitive with gasoline and put a societal pressure to use lignocellulosic biomass. Consequently, several economic studies have shown that successful fermentation of sugars derived from

lignocellulosic biomass is crucial for achieving commercial success.

Major lignocellulose resources are crop residues, forestry wastes, and energy grasses, and approximately 80% of those are what farmers already grow, but have rarely been used for production of biofuels at a commercial scale. Many different biomass feedstocks can be used for the production of fuels and chemicals (Figure 11.2). The chemical composition of agricultural waste varies among species, but biomass consists of approximately 25% lignin and 75% carbohydrate polymers, mainly cellulose and hemicellulose (Girio et al., 2010). Cellulose is a high-molecular-weight linear glucose polysaccharide, with a degree of polymerization (DP) in the range of 1000–5000 glucose molecules connected with β–1,4 glycosidic bonds. Cellulose is very strong, and its links are broken biologically only by cellulases (Figure 11.3) which can

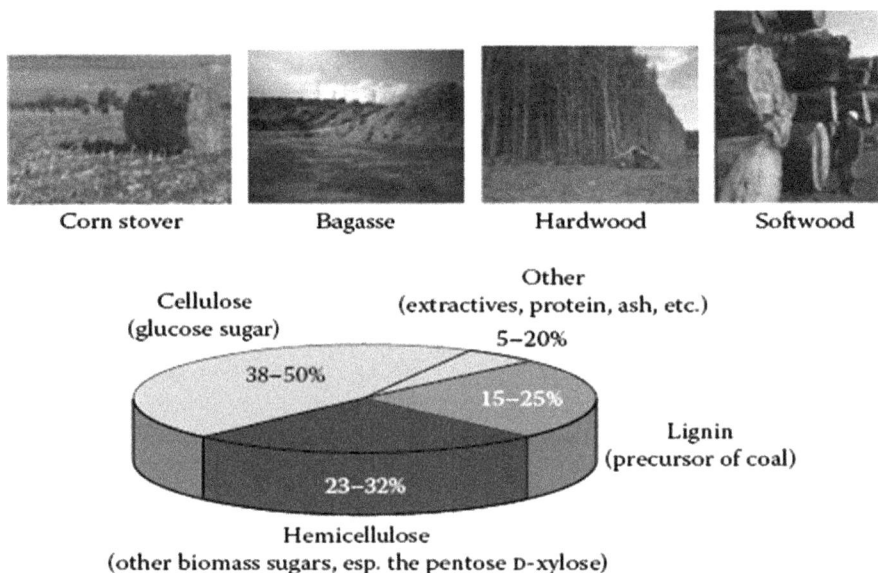

FIGURE 11.2 Examples of lignocellulosic biomass and relative macromolecular composition.

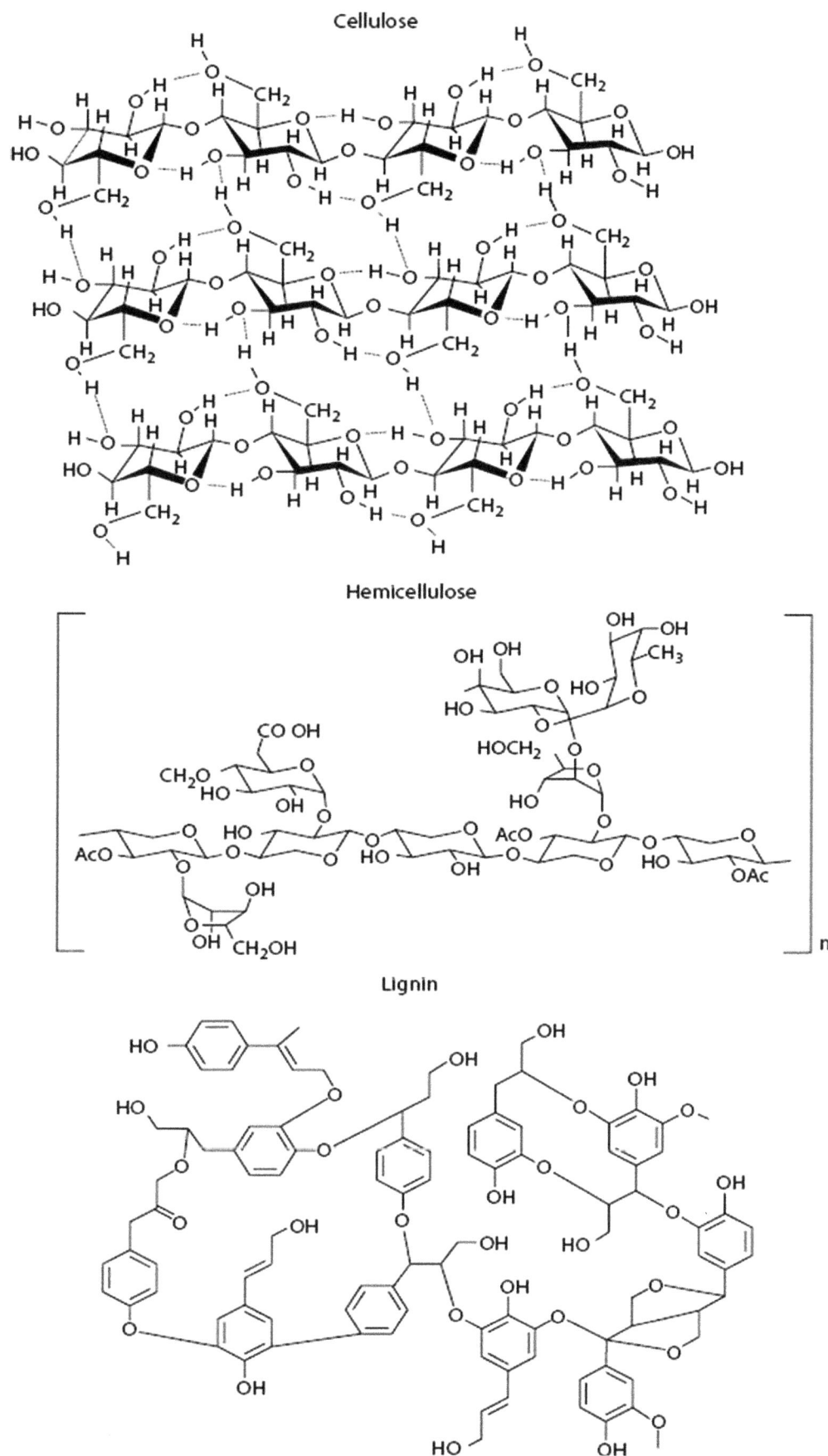

FIGURE 11.3 Lignocellulose structures and utilization.

be divided into two classes: endoglucanases (EG) and cello-biohydrolases (CBH) (Teeri et al., 1998). CBHs hydrolyze the cellulose chain from one end, whereas an EG hydrolyzes randomly along the cellulose chain. In contrast to cellulose, which is crystalline, strong, and resistant to hydrolysis, hemicellulose has a random, amorphous structure that is easily hydrolysable enzymatically or by dilute acid or base (Kuhad et al., 1997). The hemicellulose (Figure 11.3) is a rather a low-molecular-weight heteropolysaccharide (DP < 200, α–1,3 glycosidic links) with a wide variation in structure and composition. The cellulose fraction of biomass is typically high (25–60%), whereas the hemicellulose fraction is typically low (10–35%). The monomeric composition of lignocellulosic material varies depending on the biomass source (Table 11.1). In general, the carbohydrate fraction is made up primarily of glucose, with small amounts of galactose, and pentose moieties. However, the pentose fractions are rather significant: xylose 5–20% and arabinose 1–5%. Xylose is the second most abundant sugar in the hemicellulose of hardwoods and crop residues.

The conversion of cellulose and hemicellulose for the production of fuel ethanol has been studied extensively with a view to develop a technically and economically viable bioprocess. Ethanol's high octane rating and high heat of vaporization values make it more fuel-efficient than gasoline. Furthermore, ethanol is low in toxicity, volatility, and photochemical reactivity, resulting in reduced ozone emission and smog formation compared with conventional fuels. Researchers at the National Renewable Energy Laboratory (NREL) estimate that the US potentially could convert 2.45 billion ton of biomass to 270 billion gallons of ethanol each year, which is more than the current annual gasoline consumption in the US. Bioethanol, also used as a hydrogen fuel source for fuel cells, could become a vital part of the long-term solution to climate change.

The important key technologies required for the successful biological conversion of lignocellulosic biomass to ethanol have been extensively reviewed (Lee, 1997; Chandrakant and Bisaria, 1998; Gong et al., 1999; Girio et al., 2010; Olson et al., 2015). Microbial conversion of the sugar residues present in wastepaper and yard trash from US landfills alone could provide more than 400 billion liters of ethanol (Lynd et al., 1999), 10 times the corn-derived ethanol burned annually as a 10% blend with gasoline (Keim and Venkatasubramanian, 1989). Such technologies are now being realized by a number of companies: 1) INEOS Bio has started its $130 million Indian River Bioenergy Center near Vero Beach, Florida, to converts yard and wood waste to eight million gallon cellulosic ethanol; 2) POET-DSM's Project LIBERTY in Emmetsburg, Iowa, is converting corn stover into cellulosic ethanol and they are also planning to produce their hydrolyzing enzymes on-site; 3) Abengoa Bioenergy's biorefinery in Hugoton, Kansas, is converting agricultural waste into cellulosic ethanol and renewable electricity (US Department of Energy, 2016).

The biological conversion of lignocellulose to ethanol (Figure 11.4) requires the following three steps:

1. Delignification to liberate cellulose and hemicellulose from their complex with lignin.
2. Depolymerization of the carbohydrate polymers (cellulose and hemicellulose) to produce free sugars.
3. Fermentation of mixed hexose and pentose sugars to produce ethanol.

The development of a delignification process should be possible if lignin-degrading microorganisms, their eco-physiological requirements, and optimal bioreactor designs are established. Some thermophilic anaerobes and recently developed recombinant bacteria have advantageous features for direct microbial conversion of cellulose to ethanol. Bioconversion of xylose, the main pentose sugar obtained on hydrolysis of hemicellulose, is essential for the economical production of ethanol; the average cost of biomass amounts is approximately $0.06/kg of sugar.

TABLE 11.1
Percentage Composition of Fermentable and non-Fermentable Components in Various Feedstocks

| | Major Fermentables | | | | | | | | |
| | Carbohydrates | | | | | | | | |
Feedstock	Glu	Man	Gal	Xyl	Ara	Pro	Total	Lig	Ash
Corn Stover	39	0.3	0.8	14.8	3.2	4	62	15.1	4.3
Wheat Straw	36.6	0.8	2.4	19.2	2.4	3	64	14.5	9.6
Rice Straw	41	1.8	0.4	14.8	4.5	0	63	9.9	12.4
Rice Hulls	36.1	3	0.1	14	2.6	0	56	19.4	20.1
Baggase Fiber	38.1	0	1.1	23.3	2.5	3	68	18.4	2.8
Newsprint	64.4	16.6	0	4.6	0.5	0	86	21	0.4
Cotton Gin Trash	20	2.1	0.1	4.6	2.3	3	32	17.6	14.8
Douglas Fir	50	12	1.3	3.4	1.1	0	68	28.3	0.2

Note: Glu, glucose; Man, mannose; Gal, galactose; Xyl, xylose; Ara, arabinose; Pro, protein; Lig, lignin. (Lee, 1997).

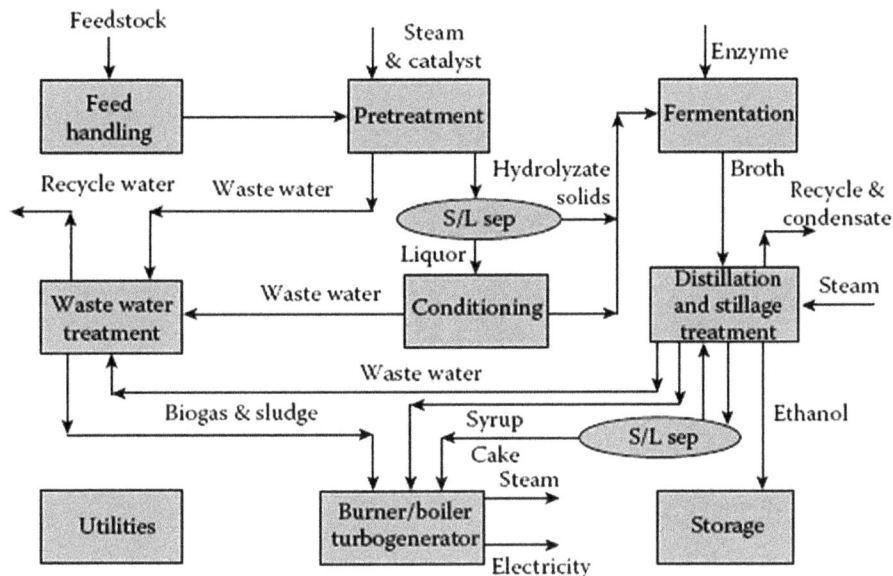

FIGURE 11.4 Conversion of lignocellulose to ethanol. Crystalline cellulose, the largest (50%) and most difficult fraction, is hydrolyzed by a combination of acid and enzymatic processes. After hydrolysis, 95–98% of the xylose and glucose are recovered and, in turn, fermented to alcohol by appropriate microorganisms.

Applied research in the area of biomass conversion to ethanol in the last 40 years has answered most of the major challenges on the road to commercialization, but, as with any new technology, there is still room for improvement. Over the past 10 to 15 years, the total cost of ethanol has dropped whereas the cost of gasoline has fluctuated and increased to the level that these two are now very comparable (Figure 11.5). Further cost reductions in ethanol production can be accomplished through the use of lignocellulosic raw materials, although many technical challenges remain unsolved. Furthermore, an efficient utilization of the hemicellulose component of lignocellulosic feedstocks offers an opportunity to reduce the cost of producing fuel ethanol significantly (Hamelinck et al., 2005). As no naturally occurring ethanol-producing organism possesses the enzyme machinery required for the fermentation of

pentoses and hexoses, there is a need to develop recombinant strains that are capable of fermenting both types of sugars. Two such examples include the introduction of ethanol genes in the bacterium *Escherichia coli* as well as the engineering of pentose-metabolizing pathways in natural ethanol producers such as the yeast *Saccharomyces cerevisiae* or the Gram-positive bacterium *Zymomonas mobilis*. *E. coli* offers several desirable attributes, such as its ability to effectively utilize C6 and C5 carbon sources, fast growth on salt media, and established genetic tools. *S. cerevisiae* and *Z. mobilis*, albeit unable to utilize C5 sugars, are considered to the best ethanol-producing microbes from C6 sugars.

Genetic improvements of the organisms have been carried out to extend their range of substrate utilization or to improve their metabolic flux toward ethanol or other desirable products

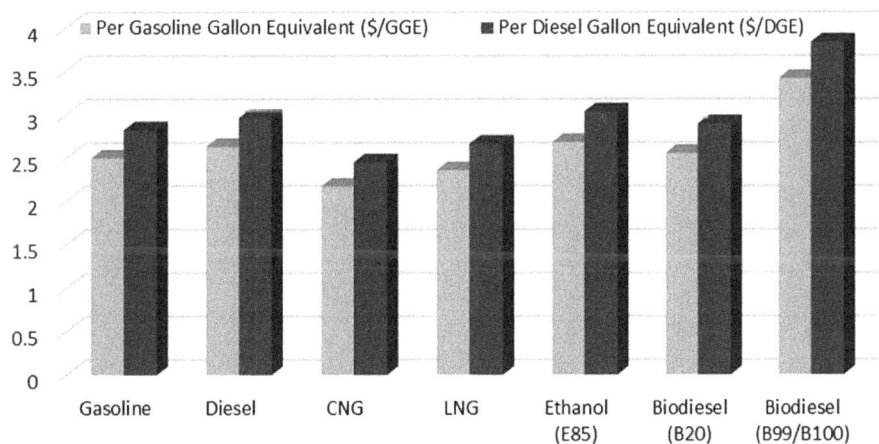

FIGURE 11.5 National average retail fuel prices on an energy-equivalent basis (for alternative fuels, prices on an energy-equivalent basis, i.e., $/GGE or $/DGE, are generally higher than the prices per gallon, due to their lower energy content). [Clean Cities Alternative Fuel Price Report (Jan. 2018), US Department of Energy, Energy Efficiency and Renewable Energy].

(e.g., organic acids). These contributions represent real significant advancements in the field for the conversion of cellulose and hemicellulose sugars to ethanol and other platform chemicals. Furthermore, the bioconversion of lignocellulosic materials to platform chemicals can also be successfully developed and optimized by applying recent innovations in synthetic and system biology to solve the key problems, such as xylose fermentation. This is because the efficient fermentation of xylose and other hemicellulose constituents is essential for the development of an economically viable process to produce fuels and chemicals from biomass.

11.2 PENTOSE FERMENTATION

Glucose and xylose are predominant sugars in most of the lignocellulosic materials. Ideally the production organism should be able to ferment all carbohydrates present in the hydrolysate as well as withstand its inhibitors which are present. A large number of microorganisms can ferment glucose efficiently but only a limited number of microorganisms can ferment xylose. However, pentose-fermenting microorganisms are found among bacteria, yeasts, and fungi, with the yeasts *Pichia stipitis*, *Candida shehatae*, and *Pachysolen tannophilus* being the most promising naturally occurring microorganisms. Yeasts produce ethanol efficiently from hexoses by the pyruvate decarboxylase-alcohol dehydrogenase (PDC-ADH) system. However, during xylose fermentation the by-product xylitol accumulates, thereby reducing the yield of ethanol. Furthermore, yeasts are reported to ferment L-arabinose only very weakly. So far only a small number of bacterial species are known that do possess the important PDC activity, i.e., *Zymomonas mobilis*, *Zymobacter palmae*, *Sarcina ventriculi*, *Acetobacter pasteurianus*, *Gluconobacter oxydans*, and *G. diazotrophicus*. While among these *Z. mobilis* has the most active PDC-ADH system for ethanol production, none of them is capable of dissimilating pentose sugars.

The microorganisms, which can ferment xylose, they in general can metabolize xylose to xylulose through two separate routes (Figure 11.6). The one-step pathway, catalyzed by xylose isomerase (XI, EC 5.3.1.5), is typical in bacteria, whereas the two-step reaction, involving xylose reductase (XR) and xylitol dehydrogenase (XDH), is usually found in yeast. Xylulose is then subsequently phosphorylated with xylulokinase (XK) to xylulose-5-phosphate that can be further catabolized via the pentose phosphate pathway and the Embden–Meyerhof–Parnas (EMP) pathway or the Entner–Doudoroff (ED) pathway in organisms such as *Z. mobilis*.

Empowered with the modern tools of genetic engineering and high throughput screening, several groups have been pursuing the construction of yeast or bacterial organisms that can efficiently convert most of the sugars present in biomass derived hydrolysates to useful products. The approaches can be divided in two groups:

1. Engineering organisms to expand their substrate spectrum
2. Engineering organisms to enhance their abilities of converting key intermediates of central carbon metabolism (e.g., pyruvate) to useful compounds such as ethanol, lactate, or succinate

The first approach has been focused on good ethanologenic organisms, such as *S. cerevisiae* or *Z. mobilis*, with the aim of introducing the pathways for xylose or arabinose metabolism. The second approach starts with organisms that have a wide sugar substrate range (C6 and C5), such as *E. coli*, and introduce pathways for converting these sugars to various fermentation products, including ethanol, lactate, acetate, pyruvate, or succinate.

11.3 GENETICALLY ENGINEERED BACTERIA

Earlier studies were focused in redirecting the fermentative metabolism to ethanol in bacteria, such as *Erwinia chrysanthemi* (Tolan and Finn, 1987a), *Klebsiella planticola* (Tolan and Finn, 1987b), and *E. coli* (Yomano et al., 1998; Ingram et al., 1999). More recently, the focus is also shifting toward next-generation biofuels (Dellomonaco et al., 2010; Liu and Khosla, 2010; Dunlop, 2011). The first generation of the recombinant organisms depended on the cloning and overexpression of the ethanol pathway genes encoding pyruvate decarboxylase (PDC) and alcohol dehydrogenase (ADH) activities. The latter enzyme enhanced the endogenous ADH activity for the efficient reduction of acetaldehyde to ethanol and oxidation of NADH (Figure 11.7). This is important because of the reason that ethanol is just one of the several fermentation products of these bacteria and a deficiency in ADH activity together with NADH accumulation would contribute to the formation of various unwanted by-products.

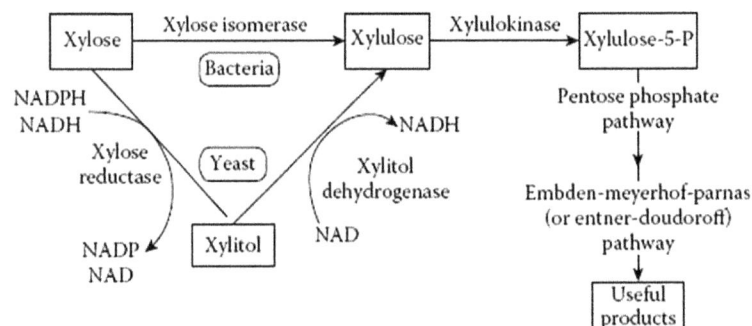

FIGURE 11.6 Xylose utilizing pathways in bacteria and yeast.

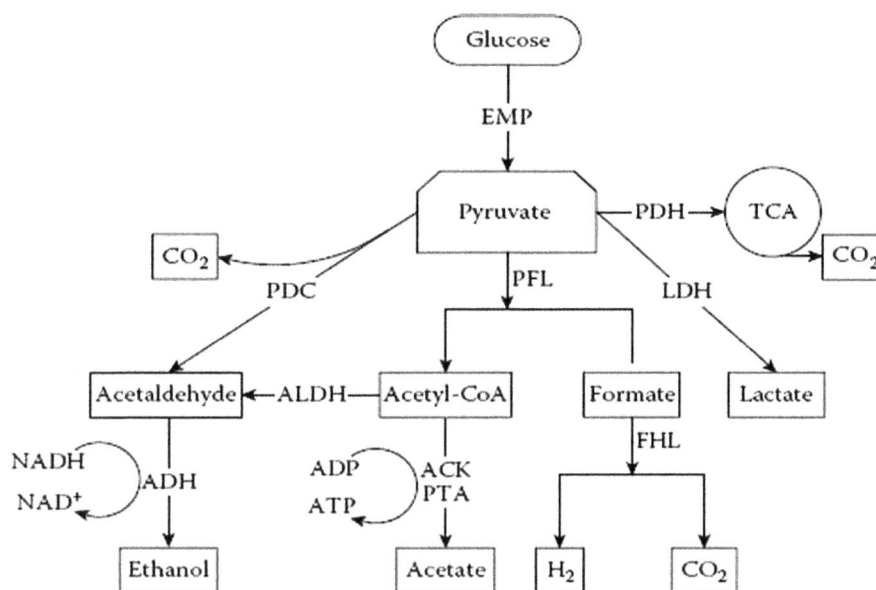

FIGURE 11.7 Competing pathways at the branch point of pyruvate. ACK: acetate kinase; PTA: phosphotransacetylase; FHL: formate hydrogen lyase.

11.3.1 *Escherichia Coli*

E. coli is a facultative anaerobe and is an attractive host organism for the conversion of renewable resources to ethanol and other useful products for several reasons:

- It can ferment efficiently a wide range of carbon substrates including C5 sugars and especially those, which are obtained from lignocellulosic material: xylose, mannose, arabinose, and galactose.
- It can sustain high metabolic fluxes aerobically as well as anaerobically.
- It has a reasonable ethanol tolerance (at least up to 50 g/L).

The wild-type *E. coli* ferment sugars and produces ethanol and a mixture of acids: lactic, acetic, formic, and succinic acid (Riondent et al., 2000). To redirect its carbon flux mainly to ethanol, a novel ethanol pathway was engineered in this microorganism. A PET operon was constructed and expressed: the ethanol pathway genes of *Z. mobilis* (*pdc* and *adhB*) coding for PDC and ADH enzymes were assembled into an artificial operon to produce a portable genetic element for ethanol production (the PET operon) and overexpressed in *E. coli*. The novel PDC enzyme competed with native lactate dehydrogenase (LDH) enzyme for pyruvate metabolism (Figure 11.7) and resulted in the enhanced metabolic flux toward ethanol with a concomitant decrease via the LDH pathway. In a study the recombinant strain produced near-maximum theoretical yields of ethanol when it was grown on a mixture of sugars typically present in hemicellulose hydrolysates, and the organism sequentially utilized the sugars in a strict order: glucose was followed by arabinose and xylose was the last to be utilized (Takahashi et al., 1994).

Depending on the growth conditions, wild-type *E. coli* can metabolize pyruvate (Figure 11.7) through LDH, PDH and pyruvate formate lyase (PFL) enzymes (K_m = 7.2, 0.4 and 2.0 mM, respectively; Table 11.2). Under anaerobic conditions, LDH and PFL pathways initiate pyruvate metabolism. On the other hand, under aerobic conditions the PDH pathway is the main route of pyruvate metabolism to give acetyl CoA and carbon dioxide. The *Z. mobilis* PDC-ADH pathway also converts pyruvate to ethanol under anaerobic conditions. Based on the K_m values (Table 11.2), while the *Z. mobilis* PDC enzyme has similar affinity for pyruvate as that of the PDH enzyme, it has much higher affinity for pyruvate as compared to those of the LDH and PFL enzymes (as reflected by its lower K_m value). Therefore, under anaerobic conditions, PDC is the predominant enzyme in the recombinant *E. coli* for the pyruvate metabolism. Thus, in the recombinant strain, pyruvate is predominantly converted to acetaldehyde which is subsequently converted to ethanol by the ADH enzyme. Furthermore, *Z. mobilis* ADH II has significantly higher affinity for NADH than that of the native ADH enzyme, which means that it can efficiently reduce the acetaldehyde to ethanol using NADH cofactor.

As mentioned above, the recombinant *E. coli* containing the *Z. mobilis* PDC-ADH can favorably compete with the native enzymes for pyruvate metabolism and ethanol production (Table 11.3); the University of Florida was awarded US Patent No. 5,000,000 for the ingenious microbe created at its Institute of Food and Agricultural Sciences. The typical final ethanol concentration by the recombinant strain was shown to be in excess of 50 g/L, with product yields on glucose approaching a theoretical maximum, i.e., 0.5 g of ethanol/g of glucose (glucose → 2 ethanol + 2CO$_2$; Ohta et al., 1991). The volumetric and specific ethanol productivities with xylose in simple batch fermentations were 0.6 g of ethanol per liter per hour and 1.3 g of ethanol per gram cell dry weight per

TABLE 11.2

Comparison of Apparent K_m Values for Pyruvate for Selected *E. coli* and *Z. mobilis* Pyruvate-Acting Enzymes

Organism	Enzyme	K_m	
		Pyruvate	NADH
E. coli	PDH	0.4 mM	0.18 mM
	LDH	7.2 mM	> 0.5 mM
	PFL	2.0 mM	—
	ALDH	—	50 µM
	NADH-OX	—	50 µM
Z. mobilis	PDC	0.4 mM	
	ADH II		12 µM

Note: ALDH, aldehyde dehydrogenase; NADH-OX, NADH oxidase; ADH II, alcohol dehydrogenase II.

TABLE 11.3

Comparison of Fermentation Products During Aerobic and Anaerobic Growth of Wild-Type and Recombinant *E. coli*

Growth	Plasmid	Fermentation Product (mM)			
		Ethanol	Lactate	Acetate	Succinate
Aerobic	None	0	0.6	55	0.2
Anaerobic	None	0.4	22	7	0.9
Aerobic	PET plasmid	337	1.1	17	4.9
Anaerobic	PET plasmid	482	10	1.2	5

PET plasmid, PLO1308-10 (Ingram. and Conway, 1988).

hour, respectively. Further ethanol yield can be improved by eliminating the competing pathways for pyruvate (e.g., phosphotransacetylase and acetate kinase or PFL or LDH). These improvements have resulted in volumetric productivities exceeding 2 g of ethanol/L-h, which can make the process economically competitive (Luli and Ingram, 2005).

BC International in Alachua FL, holding the exclusive rights to use and license the genetically engineered *E. coli* as a production organism, successfully tested several cellulosic feedstocks (e.g., sugarcane, rice straw, rice hulls, softwood forest thinning, and pulp mill sludge), and waste from the sugarcane industry and these feedstocks can be easily used for ethanol production in commercial bioethanol plants.

While various groups are working on engineering *E. coli* for the production of ethanol, there are other groups are which are engineering the microorganisms to produce other primary metabolic products and next-generation biofuels. *E. coli* has been engineered as a homofermentative producer of L-Lactic acid as a sustainable source. The optical isomers, D- and L-Lactic acid monomers are used for the production of biodegradable polylactic acid (PLA). The natural homolactate bacteria, which are used in lactic acid production, are not able

to ferment C5 sugars and they also require complex media components, such as oligopeptides and amino acids for their growth. The lactic acid produced by the engineered *E. coli* is not only a sustainable source, but it also potentially eliminates the need for complex media components. Deleting the competing pathways with LDH, on the one hand, and metabolic engineering of the LDH pathway, on the other, can improve lactic acid production. For lactic acid production, PFL is the main competing pathway under fermentative conditions in the wild type *E. coli* for pyruvate catabolism (Figure 11.7). In one case, all the genes encoding the enzymes of PFL and subsequent pathways as well as the LDH pathway were inactivated by five chromosomal deletions (*focA-pflB*, *frdBC*, *adhE*, *ackA*, and *ldhA* encoding formate transporter-pyruvate formate lyase, fumarate reductase, alcohol dehydrogenase, acetate kinase, and lactate dehydrogenase, respectively). The production of lactic acid was then recovered by cloning the *ldhL* gene of *Pediococcus acidilactici* encoding l-LDH enzyme in the knockout strain. The resulting strain produced L-Lactic acid in M9 mineral salts medium containing glucose or xylose with a yield of 93–95%, a purity of 98% (Zhou et al., 2003). Furthermore, the optical purity of the strain was improved to near 100% by deleting the *msg* gene that encoded the first enzyme of methylglyoxal pathway which was responsible for overflow catabolic flux of carbon to nonspecific lactic acid (Grabar et al., 2006). In a similar fashion, a D-Lactic acid strain of *E. coli* has also been constructed. D- and L-Lactic acid can potentially be combined to generate PLA stereocomplexes with unique and desirable physical properties. The D-Lactic acid production yield by these new strains also approached the theoretical maximum of 2 moles/mole glucose and the chemical purity of the isomers was close to 98% and the optical purity exceeded 99% (Zhou et al. 2003).

Pyruvate is used as raw material for various chemicals, such as amino acids, nutraceutical, a food additive and supplement. *E. coli* was engineered for the production of pyruvate, for example, an *E. coli* strain, TC44 was constructed for the production of pyruvate from glucose by combining the genetic changes Δ*atpFH*, Δ*adhE*, Δ*sucA*, which were aimed at minimizing ATP yield, cell growth, and respiration. This was combined with gene deletions to eliminate acetate production—*poxB::FRT* Δ*ackA*—and other by-products—Δ*focA-pflB*, Δ*frdBC*, Δ*ldhA*, Δ*adhE*. In mineral salts glucose medium, strain TC44 converted glucose to pyruvate with a yield of 0.75 g of pyruvate per gram of glucose (77.9% of theoretical yield) at a rate of 1.2 g of pyruvate/L per hour, and a maximum pyruvate titer of approximately 0.75 M. According to the authors, the efficiency of pyruvate production by strain TC44 is equal to or better than previously reported figures for other biocatalysts including yeast or bacteria (Causey et al., 2004).

Succinate is a four-carbon dicarboxylic acid, which has applications in food, pharmaceutical and chemical industries. Wild type *E. coli* only produces a small amount of succinic acid, but engineered strains can produce it in a significant amount (Thakker et al., 2012). Early attempts to engineer a succinate-overproducing *E. coli* involved the deletion of the competing pathways, i.e., PFL and LDH. Such an organism can accumulate high amounts of succinate under anaerobic conditions (Vemuri et al., 2002). Recent genetic engineering approaches have been applied with an aim of generating strains that can produce succinate under aerobic conditions. The aerobic conditions offer advantages over anaerobic fermentation in terms of faster biomass generation, carbon throughput, and product formation, albeit it introduces a significant capital and operating cost. Genetic manipulations were performed on two aerobic succinate-producing systems to increase their succinate yield and productivity. One of the aerobic succinate production systems includes five gene deletions—$\Delta sdhAB$, Δicd, $\Delta iclR$, $\Delta poxB$, and Δ ($ackA$-pta)—resulting in a strain with a highly active glyoxylate cycle. A second variation of the above includes four of the five above mutations—$\Delta sdhAB$, $\Delta iclR$, $\Delta poxB$, and Δ ($ackA$-pta)—having two routes for succinate production. One is the glyoxylate cycle and the other is the oxidative branch of the TCA cycle. Furthermore, inactivation of $ptsG$ and overexpression of a mutant sorghum $pepc$ in these two production systems resulted in strains having a succinate yield of 1.0 mol/mol glucose (close to maximum theoretical). Furthermore, the two-route production system with $ptsG$ inactivation and $pepc$ overexpression demonstrated substantially higher succinate productivity than the previous system (Lin et al., 2005). Additionally, by using metabolic engineering and directed evolution techniques, *E. coli* strain have been developed which can produce succinic acid from xylose (Khunnonkwao et al., 2018) and hemicellulose hydrolysate (Bao et al., 2014).

11.3.2 Klebsiella Oxytoca

In the early 1990s, the control of expression of the *pdc* and *adh* genes in *Z. mobilis* and *Klebsiella oxytoca* was investigated (Ohta et al., 1991). The wild-type organism has the capability to transport and metabolize cellobiose, thus minimizing the need for extracellular additions of cellobiase. As compared with *E. coli*, in *Klebsiella* strains, two additional fermentation pathways are present that convert pyruvate to succinate and butanediol. As in the case of *E. coli*, it was possible to divert more than 90% of the carbon flow from sugar catabolism away from the native fermentative pathways and toward ethanol. Overexpression of recombinant PDC alone produced only about twice the ethanol level of the parental strain. However, when PDC and ADH were elevated in *K. oxytoca* M5A1, ethanol production was very rapid and efficient: volumetric productivities more than 2.0 g/L per hour, and a yield of 0.5 g of ethanol per gram of sugar with a final ethanol of 45 g/L for glucose and xylose were obtained.

The development of methods to reduce costs associated with the solubilization of cellulose is essential for the utilization of lignocellulose as a renewable feedstock for fuels and chemicals. One promising approach is the genetic engineering of ethanol-producing microorganisms that also produce cellulase enzymes during fermentation. Efforts have also been focused with this organism on enabling its cellulose conversion to ethanol without addition of expensive cellulase enzymes. A derivative of *K. oxytoca* M5A1-containing chromosomally integrated genes for ethanol production from *Z. mobilis* (*pdc*, *adhB*) and EG genes from *E. chrysanthemi* (*celY*, *celZ*) produced over 20,000 U endoglucanase/L activity during fermentation. Because this organism has the native ability to metabolize cellobiose and cellotriose, this strain was able to ferment amorphous cellulose to ethanol without externally added cellulases with an efficiency of 58–76% of the theoretical yield (Zhou and Ingram, 2001).

2,3-BDO is a raw material building block for chemical industry and can be upgraded in high yields to produce gasoline, diesel, and jet fuel. *K. oxytoca* has been successfully engineered to produce 2,3-butanediol (2,3-BDO) from galactose and glucose mixture by introducing and expressing *E. coli galP* gene (encoding galactose permease). Like *E. coli*, *K. oxytoca* mainly catabolizes sugars through the Embden–Meyerhof–Parnas (EMP) pathway of glycolysis but it can also channel the carbon through methyglyoxal pathway which produces methylglyoxal intermediate that is toxic and inhibits sugar metabolism. Hence, the sugar consumption of *K. oxytoca* was improved by deleting its *mgsA* gene (encoding methylglyoxal synthase). The resultant strain produced 2,3-BDO with the volumetric productivity of 0.3 g/l-h (Park et al., 2017).

11.3.3 Zymomonas Mobilis

Z. mobilis is a bacterium that has been used as a natural fermentative agent in alcoholic beverage production and has been shown to have ethanol productivity superior to that of yeast. Overall, it demonstrates many of the desirable traits sought in an ideal biocatalyst for ethanol, such as high ethanol yield, selectivity, and specific productivity, as well as low pH and high ethanol tolerance. In glucose medium, *Z. mobilis* can achieve ethanol levels of at least 12% (w/v) at yields of up to 97% of the theoretical value. When compared with yeast, *Z. mobilis* exhibits 5–10% higher yield and up to 5-fold greater volumetric productivities. The notably high yield of this microbe is attributed to reduced biomass formation during fermentation, apparently limited by ATP availability. This is because of the reason that *Z. mobilis* does not use the common Embden–Meyerhof–Parnas (EMP) pathway of glycolysis for the glucose fermentation but it uses Entner–Doudoroff (ED) pathway instead. The ED pathway produces only half the amount of the ATP than that of the EMP pathway and hence, diverts most of the carbon flux toward ethanol and much less toward biomass formation.

In fact, *Zymomonas* is the only genus identified to date that exclusively utilizes the ED pathway anaerobically. The stoichiometry of ethanol production in this recombinant organism can be summarized as follows [neglecting the NAD(P)H balances]:

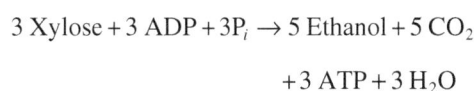

$$3\,\text{Xylose} + 3\,\text{ADP} + 3\text{P}_i \rightarrow 5\,\text{Ethanol} + 5\,\text{CO}_2$$

$$+\,3\,\text{ATP} + 3\,\text{H}_2\text{O}$$

Thus, the theoretical yield on ethanol is 0.51 g of ethanol/g of xylose (1.67 mol mol^{-1}). Because *Z. mobilis* uses ED pathway, the metabolically engineered pathway yields only 1 mol of ATP from 1 mol of xylose, compared with the 5/3 moles typically produced through a combination of the pentose phosphate and EMP pathways. When converting glucose to ethanol, this organism produces only 1 mol of ATP per mole of glucose through the ED pathway compared with 2 moles produced via the more common EMP pathway. This energy limitation results in a lower biomass formation, and thus a more efficient conversion of substrate to product, as mentioned above.

Furthermore, in *Z. mobilis* glucose can readily cross the cell membrane by facilitated diffusion, and it is efficiently converted to ethanol by an overactive PDC-ADH pathway. Moreover, this bacterium is generally recognized as a safe (GRAS) organism and its biomass is suitable for use as an animal feed.

The main drawback of this microorganism is that it can utilize glucose, fructose, and sucrose but it is unable to ferment the widely available pentose sugars. This led Zhang

and his coworkers to attempt to introduce a pathway for pentose metabolism into *Z. mobilis* (Zhang et al., 1995). Early attempts by other groups using the *xylA* and *xylB* genes encoding xylose isomerase and xylulose kinase (XK) (Figure 11.8) from either *Klebsiella* or *Xanthomonas* were met with limited success, despite the functional expression of these genes in *Z. mobilis*. It soon became evident that such failures were due to the absence of detectable levels of transketolase and transaldolase activities in *Z. mobilis*, which were necessary to complete a functional pentose metabolic pathway (Figure 11.9). When a transketolase (TK) gene (*tktA*) from *E. coli* was cloned and expressed in *Z. mobilis*, a small conversion of xylose to CO_2 and ethanol occurred (Feldmann et al., 1992). But as a result of this cloning, the strain also intracellularly accumulated a significant amount of sedoheptulose-7-phosphate which had to be further metabolized. Therefore, the next step was to introduce the transaldolase (TA) activity. To achieve this, a sophisticated cloning technique was applied for the construction of a chimeric shuttle vector (pZB5) that carried two independent operons: the first encoding the *E. coli xylA* and *xylB* genes and the second expressing

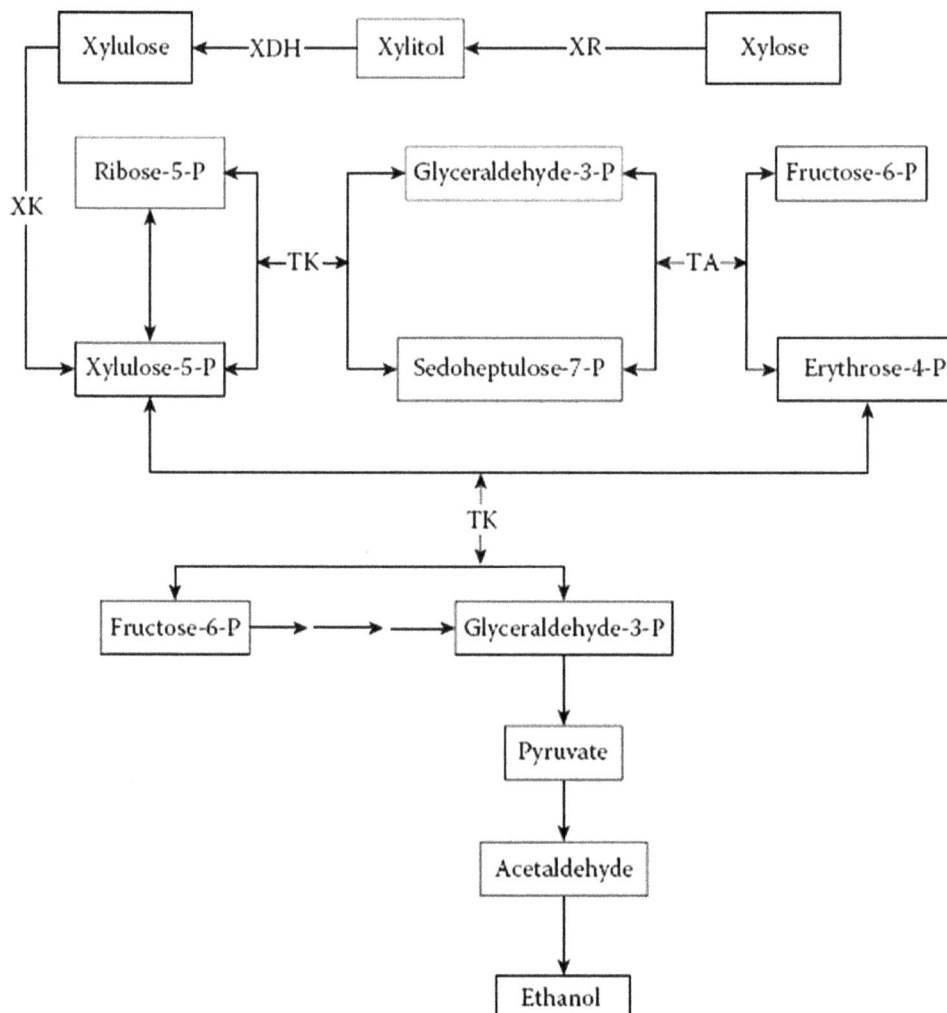

FIGURE 11.8 Ethanol productions from pentose sugars in metabolically engineered *Z. mobilis*. TK, transketolase; TA, transaldolase.

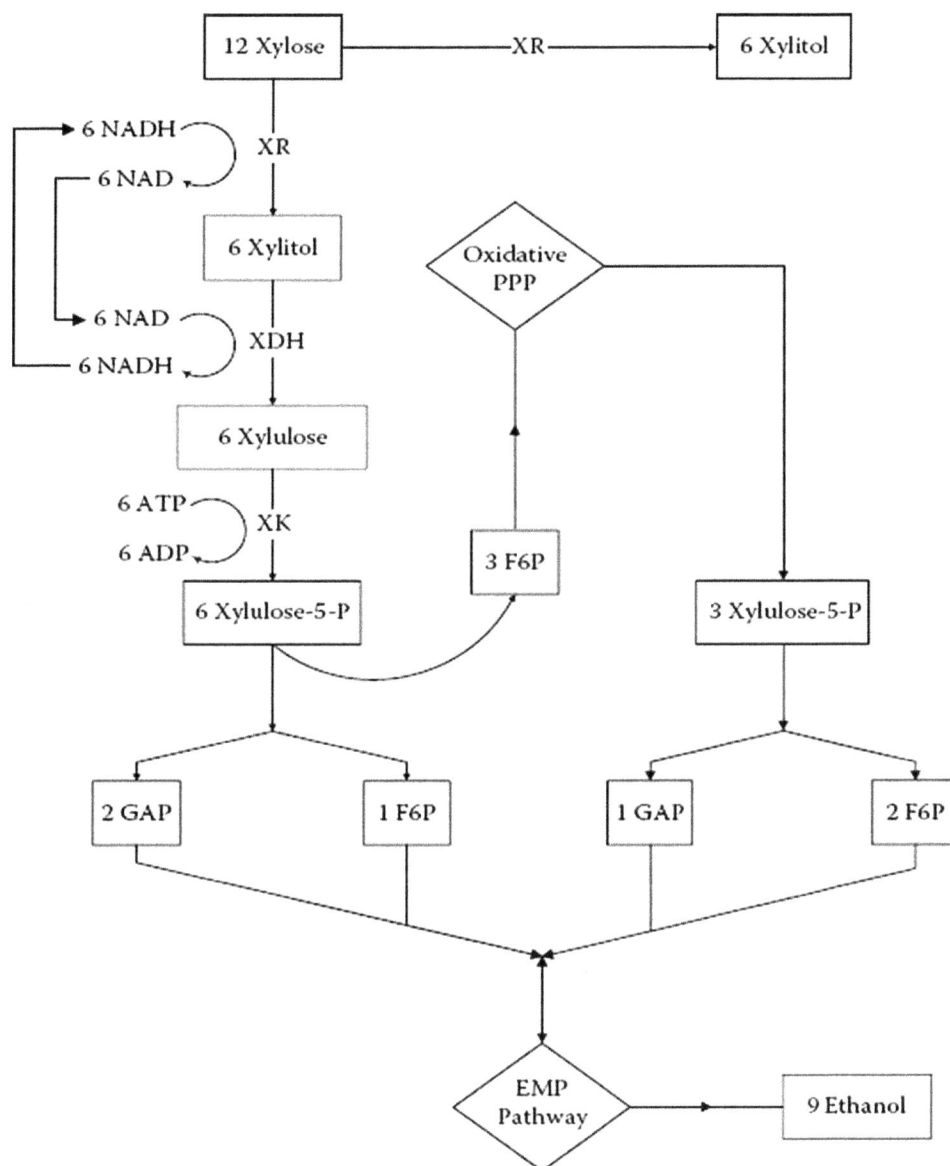

FIGURE 11.9 Anaerobic xylose utilization and cofactor regeneration in recombinant *S. cerevisiae*.

transketolase (*tktA*) and transaldolase (*tal*) again from *E. coli*. The operons were expressed successfully in *Z. mobilis* CP4. The recombinant strain was capable of growing on xylose as the sole carbon source, as well efficiently converting glucose and xylose to ethanol with 86% and 94% of their theoretical respective yields.

Subsequently the same group reported the construction of another *Z. mobilis* strain with a yet expanded substrate fermentation range to include other pentose sugar, L-arabinose, which is commonly found in agricultural residues and other lignocellulosic biomass (Deanda et al., 1996). Five genes encoding L-arabinose isomerase (*araA*), L-ribulokinase (*araB*), L-ribulose-5-phosphate-4-epimerase (*araD*), transaldolase (*talB*), and transketolase (*tktA*) were isolated from *E. coli* and introduced into *Z. mobilis* under the control of constitutive promoters. The engineered strain grew well and produced ethanol from L-arabinose as a sole carbon source

at 98% of the maximum theoretical ethanol yield, indicating that arabinose was metabolized almost exclusively to ethanol as the sole fermentation product. It was anticipated that this microorganism along with the previously developed xylose-fermenting *Z. mobilis* (Zhang et al., 1995) as a mixed culture will efficiently ferment hexose and pentose sugars of lignocellulosic feedstocks to ethanol.

Iogen Corporation of Ottawa, Canada, built a 50-ton/day biomass-to-ethanol production plant adjacent to its enzyme production facility and in 2012; it demonstrated the cellulosic ethanol production of 2.1 million litres (561,000 gallons). Furthermore, in 2015 Iogen officially launched the Costa Pinto mill, the first commercial facility employing its cellulosic ethanol technology. Earlier, the University of Toronto tested the C6/C5 co-fermentation characteristics of the NREL's metabolically engineered *Z. mobilis* using Iogen's biomass hydrolysates (Lawford et al., 2001). In this study, the biomass feedstock was an agricultural waste, namely oat hulls, which was hydrolyzed in a proprietary two-stage process involving pretreatment with dilute sulfuric acid at 200–250°C, followed by cellulase hydrolysis. The oat hull hydrolysate contained glucose, xylose, and arabinose in a mass ratio of approximately 8:3:0.5. This work examined the growth and fermentation performance of xylose-utilizing recombinant *Z. mobilis* cultures CP4:pZB5, and a hardwood prehydrolysate-adapted variant of 39676:pZB4L. In pH-stat batch fermentations with unconditioned 6% (w/v) glucose, 3% xylose, and 0.75% acetic acid, ZM4:pZB5 gave the best performance with a fermentation time of 30 h with a volumetric productivity of 1.4 g/L h. On the basis of the available glucose and xylose, the process ethanol yield for both strains was 0.47 g/g (92% maximum theoretical). Acetic acid tolerance appeared to be a major determining factor in successful co-fermentation.

Z. mobilis has been engineered to produce advanced biofuels, such as 2,3-butanediol (2,3-BDO) which is an economically important platform bulk chemical. It can be used as biosynthetic building blocks for synthetic rubber, solvents, and food additives. For example, dehydration of 2,3-BDO produces methyl ethyl ketone, which can be used as a liquid fuel additive; while its deoxydehydrated form 1,3-butadiene is an important building block for synthetic rubber. Furthermore, 1,3-butadiene can also be oligomerized in high yields to gasoline, diesel, and jet fuel. *Z. mobilis* has a great potential to produce second-generation biofuels because its substrates fermentation range has been expanded, as mentioned above and it is about 2-fold more tolerant to 2,3-BDO levels than those of ethanol. Early studies have shown that three genes of 2,3-BDO pathway, *als, aldC, and bdh* encoding acetolactate synthase, acetolactate decarboxylase, and butanediol dehydrogenase respectively from *Bacillus licheniformis* and *Enterobacter cloacae* could be successfully expressed in *E. coli* to produce 2,3-BDO (Xu et al., 2014). To engineer *Z. mobilis* to produce 2,3-BDO, synthetic biology approach was applied. First a minimized shuttle vector, pEZ15Asp was constructed which could replicate in *E. coli* as well as in *Z. mobilis* and contained only the essential elements, i.e., origins of replication for both bacteria, an antibiotic marker of spectinomycin resistance gene (*add*A), and multiple cloning sites. Then the DNA sequences of three genes, *als, aldC, and bdh* from *B. licheniformis* and *E. cloacae* were obtained. The sequences were codon optimized for *Z. mobilis*. The sequences were then synthesized and assembled in different combinations (two genes or three genes) using the Biobrick-based metabolic pathway engineering strategy and cloned into the shuttle vector pEZ15Asp (Yang et al., 2016). The results indicated that the three-gene pathways produced more 2,3-BDO than did the two-gene pathways and maximum 5 g/l of 2,3-BDO was produced. The presence of *als* gene was necessary to divert significant amount of carbon flow away from ethanol toward 2,3-BDO production. Although native ethanol pathway and the engineered butanediol pathway compete for pyruvate metabolism, attempts to inactivate *pdc* gene of the ethanol pathway met with little success possibly because an active *pdc* gene was essential for *Z. mobilis*.

11.4 GENETICALLY ENGINEERED YEAST

Yeasts produce ethanol efficiently from hexoses by the PDC-ADH pathway. The most commonly used ethanol producer, *S. cerevisiae*, has the intrinsic limitation of not being able to ferment pentose sugars such as xylose or arabinose. Although certain types of yeast such as *P. tannophilus*, *P. stipitis*, or *C. shehatae* are xylose-fermenting yeasts, they have poor ethanol yields on the pentose sugars (Bruinenberg et al., 1984) and low ethanol tolerance compared with the common glucose-fermenting yeasts such as *S. cerevisiae*.

11.4.1 SACCHAROMYCES CEREVISIAE

Saccharomyces spp. are the safest and most effective microorganisms for fermenting sugars to ethanol and traditionally have been used in industry to ferment glucose (or hexose sugar)-based agricultural products to ethanol. Cellulosic biomass, which includes agriculture residues, paper wastes, wood chips, and so on, is an ideal inexpensive, renewable, abundantly available source of sugars for fermentation to ethanol, particularly ethanol used as a liquid fuel for transportation. However, *Saccharomyces* spp. are not suitable for fermenting sugars derived from cellulosic biomass because they are not able to ferment xylose to ethanol or to use this pentose sugar for aerobic growth. Although *Saccharomyces* spp. are not able to metabolize xylose aerobically or anaerobically, there are other yeasts, such as *Pichia stipitis* and *Candida shehatae*, which are able to use xylose for aerobic growth and ferment it to ethanol. However, these naturally occurring xylose-fermenting yeasts are not effective fermentative microorganisms, and they have a relatively low ethanol tolerance.

Many attempts to introduce the one-step pathway by cloning the gene coding for xylose isomerase from *E. coli* or *Bacillus subtilis* in *S. cerevisiae* were unsuccessful because of the inactivity of the heterologous protein in the recombinant host cell (Sarthy et al., 1987). Subsequently, the *Thermus thermophilus xylA* gene encoding xylose (or glucose) isomerase was cloned and expressed in *S. cerevisiae* under the control of the yeast *PGK1* promoter (Walfridsson et al., 1996). The recombinant xylose isomerase showed the highest activity at 85°C with a specific activity of 1.0 U/mg protein. This study also demonstrated a new functional, yet low-throughput, metabolic pathway in *S. cerevisiae* with ethanol formation during oxygen-limited xylose fermentation.

Typically, microorganisms, during sugar fermentation, first produce oxidized intermediate metabolite pyruvate and reduced coenzyme NADH and then they reduce pyruvate to various products, such as ethanol, lactate, while oxidizing NADH back to NAD⁺. On the other hand, NADPH is mainly formed in pentose phosphate pathway (Cordova and Antoniewicz, 2016) from NADP and it is oxidized back to NADP in anabolic metabolic reactions and hence its major role is in biomass synthesis. Thus, the level of NADH is dominant during the sugar fermentation especially at higher sugar concentration. Therefore, for an efficient fermentation process, it is necessary that any engineered enzyme for sugar catabolism should have a higher affinity for NADH coenzyme as compared to that of NADPH.

In most yeasts and fungi, XR and XDH enzyme reactions are dependent on NADPH and NAD$^+$ coenzyme, respectively (Figure 11.9), which cause an imbalance in the levels of oxidized and reduced coenzymes. This imbalance problem is responsible for defects in xylose fermentation, accumulation of xylitol and reduction in ethanol yield Krahulec et al., 2012). However, examples of yeast XRs exist that have dual coenzyme specificity (i.e., NADPH and NADH), such as those from *P. stipitis* and *C. shehatae*. Such a type of enzyme has the advantage of preventing imbalances of the NAD$^+$/NADH redox system, especially under oxygen-limiting conditions (Granstrom et al., 2000).

The first step in yeast xylose metabolism is carried out by xylose (aldose) reductase. The gene coding for this enzyme has been given the designation *XYL1*. In most yeasts and fungi, this enzyme has cofactor specificity for NADPH, but in *P. stipitis*, the enzyme shows 70% as much activity with NADH as with NADPH (Verduyn et al., 1985). The *P. stipitis XYL1* gene has been cloned independently by a number of groups (Amore et al., 1991; Hallborn et al., 1991; Takuma et al., 1991; Jin and Jeffries, 2003). Several XR genes from various organisms have since been cloned, including those from *Kluyveromyces lactis* (Billard et al., 1995), *P. tannophilus* (Bolen et al., 1996), and even *S. cerevisiae* (Toivari et al., 2004). However, the relative affinity of various XRs for NADH and NADPH vary widely. For the most part, this enzyme has a preference for NADPH. One notable exception is the XR enzyme from *Candida boidinii* that was reported to have a higher activity with NADH than with NADPH (Vandeska et al., 1995).

XDH (XYL2) encodes for the second step in xylose metabolism, which, unlike XR, is almost always specific for NAD$^+$. Attempts have been made to modify the XR cofactor specificity (Metzger and Hollenberg, 1995; Leitgeb et al., 2005). The mutation D207→G and the double mutation D207→G and D210→G within the binding domain (GXGXXG) increased the apparent K_m for NAD$^+$ 9-fold and decreased the XDH activity. The introduction of the potential NADP-recognition sequence (GSRPVC) of the ADH from *Thermoanaerobium brockii* into the XDH allowed the mutant enzyme to use NAD$^+$ and NADP as a cofactor with equal affinity. The mutated *XYL2* gene could still mediate growth of *S. cerevisiae* transformants on xylose minimal-medium plates when expressed together with the *XYL1*. More, the gene coding for a *S. cerevisiae* XDH enzyme was also discovered (Aristidou et al., 2000; Toivari et al., 2004).

Several laboratories have attempted to engineer a xylose-fermenting *S. cerevisiae* through the expression of *XYL1* or both *XYL1* and *XYL2* (Costa et al., 2017). Expression of *XYL1* alone has not proven sufficient to enable *S. cerevisiae* to ferment or even to grow on xylose but, in the presence of glucose, *S. cerevisiae* strains expressing *XYL1* will produce primarily xylitol from xylose (Hallborn et al., 1991; Meinander et al., 1994). Production of xylitol appears to be a consequence of

redox imbalance in the cell and is affected by glycerol production (Meinander et al., 1996).

Expression of *XYL1* and *XYL2* in *S. cerevisiae* has proven to be more successful as compared to that of xylose isomerase (XI) and yielded faster xylose assimilation and ethanol production (Karhumaa et al., 2007). Kötter and Ciriacy (1993) studied xylose fermentation in *S. cerevisiae* more extensively and compared it with that of *P. stipitis* (Amore et al., 1991). In the absence of respiration, *S. cerevisiae* transformed with *XYL1* and *XYL2* converts approximately half of the xylose present in the medium into xylitol and ethanol in roughly equimolar amounts while *P. stipitis* produces only ethanol. They proposed, as had Hahn-Hägerdal et al. (2006), that in *S. cerevisiae*, ethanol production is limited by cofactor imbalance. Additional limitations of xylose utilization in *S. cerevisiae* were also attributed to the inefficient capacity of the nonoxidative pentose phosphate pathway, as indicated by the accumulation of sedoheptulose-7-phosphate (Senac and Hahn-Hägerdal, 1989; Senac and Hahn-Hägerdal, 1991).

Tantirungkij et al. (1993) took the approach one step further by subcloning *P. stipitis XYL1* into *Saccharomyces cerevisiae* under the control of the enolase promoter on a multicopy vector. This achieved 2–3 times more *XYL1* expression as was observed in *P. stipitis*. *XYL2* was also cloned and co-expressed in *S. cerevisiae* at approximately twice the level than that was achieved in induced *P. stipitis*. Despite these higher levels of expression, only low levels of ethanol (~5 g/L) were observed under optimal conditions after 100 h. These researchers also selected mutants of *S. cerevisiae* carrying *XYL1* and *XYL2* that exhibited rapid growth on xylose medium (Tantirungkij et al., 1994). The fastest growing strain showed a lower activity of XR but a higher ratio of XDH to XR activity. Southern hybridization showed that the vector carrying the two genes had integrated into the genome resulting in increased stability of the cloned genes. As a result, the yield and production rate of ethanol increased 1.6 to 2.7-fold, however, the maximum concentration of ethanol did not exceed 7 g/L after 144 hours.

The effect of the relative levels of expression of the *XYL1* and *XYL2* genes from *P. stipitis* in *S. cerevisiae* has also been investigated (Walfridsson et al., 1997). These two genes were placed in different directions under the control of the ADH I (*ADHI*) and phosphoglycerate kinase (*PGK*) promoters and inserted into the *E. coli*-yeast shuttle plasmid YEp24. Different recombinant *S. cerevisiae* strains were constructed with different specific activities of XR and XDH. The highest XR or XDH activities were obtained when the expressed gene was under the control of the *PGK* promoter. The XR/XDH ratio (i.e., the ratio of specific enzyme activities of XR and XDH) in these recombinant *S. cerevisiae* strains widely varied from 0.06 to 17.5. To enhance xylose utilization, in the *XYL1*- and *XYL2*-containing *S. cerevisiae* strains, the native *TKL1* gene encoding transketolase and the *TAL1* gene encoding transaldolase were also overexpressed, which resulted in good growth on xylose plate medium. Fermentation of the recombinant *S. cerevisiae* strains containing *XYL1*, *XYL2*, *TKL1*, and *TAL1* was studied with mixtures of glucose and xylose. A strain with higher XR:XDH ratio of 17.5 formed 0.82 g xylitol/g consumed xylose. Also, a strain with an XR:XDH ratio of 5.0 formed 0.58 g xylitol/g xylose. On the other hand, a strain with a lower XR:XDH ratio of 0.06 formed no xylitol and less glycerol and acetic acid and produced more ethanol than the other strains.

Ho and Chang (1989) have reported the construction of a recombinant *Saccharomyces* strain expressing the genes for the three xylose metabolizing enzymes: the XR (*XYL1*) and XDH (*XYL2*) genes from *P. stipitis* and XK (*XYL3*) from *S. cerevisiae*. Cloning of the *XYL3* (xylulose kinase) gene from *P. tannophilus* was first reported in 1987 (Stevis et al., 1987). Cloning of *S. cerevisiae XYL3* by complementation of a XK-deficient mutant of *E. coli* was first reported in 1988 (Rodriguez-Peña et al., 1998), and its role in xylose utilization by *S. cerevisiae* was established soon thereafter (Ho et al., 1998). Ho's group developed recombinant plasmids that could transform *Saccharomyces* spp. into xylose-fermenting yeasts. These plasmids, designated pLNH31, –32, –33, and –34, are 2-μm-based high-copy-number yeast-*E. coli* shuttle plasmids. In addition to the geneticin resistance and ampicillin resistance markers, these plasmids also contain three xylose-metabolizing genes: the *P. stipitis* genes for XR (*PsXYL1*) and XDH (*PsXYL2*) and the *S. cerevisiae* gene for XK (*XYL3*). The parental yeast strain *Saccharomyces*, 1400 is a fusion product of *Saccharomyces diastaticus* and *Saccharomyces uvarum*. It exhibits high ethanol and temperature tolerance and a high fermentation rate. Overexpression of *XYL3* in the *Saccharomyces*, 1400 strain, along with *XYL1* and *XYL2* expression, results in the production of approximately 47 g/L of ethanol with 84% of theoretical yield from a 1:1 glucose/xylose mixture (Ho et al., 1998).

Researchers at the Finnish Technical Research Institute (VTT) have addressed the redox imbalance of the XR and XDH reactions by introducing artificial transhydrogenase cycles in xylose-utilizing *S. cerevisiae* (Aristidou et al., 1999). This work was based on simultaneous expression of dehydrogenase enzymes, such as glutamate dehydrogenase (GDH) having different cofactor specificities (e.g., the yeast *GDH1*, *GDH2*), in combination with enzymes that are driven by ATP (e.g., the malic enzyme), which helped in overcoming the intrinsic limitations due to the physiological redox cofactor concentrations. Results from such genetically engineered organisms have been encouraging in terms of improving xylose utilization rates and ethanol productivities.

While in yeasts and fungi, the XR and XDH enzymes convert xylose to xylulose, in bacteria a single enzyme, xylose isomerase (XI), performs this function. Earlier attempts of expressing bacterial genes encoding XI pathway in *S. cerevisiae* were unsuccessful possibly because of the difficulties in expressing a bacterial gene functionally in yeast (Sarthy et al., 1987). Also, the novel XIs, which were sourced from a soil metagenomics library based on *E. coli* protein sequences, could not perform in *S. cerevisiae* as well as in *E. coli* (Parachin and Gorwa-Grauslund, 2011). However, discovering and expressing bacterial XI coding genes of *Thermus thermophilus* (Walfridsson et al., 1996), *Clostridium phytofermentans* (Brat et al., 2009), *Bacteroides stercoris* (Ha et al., 2011),

and fungal coding genes (Madhavan et al., 2009) enabled successful xylose fermentation by the engineered *S. cerevisiae*. Moreover, enhanced gene expression and codon optimization improved the XI activity in *S. cerevisiae* (Zhou et al., 2012). In one example, expression of the codon optimized *C. phytofermentans* XI coding gene sequence, which was based on the codon usage of glycolytic pathway genes, resulted in a 46% increase in specific growth rate on xylose in *S. cerevisiae* (Brat et al., 2009).

In addition to gene cloning, directed evolution is also a very important strategy to improve the enzyme activity. For example, after three rounds of directed evolution, Lee et al. (2012) isolated XI enzyme which had six amino acid substitutions (E15D, E144G, E129D, A177T, T242S, and V433I) and the mutated XI exhibited a 77% higher *Vmax* as compared to that of the wild type.

S. cerevisiae has not only been engineered for xylose fermentation; research is being carried out to manipulate it to produce other fuels and chemicals from lignocellulosic materials (Kwak and Jin 2017).

11.4.2 Pichia Stipitis

P. stipitis has received a significant attention because it has a natural tendency to convert xylose to ethanol to a level that depends on its cellular redox state. Respiratory and fermentative pathways coexist to support growth and product formation in this microorganism. *P. stipitis* grows rapidly without ethanol production under fully aerobic conditions, and it ferments glucose or xylose under oxygen-limited conditions, but it stops growing within one generation under anaerobic conditions. Genetic engineering techniques have been developed and applied for its efficient xylose utilization and ethanol production.

Expression of *S. cerevisiae URA1* (*ScURA1*) gene in *P. stipitis* enabled rapid anaerobic growth in minimal defined medium containing glucose when essential lipids were present. *ScURA1* encodes a dihydroorotate dehydrogenase that utilizes fumarate as an alternative electron acceptor to confer anaerobic growth. These *P. stipitis* transformants grew anaerobically and produced 32 g/L ethanol from 78 g/L glucose. The cells produced even more ethanol at faster rate after two anaerobic serial subcultures. In contrast, control strains without *ScURA1* were incapable of growing anaerobically and showed only limited fermentation. *P. stipitis* cells bearing *ScURA1* were viable in anaerobic xylose medium for long periods, and supplemental glucose allowed cell growth, but xylose alone could not support anaerobic growth even after serial anaerobic subculturing. These data imply that *P. stipitis* can grow anaerobically using metabolic energy generated through fermentation, but that it exhibits fundamental differences in cofactor selection and electron transport with glucose and xylose metabolism (Shi and Jeffries, 1998).

The *P. stipitis* XR gene (*XYL1*) was inserted into an autonomous plasmid that *P. stipitis* maintains in multicopy form (Dahn et al., 1996). The plasmid pXOR with the *XYL1* insert or a control plasmid pJM6 without *XYL1* was introduced into

P. stipitis. When grown on xylose under aerobic conditions, the strain with pXOR had up to 1.8-fold higher xylose reductase (XR) activity than the control strain. Oxygen limitation led to higher xylose (aldose) reductase (XOR) activity in experimental and control strains grown on xylose. However, the XOR activities of the two strains grown on xylose were similar under oxygen limitation. When grown on glucose under aerobic or oxygen-limited conditions, the experimental strain (with *URA1* plasmid) had XOR activity up to ten times higher than that of the control strain (plasmid without *URA1*). Ethanol production was not improved, but rather it decreased with the introduction of pXOR compared with the control, and this was attributed to nonspecific effects of the plasmid.

Jeffries's group also studied the expression of the genes encoding group I ADHs (*PsADH1* and *PsADH2*) in the xylose-fermenting yeast *P. stipitis* CBS 6054. The cells expressed PsADH1 approximately ten times higher under oxygen-limited conditions than under fully aerobic conditions when cultivated on xylose. Transcripts of PsADH2 were not detectable under either aeration condition. The *PsADH1::lacZ* fusion was used to monitor *PsADH1* expression and it was found that expression increased as oxygen decreased. The mRNA transcript of *PsADH1* was found to be repressed approximately 10-fold in cells grown in the presence of heme under oxygen-limited conditions. Concomitantly with the induction of PsADH1, PsCYC1 expression was repressed. These results indicate that oxygen availability regulates *PsADH1* expression and that regulation may be mediated by heme. The regulation of *PsADH2* expression was also examined in other genetic backgrounds. Disruption of *PsADH1* dramatically increased *PsADH2* expression on non-fermentable carbon sources under fully aerobic conditions, indicating that the expression of *PsADH2* is subject to feedback regulation under these conditions.

P. stipitis was randomly mutated with UV-irradiation followed by chemical mutagenesis with ethidium bromide, to assess its xylose fermentation and ethanol production capabilities. One mutant PSUV9 was able to produce 50% more ethanol from wheat straw hydrolysate than that of the parent strain (Koti et al., 2016). Moreover, the mutant strain was found consistently stable for 19 cycles in the hydrolysates for the ethanol production.

11.4.3 *Pichia Pastoris*

P. pastoris is a methylotrophic yeast which is generally regarded as safe (GRAS) and it grows on simple and inexpensive carbon substrate, such as methanol, at relatively high growth rates. Gene manipulation techniques have been applied to enable this yeast for xylose fermentation. A XR gene (*xyl1*) of *Candida guilliermondii* ATCC 20118 was cloned in *P. pastoris* and characterized (Handumrongkul et al., 1998). The amino acid sequence of *C. guilliermondii* XR showed 70.4% similarity to that of *P. stipitis*. The gene was placed under the control of an alcohol oxidase promoter (*AOX1*) and integrated into the genome of *P. pastoris*. The presence of methanol in the medium induced the expression

of the XR, which preferentially utilized NADPH as a cofactor. The recombinant strain was able to ferment xylose and produce 7.8 g/L xylitol under aerobic conditions.

P. pastoris has also been engineered to produce the advanced biofuel, chiral 2,3-butanediol [(2R,3R)-2,3-BDO] by introducing a synthetic metabolic pathway (Yang and Zhang, 2018). The pathway requires three enzymes to convert the intermediate metabolite, pyruvate to 2,3-BDO: The enzyme α-acetolactate synthase converts pyruvate to α-acetolactate which is then converted to R-acetoin by α-acetolactate decarboxylase enzyme and finally the enzyme (2R,3R)-2,3-BD dehydrogenase converts R-acetoin to 2,3-BDO. All three genes encoding these enzymes were introduced in *P. pastoris*. The *alsS* and *alsD* gene sequences of *B. subtilis*, encoding α-acetolactate synthase and α-acetolactate decarboxylase respectively, were codon-optimized and synthesized. Then, *S. cerevisiae BDH1* gene, encoding (2R,3R)-2,3-BD dehydrogenase, was amplified by cloning. All three genes of the pathway were then re-assembled in a cassette and cloned in the pGAPZαA plasmid under its constitutive promoter, P$_{GAP}$. The resulting plasmid vector was used to integrate the whole pathway genes including the GAP promoter in the genome of *P. pastoris*. The recombinant strain produced 12 g/L of chiral 2,3-BDO with the optical purity of more than 99% in a medium with 40 g/L glucose as carbon source in shake flask fermentation. Furthermore, under optimum cultivation conditions the strain produced 45 g/L of 2,3-BDO, with the potential of increasing the production to 74.5 g/L as the statistical model predicted.

11.4.4 Fungal Xylose Isomerase in Yeast

A significant breakthrough in xylose conversion to ethanol or other fermentation products by yeast came about as a result of completely independent and parallel efforts at NatureWorks LLC (Minneapolis, MN) and Delft Technical University (The Netherlands). All previous efforts to overexpress xylose isomerase (XI) in yeast focused on bacterial XI genes, and their success was limited. However, none of these bacterial XIs resulted in ethanol titers, rates, or yields that could be considered as commercially relevant. The most notable of these attempts was the overexpression of the original or mutated XI gene from *T. thermophilus* (Walfridsson et al., 1996).

This breakthrough came about as a result of identifying the first fungal XI gene, which was isolated from the anaerobic fungus *Piromyces sp.* strain E2 (Xarhangi et al., 2003). This organism metabolizes xylose via XI and d-XK as was shown by enzymatic and molecular analyses; this resembles the situation in bacteria. An early attempt to introduce the *Piromyces sp.* E2 XI gene into *S. cerevisiae* resulted in good XI activity. However, slow growth on xylose did not result in reported ethanol production (Kuyper et al., 2003). It was subsequently reported that expression of this particular gene in yeast, including *S. cerevisiae* (Kuyper et al., 2004, 2005a, 2005b) and nonconventional yeast such as *Kluyveromyces sp.* or *Candida sp.* (Rajgarhia et al., 2004), in conjunction with other targeted-or-not genetic changes of the xylose pathway, resulted in engineered yeast able to convert xylose to ethanol at high rates and yields. Since the isolation of the *Piromyces* XI gene, additional homologous genes have also been isolated from other anaerobic fungi (e.g., *Cyllamyces aberensis*) and successfully expressed in yeasts (Rajgarhia et al., 2004). Interestingly, XI isolated from such anaerobic yeasts turned out to be very homologous to the XI gene of the anaerobic Gram-positive bacterium *Bacteroides thetaiotaomicron*, which is a dominant member of human distal intestinal microbiota.

A *Saccharomyces* strain expressing the *Piromyces* XI together with some additional genetic modifications of the xylose pathway was reported to have good anaerobic growth and fermentation on xylose. In addition to XI, other overexpressed enzymes included XK, ribulose 5-phosphate isomerase, ribulose 5-phosphate epimerase, transketolase, and transaldolase. Furthermore, the *GRE3* gene encoding aldose reductase was deleted to further minimize xylitol production. During the growth on xylose, xylulose formation was absent and xylitol production was negligible. The specific xylose consumption rate in anaerobic xylose cultures was 1.1 g xylose per gram biomass per hour (Kuyper et al., 2005a). Further improvements were achieved through evolutionary engineering in xylose-limited chemostat cultures followed by selection in anaerobic cultivation in automated sequencing-batch reactors on glucose-xylose mixtures. A final single-strain isolate, RWB 218, rapidly consumed glucose-xylose mixtures anaerobically, in synthetic medium, with a specific rate of xylose consumption exceeding 0.9 g/g cells-h and a corresponding ethanol specific productivity of approximately 0.5 g/g-h. When the kinetics of zero trans-influx of glucose and xylose of RWB 218 strain were compared with that of the initial strain, a 2-fold higher capacity (V_{max}) and an improved K_m for xylose was measured in the selected strain (Kuyper et al., 2005b).

While the above discussion proposes that the optimal means for fermenting xylose into ethanol would use XI instead of the XR/XDH pathway, a comparison of the best publicly available yeast strains engineered to use XR/XDH or XI has shown that, regardless of the growth conditions, the strain using XR/XDH has a substantially higher ethanol productivity compared with the XI strain and the XI strain has better ethanol yields under nearly all conditions tested (Li et al., 2016). In another study, Hector et al. (2013) engineered a yeast strain with XI from a bacterium, *Prevotella ruminicola* TC2-24 and adapted for aerobic and fermentative growth by serial transfers of D-xylose cultures under aerobic and followed by microaerobic conditions. The evolved strain showed a specific growth rate of 0.23 h^{-1} on D-xylose medium, which was comparable with the best reported results for analogous engineered *S. cerevisiae* strains including those expressing the *Piromyces* sp. E2 XI. When used to ferment D-xylose, the adapted strain produced 13.6 g/L ethanol in 91 h with a metabolic yield of 83% of theoretical. From analysis of the *P. ruminicola* XI, it was found that the enzyme possessed a *Vmax* of 0.81 μmole/min/mg of protein and a *Km* of 34 mM, which was comparable with that in *S. cerevisiae* engineered with *Piromyces* XI (*Km* from 20 to 90 mM).

These studies reveal that a great deal of progress has been made for the effective conversion of biomass sugars to useful products such as ethanol or organic acids. Nevertheless, a significant amount of technological development is still necessary before having a technology that can be industrially implemented, and this includes aspects such as enhancing tolerance to hydrolysate inhibitors, genetic stability (especially in continuous processes), and ability to utilize all sugars present in hydrolysates that includes glucose and xylose as well as mannose, galactose, and arabinose.

11.5 MICROBES PRODUCING ETHANOL FROM LIGNOCELLULOSE

It is desirable that the microbes, which can convert lignocellulosic sugar to ethanol, also has a machinery to depolymerize cellulose, hemicellulose, and associated carbohydrates. Such process in which biomass conversion to fermentable sugars and product(s) formation is performed in a single step by the selected microorganism(s) is known as consolidated bioprocessing (CBP) method. The CBP of renewable feedstocks to fine chemicals requires complex metabolic processes and it is generally carried out by a single genetically engineered microorganism (Ohta et al., 1991; Beall and Ingram, 1993); however, it can also be executed by a synthetic consortium of microorganisms (Liu et al., 2018; Shahab et al., 2018).

Many plant pathogenic bacteria (soft-rot bacteria), such as *Erwinia carotovora* and *E. chrysanthemi*, have evolved sophisticated systems of hydrolases and lyases that aid the solubilization of lignocellulose and allow them to break up and penetrate plant tissue (Brencic and Winans, 2005). Genetic engineering of these bacteria for ethanol production represents an attractive alternative to the solubilization of lignocellulosic biomass by chemical or enzymatic means. *E. carotovora* SR38 and *E. chrysanthemi* EC16 were genetically engineered with the PET operon and shown to produce ethanol and CO_2 efficiently as primary fermentation products from cellobiose and glucose (Beall and Ingram, 1993). Both ethanologenic *Erwinia* strains produced approximately 50 g/L ethanol from 100 g/L cellobiose in less than 48 h with a maximum volumetric productivity of 1.5 g/L of ethanol per hour. This rate is over twice that reported for the cellobiose-utilizing yeast, *Brettanomyces custersii*, in batch culture.

Along similar lines, the incorporation of saccharifying traits into ethanol-producing microorganisms was also attempted. The gene encoding for the xylanase enzyme (*xynZ*) from the thermophilic bacterium *Clostridium thermocellum* was expressed at high cytoplasmic levels in ethanologenic strains of *E. coli* KO11 and *Klebsiella oxytoca* M5A1(pLOI555) (Ohta et al., 1991). This is a temperature-stable enzyme that depolymerizes xylan to its primary monomer (99%) xylose. To increase the amount of xylanase in the medium and facilitate xylan hydrolysis, a two-stage cyclical process was used for the fermentation of polymeric feedstocks to ethanol by a single, genetically engineered microorganism. Cells containing xylanase were harvested and added to a xylan solution at 60°C, thereby lysing and releasing xylanase for saccharification. After cooling to 30°C, the hydrolysate was fermented to ethanol while replenishing the supply of xylanase for the subsequent saccharification. *K. oxytoca* was found to be a superior strain for such an application because, in addition to xylose (metabolizable by *E. coli*), it can also consume xylobiose and xylotriose. Although the maximum theoretical yield of M5A1(pLOI555) is in excess of 48 g/L ethanol from 100 g/L xylose, approximately one-third of that was achieved in this process because xylotetrose and longer oligomers remained un-metabolized by this strain. The yield appeared to be limited by the digestibility of commercial xylan rather than by the lack of sufficient xylanase activity or by ethanol toxicity.

Shahab et al. (2018) designed a synthetic consortium of the aerobic *Trichoderma reesei* for the secretion of cellulolytic enzymes with facultative anaerobic lactic acid bacteria and co-cultivated them and achieved 34.7 g/L of lactic acid from 5% (w/w) microcrystalline cellulose. However, they faced some challenges in converting pretreated lignocellulosic biomass to lactic acid because of the presence of inhibitors, such as acetic acid, and due to carbon catabolite repression. Nevertheless, in the CBP consortium, they observed simultaneous consumption of hexoses and pentoses and with metabolic cross feeding they managed *in situ* degradation of acetic acid. As a result, they obtained superior lactic acid purities with 85.2% of the theoretical maximum.

11.6 PRODUCTION OF ETHANOL FROM CELLULOSE: AN INDUSTRIAL PERSPECTIVE

The development of commercial processes for the conversion of lignocellulosic biomass to bioethanol was greatly influenced by the realization that "first-generation" bioethanol (made from sugarcane, maize, or wheat) was in direct competition with food production and supplies on one hand and carbon footprint on the other (Inderwildi and King, 2009). In contrast, bioethanol from agricultural and food-processing wastes, which represent approximately 80% of the dry weight of crops, is free from such constraints. The majority of such research and development projects have focused on cellulose as a substrate, which forms 38–50% of agricultural wastes (Figure 11.3). The aims and achievements of this program have been detailed in a U.S. DOE report, "Breaking the Biological Barriers to Cellulosic Ethanol" (https://genomicscience.energy.gov/biofuels/2005workshop/2005low_lignocellulosic.pdf/). This document recognizes that the core barrier is the intrinsic recalcitrance of cellulosic biomass to processing to ethanol. While biomass is composed of nature's most readily available energy source, sugars, these are locked in a complex polymer exquisitely evolved to resist biological and chemical degradation. To overcome such a barrier, the steps described in the following sections are necessary.

11.6.1 REMOVAL OF THE LIGNIN THAT WATERPROOFS THE CELLULOSIC FIBERS

In itself this is already a mature technology because it is the basis of the paper-pulping industry, which is designed to produce delignified cellulose fibers from wood chips. However, paper pulping is an energy-intensive process, as illustrated by the Kraft pulping process in which strong alkali is used to wash away the lignin and hemicelluloses as "black liquor," which is then burned at high temperature to recover the alkali. This is the most costly step in the process and has a huge carbon footprint because 50–60% of the wood is converted to atmospheric CO_2, so Kraft pulping is not feasible for green bioethanol production. Use of waste paper might be a cheap route to cellulosic bioethanol production, but it would directly compete with paper recycling.

The Alcell paper-pulping process (Pye and Lora, 1991) is potentially less polluting because it uses hot ethanol to extract the lignin, leaving behind cellulose and hemicellulose. However, it has not been commercialized because of the associated fire risks. Furthermore, additional energy is required to redistill the ethanol, but the residual lignin can be used to fuel that step.

11.6.2 CONVERSION OF THE CELLULOSE FIBERS TO GLUCOSE

Cellulose fibers consist mostly of parallel β–1:4-linked glucose polymers that are linear and rigid because they are extensively crosslinked by hydrogen bonds and thereby arranged in a crystalline array that is impervious to water. Cellulose is therefore resistant to hydrolysis by mild acids or alkalis; thus, strong acids and/or high temperatures are required to convert it into fermentable glucose. Such processes have not found favor because strong acids must be recovered and dilute acids at high temperature cause "browning" reactions and the release of inhibitors, which reduce the glucose yield and can inhibit subsequent ethanol fermentations. Consequently, enzyme hydrolysis by a mixture of cellulases and cellobiases has been the preferred strategy.

However, cellulases can attack the loose ends (exo-β–1:4–glucanases) or intermittent flexible regions of amorphous cellulose at the surface of the fibers (endo-β–1:4–glucanases). Both classes of enzyme produce cellobiose, which yeasts cannot ferment, so addition of cellobiase enzyme is needed to convert this disaccharide into glucose. The overall hydrolysis rate is slow and dictated by the nature of the preceding delignification step because harsh treatments will increase the number of amorphous sites or loose ends available for cellulase attack. The "rate-limiting" step is the adsorption of the enzymes to the limited number of target sites, so one cannot increase the hydrolysis rate by adding more enzymes as in conventional enzyme reactions (Zhang et al., 2018). Hence, cellulose hydrolysis is still very slow and many large reactors are required. Nevertheless, a compensating advantage is that hydrolysis and fermentation can be carried out together in the fermentation broth in a simultaneous saccharification and fermentation (SSF) process. This also eliminates the product inhibition of cellulases by glucose, thus making SSF the preferred strategy of hydrolysis. However, cost-effective delignification remains a major problem in cellulosic bioethanol production.

Steam-explosion treatment of wood chips can cause many folds increase in the susceptibility of the chips to cellulase hydrolysis. The chips are permeated for a brief period with high-pressure steam in a sealed container, which causes an explosive decompression in the fibers when the pressure is released. This solubilizes and partly degrades the hemicellulose matrix and causes extensive depolymerization of the lignin, thus making it more readily extractable by dilute alkalis or by ethanol as in the Alcell process described in Section 11.6.1.

A sophisticated cost analysis of a hypothetical SSF process that is based on available evidence was reported by NREL in 1999 (Wooley et al., 1999). The steps in the model process are shown in Figure 11.10. The total cost was calculated on the basis of the following assumptions:

- *Feedstock*: Yellow aspen sawdust or corn stover at an assumed cost of $27.5/dry t.
- *Plant size*: 2000 t feedstock/day (2 weeks/year downtime). Annual bioethanol production is approximately 200 million L.
- *Products*: Ethanol and electricity generated from combustion of the lignin-rich residues.
- As we shall see, these neglect the animal feed value of dried distillers grains (DDGS) from the fermentation residues, which could be considerable.
- *Pretreatment*: After size reduction, the feedstock is pretreated with dilute acid at high temperature in a countercurrent reactor similar to pulp digesters. This separates the soluble hemicellulose hydrolysate and inhibitors such as acetic acid from the lignocellulosic solid phase, which is thereby rendered more amenable to cellulase hydrolysis. The acid and inhibitors are removed by "overliming" and ion exchange, and the hemicellulosic sugars are utilized for enzyme production.
- *SSCF fermentation*: The enzymes are combined with the pretreated lignocellulose for SSF fermentations with a genetically engineered *Zymomonas* strain that gives 92% theoretical yield of ethanol from glucose and 85% from xylose but produces relatively low concentrations of ethanol. The lignin-rich residues are dewatered and burned to produce steam for the plant.

- *Distillation*: The broth ethanol concentration is less than 2% w/v, so considerable energy is required to distill it to 95% ethanol in a traditional two-stage system. A vapor-phase molecular sieve is used to remove the remaining water.
- *Capital costs*: The total project investment (1997) is approximately $240 million.
- *Ethanol production cost*: $1.44/U.S. gal. ($0.37/L)

While the above model predicted the ethanol cost of $0.37/L, it is not very different than the cellulosic ethanol price of $0.40/L given by a recent model (Shah and Darr, 2016). Furthermore, there are political and environmental considerations that justify the production of cellulosic bioethanol, particularly when low-cost or polluting feedstocks can be used. For example, a Danish company (Nielsen et al., 2002) has developed an integrated biomass utilization system (IBUS) that features new technology for continuous pretreatment and liquefaction of lignocellulosic biomass in which the biomass is converted using only steam and enzymes. High-pressure steam pretreatment solubilizes most of the hemicelluloses and converts the lignin into a high-quality particulate solid biofuel. The residual cellulose fibers are then readily hydrolyzed to glucose by commercial cellulases. Enzyme hydrolysis and yeast fermentation are combined in a SSF step that gives high ethanol yields from the cellulose consumed, albeit slowly. The process is energy efficient because of very high dry matter content in all process steps and by integration with a lignin-fueled power plant that provides all of the process energy plus a surplus of heat and power. The soluble hemicellulose is currently used as feed molasses but could be used in the future for additional ethanol production.

11.7 ETHANOL FROM HEMICELLULOSIC WASTES

An alternative way to process agricultural wastes is to make ethanol only from the hemicelluloses, which form 23–32% of the dry weight (Figure 11.2) and use the lignocellulosic fibers as boiler fuel, animal feed, or for packaging. Hemicelluloses are a family of various β-linked pentose and hexose sugars, some acetylated. They form a hydrophilic matrix that surrounds the lignin-coated cellulose fibers and allows transport of nutrients throughout the plant. Because

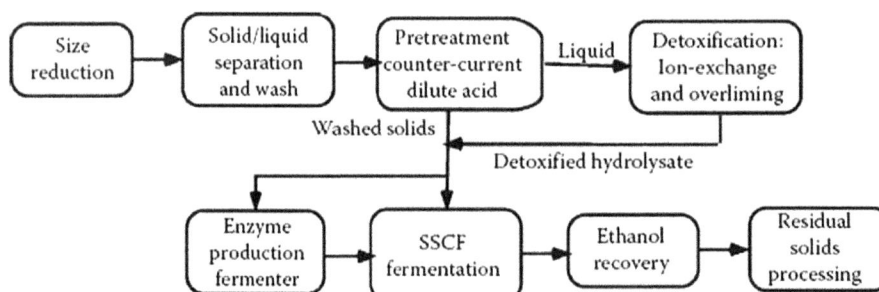

FIGURE 11.10 Flow diagram for an NREL model process.

the loose matrix is not stabilized by rigid hydrogen bonds, they are readily hydrolyzed to fermentable mono- and disaccharides by cheap and abundant microbial hemicellulases or by a short treatment with dilute acids at relatively low temperatures (Girio et al., 2010). For example, 83% of the xylans in sugarcane bagasse are hydrolyzed to xylose after treatment with 0.1% w/w sulfuric acid at 140°C for 20 min. (Pessoa et al., 1997). Such treatment increases the dry weight of the residual lignocellulosic fibers, which therefore have enhanced fuel value.

Although hemicelluloses are abundant in all agricultural and food-processing wastes (e.g., brewers or distillers spent grains), currently they have application as a low-value component of animal feed. They could in principle be converted easily to feedstocks for production of cheap bioethanol, but yeasts cannot ferment the C5 sugars that are their major component. Many soil microorganisms can hydrolyze and utilize all of the hemicelluloses but produce lactic acid as the main anaerobic product (as in silage). Early attempts to select or engineer such microorganisms to produce high yields of ethanol from mixtures of C5 and C6 sugars are discussed in Sections 11.2–11.4, but hardly any of them has provided a commercial fermentation process that competes with current yeast bioethanol fermentations. Although the feedstock costs are greatly reduced, the process costs are much higher because:

- Low ethanol yield is the most important factor. With conventional sugars, yeasts or *Zymomonas* strains can achieve 0.44–0.47 g ethanol per gram of sugars, which approaches the maximum theoretical yield of 0.51 g per gram of sugars. Some xylose-utilizing yeast strains can produce over 0.4 g ethanol per gram of xylose, but none can use the full range of hemicellulosic sugars.
- Volumetric productivity is the next biggest factor that directly affects plant capital and labor costs. Yeast fermentations are carried out on 12–20% g/L sugars in large-batch fermenters at 20–32°C over 1–3 days to produce 6–12% v/v ethanol. Hence, average volumetric productivity is approximately 0.16 L ethanol per liter of broth. All previously described xylose-utilizing strains fail this test.
- Energy costs: In current yeast fermentations, the 6–10% v/v ethanol is concentrated in distillation columns to 95% v/v. The required distillation energy is inversely proportional to the ethanol concentration but rises exponentially below 4% w/v; few novel strains reach this minimum requirement.

11.8 ETHANOL PRODUCTION BY THERMOPHILIC *BACILLI*

When fresh wet grass is put on compost heaps, there can be a very rapid rise in temperature of up to 70°C. This is caused by a class of thermophilic *Bacilli*, now classified as *Geobacilli*, which are capable of fermenting all of the hemicellulosic

sugars, including C5 and C6 sugars, very rapidly to produce mainly L-lactate with only minor amounts of acetate, formate, and ethanol. In the 1980s, a group at Imperial College London started working on these bacilli. Based on the growth properties, one of the facultative anaerobic bacilli, LLD-R was chosen and its lactate deficient mutants were isolated. These mutants, LLD-15 (Payton and Hartley, 1985), LLD-16 and T13 (Javed. 1993), which lacked lactate dehydrogenase (LDH) activity and were unable to produce lactate, could divert their anaerobic sugar metabolism from lactate to ethanol production.

Like most facultative anaerobes, these thermophiles metabolize sugars to pyruvate by Embden–Meyerhof–Parnas (EMP) pathway of glycolysis aerobically as well as anaerobically. However, a remarkable and unique feature of these thermophilic *Bacilli* is that their sugar uptake and glycolytic pathway (from sugars to pyruvate) do not appear to be regulated. Sugar uptake continues until all of the transport permeases or intracellular cofactors are saturated. The resulting large flux of sugars via glycolysis produces huge amounts of intracellular pyruvate, NADH, and ATP. Because the intracellular levels of cofactors, such as NAD^+ and ADP are limited in the cell, the cell maintains the redox balance and the adenylate energy charge, respectively, through the recycling of excess NADH and ATP.

Moreover, like most microorganisms, these thermophiles grow aerobically by the pyruvate dehydrogenase (PDH) pathway to catabolize pyruvate. The PDH is an enzyme complex that converts pyruvate to acetyl-CoA with concomitant production of carbon dioxide and reduction of NAD^+ to NADH. The NADH thus formed is reoxidized back to NAD^+ by membrane-bound electron transport chain (ETC) to produce a large amount of ATP, which is needed for cellular growth and other energy requirements as well as for raising the cellular and the surrounding temperature. As the temperature rises, it causes a reduction in oxygen solubility in the growth medium and anaerobic fermentation is observed.

As illustrated in Figure 11.11, under anaerobic conditions these thermophiles grow via pyruvate formate lyase

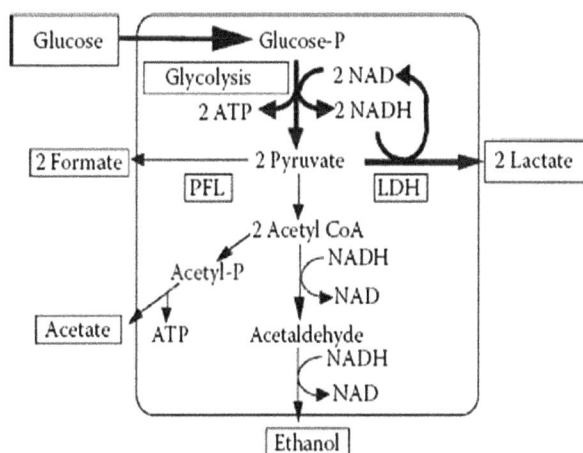

FIGURE 11.11 A schematic diagram for the anaerobic metabolism of sugar in thermophilic *Bacilli*. The PFL growth pathway is on the left and the LDH overflow pathway is on the right.

(PFL) pathway and convert pyruvate to acetyl-CoA and formate (which is excreted). Figure 11.11 also shows that during glycolysis 2 NADH molecules are produced which can only reduce half of the acetyl-CoA to acetaldehyde and then to ethanol. Hence, the other half of the acetyl-CoA is converted by phosphotransacetylase to acetyl-phosphate and then by acetate kinase to CoA and acetate (which is excreted). This final step also produces one ATP molecule, which helps to increase the cellular energy stores.

In the wild-type strain, under anaerobic conditions as sugar input and glycolytic flux increase, the intracellular pools of pyruvate and NADH continue to increase which are conveniently recycled via l-LDH (lactate dehydrogenase) enzyme to lactate (that is excreted), and NAD$^+$ respectively. Furthermore, it appears that as the sugar flux through the glycolysis increases, the pyruvate flux through the PFL pathway decreases or almost stops but it continues to rise through the LDH pathway until the cellular lactate level reaches the toxic limit when the cell growth is impaired. This major anaerobic "overflow" pathway of the LDH cannot provide growth but it provides the cell a way to produces such an excess amount of ATP through the glycolytic pathway that is far beyond than that is required for cell growth, so the excess is hydrolyzed to ADP and helps to raise the temperature.

In the native environment of these microorganisms, such as compost heap, the excess energy produces heat (as explained above), which rapidly raises the sap temperature above 60°C, where competing hemophilic microorganisms cannot grow, so thereafter the thermophiles form a predominant niche community. After the sap sugars are exhausted, these microorganisms secrete enzymes into the plant cell wall that hydrolyze the hemicellulose and amorphous cellulose to provide a further supply of C5 and C6 sugars. When the temperature eventually drops, the thermophiles sporulate and

slower-growing mesophiles may take over until a new load of fresh biomass arrives. The thermophile spores then quickly germinate and once again take over. This vignette of life in a compost heap is useful to help to understand the challenges that have been encountered in trying to develop commercial ethanol fermentations with mutant strains, which lack LDH pathway.

Since the LDH pathway is the dominant metabolic pathway for product formation in the wild type LLD-R strain, it was envisaged that the mutants lacking LDH activity, would produce ethanol as one of the main products. Hence, the spontaneous mutants of LLD-R, such as LLD-15 (Payton and Hartley, 1985), and T-13 (Javed. 1993), which could not produce lactate, were isolated.

According to the metabolic pathways, it was expected that the mutant LLD-15, that lacked LDH pathway, would ferment sugar via EMP pathway to pyruvate which would then by PFL pathway produce formate, acetate and ethanol in the molar ratios of 2:1:1 respectively. However, on the contrary the ratios were different and were favored toward ethanol. As shown in Figure 11.12, different batch fermentations of LLD-15 strain in sucrose media yielded significantly more ethanol than expected from the PFL pathway alone (Hartley and Shama, 1987). Hence, it was clear that in these microorganisms an alternative route existed for ethanol production. Up to that time, it was assumed that PDH was inactive anaerobically, but these results had shown that in the mutant strain, an anaerobic PDH pathway appeared to act as an alternative to the LDH overflow pathway to convert the excess pyruvate to ethanol and CO_2. The results have shown that the PFL flux and the PDH flux are approximately equal at pH 7.0, but at pH 6.2, the PFL flux and growth rate declines considerably (Asanuma and Hino, 2000), so the PDH overflow flux predominates, giving ethanol yields commensurate with those from yeast

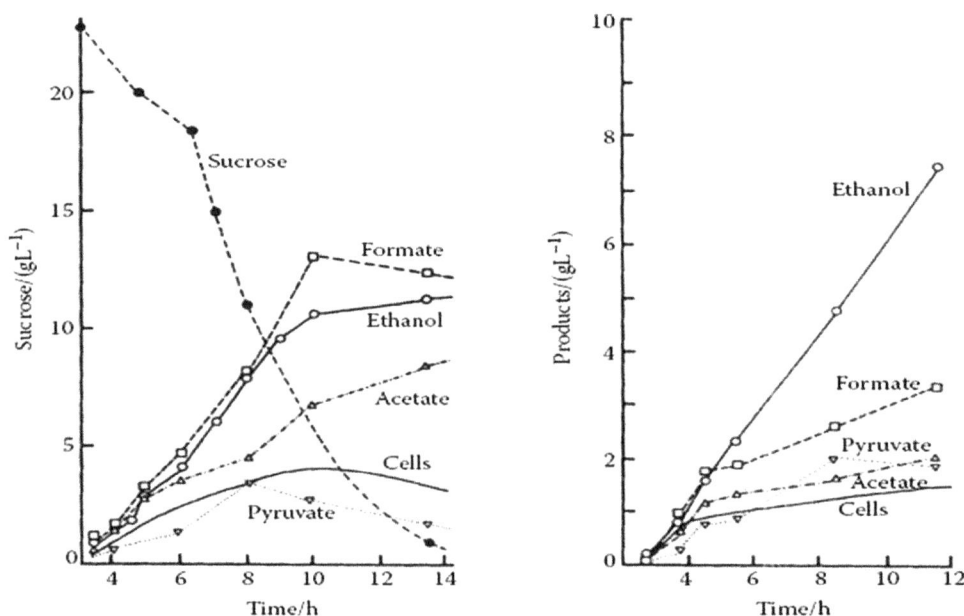

FIGURE 11.12 Anaerobic batch fermentations by strain LLD-15 on sucrose (2.35 g/L), tryptone (2.0 g/L), and yeast extract (1.0 g/L) at 60°C, pH 7.0 (left) or pH 6.2 (right).

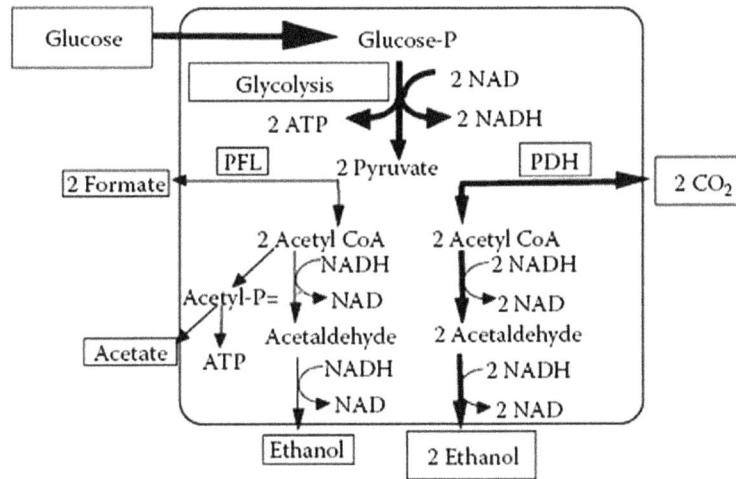

FIGURE 11.13 Anaerobic sugar metabolism in LLD-15 strain. The PFL growth pathway is on the left and the anaerobic PDH overflow pathway is on the right.

fermentations. Considering these mutant cells having lost the major pyruvate catabolic pathway (i.e., LDH), and having limited PFL flux while having an un-regulated glycolytic flux that continued very rapidly to convert sugars to pyruvate, it was logical to conclude that they had a PDH pathway which was operational aerobically as well as anaerobically.

Thus, to maintain the metabolic flux in the mutant strains, the anaerobic PDH pathway which substitutes as an overflow pathway, converts the excess pyruvate to acetyl-CoA and CO_2 with concomitant reduction of NAD^+ to NADH. This NADH and the one arising from the unregulated glycolysis is oxidized back to NAD^+ when the acetyl-CoA is reduced to acetaldehyde via acetaldehyde dehydrogenase and acetaldehyde to ethanol by alcohol dehydrogenase.

Figure 11.13 shows that maximum ethanol yields can be given by slow or nongrowing cells, so Hartley (1988) conceived a continuous "closed system" fermentation to

achieve this, as illustrated in Figure 11.14. The anaerobic production fermenter (F1) is operated at 65°C and pH 6.2, at which the growth by the PFL pathway is minimal, so this is fed continuously by fresh cells from an aerobic growth fermenter (F2). After stripping off the ethanol under mild vacuum, the spent cells are centrifuged and recycled to F1 to maintain high volumetric productivity and partly to the aerobic growth module, F2. The stillage consisting of cell debris, residual sugars, acetate, formate, residual sugars, and nutrients is fed to the aerobic growth fermenter (F2), which is operated with vigorous aeration at 65°C and pH 7. This converts most of the stillage into fresh cells, most of which are recycled to the production fermenter to maintain cell viability. The residual F2 broth containing residual cells and stillage is bled off for use as animal feed. The sugar feed rate and the F2 bleed rate essentially controls the productivity of the whole system.

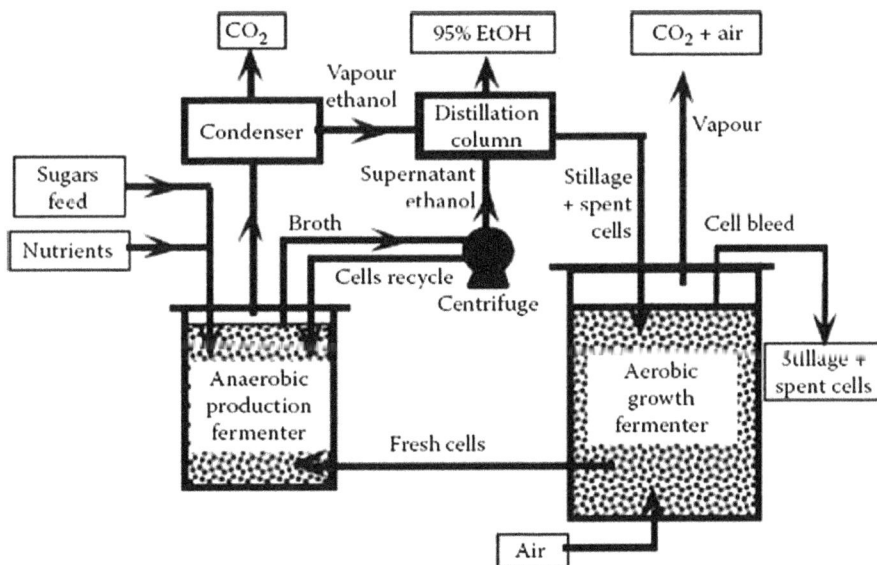

FIGURE 11.14 The model closed system for thermophilic ethanol production. (Hartley, B.S., International Patent Application PCT/GB88/004, 1988.)

However, the final objective with the above process model could not be achieved for the reason that the spontaneous LDH coding gene (*ldh*) mutation in LLD-15 and T-13 (or TN) strains proved to be unstable and these strains reverted back to wild type at high frequency. The mutation was found to arise from the insertion of an indigenous 3.2 kb transposon or insertion element (IE) into a "hot spot" within the 957 bp of the *ldh* gene at 350 bp downstream of its start codon (Javed et al., 2012). The IE element can also jump out from the hot spot at high frequency, causing the reversion to wild-type cells that can grow quickly in the production fermenter and displace the slow-growing mutant cells so as to cause production of lactic acid instead of ethanol. This phenomenon is illustrated in continuous cultures studies of strain LLD-15 fed with 2% w/v sucrose at pH 7.0 in a synthetic medium (San Martin et al., 1993). As can be seen in Figure 11.15, the PFL and PDH fluxes are approximately equal at this pH and continuous cultures can be maintained for relatively longer periods at the dilution rates below 0.2 h^{-1}. However, at higher dilution rates, pyruvate secretion and reversion to lactate production is seen as soon as the sugar uptake rate exceeds 4.2 g sucrose per gram of cells per hour. Higher dilution rates and sugar uptake rates both cause a significant increase in the intracellular pyruvate levels that is reflected by its excretion into the medium. Hence, it is rationale to presume that the higher levels of pyruvate trigger some cellular changes, which cause the IE element to jump out from the *ldh* gene and make it functional and thus mutant strain reverts to wild-type lactate producer.

An obvious way to prevent reversion was to delete the *ldh* gene, but conventional genetic engineering techniques developed for mesophiles such as *E. coli* were not readily applicable to thermophiles that grow very slowly below 60°C. Therefore, suitable techniques were eventually developed (Green et al., 2003; Green et al., 2010; Baghaei-Yazdi et al., 2006) and a stable *ldh*-deleted strain, TN-T9 was constructed (Javed et al., 2012).

The obvious concern with the stable strain was that the PDH overflow pathway became saturated which resulted in cell death which was solved by growing the strain in a chemostat culture for many generations. Furthermore, although the strain performed well in the closed system at the low sugar concentrations, it died rapidly at the high concentrations of sugar required for commercial production of bioethanol. However, this problem could also be tackled by carrying out the directed evolution experiments.

An alternative strategy, of solving the problems of increasing the pyruvate catabolic flux and growing these strains at higher input sugar, was by introducing the PDC (*pdc*) gene from *Z. mobilis* (Green et al., 2012). However, the *Zymomonas* enzyme was predictably not effective at the high temperatures necessary for the thermophilic growth. In another study, Van Zyl et al. (2014) claimed the expression of the PDC encoding gene of an aerobic bacterium, *Gluconobacter oxydans* in a *Geobacillus thermoglucosidasius* strain and achieved an ethanol yield of about 0.35 g ethanol/g glucose but the enzyme was functional only at a lower temperature of 45°C which was far below the optimum growth temperature of most *Geobacilli*.

While Javed and associates constructed the stable *ldh*-negative strain, TN-T9 by using an insertion element (Javed et al., 2012), Cripps et al. (2009) constructed an *ldh*-negative mutant of another thermophilic bacillus, *Geobacillus thermoglucosidasius* NCIMB 11955 using homologous recombination technique. Furthermore, to avoid by-product formation, such as formate, they also knocked out PFL encoding gene (*pfl*). However, such mutants could have very low pyruvate catabolic flux, therefore, they upregulated the PDH encoding gene to improve the flux in the mutant strain and the resulting strain TM242 was capable of rapidly metabolising sugars.

It appears that the PDH pathway plays a key role in the fermentation of sugars in *Geobacilli*, especially when they have

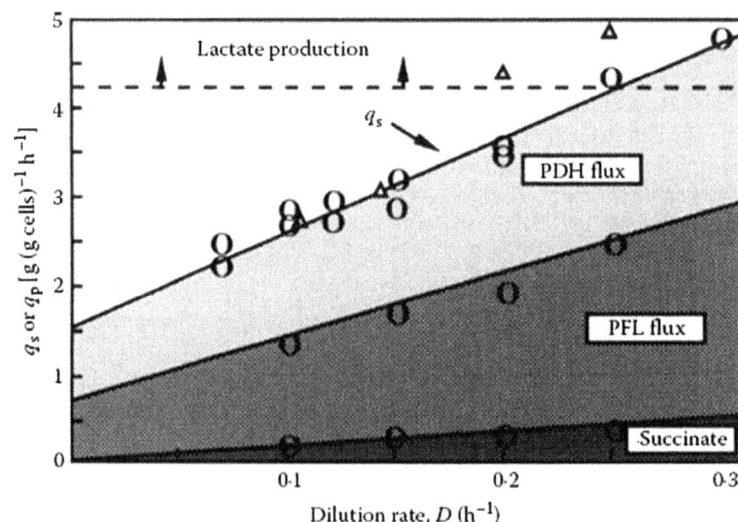

FIGURE 11.15   The specific sucrose consumption rate at 70°C, pH 7.0. (qs = g. sucrose per g. cells per hour) for strain LLD-15 at 10 g. (O) and 15 g, (Δ) input sucrose per l. The hatched areas under the q_s line (arrowed) represent the proportion of the total pyruvate flux which goes via the PDH pathway, PFL pathway or to succinate. When q_s exceeds about 4.2 g sucrose (g cells)-' h-', LLD-R revertants take over the cultures.

inactive LDH pathway. The explanation for this phenomenon can be seen in the structure and mechanism of the PDH complex (Berman et al., 1981) as illustrated in Figure 11.16. It consists of three types of subunits that catalyze consecutive steps in the reaction sequence. The E2 component forms the internal core of the complex and comprises 24 polypeptide chains arranged with octahedral symmetry. The 24 E1 components and 12 E3 components lie above this core and catalyze other steps in the reaction sequence, as schematically illustrated in Figure 11.16a. Each E2 chain contains two lipoyl-lysine residues that lie at the surface of the cluster and act as "swinging arms" that convey the reaction intermediates from one active site to another. This is normally the rate-limiting step in PDH activity.

However, as sugar feed rates increase, glycolysis and pyruvate production rates increase in parallel. They eventually exceed the maximum PDH flux rate, which is dictated by the rate-limiting E2 acetyl migration step. Thereafter, intracellular pyruvate accumulates and is in part excreted. However, unfettered glycolysis continues to consume most of the available NAD^+ until its concentration falls below the NAD^+ binding constant for the E3 lipoate dehydrogenase. At this critical point, the whole PDH flux begins to decrease dramatically. The decrease in NAD^+ levels then accelerates and the cells suffer catastrophic redox imbalance, which leads to a shortage of ATP insufficient to maintain the cell membrane potential, and instant redox death results.

By analyzing the reasons for this phenomenon, several avenues can be proposed, to avoid the redox death problem, which is necessary for a success process.

- The first approach uses a fed-batch fermentation that is controlled to maintain residual sugar levels below the critical point that leads to redox death. These

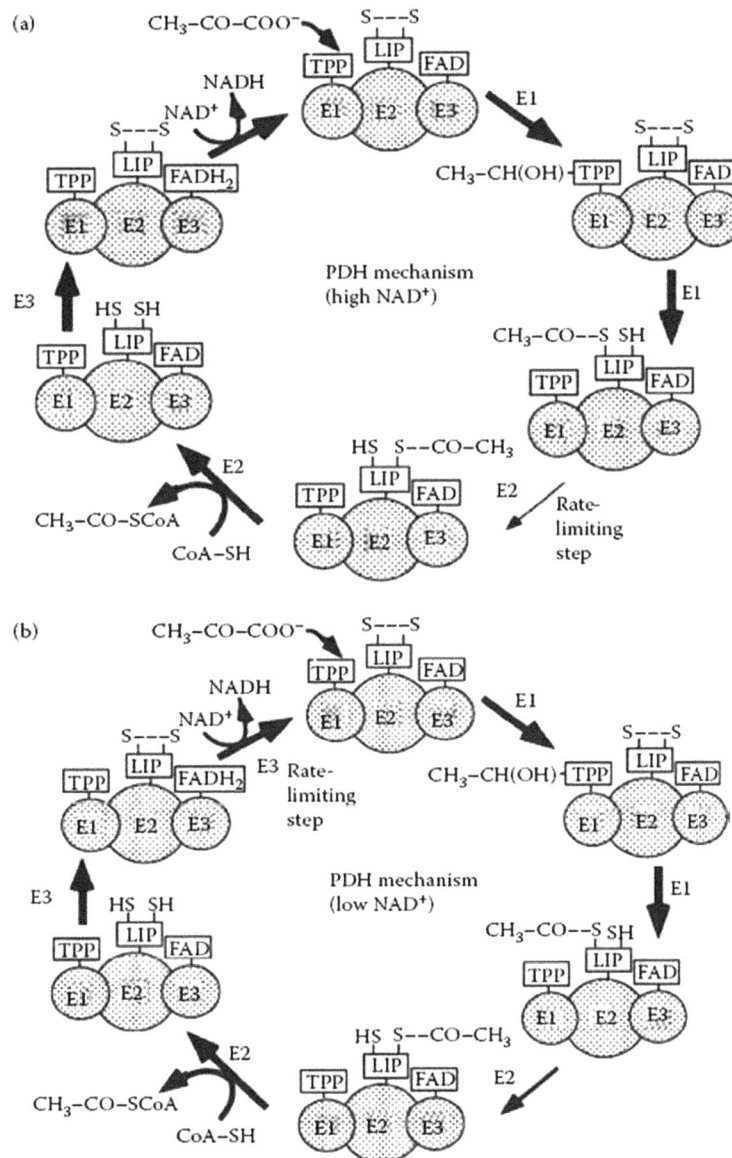

FIGURE 11.16 Mechanism of pyruvate dehydrogenase highlighting the change in the "rate-controlling" step as the metabolic environment changes from (a) NADH-rich to (b) NADH-lacking.

have been shown to produce ethanol in high yield from mixed C5 and C6 sugars at acid pH but are relatively slow in absence of a vigorous PFL growth pathway.

- A second solution to avoid redox death uses a novel pathway for ethanol production (Javed and Baghaei-Yazdi, 2006). A synthetic gene encoding a formate dehydrogenase (FDH) is introduced into the thermophile in place of the LDH gene. As illustrated in Figure 11.17, this creates an artificial PFL-FDH pathway, not found in nature that produces only ethanol and CO_2. The PDH overflow pathway also produces only ethanol and CO_2, so ethanol yields exceeding those in yeast fermentations are expected. Another important advantage is that ethanol production will be optimal when the cells are growing most vigorously at neutral pH, so yields will be close to theoretical in any type of fermentation system. However, the artificial FDH is active only below 60°C, whereas the desired growth temperature is 70°C.

- An alternative solution to the redox death problem is to supplement the sugar feedstock with a co-substrate that provides additional intracellular reducing power. Glycerol is such a co-feedstock and is currently cheap and available on an increasingly large scale as a by-product of biodiesel production containing 79% glycerol (Loaces et al., 2016). Figure 11.17 shows that the extra NADH arising from anaerobic glycerol metabolism can be used to reduce the excess acetyl-CoA arising from the PFL growth pathway, so again ethanol and CO_2 are the only products (Javed et al., 2011).

Moreover, acetate is the major component in acid hydrolysates of biomass which is toxic to the cells at acidic pH because it acts as an "uncoupling agent" that destroys the cell membrane potential. Figure 11.18 shows that the same concept applies to a mixed fermentation of glycerol and acetate alone because transported acetate in the cell is converted to acetyl-CoA which can then be reduced to ethanol using the excess NADH arising from glycerol metabolism. Hence, ethanol yields from the crude hydrolysates of biomass are potentially much higher than from conventional yeast fermentations.

However there appears a problem in the regulation of glycerol metabolism in these thermophile strains, and probably more understanding about this is required to realize this approach. Microorganisms, such as *E. coli* ferment glycerol aerobically and anaerobically by different pathways, as illustrated in Figure 11.19 (Durnin et al., 2009). The aerobic pathway (Figure 11.19) (a) involves phosphorylation by glycerol kinase (GK) encoded by the *glpK* gene followed by a membrane-bound flavoprotein, glycerophosphate dehydrogenase, which forms reduced ubiquinone that is reoxidized by the ETC to provide an adequate supply of ATP for growth. The resulting dihydroxyacetone phosphate is isomerized to phosphoglyceraldehyde and metabolized by glycolysis.

The anaerobic pathway (Figure 11.19 b) uses glycerol dehydrogenase (glyDH) encoded by gene *gldA* to form NADH and dihydroxyacetone, which is then phosphorylated by dihydroxyacetone kinase (DHAK) using phosphoenolpyruvate as a phosphate donor. The resulting dihydroxyacetone phosphate can then be isomerized to phosphoglyceraldehyde and metabolized by glycolysis. In some cases, this anaerobic pathway leads to the formation of 1,2-propanediol, or a mixture of 1-propanol and ethanol (Srirangan et al., 2014) to maintain the redox balance.

Nevertheless, Durnin et al. (2009) have shown that modest yields of ethanol can be obtained from glycerol by a microaerobic process in which growth and redox balance are maintained by a minimal flux through the pathway shown in Figure 11.19(a). The excess pyruvate is then guided into ethanol production. This process differs from that proposed by Baghaei-Yazdi et al. (2009) in Figure 11.19(b) that uses sugars metabolized by the anaerobic PFL pathway for growth and a combination of the PFL and glycerol pathways to maintain

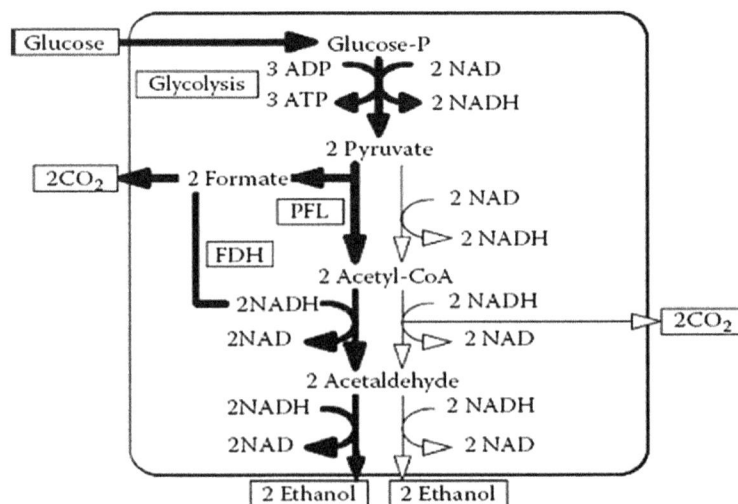

FIGURE 11.17 Anaerobic pathways for combined glycerol and sugar utilization. The glycerol pathway involves GK and glycerol phosphate dehydrogenase (GPD) followed by glycolytic enzymes.

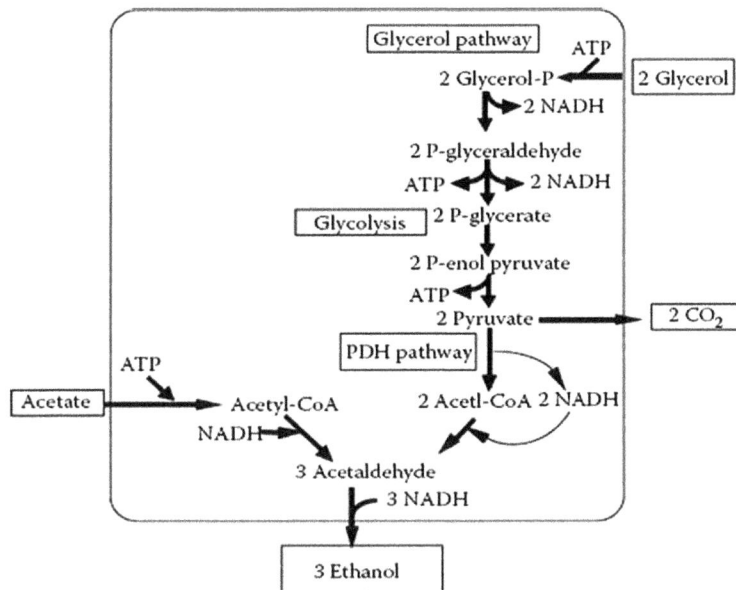

FIGURE 11.18 Metabolic routes for (a) acetate and (b) glycerol metabolism.

the redox balance. Thereby all of the sugars and glycerol can be converted to ethanol in yields greater than those provided by conventional yeast fermentations. However, under anaerobic conditions, the glycerol pathway flux is very slow in these *Geobacillus* strains.

There also appears a general challenge in the regulation of glycerol uptake. In *E. coli*, a membrane-bound glycerol transport facilitator protein (permease) encoded by the *glpF* is required for rapid glycerol uptake by the aerobic pathway (Weissenborn et al., 1992). The *glpF* gene is in an operon followed by the *glpK* gene for GK, which is preceded by a *glpR* gene that encodes a 30-kDa tetrameric repressor protein that is induced by glycerol phosphate. The glycerophosphate dehydrogenase gene *glpD* is in a different operon that is also

repressed by *glpR*. Within the thermophile genome there are genes, which are homologous to these *E. coli* genes and it is envisaged that deleting the putative *glpR* homologue will provide a strain that can constitutively metabolize sugars and glycerol.

The comparison of a genome scale model of *Geobacillus thermoglucosidasius* strain with that of the *E. coli* shows that it is capable of producing a wider range of products, in the yields, which are comparable with those of the *E. coli* (Ahmad et al., 2017). Hence, a range of techniques has been developed to manipulate *Geobacillus* strains to exploit their biorefinery potential (Hussein et al., 2015). A number of synthetic ribosome binding sites have been constructed (Pogrebnyakov et al., 2017), and modular shuttle plasmids have been developed

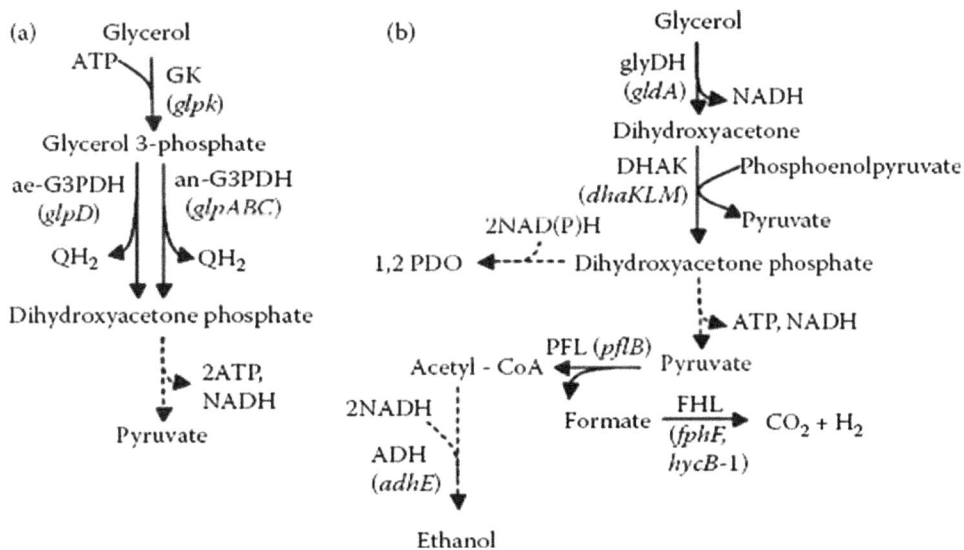

FIGURE 11.19 Anaerobic pathways for the co-utilization of glycerol and its conversion to pyruvate (a) and alcohol (b) by microorganisms. The glycerol pathway involves glycerol kinase (GK) and glycerol phosphate dehydrogenase (GPD) followed by glycolytic enzymes (Durnin et al. *Biotechnology and Bioengineering*, 103, 143–161, 2009).

which can replicate in *E. coli* and *Geobacilli* and contain two origins of replication, two selectable markers and three reporter genes (Reeve et al., 2016). The mutant selection strategies for *Geobacillus* strains have been improved, for example, a positive selection strategy has been devised (Suzuki et al., 2012) by inactivating the uracil pathway genes (*pyrF* and *pyrR*) and can be selected on agar plate medium containing fluoroorotic acid (FOA), because such uracil auxotroph mutants become resistant to the toxic FOA. Bacon et al. (2017) described another technique of counter selection in which the β-glucosidase encoding gene (*Bgl*) and a synthetic substrate X-Glu (5-bromo-4-chloro-3-indolyl-β-d-glucopyranoside) that is toxic to the cells and reduces the colony size significantly unless the *Bgl* gene is expressed to cleave the toxic X-Glu. It is therefore anticipated that because of these advancements, the development of the commercial thermophilic bioprocesses (including the above glycerol + acetate to ethanol process) will be faster to produce ethanol and fine chemicals from hemicellulosic sugars as well as from glycerol which is the major by-products of existing biodiesel plants. It is therefore timely to consider the feedstocks that may become available for these fermentation processes.

Furthermore, the research is continued to develop technologies to produce fuel and chemicals (Hussein, et al. 2015) from cellulosic and hemicellulosic sugars and glycerol feedstocks using a range of thermophilic microorganisms: facultative anaerobes like *Geobacillus thermoglucosidasius*, strict anaerobes like *Thermoanaerobacterium saccharolyticum*, *Thermoanaerobacter ethanolicus*, *Thermoanaerobacter mathranii*, *Clostridium thermocellum*, and *Caldicellulosiruptor bescii*, anaerobic archaeon like *Pyrococcus furiosus*, and methylotrophic yeasts like *Ogataea polymorpha* (Blumer-Schuette et al., 2014; Olson et al., 2017).

11.9 HEMICELLULOSIC FEEDSTOCKS FOR THERMOPHILIC ETHANOL FERMENTATION PROCESSES

Optimal feedstocks are lignocellulosic residues arising from processing of already harvested crops in which all agricultural and transport costs are borne by the primary product. Such residues are generally dried and burned as plant fuel or sold as low-grade animal feed, but they are sometimes composted and returned to the soil. Mild acid or enzyme hydrolysis releases most of the hemicellulosic sugars for bioethanol production and can actually add value to the insoluble lignocellulosic fibers by reducing drying costs and/or supplementing the animal feed with protein-rich spent thermophile cells, which resemble fishmeal in containing lysine and methionine and as such have similar animal feed value. Hence, the cost of the C5 and C6 sugar feedstock is essentially only the hydrolysis cost. Some examples are discussed in Sections 11.9.1–11.9.7.

11.9.1 SUGARCANE

World sugarcane production in 2017 was reported to be in excess of 1918 million tons, with Brazil and India as the major producers. The cane is shredded, crushed, and washed to yield 10–15% v/v cane juice, which is evaporated and crystallized to yield granulated sugar and residual molasses (20–30% sucrose and 15–25% glucose and fructose), which are mostly fermented to rum. Alternatively, the cane juice is fermented directly to ethanol, as described in Section 11.1.

Bagasse is the residual crushed cane that contains mainly the lignocellulosic rind and significant amounts of hemicelluloses and residual cane sugars arising from the pith (Table 11.1). The bagasse is normally air-dried, which allows fermentation of the residual sugars by adventitious microorganisms, including thermophiles, to increase calorific value.

The Tilby process (www.tilbytechnologies.com/technology) could greatly increase bioethanol yield from sugarcane. Instead of crushing the cane, it is chopped mechanically into 30-cm billets that are sliced down the middle to expose the central pith core. This is mechanically scraped out to leave the dry external rind that contains all of the lignocellulosic fibers (6% of the dry weight of cane). These fibers can yield various useful products such as building board, pulp, or paper, but they also have high calorific value sufficient to fuel the whole bioethanol plant. The pith contains the cane juice (~17% dry weight) sugars, and hemicellulosic fibers (~7% dry weight) that could be hydrolyzed by enzymes and/or mild acid to increase the total sugar concentration by approximately 40%. Thermophilic fermentations of the juice and the hemicellulosic sugars would slash the cost of cane bioethanol, already the lowest, to below that of gasoline.

11.9.2 SUGAR BEET

Approximately 30% of world sugar production is derived from sugar beet, mainly in Europe. Yields per hectare are lower than for cane sugar, so this crop has not found wide favor as a source of first-generation bioethanol. However, the whole crop is rich in hemicelluloses and pectin that could potentially greatly lower the production cost. Only the roots are currently harvested and the tops are used for fodder or ploughed in. These could be combined with the beet pulp and vinasse that are major by-products of beet sugar production to produce yet more bioethanol from thermophilic fermentations. Because that bioethanol would be produced from agricultural wastes it would not carry a significant carbon footprint.

11.9.3 BREWERS AND DISTILLERS SPENT GRAINS

These are residues from conventional yeast fermentations of barley, wheat, or maize that are predominantly marketed as animal feed. Large breweries and conventional bioethanol plants produce a large volume of such residues that are rich in hemicelluloses. For example, wheat bran contains 34% w/w of residual starch and fibers that contain 40% hemicelluloses, 13.5% crude protein, and only 5% lignin. Palmarois-Adrados et al. (2005) have shown that a combination of mild acid and enzymic hydrolysis can convert 80% of the total carbohydrate to sugars that could all be converted to ethanol by the thermophilic fermentation process. Moreover, as described above, the animal feed value of the fibrous residues remains similar to that of the original bran, so feedstock costs will be essentially only the hydrolysis costs.

11.9.4 Oilseed Rape

Oilseed rape is increasingly grown for biodiesel production, but the straw is left to rot uselessly in the field. Unlike cereal straws it is low in lignin, so the hemicelluloses are easily hydrolyzed to fermentable sugars by a combination of mild acid and steam. A process devised by a Danish research group can harvest fresh whole-crop oilseed rape efficiently, and the seeds could be separated, dried, and crushed at the factory, saving combine-harvesting costs. They would then be transesterified with ethanol to produce biodiesel, glycerol, and rapeseed meal as major by-products. The latter is currently used as low-grade animal feed but could be combined with the fresh straw for acid/steam hydrolysis to produce C5 and C6 sugars suitable for the glycerol/sugar fermentation described in Section 11.8.

There is already evidence that straw can be used for cardboard or paper production (Leponiemi, 2011), so it is probable that the residual fibers from this process would be even more useful. Hence, the whole crop could be converted into biodiesel plus (bioethanol, renewable packaging, and high-protein animal feed).

11.9.5 Palm-Oil Residues

Huge volumes are available in tropical countries and could be another major feedstock for bioethanol production. The main by-products from oil extraction factories are lignocellulosic residues (~6 ton per ton of palm oil) that include empty fruit bunches, palm kernels, and oil presscake that all have high hemicellulosic content. The oil and presscake are increasingly being exported for biodiesel and animal feed production, respectively, but this has engendered a "fuel versus food" debate. It would make economic and environmental sense to combine the oil extraction with biodiesel production on-site and to use the palm-oil residues and glycerol to produce bioethanol and animal feed by the thermophilic fermentation process described in Section 11.8. A portion of the ethanol could be used to enhance oil extraction from the kernels and presscake and to recycle the extract for use in transesterification in the biodiesel process.

11.9.6 Paper-Pulping Residues

As discussed in previous sections, production of bioethanol from cellulose is not an attractive economic or environmental option, but bioethanol from paper-pulping residues could be both. Removal of xylans to produce top-grade paper pulps by enzyme or mild alkali treatment after mechanical paper pulping is already a commercial practice, but the solubilized pentose polymers (xylans) are not commercially utilized, thus setting the target for thermophilic ethanol fermentations.

11.9.7 Straw Pretreatment

Cereal straws contain large amounts of potassium chloride that sublimes at high temperatures and attacks the stainless steel tubing in boilers that are used for efficient electricity production. Pretreatments to remove the potassium chloride can be combined with hydrolysis to solubilize hemicelluloses so as to increase the calorific value of the straw. As discussed in Section 11.6.2, steam explosion followed by cellulase and cellobiase hydrolysis can give high yields of fermentable sugars (Galbe and Zacchi, 2002). If hemicellulase and cellulase hydrolysis were used instead, the lignin content and calorific value of the residues would increase, and over half of the straw could be converted to ethanol in thermophilic fermentations.

SUMMARY

The major component, cellulose, can be converted to glucose, which is easily fermented to ethanol by yeasts, but this requires expensive and energy-intensive pretreatment to separate the cellulose fibers from their lignin coat. In contrast, hemicelluloses can easily be hydrolyzed to a mixture of glucose, xylose, and arabinose, but yeasts cannot efficiently ferment the latter.

The initial enthusiasm for biofuels as a sustainable alternative to gasoline and an answer to global warming has died down with the realization that first-generation biofuels compete for limited land and water resources with the food production required by an expanding world population. However, that criticism does not apply to "second-generation" bioethanol derived from the relatively useless lignocellulosic residues that are 80% of what farmers grow, because the food production already carries the agronomic carbon footprint.

Lignocellulosic materials (cellulose, hemicellulose, and lignin) are the most abundant renewable organic resource on earth, so the development of new processes for their conversion to useful products has been recognized as an urgent need. Until recently, cellulose pretreatment, hydrolysis, and fermentation have been the major focus for research and development, but the resulting bioethanol is still not competitive with gasoline.

Unfortunately, the emphasis on cellulose utilization has also cast a cloud over second-generation biofuels because the energy inputs required to break down the lignocellulose scarcely match the energy output of the resulting bioethanol. Wood is already a sustainable biofuel for much of the world's population, so efficient use of lignocellulosic residues for heat and electricity production remains a sensible target. But none of these criticisms apply to the production of bioethanol from hemicelluloses, as illustrated in this chapter.

A good deal of attention has therefore been focused on the developments of genetically modified strains that are capable of efficiently utilizing all of the hemicellulose sugars. Because powerful genetic engineering tools have been developed for enteric bacteria such as *E. coli*, it is natural that these were first choice for developing strains to ferment hemicellulosic sugars such as xylose. However, they are far from ideal production hosts because they grow naturally at 37°C° in a semiaerobic environment on various feedstocks produced by the action of intestinal enzymes.

Pentoses are not among these, so foreign genes must be introduced to import and metabolize them. Also, sugar uptake is regulated in these microorganisms, so utilization of mixed sugars will be relatively slow and stepwise. Unlike yeasts, they are not ethanol tolerant, so they can produce only low ethanol concentrations, *Zymomonas* species are an exception to some of these criticisms, but they require considerable genetic manipulation for pentose fermentations.

In our opinion, the future lies in ethanol fermentations by thermophilic *Bacilli* such as *Geobacillus*. As described previously, they are naturally selected to ferment a wide range of C5 and C6 sugars extremely rapidly by relying on unregulated sugar uptake. In consequence, they squander energy to raise the ambient temperature to 70–75°C and thereby take over competing mesophiles.

Because oxygen solubility is very low at these temperatures, they are well adapted to anaerobic or microaerobic environments. Gas sparging or mild vacuum can be used to keep the broth ethanol concentration safely below the inhibitory levels. No such fermentation system is yet commercialized, but the powerful tools are now available from a combination of protein engineering, genetic engineering, and metabolic engineering combined, with novel fermentation systems that make it almost certain that commercialization is possible in the near future.

The inherent advantages of such thermophilic fermentations guarantee that they will then slash the costs of bioethanol for the following reasons:

- Reduced feedstock costs from cheaper processing of agricultural and municipal wastes.
- Reduced capital costs from rapid continuous fermentations with high volumetric productivity.
- Reduced energy costs from elimination of cooling water.
- Reduced distillation costs from continuous ethanol removal as a 20–30% w/v vapor.
- The by-product is spent cells with high animal feed value.

Given such commercial profitability and abundant waste resources, hemicellulosic bioethanol will make a major contribution to replacement of gasoline as a transport fuel and to reduction of global warming.

The application of metabolic engineering also offers the potential to revolutionize the chemical industry. It is estimated that in very near future, integrated biorefineries will play a role comparable to today's petrochemical manufacturers. Biorefineries will use row crops, energy crops, and agricultural waste as inputs to extract oil and starch for food, protein for feed, lignin for combustion or chemical conversion, cellulose for conversion into fermentable sugars, and other by-products.

One can conclude that the future of integrated biorefineries is bright and that the role of metabolic engineering is only limited by our imagination.

REFERENCES

Ahmad, A., H.B. Hartman, S. Krishnakumar, D.A. Fell, M.G. Poolman, and S. Srivastava. 2017. A genome scale model of *Geobacillus thermoglucosidasius* (C56-YS93) reveals its biotechnological potential on rice straw hydrolysate. *J Biotechnol* 251:30–37.

Amore, R., P. Kötter, C. Kuster, M. Ciriacy, and C.P. Hollenberg. 1991. Cloning and expression in *Saccharomyces cerevisiae* of the NAD(P)H-dependent xylose reductase-encoding gene (*XYL1*) from the xylose-assimilating yeast *Pichia stipitis*. *Gene* 109:89–107.

Asanuma, N., and T. Hino. 2000. Effects of pH and energy supply on activity and amount of pyruvate formate-lyase in *Streptococcus bovis*. *Appl Environ Microbiol* 66:3773–3777

Aristidou, A., J. Londesborough, M. Penttilä, P. Richard, L. Ruohonen, H. Soderlund, A. Teleman, and M. Toivari. 1999. Transformed micro-organisms with improved properties. Patent Application PCT/FI99/00185, WO 99/46363.

Aristidou, A., P. Richard, L. Ruohonen, M. Toivari, J. Londesborough, and M. Penttilä. 2000. Redox balance in fermenting yeast. *Monograph European Brewing Convention* 28:161–170.

Bacon, L.F., C. Hamley-Bennett, M.J. Danson, and D.J. Leak. 2017. Development of an efficient technique for gene deletion and allelic exchange in *Geobacillus* spp. *Microb Cell Fact* 16:58–66.

Baghaei-Yazdi, N., F. Cusdin, E.M. Green, and M. Javed. 2006. Modification of bacteria. *International Patent Application* 664076.

Baghaei-Yazdi, N.B., M. Javed, and B.S. Hartley. 2009. Enhancement of ethanol production. *International Patent Application*. WO 2009/10145 A1.

Bao, H., R. Liu, L. Liang, Y. Jiang, M. Jiang, J. Ma, K. Chen, H. Jia, P. Wei, and P. Ouyang. 2014. Succinic acid production from hemicellulose hydrolysate by an *Escherichia coli mutant* obtained by atmospheric and room temperature plasma and adaptive evolution. *Enzyme Microb Tech* 66:10–15

Beall, D.S., and L.O. Ingram. 1993. Genetic engineering of soft-rot bacteria for ethanol production from lignocellulose. *J Ind Microbiol* 11:151–155.

Berman, J.N., G.X. Chen, G. Hale, and R.N. Perham. 1981. Lipoic acid residues in a take-over mechanism for the pyruvate dehydrogenase multienzyme complex of *Escherichia coli*. *Biochem J* 199:513–520.

Billard, P., S. Menart, R. Fleer, and M. Bolotin-Fukuhara. 1995. Isolation and characterization of the gene encoding xylose reductase from *Kluyveromyces lactis*. *Gene* 162:93–97.

Blumer-Schuette, S.E., S.D. Brown, K.B. Sande, E.A. Bayer, I. Kataeva, J.V. Zurawski, J.M. Conway, M.W.W. Adams, and R.M. Kelly. 2014. Thermophilic lignocellulose deconstruction. *FEMS Microbiol Rev* 38:393–448.

Bolen, P.L., G.T. Hayman, and H.S. Shepherd. 1996. Sequence and analysis of an aldose (xylose) reductase gene from the xylose-fermenting yeast *Pachysolen tannophilus*. *Yeast* 12:1367–1375.

Brat, D., E. Boles, and B. Wiedemann. 2009. Functional expression of a bacterial xylose isomerase in *Saccharomyces cerevisiae*. *Appl Environ Microbiol* 75:2304–2311.

Brencic, A., and S.C. Winans. 2005. Detection of and response to signals involved in host-microbe interactions by plant-associated bacteria. *Microbiol Mol Biol Rev* 69:155–194.

Bruinenberg, P.M., P.H.M. De Bot, J.P. Van Dijken, and W.A. Scheffers. 1984. NADH-linked aldose reductase: The key to anaerobic alcoholic fermentation of xylose by yeasts. *Appl Microbiol Biotechnol* 19:256–260.

Causey, T.B., K.T. Shanmugam, L.P. Yomano, and L.O. Ingram. 2004. Engineering *Escherichia coli* for efficient conversion of glucose to pyruvate. *Proc Natl Acad Sci USA* 101:2235–2240.

Chandrakant, P., and V.S. Bisaria. 1998. Simultaneous bioconversion of cellulose and hemicellulose to ethanol. *Crit Rev Biotechnol* 18:295–331.

Cordova, L.T., and M.R. Antoniewicz. 2016. C-13 metabolic flux analysis of the extremely thermophilic, fast growing, xylose-utilizing *Geobacillus* strain LC300. *Metab Eng* 33:148–157.

Costa, C.E., A. Romaní, J.T. Cunha, B. Johansson, and L. Domingues. 2017. Integrated approach for selecting efficient *Saccharomyces cerevisiae* for industrial lignocellulosic fermentations: Importance of yeast chassis linked to process conditions. *Bioresource Technol* 227:24–34.

Cripps, R.E., K. Eley, D.J. Leak, B. Rudd, M. Taylor, M. Todd, S. Boakes, S. Martin, and T. Atkinson. 2009. Metabolic engineering of *Geobacillus thermoglucosidasius* for high yield ethanol production. *Metab Eng* 11:398–408.

Dahn, K., B. Davis, P. Pittman, W. Kenealy, and T. Jeffries. 1996. Increased xylose reductase activity in the xylose-fermenting yeast *Pichia stipitis* by overexpression of *XYL1*. *Appl Biochem Biotechnol* 57–58:267–276.

Deanda, K., M. Zhang, C. Eddy, and S. Picataggio. 1996. Development of an arabinose-fermenting *Zymomonas mobilis* strain by metabolic pathway engineering. *Appl Environ Microbiol* 62:4465–4470.

Dellomonaco, C., F. Fava, and R. Gonzalez. 2010. The path to next generation biofuels: Successes and challenges in the era of synthetic biology. *Microb Cell Fact* 9:3.

Dunlop M.J. 2011. Engineering microbes for tolerance to next-generation biofuels. *Biotechnol Biofuel* 4:32

Durnin, G., J. Clomburg, Z. Yeates, J.J. Pedro, P.J.J Alvarez, K. Zygourakis, P. Campbell, and R. Gonzalez. 2009. Understanding and harnessing the microaerobic metabolism of glycerol in *Escherichia coli*. *Biotechnol Bioengin* 103:143–161.

Feldmann, S., H. Sahm, and G.A. Sprenger. 1992. Pentose metabolism in *Zymomonas mobilis* wild type and recombinant strains. *Appl Microbiol Biotechnol* 38:354–361.

Galbe, M., and G. Zacchi. 2002. A review of the production of ethanol from softwood. *Appl Microbiol Biotechnol* 6:618–628.

Girio, F.M., C. Fonseca, F. Carvalheiro, L.C. Duarte, S. Marques, and R. Bogel-Lukasik. 2010. Hemicelluloses for fuel ethanol: A review. *Bioresource Technol* 101:4775–4800.

Gong, C.S., N.J. Cao, J. Du, and G.T. Tsao. 1999. Ethanol production from renewable resources. *Adv Biochem Eng Biotechnol* 65:207–241.

Grabar T.B., S. Zhou, K.T. Shanmugam, L.P. Yomano, and L.O. Ingram. 2006. Methylglyoxal bypass identified as source of chiral contamination in l(+) and d(−) lactate fermentations by recombinant *Escherichia coli*. *Biotechnol Lett* 28:1527–1535.

Granstrom, T.B., A.A. Aristidou, J. Jokela, and M. Leisola. 2000. Growth characteristics and metabolic flux analysis of *Candida milleri*. *Biotechnol Bioeng* 70:197–207.

Green, E., M. Javed, and N. Bhaghei-Yazdi. 2010. Ethanol production. US Patent No.: US 7,691,620.

Green, E., M. Javed, and N. Bhaghei-Yazdi. 2012. Ethanol production. US Patent No.: US 8,192,977.

Green, E.M., F.H. Cusdin, N. Bhaghei-Yazdi, and. M. Javed. 2003. Modification of bacteria. US Patent No.: US 6,664,076.

Ha, S.J., S.R. Kim, J.H. Choi, M.S. Park, and Y.S. Jin. 2011. Xylitol does not inhibit xylose fermentation by engineered *Saccharomyces cerevisiae* expressing *xyl*A as severely as it inhibits xylose isomerase reaction *in vitro*. *Appl Microbiol Biotechnol* 92:77–84.

Hahn-Hägerdal B., M. Galbe, M.F. Gorwa-Grauslund, G. Liden, and G. Zacchi. 2006. Bio-ethanol—the fuel of tomorrow from the residues of today. *Trend Biotechnol* 24:549–556.

Hallborn, J., M. Walfridsson, U. Airaksinen, H. Ojamo, B. Hahn-Hägerdal, M. Penttilä, and S. Keranen. 1991. Xylitol production by recombinant *Saccharomyces cerevisiae*. *Biotechnology* 9:1090–1095.

Hamelinck, C.N., G.V. Hooijdonk, and A.P.C. Faaij. 2005. Ethanol from lignocellulosic biomass: Technoeconomic performance in short-, middle- and long-term. *Biomass Bioenergy* 28:384–410.

Handumrongkul, C., D.P. Ma, and J.L. Silva. 1998. Cloning and expression of *Candida guilliermondii* xylose reductase gene (*xyl1*) in *Pichia pastoris*. *Appl Microbiol Biotechnol* 49:399–404.

Hartley, B.S., and G. Shama. 1987. Novel ethanol fermentations from sugar cane and straw. *Phil Trans Roy Soc Lond* A321:555–568.

Hartley, B.S. 1988. Thermophilic ethanol production. *International Patent Application* PCT/GB88/004.

Hector, R.E., B.S. Dien, M.A. Cotta, and J.A. Mertens. 2013. Growth and fermentation of D-xylose by *Saccharomyces cerevisiae* expressing a novel D-xylose isomerase originating from the bacterium *Prevotella ruminicola* TC2-24. *Biotechnol Biofuel* 6:84–96.

Ho, N.W.Y., and S.F. Chang. 1989. Cloning of yeast xylulokinase gene by complementation of *E. coli* and yeast mutations. *Enzyme Microbiol Technol* 11:417–421.

Ho, N.W.Y., Z. Chen, and A.P. Brainard. 1998. Genetically engineered *Saccharomyces* yeast capable of effective cofermentation of glucose and xylose. *Appl Environ Microbiol* 64:1852–1859.

Hussein, A.H., B.K. Lisowska, and D.J. Leak. 2015. The genus *Geobacillus* and their biotechnological potential. *Adv Appl Microbiol* 92:1–48

Inderwildi, O.R., and D.A. King. 2009. "Quo Vadis Biofuels." *Energy Environ Sci* 2, 343–346.

Ingram, L.O., H.C. Aldrich, A.C.C. Borges, T.B. Causey, A. Martinez, F. Morales, A. Saleh, S.A. Underwood, L.P. Yomano, S.W. York, J. Zaldivar, and S. Zhou. 1999. Enteric bacterial catalysts for fuel ethanol production. *Biotechnol Prog* 15:855–866.

Ingram, L.O. and T. Conway. Click here to enter text.1988. Expression of different levels of ethanologenic enzymes from *Zymomonas mobilis* in recombinant strains of *Escherichia coli*. *Appl Environ Microbiol* 54:397–404.

Javed, M. 1993. Strain improvement of a thermophilic *Bacillus* for ethanol production. PhD thesis, Imperial College London.

Javed M., and N. Baghaei-Yazdi. 2006. Enhancement of microbial ethanol production. Patent Publication Number 20,090,226,992.

Javed M., B.S. Hartley, and N. Baghaei-Yazdi. 2011. Increased ethanol production by bacterial cells. Patent Publication Number US 2011/0020890 A1.

Javed, M., F.H. Cusdin, P.I. Milner, and E. Green. 2012. Ethanol production in *bacillus*. US Patent No.: US 8,097,460.

Jin, Y.S., and T.W. Jeffries. 2003. Changing flux of xylose metabolites by altering expression of xylose reductase and xylitol dehydrogenase in recombinant *Saccharomyces cerevisiae*. *Appl Biochem Biotechnol* 106:277–286.

Khunnonkwao P., S.S. Jantama, S. Kanchanatawee, and K. Jantama. 2018. Re-engineering Escherichia coli KJ122 to enhance the utilization of xylose and xylose/glucose mixture for efficient succinate production in mineral salt medium. *Appl Microbiol Biotechnol.* 102(1):127–141.

Koti, S., S.P. Govumoni, J. Gentela, and L.V. Rao. 2016. Enhanced bioethanol production from wheat straw hemicellulose by mutant strains of pentose fermenting organisms *Pichia stipitis* and *Candida shehatae. SpringerPlus* 5:1545

Krahulec, S, M. Klimacek, and B. Nidetzky. 2012. Analysis and prediction of the physiological effects of altered coenzyme specificity in xylose reductase and xylitol dehydrogenase during xylose fermentation by *Saccharomyces cerevisiae. J Biotechnol* 158:192–202.

Keim, C.R., and K. Venkatasubramanian. 1989. Economics of current biotechnological methods of producing ethanol. *Trends Biotechnol* 7:22–29.

Kötter, P., and M. Ciriacy. 1993. Xylose fermentation by *Saccharomyces cerevisiae. Appl Microbiol Biotechnol* 38: 776–783.

Kuhad, R.C., A. Singh, and K.E. Eriksson. 1997. Micro-organisms and enzymes involved in the degradation of plant fiber cell walls. *Adv Biochem Eng Biotechnol* 57:45–125.

Kuyper, M., H.R. Harhangi, A.K. Stave, A.A. Winkler, M.S.M. Jetten, W.T.A.M. De Laat, J.J.J. Den Ridder, H.J.M. Op den Camp, J.P. Van Dijken, and J.T. Pronk. 2003. High-level functional expression of a fungal xylose isomerase: The key to efficient ethanolic fermentation of xylose by *Saccharomyces cerevisiae? FEMS Yeast Res* 4:69–78.

Kuyper, M., M.M.P. Hartog, M.J. Toirkens, M.J.H. Almering, A.A. Winkler, J.P. van Dijken, and J.T. Pronk. 2005a. Metabolic engineering of a xylose-isomerase-expressing *Saccharomyces cerevisiae* strain for rapid anaerobic xylose fermentation. *FEMS Yeast Res* 5:399–409.

Kuyper, M., M.J. Toirkens, J.A. Diderich, A.A. Winkler, J.P. van Dijken, and J.T. Pronk. 2005b. Evolutionary engineering of mixed-sugar utilization by a xylose-fermenting .Saccharomyces cerevisiae strain. *FEMS Yeast Res* 5:925–934.

Kuyper, M., A.A. Winkler, J.P. van Dijken, and J.T. Pronk. 2004. Minimal metabolic engineering of *Saccharomyces cerevisiae* for efficient anaerobic xylose fermentation: A proof of principle. *FEMS Yeast Res* 4:655–664.

Kwak, S., and Y.S. Jin. 2017. Production of fuels and chemicals from xylose by engineered *Saccharomyces cerevisiae*: a review and perspective. *Microb Cell Fact* 16:82–97.

Loaces, I., C. Rodríguez, V. Amarelle, E. Fabiano, and F. Noya. 2016. Improved glycerol to ethanol conversion by *E. coli* using a metagenomic fragment isolated from an anaerobic reactor. *J Ind Microbiol Biotechnol* 43:1405–1416.

Lawford, H.G., J.D. Rousseau, and J.S. Tolan. 2001. Comparative ethanol productivities of different *Zymomonas* recombinants fermenting oat hull hydrolysate. *Appl Biochem Biotechnol* 91–93:133–146.

Lee, J. 1997. Biological conversion of lignocellulosic biomass to ethanol. *J Biotechnol* 56:1–24.

Lee, S.M., T. Jellison, and H.S. Alper. 2012. Directed evolution of xylose isomerase for improved xylose catabolism and fermentation in the yeast *Saccharomyces cerevisiae. Appl Environ Microbiol* 78:5708–5716.

Leitgeb, S., B. Petschacher, D.K. Wilson, and B. Nidetzky. 2005. Fine tuning of coenzyme specificity in family 2 aldo-keto reductases revealed by crystal structures of the Lys-274→Arg mutant of *Candida tenuis* xylose reductase (*AKR2B5*) bound to NAD$^+$ and NADP$^+$. *FEBS Lett* 579:763–7.

Leponiemi, A. 2011. *Fibres and energy from wheat straw by simple practice.* Espoo: VTT Publications 767.

Lerouge, P., M. Cabanes-Macheteau, C. Rayon, A.C. Fischette-Laine, V. Gomord, and L. Faye. 1998. N-glycoprotein biosynthesis in plants: Recent developments and future trends. *Plant Mol Biol* 38:31–48.

Li, X., A. Park, R. Estrela, S.R. Kim, Y.S. Jin, and J.H.D. Cate. 2016. Comparison of xylose fermentation by two high-performance engineered strains of *Saccharomyces cerevisiae. Biotechnol Rep* 9:53–56.

Lin, H., G.N. Bennett, and K.Y. San. 2005. Metabolic engineering of aerobic succinate production systems in *Escherichia coli* to improve process productivity and achieve the maximum theoretical succinate yield. *Metab Eng* 7:116–127.

Liu, T., and C. Khosla. 2010. Genetic engineering of *Escherichia coli* for biofuel production. *Ann Rev Genet* 44:53–69.

Liu, H., J. Sun, J.S. Chang, and P. Shukla. 2018. Engineering microbes for direct fermentation of cellulose to bioethanol. *Crit Rev Biotechnol* 10:1–17.

Luli, G., and L. Ingram. 2005. *UF/IFAS Researcher's Biomass-to-Ethanol Technology Could Help Replace Half of Auto Fuel in U.S.* University of Florida News.

Lynd, L.R., C.E. Wyman, and T.U. Gerngross. 1999. Biocommodity engineering. *Biotechnol Prog* 15:777–793.

Madhavan, A., S. Tamalampudi, K. Ushida, D. Kanai, S. Katahira, and A. Srivastava. 2009. Xylose isomerase from polycentric fungus *Orpinomyces*: Gene sequencing, cloning, and expression in *Saccharomyces cerevisiae* for bioconversion of xylose to ethanol. *Appl Microbiol Biotechnol* 82:1067–1078.

Meinander, N., B. Hahn-Hägerdal, M. Linko, P. Linko, and H. Ojamo. 1994. Fed-batch xylitol production with recombinant *XYL1* expressing *Saccharomyces cerevisiae* using ethanol as a co-substrate. *Appl Microbiol Biotechnol* 42:334–339.

Meinander, N., G. Zacchi, and B. Hahn-Hägerdal. 1996. A heterologous reductase affects the redox balance of recombinant *Saccharomyces cerevisiae. Microbiology* 142:165–172.

Metzger, M.H., and C.P. Hollenberg. 1995. Amino acid substitutions in the yeast *Pichia stipitis* xylitol dehydrogenase coenzyme-binding domain affect the coenzyme specificity. *Eur J Biochem* 228:50–54.

Nielsen, C., J. Larsen, F. Iversen, C. Morgen, and B.H. Christensen. 2002. Integrated Biomass Utlisation System (IBUS) for co-production of electricity and bioethanol. *EU Contract Report* ENK6-CT-2002–00650.

Ohta, K., D.S. Beall, J.P. Mejia, K.T. Shanmugam, and L.O. Ingram. 1991. Metabolic engineering of *Klebsiella oxytoca* M5A1 for ethanol production from xylose and glucose. *Appl Environ Microbiol* 57:2810–2815.

Olson, D.G., R. Sparling, and L.R. Lynd. 2015. Ethanol production by engineered thermophiles. *Curr Opinion Biotechnol* 33:130–141

Palmarois-Adrados, B., P. Choteborska, M. Gelbe, and G. Zacchi. 2005. Ethanol production from non-starch carbohydrates of wheat bran. *Bioresource Technol* 96:843–850.

Park, J.H, M.A. Choi, Y.J. Kim, Y. Kim, Y.K. Chang, and K.J. Jeong. 2017. Engineering of *Klebsiella oxytoca* for production of 2,3-butanediol via simultaneous utilization of sugars from a *Golenkinia* sp. hydrolysate. *Bioresource Tech* 245:1386–1392.

Pessoa, A., I.M. Mancilha, and S. Sato. 1997. Acid hydrolysis of hemicellulose from. Sugarcane bagasse. *Brazilian J Chem Eng* 14:3.

Parachin, N.S., and M.F. Gorwa-Grauslund. 2011. Isolation of xylose isomerases by sequence- and function-based screening from a soil metagenomic library. *Biotechnol Biofuel* 4:9–19.

Payton, M.A., and B.S. Hartley. 1985. Mutants of *Bacillus stearothermophilus* lacking NAD-linked L-lactate dehydrogenase. *FEMS Microbiol Lett* 26:333–3336.

Pogrebnyakov, C.B. Jendresen, and A.T. Nielsen. 2017. Genetic toolbox for controlled expression of functional proteins in *Geobacillus* spp. *PLoS One* 12:1–15.

Pye, E.K., and Lora, J.H., (1991). The Alcell process, a proven alternative to kraft pulping. *Tappi J* 74:113–118.

Rajgarhia, V., K. Koivuranta, M. Penttilä, M. Ilmen, P. Suominen, A. Aristidou, C. Miller, S. Olson, and L. Ruohonen. 2004. Genetically modified yeast species and fermentation processes using genetically modified yeast. Patent Application WO 2004099381.

Reeve, B., E. Martinez-Klimova, J. de Jonghe, D.J. Leak, and T. Ellis. 2016. The *Geobacillus* plasmid set: A modular toolkit for thermophile engineering. *ACS Synth Biol* 5:1342–1347.

Riondent, C., R. Cachon, Y. Wache, G. Alcaraz, and C. Divies. 2000. Extracellular oxidoreduction potential modifies carbon and electron flow in *Escherichia coli*. *J Bacteriol* 182:620–626.

Rodriguez-Peña, J.M., V.J. Id, J. Arroyo, and C. Nombela. 1998. The *YGR194c* (*XKS1*) gene encodes the xylulokinase from the budding yeast *Saccharomyces cerevisiae*. *FEMS Microbiol Lett* 162:155–160.

San Martin, R., D. Busshell, D. Leak, and B.S. Hartley. 1993. Ethanolic fermentation of lignocellulose hydrolysates. *J Gen Microbiol* 139:1033–1040.

Sarthy, A.V., B.L. McConaughy, Z. Lobo, J.A. Sundstrom, C.E. Furlong, and B.D. Hall. 1987. Expression of the *Escherichia coli* xylose isomerase gene in *Saccharomyces cerevisiae*. *Appl Environ Microbiol* 53:1996–2000.

Senac, T., and B. Hahn-Hägerdal. 1989. Intermediary metabolite concentrations in xylulose- and glucose-fermenting *Saccharomyces cerevisiae* cells. *Appl Environ Microbiol* 56:120–126.

Senac, T., and B. Hahn-Hägerdal. 1991. Effects of increased transaldolase activity on D-xylulose and D-glucose metabolism in *Saccharomyces cerevisiae* cell extracts. *Appl Environ Microbiol* 57:1701–1706.

Shah, A., and M. Darr. 2016. A techno-economic analysis of the corn stover feedstock supply system for cellulosic biorefineries. *Biofuel Bioprod Bioref* 10:542–59.

Shahab, R.L., J.S. Luterbacher, S. Brethauer, and M.H. Studer. 2018. Consolidated bioprocessing of lignocellulosic biomass to lactic acid by a synthetic fungal-bacterial consortium. *Biotechnol Bioeng* 115:1207–1215.

Shi, N.Q., and T.W. Jeffries. 1998. Anaerobic growth and improved fermentation of *Pichia stipitis* bearing a *URA1* gene from *Saccharomyces cerevisiae*. *Appl Microbiol Biotechnol* 50:339–345.

Srirangan, K., X. Liu, A. Westbrook, L. Akawi, M.E. Pyne, M. Moo-Young, and C.P Chou. 2014. Biochemical, genetic, and metabolic engineering strategies to enhance coproduction of 1-propanol and ethanol in engineered *Escherichia coli*. *Appl Microbiol Biotechnol* 98:9499–9515.

Stevis, P.A., J.J. Hang, and N.W.Y. Ho. 1987. Cloning of the *Pachysolen tannophilus* xylulokinase gene by complementation in *Escherichia coli*. *Appl Environ Microbiol* 53:2975–2977.

Suzuki, H., A. Murakami, and K. Yoshida. 2012. Counterselection system for *Geobacillus kaustophilus* HTA426 through disruption of *pyrF* and *pyrR*. *App Environ Microbiol* 78:7376–7383.

Takahashi, D.F., M.L. Carvalhal, and F. Alterhum. 1994. Ethanol production from pentoses and hexoses by recombinant *Escherichia coli*. *Biotechnol Lett* 16:747–750.

Takuma, S., N. Nakashima, M. Tantirungkij, S. Kinoshita, H. Okada, T. Seki, and T. Yoshida. 1991. Isolation of xylose reductase gene of *Pichia stipitis* and its expression in *Saccharomyces cerevisiae*. *Appl Biochem Biotechnol* 28–29:327–340.

Tantirungkij, M., T. Izuishi, T. Seki, and T. Yoshida. 1994. Fed-batch fermentation of xylose by a fast growing mutant of xylose-assimilating recombinant *Saccharomyces cerevisiae*. *Appl Microbiol Biotechnol* 41:8–12.

Tantirungkij, M., N. Nakashima, T. Seki, and T. Yoshida. 1993. Construction of xylose-assimilating *Saccharomyces cerevisiae*. *J Ferment Bioeng* 75:83–86.

Teeri, T.T., A. Koivula, M. Linder, G. Wohlfahrt, C. Divne, and T.A. Jones. 1998. *Trichoderma reesei* cellobiohydrolases: Why so efficient on crystalline cellulose? *Biochem Soc Trans* 26:173–178.

Thakker, C., I. Martínez, K.-Y. San, and G. N. Bennett. 2012. Succinate production in *Escherichia coli*. *Biotechnol J* 7:213–224.

Toivari, M.H., L. Salusjärvi, L. Ruohonen, and M. Penttila. 2004. Endogenous xylose pathway in *Saccharomyces cerevisiae*. *Appl Environ Microbiol* 70:3681–3686.

Tolan, J.S., and R.K. Finn. 1987a. Fermentation of D-xylose and L-arabinose to ethanol by *Erwinia chrysanthemi*. *Appl Environ Microbiol* 53:2033–2038.

Tolan, J.S., and R.K. Finn. 1987b. Fermentation of D-xylose to ethanol by genetically modified *Klebsiella planticola*. *Appl Environ Microbiol* 53:2039–2044.

US Department of Energy. 2016. *2016 Billion-Ton Report: Advancing Domestic Resources for a Thriving Bioeconomy.* www.energy.gov/sites/prod/files/2016/12/f34/2016_billion_ton_report_12.2.16_0.pdf.

Vandeska, E., S. Kuzmanova, and T. W. Jeffries. 1995. Xylitol formation and key enzyme activities in *Candida boidinii* under different oxygen transfer rates. *J Ferment Bioeng* 80:513–516.

Van Zyl, L.J., M.P. Taylor, K. Eley, M. Tuffin, and D.A. Cowan. 2014. Engineering pyruvate decarboxylase-mediated ethanol production in the thermophilic host *Geobacillus thermoglucosidasius*. *Appl Microbiol Biotechnol* 98:1247–1259.

Vemuri, G.N., M.A. Eiteman, and E. Altman. 2002. Succinate production in dual-phase *Escherichia coli* fermentations depends on the time of transition from aerobic to anaerobic conditions. *J Indust Microbiol Biotechnol* 28:325–332.

Verduyn, C., R. Van Kleef, J. Frank, H. Schreuder, J.P. Van Dijken, and W.A. Scheffers. 1985. Properties of the NAD(P)H-dependent xylose reductase from the xylose-fermenting yeast *Pichia stipitis*. *Biochem J* 226:669–677.

Walfridsson, M., M. Anderlund, X. Bao, and B. Hahn-Hägerdal. 1997. Expression of different levels of enzymes from the *Pichia stipitis XYL1* and *XYL2* genes in *Saccharomyces cerevisiae* and its effects on product formation during xylose utilisation. *Appl Microbiol Biotechnol* 48:218–224.

Walfridsson, M., X. Bao, M. Anderlund, G. Lilius, L. Bulow, and B. Hahn-Hägerdal. 1996. Ethanolic fermentation of xylose with *Saccharomyces cerevisiae* harboring the *Thermus thermophilus xylA* gene, which expresses an active xylose (glucose) isomerase. *Appl Environ Microbiol* 62:4648–4651.

Weissenborn, D.L., N. Wittekindt, and T.J. Larson. 1992. Structure and regulation of the *glpFK* operon encoding glycerol diffusion facilitator and glycerol kinase of *Escherichia coli* K-12. *J Biol Chem* 267:6162–31.

Wooley, R., M. Ruth, J. Sheehan, K. Ibsen, H. Majdeski, and A. Galvez. 1999. Lignocellulosic biomass to ethanol process design and economics. Technical Report Number: NREL/TP-580-26157

Xarhangi, H.R., A.S. Akhmanova, R. Emmens, C. van der Drift, W.T.A.M. de Laat, J.P. van Dijken, M.S.M. Jetten, J.T. Pronk, and H.J.M. Op den Camp. 2003. Xylose metabolism in the anaerobic fungus *Piromyces* sp. strain E2 follows the bacterial pathway. *Arch Microbiol* 180:134–141.

Xu, Y, H. Chu, C. Gao, F. Tao, Z. Zhou, K. Li, L. Li, C. Ma, and P. Xu. 2014, Systematic metabolic engineering of *Escherichia coli* for high-yield production of fuel bio-chemical 2,3-butanediol. *Metab Eng* 23:22–33.

Yang, S, A Mohagheghi, M.A. Franden, Y.C. Chou, X. Chen, N. Dowe, M.E. Himmel, and M. Zhang. 2016. Metabolic engineering of *Zymomonas mobilis* for 2,3-butanediol production from lignocellulosic biomass sugars *Biotechnol Biofuel* 9:189–204.

Yang, Z., and Z. Zhang. 2018. Production of (2R, 3R)-2,3-butanediol using engineered *Pichia pastoris*: Strain construction, characterization and fermentation. *Biotechnol Biofuel* 11:35–51.

Yomano, L.P., S.W. York, and L.O. Ingram. 1998. Isolation and characterization of ethanol-tolerant mutants of *Escherichia coli* KO11 for fuel ethanol production. *J Ind Microbiol Biotechnol* 20:132–138.

Zhang, K.D., W. Li, Y.F. Wang, Y.L. Zheng, F.C. Tan, X.O. Ma, L.S. Yao, E.A. Bayer, L.S. Wang, and F.L. Li. 2018. Processive degradation of crystalline cellulose by a multimodular endo-glucanase via a wirewalking mode. *Biomacromolecules* doi:10.1021/acs.biomac.8b00340. [Epub ahead of print].

Zhang, M., C. Eddy, K. Deanda, M. Finkelstein, and S. Picataggio. 1995. Metabolic engineering of a pentose metabolism pathway in ethanologenic *Zymomonas mobilis*. *Science* 267:240–243.

Zhou, H., J.S. Cheng, B.L. Wang, G.R. Fink, and G. Stephanopoulos. 2012. Xylose isomerase overexpression along with engineering of the pentose phosphate pathway and evolutionary engineering enable rapid xylose utilization and ethanol production by *Saccharomyces cerevisiae*. *Metab Eng* 14:611–622.

Zhou, S., T. B. Causey, A. Hasona, K.T. Shanmugam, and L.O. Ingram. 2003. Production of optically pure D-Lactic acid in mineral salts medium by metabolically engineered *Escherichia coli* W3110. *Appl Environ Microbiol* 69:399–407.

Zhou, S., and L.O. Ingram. 2001. Simultaneous saccharification and fermentation of amorphous cellulose to ethanol by recombinant *Klebsiella oxytoca* SZ21 without supplemental cellulase. *Biotechnol Lett* 23:1455–1462.

12 Biorefineries
Industrial Perspectives and Challenges in Primary and Secondary Metabolism

David Mousdale

CONTENTS

> "When you have eliminated the impossible, whatever remains, however improbable, must be the truth."
>
> **Arthur Conan Doyle *"The Sign of the Four"* (1890)**

12.1 INTRODUCTION: BIOFUELS APPROACHING 2020—THE GLOBAL BIOECONOMY

"Bioeconomy" is (as noted by the Organization for Economic Co-operation and Development, OECD) an ambiguous term but the OECD has offered a broad and uncontentious definition: "the aggregate set of economic operations in a society that use the latent value incumbent in biological products and processes to capture new growth and welfare benefits for citizens and nations" (Organization for Economic Co-operation and Development, 2006). The major sector of the global bioeconomy, in terms of publicity and economic/commercial impact, remains that of biofuels. Although many of the statistics advanced for the global biofuels industry—billions of liters or millions of barrels of oil equivalent produced per year—are often highly impressive, the economic reality is that biofuels remain a niche industry (Mousdale, 2014). This conclusion is fully supported by recent releases of updated energy indicators:

- The International Energy Agency (IEA) estimated that biofuels accounted for only 2.8% of all liquid fuel usage for transport in 2016 (International Energy Agency, 2016).
- BP estimated that the use of liquid biofuels was equivalent to 1.9% of total oil consumption in 2016 (BP, 2017).

In, 2018, 22 years after the publication of the most influential treatise on biofuels (Wyman, 1996), this disappointing lack of market penetration of biofuels as global phenomena must, therefore, be considered in order to place the undoubted and considerable ingenuity of microbial biochemists and genetic engineers over several decades (Aristidou et al., 2012) in a clear technological perspective.

As indicators of the health of the global bioeconomy, biofuels and their various synthetic technologies and production routes are invaluable but any analysis leads to the inescapable conclusion that both biofuels and the broader bioeconomy are constrained not by scientific advances but by radically divergent national and international policies on the use of conventional fossil fuels and of fossil fuels as feedstocks to the petrochemical industry, primarily based on short-term sales costs.

The production costs of biofuels have never ceased to be important foreground considerations in techno-economic analyses of biofuels (Putsche and Sandor, 1996; Schubert, 2006; Mousdale, 2010; Lane, 2015). The IEA presented in 2007 data for the minimum cost of oil (approximately $70 per barrel) for fuel ethanol to be cost competitive; this has proved to be of major and continuing interest to commentators on biofuels economics and policies (International Energy Agency, 2007). In retrospect from post-2014, the extensive scientific work on ethanol as a biofuel in 1996 was visionary in nature because oil prices had been much lower than the $70 per barrel guideline and remained sub-threshold until 2006 (Figure 12.1). Between, 2006 and 2014, whether in nominal or inflation-adjusted terms, oil prices were at levels which greatly promoted research into biofuels and other alternative forms of energy.

The upward trend in oil prices from 2000 helped to increase commercial interest and investment in biofuels, which remained high in this era despite the short-lived dip in oil prices in the post-2007 recession. In contrast, the sudden collapse in oil prices in 2014 had (and continues to have) a major impact on business confidence in the biofuels sector (Lane, 2015). A major casualty of this reformatting of international energy supply, demand, and costs was the world-leading producer of algal biofuels and its transfer of interest away from biofuels to other bioproducts (Fehrenbacher, 2016). The

continued low oil price but the survival of the oil gas and shale industry in North America has, however, questioned the logic of further development of "uneconomic" alternative fuels.

The IEA has made several projections of biofuels usage up to the year 2040 modeling different sets of assumptions (International Energy Agency, 2016). In the scenario in which current energy policies (national, international, and regional) continue to be pursued, biofuels use reaches a 4% share of total transport fuels. In another and radically different scenario, this figure jumps to 18% but supporting measures for this require a very different and environmentally driven mindset.

12.1.1 Environmental and Geopolitical Drivers for Biofuels Use

The "450 Scenario" envisioned by the IEA is a pathway to limit long-term global warming to 2°C above pre-industrial levels, i.e., place stringent limits on how high atmospheric CO_2 concentrations are allowed to rise (< 450 ppm). To achieve this, a range of specific policies is invoked, including carbon price, the removal of subsidies for fossil fuels and their derivatives, greatly improved efficiencies of medium- and heavy-duty vehicles, internationally approved and imposed CO_2 emission limits for new cars and aircraft and support for renewable technologies (International Energy Agency, 2016).

An essential feature of the theoretical underpinning for this scenario is the requirement to deliberately *not* utilize known reserves of fossil fuels (McGlade and Ekins, 2014, 2015). The extent of this non-use is very high, reaching 100% of known reserves in some cases; the negative impacts of such policies on national economies are variable but are expected to be highest in those rapidly developing economically, most obviously in China and India.

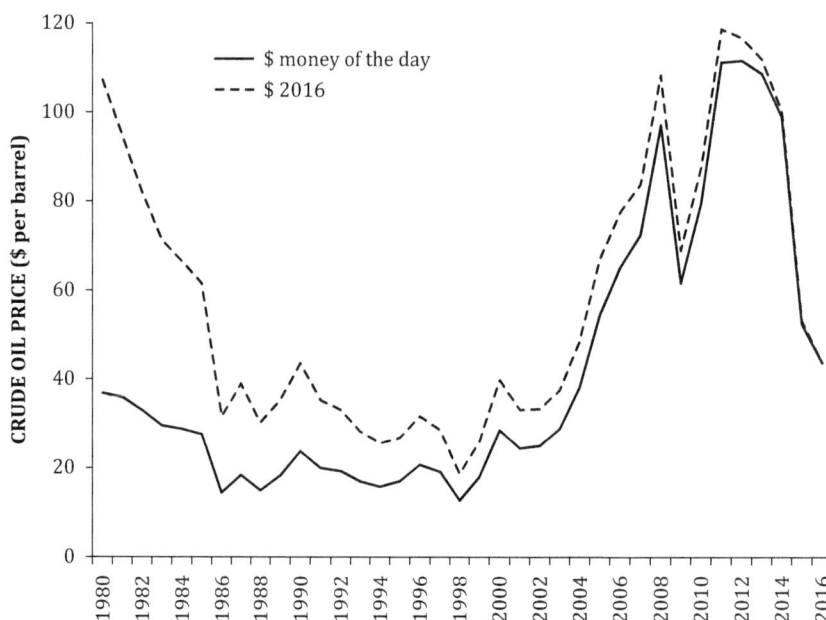

FIGURE 12.1   Oil price trends 1980–2016. (Data from the BP Statistical Review of World Energy, 2017, underpinning data).

This should (logically) have acted as a driver for biofuels to substitute conventional fossil fuel use but three counteracting factors have become important:

1. The perceived conflict between food production and land use for growing crops suitable for biofuels production.
2. The environmentally damaging effects of massive first-generation biofuels production.
3. The rapid move to national policies aiming at completely replacing liquid fuel-dependent vehicles by electric vehicles.

"Food versus fuel" was one of several counter-arguments adduced by opponents of large-scale biofuels production, other major arguments being that biofuels per se had no or insignificant beneficial effects on greenhouse gas emissions and that bioenergy crop plantations were inherently biologically unsustainable (Pimental et al., 2009).

Intertwined with his concern was the realization that large-scale replanting of land for bioenergy crops caused indirect land-use changes that exacerbated greenhouse gas emissions (Lapola et al., 2010; Di Lucia et al., 2012). This persuaded the European Union to abandon support for first-generation biofuels in favor of "advanced" biofuels produced from plant material with no alternative use as food for human consumption.

National policies that set time limits for the production and sale of conventional petrol/gasoline and diesel vehicles are more recent but include those announced for the United Kingdom and France (2040), The Netherlands, Germany, and India (2030), Norway (2025, provisional), together with plans to encourage electric vehicle use (Gray, 2017).

The cumulative effect of these factors would be to greatly reduce the market for biofuels, potentially resulting in a commodity lifespan of only 30–50 years—of the same order as magnitude as 20th century electronics items such as VHS recorders, laser discs and digital audio tapes. The environmental and geopolitical drivers thus become inhibitors for future biofuels development. That this "obsolescence" need not be fatal for further R&D into biofuels will, however, be examined in this chapter, which aims to place accumulated experience of biofuels production and products within the broader and dynamically evolving bioeconomy of the 21st century.

12.1.2 Linking Biofuels to Fine Chemicals in Biomanufacturing

The large-scale production of the so-called first-generation biofuels (ethanol and methyl esters of long-chain fatty acids) is a mature and global industrial sector using mostly food crops (corn and sugarcane) for ethanol and a widening portfolio of plant and animal fats and oils for the methyl esters that constitute biodiesel. In addition, a series of other "advanced biofuels" molecules have established production routes: n-butanol, isobutanol, 4-methyl-pentanol, n-pentanol, and so on (Meadows et al., 2018). All these chemical entities can be intermediates in the biosynthesis of higher-value bioproducts; in other words, decades of biotechnological and bioengineering experience with biofuels are not to be discarded but should be seen as a platform for further technological R&D in biorefineries for a wide spectrum of industrial, agricultural and consumer chemicals.

The essential link is that biofuels as molecules can function as feedstocks for other and sequential processes. This was first developed a conceptual biorefinery to produce ethylene and ethanolamine (via ethylene oxide) from sugarcane-derived ethanol (Bruscino, 2009). This pioneering case study modeled a biorefinery with raw sugarcane as the feedstock and the factory gate production of 5.0×10^4 tons per year ethanolamine and 6.0×10^4 tons per year ethylene (Figure 12.2).

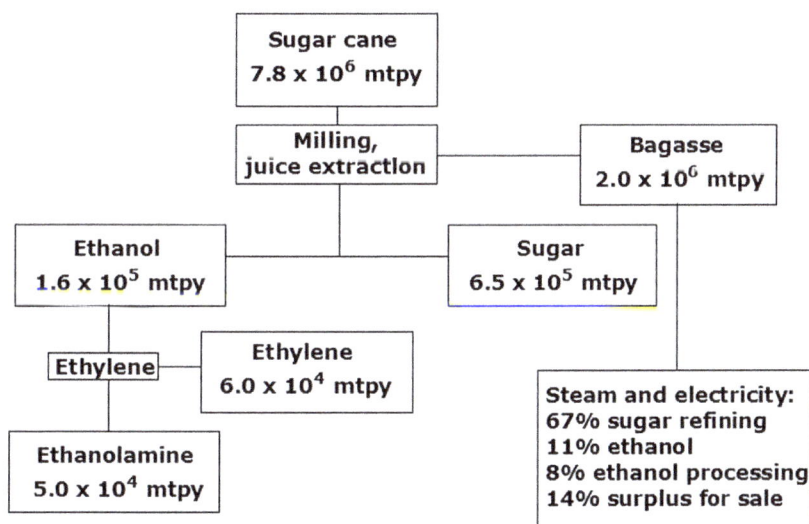

FIGURE 12.2 Outline design scheme for a biorefinery processing approximately 8 million metric tons per year (mtpy) of sugarcane (Redrawn from data in Bruscino (2009).

The biotechnological input is the fermentation of sugars to ethanol; after the dehydration of ethanol to ethylene, other C2 derivatives could be manufactured by standard process chemistry. In effect, this model added ethanol-processing chemistry to a sugarcane refinery and ethanol production unit, utilizing the sugarcane biomass waste stream (bagasse) as a heat and power substrate.

India Glycols Limited has produced "green" ethylene glycols via ethanol using this integrated technology since the 1980s with sugarcane molasses as the feedstock (www.indiaglycols.com). The same basic design as outlined in Figure 12.2 could focus on ethylene as the sole product and Braskem in Brazil began to manufacture "green" polyethylene from sugar-derived ethanol in 2010 with annual production presently reaching 2.0×10^5 ton (http://green-polyethylene.com).

These examples illustrate the industrial biotechnology of *primary* products in biorefineries, i.e., utilizing metabolic pathways to produce high volume/low to medium cost chemicals that are essential features of cellular processes. Establishing biorefineries as production units for primary products is an essential first step in the development of a functioning bioeconomy but—as discussed in section 12.4—secondary products (Zhang et al., 2010), i.e., those derived from primary metabolites and considered non-essential for cellular activities in microbial life cycles, are equally potential outputs from future biorefineries; these products include, enzymes, recombinant proteins, antimicrobials, pigments, enzyme inhibitors, "rare" sugars and novel medicines.

12.2 FIRST-GENERATION BIOREFINERIES: PRIMARY METABOLITES AND BIOENERGY

12.2.1 Biorefinery: An Evolving Etymology

While ethanol and biodiesel were essentially "drop-in" products to conventional petrofuels with current blending levels in the 2–10% (v/v) range (Mousdale, 2014), far more ambitious uses of plant biomass as feedstock sources were advanced from at least the 1980s, generally using the word "biorefinery" as a metaphor for conceptually mirroring many long-established crude oil refineries with their multiplicity of products, i.e., conventional liquid and gaseous fuels as well as feedstock chemicals for the global petrochemical industry (Levy et al., 1981; Kamm et al., 2006).

Radically different concepts of a biorefinery—broadly differentiable as those grounded in either the food industry or in process chemistry—have continued to evolve new variants as academic, commercial, and industrial interest has intensified and a wide range of feedstocks has come into consideration (Schieb et al., 2015; Pandey et al., 2015). This has resulted in a degree of degradation of the term and its confusion with a closely related one, "biomass refining," which aims to tap the enormous store of preformed chemicals represented by plant biomass, especially lignocellulosic biomass (Bungay, 1982; Esposito and Antonietti, 2015; Rouilly and Vaca-Garcia, 2015).

The attempt to be all encompassing is best summed up by the Bioenergy Program of the IEA; Task 42 Biorefining addresses "the sustainable and synergetic processing of biomass into marketable food and feed ingredients, chemicals, materials and energy (fuels, power, heat)" (van Ree and van Zealand, 2014).

In the background, however, is another layer of meaning to the biorefinery concept: product replacement, "bio" for "synthetic" (= "plastic") as part of a circular bio-economy in which sustainable management of natural resources, sustainable bio-production as well as mitigation of global climate change are prominent features (D'Amato et al., 2017). Models of climate change mitigation by carbon taxation, carbon capture and storage and increased energy efficiency converge on the absolute requirement to develop competitively priced alternative sources of energy such as solar and wind power (Grimaud et al., 2011; Mathews, 2014; Marcucci and Fragkos, 2015). The integration of these various economic and environmental factors leads to the conclusion that directed technical change to replace petrochemical derivatives in industrial and consumer products by the outputs of biorefineries is a technology that is (or will eventually be) inescapable in the circular bio-economy in which waste is minimized and man-made non-degradable polymers (highly polluting to the marine environment) are progressively eliminated.

The Circular Economy is highly environmentalist by design and emphasizes restorative and regenerative features in an approach to minimize waste and efficiently use global finite resources; in this context, biorefining is seen as a key enabling technology that can incorporate the entire resource base from raw biomass, by-products, post-consumer residues, water, and carbon as photosynthate from terrestrial and marine organisms (IEA Bioenergy, 2017). European initiatives in biorefineries—as discussed in section 12.2.2—have been highly influenced by Circular Economy considerations.

12.2.2 Biorefineries: A Global Stocktaking

In 2014, IEA Bioenergy's Task 42 Biorefining listed biorefinery facilities in the participating countries of this initiative, Europe, North America, and Australasia (van Ree and van Zealand, 2014). Stripping out service units with no defined in-house products, this list is combined with examples of biofuel-driven biorefineries taken from an earlier IEA Bioenergy publication (Jungmeier et al., 2013) in Table 12.1. The combined list is heavily populated with ethanol producers from non-food plant sources although a conventional rapeseed oil biodiesel plant, an algal phytochemical extraction unit and a wood-based biomethane plant are classed by the IEA as biorefineries. Also included are demonstration plants for Fischer-Tropsch (FT) biofuels (see section 12.3.2)

Ethanol is a prominent output from these small-scale production units, demonstrating again the important link between ethanol as a biofuel and as a major bioeconomy chemical. Indeed, some examples from Table 12.1 can be viewed as fitting easily with the linear model of the biorefinery concept as illustrated by Figure 12.2, in which the substitution of an

TABLE 12.1

Biorefinery Operations Recognized by IEA Bioenergy Task 42 Biorefining (Jungmeier et al., 2013; van Ree and van Zealand, 2014)

Location (ISO Code)	Scale[a]	Feedstock	Organic Outputs
Victoria (AU)	C	Animal fats, vegetable oil	Biodiesel, biogas, liquid burner fuel, glycerol
Sydney (AU)	P	Sugarcane bagasse, corn stover	Ethanol, yeast, green coal
Bomaderry (AU)	C	Wheat, wheat flour	Ethanol, glucose starch, gluten, animal feed
Vienna (AT)	P	Wood chips, straw	Upgradable diesel, pyrolysis oil, biochar
Pischelsdorf (AT)	C	Wheat	Ethanol, animal feed
Bruck a/d Leitha (AT)	D	Microalgae	Omega-fatty acids, astaxanthin, algal oil
Pöls (AT)	C	Wood	Tall oil, turpentine
Alberta (CA)	C	Wood	Methanol
Ontario (CA)	P	Crop/tree residues	Ethanol, acetic acid, animal feed
Alberta (CA)	C	Municipal solid waste	Ethanol, methanol
Ballerup (DK)	P	Crop residues	Ethanol
Maabjerg (DK)	C	Waste streams,	Biomethane, ethanol, fertilizer
Kalundborg (DK)	D	Agricultural residues	Ethanol, animal feed
Karlsruhe (DE)	P	Agricultural residues	Gasoline (via syngas)
Brensbach (DE)	C	Grass	Cellulosic composites
Leuna (DE)	P	Isobutene (maize)	*Iso*octane
Dundalk (IE)	Cl	Straw. lactose whey, brewers grain	Lactic acid, ethyl lactate
Porto Torres (IT)	C	Non-food crop oils	Biolubricants, glycerol, acids
Crescentino (IT)	D	Non-food plants, straw	Ethanol, animal feed
Amsterdam (NL)	P	Starch and sugars	Bioplastics, levulinic acid
Wageningen (NL)	P	Microalgae	Biochemicals
Kinleith (NZ)	C	Wood	Tall oil, red oil, turpentine
Edgecombe (NZ)	C	Milk whey/permeate	Ethanol
Glenbrook (NZ)	P	Wood wastes	Ethanol
Vero Beach (US)	C	Vegetable and wood wastes	Ethanol
St Joseph (US)	P	Energy crops	Ethanol, animal feeds
Boardman (US)	C	Poplar, corn stover, wheat straw	Ethanol, ethyl acetate
Sternberg (DE)	C	Rapeseed oil	Biodiesel, glycerol
Dornsjö (SE)	D	Grass, manure	Biomethane, lactic acid, amino acids
Güssing (AT)	D	Wood	Biomethane, hydrogen
Sarpsborg (NO)	D	Woodchips, saw mill residues	Ethanol, lignin composites
Piteå (SE)	C	Woodchips	Biodimethylether

[a] C, commercial; D, demonstration; P, pilot.

established commercial product (for example, petrochemically derived polyethylene or food crop-derived ethanol) is the major aim.

There is, however, a strong Circular Bioeconomy component in the facilities listed in Table 12.1 and a radically different approach, i.e., product *replacement* in a sustainable bio-based economy has become an important factor in leveraging investment and (inter)national funding for biorefinery R&D. This new approach foresees the introduction of biopolymers to replace synthetic plastics and of biochemical building blocks as the starting points for novel consumer chemicals. The intellectual underpinning for this is that plant and microbial biology already provides chemicals that can be marketed as consumer products, for example biosurfactants (Daniel et al., 1998; Müller et al., 2012; Varvaresou and Iakovou, 2015) or (more generally) functional molecules that would be difficult to synthesize chemically from petrochemicals derivatives (Clark et al., 2012; Dusselier et al., 2014).

Such a product-replacing process is run by Avantium YXY Fuels & Chemicals (The Netherlands) to produce furan 2,5-dicarboxylic acid (FDCA) as the precursor for a recyclable polymer (polyethylene-furanoate) to replace polyethylene terephthalate in drinks bottles and synthetic fibers (www.avantium.com). FDCA is one of a family of compounds derived from 5-hydroxymethylfurfural (HMF), a dehydration product of fructose formed in acidic conditions; other related products include the potential liquid biofuels 2-methylfuran and 2,5-dimethylfuran (Román-Leshkov et al., 2007).

HMF is a well-characterized by-product of the thermochemical processing of lignocellulosic biomass (Zhang et al., 2015). A flavin-dependent oxidase catalyzes the conversion of HMF to FDCA in four consecutive oxidative steps at high yield and at ambient temperature and pressure (Dijkman et al., 2014). Any biological source of glucose can, however, be used as a feedstock for FDCA production if glucose isomerase

(EC 5.3.1.9) is used to catalyze the isomerization of glucose to fructose (Schmidt and Dauenhauer, 2007).

Avantium also offers levulinic acid, with FDCA an entry in a small but growing list of biochemical building blocks viewed as potential replacements for petrochemicals derivatives in polymers and/or novel biopolymers (Adkins et al., 2012; Erickson et al., 2012):

- (C3) glycerol, 3-hydroxypropionic acid, lactic acid, 1,3-propanediol
- (C4) succinic acid, fumaric acid, malic acid, aspartic acid, 3-hydroxybutyrolactone, 3-hydroxybutyric acid, 4-hydroxybutyric acid, 1,3-butanediol, 2,3-butanediol, putrescine
- (C5) FDCA, levulinic acid, gluconic acid, itaconic acid, xylitol, arabinitol, 3-hydroxyvaleric acid, 5-aminovaleric acid, cadaverine
- (C6) sorbitol, glucaric acid, adipic acid, 1,6-hexanediamine, 6-aminocaproic acid
- (C8) styrene, 4-hydroxystyrene

Some of these compounds have independent uses without polymerization (for example, xylitol as a zero-calorie sweetener), some are products of metabolic engineering for novel pathways and most have been commercialized or are close to small-scale commercialization (Choi et al., 2015).

Glycerol is a particularly versatile chemical and has global markets in personal care products, foods and beverages and in pharmaceuticals (Erickson et al., 2012). Chemocatalytic routes for the conversion of glycerol into citric acid, lactic acid, 1,3-dihydroxyacetone, 1,3-propanediol, dichloro-2-propanol, acetaldehyde, acrolein and formaldehyde are known (Pathak et al., 2010; Bagheri et al., 2015). Glycerol is a carbon source for many microbes and could support potential niche markets for the much researched but commercially unsuccessful hydroxyalkanoate biopolymers (Canadas et al., 2014).

Similarly, succinic acid (1,4-butanedicarboxylic acid) finds applications in polyurethanes, paints and coatings, adhesives, sealants, artificial leathers, food additives, cosmetics and personal care products, biodegradable plastics, nylons, industrial lubricants, phthalate-free plasticizers, pigments, pharmaceuticals, and metal plating (www.bio-amber.com; www.myriant.com).

Industrial-scale production of succinic acid (produced chemically from butane-1,4-diol or directly from ethylene by carbonylation) by fermentation was achieved after 2000 to facilitate future commercialization (Jansen and van Gulik, 2014). Different microbial (predominantly bacterial) host platforms have been developed, including *E.coli* (Agarwal et al., 2006), *Actinobacillus succinogenes* (van Heerden and Nicol, 2013), *Anaerobiospirillum succiniciproducens* (Lee et al. 2000), *Mannheimia succiniciproducens* (Lee et al., 2002) and also yeast species (Ito et al., 2014; Jost et al., 2015). Succinity GmbH is a joint venture between BASF (Germany) and Corbion (The Netherlands) to commercialize bio-based

succinic acid and polybutylene succinate as a copolymer of butan-1,4-diol and succinic acid (www.succinity.com).

A defining feature of a biorefinery—as the term is understood in the context of the sustainable bio-economy (van Ree and van Zealand, 2014)—is production of market-ready chemicals from non-food resources (Lynd et al., 1999; Zeng and Biebl, 2002).

Even more important for feedstock supply is the capability of producing microbial species in fermentations to utilize a variety of substrates, both during the year (seasonal availability) and from year to year (fluctuations in crop yield). An example of the intensity with which this issue is being experimentally investigated is given by that of biologically derived succinic acid, which can be produced from a wide range of non-food biomass streams (Table 12.2).

Large-scale biorefineries are estimated to require start-up costs in excess of $500 million in North America or approximately €500 million in Europe (Brown et al., 2013; Lane, 2015; Rizwan et al., 2015). It is not surprising, therefore, that funding applications for pilot and demonstration facilities have relied heavily on Circular Economy arguments to partially offset financial considerations.

Central to "circular" economic modeling is the minimization of non-renewable energy use and greenhouse gas emissions in life cycle analyses, i.e., from "cradle-to-factory gate" (Subhadra, 2010; Sacramento-Rivero, 2012; Cok et al., 2014). Techno-economic analysis of a conceptual biorefinery based on these principles and using wood chips, acid hydrolysis to release a hexose/pentose sugar stream, fermentation to ethanol, 45% utilization of lignin residues for heat and power generation by combustion, and 55% for the extraction of phenolic residues claimed approximately 90% reductions in fossil fuel usage and greenhouse gas emissions; the conventional reference system was oil-based gasoline and phenols as petrochemicals derivatives with heavy oil fractions and natural gas supplying heat and power (Jungmeier et al., 2013).

The ability to sell excess power and electricity to national grids has been an important feature of economics analyses of such biorefineries; if, for example, all the lignin by-product of woody biomass chemical processing could be sold at its energy price, the minimum economic selling price of the ethanol output would be reduced by $0.50 per gallon (Klein-Marcuschamer et al., 2011). Significantly, however, the total energy demand for this type of biorefinery model was 46% higher than in the fossil fuel reference model (Jungmeier et al., 2013). The extra energy demand was envisaged to be mostly (>90%) met by biomass resources and for this reason the feasibility of the provision of sufficiently large amounts of biomass for biorefinery operations has become a significant issue in the bioenergy and biorefinery debate, particularly for land requirements and transportation costs (Argo et al., 2013; Lin et al., 2013).

Based on these arguments and the examples in Table 12.1, a generic outline scheme for the biorefinery of the Circular Bioeconomy is depicted in Figure 12.3.

TABLE 12.2

Non-Food Crop Feedstocks Demonstrated to be Utilizable for the Microbial Bio-Production of Succinic Acid

Feedstock	Producing Species	Reference
Cane molasses, corn steep liquor	E. coli	Agarwal et al. (2006) J Appl Microbiol 100:1348–1354
Industrial hemp	Actinobacillus succinogenes	Gunnarsson et al. (2015) Bioresour Technol 182:58–66.
Cassava bagasse hydrolysate	Corynebacterium glutamicum	Shi et al. (2014) Bioresour Technol 174:190–197.
Wood hydrolysate	Mannheimia succiniciproducens	Kim et al. (2004) Enz Microb Technol 35:648–53.
Carob pods	Actinobacillus succinogenes	Carvalho et al. (2014) Bioresour Technol 170:491–498.
Restaurant food waste	E. coli	Sun et al. (2014) Appl Biochem Biotechnol 174:1822–1833
Sugarcane bagasse hydrolysate	E. coli	Liu et al. (2013) Bioresour Technol 149:84–89.
Corn cob hydrolysate	Corynebacterium glutamicum	Wang et al. (2014) Appl Biochem Biotechnol 172:340–350.
Rice husk pyrolysis oil	E. coli	Wang et al. (2013). Biotechnol Biofuels 6:74.
Distillers' grain from spirits	Actinobacillus succinogenes	Zhou and Zheng (2013) Biotechnol Lett 35:679–684.
Beechwood xylan hydrolysate	E. coli	Zheng et al. (2012). Microb Cell Fact 11:37.
Orange peel, wheat straw	Fibrobacter succinogenes	Li et al. (2010) Appl Microbiol Biotechnol 88:671–678.
Wheat bran	Actinobacillus succinogenes	Dorado et al. (2009) J Biotechnol 143:51–9.
Rice straw	Actinobacillus succinogenes	Zheng et al. (2009) Bioresour Technol 100:2425–2429.
Cheese whey	Actinobacillus succinogenes	Wan et al. (2008) Appl Biochem Biotechnol 145:111–119.
Waste bread	Actinobacillus succinogenes	Leung et al. (2012). Biochem Eng J 65: 10–15.
Rapeseed meal	Actinobacillus succinogenes	Chen et al. (2011). Enzyme Microb Technol 48:339–344.
Arundo donax	Basfia succiniciproducens	Cimini et al. (2016) Bioresour Technol 222:355–360.
Biodiesel glycerol coproduct	Yarrowia lipolytica	Gao et al. (2016) Biotechnol Biofuels 9:179.
Agave tequilana bagasse	Actinobacillus succinogenes	Corona-González et al. (2016) Bioresour Technol 205:15–23
Red alga hydrolysate	E. coli	Olajuyin et al. (2016) Bioresour Technol 214:653–659
Corn stover hydrolysate	Basfia succiniciproducens	Salvachúa et al. (2016). Bioresour Technol 214:558–566
Duckweed hydrolysate	Actinobacillus succinogenes	Shen J et al. (2018). Bioresour Technol 250:35–42.

The central features are:

1. The facility processes lignocellulosic biomass gathered locally to minimize fossil fuel use in transportation (Kim and Dale, 2015)
2. Biomass processing yields sugar streams that are separated into fermentations and/or chemocatalytic routes to biofuels and building block biochemicals.
3. By-products of economic value (for example, animal feeds) may be economically important to the overall scheme.
4. Lignin is dried and combusted as a source of heat and power for the facility and, if energy is

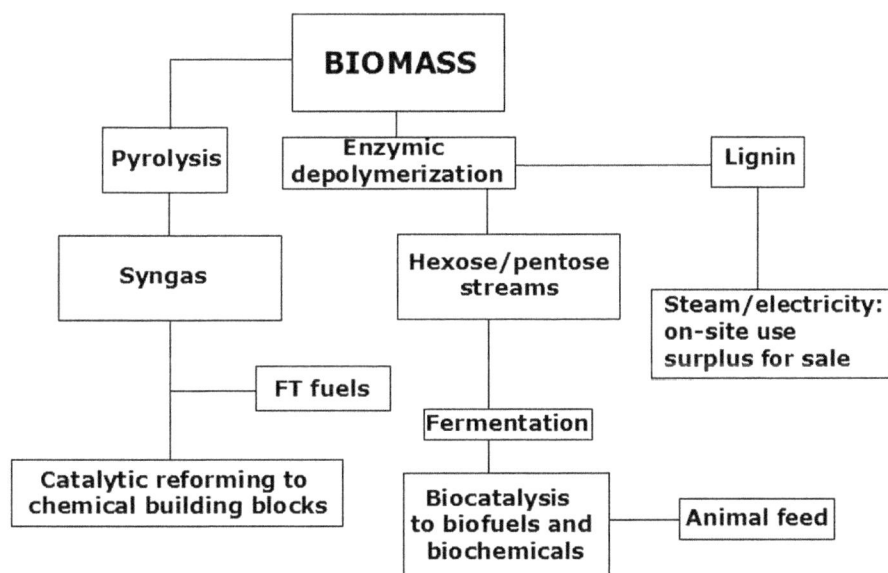

FIGURE 12.3 Conceptual biorefinery scheme based on plant biomass processed thermochemically or enzymically to yield biofuels, biochemicals, feed coproducts, and heat/power.

generated in excess, for sale to local or national utilities.

The major difference between this and the chemical (linear) biorefinery in Figure 12.2 is the use of a non-food renewable feedstock; any number of microbial fermentations or chemical conversions of the sucrose, glucose, and fructose from refined sugarcane and from sugarcane molasses could be incorporated into Figure 12.2.

Additionally, the generation of a range of outputs more chemically complex than hydrocarbons or their simple derivatives is important. The choice of how many outputs actually used is inevitably an economic one because even for simple products such as succinic acid downstream processing costs in a fermentation facility account for >50% of the total operating costs (Chen and Patel, 2012). In practical terms, a multiplicity of outputs is desirable to even out fluctuations in market prices for individual items but this flexibility implies a much greater capital expenditure.

As with linear biorefineries aiming at product substitution, costs have proved to be the major obstacle to developing renewable biomass-to-chemicals biorefineries. Using cellulosic ethanol as a benchmark, large units (>5 × 10⁷ gallons annual output) require a capital investment of $7–10 per gallon installed capacity (Chen and Patel, 2012; Lane, 2015) whereas small demonstration units suffer severely from the negative effects of scale: $55–73 per gallon installed capacity (100 ton per day dry biomass, thermochemical process) or $82–163 per gallon installed capacity at 10 ton per day with a biological process (Dutta et al., 2011). Moreover, the implementation of any terrestrial biorefinery strategy on an industrial scale ignores the highly contentious issue of how much biomass could sustainably be generated using short growth cycle trees or other bioenergy crops in the face of soil erosion, plant productivity limits, water supply, fossil fuel usage, and nutrient supply (Patzek and Pimentel, 2005).

12.3 SECOND-GENERATION BIOREFINERIES: CHEMICALS AND BIOENERGY

12.3.1 MODELS FOR ADVANCED BIOREFINERIES

The major economic challenges for biorefineries beyond the demonstration and pilot scales have resulted in three radically different models for biorefineries. The first is driven by the aim to decarbonize the global economy for environmental reasons (McGlade and Ekins, 2014, 2015; International Energy Agency, 2016) and addresses the question of the complete replacement of petrofuels and petrochemical feedstock from renewable plant biomass sources. The second looks to tap the vast resources of marine resources to overcome the logistical hurdles of transporting large but unpredictable quantities of biomass from primary producing zones to centralized terrestrial biorefinery sites (Baghel et al., 2015). The third seeks to directly utilize CO_2 in the atmosphere or from industrial processes as the carbon substrate.

12.3.2 THE PETROCHEMICAL MOLECULAR WORLD

It is a commonplace assertion that petrochemicals underlie much of the materials of modern life, from detergents, chewing gum and engine lubricants to textiles, organic chemicals, synthetic rubber, vehicle fuels, printing inks and unbreakable glass. This vast range of products contributes to an estimated worldwide annual production of 4×10^8 tons of petrochemicals derivatives (Bender, 2014). The scale of these figures provides a useful benchmark for the worldwide numerical scales of production future biorefineries will be required to meet as industrial realities.

Vacuum fractional distillation of crude oil yields the different fractions with increasing number of carbon atoms per molecule that make up, successively, light alkanes and olefin gases (C1–C5), naphtha (C6–C11), gasoline (C7–C9), kerosene/paraffin oil (C11–C18), diesel oils (C11–C18), lubrication oil (C18–C25), and fuel oil (C20-C27). Subsequent steam cracking and catalytic reforming generate the small-molecule "platform" petrochemicals:

- (C1) methane (also from natural gas) and, by steam reforming, methanol
- (C2) ethylene (ethene)
- (C3) propylene (propene)
- (C4) the "raffinate" stream of n-butenes, isobutene and butadiene
- (C6-C8) aromatics, mostly benzene, toluene, and xylenes.

From these simple hydrocarbons and aromatics, a group of 20–30 derivatives form the precursors for a wide spectrum of addition and condensation polymers and resins (including synthetic rubbers, polyvinylchloride plastics, nylon, polyester fabrics, polyurethane foams, solvents, plasticizers, fuel ether (methyl*tert*butylether), and numerous chemical intermediates) (Figure 12.4).

The scientific and technical challenge is to elucidate biotechnological, enzymic, chemocatalytic or thermochemical routes from biomass to substitute for crude oil and natural gas as feedstock.

A natural starting is ethanol, which can be catalytically dehydrated to yield not only ethylene but a range of higher hydrocarbons (Costa et al., 1985; Gayubo et al., 2001). A more important primary chemical intermediate is syngas, which can be made by pyrolytic treatments of plant material and is composed of a CO, H_2, and CO_2 in varying proportions (Asadullah et al., 2002; Lv et al., 2007). Biomass-derived syngas has proven to be a highly versatile substrate. One route to upgrading the gas mixture is via the Fischer-Tropsch (FT) reaction with iron or cobalt-based catalysts:

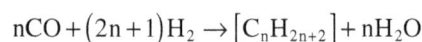

$$nCO + (2n+1)H_2 \rightarrow [C_nH_{2n+2}] + nH_2O$$

The resulting hydrocarbons (represented in the above equation as alkanes) can include the various fuel fractions derived from crude oil, i.e., gasoline, diesel, and kerosene, and so

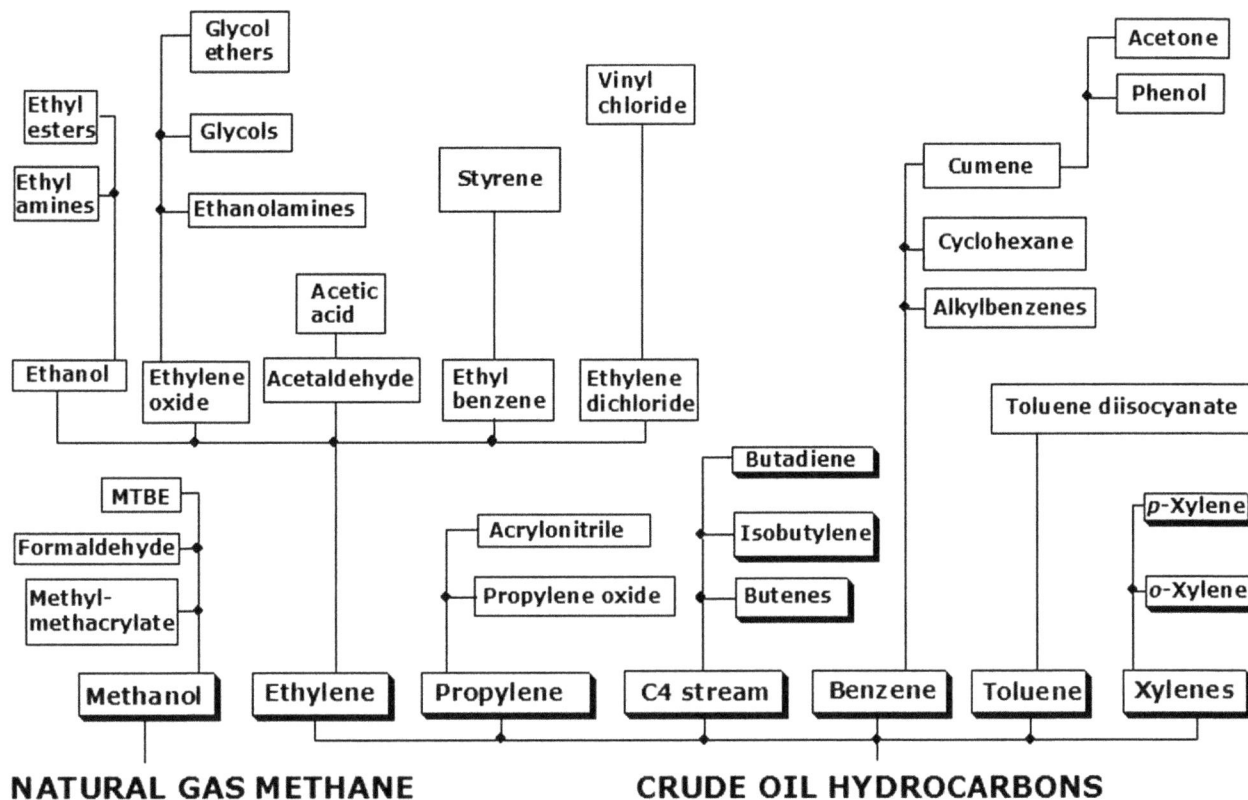

FIGURE 12.4 Petrochemicals derivatives in a conventional petro-refinery: Formation of methanol and C_2-C_8 alkenes and aromatics as chemical building blocks and monomers.

on (Schulz, 1999; Diehl and Khodakov, 2009). Any biomass feedstock is in principle acceptable, although process economics will determine the practical choices made for commercial large-scale processes (Baliban et al., 2013a,b; Petersen et al., 2015).

Pyrolysis of plant biomass also yields solid charcoal ("biochar") and liquid products conventionally described as "bio-oil." The Catalytically upgrading bio-oil in the vapor phase at 500°C uses zeolite (HZSM-5) and, for higher conversion efficiencies, modified zeolite (Ni/HZSM-5) catalysts; products with high hydrocarbon contents were generated (Cheng et al., 2015); this processing is upstream of subsequent fractionation steps to generate hydrocarbon fuel grades. Similarly, the major means of processing lignocellulosic biomass for paper making is Kraft pulping, in which alkali treatment in the presence of a Na_2S catalyst separates the lignin and hemicellulose from cellulose and yields "black liquor" as a source of semi-purified lignin by precipitation. Softwood Kraft lignin can be converted at 600°C in the presence of zeolite catalysts by removal of the hydroxyl and carboxylic acid functionalities into a gasoline-compatible product, i.e., a mixture of mostly hexanes, heptanes, and octanes (Ben and Ragauskas, 2011).

The trihydric alcohol glycerol is an obligatory by-product of biodiesel manufacture as glycerol tri-esters are transesterified with methanol to yield methyl esters of fatty acids (Pagliaro et al., 2007). In industrial biotechnology, microbial fermentation bioprocesses yield glycerol directly from glucose at up to 50% carbon conversion efficiency (Overkamp et al., 2002). Via syngas and Fischer-Tropsch conversion, octanes can be produced as liquid biofuels from glycerol (Soares et al., 2006).

12.3.2.1 Ethylene

The hybrid biotechnological-chemocatalytic route via ethanol and the dehydration of ethanol has been discussed in section 12.1.2. To make "bioethylene" a genuine backstop technology—a technology substituting a product from an exhaustible resource by using more abundant alternative production inputs—would require improvements in its cost competitiveness and advances in three areas of bioengineering:

1. The development of more efficient catalysts for the dehydration of ethanol to ethylene with enhanced stability, the ability to catalyze the endothermic dehydration reaction at a lower temperature and to exhibit greater selectivity over the formation of higher hydrocarbons (Chen et al., 2007; Bi et al., 2010; Matachowski et al., 2012).

2. Improved energy efficiency in ethanol-to-ethylene production sites (Arvidsson and Lundin, 2011).

3. Integration of more advanced forms of ethanol production, for example using algal biotechnology with its more area-intensive route to ethanol in photobioreactors sited on non-arable land (Li et al., 2014; Singh and Singh, 2014; Wijffels et al., 2014).

The initial form in which ethanol in produced by microbial fermentation is a dilute aqueous solution (approximately 10% v/v) and distillation only yields an azeotropic mixture with an irreducible minimum water content of 4.4% (v/v). Transforming hydrous ethanol to anhydrous ethanol is a significant energy and cost burden; catalysts capable of dehydrating hydrous ethanol would be an advantage and several catalysts are known to function in the presence of water (Phillips and Datta, 1997; Gurgul et al., 2011). The detailed mechanistic pathway of the dehydration reaction may even differ in the presence of water (Meeprasert et al., 2009).

A direct thermochemical transformation of biomass to ethylene offers an interesting alternative. Ethylene, methane and ethane are minor constituents in the syngas production of CO and H_2 mixtures but their concentrations increase at higher treatment temperatures (Di Blasi et al., 1999). Techno-economic evaluation of European biomass (wood chip) gasifiers for bio-synthetic natural gas has led to the conclusion that in some process variants more energy is present in light olefins and aromatics than in CO and H_2 and that, with large gasifiers, cryogenic separation of ethylene would be economically possible (Rabou and van der Drift, 2011). More ambitiously, an outline design in 2011 envisioned a 7.5×10^5 ton per year wood chip gasification facility with a 24% mass conversion to ethylene and propylene (Gay et al., 2011). This process design extrapolated known technologies for syngas production, catalytic conversion to dimethylether and methanol and onward to yield light olefins (Jin et al., 2004; Hu et al., 2005; Zhao et al., 2006).

Some bacterial species also form ethylene from a carboxylic acid and an amino acid with a complex stoichiometry (Eckert et al., 2014):

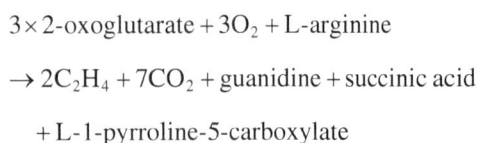

$$3 \times 2\text{-oxoglutarate} + 3O_2 + \text{L-arginine}$$

$$\rightarrow 2C_2H_4 + 7CO_2 + \text{guanidine} + \text{succinic acid}$$

$$+ \text{L-1-pyrroline-5-carboxylate}$$

Both 2-oxoglutarate and L-arginine are prominent in central metabolism but the mechanism of the ethylene-forming enzyme (EFE) and its regulation are speculative. Sequence analyses of EFE and EFE homolog genes in *Pseudomonas syringae*, *Penicillium digitatum* and *P. chrysogenum* position EFE in a subclass of 2-oxoglutarate/Fe(II) dependent dioxygenases (Johansson et al., 2014). That EFE is a single protein encoded by a unique gene suggests that it represents a highly manipulatable target for genetic engineering, metabolic engineering, and overexpression in easily cultivated microorganisms. An alternative platform has been achieved by expressing higher plant EFEs in the cyanobacterium *Synechoccus elongatus*; using this photosynthetic microbe, bioethylene could potentially be produced from atmospheric CO_2 or waste CO_2 streams (Jindou et al., 2014).

12.3.2.2 Methane and Methanol

The biochemical C1 pool entry point of methane (Figure 12.4) is by far the easiest petrochemical feedstock compound to substitute because of the near omnipresence of anaerobic digesters in the global ecosystem. Consortia of hydrolyzing, digestive, acetogenic, homoacetogenic, sulfate-reducing and methanogenic bacteria reduce biomass residues and food, animal and human waste products to the cheapest and most universally available biofuel, biogas, containing up to 75% methane (Weimer et al., 2009). The mixed cultures of methanogens present can accept many different substrates, including carboxylic acids, organic amines and mercaptans, CO and H_2 to generate biogas. If impurities are removed from the methane during upgrading—up to 50% of biogas may be CO_2, up to 7% water, with much smaller amounts of H_2, O_2, N_2, CO, and H_2S—the resulting gaseous product "(biomethane") is fully compatible with conventional natural gas (IEA Bioenergy, 2014).

The European Union has the installed capacity to generate over 500 million m^3 of biomethane annually, although this represents only 0.3% of natural gas imports. An alternative route for biomethane production is from biomass gasification but this has not yet achieved full commercialization (da Costa Gomez, 2013). The major practical obstacle to biomethane in industrial syntheses is that biogas and biomethane are too attractive as primary fuels and their value as sources of renewable carbon is outweighed by their ease of use as local fuels, transportation fuels and to synergize with established pipelines of natural gas (Schobert, 2013; Batstone and Virdis, 2014). Nevertheless, the potential of biomethane as a substitute petrochemical feedstock is evident and, from relatively low-technology roots, methane by anaerobic digestion has moved into contemporary bioindustry as microwave pretreatment to improve biomethane production kinetics has extended this process to more resilient plant biomass feedstocks (www.advancedmicrowavetechnologies.com).

The first true petrochemicals derivative derived from methane is methanol (Figure 12.4). The current global production capacity for methanol is approximately 1×10^8 tons and natural gas is the dominant feedstock (www.methanol.org). That syngas mixtures containing CO, CO_2 and H_2 can be processed to methanol has been known and exploited industrially since 1913 (Ren et al., 2003):

$$CO + 2H_2 \rightarrow CH_3OH$$

Early processes used ZnO catalysts at 400°C and various formulas and structures are known, for example $Cu/ZnO/Al_2O_3$, $Cu/ZnO/Cr_2O_3$ and $CuZnOZrO_2$ (Yang et al., 2006). Direct reduction of CO_2 (inevitably present in raw syngas) to methanol can also be performed using (among others) NiGa and Ni_5Ga_3 (Sharafutdinov et al., 2014), Pd_4/In_2O_3(Ye et al., 2014) and Mo_6S_8 (Liu et al., 2010) catalysts.

A techno-economic analysis of biomass-to-methanol technologies showed that feedstock costs were crucial for commercialization, representing up to 62% of production costs (Hamelinck and Faaij, 2002). Start-up companies using waste streams have been announced or have started operating in Europe and Canada: BioMCN (using crude glycerol from biodiesel manufacture, www.biomcn.eu), VärmlandsMetanol AB (forestry residues, www.varmlandsmetanol.se) and Enerkem (non-recyclable municipal solid waste, http://enerkem.com).

12.3.2.3 Propylene

Interconversion of propylene and ethylene/but-2-ene were first used commercially in the 1960s and can be run in reverse for the metathesis reaction to form propylene (Banks and Kukes, 1985; Mol, 1999). More recently, direct production of propylene from ethylene has been achieved and may involve a similar sequence of reactions: dimerization of ethylene to but-1-ene, isomerization to but-2-ene and the metathesis step (C2 + C4 → 2 x C3) to propylene (Iwamoto, 2011)

Given a biological (fermentation) route via ethanol to ethylene, "green polypropylene" is a straightforward extrapolation and Braskem in Brazil has installed such a production unit (www.braskem.com). Catalytic conversion of methanol to propylene is also an established industrial process (Patcas, 2005; Olsbye et al., 2012). In contrast, gasification of biomass to syngas and thence to propylene is an experimental approach (Gay et al., 2011). Another demonstrated chemical route is by the hydro-deoxygenation of glycerol generated as a by-product of biodiesel production (Zacharopoulou et al., 2015).

Two biotechnological routes have also been explored, one using a propionic acid fermentation with a *Propionibacterium* and subsequent chemical hydrogenation and dehydration (Rodriguez et al., 2014), a second with an engineered *Escherichia coli* producing propan-2-ol with a high molar yield for subsequent dehydration to propylene (Hanai et al., 2007).

12.3.2.4 C4 Units

Biomass-derived ethylene can be transformed to butenes in catalyzed gas-phase reactions (Al-Sa'doun, 1993; Scholz et al., 2014). Since facile routes to the formation of butenes from butanols are well known (West et al., 2009; Zhang et al., 2010; Choi et al., 2013), a direct biological route can use the fermentation of sugars or other carbon sources from biomass to support fermentations for the production of *n*-butanol (butan-1-ol) as the butene precursor.

Fermentations to produce *n*-butanol originally used strictly anaerobic clostridial species; this had the disadvantage of multiple products, including acetone, ethanol, and carboxylic acids (Holt et al., 1988). Genetic engineering and synthetic biology have subsequently moved the pathway to other and more industrially friendly microbes, notably the yeast *Saccharomyces cerevisiae* (Krivoruchko et al., 2013; Sakuragi et al., 2015) and *E. coli* (Atsumi et al., 2008). Nevertheless, the highest titers are obtained with Clostridia and these bacteria can biomanufacture butanol from a wide range of feedstocks, including agricultural residues, the thin stillage resulting from ethanol fermentations, domestic organic waste, cheese whey and pulping liquors (Claassen et al., 2000; Chen et al., 2013; Komonkiat and Cheirsilp, 2013; Shukor et al., 2014; Becerra et al., 2015; Kudahettige-Nilsson et al., 2015; Su et al., 2015). For these reasons, European Union policy makers actively promote biobutanol as an advanced biofuel.

*Iso*butanol (2-methylpropan-1-ol) has long been known as a minor by-product of potable ethanol fermentations as a component of "fusel oils" but metabolic engineering has produced microbial strains with much higher production rates for *iso*butanol: *E. coli* (Smith and Liao, 2011), *S. cerevisiae* (Matsuda et al., 2013; Park et al., 2014), *Corynebacterium glutamicum* (Smith et al., 2010; Yamamoto, et al. 2013), *Clostridium* spp. (Higashide et al., 2011; Gak et al., 2014), a photosynthetic unicellular alga (Li et al., 2014), *Bacillus subtilis* (Qi et al., 2014), and the metal-reducing bacterium *Shewanella oneidensis* (Jeon et al., 2015).

*Iso*butanol is dehydrated to isobutylene (*iso*butene) over alumina catalysts (Taylor et al., 2010). Direct fermentative production of isobutylene was first demonstrated 40 years ago (van Leeuwen et al., 2012) and the whole-cell biotransformation by *E. coli* of 3-hydroxy-3-methylbutyrate employing a heterologously expressed yeast diphosphomevalonate decarboxylase gene product has been achieved (Gogerty and Bobik, 2010).

Biobutanol production began to be implemented by retrofitting North American ethanol units in 2013 by ButamaxTM Advanced Biofuels (www.butamax.com). Gevo (www.gevo.com) announced its first sales of gasoline blended with *iso*butanol as a fuel oxygenate in 2015. A proposal for a biorefinery producing liquid fuels, solvents and a diesel additive with *iso*butanol as the platform chemical appeared in 2014 (Posada Duque et al., 2014).

2,3-Butanediol is converted to 1,3-butadiene by Sc_2O_3 or Sc_2O_3/Al_2O_3 at 320-410 °C (Duan et al., 2014) The direct microbial production of 2,3-butanediol is an area of intense research effort upscaling highly productive bioprocesses using yeast (Bao et al., 2015) and bacteria (Adlakha and Yazdani, 2015; Bai et al., 2015; Ji et al., 2015; Lee et al., 2015).

12.3.2.5 Benzenoid Aromatics

Higher plant biomass sources—in particular, lignocellulosics—possess abundant stores of benzenoid metabolites and polymeric structures derived from aromatic precursors. Lignin is the second most abundant natural polymer in the biosphere (second only to cellulose) but, like most biosynthesized aromatic structures, it is invariably more oxidized than the simple aromatics in petrochemicals derivatives (Shahzadi et al., 2014). Chemical processing of lignocellulosic biomass via pyrolysis followed by hydroprocessing with zeolite catalysts produces simple aromatics (C6-C8) and simple alkenes (C2–C4) at over 60% conversion (Vispute et al., 2010).

Benzene, toluene and xylenes are the principal aromatics in the upgraded pyrolysis oil, reducing the pore size of the catalyst by the chemical liquid deposition of silica results in a greater selectivity for *p*-xylene as the end product (Cheng et al., 2012). Cycloaddition of biomass-derived ethylene with dimethylfuran (formed from fructose by dehydration) and subsequent product dehydration yields *p*-xylene with high selectivity using a zeolite catalyst at 300°C (Williams et al., 2012).

Anellotech (http://anellotech.com/) has developed a technology to process wood, sawdust, corn stover or sugar cane bagasse to a mixture of benzene, toluene, and xylenes as a drop-in product for further separation in petrochemical infrastructures.

12.3.2.6 Complete Substitution of Petrochemical Derivatives

A complete portfolio of technologies—thermochemical, chemocatalytic and biotechnological—already exists for the production of the key petrochemicals derivatives from renewable biomass sources (Figure 12.5).

Start-up companies and new ventures are seeking funding for this "green" landscape of bulk chemicals. The processes are linear in design with an accessible biomass resource stream and with the commercial intention to provide drop-in substitutes for existing petrochemicals derivatives. Implicitly, the mingling of conventional and renewables streams is anticipated to gradually implement this product substitution in the coming decades; the established approach for this is the hydrotreatment of vegetable oils for combining with heavy gas-oil mixtures in conventional oil refineries (Huber et al., 2007; Mikkonen, 2008; Hilbers et al., 2015).

12.3.3 MARINE (ALGAL) BIOREFINERIES

The impetus to develop marine biorefineries based on microalgae was the high fat content of microalgal species, which was ideal for the bioproduction of the long-chain fatty acids for biodiesel (Singh and Gu, 2010; de Jaeger et al., 2014). It was rapidly appreciated that these species elaborated a range of compounds that could collectively form the fine chemical and biochemical outputs

of a coastal or estuarine biorefinery model (Ruiz et al., 2016). Among these higher-value products are:

- "Functional foods" such as carotenoids, β-glucans, and ω-3 and poly-unsaturated fatty acids
- Compounds with nutraceutical health-promoting properties, for example the carotenoid fuxcoxanthin (Hosokawa et al., 2010).
- Natural pigments (Gouveia et al., 2007).
- Antioxidants for cosmetics
- Protein fractions for hydrolysis as human animal and fish food supplements (Bleakley and Hayes, 2016).

A multi-valorization product chain could be envisaged in which multiple products could be flexibly scheduled to meet fluctuating market demand. Furthermore, two technical advantages of microalgal-based biorefineries have been heavily promoted and publicized:

1. Multiple growth environments on-site with widely differing technical requirements, the extremes being closed-loop photobioreactors and open-pond systems.
2. The ability to utilize waste CO_2 streams from, for example, geographically proximate cement factories, as well as atmospheric CO_2.

Algal biorefineries have been intensively funded by the European Commission's R&D agencies. Illustrative examples of such projects are:

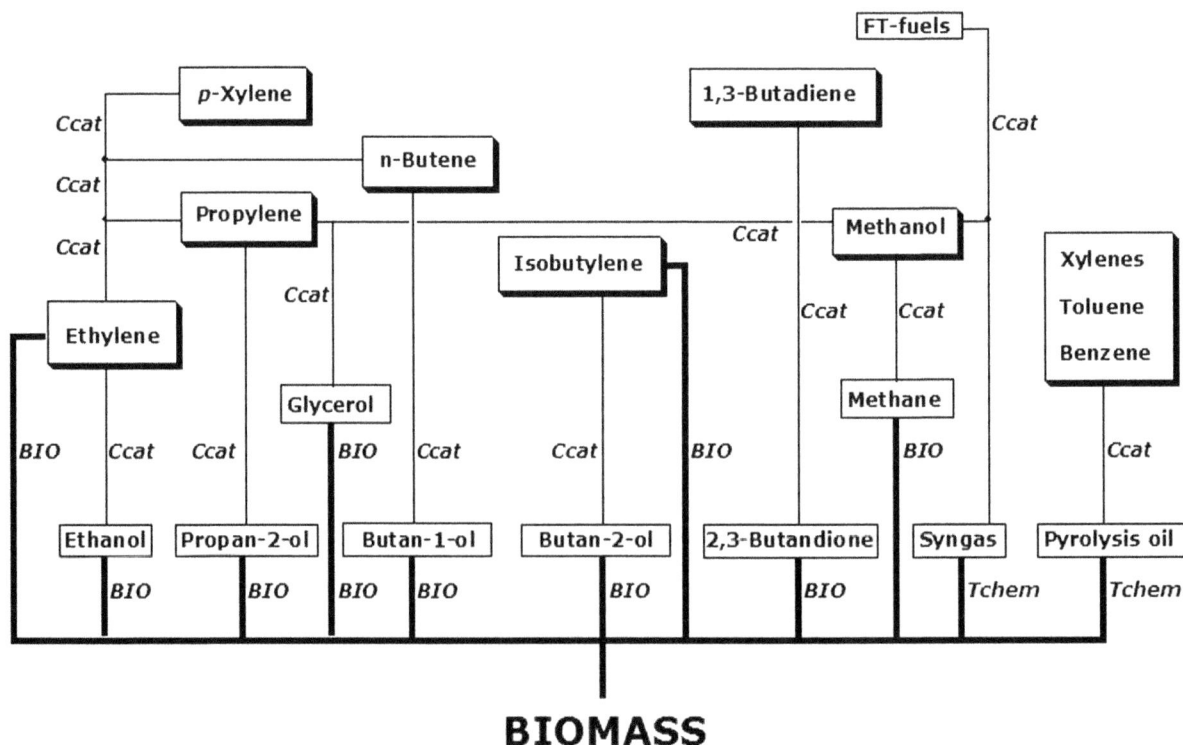

FIGURE 12.5 Portfolio of biocatalytic (*BIO*), thermochemical (*Tchem*) and chemocatalytic (*Ccat*) routes demonstrated for the formation from biomass of C1-C8 chemical building blocks as drop-in products for synthetic chemistry.

- Bisigodos (www.bisigodos.eu); high added-value chemicals and bioresins from algal biorefineries produced from CO_2 provided by industrial emissions (cement, steel factory, thermal power plants, etc.); products include amino acids, novel anti-corrosive coatings, surfactants, and biobased resins for aqueous inks.
- D-Factory (www.d-factoryalgae.eu); food and feed supplements and additives (colorants, sources of nutritional antioxidants and pro-Vitamin A, fiber, protein, and starch-rich extracts) and pharmaceuticals and cosmeceuticals (bioactives, excipients, and emulsifiers).
- Miracles (http://miraclesproject.eu); specialty products from microalgae for application in food, aquaculture, and non-food products.
- Marisurf (http://www.marisurf.eu); marine microbes as sources of novel surfactant materials,

These projects include life cycle analysis projections and other actions to minimize carbon emissions to harmonize with Circular Bioeconomy principles. Major targets include reducing unfavorable economics for biorefinery processes and adapting downstream processing to multiple (both small-molecule and macromolecular) products.

Macroalgae are less suitable as biorefinery producing organisms because of their cellular compositions and relatively slow growth rates but could be platforms for locally based production of biofertilizers, biofuels and methane by anaerobic digestion (Ingle et al., 2018).

12.3.4 CO_2 as the Ultimate Biorefinery Substrate

Atmospheric CO_2 remains the most abundant gaseous carbon source available to photosynthetic and non-photosynthetic microbes. Many organisms on Earth—including *Homo sapiens*—can "fix" CO_2 in single enzyme-catalyzed reactions but only photosynthetic organisms and some highly autotrophic microbial species can genuinely conserve the fixed carbon in their cellular biochemistry. Enzymes such as pyruvate carboxylase (E.C. 6.4.1.1) are essential in humans to replenish the intermediates of the tricarboxylic acid (citric) cycle used for the biosynthesis of amino acids, etc., the "anaplerotic" pathways, which are combined with "cataplerotic" reactions (the extraction of central metabolites for synthetic pathways) in gluconeogenesis, fatty acid synthesis and glycerogenesis (Owen et al., 2002).

Electrochemical reduction of CO_2 into hydrocarbons (predominantly methane and ethylene) with copper electrodes has been demonstrated (Peterson et al., 2010). However, non-photosynthetic microbial species have been recognized that can directly assimilate syngas components (CO, CO_2 and H_2) and form compounds of higher value to the biosphere (Wilkins and Atiyeh, 2011; Daniell et al., 2012; Ramió-Pujol et al., 2015; Phillips et al., 2017). This remarkable group of microbes utilizes the Wood-Ljungdahl pathway with CO and CO_2 as the carbon substrates (Figure 12.6).

Bifunctional carbon monoxide dehydrogenase acetyl-CoA synthase (CODH-ACS) combines a CO with a methyl group-carrying cofactor to generate acetyl-coenzyme A (acetyl-CoA); reducing power is generated from H_2 and CO oxidation steps. From acetyl-CoA, a range of C2, C3 and C4 compounds can be synthesized to feed into general metabolic biochemistry (Liew et al., 2016). With the genetic manipulation of acetogenic bacteria being possible, investigation of a wider spectrum of value-added products is now a significant R&D sector (Daniell et al., 2016).

Novel microbial bioreactor systems have been developed to synthesize ethanol from syngas (Handler et al., 2015) and the process is being commercialized (www.lanzatech.com). A European Union-funded project involving academic

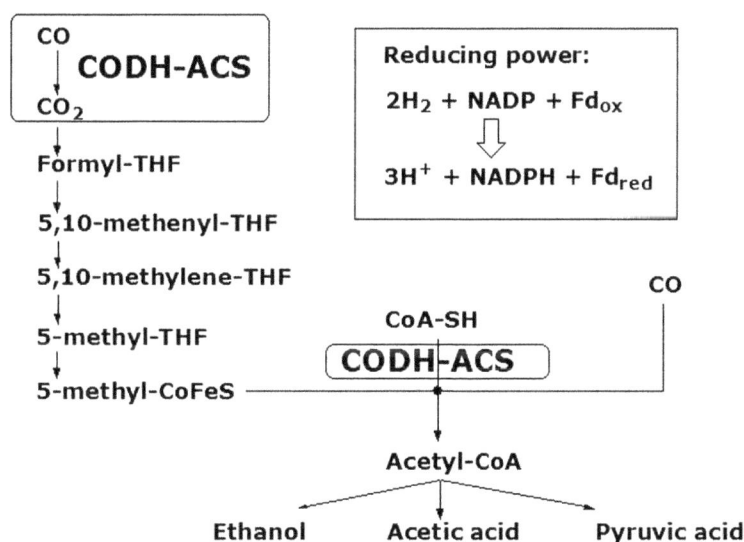

FIGURE 12.6 Wood-Ljungdahl pathway of syngas gas utilization by the acetogen *Clostridium autoethanogenum*. CODH-ACS, carbon monoxide dehydrogenase acetyl-CoA; THF, tetrahydrofolate; CoFeS, corrinoid iron-sulfur-containing protein; Fd, ferredoxin. Redrawn from Liew et al. (2016).

researchers and private industry is CELBICON (cost-effective CO_2 conversion into chemicals via combination of Capture, ELectrochemical and BIochemical CONversion technologies) aiming to use atmospheric CO_2 via electrochemically generated syngas to support microbial fermentations for polyhydroxyalkanoates and methane while a second electrochemical conversion of CO_2 to C1 compounds (i.e., entirely bypassing the need for the biology of the Wood-Ljungdahl pathway) provides substrates for the fermentative production of lactic acid, isoprene and mono-terpenoids (www.celbicon. org).

Heterotrophic microbial fixation of CO_2 has been demonstrated or postulated with various cyclic combinations of mono- and dicarboxylic acids (Menendez et al., 1999; Huber et al., 2008; Hawkins et al., 2015). Figure 12.7 outlines one of these cyclic pathways, with seven C3 and eight C5 intermediates. Included in this molecular shuffling with its twin carboxylation steps are four examples of "biobuilding block" biochemicals: succinic, malic, fumaric, and 4-hydroxybutyric acids. Metabolic engineering could, with appropriate use of anaplerotic pathways, guide different major outputs from the same basic genetic toolbox.

12.4 SECONDARY METABOLISM IN THE BIOREFINERY CONCEPT

What role do secondary metabolites play in the biorefinery? The "traditional" biomanufacturing sector (White

Biotechnology) will be substantially unchanged and focused on highly regulated fermentations and cell culture bioprocesses for antibiotics and recombinant proteins. Enzymes, amino acids, animal growth promotors, carboxylic acids and vitamins will be major markets while specialized biotechnologies for the production of novel bioactive and newly discovered enzymes with desirable traits for the synthetic chemist will continue to be researched.

The biorefinery offers, however, a base for a broad range of microbial processes exemplified by biosurfactants (Müller et al., 2012) or L-hexoses and other "rare" sugars (Izumori, 2002). While commercial scale production remains elusive, yields and productivities will be sidelined but these issues will eventually need to be addressed and the problems of process scale-up and optimization that beset all industrial-scale fermentations will be multiply rediscovered in facilities without the central R&D resources of multinational chemical and pharmaceutical companies. In the biorefinery era, therefore, generic solutions to microbial physiology in stirred-tank reactors will be increasingly attractive and two major areas of attention are growth control and productivity management.

12.4.1 GROWTH CONTROL AND MANAGEMENT

Microbiologists have long appreciated the much higher growth rates for microbial cultures attainable in "complex" media, i.e., those replete with the preformed building blocks of amino

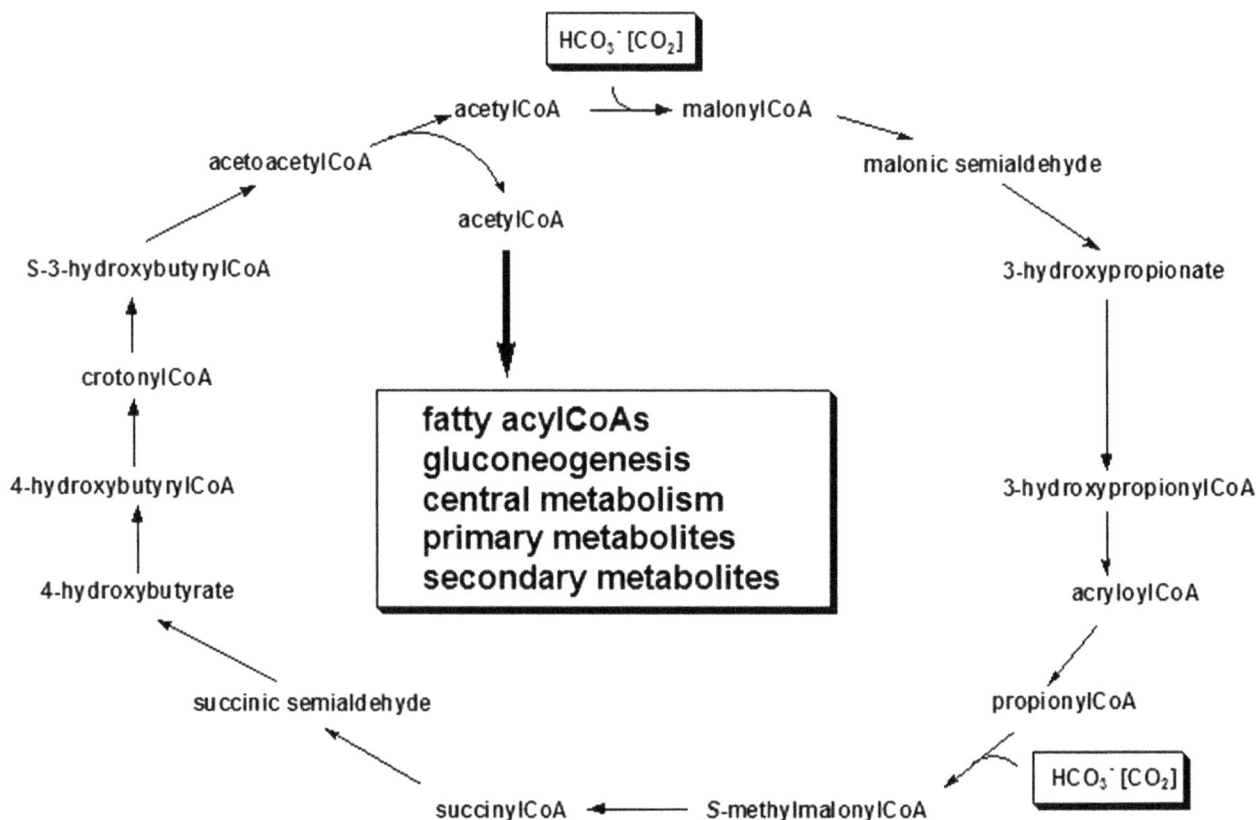

FIGURE 12.7 A heterotrophic reaction cycle for the assimilation of CO_2 via carboxylation reactions to form C2 precursors for cell growth and product formation. Redrawn from Huber et al. (2008).

acids, peptides, and vitamins, over the defined media which are more useful for laboratory studies. Understanding how precise is the eventual limitation to rapid growth in complex media in industrial fermentations has been speculative but secondary metabolite fermentations have the crucial underlying tendency to balance a phase of rapid growth with a subsequent deceleration of growth that initiates secondary product formation. The starting batched medium has, therefore, an inbuilt factor whose exhaustion limits the phase of most rapid growth in the early stages of the fermentation. An example from an industrial process is show in Figure 12.8.

In this case, a protein mixture supplied the nitrogen sources required for cell multiplication; detailed modeling of known cellular constituents, including amino acids required for growth and amino acids supplied (as protein) in the medium (Figure 12.9) showed that the critical transition was that when the amino acids required for further rapid growth could no longer be supplied by the medium and could only be formed by de novo biosynthesis in the producing organism.

This represented the primary signal that triggered the change to slower growth rates and increasing rates of product formation (a polyketide) that occurred reproducibly at 16–17 hours into the fermentation (Figure 12.8). Surprisingly, the amino acid limitation was most evident with the "common" amino acids, alanine, aspartic acid, and serine rather than the metabolically more "costly" amino acids to biosynthesize (for example, aromatic amino acids and tryptophan).

In future biorefineries, nitrogen sources may be locally sourced (for example, plant seed proteins) and their exact compositions will be variable. Adapting fermentation media for seasonal campaigns of production will require a focused and detailed analysis of growth kinetics.

12.4.2 Productivity and the Role of Artificial Intelligence

The large number of biochemical pathways required for the assimilation of preformed nutrients and the biosynthesis of others constitute the essential primary routes involved in cellular growth and division. Once secondary product formation has commenced, however, a much-restricted selection is made from the microbial genome; "crucial metabolism" is the summation of these comparatively few genes and pathways that focus metabolic flux into secondary product formation.

The analysis of gene expression in transcriptomics is the contemporary approach to understanding the temporal sequence of gene expression during the transition from unproductive rapid growth to productive modest (or zero) growth in a microbial fermentation (Yang et al., 2015). Most importantly, the aim is to explain—and (optimistically) to actively manage—productivity, which typically follows a sigmoidal progress curve with time-limited peak productivity. Broadly, three main scenarios can be encountered in industrial practice:

1. Maximal productivity is associated with a critical range of specific growth rates—this can defined only by chemostat studies in the laboratory—and pathway gene expression is maximal for only a restricted time within the process (but could be lengthened by judicious feeding of the growth-limiting substrate).

2. Maximum productivity is restricted by an inappropriate feeding strategy; an example could be the use of a reducing sugar assay for glucose fed in a molasses mixture that also contains non-utilizable di- and tri-saccharides that accumulate inside the fermentor and, in effect, shut down feeding of glucose that could still be metabolized.

3. Alternative pathways begin to be expressed during the middle-to-later stages of the period of highest productivity. This is the most difficult option to identify and requires detailed metabolic analysis. Figure 12.10 illustrates a polyketide fermentation where competing pathways to previously unrecognized products became quantitatively important at the time when peak productivity was achieved and subsequently lost.

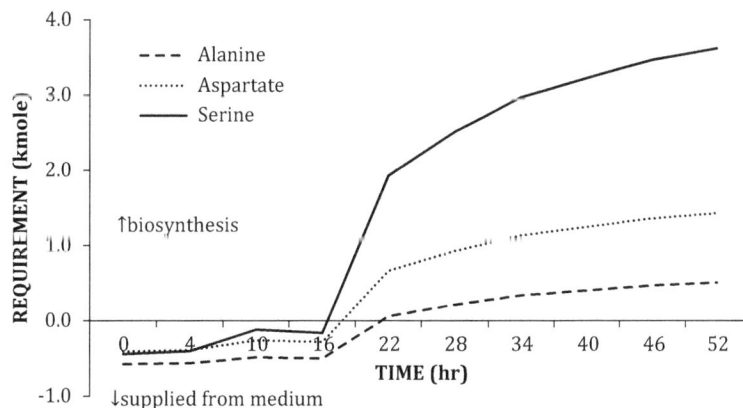

FIGURE 12.8 Growth limitation imposed by availability of amino acids in a medium for the industrial-scale fermentation of a polyketide secondary metabolite.

triglycerides, oleate

FERMENTATION
BROTH

thr   leu   lys   gln   glu

CELL

arg

teichoic
acid   0.012   fatty acids

cell
protein   0.192   lys

thr   0.206

teichoic   leu   0.306   lys

acid   cell
protein

0.409   cell
protein   0.156   gln

glycerol   cell lipids   glu   0.159
0.024   peptidoglycan   0.111   cell
protein

cell wall polymers   SMP   0.003   cell polymers   cell lipids
0.492   0.459   0.734

arg   0.145   cell
protein

pro   pro

0.174

phe,   phe,
tyr   tyr

ile   ile

carbohydrates   Glucose 6P   GAP   pyruvate/PEP   AcCoA   CIT   2-oxoglutarate

DNA   RNA   CO2   CO2

0.197   cell
protein

peptidoglycan   0.209   OAA   SMP   CO2   0.329

cell lipids   SUCCoA

cell
protein   DNA   cell
protein   0.557   ala   DNA   0.058   FUM

0.329   ser   0.189   RNA   0.552   asp   asn

gly   CO2   0.024   0.054   cell
protein   0.163   0.125   0.162   val   val

0.107   trp, his   0.069   0.091   peptidoglycan   cell
protein

peptidoglycan   RNA   2.531   cys, met   phe, tyr
0.098   0.066

gly   ser   ala   asp   asn

trp, his   cys, met   phe, tyr

SMP = small metabolite pool (intracellular)   2.531: data units mmol (g dry mass)$^{-1}$

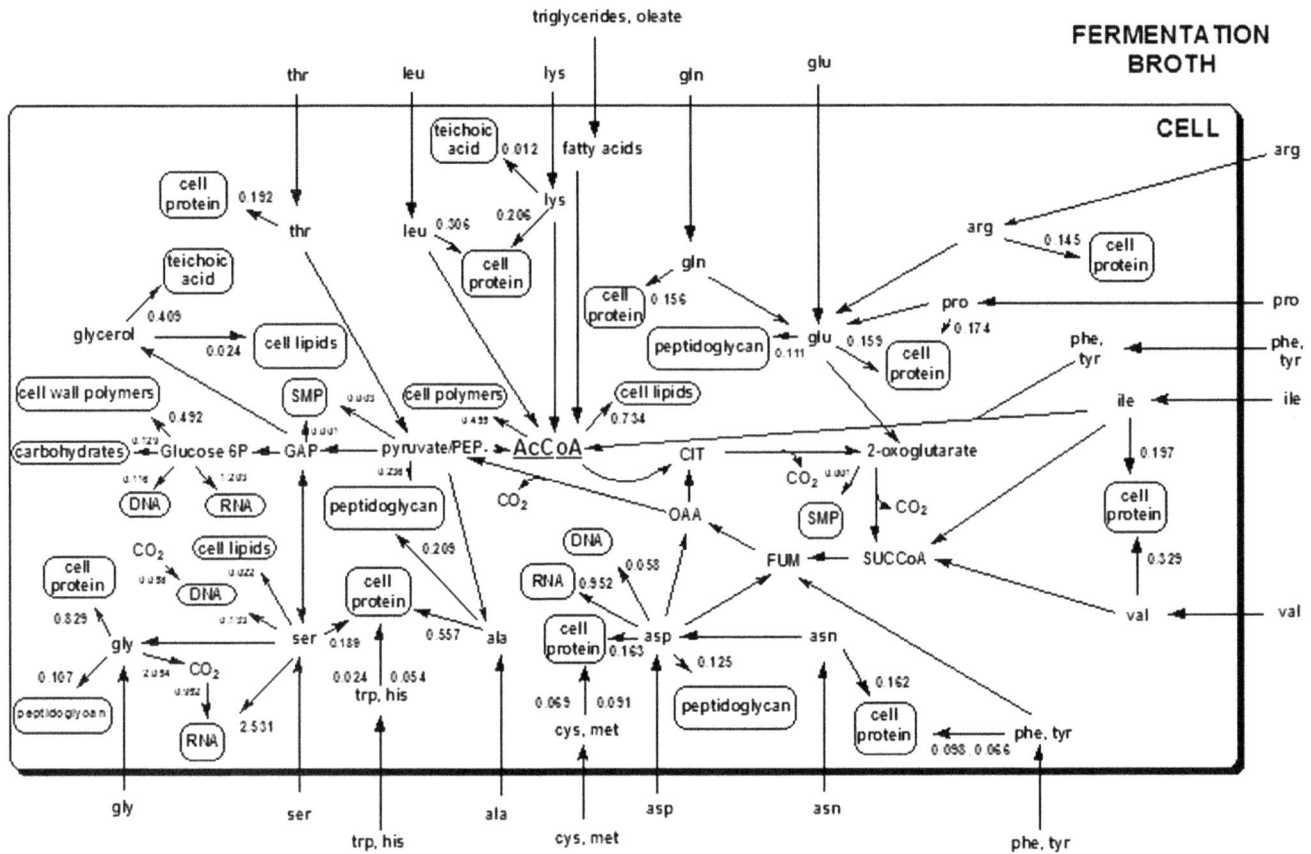

FIGURE 12.9   Growth limitation imposed by availability of amino acids in a medium for the industrial-scale fermentation of a polyketide secondary metabolite.

Industrial biomanufacturing processes generate very large amounts of primary data, in excess of 1×10^6 data points over the course of 4–7 days: pH, temperature airflow, back pressure, concentrations in exit gas streams, and so on. This is a Big Data exercise that is highly amenable to machine learning but has received remarkably little attention from biochemical engineers (Steyer et al., 1993). With Fog Computing able to link continuous data streams to both local servers and distant consolidated (Cloud Computing) resources (Bonomi et al., 2014), the in-depth analysis of massive quantities of collected on-line information and of the relationships they reveal to key events in a fermentation detected by in-line or *in situ* sensors will become a more routine example of data processing.

SUMMARY

This chapter has traced technological roadmaps for industrial-scale biorefineries that build on established pilot and demonstration facilities or which envisage radically different approaches to extensive product substitution and replacement from biomass or CO_2 as the carbon substrates.

Linear biorefineries with relatively little emphasis placed on biomass resource choices (i.e., food or non-food crops) are already functioning with the explicit aim of drop-in products for established petrochemicals derivatives markets. More advanced sites with entirely renewal/sustainable feedstock and products that are chemically more complex and oxygenated than hydrocarbons are rare. Neither model is likely to prosper in a climate of low crude oil prices that reduce or even eliminate financial returns on investment (Mousdale, 2014; Brownbridge et al., 2014). Carbon taxation, national subsidies for bio-commodity chemical production to support public acceptance of "green" alternatives in the face of negative effects on gross domestic products, public-private investment strategies for biorefinery construction, tax incentives, and a 30–50 year commitment to technology development for the decarbonization of world energy markets may all be mandatory to foster entrepreneurial confidence and installation (Luderer et al., 2009; Jenkins, 2014).

Central to any analysis of biorefineries is the concept of "backstop" technologies that will become more affordable as the costs of the non-renewable resources themselves become unaffordable. Physical renewable energy sources provide an exemplary case of such backstops coming to maturity in the combination of fossil reserves that are perceived as either exhaustible or environmentally unusable or too expensive because of some form of carbon taxation added to production

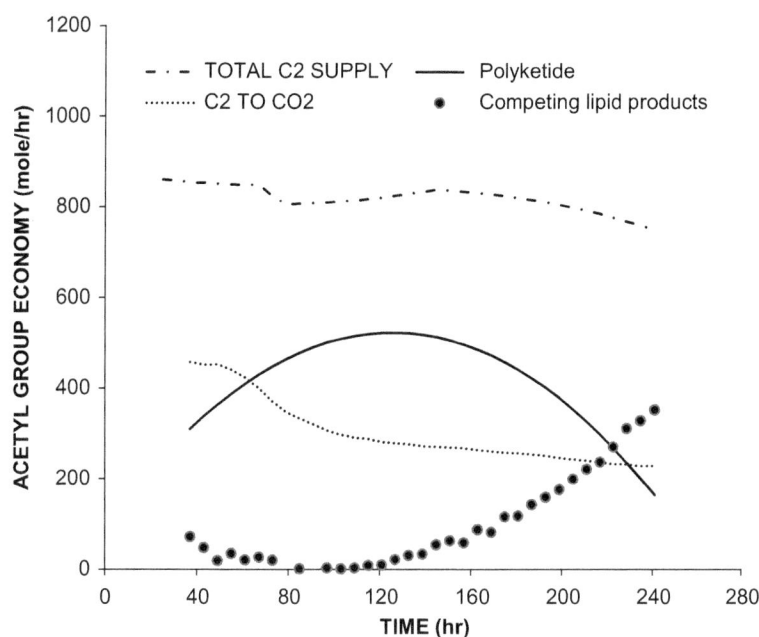

FIGURE 12.10 Metabolic incorporations of acetyl (C2) units into a polyketide, CO_2 and competing lipid products.

and distribution costs. The combination of biotechnological, thermochemical, and chemocatalytic developments presented in this review support the credibility of backstop technologies for carbon product substitution or replacement.

Increasing numbers of novel fermentations for secondary products are likely with an expansion of biorefinery operations. The concept of "crucial metabolism" is proposed to identify the major quantitative pathways for carbon and other fluxes during product formation and their relationship to transcriptomics (gene expression) and growth kinetics.

Despite major uncertainties regarding future conventional fuel costs and security of supply, accelerated R&D efforts to bring existing and future demonstrable biomass-to-fuels and biomass-to-commodity chemicals can—to optimally follow a rapid approach path model (Tsur and Zemel, 2003)—bring new production processes to commercial viability in a timescale consistent with a major deceleration in fossil reserve extraction and use (McGlade and Ekins, 2014, 2015).

Interested readers can follow developments in the innovative aspects of the various technologies in biorefineries via the private-company and funded project websites included in this text, which provide the means of continuously updating key areas of scientific knowledge and their applications in biorefinery operations. To what extent these transform the industrial world, however, is not a scientific or technological question but is determined by geopolitical change: if the Paris Climate Accord of 2015 is to be ignored (or abandoned) in a continuing era of cheap oil from both from conventional oil fields and shale oil reserves, biorefinery operations will service only niche markets—such as biofuels and biosurfactants—and specialist commercial companies and funded research programs will remain on the fringes of 21st century society.

REFERENCES

Adkins. J., S. Pugh, R. McKenna and D.R. Nielsen. 2012. Engineering microbial chemical factories to produce renewable "biomonomers." *Front Microbiol* 3:313.

Adlakha, N. and S.S. Yazdani. 2015. Efficient production of (R,R)-2,3-butanediol from cellulosic hydrolysate using *Paenibacillus polymyxa* ICGEB2008. *J Ind Microbiol Biotechnol* 42:21–28.

Agarwal. L., J. Isar, G.K. Meghwanshi and R.K. Saxena. 2006. A cost effective fermentative production of succinic acid from cane molasses and corn steep liquor by *Escherichia coli*. *J Appl Microbiol* 100:1348–1354.

Al-Sa'doun, A.W. 1993. Dimerization of ethylene to butene-1 catalyzed by Ti(OR')4-AlR3. *App Catal A: Gen* 105:1–40.

Argo, A.M., E.C.D. Tan, D. Inman, M.H. Langholtz, L.M. Eaton, J.J. Jacobson, C.T. Wright, D.J. Muth, M.M. Wu, Y.W. Chiu and R.L. Graham. 2013. Investigation of biochemical biorefinery sizing and environmental sustainability impacts for conventional bale system and advanced uniform biomass logistics designs. *Biofuels Bioprod Biorefin* 7:282–302.

Aristidou, A.A., N. Baghaei-Yazdi, M. Javed and B.S. Hartley. 2012. Conversion of renewable resources to biofuels and fine chemicals: current trends and future prospects, in El-Mansi, E.M.T., Bryce, C.F.A., Dahlou, B., Sanchez, S., Demain, A.L. and Allman, A.R. (eds.) *Fermentation Microbiology and Biotechnology*, 3rd ed., Boca Raton, New York and London: CRC Press, 225–61.

Arvidsson, M. and B. Lundin. 2011. Process integration study of a biorefinery producing ethylene from lignocellulosic feedstock for a chemical cluster. http://publications.lib.chalmers.se/records/fulltext/140886.pdf.

Asadullah, M., S.I. Ito, K. Kunimori, M. Yamada and K. Tomishige. 2002. Biomass gasification to hydrogen and syngas at low temperature: Novel catalytic system using fluidized-bed reactor. *J Catal* 208:255–259.

Atsumi, S., A.F. Cann, M.R. Connor, C.R. Shen, K.M. Smith, M.P. Brynildsen, K.J. Chou, T. Hanai and J.C. Liao. 2008. Metabolic engineering of *Escherichia coli* for 1-butanol production. *Metab Eng* 10:305–311.

Baghel, R.S., N. Trivedi, V. Gupta, A. Neori, C.R.K. Reddy, A. Lalid and B.B. Jhaab. (2015). Biorefining of marine macroalgal biomass for production of biofuel and commodity chemicals. *Green Chem* 17:2436–2443.

Bagheri, S., N.M. Julkapli and W.A. Yehye. 2015. Catalytic conversion of biodiesel derived raw glycerol to value added products. *Renew Sust Energ Rev* 41:113–127.

Bai, F., L. Dai, J. Fan, N. Truong, B. Rao, L. Zhang and Y. Shen. 2015. Engineered *Serratia marcescens* for efficient (3*R*)-acetoin and (2*R*,3*R*)-2,3-butanediol production. *J Ind Microbiol Biotechnol* 42:779–786.

Baliban, R.C., J.A. Elia and C.A. Floudas. 2013a. Biomass to liquid transportation fuels (BTL) systems: process synthesis and global optimization framework. *Energy Environ Sci* 6:267–287.

Baliban, R.C., J.A Elia, C.A. Floudas, X. Xiao, Z. Zhang, J. Li, H. Cao, J. Ma, Y. Qiao and X. Hu. 2013b. Thermochemical conversion of duckweed biomass to gasoline, diesel, and jet fuel: Process synthesis and global optimization. *Ind Eng Chem Res* 52:11436–11450.

Banks, R.L. and S.G. Kukes. 1985. New developments and concepts in enhancing activities of heterogeneous metathesis catalysts. *J Mol Catal* 28:117–131.

Bao, T., X. Zhang, X. Zhao, Z. Rao, T. Yang and S. Yang. 2015. Regulation of the NADH pool and NADH/NADPH ratio redistributes acetoin and 2,3-butanediol proportion in *Bacillus subtilis*. *Biotechnol J* 10:1298–1306.

Batstone DJ. and B. Virdis. 2014. The role of anaerobic digestion in the emerging energy economy. *Curr Opin Biotechnol* 27:142–9.

Becerra, M., M.E. Cerdán and M.I. González-Siso. 2015. Biobutanol from cheese whey. *Microb Cell Fact* 14:27.

Ben H. and A.J. Ragauskas. 2011. NMR characterization of pyrolysis oils from Kraft lignin. *Energy Fuels* 25:2322–2332.

Bender, M. 2014. An overview of industrial process for the production of olefins – C4 hydrocarbons. *ChemBioEng Rev* 1:136–147.

Bi, J., X. Guo, M. Liu and X. Wang. 2010. High effective dehydration of bio-ethanol into ethylene over nanoscale HZSM-5 zeolite catalysts. *Catal Today* 149:143–147.

Bleakley, S. and M. Hayes. 2017. Algal proteins: Extraction, application, and challenges concerning production. *Foods* 6, 33.

Bonomi, F., R. Milito, P. Natarajan and J. Zhu. 2014. Fog computing: A platform for internet of things and analytics, in Bessis, N. and Dobre, C. (eds) *Big Data and Internet of Things: A Roadmap for Smart Environments*. Cham (Switzerland) Springer International, pp. 169–186.

BP plc. 2017. BP Statistical Review of World Energy June 2017. www.bp.com/content/dam/bp/en/corporate/pdf/energy-econo mics/statistical-review-2017/bp-statistical-review-of-world-energy-2017-full-report.pdf.

Brown, T.R., R. Thilakaratne, R.C. Brown and G. Hu. 2013. Regional differences in the economic feasibility of advanced biorefineries: Fast pyrolysis and hydroprocessing. *Energy Policy* 57:234–243.

Brownbridge, G., P. Azadi, A. Smallbone, A. Bhave, B. Taylor and M. Kraft. 2014. The future viability of algae-derived biodiesel under economic and technical uncertainties. *Bioresour Technol* 151:166–173.

Bruscino, M. 2009 Biorefineries: Fact or fiction? *Hydrocarbon Processing* 88(8):65–70.

Bungay, H.R. 1982. Biomass refining. *Science* 218:643–646.

Canadas, R.F., J.M.B.T. Cavalheiro, J.D.T Guerreiro, M.C., de Almeida, E. Polle, C.L da Silva, M.M. da Fonseca and F.C. Ferreira. 2014. Polyhydroxyalkanoates: waste glycerol upgrade into electrospun fibrous scaffolds for stem cells culture. *Int J Biol Macromol* 71:131–140.

Chen, G., S. Li, F. Jiao and Q. Yuan. 2007. Catalytic dehydration of bioethanol to ethylene over TiO_2/γ-Al_2O_3 catalysts in microchannel reactors. *Catal Today* 125:111–119.

Chen, G.Q. and M.K. Patel. 2012. Plastics derived from biological sources: Present and future: A technical and environmental review. *Chem Rev* 112:2082–2099.

Chen, W.H., Y.C. Chen and J.G. Lin. 2013. Evaluation of biobutanol production from non-pretreated rice straw hydrolysate under non-sterile environmental conditions. *Bioresour Technol* 135:262–268.

Cheng, Y.T., Z. Wang, C.J. Gilbert, W. Fan and G.W. Huber. 2012. Production of *p*-xylene from biomass by catalytic fast pyrolysis using ZSM-5 catalysts with reduced pore openings. *Angew Chem Int Ed* 51:11097–11100.

Cheng, Z., W. Lin, X. Zhao, Y. Huang, D. Raynie, C. Qiu, J. Kiratu and Y. Yu. 2015. Directly catalytic upgrading bio-oil vapor produced by prairie cordgrass pyrolysis over Ni/HZSM-5 using a two stage reactor. *AIMS Energy* 3:227–240.

Choi, H., J.H. Bae, D.H. Kim, Y.K. Park and J.K. Jeon. 2013. Butanol dehydration over V_2O_5-TiO_2/MCM-41 catalysts prepared via liquid phase atomic layer deposition. *Materials* 6:1718–1729.

Choi, S., C.W. Song, J.H. Shin and S.Y. Lee. 2015. Biorefineries for the production of top building block chemicals and their derivatives. *Metab Eng* 28:223–239.

Clark, J.H., R. Luque and A.S. Matharu. 2012. Green chemistry, biofuels, and biorefinery. *Ann Rev Chem Biomol Eng* 3:183–207.

Claassen, P.A.M., M.A.W. Budde and A.M. López-Contreras. 2000. Acetone, butanol and ethanol production from domestic organic waste by solventogenic clostridia. *J Mol Microbiol Biotechnol* 2:39–44.

Cok, B., I. Tsiropoulos, A.L. Roes and M.K. Patel. 2014. Succinic acid production derived from carbohydrates: An energy and greenhouse gas assessment of a platform chemical toward a bio-based economy. *Biofuels Bioprod Biorefin* 8:16–29.

Costa, E., A. Uguina, J. Aguado and P.J. Hernandez. 1985. Ethanol to gasoline process: Effect of variables, mechanism, and kinetics. *Ind Eng Chem Proc Design Dev* 24:239–244.

da Costa Gomez, C. 2013. Biogas as an energy option: An overview, in Wellinger, A., Murphy, J.P. and Baxter, D. (eds), *The Biogas Handbook: Science, Production and Applications*. Sawston: Woodhead Publishing, pp. 1–16.

D'Amato, D., N. Droste, B. Allen, M. Kettunen, K. Lähtinen, J. Korhonen, P. Leskinen, B.D. Matthies and A. Toppinen. 2017. Green, circular, bio economy: A comparative analysis of sustainability avenues. *J Cleaner Prod* 168:716–734.

Daniel, H.J., M. Reuss and C. Syldatk. 1998. Production of sophorolipids in high concentration from deproteinized whey and rapeseed oil in a two stage fed batch process using *Candida bombicola* ATCC 22214 and *Cryptococcus curvatus* ATCC 20509. *Biotechnol Lett* 20:1153–1156.

Daniell, J., M. Köpke and S. D. Simpson. 2012. Commercial biomass syngas fermentation. *Energies* 5:5372–5417.

Daniell, J., S. Nagaraju, F. Burton, M. Köpke and S.D. Simpson. 2016. Low-carbon fuel and chemical production by anaerobic gas fermentation. *Adv Biochem Eng Biotechnol* 156:293–321.

de Jaeger L., R.E.M Verbeek, R.B. Draaisma, D.E. Martens, J. Springer, G.G. Eggink and R.H. Wijffels. 2014. Superior triacylglycerol (TAG) accumulation in starchless mutants of *Scenedesmus obliquus*: (I) mutant generation and characterization. *Biotechnol Biofuels* 7:69.

Di Blasi, C., G. Signorelli, C. Di Russo and G. Rea. 1999 Product distribution from pyrolysis of wood and agricultural residues. *Ind Eng Chem Res* 38:2216–2224.

Diehl, F. and A.Y. Khodakov. 2009. Promotion of cobalt Fischer-Tropsch catalysts with noble metals: A review. *Oil Gas Sci Technol* 64:11–24.

Dijkman, W.P., D.E. Groothuis and M.W. Fraaije. 2014. Enzyme-catalyzed oxidation of 5-hydroxymethylfurfural to furan-2,5-dicarboxylic acid. *Angew Chem Int Ed* 53:6515–6518.

Di Lucia, L., S. Ahlgren and K. Ericsson. 2012. The dilemma of indirect land-use changes in EU biofuel policy – An empirical study of policy-making in the context of scientific uncertainty. *Environ Sci Policy* 16:9–19.

Duan, H., Y. Yamada and S. Sato. 2015. Efficient production of 1,3-butadiene in the catalytic dehydration of 2,3-butanediol. *Appl Catal A: Gen* 491:163–169.

Dusselier, M., M. Mascal and B.F. Sels. 2014. Top chemical opportunities from carbohydrate biomass: A chemist's view of the biorefinery. *Top Curr Chem* 353:1–40.

Dutta, A., M. Talmadge, J. Hensley, M. Worley, D. Dudgeon, D. Barton, P. Groenendijk, D. Ferrari, B. Stears, E.M. Searcy, C.T. Wright and J.R. Hess. 2011. Process design and economics for conversion of lignocellulosic biomass to ethanol. Thermochemical pathway by indirect gasification and mixed alcohol synthesis. www.nrel.gov/docs/fy11osti/51400.pdf.

Eckert, C., W. Xu, W. Xiong, S. Lynch, J. Ungerer, L. Tao, R. Gill, P.C. Maness and J. Yu. 2014. Ethylene-forming enzyme and bioethylene production. *Biotech Biofuels* 7:33.

Erickson, B., J.E. Nelson and P. Winters. 2012. Perspective on opportunities in industrial biotechnology in renewable chemicals. *Biotechnol J* 7:176–185.

Esposito, D. and M. Antonietti. 2015. Redefining biorefinery: The search for unconventional building blocks for materials. *Chem Soc Rev* 44:5821.

Fehrenbacher, K. 2016. Solazyme Ditches Biofuels (& Name) in a World of Cheap Oil. http://fortune.com/2016/03/16/solazyme-terravia-ditches-biofuels/.

Gak, E., M. Tyurin and M. Kiriukhi. 2014. Genome tailoring powered production of isobutanol in continuous CO_2/H_2 blend fermentation using engineered acetogen biocatalyst. *J Ind Microbiol Biotechnol* 41:763–781.

Gay, M., B. Pope and J. Wharton. 2011. Propylene from biomass. https://repository.upenn.edu/cgi/viewcontent.cgi?article=1028&context=cbe_sdr.

Gayubo, A.G., A.M. Tarrío, A.T. Aguayo, M. Olazar and J. Bilbao. 2001. Kinetic modelling of the transformation of aqueous ethanol into hydrocarbons on a HZSM-5 zeolite. *Ind Eng Chem Res* 40:3467–3474.

Gogerty, D.S. and T.A. Bobik. 2010. Formation of isobutene from 3-hydroxy-3-methylbutyrate by diphosphomevalonate decarboxylase. *Appl Environ Microbiol* 76:8004–8010.

Gouveia, L., B.P. Nobre, F.M. Marcelo, S. Mrejen, M.T. Cardoso, A.F. Palavra and R.L. Mendes. 2007. Functional food oil coloured by pigments extracted from microalgae with supercritical CO_2. *Food Chem* 101:717–723.

Gray, A. 2017. Countries are announcing plans to phase out petrol and diesel cars. Is yours on the list? www.weforum.org/agenda/2017/09/countries-are-announcing-plans-to-phase-out-petrol-and-diesel-cars-is-yours-on-the-list.

Grimaud, A., G. Lafforgue and B. Magné. 2011. Climate change mitigation options and directed technical change: A decentralized equilibrium analysis. *Res Energy Econ* 33:938–962.

Gurgul, J., M. Zimowska, D. Mucha, R.P. Socha and L. Matachowski. 2011. The influence of surface composition of $Ag_3PW_{12}O_{40}$ and $Ag_3PMo_{12}O_{40}$ salts on their catalytic activity in dehydration of ethanol. *J Mol Catal A: Chem* 351:1–10.

Hamelinck, C.N. and A.P.C. Faaij. 2002. Future prospects for production of methanol and hydrogen from biomass. *J Power Sources* 111:1–22.

Hanai, T., S. Atsumi and J.C. Liao. 2007. Engineered synthetic pathway for isopropanol production in *Escherichia coli*. *Appl Env Microbiol* 73:7814–7818.

Handler, R.M., D.R. Shonnard, E.M. Griffing, A. Lai and I. Palou-Rivera. 2015. Life cycle assessments of ethanol production via gas fermentation: Anticipated greenhouse gas emissions for cellulosic and waste gas feedstocks. *Ind Eng Chem Res* 55:3253–3261.

Hawkins, A.B., H. Lian, B.M. Zelde, A.J. Loder, G.L. Lipscomb, G.J. Schut, M.W. Keller, M.W.W. Adams and R.M. Kelly. 2015. Bioprocessing analysis of *Pyrococcus furiosus* strains engineered for CO_2-based 3-hydroxypropionate production. *Biotechnol Bioeng* 112:1533–1543.

Higashide, W., Y. Li, Y. Yang and J.C. Liao. 2011. Metabolic engineering of *Clostridium cellulolyticum* for production of isobutanol from cellulose. *Appl Environ Microbiol* 77:2727–2733.

Hilbers, T.J., L.M.J. Sprakel, L.B.J. van den Enk, B. Zaalberg, H. van den Berg and L.G.V. van der Ham. 2015. Green diesel from hydrotreated vegetable oil process design study. *Chem Eng Technol* 38:651–657.

Holt, R.A., A.J. Cairns and J.G. Morris. 1988. Production of butanol by *Clostridium puniceum* in batch and continuous culture. *Appl Microbiol Biotech* 27:319–324.

Hosokawa, M., T. Miyashita, S. Nishikawa, S. Emi, T. Tsukui, F. Beppu, T. Okada and K. Miyashita. 2010. Fucoxanthin regulates adipocytokine mRNA expression in white adipose tissue of diabetic/obese KK-Ay mice. *Arch Biochem Biophys* 504:17–25.

Hu, J., Y. Wang, C. Cao, D.C. Elliott, D.J. Stevens and J.F. White. 2005. Conversion of biomass syngas to DME using a microchannel reactor. *Ind Chem Eng* 44:1722–1727.

Huber, G.W., P. O'Connor and A. Corm. 2007. Processing biomass in conventional oil refineries: Production of high quality diesel by hydrotreating vegetable oils in heavy vacuum oil mixtures. *Appl Catal A: Gen* 329:120–129.

Huber, H., M. Gallenberger, U. Jahn, E. Eylert, I.A. Berg, D. Kockelkorn, W. Eisenreich and G. Fuchs. 2008. A dicarboxylate/4-hydroxybutyrate autotrophic carbon assimilation cycle in the hyperthermophilic Archaeum *Ignicoccus hospitalis*. *Proc Natl Acad Sci USA* 105:7851–7856.

International Energy Agency (2007). Biofuel Production (IEA Energy Technology Essentials, ETE02). www.iea.org/publications/freepublications/publication/essentials2.pdf.

International Energy Agency (2016). World Energy Outlook 2016. www.iea.org/media/publications/weo/WEO2016Chapter1.pdf.

IEA Bioenergy Task 40 and Task 37 (2014). Biomethane – status and factors affecting market development and trade. www.ieabioenergy.com/publications/biomethane-status-and-factors-affecting-market-development-and-trade/.

IEA Bioenergy. 2017. Annual report 2017. http://task42.ieabioenergy.com/wp-content/uploads/2018/04/IEA-Bioenergy-Annual-Report-2017.pdf.

Ingle, K., E. Vitkin, A. Robin, Z. Yakhini, D. Mishori and A. Golberg. 2018. Macroalgae biorefinery from *Kappaphycus alvarezii*: Conversion modeling and performance prediction for India and Philippines as examples. *BioEnergy Res* 11:22–32.

Ito, Y., T. Hirasawa and H. Shimizu. 2014. Metabolic engineering of *Saccharomyces cerevisiae* to improve succinic acid production based on metabolic profiling. *Biosci Biotechnol Biochem* 78:151–9.

Iwamoto, M. 2011. One step formation of propene from ethene or ethanol through metathesis on nickel ion-loaded silica. *Molecules* 16:7844–7863.

Izumori, K. 2002. Bioproduction strategies for rare hexose sugars. *Naturwissenschaften* 89:120–124.

Jansen, M.L. and V.M. van Gulik. 2014. Towards large scale fermentative production of succinic acid. *Curr Opin Biotechnol* 30:190–197.

Jenkins, J.D. 2014. Political economy constraints on carbon pricing policies: What are the implications for economic efficiency, environmental efficacy, and climate policy design? *Energy Policy* 69:467–477.

Jeon, J.M., H. Park, H.M. Seo, J.H. Kim, S.K. Bhatia, G. Sathiyanarayanan, H.S. Song, S.H. Park, K.Y. Choi, B.I. Sang and Y.H. Yang. 2015. Isobutanol production from an engineered *Shewanella oneidensis* MR-1. *Bioproc Biosyst Eng* 38:2147–2154.

Ji, X.J., L.G. Liu, M.Q. Shen, Z.K. Nie, Y.J. Tong and H. Huang. 2015. Constructing a synthetic metabolic pathway in *Escherichia coli* to produce the enantiomerically pure (R, R)-2,3-butanediol. *Biotechnol Bioeng* 112:1056–1059.

Jin, Y., S. Asaoka, X. Li, K. Asami and K. Fujimoto. 2004. Synthesis of liquefied petroleum gas via methanol and/or dimethyl ether from natural gas (part 1) Catalysts and reaction behaviors associated with methanol and/or dimethyl ether conversion. *J Japan Petrol Inst* 47:394–402.

Jindou, S., Y. Ito, N. Mito, K. Uematsu, A. Hosoda and H. Tamura. 2014. Engineered platform for bioethylene production by a cyanobacterium expressing a chimeric complex of plant enzymes. *ACS Synth Biol* 18:487–496.

Johansson, N., K.O. Persson, C. Larsson and J. Norbeck. 2014. Comparative sequence analysis and mutagenesis of ethylene forming enzyme (EFE) 2-oxoglutarate/Fe(II)-dependent dioxygenase homologs. *BMC Biochem* 15:22

Jost, B., M. Holz, A. Aurich, G. Barth, T. Bley and R.A. Müller. 2015. The influence of oxygen limitation for the production of succinic acid with recombinant strains of *Yarrowia lipolytica*. *Appl Microbiol Biotechnol* 99:1675–1686.

Jungmeier, G., M. Hingsamer and R. van Ree. 2013. Biofuel-driven Biorefineries. www.nachhaltigwirtschaften.at/resources/iea_pdf/iea_task_42_biofuel_driven_biorefineries_lr.pdf.

Kamm, B., M. Kamm, P.R. Gruber and S. Kromus. 2006. Biorefinery systems–an overview, in Kamm, B., Gruber, P.R. and Kamm, M. (eds.) *Biorefineries –Industrial Processes and Products*, Volume 1, Weinheim: Wiley-VCH, 1–40.

Kim, S. and D.E. Dale. 2015. All biomass is local: The cost, volume produced, and global warming impact of cellulosic biofuels depend strongly on logistics and local conditions. *Biofuels Bioprod Biorefin* 9:422–434.

Klein-Marcuschamer, D., B. A. Simmons and H.W. Blanch. 2011. Techno-economic analysis of a lignocellulosic ethanol biorefinery with ionic liquid pre-treatment. *Biofuels Bioprod Biorefin* 5:562–569.

Komonkiat, I. and B. Cheirsilp. 2013. Felled oil palm trunk as a renewable source for biobutanol production by *Clostridium* spp. *Bioresour Technol* 146:200–207.

Krivoruchko, A., C. Serrano-Amatriain, Y. Chen, V. Siewers and J. Nielsen. 2013. Improving biobutanol production in engineered *Saccharomyces cerevisiae* by manipulation of acetyl-CoA metabolism. *J Ind Microbiol Biotechnol* 40:1051–1056.

Kudahettige-Nilsson, R.L., J. Helmerius, R.T. Nilsson, M. Sjöblom, D.B. Hodge and U. Rova. 2015. Biobutanol production by *Clostridium acetobutylicum* using xylose recovered from birch Kraft black liquor. *Bioresour Technol* 176:71–79.

Lane, J. 2015. Biorefinery 2015 – transformations in biofuels costs, financing. www.biofuelsdigest.com/bdigest/2013/03/25/biorefinery-2015-transformations-in-biofuels-costs-financing.

Lapola, D.M., R. Schaldach, J. Alcamo, A. Bondeau, J. Koch, C. Koelking and J.A. Priesse. 2010. Indirect land-use changes can overcome carbon savings from biofuels in Brazil. *Proc Natl Acad Sci U S A* 107:3388–3393.

Lee, P.C., W.G. Lee, S.Y. Lee, H.N. Chang and Y.K. Chang. 2000. Fermentative fermentation of succinic acid from glucose and corn steep liquor by *Anaerobiospirillum succiniciproducens*. *Biotechnol Bioproc Eng* 5:379–381.

Lee, P.C., S.Y. Lee, S.H. Hong and H.N. Chang. 2002. Isolation and characterization of a new succinic acid-producing bacterium, *Mannheimia succiniciproducens* MBEL55E from bovine rumen. *Appl Microbiol Biotechnol* 58:663–668.

Lee, S., B. Kim, J. Yang, D. Jeong, S. Park and J. Lee. 2015. A non-pathogenic and optically high concentrated (R, R)-2,3-butanediol biosynthesizing *Klebsiella* strain. *J Biotechnol* 209:7–13.

Levy, P.F., J.E. Sanderson, R.G. Kispert and D.L. Wise. 1981. Biorefining of biomass to liquid fuels and organic chemicals. *Enz Microb Tech* 3:207–215.

Li. K., S. Liu and X. Liu. 2014. An overview of algae bioethanol production. *Int J Energ Res* 38:965–977.

Li, X., C.R. Shen and J.C. Liao. 2014. Isobutanol production as an alternative metabolic sink to rescue the growth deficiency of the glycogen mutant of *Synechococcus elongatus* PCC 7942. *Photosyn Res* 120:301–310.

Liew, F., M.E. Martin, R.C. Tappel, B.D. Heijstra, C. Mihalcea and M. Köpke. 2016. Gas fermentation—a flexible platform for commercial scale production of low-carbon-fuels and chemicals from waste and renewable feedstocks. *Front Microbiol* 7:694.

Lin, T., L.F. Rodríguez, Y.N. Shastri, A.C. Hansen and K.C. Ting. 2013. GIS-enabled biomass-ethanol supply chain optimization: model development and *Miscanthus* application. *Biofuels Bioprod Biorefin* 7:314–333.

Liu, P., Y.M. Choi, Y. Yang and M.G. White. 2010. Methanol synthesis from H_2 and CO_2 on a Mo_6S_8 cluster: A density functional study. *J Phys Chem A* 114:3888–3895.

Luderer, G., V. Bosetti, J. Steckel, H. Waisman, N. Bauer, E. Decian, M. Leimbach, O. Sassi and M. Tavoni. 2009. The Economics of Decarbonization – Results from the RECIPE model intercomparison. RECIPE Background Paper. www.pik-potsdam.de/recipe.

Lv, P., Z. Yuan, C. Wu, L. Ma, Y. Chen and N. Tsubaki. 2007. Biosyngas production from biomass catalytic gasification. *Energy Conver Man* 48:1132–1139.

Lynd, L.R., C.E. Wyman and T.U. Gerngross. 1999. Biocommodity engineering. *Biotechnol Prog* 15:777–793.

Marcucci, A. and P. Fragkos. 2015. Drivers of regional decarbonization through 2100: A multi-model decomposition analysis. *Energy Econ* 51:111–124.

Matachowski, L., M. Zimowska, D. Mucha and T. Machej. 2012. Ecofriendly production of ethylene by dehydration of ethanol over $Ag_3PW_{12}O_{40}$ salt in nitrogen and air atmospheres. *Appl Catal B:Environ* 123-124:448–456.

Mathews, J.A. 2014. *Greening of Capitalism: How Asia Is Driving the Next Great Transformation*, Redwood City: Stanford University Press.

Matsuda, F., J. Ishii, T. Kondo, K. Ida, H. Tezuka and A. Kondo. 2013. Increased isobutanol production in *Saccharomyces cerevisiae* by eliminating competing pathways and resolving cofactor imbalance. *Microb Cell Fact* 12:119.

McGlade, C. and P. Ekins. 2014. Un-burnable oil: An examination of oil resource utilisation in a decarbonised energy system. *Energy Policy* 64:102–112.

McGlade, C. and P. Ekins. 2015. The geographical distribution of fossil fuels unused when limiting global warming to 2°C. *Nature* 517:187–190.

Meadows, C.W., A. Kang and T.S. Lee. 2018. Metabolic engineering for advanced biofuels production and recent advances toward commercialization. *Biotechnol J* 13, 1600433.

Meeprasert, J., S. Choomwattana, P. Pantu and J. Limtrakul. 2009. Dehydration of ethanol into ethylene over H-MOR: A quantum chemical investigation of possible reaction mechanisms in the presence of water. *Technical Proceedings of the 2009 NSTI Nanotechnology Conference and Expo*, Houston, pp. 288–291.

Menendez, C., Z. Bauer, H. Huber, N. Gad'on, K.O. Stetter and G. Fuchs. 1999. Presence of acetyl coenzyme A (CoA) carboxylase and propionyl-CoA carboxylase in autotrophic *Crenarchaeota* and indication for operation of a 3-hydroxypropionate cycle in autotrophic carbon fixation. *J Bacteriol* 181:1088–1098.

Mikkonen, S. 2008. Second-generation renewable diesel offers advantages. *Hydrocarbon Processing* 87(2):63–66.

Mol, J.C. 1999. Olefin metathesis over supported rhenium oxide catalysts. *Catal Today* 51:289–299.

Mousdale, D. 2010. *Introduction to Biofuels*. Boca Raton: CRC Press, pp. 217–263.

Mousdale, D. 2014. Ten top indicators for liquid biofuel use in the next decade. *Biofuels Bioprod Biorefin* 8:302–305.

Müller, M.M., J.H. Kügler, M. Henkel, M. Gerlitzki, B. Hörmann, M. Pöhnlein, C. Syldatk and R. Hausmann. 2012. Rhamnolipids–next generation surfactants? *J Biotechnol* 162:366v80.

Organisation for Economic Co-operation and Development. 2006. *The Bioeconomy to 2030: Designing a Policy Agenda*. Paris: OECD Publications. www.oecd.org/sti/biotech/34823102.pdf.

Olsbye, U., S. Svelle, M. Bjørgen, P. Beato, T.V. Janssens, F. Joensen, S. Bordiga and K.P. Lillerud. 2012. Conversion of methanol to hydrocarbons: How zeolite cavity and pore size controls product selectivity. *Angew Chem Int Ed* 51:5810–5831.

Overkamp, KM, B.M. Bakke, P. Kötter, M.A.H. Luttik, J.P. van Dijken and J.T. Pronk. 2002. Metabolic engineering of glycerol production in *Saccharomyces cerevisiae. Appl Environ Microbiol* 68:2814–2821.

Owen, O.E., S.C. Kalhan and R.W. Hanson. 2002. The key role of anaplerosis and cataplerosis for citric acid cycle function. *J Biol Chem* 277:30409–30412.

Pagliaro, M, R. Ciriminna, H. Kimur, M. Rossi and C. Della Pina. 2007. From glycerol to value-added products. *Angew Chem Int Ed* 46:4434–4440.

Pandey, P., R. Hofer, C. Larroche, M. Taherzadeh and M. Nampoothiri. 2015. *Industrial Biorefineries and White Biotechnology*. Amsterdam: Elsevier.

Park, S.H., S. Kim and J.S. Hahn. 2014. Metabolic engineering of *Saccharomyces cerevisiae* for the production of isobutanol and 3-methyl-1-butanol. *Appl Microbiol Biotechnol* 98:9139–9147.

Patcas, F.C. 2005. The methanol-to-olefins conversion over zeolite-coated ceramic foams. *J Catal* 231:194–200.

Pathak, K., K. Mohan Reddy, N.N. Bakhshi and A.K. Dalai. 2010. Catalytic conversion of glycerol to value added liquid products. *Appl Catal A: Gen* 372:224–238.

Patzek, T.W. and D. Pimentel. 2005. Thermodynamics of energy production from biomass. *Crit Rev Plant Sci* 24:327–364.

Peterson, A.A., F. Abild-Pedersen, F. Studt, J. Rossmeisl and J.K. Nørskov. 2010. How copper catalyzes the electroreduction of carbon dioxide into hydrocarbon fuels. *Energy Environ Sci* 3:1311–1315

Petersen, A.M., S.X. Farzad and J.F. Görgens. 2015. Techno-economic assessment of integrating methanol or Fischer-Tropsch synthesis in a South African sugar mill. *Bioresour Technol* 183:141–152.

Phillips, C.B. and R. Datta. 1997. Production of ethylene from hydrous ethanol on H-ZSM-5 under mild conditions. *Ind Eng Chem Res* 36:4466–4475.

Phillips, J.R., R. L. Huhnke and H. K. Atiyeh. 2017. Syngas fermentation: A microbial conversion process of gaseous substrates to various products. *Fermentation* 3(2):28.

Pimentel, D., A. Marklein, M.A. Toth, M.N. Karpoff, G.S. Paul, R. McCormack J. Kyriazis and T. Krueger. 2009. Food versus biofuels: Environmental and economic costs. *Hum Ecol* 37:1.

Posada Duque, J.A., H. Zirkzee, E.W. van Hellemond, A. Lopez-Contreras, J.W. van Hal, and A.J.J. Straathof. 2014. A Biorefinery in Rotterdam with Isobutanol as Platform? www.ecn.nl/publications/PdfFetch.aspx?nr=ECN-V--14-004.

Putsche, V. and D. Sandor. (1996). Strategic, economic and environmental; issues for transportation fuels, in Wyman, C.E., (ed.). *Handbook on Bioethanol: Production and Utilization*. Washington, DC and London: Taylor and Francis, pp.19–35.

Qi, H., S. Li, S. Zhao and D. Huang. 2014. Model-driven redox pathway manipulation for improved isobutanol production in *Bacillus subtilis* complemented with experimental validation and metabolic profiling analysis. *PLoS ONE* 9(4): e93815.

Rabou, L.P.L.M. and A. van der Drift. 2011. Benzene and ethylene in Bio-SNG production: Nuisance, fuel or valuable products? *Proceedings of the International Conference on Polygeneration Strategies 11, Vienna*, Hofbauer, H. and Fuchs, M. (eds) Vienna University of Technology, pp. 157–162.

Ramió-Pujol, S., R. Ganigué, L. Bañeras and J. Colprim. 2015. How can alcohol production be improved in carboxydotrophic clostridia? *Process Biochem* 50:1047–1055.

Ren, F., H. Li, D. Wang, J. Wang, W. Pan and J. Li. 2003. Methanol synthesis from syngas in a slurry reactor. *Am Chem Soc Div Fuel Chem* 48:921–922.

Rizwan, M., J.H. Lee and R. Gani. 2015. Optimal design of microalgae-based biorefinery: Economics, opportunities and challenge. *Appl Energy* 150:69–79.

Rodriguez, B.A., C.S. Stowers, V. Phama and B.M. Cox. 2014. The production of propionic acid, propanol and propylene via sugar fermentation: An industrial perspective on the progress, technical challenges and future outlook. *Green Chem* 16:1066–1076.

Román-Leshkov, Y., C.J. Barrett, Z.Y. Liu and J.A. Dumesic. 2007. Production of dimethylfuran for liquid fuels from biomass-derived carbohydrates. *Nature* 447:982–985.

Rouilly, A. and C. Vaca-Garcia. 2015. Bio-based Materials, in: Clark, J.H. and Deswarte, F. (eds.) *Introduction to Chemicals from Biomass*, 2nd ed., Hoboken: Wiley-Blackwell, pp. 205–248.

Ruiz J., G. Olivieri, J. de Vree, R. Bosma, P. Willems, J.H. Reith, M.H.M. Eppink, D.M.M. Kleinegris, R.H. Wijffels and M.J. Barbosa. 2016. Towards industrial products from microalgae. *Energy Environ Sci* 24:405–413.

Sacramento-Rivero, J.C. 2012. A methodology for evaluating the sustainability of biorefineries: Framework and indicators. *Biofuels Bioprod Biorefin* 6:32–44.

Sakuragi, H., H. Morisaka, K. Kuroda and M. Ucda. 2015. Enhanced butanol production by eukaryotic *Saccharomyces cerevisiae* engineered to contain an improved pathway. *Biosci Biotechnol Biochem* 9:314–320.

Schieb, P.A., H. Lescieux-Katir, M. Thénot and B. Clément-Larosière. 2015. *Biorefinery 2030: Future Prospects for the Bioeconomy*, Berlin: Springer.

Schmidt, L.D. and P.J. Dauenhauer. 2007. Chemical engineering: Hybrid routes to biofuels. *Nature* 447:914–915.

Schobert, H.H. 2013. *The Chemistry of Hydrocarbon Fuels* (revised edition), Oxford: Butterworth-Heinemann.

Scholz, J., V. Hager, X. Wang, F.T.U. Kohler, M. Sternber, M. Haumann, N. Szesni, K. Meyer and P. Wasserscheid. 2014. Ethylene to 2-butene in a continuous gas phase reaction using SILP-type cationic nickel catalysts. *Chem Cat Comm* 6:162–169.

Schubert, C. 2006. Can biofuels finally take center stage? *Nat. Biotechnol.* 24:777.

Schulz, H. 1999. Short history and present trends of FT synthesis. *Appl Catalysis A: Gen* 186:3–12.

Shahzadi, T., S. Mehmood, M. Irshad, Z. Anwar, A. Afroz, N. Zeeshan, U. Rashid and K. Sughra. 2014. Advances in lignocellulosic biotechnology: A brief review on lignocellulosic biomass and cellulases. *Adv Biosci Biotechnol* 5:246–251.

Sharafutdinov, I., C.F. Elkjær, H.W. Pereira de Carvalho, D. Gardini, G.L. Chiarello, C.D. Damsgaard, J.B. Wagner, J.-D. Grunwaldt, S. Dahl and Chorkendorff, I. 2014. Intermetallic compounds of Ni and Ga as catalysts for the synthesis of methanol. *J Catal* 320:77–88.

Shukor, H., N.K.N. Al-Shorgani, P. Abdeshahian, A.A. Hamid, N. Anuar, N.A. Rahman and M.S. Kalil. 2014. Production of butanol by *Clostridium saccharoperbutylacetonicum* N1-4 from palm kernel cake in acetone-butanol-ethanol fermentation using an empirical model. *Bioresour Technol* 170:565–573.

Singh, A.K. and M.P. Singh. 2014. Importance of algae as a potential source of biofuel. *Cell Mol Biol* 60:106–109.

Singh, S. and S. Gu, 2010. Commercialization potential of microalgae for biofuels production. *Renew Sus Energy Rev* 14:2596–2610.

Smith, K.M., K.M. Cho and J.C. Liao. 2010. Engineering *Corynebacterium glutamicum* for isobutanol production. *Appl Microbiol Biotechnol* 87:1045–1055.

Smith, K.M. and J.C. Liao. 2011. An evolutionary strategy for isobutanol production strain development in *Escherichia coli*. *Metab Eng* 13:674–681.

Soares, R.R., D.A. Simonetti and J.A. Dumesic. 2006. Glycerol as a source for fuels and chemicals by low-temperature catalytic processing. *Angew Chem Int Ed* 45:3982–3985.

Steyer, J.P., I. Queinnec and D. Simoes. 1993. Biotech: A real-time application of artificial intelligence for fermentation processes. *Control Eng Practice* 1:315–321.

Su, H., G. Liu, H. Mingxiong and F. Tan. 2015. A biorefining process: Sequential, combinational lignocellulose pretreatment procedure for improving biobutanol production from sugarcane bagasse. *Bioresour Technol* 187:149–160.

Subhadra, B.G. 2010. Sustainability of algal biofuel production using integrated renewable energy park (IREP) and algal biorefinery approach. *Energy Policy* 38:5892–5901.

Taylor, J.D., M.M. Jenni and M.W. Peters. 2010. Dehydration of fermented isobutanol for the production of renewable chemicals and fuels. *Topics Catal* 53:1224–1230.

Tsur, Y. and A. Zemel. 2003. Optimal transition to backstop substitutes for nonrenewable resources. *J Econ Dynamics Control* 27:551–572.

van Heerden, C.D. and W. Nicol. 2013. Continuous succinic acid fermentation by *Actinobacillus succinogenes*. *Biochem Eng J* 73:5–11.

van Leeuwen, B.N.M., A.M. van der Wulp, I. Duijnstee, A.J. van Maris and A.J. Straathof. 2012. Fermentative production of isobutene. *Appl Microbiol Biotechnol* 93:1377–1387.

van Ree, R. and A. van Zealand. 2014. IEA Bioenergy Program Task 42 Biorefining, www.ieabioenergy.com/wp-content/uploads/2014/09/IEA-Bioenergy-Task42-Biorefining-Brochure-SEP2014_LR.pdf.

Varvaresou, A. and K. Iakovou. 2015. Biosurfactants in cosmetics and biopharmaceuticals. *Lett Appl Microbiol* 61:214–223.

Vispute, T.P., H. Zhang, A. Sanna, R. Xiao and G.W. Huber. 2010. Renewable chemical commodity feedstocks from integrated catalytic processing of pyrolysis oils. *Science* 330:1222–1227.

Weimer, P.J., J.B. Russell and R.E. Muck. 2009. Lessons from the cow: What the ruminant animal can teach us about consolidated bioprocessing of cellulosic biomass, *Bioresource Technol* 100:5323–5331.

West, R.M., D.J. Braden and J.A. Dumesic. 2009. Dehydration of butanol to butene over solid acid catalysts in high water environments. *J Catal* 262:134–143.

Wijffels, R.H., O. Kruse and K.J. Hellingwerf. 2014. Potential of industrial biotechnology with cyanobacteria and eukaryotic microalgae. *Curr Opin Biotechnol* 24:405–413.

Wilkins, M.R. and H.K. Atiyeh. 2011. Microbial production of ethanol from carbon monoxide. *Curr Opin Biotechnol* 22:326–330.

Williams, C.L., C.C. Chang, P. Do, N. Nikbin, S. Caratzoulas, D.G. Vlachos, R.F. Lobo, W. Fan and P.J. Dauenhauer. 2012. Cycloaddition of biomass-derived furans for catalytic production of renewable *p*-xylene. *ACS Catal* 2:935–939.

Wyman, C.E., (ed.). 1996. *Handbook on Bioethanol: Production and Utilization*. Washington, DC and London: Taylor and Francis.

Yamamoto, S., M. Suda, S. Niimi, M. Inui and H. Yukawa. 2013. Strain optimization for efficient isobutanol production using *Corynebacterium glutamicum* under oxygen deprivation. *Biotechnol Bioeng* 110:2938–2948.

Yang, C., Z. Ma, N. Zhao, W. Wei, T. Hu and Y. Sun. 2006. Methanol synthesis from CO_2-rich syngas over a ZrO_2 doped CuZnO catalyst. *Catal Today* 115:222–227.

Yang, Y., B. Liu, X. Du, P. Li, B. Liang, X. Cheng, L. Du, D. Huang, L. Wang and S. Wang. 2015. Complete genome sequence and transcriptomics analyses reveal pigment biosynthesis and regulatory mechanisms in an industrial strain, *Monascus purpureus* YY-1. *Sci Rep* 5: 8331

Ye, J., C.J. Liu, D. Mei and Q. Ge. 2014. Methanol synthesis from CO_2 hydrogenation over a Pd_4/In_2O_3 model catalyst: A combined DFT and kinetic study. *J Catal* 317:44–53.

Zacharopoulou, V., E.S. Vasiliadou and A.A. Lemonidou. 2015. One-step propylene formation from bio-glycerol over molybdena-based catalysts. *Green Chem* 17:903–912.

Zeng, A.P. and H. Biebl. 2002. Bulk chemicals from biotechnology: The case of 1,3-propanediol production and the new trends. *Adv Biochem Eng/Biotechnol* 74:239–259.

Zhang, J., J. Li, Y. Tang, L. Lin and M. Long. 2015. Advances in catalytic production of bio-based polyester monomer 2,5-furandicarboxylic acid derived from lignocellulosic biomass. *Carbohydr Polym* 130:420–428.

Zhang, W., I.S, Hunter and R. Tham. 2010. Microbial and plant cell synthesis of secondary metabolites and strain improvement, in El-Mansi, E.M.T., Bryce, C.F.A., Dahlou, B., Sanchez, S., Demain, A.L. and Allman, A.R. (eds.) *Fermentation Microbiology and Biotechnology*, 3rd ed., Boca Raton, New York and London: CRC Press, 101–135.

Zhang, D., R. Al-Hajri, S.A.I. Barri and D. Chadwick. 2010. One-step dehydration and isomerisation of *n*-butanol to iso-butene over zeolite catalysts. *Chem Commun* 46:4088–4090.

Zhao, T.S., T. Takemoto and N. Tsubaki. 2006. Direct synthesis of propylene and light olefins from dimethyl ether catalyzed by modified H-ZSM-5. *Catal Commun* 7:647–650.

13 Solid-State Fermentation
Current Trends and Future Prospects

Reeta Rani Singhania, Anil Kumar Patel, Lalitha Devi Gottumukkala,
Kuniparambil Rajasree, Carlos Ricardo Soccol, and Ashok Pandey

CONTENTS

"Live with purpose. Don't let people or things around you get you down."

Albert Einstein

13.1 INTRODUCTION

Solid-state fermentation (SSF) is defined as a fermentation process that utilizes a solid matrix containing enough moisture to support microbial activities without additional free water. The solid matrix could be the source of nutrients or simply a supporting material impregnated with all the nutrients that are required for microbial growth (Pandey, 1992, 1994; Singhania et al., 2009, 2017; Thomas et al., 2013). SSF resembles the natural habitat of microorganisms and has proved to be a useful tool in the production of value-added products.

SSF processes (e.g., composting and ensiling the preservation of green fodder in pits) recycle agricultural wastes and play a major role in the utilization of renewable resources. In addition, SSF is one of the oldest fermentation processes known to have been practiced by mankind; it has been associated with the production of traditional fermented foods such as Indonesian "tempeh" and Indian "ragi." The use of *Aspergillus oryzae* in the making the "koji" and the use of *Penicillium roquefortii* in the manufacture of cheese in Europe is another common example. In addition, the manufacture of soya sauce in Asia, and bread-making in Egypt are well-established technologies (Pandey, 1992;).

In the 1970s, SSF rose to prominence with the discovery of mycotoxins as a product of SSF fungal metabolism. as well as the realization that SSF can be used effectively to produce protein-rich animal feed. In addition, SSF is being used for production of high value–added products such as enzymes, organic acids, bio-pesticides, flavor enhancers, and biofuels (Table 13.1).

In recent years, SSF has emerged as an attractive alternative to submerged fermentation (SmF), because it affords a sustainable and more economical process than SmF. This is primarily because SSF utilizes low-cost agricultural residues as a feedstock and does not require sophisticated fermentors, nor does it require heavy investments. SSF has continued to build up credibility in biotech industries due to its versatile applications in the production of biologically active secondary metabolites as well as feedstock, biofuel, fine chemicals, and pharmaceuticals. SSF has emerged as a viable technology for the bioremediation and biodegradation of hazardous compounds as well as the recycling of agriculture residues and biomass conservation. Agro-waste peanut shell in natural state proved to be a promising adsorbent for dye removal from aqueous solution. Fungal solid-state fermentation produced beneficial laccase enzyme and simultaneously changed physicochemical properties of peanut shell further improving its adsorbent capacity (Liu et al., 2018).

Even today SSF plays important role in food industry and research continues exploring the technology for further improving the food products. Okara (soybean pulp), a by-product of the soy food industry and of which significant amounts are produced, is currently of no useful use due to its fibrous texture in the mouth, low digestibility, and grassy off-odor. Interestingly, however, the composition of Okara and its nutritional value as well as its organoleptic qualities (flavor, odor, etc.) changes favorably following fermentation using *Rhizopus oligosporus* and *Yarrowia lipolytica* (Vong et al., 2018). Similarly, SSF of oats using *Monascus anka* yielded a product with higher nutritional value, which is due to optimization of the production of phenolic compounds (Bei et al., 2018).

Another example of the usability of SSF for industrial-scale applications is in the bioremediation of contaminated land and shorelines. This is especially useful for spills of hydrocarbons, e.g., keerosene, where landfall on a beach allows immobilization and concentration of the contaminant. The use a laboratory-scale SSF allows the determination of the appropriate "cocktail of microbes" for optimum remediation, the rate of remediation in conditions equivalent to those found on-site (using carbon dioxide evolution), and the influence of mixing methodologies and rates. As mycelial threads are easily broken by gentle continuous rotation, occasional turning over of the solid substrate only once or twice per day may be the most effective treatment.

An extension of the SSF is to combine it with a second, liquid culture in the same vessel after an initial phase on solid medium. This a way of combining two distinct process steps as a cascade in one vessel, with different operating conditions applying in each phase. The first stage mixes an enzyme with a solid substrate to release hemicelluloses from, e.g., agricultural waste. For this to be effective the bioreactor needs special modifications in terms of a powerful, high-torque motor, temperature control in a water jacket adapted to the relatively poor heat transfer in solids and special impellor designs for mixing of solids, e.g., anchor or helical. The second phase is a conventional anaerobic fermentation using yeast to produce

TABLE 13.1
Applications of SSF in Different Sectors

Economical sector	Application	Products and the Microorganism Involved
Agro-food industry	Traditional food fermentations	Koji, fermented cheese
	Mushroom production and spawn	*Agaricus*, *Pleurotus*
	Bioconversion by-products	Sugar pulp bagasse, coffee pulp silage composting, detoxication
	Food additives	Flavors dyestuffs, essential fat, and organic acids
Agriculture	Biocontrol, bioinsecticide	*Beauveria*, *Metarhizium*, *Trichoderma*
	Plant growth hormones	*Gibberellins*, *Rhizobium*
	mycorhization, wild mushroom	Plant inocultion
Industrial	Enzymes production	Amylases, cellulases proteases, pectinases, xylanases
Fermentation	Antibiotic production	Penicillin, feed, and probiotics
	Organic acid production	Citric acid, fumaric acid, gallic acid, lactic acid
	Ethanol production	*Saccharomyces* sp. starch malting and brewing
	Fungal metabolites	Hormones alkaloids

Source: Adapted from Raimbault 1998. *Elect J Biotech* 1:11–15. With permission.

alcohol as a biofuel. In this case, the gentle mixing of liquids by the same impellor(s) as used for the solid substrate is of little importance since the process does not require significant oxygen transfer. An example of a system based on some of these principles is provided in Section 13.7.4.

13.2 SUITABILITY OF MICROORGANISMS FOR SSF PROCESSES

The ability of a given microorganism to grow on solid substrate depends on

- The capacity of the organism to adhere and penetrate the substrate
- The ability of the organism to assimilate complex substrates
- The level of water-activity requirement

Many bacteria and fungi are capable of growing on solid substrates (Table 13.2). However, filamentous fungi are better adapted to solid substrates because hyphal growth allows the fungi to penetrate the substrates. In addition, filamentous fungi possess a good tolerance to low water activity (A_w) and high osmotic pressure (Chundakkadu, 2005).

However, bacteria and yeasts have also been used in traditional SSF processes (e.g., the production of enzymes, composting, ensiling, and food manufacturing). Yeasts have been mainly used for ethanol production and protein enrichment of agricultural residues. Measurement of the specific growth rate of a given microorganism is necessary for the understanding of fermentation kinetics; this aspect was dealt with in Chapter 2 and more extensively in Chapter 3 of this edition.

13.3 BIOMASS MEASUREMENT

Biomass is a fundamental parameter in the characterization of microbial growth. Direct measurement of biomass is not possible in SSF because microbial biomass cannot be separated from the substrate. The consumption of oxygen and the liberation of carbon dioxide (CO_2) are directly associated with cellular metabolism and as such can be used as indicators of microbial growth. Similarly, the excretion of extracellular enzymes and the production of primary metabolites can also be used as indicators of microbial growth.

Because direct measurement of biomass is difficult; the following global stoichiometric equation is used:

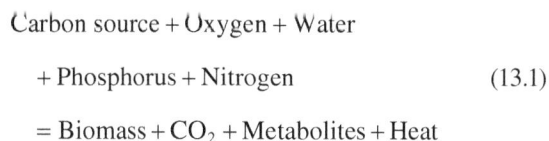

$$\text{Carbon source} + \text{Oxygen} + \text{Water}$$
$$+ \text{Phosphorus} + \text{Nitrogen} \qquad (13.1)$$
$$= \text{Biomass} + CO_2 + \text{Metabolites} + \text{Heat}$$

Measuring any one of the above components allows for determining the evolution of others, if all of the coefficients remain constant. It has to be remembered, however, that SSF is at a disadvantage, because its bioprocess variables are not easily measurable (Pandey, 1994; Ramachandran et al., 2007).

13.3.1 BIOMASS COMPONENTS

Specific components in biomass can be measured to estimate the growth of the microorganisms. It is necessary to determine the particular component of the cell or mycelium that is not present in the solid substrate.

13.3.1.1 Protein Content

Protein content of the biomass can be measured by determining the nitrogen content by the Kjeldahl method. Accurate measurement of biomass can be obtained by an amino-acid analyzer. The principal problem in determining protein content is which part of the protein present in the substrate is not consumed or transformed. This method of biomass measurement is reliable if the solid substrate has no protein or high protein content.

13.3.1.2 Glucosamine Content

Glucosamine is present as acetylglucosamine monomers in the chitinous cell wall of fungi. Chitin is an insoluble polymer present in the mycelium and it should be depolymerized first for glucosamine determination. Glucosamine determination has a disadvantage of a lengthy procedure, which takes approximately 24 h. Interference with this method may occur when using complex agricultural substrates containing glucosamine in glycoproteins.

13.3.1.3 Nucleic Acid Determination

DNA content in the medium depends on the growth of the biomass. DNA contents increase during early growth and level off as the stationary phase is approached. Methods based on DNA and RNA determination are reliable only if the substrate has little nucleic acid and no interfering chemical present.

13.3.1.4 Ergosterol Measurement

Ergosterol is the predominant sterol in fungi. It is easy to determine ergosterol because it can be separated from other sterols endogenous to the solid substrate by using high-performance liquid chromatography (HPLC) and can be quantified simply by spectrophotometry. Ergosterol determination is an unreliable method to follow the growth of the microorganism as it varies with culture conditions, aeration, and substrate composition.

13.3.1.5 Physical Measurement of Biomass

Biomass can be quantified directly by measuring electrical conductivity or indirectly by measuring the drop in pressure. During the early condiophore stage, the pressure drops drastically and a break point can be easily observed.

13.4 FACTORS AFFECTING SSF

The SSF process (Figure 13.1) can be affected by various factors, and depending upon the nature of substrates and the microorganisms used, they may vary from process to process. These factors can be broadly divided into biological, physicochemical, and environmental factors. Because the microorganisms

TABLE 13.2

Examples of Products Produced by the SSF Process Highlighting the Primary Substrate and the Microorganism Used

Product	Substrate	Organism
Cellulase	Rice straw, wheat bran, wheat straw	*Aspergillus niger*, *Trichoderma reesei Humicola fasiolense*, *Penicillium* sp.
Xylanase	Rice straw, sorghum flour, rice bran	*Aspergillus niger Trichoderma harzianum Bacillus* sp. JB-99
Aroma	Cassava, soybean, amaranth grain	*Rhizopus oryzae*
Neutral protease	Wheat bran	*Aspergillus oryzae* NRRL 1808
Pectinase	Grape pomace	*Aspergillus awamori*
α-Amylase	Wheat bran, coconut oil cake	*Aspergillus* sp. *Bacillus subtilis* *Aspergillus oryzae*
Lipase	Wheat bran + olive oil	*Aspergillus niger*
Phytase	Wheat bran + soy meal coconut oil cake	*Aspergillus niger Rhizopus oligosporous Mucor recemosus*
Chitinase	Wheat bran + chitin + yeast extract, wheat bran + chitin	*Trichoderma harzianum*, *P. chrysogenum*
Alkaline protease	Wheat bran + soy protein	*Pencillium* spp.
Lipase	Soy cake, sugarcane bagasse	*P. simplicissimum, Rhizopus homothallicus*
Alkaline protease	Green gram husk	*Bacillus* sp.
Pectinase	Wheat bran + polygalacturonic acid	*Bacillus* sp.
Aflatoxin	Cassava, rice, maize, pea nuts	*Aspergillus niger* *Aspergillus panasitus*
Antibiotic	Wheat, corn	*Alternaria brassicola*
Bacterial endotoxins	Coconut	*Bacillus thurungiensis*
Cephalosporin	Rice grains, barley	*Streptomyces claValigerus*
Gibberlic acid	Wheat bran	*Fusarium moniliforme*, *Gibberlla fujikuroi*
Mycotoxins	Corn, wheat, oats	*Aspergillus flavus*
Penicillin	Bagasse	*Penicillium chrysogenum*
Surfactin	Soya	*Bacillis subtilis*
Tetracyclines	Sweet potato	*Aspergillus*
Zearalenone	Corn	*Fusarium moniliforme*
Citric acid	Pineapple waste	*Aspergillus niger* DS-1
γ-Linolenic acid	Rice bran and soya bean meal	*Mucor rouxii*
Mycophenolic acid	Pearl barley	*Penicillium brevicompactum*
Gluconic acid	Sugarcane molasses	*Aspergillus niger* ARNU-4
Lipopeptides + poly-γ-glutamic acid	Soybean and sweet potatoes	*Bacillus subtillus*
Cephalosporin	Rice grains, barley	*Streptomyces claValigerus*
Gibberlic acid	Wheat bran	*Fusarium moniliforme*, *Gibberlla fujikuroi*
Penicillin	Bagasse	*Penicillium chrysogenum*
Surfactin	Soya	*Bacillis subtilis*
Tetracyclines	Sweet potato	*Aspergillus*
Zearalenone	Corn	*Fusarium moniliforme*
Citric acid	Pineapple waste	*Aspergillus niger* DS-1
Gamma-linolenic acid	Rice bran and soya bean meal	*Mucor rouxii*
Lactic acid	Cassava bagasse	*Lactobacillus delbrueckii*
Mycophenolic acid	Pearl barley	*Penicilliu brevicompactum*
Mevastatin	Wheat bran, ragi floor	*Penicillium citrinum* NCIM 768, *Aspergillus terreus*

predominately used for SSF are fungal species, the main focus of this discussion is also based on fungal species.

13.4.1 Inoculum Type

Fermentation using SSF is generally initiated using a mycelial or spore inoculum. There are several advantages of using spore inoculum, including homogeneity, prolonged shelf life and ease of manipulation. However, in wheat straw fermentation by *Chaetomium cellulolyticum*, the use of mycelial inoculum is more effective in yielding high level of proteins, presumable due to instant availability of hydrolytic (cellulase and hemicellulase) enzymes (Abdullah et al., 1985). Another important factor is the density of the inoculum, as a higher density minimizes contamination with undesired organisms (Singh et al., 2017).

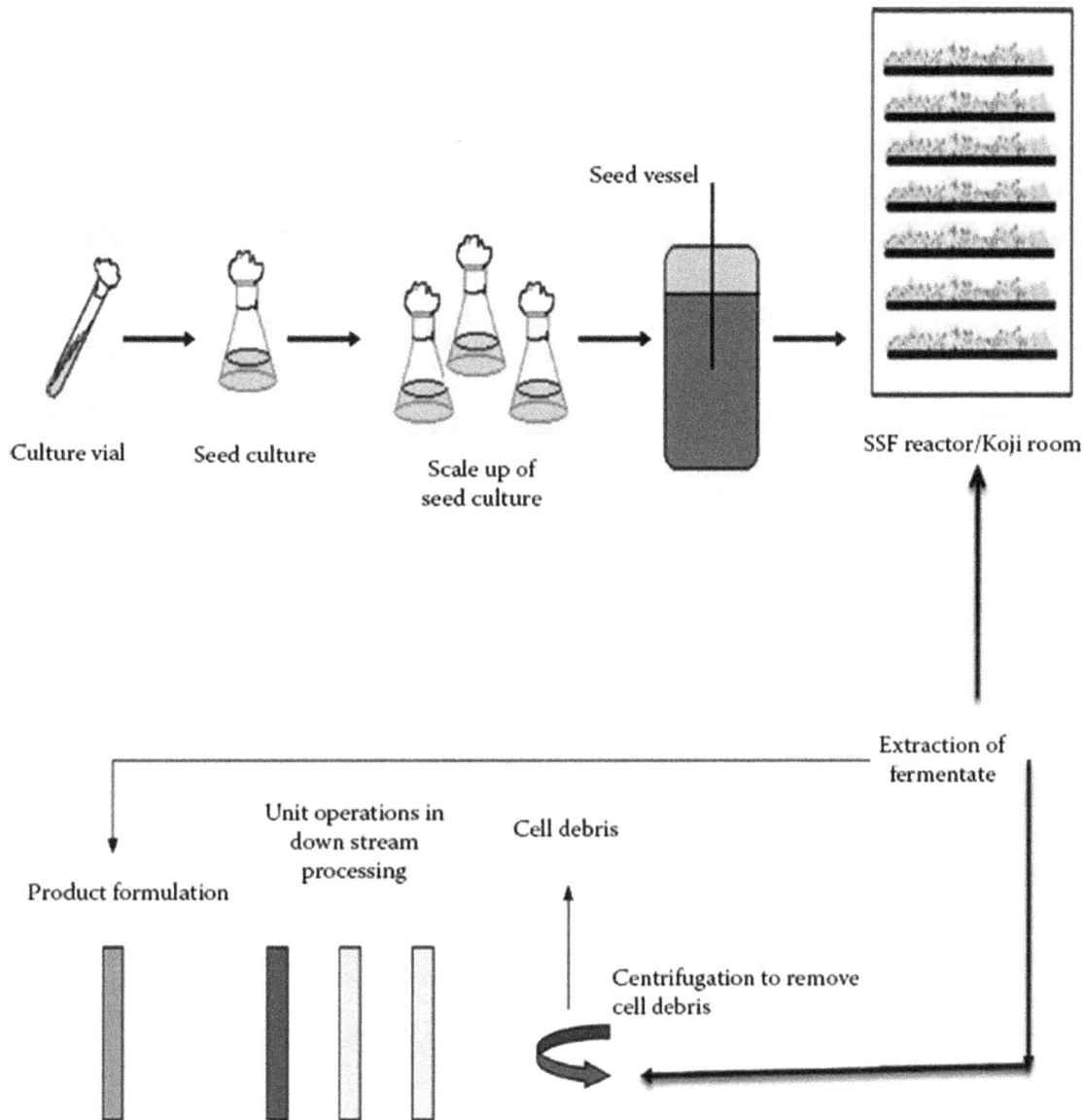

FIGURE 13.1 A schematic representation of the steps involved in a typical SSF process.

13.4.2 Moisture and Water Activity

Moisture and water content of the substrate have a critical role in the SSF process. An optimal moisture level is needed for microbial growth. Lower moisture content causes reduction in solubility of nutrients of the substrate whereas higher moisture levels can cause a reduction in enzyme yield by reduction in porosity (inter-particle spaces) of the solid matrix, thus interfering with oxygen transfer. Evaporation and microbial growth reduces the water level of the substrate during fermentation and this leads to the loss of humidity.

Water activity (A_w) is the accessible or available water for the growth of the microorganism. It gives the amount of unbounded water available in the immediate surroundings of the microorganism. It is closely related, but not equal to, the water content. The definition of the water activity is:

$$A_w = P_s/P_o \qquad (13.2)$$

where P_s is the equilibrium vapor pressure of water within the solid substrate and P_o is the vapor pressure of pure water.

In general, fungal and yeast cultures have been found to be more effective than bacteria in SSF processes. This has been essentially based on the theoretical concept of water activity, as fungi and yeast have lower water activity requirements, typically around 0.5-06 A_w. Bacterial cultures have higher water-activity requirement (around 0.8-0.9 A_w), which makes them less suitable for SSF processes. However, it must be remembered that the choice of microorganism for use in the SSF is dictated by the nature of the substrate to be fermented.

13.4.3 Temperature

Temperature is one of the most important factors affecting the SSF process. Growth of microorganisms, production of enzymes, secondary metabolite synthesis, and so on, depend on temperature. Fungi can grow under a wide range of temperatures; however, the optimal temperature for growth and product formation varies greatly. When the growth rate and metabolic activity increase, the temperature also increases inside of the fermentor. If there is not an efficient system to remove the heat generated, this will adversely affect the product formation and growth. Aeration and agitation are the two main conventional methods used to control the temperature of the substrate. In aeration, water-saturated air is used to control the temperature, so that it can reduce the water activity of the substrate.

13.4.4 Hydrogen Ion Concentration (pH)

Because the substrate used in SSF itself having a buffering effect due to its complex chemical composition, though it is difficult to monitor the changes. Filamentous fungi and yeasts have a broad range of pH for growth that can be exploited to prevent bacterial contamination by using a low pH. To control the pH in SSF, ammonium salts have been used in combination with urea or nitrate salts to neutralize the effect of acidification and alkalization.

13.4.5 Substrates

Substrates that are meant for solid support in SSF have a common basic macromolecular structure such as cellulose, starch, pectin, lignocellulose, fibers, and so on. Generally, this matrix can also serve as a carbon and energy source. Based on the nature of the substrate, certain preparative and pretreatment steps are necessary to convert the raw substrate into a suitable form:

- Size reduction by grinding, milling, rasping, or chopping;
- Physical, chemical, or enzymatic hydrolysis of polymers to increase substrate availability by the fungus;
- Supplementation with nutrients (phosphorus, nitrogen, salts) and setting the pH and moisture content with a mineral solution; and
- Cooking or vapor treatment for macromolecular structure pre-degradation and elimination of major contaminants.

Several factors influence the selection of substrates for SSF. Three crucial factors are cost, availability, and heterogeneity of the substrates. Based on the nature of the substrate, two types of SSF systems exist. The most common is having natural substrates that can act as a carbon and energy source. Less common is inert material supplemented with a mineral medium, which can give only solid support. As already mentioned, agro-industrial residues are the best choice as a solid substrate for the growth of microorganisms and include sugar cane bagasse, wheat bran, rice bran, wheat and rice straw, coconut coir pith, tea and coffee waste, cassava waste and various pulps, and so on. One of the major drawbacks of using these natural substrates as a solid support is that during the growth of microorganisms, physical and geometrical changes occur in the structure of the substrates that eventually lead to a reduction in the heat and mass transfer. This problem can be solved by using an inert support as a solid substrate, so that it can maintain the physical structure throughout. This will give a proper control over heat and mass transfer.

13.4.5.1 Particle Size

Particle size is an important parameter for helping gaseous exchange as well as heat and mass transfer between particles. It affects the surface area-to-volume ratio of the particle that is initially accessible to the microorganism and the packing density within the surface mass. The size of the substrate determines the void space, which is occupied by air. Because the rate of oxygen transfer into the void space affects growth, the substrate should contain particles of suitable size to enhance mass transfer (Singhania et al., 2009). Smaller substrate particles would generally provide larger surface area for microbial action, but particles which are too small may result in substrate agglomeration, which may interfere with microbial respiration/aeration and thus result in poor growth. Smaller particle size is also advantageous for heat transfer and exchange of oxygen and CO_2 between the air and the solid surface. At the same time, larger particles also provide better respiration/aeration efficiency but provide limited surface for microbial action (Mithell et al., 1992).

13.4.5.2 Aeration and Agitation

Aeration and agitation play a significant role in the SSF process because they face two fundamental problems, i.e., oxygen demand in the aerobic process plus heat and mass transport phenomena in a heterogeneous system. Oxygen demand in SSF can be satisfied with relatively low aeration levels. Agitation may also promote or prevent aggregate formation of the fermenting mass depending on the nature of the solids (Lonsane et al., 1992). More is discussed in Section 13.5.3.

13.5 SCALE-UP

After standardization of unit operations on a laboratory scale, scale-up studies should be done for setting up a pilot plant. Scale-up is a crucial link in transferring laboratory-scale processes to a commercial production scale. For a process or a product to be commercialized, it must undergo studies on four different levels of scale: flask level (50–100 g), laboratory fermentor level (5–20 kg), pilot fermentor level (50–5000 kg), and production fermentor level (25–1000 t) (Lonsane et al., 1992). During scale-up, several problems are encountered because of the size of fermentation and process productivity. Scale-up using any type of SSF bioreactor is a complicated process because of intense heat generation in the system and heterogeneity. A proper bioreactor should be selected or designed for efficient heat removal, moisture balance, aeration, agitation, and so on, but because of the heterogeneity of the system, difficulties will be encountered in achieving proper mass transfer and diffusion with an increase in scale. There are several major and minor areas to be studied for a proper SSF scale-up. Important ones will be discussed below.

13.5.1 Large-Scale Inoculum Development

Generally, in large-scale SSF, high inoculum density is used to reduce the risk of contamination and to produce the desired level of product in a shorter time. In large-scale production, inoculum generation itself becomes a distinct unit operation. If the inoculum has to be generated in liquid medium, large-scale fermentors should be used. In many cases, a spore inoculum is preferred over an inoculum generated in liquid medium because of the ease of uniform mixing of spores with a moist solid substrate. Generating spore inocula, especially in fungal cultures, takes more time, and the chance of mutation is high because of the formation of spores in earlier inoculum development stages and their germination in the production stage (Reusser et al., 1961). To reduce the potential chance of mutation, a minimal amount of sub culturing should be done during inoculum development.

13.5.2 Medium Sterilization

Batch sterilization is preferred over continuous sterilization because there are difficulties due to the physical nature of moist solid medium in SSF systems. Solid substrates such as wheat bran are modified and made more amenable to microbial growth during autoclaving. It is essential that each solid particle is heated to 121°C for 60 min in large-scale sterilization.

13.5.3 Aeration and Agitation

Aeration essentially has two functions: oxygen supply (most important) and removal of CO_2, heat, and volatile compounds. Oxygen uptake by the microorganisms present in the SSF system depends on different parameters such as porosity of the moist substrate, bed depth, perforations in the culture vessel, forced aeration, and mixing. Flow rate and air pressure matter if the aeration is forced into the culture vessel. Gaseous diffusion increases with an increase in pore size and decreases with a reduction of diameter. The rate of aeration must be determined by the growth requirements of the organism and heat evolution. In the case of a static fermentor, gaseous diffusion is difficult, causing the hyphae which penetrate into the moist solid particles to encounter problems in oxygen uptake. It is always good to combine aeration with agitation for efficient distribution of oxygen throughout the SSF system and proper mass transfer. Agitation also helps in efficient nutrient mixing, which is important for bacteria and yeast. Agitation may also prevent or promote aggregation of solid mass depending on the nature of the solid. Intermittent mixing and low agitation speed help to prevent damage to mycelia and reduce aggregation of solids. In certain conditions, such as tray fermentation, agitation is not possible, and these are usually kept as static systems.

13.5.4 Heat Removal and Moisture Balance

Heat generated is proportional to the metabolic activity in the system. Generally, there is an increase in temperature in the center of the bed which may be 17–20°C higher when compared with the set value. It is very difficult to control the temperature of the system just by placing it in a temperature-controlled room or by passing cooling water through the jacket of the fermentor. Heat removal through conduction and convection is not effective because of a low or no rate of agitation and poor thermal conductivity of solid substrates. Evaporative cooling is beneficial but has the disadvantage of moisture loss (Narahara et al., 1984). Moisture of the substrate should also be maintained properly for better productivity. This can be achieved with a high initial moisture content of the medium and a humidified atmosphere in the fermentor (90–98%) (Ahmed et al., 1957).

13.5.5 pH Control

In large-scale solid-state fermentation, it is often difficult to have a continuous monitor of pH change and hence it is advised to use buffers in the medium (see Section 13.4.4 for more detail). In open systems, such as tray fermentation, aseptic operation is not possible and, therefore, pH and moisture play a crucial role in reducing the risk of contamination.

13.5.6 Product Recovery

Recovery of product from solid substrates is a highly complex and labor-intensive process. Product should be leached out from the solid substrate using a recognized solvent system. Efficiency of product recovery depends on temperature, pH, incubation time, agitation speed, type of solvent, miscibility of the product in the solvent, ratio of solids to solvent, and so on. There are several types of leaching techniques, including percolation, multiple-contact countercurrent leaching, pulsed plug-flow extraction in column, hydraulic pressuring, and supercritical fluid extraction. An economically feasible leaching technique should be selected based on the fermentation product.

13.6 MODELING IN SSF

The concept of modeling in SSF is the search for mathematical expressions that represent the system under consideration. The objective of these expressions is to establish the relationships or functions between two different variables that characterize the system. This is to ascertain the validity of the system, to establish different parameters that characterize a specific process, and to find appropriate mechanisms for development and control of the process (Pandey et al., 2001). Modeling of bioreactors used in SSF processes can play a crucial role in the analysis, design, and development of bioprocesses based on phenomena that encompass various applications ranging from the production of enzymes to the treatment of agro-industrial residues. Existing models of SSF processes describe coupled substrate conversion and diffusion, along with the consequent microbial growth. These models neglect many of the significant phenomena that are known to influence SSF. Consequently, the available models fail to explain the generation of numerous products that form during any SSF process and the outcome of the process in terms of the characteristics of the final product.

Information regarding modeling in SSF has been limited because of the unavailability of a suitable method for direct measurement of microbial growth because of the difficulty in separating microorganisms from the substrate and determination of the rate of substrate utilization. Among several approaches to tackling this problem, an important one has been to use a synthetic model substrate. It is well known that the fermentation kinetics is extremely sensitive to the variation in ambient and internal gas compositions. Therefore, the cellular growth of the microorganisms can be determined by measuring the change in gaseous compositions inside of the bioreactor.

The importance of models in SSF, as in any chemical or biochemical process, lies in the establishment of the parameter values which can explain a particular process; also providing the basis to evaluate the process, including the way in which a process could be economically scaled-up, e.g., using design and control criteria.

The major challenges to overcome in scale-up SSF processes include the heat accumulation and heterogeneous nature of the substrate, comprising a three-phase gas-liquid-solid multiphase system during the fermentation. The developments of models in SSF are focused on:

- representation of the microbial activity (kinetic patterns and thermodynamic concerns);
- studies of the problems of heat and mass transfer in the three-phase systems and the connection between them; and
- the selection of the best type of the fermentor.

In order to achieve these goals, it is necessary to understand and estimate the heat and mass transfer parameters, which would be helpful in developing the mathematical models considered as the key for scale-up data.

In last few years, numerous studies have been performed in this direction and a lot of modeling data has been generated on SSF bioreactor studies. Wang et al. (2009) developed a mathematical model of a rotating drum bioreactor used for the production of ethanol, considering the radial temperature distribution in the substrate bed (sorghum stalk). The model was a good fit compared with the experimental data, which showed that this mathematical model is a powerful tool to investigate the design and scale-up of SSF processes. Modeling of SSF bioprocesses for different products is possible, e.g., see the example given by Thomas et al. (2013).

13.7 TYPES OF SSF BIOREACTORS

SSF bioreactor design largely depends on the solid matrix. Some of the important criteria while designing a SSF include perfecting control systems for temperature, airflow, and humidity; maintenance of aseptic conditions for preventing contamination; maintenance of homogenous water activity, temperature distribution, and composition so that microbes can grow uniformly; plus the quick and efficient removal of harmful metabolites for labor saving and easy handling. SSF processes could be operated in batch, fed-batch, or continuous modes, although batch processes are the most common.

Two categories of bioreactors exist for SSF processes: (1) laboratory-scale bioreactors that use quantities of dry solid medium from a few grams up to few kilograms, and (2) pilot- and industrial-scale bioreactors where several kilograms up to several tons are used. In laboratory-scale bioreactors, Petri dishes, jars, wide-mouth Erlenmeyer flasks, Roux bottles, and roller bottles can be used as fermentors for SSF. In laboratory-scale bioreactors, maintenance of completely aseptic conditions is possible to a certain extent; however, aeration and agitation are not possible. Several industrial-level SSF bioreactors have been designed, but all of them invariably face the problem of heat generation and its removal along with handling difficulties.

13.7.1 SHALLOW-TRAY FERMENTOR

This is one of the technologies used in China and other countries for SSF. The shallow tray can be made up of wood, bamboo, metal, or plastic. The bottom of the tray is made up of sieve plate or wire mesh to help airflow. The tray is 30–50 mm deep. A suitable space is left between two trays. The shallow-tray fermentor needs to be kept in a sterile environment to prevent contamination. Low investments and simplicity in construction are the two main advantages of using shallow-tray fermentors. Figure 13.2 shows a schematic representation of the shallow-tray fermentor. Air saturated with water vapor was blown into the bottom of a culture vessel and the exhaust air was withdrawn at the top.

13.7.2 COLUMN FERMENTORS

A column fermentor is a fixed-bed reactor. Solid medium is put into the column with entries at both ends for aeration. Figure 13.3 shows a column fermentor developed by Saucedo-Castaneda and colleagues (1990). Sterile air can be supplied by a radial or axial gradient method. The water activity is maintained by humidified air. The temperature for the solid fermentation is monitored and controlled by recycling water in the jacket from an isothermal bath. In a fixed-bed column fermentor, the oxygen transfer and CO_2 dissipation are improved by forced convection. However, it is difficult to regulate the water activity and temperature in the fixed bed. In addition, handling the solid materials in a column fermentor is a huge problem and the scale-up of a column fermentor is troublesome.

13.7.3 ROTATING DRUM BIOREACTORS

Drum bioreactors are mainly of two types: continuously rotating and discontinuously rotating drums. Continuously rotating drum reactors are mainly used in laboratory-scale and pre-pilot scale bioreactors. They consist of a rotating drum that is

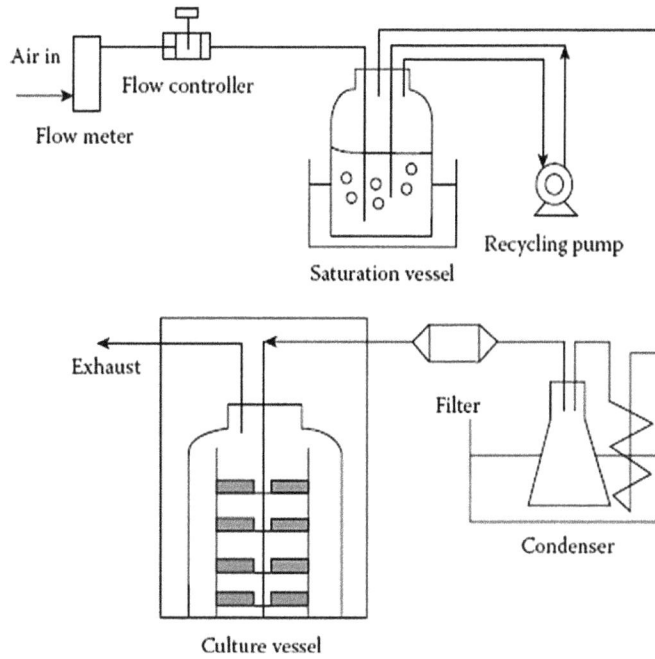

FIGURE 13.2 Shallow-tray fermentor used for solid-state cultivation.

meant for thorough mixing of the substrate particles. When the rotation rate of the drum is increased, it can affect the mycelial growth, presumably because of shear effects (Fung and Mitchell, 1995). The largest bioreactor cited in the literature was a stainless steel rotating drum (Ø 56 and 90 cm long) that used 10 kg of steamed wheat bran as substrate (Stuart et al., 1998) for kinetic studies of *Rhizopus*. In the case of discontinuously rotating drum reactors, the operating principle is intermittent mixing and static conditions, which help to reduce the rotation rate. During the static period, the reactor is similar to a tray reactor. In addition to the above, rocking drum bioreactors, gas solid fluidized-bed reactors, continuously stirred aerated bed reactors, and so on, are also available, having continuous mixing and forced aeration. This can overcome some limitations, such as the oxygen mass transfer being enhanced and overheating is prevented. The heterogeneity of the system

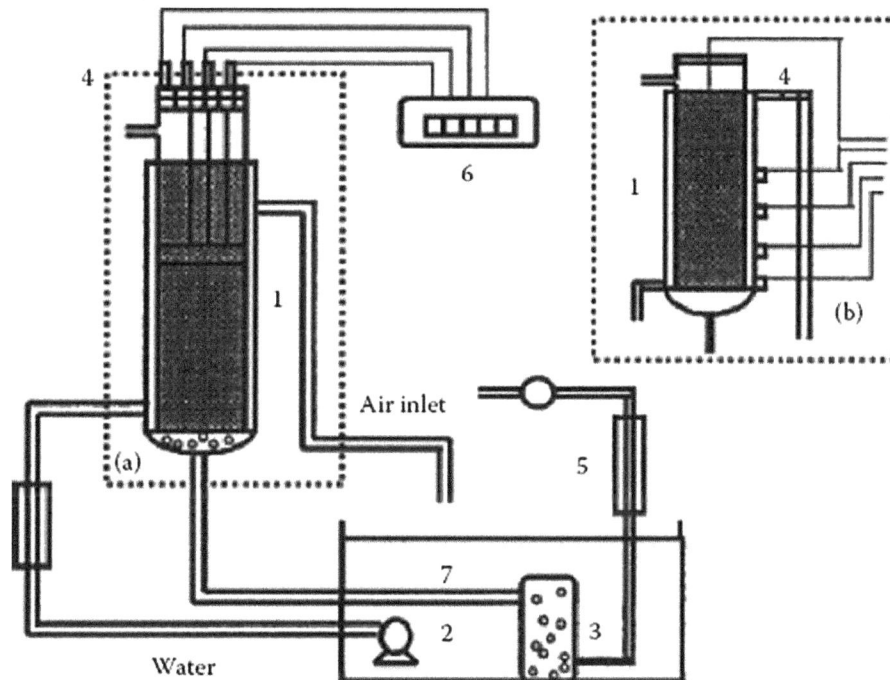

FIGURE 13.3 Column fermentor for SSF: (a) radial gradient and (b) axial gradient. Column fermentor (volume = 1 L, radius = 6 cm, length = 35 cm): 1, jacket fermentor; 2, water pump; 3, humidifier; 4, thermocouples; 5, pressure and airflow controls; 6, temperature display device; 7, water bath.

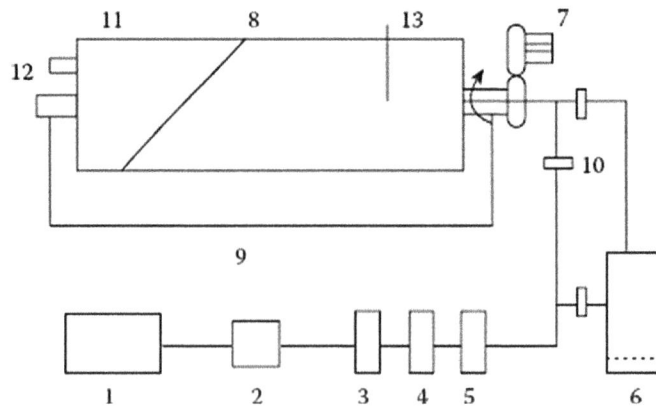

FIGURE 13.4 Rotation drum bioreactor: 1, air compressor; 2, pressure regulation valve; 3, oil separator; 4, air filter; 5, air heater; 6, atomizer; 7, rotating device; 8, rotary-drum fermentor; 9, fermentor stand; 10, gas valve; 11, fermentor cover, 12, air outlet; 13, thermistor (Pandey, A., *Process Biochem*, 27:109–17, 1992).

is also reduced to a large extent, compared with static trays or a packed-bed fermentor. Figure 13.4 shows a rotating drum bioreactor used by Tao et al. (1997) for cellulase production. However, during scale-up major limitations with all of the SSF bioreactors include the maintenance of a sterile environment, irregular growth and heat removal, which is another challenge toward engineering of SSF bioreactors.

13.7.4 RECENT ADVANCES IN SSF BIOREACTOR, AN EXAMPLE

Chen and Li, (2013) recently developed an industrial bio-process for non-isothermal simultaneous solid-state scarifi-cation, fermentation, and separation for ethanol production (NSSSFS), in which the enzymatic scarification and fermen-tation proceeded at around 50°C and 37°C, respectively, and were coupled together as a cascade based on the hydrolyza-tion. The glucose produced from the enzymatic scarification was consumed by yeast and the resulting ethanol was sepa-rated online by CO_2 gas stripping. This new system affords higher solids substrate loading and increased the efficiency of carbon conversion to ethanol production. The NSSSFS could, therefore, be used to resolve the inherent problems of simul-taneous scarification and fermentation during the course of ethanol at the industrial level.

13.8 CELLULASE PRODUCTION USING SSF AND SmF: A COMPARITIVE ANALYSIS

Cellulases, along with amylases and proteases, play a central role in creating sustainable bioprocess for the bioconversion of renewable resources into biofuels and other desirable end products. Among the many microorganisms that can produce cellulases, filamentous fungi are the most efficient and, as such, have been selected for the commercial production of cel-lulases. As a soil microorganism, filamentous fungi are well adapted to solid-state fermentation as it imitates their natural habitat. Industrialists have argued that cellulase production by SSF is more economical than submerged fermentation (SmF),

as it utilizes cheaper substrates and requires inexpensive infrastructure (Tengerdy R P, 1996; Singhania et al., 2010). Earlier reports concluded that cellulase production by SSF is ten times cheaper than cellulase produced by SmF (Tengerdy R P, 1996). The use of SSF also facilitates the accumulation of high concentrations of diverse biomass hydrolyzing enzymes and this, in turn, leads to higher yield of desirable end prod-ucts (Prajapati et al., 2018). Wheat bran, pretreated rice straw, wheat straw, and other lignocellulosic material have been used as solid substrates for cellulase production with a good deal of success. It has also been debated that cellulase produced on biomass for hydrolyzation at a later phase is superior, since it is expected to induce all the enzymes in the cellulolytic cock-tail required to hydrolyze that specific biomass. Singhania et al. (2015) have adopted SSF using a combination of wheat bran and avicel (4:1) as co-substrates to induce the production of cellulases. Upon the completion of fermentation, whole fermented matter was used as a cellulose-rich inoculum for the hydrolyzation of renewable resources. Using whole fer-mented matter as an inoculum has the added advantage of reducing the operating cost as well as avoiding the dilution of the hydrolytic capacity of the enzymes. This in turn avoids the need for harvesting, thus simplifying downstream pro-cessing, which in turn adds value to the whole process of bio-ethanol production. Sequential scarification and fermentation employing fermented matter, as well as extracted cellulose, was compared for bioethanol production. In this study, cel-lulase production, scarification of biomass and ethanol pro-duction were investigated in a single vessel and the results showed that SSF bioprocesses produce a cocktail of cellulases that are of superior quality and diverse specificities to those produce by SmF. This finding makes SSF an ideal tool for the hydrolysis renewable resources and their conversion into commercial products. A comparative analysis of cellulase activities produced by SSF (20FPU/gds) and SmF (3.1FPU/ ml) for the same fungus, revealed the superiority of SSF over the SmF by these criteria (Singhania et al., 2014, 2015). In the case of the requirement for higher solid loading, the amount of liquid which can be added is restricted, hence the need to concentrate the cellulase obtained by SmF, thus increasing the cost of production, In contrast, 1 g of SSF fermented matter contains 20 FPU of cellulase and can be used directly. Although direct comparison of SSF and SmF is not possible, SSF has the added advantage of allowing the accumulation of products to a much higher concentration.

Despite the many advantages of SSF, the commercial production of cellulases continues to be produced through SmF; principally due to the lack of adequate automation and the need to create biosensors for online monitoring of cellulases production in SSF bioreactors.

13.9 CHALLENGES IN SSF

SSF bioprocess faces a wide range of challenges, some of which are associated with its operation, while others are associated with fermentation monitoring and product harvesting. The substrates used in SSF differ greatly in composition, chemical nature, mechanical properties, particle size (including inter- and intra-particle spaces), water retention capacity, surface area, and so on. These factors affect the overall process design and product development.

Successful development of SSF bioprocesses is influenced by a number of factors, including technical and economic considerations. While several of these are of generic in nature, they still hold a significant impact and need to be considered in a holistic manner. These include the selection of microorganism and substrate, optimum physical-chemical and biological process parameters along with purification of the desired products. These are all ongoing challenges for SSF. Thus, optimization of downstream processing is considered to be the most important challenge in SSF. Separation of product from the solids is a technically and economically challenging process, and, although there is much advent in biochemical engineering, SSF is generally being used only for the production of metabolites where low levels of purity are required.

Understanding the impact of heat and mass transfer on the fermentation of a given primary feedstock in SSF is a critical aspect in fermentation and deserves more attention. It follows, control of heat and mass transfer represents a significant challenge for the bio-engineer, not only for successful design and operation of bioreactors, but also for scaling-up the SSF processes for the commercial production of bio-products. Furthermore, the heterogeneous nature of the substrate used in SSF (agro-industrial residues) represents another challenge, which has to be investigated through studying the kinetics of substrate conversion into bio-products, as well as establishing a viable model for the fermentation process itself (Pandey, 2003; Singhania et al., 2009; Ali and Zulkali, 2011). Another well-known and difficult challenge of SSF is bioreactor design for large-scale production. The design of bioreactor varies with substrate, organism, type of product, and so on. With the present advanced technology, several bioreactors are designed with the capability for online monitoring of several parameters alongside heat and mass transfer. Monitoring biomass growth and its estimation is necessary for kinetic studies of fermentation. However, in SSF, separation of biomass from solid mass is not possible and accurate measurement of biomass is almost impractical. There are several indirect methods to measure biomass and almost every method has its own drawback. In addition to these major drawbacks, there are several other minor drawbacks such as solid handling, waste management, and so on. If fermented solid is not itself the final product, waste-treatment expenses add heavily to the product cost. Various management strategies such as recycling of the spent solid as animal feed or biogas generation could be beneficial.

SUMMARY

- SSF is widely used in the production of many products, including animal feed, biofuel, and industrial and pharmaceutical chemicals.
- More recently, SSF has proved to be a great success in the production of biologically active secondary metabolites.
- In some applications, SSF has emerged as an attractive alternative to submerged fermentation.
- SSF is currently being used for bioremediation, bioleaching, bio pulping, and bio beneficiation.
- SSF stands to play a major role in the conversion of renewable resources in general and agricultural wastes into useful products.
- Although the production of cellulases using SSF is significantly cheaper than SmF, the commercial production of cellulases continue to be through SmF bioprocess, primarily because of the of lack adequate automation and online monitoring of industrial scale SSF.
- Full exploitation of SSF in bioprocess technology awaits extensive investigations and further investment to overcome the various limitations imposed on the use of SSF for the conversion of renewable resources to desirable end products.

REFERENCES

Abdullah, A.L., R.P. Tengerdy, and V.G. Murphy. 1985. Optimization of solid-state fermentation of wheat straw. *Biotechnol Bioeng* 27:20–27.

Ahamad, M.Z., B.P. Panda, S. Javed, and M. Ali. 2010. Production of mevastatin by solid-state fermentation using wheat bran as substrate. *Research Journal of Microbiology*. 5(11):1165–1169.

Ahmed, S.Y., B.K. Lonsane, N.P. Ghildyal, and S.V. Ramakrishna. 1957. Design of solid state fermentor for production of fungal metabolites on large-scale. *Biotechnol Techniques* 1:97–102.

Ali, H.K.Q., and M.M.D. Zulkali. 2011. Design aspects of bioreactors for solid-state fermentation: A review. *Chem. Biochem Eng Quart* 25:255–266.

Bei, Q., G. Chen, F. Lu, S. Wu, and Z. Wu. 2018. Enzymatic action mechanism of phenolic mobilization in oats (Avena sativa L.) during solid-state fermentation with *Monascus anka*. *Food Chem* 245:297–304.

Chen, H., and G. Li. 2013. An industrial level system with non-isothermal simultaneous solid-state saccharification, fermentation and separation for ethanol production. *Biochem Eng J* 74:121–126.

Chundakkadu, K. 2005. Solid-state fermentation systems: An overview. *Crit Rev Biotechnol* 25:1–30.

Fung, C.J., and D.A. Mitchell. 1995. Baffles increase performance of solid state fermentation in rotating drums. *Biotechnol Tech* 9:295–298.

Liu, J., Z. Wang, H. Li, C. Hu, P. Raymer, and Q. Huang. 2018. Effect of solid state fermentation of peanut shell on its dye adsorption performance. *Bioresour Technol* 249:307–314.

Lonsane, B.K., G. Saucedo-Castaneda, M. Raimbault, S. Roussos, G. Viniegra-Gonzalez, N.P. Ghildyal, M. Ramakrishna, and M.M. Krishnaiaha. 1992. Scale up strategies for solid state fermentation systems. *Process Biochem* 27:259–273.

Mitchell, D.A., Z. Targonski, J. Rogalski, and A. Leonowicz. 1992. Substrates for processes in solid substrate cultivation. H.W. Doelle, D.A. Mitchell, C.E. Rolz (eds.), *Solid Substrate Cultivation*, pp. 29-52. London: Applied Science.

Narahara, H., Y. Koyama, T. Yoshidu, P. Atthaspunna, and H.J. Taguchi. 1984. Control of water content in a solid state culture of *Aspergillus oryzae. Ferment Tech* 62:453–459.

Pandey, A. 1992. Recent process developments in solid-state fermentation. *Process Biochem* 27:109–117.

Pandey, A. 1994. Solid-state fermentation: An overview. In A. Pandey (ed.), *Solid State Fermentation*, pp. 3–10. New Delhi, India: Wiley.

Pandey, A. 2003. Solid-state fermentation, *Biochem. Eng. J.* 13:81–84.

Pandey, A., C.R. Soccol, J.A.R. Leo, and P. Nigam. 2001. *Solid-State Fermentation in Biotechnology*. New Delhi, India: Asiatech Publishers.

Prajapati, B.P., R.K. Suryawanshia, S. Agrawala, M. Ghosh, and N. Kangoa. 2018. Characterization of cellulase from *Aspergillus tubingensis* NKBP-55 for generation of fermentable sugars from agricultural residues. *Bioresour Techol* 250:733–740.

Ramachandran, S., P. Fontanille, A. Pandey, C. Larroche. 2007. Spores of *Aspergillus niger* as reservoir of glucose oxidase synthesized during solid-state fermentation and their use as catalyst in gluconic acid production. *Lett Appl Microbiol* 44:155–160.

Reusser, F., H.I. Koepseil, and G.M. Savage. 1961. Degeneration of *Streptomyces niveus* with repeated transfers. *Appl Microbiol* 9:342–345.

Saucedo-Castaneda, G., M. Guierrez Rojas, G. Bacquet, M. Raimbault, and G. Viniegra Gonzalez. 1990. Heat transfer simulation in solid substrate fermentation. *Biotechnol Bioengin* 35:802.

Singh, A., M. Adsul, N. Vaishnav, A. Mathur, and R.R. Singhania. 2017. Improved cellulase production by *Penicillium janthinellum* mutant. *Indian Journal of Experimental Biology* 55:436–440.

Singhania, R.R., A.K. Patel, C.R. Soccol, and A. Pandey. 2009. Recent advances in solid-state fermentation. *Biochem Eng J* 44:13–18.

Singhania, R. R., R. K. Sukumaran, A. K. Patel, C. Larroche A. Pandey. 2010. Advancement and comparative profiles in the production technologies using solid-state and submerged fermentation for microbial cellulases. *Enzyme Microbial Technol* 46:541–549.

Singhania, R. R., J.K. Saini, R. Saini, M. Adsul, A. Mathur, R. Gupta, and D. Tuli. 2014. Bioethanol production from wheat straw via enzymatic route employing *Penicillium janthinellum* cellulases. *Bioresource Technology* 169:490–495.

Singhania, R. R., R. Saini, M. Adsul, J.K. Saini, A. Mathur, and D.K. Tuli. 2015. An integrative process for bio-ethanol production employing SSF produced cellulase without extraction. *Biochem Eng J* 102:45–48.

Singhania, R.R., A.K. Patel, L. Thomas, and A. Pandey. 2017. Solid-state fermentation. In C. Wittmann, J.C. Liao (eds.), *Industrial Biotechnology: Products and Processes*, pp. 187–204. Hoboken: John Wiley & Sons.

Stuart, D.M., D.A. Mitchell, M.R. Johns, and J.D. Lister. 1998. Solid-state fermentation in rotating drum bioreactors: Operating variables affect performance through their effects on transport phenomena. *Biotechnol Bioeng* 63:383–391.

Tao, S., L. Beihui, and L. Zuohu. 1997. Enhanced cellulase production in fed-batch solid-state fermentation for *Trichoderna viride* SL-1. *J Chem Tech Biotechnol* 69:429–432.

Tengerdy, R.P. 1996. Cellulase production by solid substrate fermentation. *J Sci Ind Res* 55:313–316.

Thomas, L., C. Larroche, A. Pandey. 2013. Current developments in solid-state fermentation. *Biochem Eng J* 81:146–161.

Vong, W.C., X.Y. Hua, and S.Q. Liu. 2018. Solid-state fermentation with *Rhizopus oligosporus* and *Yarrowia lipolytica* improved nutritional and flavour properties of okara, LWT (2018):316–322.

Wang, E.Q., S.Z. Li, L. Tao, X. Geng, and T.C. Li. 2009. Modeling of rotating drum bioreactor for anaerobic solid-state fermentation. *Appl Energy* 87:2839–2845.

Section V

Fermentation Microbiology and Biotechnology

Tools, Monitoring, and Control of Fermentation Processes

14 Functional Genomics
Current Trends, Tools, and Future Prospects in the Fermentation and Pharmaceutical Industries

Surendra K. Chikara, Toral Joshi, and Mahmoud M. A. Moustafa

CONTENTS

"When you are face to face with a difficulty, you are up against a discovery."

Lord Kelvin

14.1 INTRODUCTION

Functional genomics is a discipline and enterprise that exploits the vast wealth of data produced by genome sequencing projects. A key feature of functional genomics is their genome-wide approach, which invariably utilizes high-throughput technologies and relies on sophisticated analytical tools.

Understanding the genome sequence and its relevance for various applications is therefore central to the development and production of new biopharmaceuticals. Current advances in bioinformatics and high-throughput technologies such as microarray analysis have revolutionized our perception and understanding of the molecular mechanisms underlying normal and abnormal (dysfunctional) biological functions. Microarray studies and other genomic techniques are also inspiring the discovery of new targets for disease treatment and control, thus aiding drug development immunotherapeutics, and gene therapy (Zhao et al., 2011; Beitelshees et al., 2017). Current trends and challenges in the field include

- Exploitation of new innovations in nanotechnology and microfluidics for the development of low-cost technologies for sequencing and genotyping as well as for the identification and confirmation of functional elements that do not encode protein (e.g., introns, promoters, **telomeres**, and regulatory and structural features);

    Telomere: A telomere is a region of repetitive DNA located at the end of chromosomes to protect them from degradation.
- *In vivo*, real-time monitoring of gene expression and functional modification of gene products in all relevant cell types using large-scale mutagenesis, small-molecule inhibitors, and knockdown approaches;
- Identifying genes and pathways that have thus far proved difficult to study biochemically and deciphering cellular phenomena related to the networking of metabolic pathways;
- Monitoring of membrane proteins, modified proteins, and regulatory proteins of low concentrations; and
- Correlation of genetic variation to human health and disease using haplotype and comprehensive variation information to provide large databases that are amenable to statistical methods.

The challenges faced in the human genome project and the need for high-throughput capacity have led to the development of next-generation sequencing technology (NGST), which has dramatically accelerated biological and biomedical research because it renders the comprehensive analysis of genomes, transcriptomes, and interactomes inexpensive, which should be helpful toward achieving the goal of personalized medicine and designer crops (Horner et al., 2010). Of pharmaceutical relevance is sequencing of multiple strains of pathogens to monitor drug resistance and pathogenicity. Resequencing of selected regions to search for human variation in population and for tumor profiling to guide cancer therapies is also of great interest. In this chapter, we review current trends and methodologies in functional genomics and assess their relevance, strengths, and weaknesses.

Completion of the human genome project signaled a new beginning for modern biology, one in which most biological and biomedical research will be conducted in a sequence-based fashion. It is now possible to deconstruct the genomic sequences of multiple organisms with the view to unravel the interrelationship between the physiology or the pathogenicity of a given organism and its genome constituent parts (genes) (Cook Degan, 1991; 56Johnson, 1992; Greenhalgh, 2005; Evans, 2010). The deconstruction of the genome to assign biological functions to genes, groups of genes, and particular gene-gene or gene-protein interactions is well underway and documented in MEDLINE (Patterson and Gabriel, 2009). These functions may be directly or indirectly the result of a gene's transcription (Mamanova et al., 2010). A computationally intensive branch of functional genomics has emerged as a result of the practical implementation of technologies to assess gene expression of thousands of genes at a time. The ability to comprehensively measure gene expression affords an excellent opportunity to further expand a target-oriented approach as opposed to the conventional hypothetical-based approach.

The need to generate, analyze, and integrate large and complex sets of molecular data has led to the development of whole-genome approaches, such as microarray technology, to expedite the process of translating molecular data into biologically meaningful information (Miyake and Matsumoto, 2005). The comparison of the strengths and weakness of next-generation sequencing and microarray techniques is illustrated in Table 14.1 (Asmann et al., 2008).

The paradigm shift from traditional single-molecule studies to whole-genome approaches requires standard statistical modeling and algorithms as well as high-level hybrid computational/statistical automated learning systems for improving

TABLE 14.1

The Strength and Weaknesses of NGS and Microarray Techniques

Microarray analysis		Massively parallel, or NGS	
Pros	**Cons**	**Pros**	**Cons**
Relatively inexpensive	High background, low sensitivity	Low background, very sensitive	Expensive
Easy sample preparation	Limited dynamic range	Large dynamic range	Complex sample preparation
Mature informatics and statistics	Not quantitative	Quantitative	Limited bioinformatics
Competitive hybridization		Massive information technology infrastructure required	
Annotation of the probes			

Source: Information from Asmann, Y.W., M.B. Wallace, and E.A. Thompson. 2008. *Gastroenterology*, 135:1466–1468, and reproduced with the kind permission of Elsevier, New York.

the understanding of complex traits. Microarray experiments provide unprecedented quantities of genome-wide data on gene expression patterns. The implementation of a successful uniform program of expression analysis requires the development of various laboratory protocols and the development of database and software tools for efficient data handling and mining (Teng and Xiao, 2009). The computational tools necessary to analyze the data are rapidly evolving. The effect of microarray measurements on biology and bioinformatics has been astonishing.

14.2 MICROARRAYS: ROLE IN FUNCTIONAL GENOMICS

Complementary DNA microarrays, oligonucleotide microarrays, or serial analysis of gene expression (SAGE) are widely used microarray tools (Antipova et al., 2002; Tuteja and Tutej, 2004). In this section, we will focus on microarrays, which are artificially constructed grids of DNA such that each holds a DNA sequence that is a reverse complement to the target RNA sequence. Although there are many protocols, the basic technique involves the incorporation of fluorescent nucleotides or a tag, which is later stained with fluorescence, into the extracted RNA. The labeled RNA is then hybridized to a microarray for a period of time, after which the excess is washed off and the microarray is scanned under laser light. All probes are designed to be hypothetically similar with regard to hybridization temperature and binding affinity for oligonucleotide microarrays (Brennan et al., 2004; Cimaglia et al., 2018), and for which each microarray measures the level of each RNA molecule in each sample, although this absolute measurement might not correlate exactly with RNA concentration in terms of micrograms per unit volume (Steibel and Rosa, 2005). With cDNA microarrays (Rogers et al., 2005; Bumgarner, 2013), in which each probe has its own hybridization characteristic, each microarray measures two samples and provides a relative measurement level for each RNA molecule. Because a complete experiment involves hundreds of microarrays, the resultant RNA expression data sets can vary greatly in size. As the cost of microarrays continues to drop,

it is clear that microarrays are becoming more integral to the drug discovery programs (Wu et al., 2005; Mustapha et al., 2017). In addition to the obvious use of functional genomics in basic research and target discovery, there are many other specific uses (e.g., biomarker determination) to find genes that correlate with and presage disease progression, and in toxicogenomics, to find gene expression patterns in tissues or organisms exposed in response to a given drug to develop early warning diagnostic predictors for the onset of disease (Wu et al., 2005). Many free and commercial software packages are available to analyze microarray data, although it is still difficult to find a single off-the-shelf software package that answers all functional genomics questions (Aburatani, 2005; Shittu et al., 2018).

The two basic applications of DNA microarrays are studying genome structure and gene expression analysis.

14.2.1 APPLICATIONS OF MICROARRAY TO THE STUDY OF GENOME STRUCTURE

Microarray methods used for the study of genome structure include

- *Comparative genomic hybridisation (CGH) arrays*: This method has been widely used for the detection of large changes in the genome (e.g., deletions, insertions, and copy number variation) (Miyake and Matsumoto, 2005; Choi et al., 2018).
- *Single-nucleotide polymorphism (SNP) arrays*: This method is useful in genotyping, designed at present to identify SNPs, other than that it also identifies loss of heterozygosity (LOH), copy number variation (CNV), allelic imbalance (AI), and uniparental dispute (UPD) (You et al., 2018).
- *Tiling arrays*: This method is used in the study and analysis of DNA-protein interactions, methyl-DNA immunoprecipitation (MeDIP-chip, epigenetic studies), chromatin hypersensitive sites localization (DNase-chip), gene annotation, and mapping (Johnson et al., 2006).

14.2.2 Applications of Microarray to the Study of Gene Expression and Profiling

Gene expression profiles are precise, and the methods of microarrays used and their applications are as follows:

- *cDNA arrays*: In this method, the probes and cDNA clones are spotted onto a glass slide using printing robots.
- *Tiling arrays*: Play a major role in discovering new transcriptionally active regions and splice variants.
- *Exon arrays*: Utilizes separate probes designed to detect individual exons and is useful in the detection of splicing isoforms.
- *CSH array*: This method is specially designed for the organisms for which the genome is not sequenced and helps to identify the gene expression for the same.

A few microarray platforms have been developed to accomplish this task, and the basic idea for each is simple: A glass slide or membrane is spotted or "arrayed" with DNA fragments or an oligonucleotide that represents specific gene coding regions. After that, purified RNA is fluorescently or radioactively labeled and hybridized to the slide or membrane. In some cases, hybridization is done simultaneously with reference RNA to facilitate comparison of data across multiple experiments. Subsequent to thorough washing, the raw data are obtained by laser scanning or autoradiographic imaging (Figure 14.1).

14.3 MICROARRAY QUALITY CONTROL

Microarray quality control (MAQC) compares performance of different microarray platforms with respect to their sensitivity, specificity, dynamic range, precision, and accuracy (Hardiman, 2004; MAQC Consortium, 2006; Kricka and Master, 2009). The study included microarrays from five major vendors—Affymetrix, Agilent, Applied BioSystems, Nimblegen, and Illumina—and is published online in *Nature Biotechnology*, 2006 (www.nature.com/nature/journal/v442/n7106/full/4421067a.html). One of the major limitations of expanding the use of microarray technology is that the values of expression generated on different platforms cannot be directly compared because of the unique labeling methods and probe sequences used. However, the results of the MAQC consortium demonstrated that most major commercial platforms can be selected with confidence (Hester et al., 2009).

14.3.1 Affymetrix

Affymetrix microarray technology is based on the hybridization of small, high-density arrays containing tens of thousands of synthetic oligonucleotides (Elo et al., 2005; Lee et al., 2018). The arrays are designed based on sequence information alone and are synthesized in situ using a combination of photolithography and oligonucleotide chemistry. RNAs present at a frequency of 1:300,000 are readily detected and quantified; the method is readily scalable to the simultaneous monitoring of tens of thousands of genes. The Affymetrix integrated GeneChip arrays include up to 500,000 unique probes corresponding to tens of thousands of gene expression measurements (Della et al., 2008; Cho et al., 2018).

Affymetrix manufactures arrays that monitor the global activities of genes in yeast, *Arabidopsis*, *Drosophila*, mice, rats, and humans. Apart from that, custom expression arrays can be designed for other model organisms, proprietary sequences, or specific subsets of known genes. For the

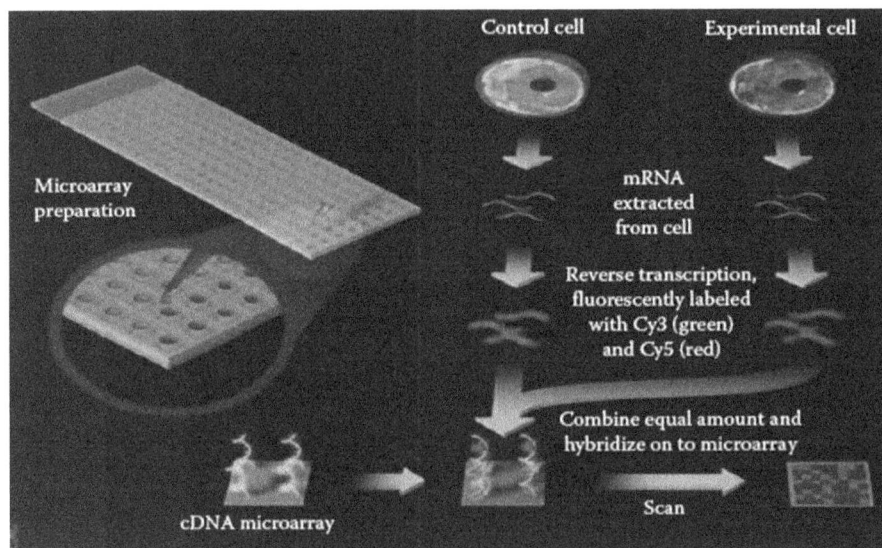

FIGURE 14.1 Microarray technology. The picture depicts extraction of mRNA from two types of cells (e.g., tumor cells and control cells) and then labels the samples with different fluorescent dyes. When washed over the microarray, these colored transcripts bind to their complementary probes, leaving a trail of informative spots: red for genes turned on in cancer cells, green for genes turned on in normal cells, yellow for genes turned on in both types of cells—with more intense color indicating higher gene activity. (Courtesy of the *Science Creative Quarterly*, Jiang Long, artist.)

starting point, the clusters are used, and sequences are further subdivided into subclusters representing distinct transcripts. This categorization process involves alignment to the genome, which reveals splicing and polyadenylation variants. The newest product for Affymetrix is Axiom for genome-wide association studies (GWAS), replication studies, and candidate gene association studies. It includes predesigned and personalized array plates with validated genomic content from the Axiom genomic database. This solution also includes complete reagent kits, data analysis tools, and a fully automated workflow.

14.3.2 NimbleGen MS 200 Microarray and Scanner

Roche NimbleGen, Inc. distinctively produces high-density arrays of long oligonucleotide probes that provide better information content and higher data quality necessary for studying the full diversity of genomic and epigenomic variation. Roche NimbleGen ensures high-definition genomics by providing scientists with cost-effective, high-throughput tools for extracting and integrating complex data on important forms of genomic and epigenomic variation not previously accessible on a genome-wide scale. This enhanced feature is made possible by Roche NimbleGen's proprietary Maskless Array Synthesis (MAS) technology, which utilizes digital light processing, and rapid, high-yield photochemistry

to synthesize long oligonucleotide, high-density DNA microarrays with extreme flexibility, thus facilitating a clearer understanding of epigenomics. Roche NimbleGen offers a complete catalog of gene expression arrays for whole-genome expression profiling. With the unique combination of long oligonucleotide probes, flexible array content, and high probe density (up to 385,000 probes per array), NimbleGen gene expression arrays enable analysis of eukaryotic and prokaryotic genomes. Built on the ultrahigh-density HD2 platform, the 12- by 135-K array format combines a multiplex option with high probe density for high-throughput projects (Figure 14.2). It allows for running of replicates within the same array that validate analysis within the same experiment. Also, conducting the array hybridization is relatively straightforward, and analysis of the results is user-friendly and does not require highly specialized informatics skills (Mueckstein et al., 2010; Reutersward, 2018).

NimbleGen microarrays enable precise, sensitive, and specific interrogation of genome-wide expression for any prokaryotic or eukaryotic sequenced and annotated genome.

14.3.3 Agilent DNA Microarray and Scanner

Agilent Technologies manufactures probes using a proprietary DNA synthesis method based on inkjet printing technology for microarrays. The method, in which layers of DNA

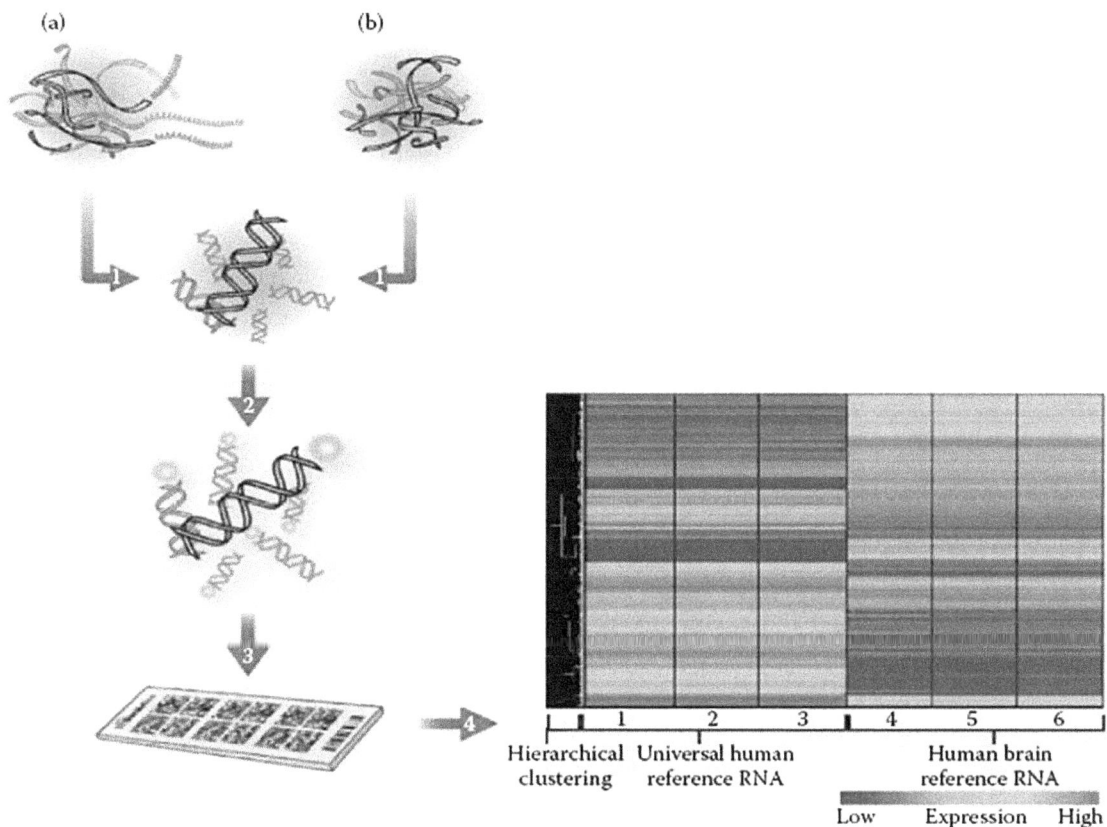

FIGURE 14.2 (1) cDNA synthesis: eukaryotic (a) or prokaryotic mRNA (b) is converted via oligo-dT or random priming, respectively, to double-stranded cDNA. (2) cDNA labeling: the cDNA is labeled with Cy3 dye. (3) Hybridization: labeled cDNA is hybridized to a NimbleGen gene expression array. (4) Data analysis: Data are extracted using NimbleScan software and analyzed for differentially expressed genes.

nucleotides are "printed" onto desired microarray feature locations to synthesize probes, circumvents many of the limitations of light-based synthesis methods and is capable of producing oligonucleotide probes of unprecedented quality and length. Light-based DNA chemistry is another common method implemented for synthesizing oligonucleotide probes for microarrays (Zahurak et al., 2007; Chow et al., 2009). Agilent has combined some of the best features of both of these approaches and has gained widespread acceptance as a result.

Agilent combines two-sample hybridization with use of long (60-mer) oligonucleotides. These arrays are hybridized with two different fluorescent samples and measurements of differential expression obtained from the relative abundance of hybridized mRNA. Agilent's methodology achieves an extraordinary 99.5% stepwise DNA synthesis yield and produces probes with very high sequence fidelity. This platform can produce accurate relative quantization of gene expression levels over a range of more than five orders of magnitude, far exceeding the capabilities of other systems.

Agilent's Feature Extraction algorithms were developed aiming to reduce systematic errors that arise from labeling bias, irregular feature morphologies, and mismatched sample concentrations and cross-hybridization. They quantify feature signals and their background, perform background subtraction and dye normalization, and calculate feature log ratios and error estimates (Lopez-Romero et al., 2010). The error estimates, which are based on an extensive error model and pixel-level statistics calculated from the feature and background for each spot, are used to generate a P value for each log ratio. The file produced by Agilent's extraction software also contains raw pixel intensity data. These intensities, whether they are mean or median values, can easily be exported to other software, such as R.

14.4 MICROARRAY APPLICATIONS

14.4.1 GENE EXPRESSION PROFILING

Gene expression profiling is the measurement of the activity (the expression) of thousands of genes at once to create a global picture of cellular function (Velculescu et al., 1995). These profiles can distinguish between cells that are actively dividing or show how the cells react to a particular treatment. Many experiments of this sort simultaneously measure an entire genome; that is, every gene present in a particular cell (Shiu and Borevitz, 2008).

In an mRNA or gene expression profiling experiment, the expression levels of thousands of genes are simultaneously monitored to study the effects of certain treatments, diseases, and developmental stages on gene expression. Gene expression profiling may become an important diagnostic test. Expression profiling provides new information about what genes do under various conditions. Overall, microarray technology produces reliable expression profiles. Apart from this the other applications are in the field of comparative genomic hybridization using oligonucleotide microarray (Barrett et al., 2004; Fallahshahroudi et al., 2018).

14.4.2 CGH

CGH provides an alternative means of genome-wide screening for CNVs (Pinkel and Albertson, 2005). First developed to detect copy number changes in solid tumors, CGH uses two genomes—a test and a control—that are differentially labeled and competitively hybridized to metaphase chromosomes. The fluorescent signal intensity of the labeled test DNA relative to that of the reference DNA can then be linearly plotted across each chromosome, allowing for the identification of copy number changes.

Unlike traditional techniques used to detect copy number gains and losses, which rely on the examination of a single target and prior knowledge of the region under investigation, CGH can be used to quickly scan an entire genome for imbalances (Hester et al., 2009; Asakawa et al., 2016). In addition, CGH does not require cells that are undergoing division. However, as with earlier cytogenetic methods, the resolution of CGH has been limited to alterations of approximately 5–10 Mb for most clinical applications.

14.4.3 DETECTION OF SNP

An SNP array is a useful tool to study the whole genome. The most important application of SNP array is in determining disease susceptibility and, consequently, in pharmacogenomics by measuring the efficacy of drug therapies specifically for the individual. Because each individual has many SNPs that together create a unique DNA sequence, SNP-based genetic linkage analysis could be performed to map disease loci and hence determine disease susceptibility genes for an individual (Rauch et al., 2004). The combination of SNP maps and high-density SNP array allows for the efficient use of SNPs as the markers for Mendelian diseases with complex traits. A SNP array can also be used to generate a virtual karyotype using specialized software to determine the copy number of each SNP on the array and then align the SNPs in chromosomal order. In addition, SNP array can be used for studying the LOH. LOH is a form of AI that can result from the complete loss of an allele or from an increase in copy number of one allele relative to the other (Leykin et al., 2005; Kamath et al., 2009; McGranahan et al., 2017). SNP array is also capable of detecting extra chromosomes due to UPD.

Karyotype [kar-ee-uh-tahyp]: A systematized arrangement of chromosomes displayed in pairs and arranged in descending order of size.

14.4.4 CHROMATIN IMMUNOPRECIPITATION

ChIP-on-chip is a technique that combines chromatin immunoprecipitation with microarray technology. ChIP-on-chip is used to investigate *in vivo* interactions between proteins and DNA. It allows for the identification of the cistrome, which is the set of DNA binding sites of a *trans*-acting factor on a genome-wide basis. Whole-genome analysis can be performed to determine the locations of binding sites for almost any protein of interest. The aim is to localize protein binding sites that may help identify functional elements in the genome. For example, in the case of a transcription factor as a protein of interest, one can determine its transcription factor binding sites throughout the genome. Other proteins allow for the identification of promoter regions, enhancers, repressors and silencing elements, insulators, boundary elements, and sequences that control DNA replication. One of the long-term goals of ChIP-on-chip is to establish a catalogue of organisms that lists all protein-DNA interactions under various physiological conditions (Von, 2008). This knowledge would ultimately help in the understanding of the machinery behind gene regulation, cell proliferation, and disease progression. Hence, ChIP-on-chip offers not only huge potential to complement our knowledge about the orchestration of the genome on the nucleotide level, but also on higher levels of information and regulation as it is propagated by research on epigenetics (Gebauer, 2004; Lambert et al., 2018).

14.4.5 Transcriptome Analysis

Whole **transcriptome** analysis is one of the most general requirements to address many applied or basic biological questions. One of the most efficient tools to carry out such analyses relies on the use of microarrays coupled to specific labeling of cDNA populations generally reflecting two states of the transcriptome. Up until recently, DNA microarray was the procedure of choice for transcriptome analysis, but it has now been superseded by high-throughput sequencing technologies (RNA-Seq), which have become an additional alternative to microarrays.

> **Transcriptome**: A transcriptome is the entire set of all RNA molecules, including mRNA, rRNA, and tRNA species that exist in a given organism under certain conditions. Because it includes all mRNA transcripts in the cell, the transcriptome reflects the genes that are actively expressed under certain conditions.

14.4.6 Application of DNA Microarray in the Pharmaceutical Industries

Natural product research is often based on ethnobotanical information, and many of the drugs used today were used in indigenous societies. The aim of ethnopharmaceutical research is better understanding of the pharmacological effects of different medicinal plants traditionally used in healthcare. Plants are regarded as a promising source of novel therapeutic agents because of their higher structural diversity as compared with standard synthetic chemistry. Plants have applications in the development of therapeutic agents as a source of bioactive compounds for possible use as drugs. There are three approaches to natural product-based drug discovery: screening of crude extracts, screening of prefractionated extracts, and screening of pure compounds. There are three main applications of DNA microarrays:

1. The pharmacodynamics for discovery of new diagnostic and prognostic indicators and biomarkers of therapeutic response; elucidation of molecular

mechanism of action of an herb, its formulations, or its phytochemical components; and identification and validation of new molecular targets for herbal drug development (Crowther, 2002; Dumbrava et al., 2018).

2. Pharmacogenomics for prediction of potential side effects of the herbal drug during preclinical activity and safety studies, identification of genes involved in conferring drug sensitivity or resistance, and prediction of patients most likely to benefit from the drug and use in general pharmacogenomics studies (Ko et al., 2016).

3. Pharmacognosy for correct botanical identification and authentication of crude plant materials, which is done as part of standardization and quality control (Crowther, 2002; Binitha et al., 2018).

14.4.7 Microarray Future Prospects

At present, the major practical applicability of DNA microarrays remains in SNP mapping, geno-typing, and pharmacogenetics. In recent years, array-based sequencing that combines target hybridization with enzymatic primer extension reactions has emerged as a powerful means to scan for all possible DNA sequence variations. Although DNA microarrays have huge potential for pharmacodynamic and toxicogenomic applications, these are still in the exploratory stage and need validation by other biological experiments (Debouck and Goodfellow, 1999). With the development of new, uniform, and more sophisticated experimental designs, data management system statistical tools, and algorithms for data analysis, DNA microarrays can be optimally used in herbal drug research. Despite the huge potential offered by microarray technology, the importance of *in vitro* biological assays, cell-line studies, and *in vivo* animal studies cannot be ignored. A comprehensive strategy integrating information from diverse scientific experiments and technologies will lead to molecular evidence-based herbal medicine.

> **Functional genomics**: Functional genomics is a discipline of biotechnology that attempts to exploit the vast wealth of data produced by genome sequencing projects. A key feature of functional genomics is its genome-wide approach, which invariably involves the use of a high-throughput approach.

Microarrays play an important role in functional genomics, and the completion of many genome sequencing projects fuels more and more projects in the area (Jares, 2006). Genomic differences such as SNPs can result in functional differences by changing particular entities (e.g., protein coding parts) or affecting gene regulation (regulatory SNPs). In short, genotypic differences need to be seen in a functional context and do not provide conclusive information on their own or in isolation.

A thorough downstream analysis of functional connections and consequences is mandatory to close the gap between mere data (sequence tags) and functional changes observed on a higher level paralleling the requirements already known from microarray analyses (Putonti, 2007). The very first step on this path from data to knowledge is to put the raw sequence reads into the coordinate system of a reference genome via genome annotation establishing the physical context.

14.5 NEXT-GENERATION DNA SEQUENCING

Next-generation sequencing (NGS) is well suited to provide all of the primary data required (i.e., the sequences) for genomics, epigenetics, transcription factor binding, and transcriptomics in virtually unlimited detail. NGS also enables the cheapest and fastest method of *de novo* sequencing and resequencing. New sequencing technologies have emerged, including 454 life sciences (Roche), Genome Analyser (Illumina, Inc.), SOLiD (Applied BioSystems), HeliScope (Helicos BioSciences), Pacific Bio, VisiGen, and Ion Torrent, and provide sequencing capability with much higher throughputs and at greatly reduced costs (Mardis, 2008a; Anderson and Schrijver, 2010; Ansorge, 2018). In Section 14.5.1, we shall highlight the strength and limitations of current and emerging techniques.

14.5.1 First-Generation Sequencers

14.5.1.1 Sanger's Sequencing Method

Sanger's method, also referred to as "dideoxy sequencing" or "chain termination," is based on the use of dideoxynucleotides (ddNTPs) in addition to the normal nucleotides (NTPs) found in DNA. ddNTPs differ from NTPs in that the hydroxyl group at the 3'-carbon atom is replaced by a hydrogen atom. Incorporation of a ddNTP will prevent the formation of a 3'-5' phosphodiester bond, which in turn brings chain extension to an end. Apart from nonspecific binding of the primer to the DNA, this method is time-consuming and excessively expensive.

14.5.2 Second-Generation Sequencers

The second-generation sequencers are those based on massive parallel analysis. "Second generation" is in reference to the various implementations of cyclic-array sequencing that has recently been realized in a commercial product. A detailed comparison of second-generation sequencers is given in Table 14.2.

14.5.2.1 Roche GS FLX (Pyrosequencing)

In pyrosequencing, incorporation of each nucleotide by DNA polymerase results in the release of pyrophosphate, which initiates a series of cascading reactions that ultimately yield light by the firefly enzyme luciferase. The amount of light produced is proportional to the number of nucleotides incorporated. In the Roche/454 approach, the library fragments are mixed with a population of agarose beads for which the surfaces are coated with oligonucleotides complementary to the 454-specific adapter sequences on the fragment library so that each bead is associated with a single fragment (Mardis, 2008a). Each of these fragment bead complexes is isolated into individual oil water micelles that also contain PCR reactants, and thermal cycling (emulsion PCR) of the micelles produces approximately 1 million copies of each DNA fragment on the surface of each bead. These amplified single molecules are then sequenced in mass. First, the beads are arrayed into a Picotiter Plate that holds a single bead in each of several hundred thousand single wells; this provides a fixed location at which each sequencing reaction can be monitored. Enzyme-containing beads that catalyze the downstream pyrosequencing reaction steps are then added to the Picotiter Plate (PTP), and the mixture is centrifuged to surround the agarose beads (Ansorge, 2009). On the instrument, the PTP acts as a flow cell into which each pure nucleotide solution is introduced in a stepwise fashion with an imaging step after each nucleotide incorporation step. The PTP is seated opposite a CCD camera that records the light emitted at each bead. The first four nucleotides (TCGA) on the adapter fragment adjacent to the sequencing primer added in library construction correspond to the sequential flow of nucleotides into the flow cell. For more details, the reader is advised to consult www.454.com (Morozova and Marra, 2008a).

The present Genome Sequencer FLX Instrument, powered by GS FLX Titanium series reagents, features a groundbreaking combination of long reads, exceptional accuracy, and high throughput. The breadth of applications includes *de novo*

TABLE 14.2
Comparison Between Second-Generation Sequencers Depending on the Read Lengths Data per Run, Cost of the Machine, and Cost of the Base per Run

	ABI 3730 XL (first released in 2005)	Roche GS FLX (first released in October 2005)	GAIIx (first released in June 2006)	5500, 5500XL, and SOLiD4 (first released in October 2007)
Read length	600–900 bp	25, 100, 300, 400, and 500 bp	18, 26, 36, 50, 75, and 100 bp	10,25,35,50, and 75 bp
Data per run	1 mbps	350–450 Mb	30–100 Gb	40–300 Gb
Machine cost	$400,000	$700,000	$850,000	800,000$
Run cost per base	$1000/Mb	$27,000/Gb	$1200/Gb	1250$/Gb
Limitation	Expensive, low throughput, labor-intensive	High cost, repetitive region	Short read assembly mapping	Short read assembly mapping
Application of choice	*De novo* sequencing of complex genome	*De novo* genome, *de novo* transcript, metagenomics	Resequencing SNP transcriptome	Resequencing SNP, miRNA transcriptome

sequencing and resequencing of whole genomes, metagenomics, and RNA analysis (Shendure and Ji, 2008; Voelkerding et al., 2009; Fakruddin et al., 2013; Weickardt et al., 2018).

14.5.2.2 Illumina Genome Analyzer

The Solexa sequencing platform was commercialized in 2006. The principle is based on sequencing-by-synthesis chemistry with novel reversible terminator nucleotides for the four bases, each labeled with a different fluorescent dye, and a special DNA polymerase enzyme able to incorporate them. DNA fragments are ligated at both ends to adapters and, after denaturation, are immobilized at one end on a solid support. The surface of the support is coated densely with the adapters and the complementary adapters (Ansorge, 2009). Each single-stranded fragment, immobilized at one end on the surface, creates a "bridge" structure by hybridizing with its free end to the complementary adapter on the surface of the support. In the mixture containing the PCR amplification reagents, the adapters on the surface act as primers for the following PCR amplification. Again, amplification is needed to obtain sufficient light signal intensity for reliable detection of the added bases. After several PCR cycles, random clusters of approximately 1000 copies of single-stranded DNA fragments (termed DNA "polonies," resembling cell colonies after polymerase amplification) are created on the surface. The reaction mixture for the sequencing reactions and DNA synthesis is supplied onto the surface and contains primers, four reversible terminator nucleotides each labeled with a different fluorescent dye, and the DNA polymerase. After incorporation into the DNA strand, the terminator nucleotide, as well as its position on the support surface, is detected and identified via its fluorescent dye by the CCD camera as shown in Figure 14.3 (Mardis, 2008b). The terminator group at the 3'-end of the base and the fluorescent dye are then

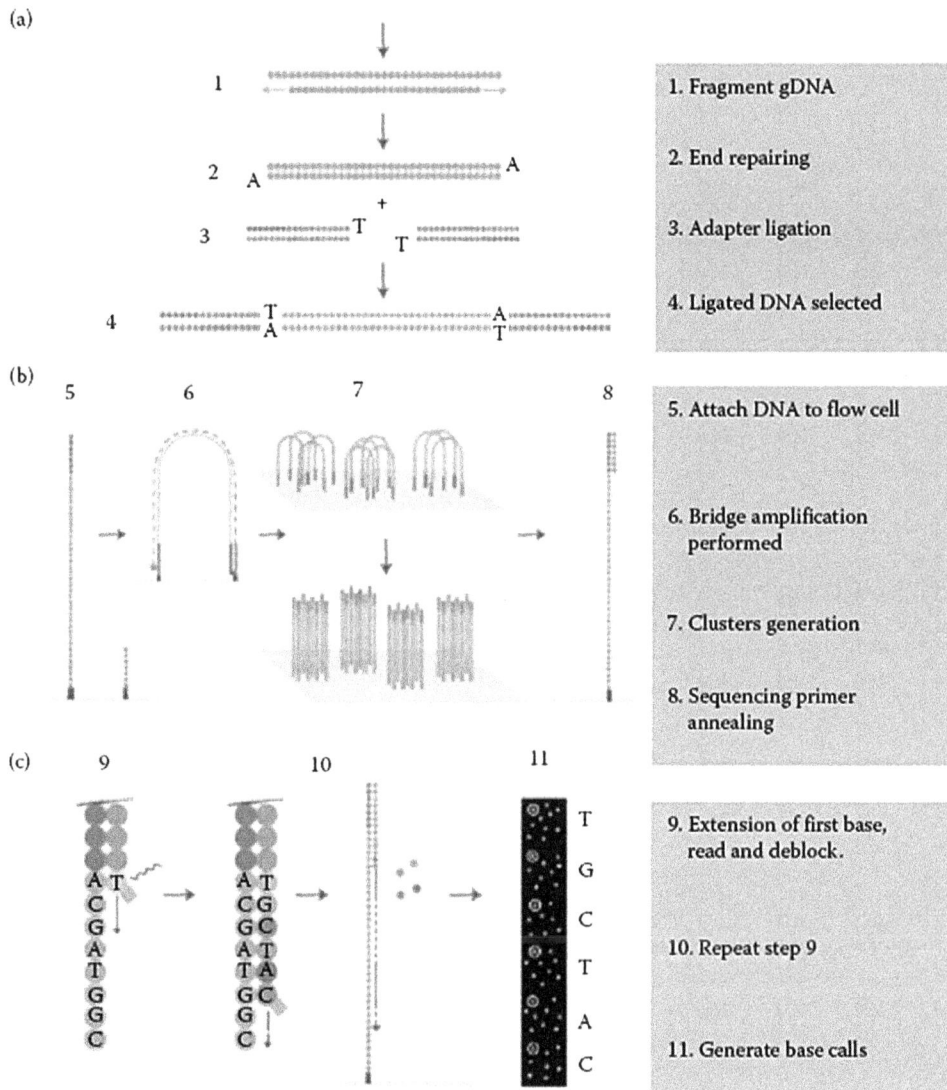

FIGURE 14.3 Outline of the Illumina Genome Analyzer workflow. Similar fragmentation and adapter ligation steps take place (a), before applying the library onto the solid surface of a flow cell. Attached DNA fragments form "bridge" molecules that are subsequently amplified via an isothermal amplification process, leading to a cluster of identical fragments that are subsequently denatured for sequencing primer annealing (b). Amplified DNA fragments are subjected to sequencing-by-synthesis using 30 blocked labeled nucleotides (c).

removed from the base and the synthesis cycle is repeated. The sequence read length achieved in the repetitive reactions is approximately 35 nucleotides. The sequence of at least 40 million polonies can be simultaneously determined in parallel, resulting in a very high sequence throughput on the order of gigabases per support. A base-calling algorithm assigns sequences and associated quality values to each read, and a quality-checking pipeline evaluates the Illumina data from each run, removing poor-quality sequences.

In, 2008, Illumina introduced an upgrade, the Genome Analyser II, that triples output compared with the previous Genome Analyzer instrument. Evidenced by a vast number of peer-reviewed publications in an ever-broadening range of applications, Illumina sequencing technology with the Genome Analyzer is a proven platform for genomic discovery and validation.

14.5.2.3 Applied Biosystems SOLiD™ 4 System

The SOLiD platform uses an adapter-ligated fragment library and an emulsion PCR approach with small magnetic beads to amplify the fragments for sequencing. Unlike other platforms, SOLiD uses DNA ligase and a unique approach as illustrated in Figure 14.4.

In this technique, DNA fragments are ligated to adapters and then bound to beads. A water droplet in oil emulsion contains the amplification reagents and only one fragment bound per bead; the emulsion PCR amplifies DNA fragments on the beads. After DNA denaturation, the beads are deposited onto a glass support surface. Libraries may be constructed by any method that gives rise to a mixture of short, adaptor-flanked fragments, although much effort with this system has been put into protocols for mate-paired tag libraries with controllable and highly flexible distance distributions. Clonal sequencing features are generated by emulsion PCR, with amplicons captured to the surface of 1-μM paramagnetic beads. After breaking the emulsion, beads bearing amplification products are selectively recovered and then immobilized to a solid planar substrate to generate a dense, disordered array. Sequencing-by-synthesis is driven by a DNA ligase rather than a polymerase. A universal primer complementary to the adaptor sequence is hybridized to the array of amplicon-bearing beads (Ansorge, 2009). Each cycle of sequencing involves the ligation of a degenerate population of fluorescently labeled octamers. The octamer mixture is structured in that the identity of specific position(s) within the octamer (e.g., base 5) correlate with the identity of the fluorescent label. After ligation, images are acquired in four channels that effectively collect data for the same base positions across all template-bearing beads. The octamer is then chemically cleaved between positions 5 and 6, removing the fluorescent label. Progressive rounds of octamer ligation enable sequencing of every fifth base (e.g., bases 5, 10, 15, and 20). Upon completing several such cycles, the extended primer is denatured to reset the system. Subsequent iterations of this process can be directed at a different set of positions (e.g., bases 4, 9, 14, and 19) by using a primer that is set back one or more bases from the adaptor-insert junction or by using different mixtures of octamers in which a different position

(e.g., base 2) is correlated with the label. An additional feature of this platform involves the use of two-base encoding, which is an error-correction scheme in which two adjacent bases (Mardis, 2008a), rather than a single base, are correlated with the label. Each base position is then queried twice such that miscalls can be more readily identified.

The Applied Biosystems SOLiD 4 system is a revolutionary genetic analysis platform (Figure 14.5) that enables parallel sequencing of clonally amplified DNA fragments linked to beads. The method has the following advantages over its counterparts:

- It enables researchers to obtain higher-quality genomes at lower cost without the purchase of a new instrument.
- *Scalable system:* The open-slide format and flexible bead densities enable increases in throughput with protocol and chemistry optimizations.
- *Superior accuracy:* Accuracy greater than 99.94% because of two-base encoding.
- *Uniform coverage:* Coverage is consistent for the sequencing.
- *High throughput:* It generates over 100 gigabases and 1.4 billion tags per run.
- *Flexible:* The independent flow cell configuration of the SOLiD analyzer enables users to run two completely independent experiments in a single run—essentially providing two instruments in one.

14.5.2.4 Ion Personal Genome Machine Sequencer

Ion Torrent has developed the world's first semiconductor-based DNA sequencing technology that directly translates chemical information into digital data. DNA sequencing is performed with all natural nucleotides and Ion Torrent has developed a DNA sequencing system that directly translates chemical signals (A, C, G, T) into digital information (0, 1) on a semiconductor chip. The result is a sequencing system that is simpler, faster, and more cost-effective and scalable than any other technology available.

The Ion Torrent Personal Genome Machine (PGM™) sequencer is a bench-top system utilizing groundbreaking and disruptive semiconductor technology that enables rapid and scalable sequencing experiments. Ion Torrent technology uses a massive, parallel array of proprietary semiconductor sensors to perform direct real-time measurement of the hydrogen ions produced during DNA replication. A high-density array of wells on the Ion Torrent semiconductor chips provides millions of individual reactors, and integrated fluidics allow reagents to flow over the sensor array. This unique combination of fluidics, micromachining, and semiconductor technology enables the direct translation of genetic information (DNA) to digital information (DNA sequence), rapidly generating large quantities of high-quality data. The PGM along with Ion Torrent semiconductor sequencing chips, Ion Torrent reagent kits, and the Torrent Server/Torrent Suite software allow Ion Torrent to deliver a cutting-edge sequencing solution that is simple, stable, and fast.

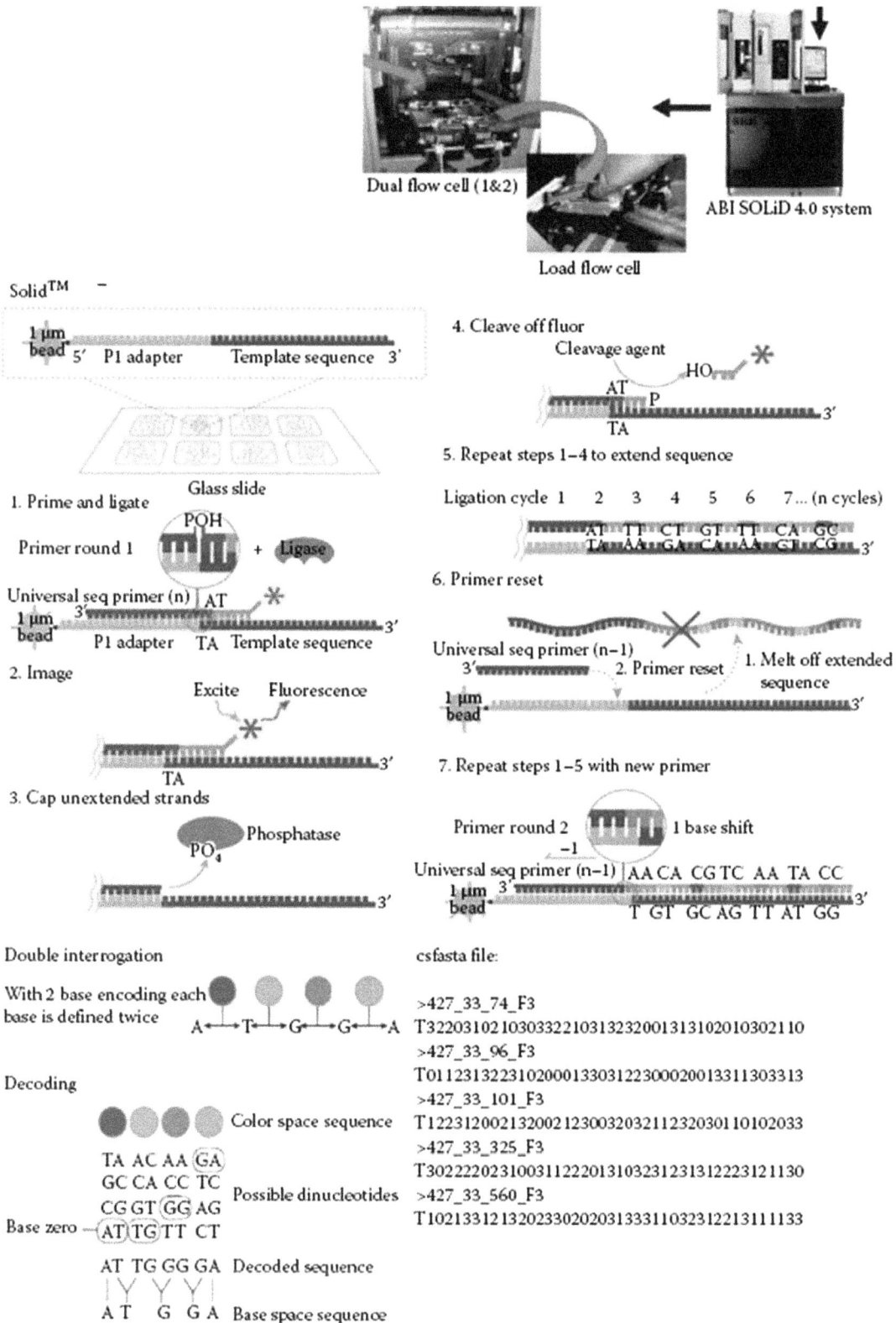

FIGURE 14.4 Sequencing-by-ligation using, the SOLiD DNA sequencing platform. Step 1–6 (a): Primers hybridize to the P1 adapter within the library template. A set of four fluorescence labeled di-base probes competes for ligation to the sequencing primer. These probes have a partly degenerated DNA sequence (indicated by n and z) and for simplicity only one probe is shown (labeling is denoted by asterisk). Specificity of the di-base probe is achieved by interrogating the first and second base in each ligation reaction (CA in this case for the complementary strand). Following ligation, the fluorescent label is enzymatically removed together with the three last bases of the octamer. Step 7 (b): Sequence determination by the SOLiD DNA sequencing platform is performed in multiple ligation cycles, using different primers, each one shorter from the previous one by a single base. The number of 5–15 ligation cycles determines the eventual read length, from primer (n) to primer (n = 4), and double interrogation and decoding of csfasta format of the read data.

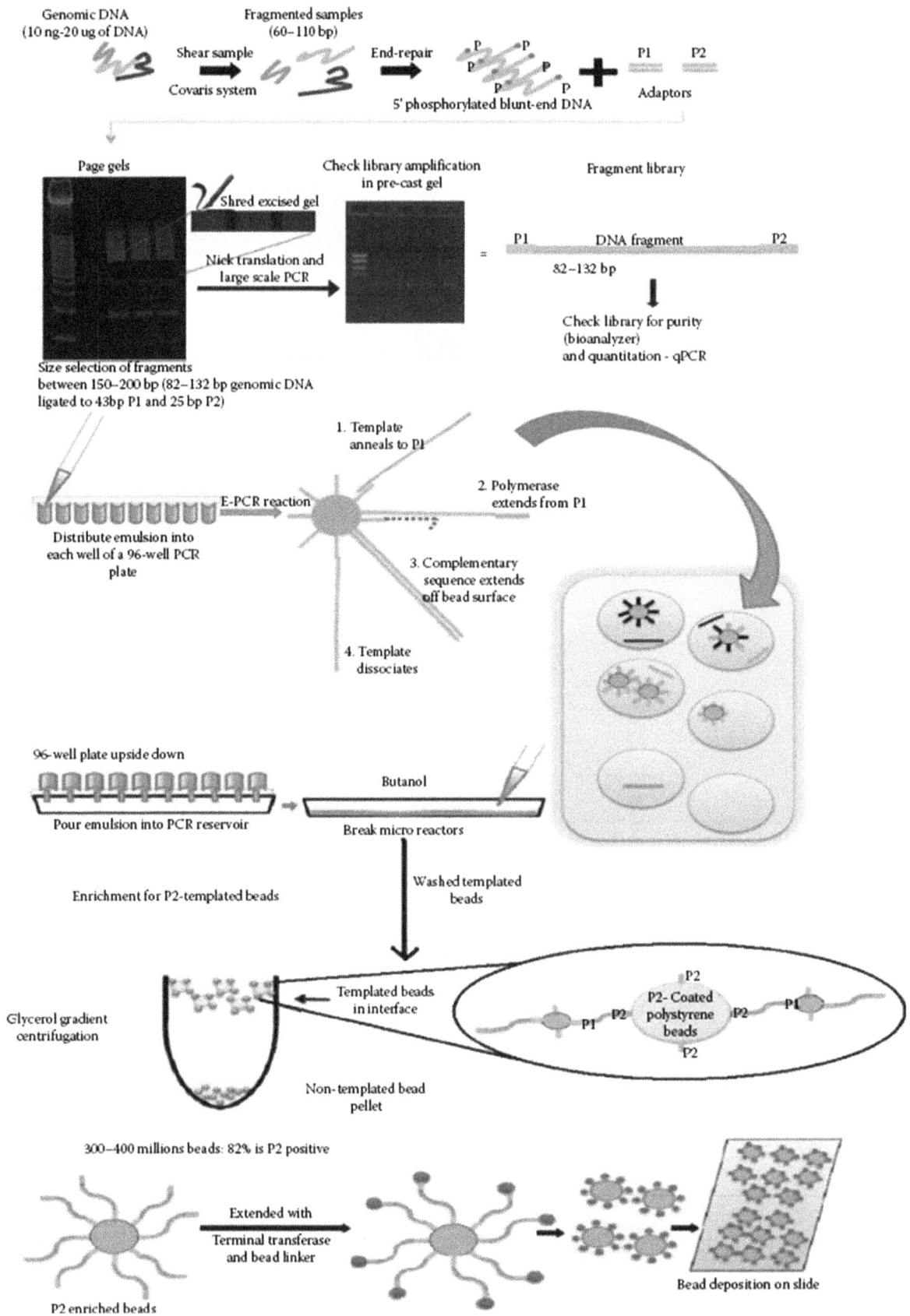

FIGURE 14.5 The process depicted fragment library construction from genomic DNA, preparation of the P2 templated beads using ePCR, and enrichment of the P2-positive beads using glycerol gradient centrifugation. The enriched beads modification and its deposition on treated glass slides for carrying out the sequencing on a SOLiD system.

The systems discussed above require the emulsion PCR amplification step of DNA fragments to make the light signal strong enough for reliable base detection by the CCD cameras (Mamanova et al., 2010). PCR amplification has revolutionized DNA analysis, but in some instances, it may introduce base sequence errors into the copied DNA strands or favor certain sequences over others, thus changing the relative frequency and abundance of various DNA fragments that existed before amplification. Ultimate miniaturization into the nanoscale and the minimal use of biochemicals would be achieved if the sequence could be determined directly from a single DNA molecule without the need for PCR amplification and its potential for distortion of abundance levels. This requires a very sensitive light detection system and a physical arrangement capable of detecting and identifying light from a single dye molecule. Techniques for the detection and analysis of single molecules have been under intensive development over past decades, and several very sensitive systems for single-photon detection have been produced and tested. Therefore, the third generation was developed for which PCR emulsion is not needed (Reis-Filho, 2009; Metzker, 2010).

14.5.3 Third-Generation Sequencers

The desired characteristics and features for the third-generation sequencers are high throughput, low cost, and long and fast read through (Buehler et al., 2010). Measurements are directly linked to the nucleotide sequence rather than capturing images that require a conversion into quantitative data for base calling. Third-generation sequencers are those based on single-molecule sequencing in addition to massive parallel analysis.

14.5.3.1 Helicos

Helicos single-molecule sequencing provides a unique view of genome biology through direct sequencing of cellular nucleic acids in an unbiased manner, providing quantitative and accurate sequence information. The simple sample preparation involves no ligation or PCR amplification, allowing for direct sequencing of targeted DNA or RNA molecules. DNA and RNA can be directly hybridized to the flow cell, eliminating many intermediary steps that can introduce sample loss or bias. One of the first techniques for sequencing from a single DNA molecule was described by the team of S. Quake and licensed by Helicos BioSciences. Helicos introduced the first commercial single-molecule DNA sequencing system in 2007. The nucleic acid fragments are hybridized to primers covalently anchored in random positions on a glass coverslip in a flow cell. The primer, polymerase enzyme, and labeled nucleotides are added to the glass support. The next base incorporated into the synthesized strand is determined by analysis of the emitted light signal in the sequencing-by-synthesis technique (Figure 14.6). This system also simultaneously analyzes many millions of single DNA fragments, resulting in sequence throughput in the gigabase range (Milos, 2008; Thompson and Steinmann, 2010). Although still in the first years of operation, the system has been tested and validated in several applications with promising results. In the homopolar regions, multiple fluorophore incorporations could decrease emissions, sometimes below the level of detection; when errors did occur, most were deletions. Helicos announced that it has recently developed a new generation of "one-base-at-a-time" nucleotides that allows for more accurate homopolymer sequencing and lower overall error rates. The latest model of

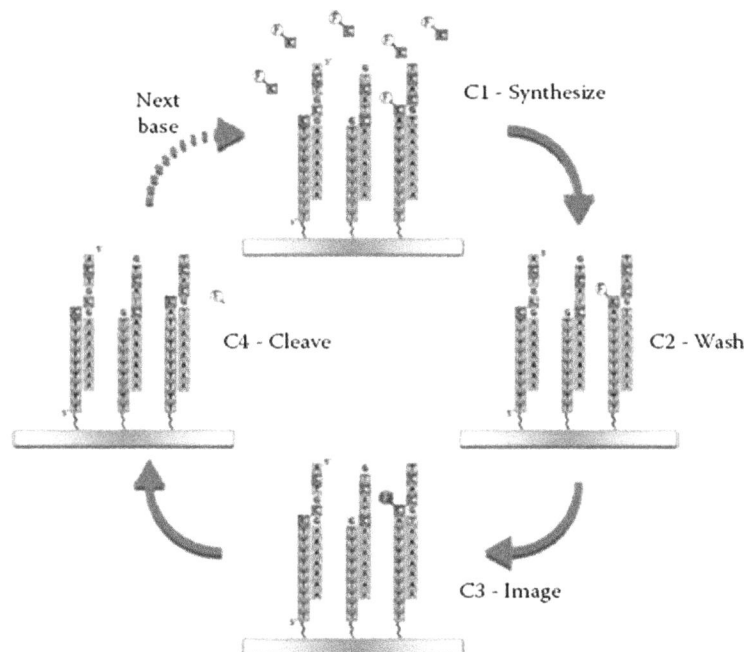

FIGURE 14.6 Helicos BioSciences' approach to single-molecule sequencing involves binding of targets to a flow cell followed by sequential rounds of nucleotide addition, imaging, and cleavage.(Courtesy of Helicos BioSciences, Cambridge, MA.)

Helicos sequencer consists of the following (Metzker, 2010; Azhikina, 2014: Ma et al., 2017):

- It utilizes an advanced fluidics and optics configuration, simultaneously performing strand synthesis and imaging to maximize run efficiency.
- It provides the simplicity and flexibility to balance numerous experimental factors associated with today's experiments, including those applications requiring high coverage, strand-counting, or both—all at a cost lower than other technologies.
- It currently outperforms any other genetic analysis technology, yet it is designed to accommodate future enhancements in accuracy and throughput without the need to upgrade the instrument.

14.5.3.2 Pacific Biosciences

Pacific Biosciences have developed a novel approach to study the synthesis and regulation of DNA, RNA, and protein. Combining recent advances in nanofabrication, biochemistry, molecular biology, surface chemistry, and optics, the powerful technology platform called "single molecule, real-time," or SMRT technology, was created. Pacific Biosciences have introduced a platform for SMRT observation of biological events named PacBio RS (Shendure and Ji, 2008). This uses the proprietary SMRT technology and maintains many of the key attributes of currently available sequencing technologies while solving many of the inherent limitations of previous technologies. The PacBio RS consists of an instrument platform that uses our consumables including the proprietary SMRT cell. PacBio RS is a user-friendly instrument that conducts monitors and analyzes single-molecule biochemical reactions in real-time. The PacBio RS uses a high numerical aperture objective lens and four single-photon-sensitive cameras to collect the light pulses emitted by fluorescence, allowing the observation of biological processes. An optimized set of algorithms is used to translate the information that is captured by the optics system. Using the recorded information, light pulses are converted into an A, C, G, or T base call with associated quality metrics. Once sequencing is started, the real-time data are delivered to the system's primary analysis pipeline, which outputs base identity and quality values (QVs). To generate a consensus sequence from the data, an assembly process aligns the different fragments from each ZMW based on common sequences. The system provides long read lengths, flexibility in experimental design, fast time to result, and is easy to use (Metzker, 2010).

14.5.3.3 VisiGen

VisiGen Biotechnologies is developing a breakthrough technology to sequence a human genome in less than a day for less than $1,000. The ability to achieve $1,000 human genome sequencing is directly related to the successful implementation of a single-molecule approach, and the ability to accomplish this feat in a day is directly related to the implementation of the approach on a massive parallel scale (Shendure and Ji, 2008). VisiGen is distinguished from other leading developers of NGS technologies in that it exploits the natural process of DNA replication in a way that enhances accuracy and minimally affects efficiency. VisiGen has engineered polymerase, the enzyme that synthesizes DNA and nucleotides—the building blocks of a DNA strand, to act as direct molecular sensors of DNA base identity in real time, effectively creating nanosequencing machines that are capable of determining the sequences of any DNA strand (Figure 14.7). The technology platform detects sequential interactions between a single polymerase and each nucleotide that the polymerase inserts into the elongating DNA strand. Importantly, before initiating sequencing activity, the nanosequencers are immobilized on a surface so that the activity of each can be monitored in parallel.

VisiGen builds nanosequences by drawing on the disciplines of single-molecule direction, fluorescent molecule chemistry, computational biochemistry, and genetic engineering of biomolecules. VisiGen's strategy involves monitoring single-pair Forster resonance energy transfer between a donor fluorophore attached to or associated with a polymerase and a color-coded acceptor fluorophore attached to the γ-phosphate of a dNTP during nucleotide incorporation and pyrophosphate release. The purpose of the donor is to excite each acceptor to produce a fluorescent signal for which the emission wavelength and intensity provide a unique signature of base identity. Working at a single-molecule level makes sequencing signal maximization and background noise minimization critical; thus, the core of VisiGen's technology exploits aspects of physics that enhance signal detection to enable real-time, single-molecule sequencing (Lü et al., 2004).

14.5.3.4 BioNanomatrix

BioNanomatrix and Complete Genomics announced in 2007 the formation of a joint venture to develop technology to sequence a human genome in 8 h for less than $100. The proposed platform will use Complete Genomics' sequencing chemistry and BioNanomatrix's nanofluidic technology. They plan to adapt DNA sequencing chemistry with linearized nanoscale DNA imaging to create a system that can read DNA sequences larger than 100,000 bases. With their design and price, they target the possible sequencing of many genomes. Complete Genomics recently presented a new method using rolling circle PCR amplification that resulted in DNA nanoballs and a modified ligation technique for fast and inexpensive sequencing of human genomes. A very different approach to single-molecule DNA sequencing using RNA polymerase (RNAP) has been presented recently (Metzker, 2010: Glyde et al., 2018; Heyduk and Heyduk, 2018).

In the planned method, RNAP is attached to one polystyrene bead whereas the distal end of a DNA fragment is attached to another bead. Each bead is placed in an optical trap, and the pair of optical traps levitates the beads. The RNAP interacts with the DNA fragment, and the transcriptional motion of RNAP along the template changes the length of the DNA between the two beads. This leads to displacement of the two beads that can be registered with precision in the Angstrom range, resulting in single-base resolution on a single DNA molecule. By aligning four displacement records, each with a lower concentration of one of the four nucleotides

Nano-sequencer components

FRET

Real-time data collection

FIGURE 14.7   VisiGen Biotechnologies are developing a fluorescence resonance energy transfer (FRET)-based approach to single-molecule sequencing. VisiGen's technology is a third-generation technology that uses the nanosequencer components. It based on single-molecule changes.

and in a role analogous to the primers used in Sanger sequencing, and using for calibration the known sequences flanking the unknown fragment to be sequenced, it is possible to deduce the sequence information. Thirty of the 32 bases were correctly identified in approximately 2 min. The technique demonstrates that the movement of a nucleic acid enzyme and the very sensitive optical trap method may allow for direct extraction of sequence information from a single DNA molecule (Ansorge, 2009: Hashmi et al., 2015).

14.6   COMPARISON OF THE SEQUENCING TECHNIQUES

The sequencing techniques provided by various companies follow different techniques, which are listed above, and all of these techniques have various advantages and disadvantages. All the characteristics are listed in Table 14.3. Other than the platform comparison, the analysis and annotation of the data generated by the sequencers are to be done. This analysis is easy to understand and interpret with the help of Bioinformatics (Morishita, 2009; Sugano, 2009). The strength of NGS lies in its diverse array of applications, which are

listed in Table 14.4 and illustrated in Figure 14.8 (Marguerat et al., 2008).

14.7   DISCOVERY OF SNP

One of the central themes in genomics is to study genome differences or variations, including SNP. In the past, human SNP discovery relied on PCR amplification of targeted regions, followed by capillary sequencing and *in silico* sequence alignment. However, this approach was impractical for other species because it was rather laborious and costly. The NGS platforms, which produce millions of short reads significantly faster and cheaper, fit in perfectly and have started to change the processes for SNP discovery (Chan, 2009; Shen et al., 2010; Arslan, 2018).

14.8   TRANSCRIPTOME ANALYSIS

14.8.1   GENE EXPRESSION: SEQUENCING THE TRANSCRIPTOME

Historically, the mRNA expression has been gauged with microarray or quantitative PCR (qPCR)-based approaches; the

TABLE 14.3
Comparison of NGS Technologies

Platform	Library/template preparation	Read length (bases)	Run time (days)	Gb per run	Pros	Cons	Biological applications
Roche/454's GS FLX Titanium	Fragment, MP/emPCR	330	0.35	0.45	Longer reads improve mapping in repetitive regions; fast run times	High reagent cost; high error rates in homopolymer repeats	Bacterial and insect genome *de novo* assemblies; medium scale (<3 Mb) exome capture; 16S in metagenomics
Illumina/Solexa's GAII	Fragment, MP/ solid-phase	75 or 100	40–49	18, 35	Currently the most widely used platform in the field	Low multiplexing capability of samples	Variant discovery by whole-genome resequencing or whole-exome capture; gene discovery in metagenomics
Life/APG's SOLiD 3	Fragment, MP/emPCR	50	7, 14	30–50	Two-base encoding provides inherent error correction	Long run times	Variant discovery by whole-genome resequencing or whole-exome capture; gene discovery in metagenomics
Polonator G.007	MP only/emPCR	26	5	12	Least-expensive platform; open source to adapt alternative NGS chemistries	Users are required to maintain and quality-control reagents; shortest NGS read lengths	Bacterial genome resequencing for variant discovery
Helicos BioSciences HeliScope	Frag, MP/single molecule	32	8	37	Non-biased Representation of templates for genome and sequence-based applications	High error rates compared with other reversible terminator chemistries	Sequence-based methods
Pacific Biosciences	Frag only/single molecule	964	N/A	N/A	Has the greatest potential for reads exceeding 1 kb	Highest error rates compared with other NGS chemistries	Full-length transcriptome sequencing; complements other resequencing efforts in discovering large structural variants and haplotype blocks

TABLE 14.4
Applications of NGS

Category	Examples of applications
Complete genome resequencing	Comprehensive polymorphism and mutation discovery in individual human genomes
Reduced representation sequencing	Large-scale polymorphism discovery
Targeted genomic resequencing	Targeted polymorphism and mutation discovery
Paired end sequencing	Discovery of inherited and acquired structural variation
Metagenomic sequencing	Discovery of infectious and commensal flora
Transcriptome sequencing	Quantification of gene expression and alternative splicing; transcript annotation; discovery of transcribed SNPs or somatic mutation
Small RNA sequencing	MiRNA profiling 64 sequencing of bisulfite-treated DNA determining patterns of cytosine methylation in genomic DNA
Chromatin immunoprecipitation sequencing	Genomic-wide mapping of protein-DNA interactions
Nuclease fragmentation and sequencing	Nucleosome positioning
Molecular barcoding	Multiplex sequencing of samples from multiple individuals

latter is most efficient and cost-effective for a genome-wide survey of gene expression levels. However, even the exquisite sensitivity of qPCR is not absolute, nor is it straightforward or reliable to evaluate novel alternative splicing isoforms using either technology (Bateman and Quackenbush, 2009). In the past, SAGE and its variants have provided a digital readout of gene expression levels using DNA sequencing. These powerful approaches have the ability to report the expression of genes at levels below the sensitivity of microarrays, but they have been limited in their application by the cost of DNA sequencing. In contrast, the rapid and inexpensive sequencing capacity offered by the NGS instruments meshes perfectly with SAGE tagging or conventional complementary DNA sequencing approaches (Morozova et al., 2009), as evidenced by several studies that used Roche/454 technology. Without doubt, the shorter read lengths offered by the Illumina and Applied Biosystems instruments will be utilized with these approaches in the future, offering the advantage of sequencing individual SAGE tags rather than requiring concatenation of the tags before sequencing. Indeed, combining the data obtained from isolating and sequencing ChiP-derived DNA bound by a transcription factor of interest to the corresponding co-isolated and sequenced mRNA population from the same cells might be imagined (Forrest and Carninci, 2009; Sigdel et al., 2018).

14.8.2 APPLICATIONS OF TRANSCRIPTOMICS ANALYSIS

Transcriptome analysis could be applied at several stages of clinical drug development (Morozova et al., 2009). A basic paradigm of clinical pharmacology is that a drug can be distinguished from a poison by its therapeutic index. As the dose of a drug increases, the pharmacologic effect increases, as do the toxicologic effects. Drugs have pharmacologic effects at doses lower than their toxicologic effects, whereas poisons have toxicologic effects at doses lower than their pharmacologic effects. When human testing commences, the pharmacokinetics is carefully studied, usually in healthy normal volunteers, and adverse effects are carefully monitored. When plasma levels are achieved that would be predicted to have appropriate pharmacologic activity, and that appear to be safe and well tolerated, the drug proceeds to testing in patients with the target disease. If a biomarker or a surrogate marker for pharmacologic activity is available, it can be used to assess dose-responses in normal volunteers and in patients. Conversely, if a biomarker for a potential toxic effect is available, this will also be assessed in the normal volunteers and in patients. Such biomarkers are extremely important in early clinical development where their utilization enables more rapid and accurate definition the potential clinical dose range, detection of pharmacologic activity, and assessment of the risk of side effects (Tanaka, 2000). Transcriptome analysis provides the opportunity to simultaneously measure many potential biomarkers. Such measurements could provide more comprehensive assessment of the pharmacologic and toxicologic effects of drugs early in clinical development. This is how transcriptome analysis has obtained its importance in the pharmaceutical sector (Gong et al., 2018).

14.8.3 SMALL RNA ANALYSIS

Small RNA (sRNA) is a functional RNA molecule that is not translated into a protein. It is also called noncoding RNA (NC-RNA), nonprotein-coding RNA (npcRNA), non-messenger RNA (nmRNA), or functional RNA (fRNA) Small NC-RNAs are typically only approximately 18–40 nucleotides in length; however, their effect on cellular processes is profound (Axtell et al., 2007). Small RNA has been shown to play critical roles in developmental timing, tumor progression, and neurogenesis.

- MicroRNA (miRNA)
- Short interfering RNA (siRNA)
- Piwi-interacting RNA (piRNA)
- siRNA RNA (rasiRNA)

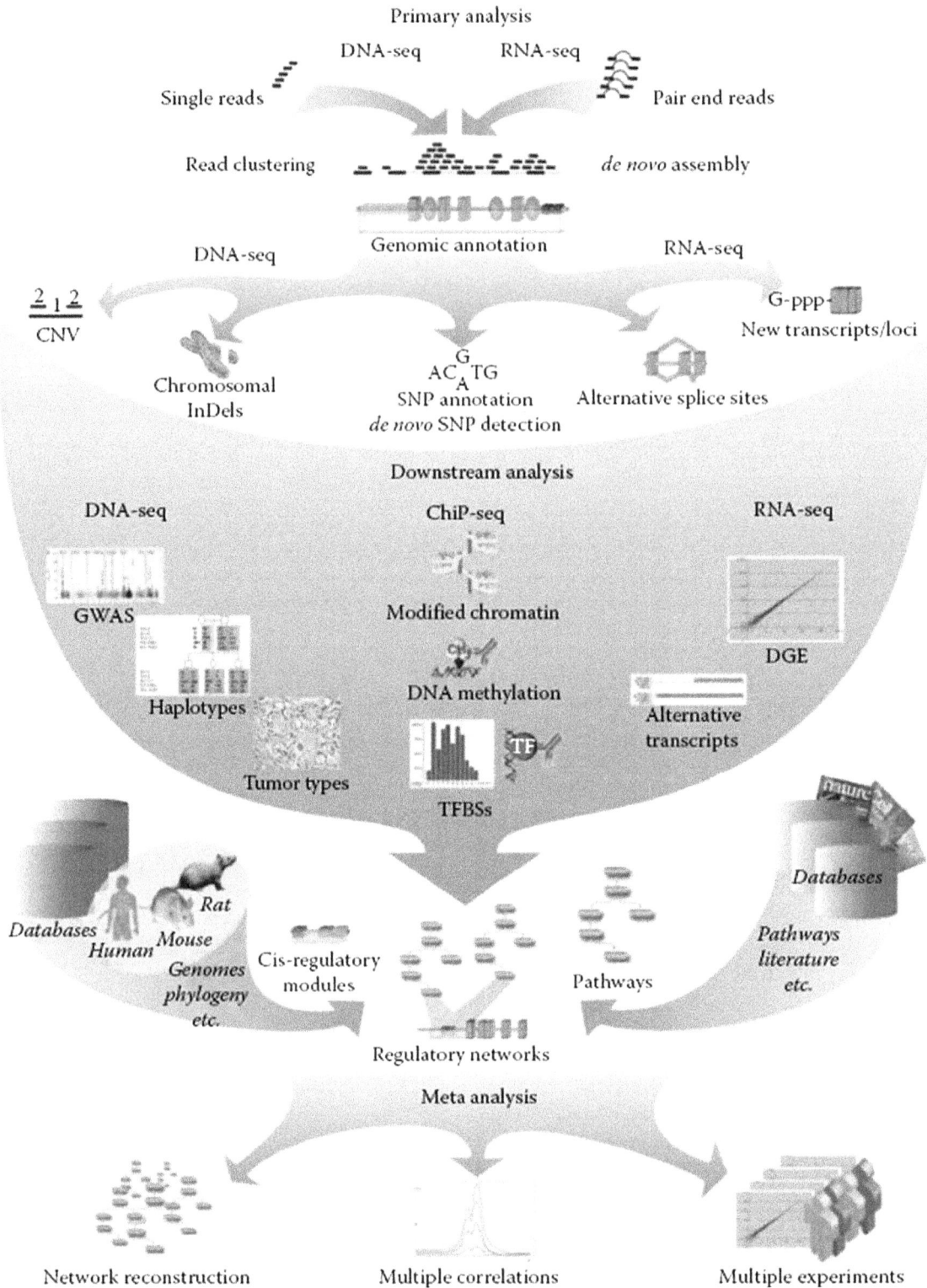

FIGURE 14.8 Overview of NGS-based analysis strategies. Primary analysis: This part describes analyses steps that are used directly on the reads: CNV, chromosomal indels are insertions or deletions, SNP annotation, alternative splice sites can be detected and new transcripts/loci are derived by direct mapping of novel exons and splice overlaps. Downstream analysis: This part differs for the three major application areas: GWAS, the definition of haplotypes and tumor typing. ChIP-Seq determines genome-wide patterns of modified chromatin, usually DNA-dependent RNA polymerases or transcription factors leading to the definition of patterns such as transcription factor binding sites. RNA-Seq determines the unknown transcripts. Finally, meta-analysis allows merging of several fields via network reconstruction, multiple correlations of various lines of evidence, and the cross examination of multiple experiments such as transcriptional profiles from several patients.

sRNA (small RNA): sRNA refers to small bacterial NC-RNAs, with the DNA coding sequences being referred to as **RNA genes** or noncoding RNA genes.

14.8.3.1 Techniques for RNA Analysis

Deep sequencing RNA analysis pipeline (DSAP) is an automated multiple task web service designed to provide a total solution to analyzing deep-sequencing small RNA data sets generated by NGS technology (Huang et al., 2010). DSAP uses a tab-delimited file as an input format that holds the unique sequence reads (tags) and their corresponding number of copies generated by the Solexa sequencing platform. The input data will go through four analysis steps in DSAP:

1. *Cleanup:* Removal of adaptors and poly-A/T/C/G/N nucleotides
2. *Clustering:* Grouping of cleaned sequence tags into unique sequence clusters
3. *NC-RNA matching:* Sequence homology mapping against a transcribed sequence library from the NC-RNA database Rfam
4. *Known miRNA matching:* Detection of known miRNAs in miRBase based on sequence homology

miRNAs: MicroRNAs are posttranscriptional regulator molecules that are capable of binding to mRNA complementary sequences. This in turn brings about a repression of translation.

siRNA (small-interfering RNA, or silencing RNA): siRNA is a class of double-stranded RNA molecules that play various roles [e.g., repression of expression of enzymes of RNAi (RNA interference) pathway].

piRNA (Piwi-interacting RNA): Small RNA molecules that are capable of forming RNA-protein complexes through interactions with Piwi proteins. piRNA are slightly larger in size than miRNA.

rasiRNA (repeat-associated small-interfering RNA (rasiRNA)): rasiRNA are a class of small RNA that associates with the Ago and Piwi proteins. rasiRNA is involved in establishing and maintaining heterochromatin structure, controlling transcripts that emerge from repeat sequences, and silencing transposons and retrotransposons

NC-RNA (noncoding RNA molecule): NC-RNA is also known as nonproteincoding RNA (npcRNA), non-messenger RNA (nmRNA), small non-messenger RNA (snmRNA), and functional RNA (fRNA).

The expression levels corresponding to matched NC-RNAs and miRNAs are summarized in multicolor clickable bar charts linked to external databases. DSAP is also capable of displaying miRNA expression levels from different jobs using a log_2-scaled color matrix. Furthermore, a cross-species comparative function is also provided to show the distribution of identified miRNAs in different species as deposited in miRBase.

14.8.3.2 Discovering NC-RNA

One of the most exciting areas of biological research in recent years has been the discovery and functional analysis of NC-RNA systems in different organisms. Perhaps the most profound impact of NGS technology has been on the discovery of novel NC-RNAs belonging to distinct classes in an extraordinarily diverse set of species. In fact, this approach has been responsible for the discovery of NC-RNA classes in organisms not previously known to possess them. These discoveries are being coupled with an ever-expanding comprehension of the functions embodied by these unique RNA species, including gene regulation by various mechanisms. In this regard, studying the roles of specific miRNAs in cancer is helping to uncover certain aspects of the disease. NC-RNA discovery is best accomplished by sequencing because the evolutionary diversity of NC-RNA gene sequences makes it difficult to predict their presence in a genome with high certainty by computational methods alone. The unique structures of the processed NC-RNAs pose difficulties for converting them into NGS libraries, but remarkable progress has already been made in characterizing these molecules. With these barriers dissolving, the high capacity and low cost of next-generation platforms ensure that discovery of NC-RNAs will continue at a rapid pace and that sequence variants with important functional impacts will also be determined. For example, because the readout from next-generation sequencers is quantitative, NC-RNA characterization will include detecting expression level changes that correlate with changes in environmental factors, with disease onset and progression, and perhaps with complex disease onset or severity. Importantly, the discovery and characterization of NC-RNAs will enhance the annotation of sequenced genomes such that, especially in model organisms and humans, the effect of mutations will become more broadly interpretable across the genome (Mardis, 2008b; Wang and Chang, 2011; Liu et al., 2015).

14.9 EPIGENETICS

Epigenetics is the study of inherited changes in phenotype (appearance) or gene expression caused by mechanisms other than changes in the underlying DNA sequence. These changes may remain through cell divisions for the remainder of the cell's life and may also last for multiple generations. However, there is no change in the underlying DNA sequence of the organism; instead, nongenetic factors cause the organism's genes to behave (or "express themselves") differently.

The molecular basis of epigenetics is complex. It involves modifications of nucleotides in a given sequence, without changing the order or the sequence itself. Additionally, the chromatin proteins associated with DNA may be activated or silenced (Bird, 2007; Suzuki and Bird, 2008). This accounts for the observation that differentiated cells in a multicellular organism express only the genes that are necessary for their own activity. Epigenetic changes are preserved when cells divide. Most epigenetic changes only occur within the course of one individual organism's lifetime; however, some epigenetic changes are inherited from one generation to another, which begs the question of whether or not epigenetic changes in an organism alter the basic structure of its DNA (Zhou et al., 2010; Feinberg, 2018).

Specific epigenetic processes include paramutation, bookmarking, imprinting, gene silencing, X chromosome inactivation, position effect, reprogramming, transvection, maternal effects, the progress of carcinogenesis, many effects of teratogens, regulation of histone modifications and heterochromatin, and technical limitations affecting parthenogenesis and cloning. Epigenetic research uses a wide range of molecular biologic techniques to further our understanding of epigenetic phenomena, including chromatin immunoprecipitation (together with its large-scale variants ChIP-on-chip and ChIP-seq), fluorescent in situ hybridization, methylation-sensitive restriction enzymes, DNA adenine methyl transferase identification, and bisulfite sequencing (Park, 2008; Rao et al., 2018).

14.10 CHROMATIN IMMUNOPRECIPITATION, CHIP-SEQ TECHNIQUE

NGS technology allowed replacement of microarrays in the mapping step with high-throughput sequencing of DNA binding sites, and their direct mapping to a reference genome in the database. The sequence of the binding site is mapped with high resolution to regions shorter than 40 bases, a resolution not achievable by microarray mapping. Moreover, the ChiP-Seq technique is not biased and allows for the identification of unknown protein binding sites, which is not the case with the ChIP-on-chip approach, in which the sequence of the DNA fragments on the microarray is predetermined (e.g., in promoter arrays, exon arrays) (Nelson et al., 2006).

The association between DNA and proteins is a fundamental biological interaction that plays a key part in regulating gene expression and controlling the availability of DNA for transcription, replication, and other processes. These interactions can be studied in a focused manner using a technique called chromatin immunoprecipitation (ChIP). ChIP entails a series of steps:

1. DNA and associated proteins are chemically crosslinked.
2. Nuclei are isolated, lysed, and the DNA is fragmented.
3. An antibody specific for the DNA binding protein (transcription factor, histone, etc.) of interest is used to selectively immunoprecipitate the associated protein-DNA complexes.
4. The chemical crosslinks between DNA and protein are reversed, and the DNA is claimed for downstream analysis.

In early applications, typical analyses examined the specific gene of interest by quantitative PCR or Southern blotting to determine if corresponding sequences were contained in the captured fragment population. Recently, genome-wide

ChIP-based studies of DNA-protein interactions became possible in sequenced genomes by using genomic DNA microarrays to assay the released fragments (Carey et al., 2009; Dey et al., 2012). Although utilized for several important studies, it has several drawbacks, including a low signal-to-noise ratio and a need for replicates to build statistical power to support putative binding sites.

14.11 METAGENOMICS

Cataloging the biodiversity found on Earth is of particular interest as we enter a critical stage in which our ecosystem is changing on an alarming scale. DNA- or RNA-based approaches for this purpose are becoming increasingly powerful as the growing number of sequenced genomes enables us to interpret partial sequences obtained by direct sampling of specific environmental niches (Mardis, 2008a). Such investigations are referred to as metagenomics. Conventionally, metagenomics are addressed by isolating DNA from an environmental sample, amplifying the collective of 16S ribosomal RNA (rRNA) genes with degenerate PCR primer sets, subcloning the PCR products that result, and classifying the taxa present according to a database of assigned 16S rRNA sequences. As an alternative, DNA (or RNA) is isolated, subcloned, and then sequenced to produce a fragment pool representative of the existing population. These sequences can then be translated *in silico* into protein fragments and compared with the existing database of annotated genome sequences to identify community members. In both approaches, deep sequencing of the population of subclones is necessary to obtain the full spectrum of taxa present and is limited by potential cloning bias that can result from the use of bacterial cloning. With aids of **metabolomics** and sampling RNA sequences from a metagenomic isolate, one can attempt to reconstruct metabolic pathways that are active in a given environment. By contrast, the rapid, inexpensive, and massive data production enabled by next-generation platforms has caused a recent explosion in metagenomic studies. These studies include previously sampled environments such as the ocean and an acid mine site, but soil and coral reefs also were studied by Roche/454 pyrosequencing (Tringe and Rubin, 2005; Wommack et al., 2008; Misra, 2018).

> **Metabolomics**: Metabolomics involves the rapid, high throughput characterization of the small-molecule metabolites found in an organism. It is closely tied to the genotype of an organism, its physiology, and its environment.

> **Metabolome**: A metabolome refers to the complete set of primary and secondary metabolites as well as activators, inhibitors, and hormones that are produced by a given organism under certain conditions. However, it is noteworthy that it is not currently possible to analyze the entire range of metabolites by a single analytical method.

A new generation of non-Sanger-based sequencing technologies have delivered on their promise of sequencing DNA at unprecedented speed, thereby enabling impressive scientific achievements and novel biological applications (Gill et al., 2006). However, before stepping into the limelight, NGS had to overcome the inertia of a field that relied on Sanger-sequencing for 30 years. Each next-generation platform is optimized for specific sequencing applications. Next-generation DNA sequencing has the potential to dramatically accelerate bio-logical and biomedical research by enabling the comprehensive analysis of genomes, **transcriptomes**, and **interactomes** to become inexpensive, routine, and widespread rather than requiring significant production-scale efforts (Wooley et al., 2010; Yang et al., 2015).

Transcriptome: The transcriptome is the set of all RNA molecules, including mRNA, rRNA, tRNA, and other non-coding RNA produced in one or a population of cells

Interactome: An interactome is the whole set of interactions that takes place between different molecules within the cell, both within a given family of molecules or between molecules belonging to different biochemical families. When discussed in terms of functional genomics, it refers to gene-gene interaction.

There are numerous applications of NGS because NGS plays a huge role in functional genomic and expressional analysis. The DNA sequencing process is followed by the assembly and the annotation of the genome. The annotation of the genome is a key step for identifying the functional genes and predicting the proteins. The genome analysis and the genome annotation are explained in Section 14.12.

14.12 GENOME ANALYSIS

14.12.1 WHOLE-GENOME SEQUENCING

Most of the methods give rather short read lengths; they have mainly been used for resequencing. In this case, it is not necessary to do a complete, independent genome assembly, but the sequence reads can be aligned to a reference genome sequence (Sundquist et al., 2007). For example, the sequence reads from a single person can be aligned to the reference human genome. However, all of the methods have been modified to produce "paired reads" in which both ends of a DNA fragment of known length are sequenced (Bentley et al., 2009). This makes it possible to do *de novo* assemblies of genomes. In whole-genome sequencing the aim is to produce a complete, continuous sequence of high quality or a fragmented draft version of the genome. Although the draft version can be produced faster and at lower cost compared with the complete sequence, only the latter can be reliably used in different analyses; for example, because a gene that is not found in the sequence is truly missing. If a certain region of the genome is of special interest it can be sequenced separately; for example, a gene known to be associated with this specific disease can be sequenced from several individuals with differing symptoms or disease outcomes to establish the connection between the genotype and phenotype.

Two major strategies—clone-by-clone and whole-genome shotgun approaches—have been used for whole-genome sequencing. The most suitable methods depend on the organism to be sequenced. For relatively small and nonrepetitive genomes the whole-genome shotgun methods is advantageous because mapping and construction of a large insert clone are avoided (Ng and Kirkness, 2010). More complex genomes such as human genomes are difficult to sequence using this method. The high levels of repetitive sequences cause difficulties in the assembly of the genome. Combination of the two methods (i.e., hybrid strategies) might be more successful when sequencing complex genomes (Huang et al., 2009; Zhao and Grant, 2010; Fuselli et al., 2018; Yue and Wang, 2018).

14.12.2 *DE NOVO* ASSEMBLY

Genome assembly refers to the process of taking many short DNA sequences, all of which were generated by a shotgun-sequencing project and putting them back together to create a representation of the original chromosomes from which the DNA originated. In a shotgun-sequencing project, the entire DNA from a source is first fractured into millions of small pieces (Pevzner et al., 2004). These pieces are then "read" by automated sequencing machines, which can read up to 900 nucleotides or bases at a time. A genome assembly algorithm works by taking all of the pieces and aligning them to one another and detecting all places where two of the short sequences, or reads, overlap. These overlapping reads can be merged together, and the process continues (Morozova, 2008a).

The assembly of a genome is a very difficult computational problem, made more difficult because many genomes contain many identical sequences, known as repeats. These repeats can be thousands of nucleotides long, and some occur in thousands of different locations, especially in the large genomes of plants and animals. The resulting genome sequence is created by combining information from the sequenced contigs and using linking information for creating scaffolds. These scaffolds are positioned along the physical map of the chromosomes creating a "golden path." The efficiency of the *de novo* genome sequence assembly processes depends heavily on the length, fold-coverage, and per-base accuracy of the sequence data. Despite substantial improvements in the quality, speed, and cost of Sanger sequencing, generating a high-quality draft *de novo* genome sequence for a eukaryotic genome remains expensive. The newly available sequencing-by-synthesis systems from Roche (454), Illumina (Genome Analyser), and ABI (SOLiD) offer greatly reduced per-base sequencing costs. Although they are attractive for generating *de novo* sequence assemblies for eukaryotes, these technologies add several complicating factors: they generate short (typically 450 base pairs for Roche/454; 50–100 base pairs for Illumina and SOLiD) reads that cannot resolve low-complexity sequence regions or distributed repetitive elements, they have system-specific error models, and they can have higher base-calling error rates (Li et al., 2010). The whole process of *de novo* sequencing is illustrated in Figure 14.9.

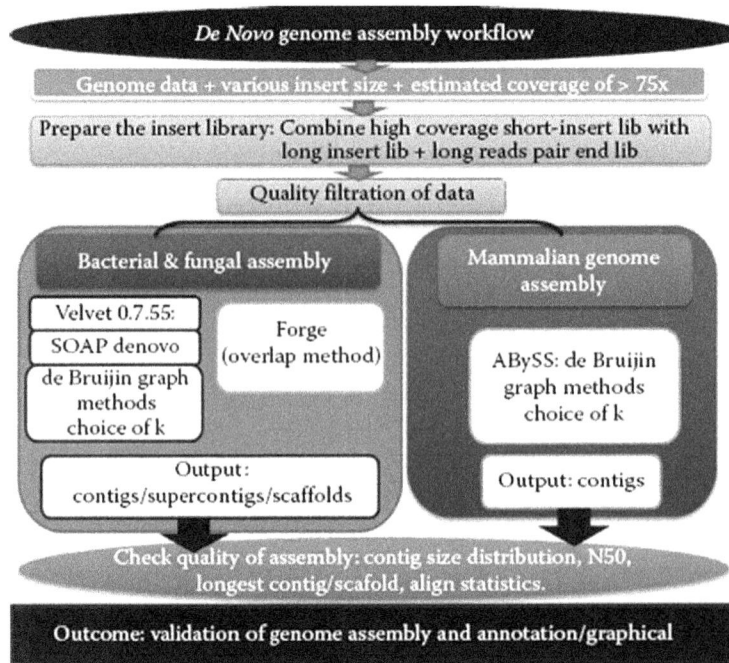

FIGURE 14.9 Schematic overview of the *de novo* genome assembly workflow for assembling the genome (Zhao et al. 2011. *Physiol. Genomics*, 43:325–45).

De novo sequencing using next-generation technologies necessitates the development of new algorithms for assembling the short and more error-prone reads that they generate. Several *de novo* assembly algorithms based on de-Bruijn graphs (EULER-SR and Velvet), hash-extension (VCAKE), and overlap layout (EDENA) and for paired-end reads (ALLPATHS) have been recently developed. These algorithms are able to assemble millions of short reads from NGS technologies into thousands of contigs with varying degrees of efficiency (Paszkiewicz and Studholme, 2010; Khan et al., 2018).

14.12.3 Annotation of Genome Assembly

Annotation of the genome is the process of attaching biological information to sequences. It consists of two main steps: Identifying the elements on the genome, a process called "gene finding," and attaching biological information to these elements (Srinivasan et al., 2005). Automatic annotation tools try to perform all of these with computational analysis, as opposed to manual annotation that involves human expertise. These approaches ideally coexist and complement each other in the same annotation pipeline (Morozova, 2008a; Dominguez et al., 2018).

The basic level of annotation is done using BLAST to find similarities and then annotating genomes on the basis of that. However, nowadays more and more information is added to the annotation platform. The additional information allows manual annotators to deconvolute discrepancies between genes that are given the same annotation. The genome annotations are of two types: structural annotation and functional annotation. Structural annotation consists of the identification of the genomic elements (1) open reading frames and their localization, (2) gene structure, (3) coding region, and (4) location of regulatory motifs. Functional annotation consists in attaching (1) biological information to genetic elements, (2) biochemical function, (3) regulatory elements, and (4) interaction expression. Genome annotation is an active area of investigation involving several different organizations in the life science community that publish the results of their efforts in publicly available biological databases accessible via the Web and other electronic means (Kawaji and Hayashizaki, 2008; Raza, 2010). The NGS and the genome annotation procedures are inseparable.

14.12.4 Genome Mapping of Next Generation Sequencing Data

Assembling reads against an existing backbone sequence, building a sequence is called "genome mapping," which is similar but not necessarily identical to the backbone sequence. The NGS technologies pose new challenges to the sequencing community (Swidan et al., 2006). In addition to the enormous amount of data that they can produce through their high throughput, the read length of the sequence tags is also much shorter than the conventional sequencing methods. Whereas Sanger sequencing can produce read lengths over 700 base pairs, the GS FLX (Roche/454 life sciences) can read average lengths of up to 400 base pairs and the Genome Analyser (Illumina) and SOLiD (Applied Bio Systems) can read tags up to 35 or 50 base pairs, respectively. Mapping those short reads to the reference genomes can pose different challenges in determining the nature of mismatches (Mychaleckyj, 2007). A mismatch can occur because of sequencing error, differences between the query sequence and the reference sequence, or existence of repetitive regions. Although read lengths have constantly been improved by different technologies over the past few years after the release, there will remain a certain cutoff read length that is practical for bioinformatics use, economical, and affordable. The region for finding a balance here is that the cost of sequencing goes up through higher read lengths because of the use of more chemical reagents. In Section 14.12.5; the emphasis will be on alignment programs developed for the Genome Analyzer and SOLiD because those two technologies produce ultrashort reads and are therefore more challenging to use with older alignment programs developed for Sanger sequencing. The drawbacks of the NGS methods are that they generate short read lengths and higher errors, which are not taken into account by sequence alignment programs that have been developed for conventional sequencing methods, such as BLAST (Shumway et al., 2010: Heather et al., 2017).

14.12.5 TARGETING RESEQUENCING

The process of targeted resequencing involves isolation of genomic regions of interest in a sample library for a focused genomic search. Leveraging superior data quality and high throughput, the Genome Analyzer is the system of choice for the most flexible and efficient targeted resequencing solution (Marguerat et al., 2008; Niedzicka et al., 2016). The isolating of genomic regions of interest through targeted resequencing of many samples enables systematic detection of common and rare variants for high-throughput sequencing at a lower cost per sample. The Genome Analyzer offers a flexible and efficient solution for targeted resequencing applications, including

- Sequencing genes or regions in very large populations
- Following up on identified genomic regions of interest in GWAS
- Focusing on genes involved in certain pathways
- Identifying signatures associated with disease prevalence
- Discovering rare variants

Despite great efforts, the scientific community still faces tremendous challenges in understanding the genetics behind complex diseases such as cancer. Systematic detection of common and rare variants through targeted resequencing of many samples will likely hold the key to advancing our knowledge of the basis of disease and the development of treatments (Li et al., 2009: Salk et al., 2018).

Targeted resequencing experiments have been limited by the high costs of Sanger sequencing and lack of a scalable, efficient method for partitioning genomic regions of interest. The Genome Analyzer is the most widely adopted and easiest to use NGS platform. Superior data quality and a wide range of read lengths have made it the system of choice for whole genome *de novo* sequencing and resequencing. In addition to these applications, the ever-increasing output of the Genome

Analyzer can be harnessed to analyze a select region of interest in the genome (Prabhu and Pe'er, 2009; Olova et al., 2018).

REFERENCES

Aburatani, H. 2005. Functional genomic analysis with microarray technology. *Nippon Rinsho* 63(Suppl 12):171–175.

Anderson, M.W., and I. Schrijver. 2010. Next generation DNA sequencing and the future of genomic medicine. *Genes* 1:38–69. doi:10.3390/genes1010038.

Ansorge, W.J. 2018. *Top 10 Contributions on Bioinformatics & Systems Biology Chapter 08 Next Generation DNA Sequencing (II): Techniques, Applications.* Copyright: © 2018. www.avidscience.com.

Ansorge, W.J. 2009. Next-generation DNA sequencing techniques. *N Biotechnol* 25:195–203.

Antipova, A.A., P. Tamayo, and T.R. Golub. 2002. A strategy for oligonucleotide microarray probe reduction. *Genome Biol* 3:RESEARCH0073.1–4.

Arslan, A. 2018. Mapping the schizophrenia genes by neuroimaging: The opportunities and the challenges. *Int J Mol Sci* 9(219):1–13. doi:10.3390/ijms19010219.

Asakawa, J., M. Kodaira, A. Miura, T. Tsuji, Y. Nakamoto, M. Imanaka, J. Kitamura, H. Cullings, M. Nishimura, Y. Shimada, and N. Nakamura. 2016. Genome-wide deletion screening with the array CGH method in mouse offspring derived from irradiated spermatogonia indicates that mutagenic responses are highly variable among genes. *Radiation Res* 186:568–576. DOI:10.1667/RR14402.1.

Asmann, Y.W., M.B. Wallace, and E.A. Thompson. 2008. Transcriptome profiling using next-generation sequencing. *Gastroenterology* 135:1466–8.

Axtell, M.J., J.A. Snyder, and D.P. Bartel. 2007. Common functions for diverse small RNAs of land plants. *Plant Cell* 1:1750–1769.

Azhikina, T.L. 2014. Sequencing strategies and tactics in DNA and RNA analysis. *Encyclopedia Anal Chem* DOI:10.1002/9780470027318.a1429.pub2.

Barrett, M.T., A. Scheffer, A. Ben-Dor, N. Sampas, D. Lipson, R. Kincaid, P. Tsang, B. Curry, K. Baird, P.S. Meltzer, Z. Yakhini, L. Bruhn, and S. Laderman. 2004. Comparative genomic hybridization using oligonucleotide microarrays and total genomic DNA. *Proc Natl Acad Sci USA* 101:17765–17770.

Bateman, A., and J. Quackenbush. 2009. Bioinformatics for next generation sequencing. *Bioinformatics* 25:429.

Beitelshees, M., A. Hill, P. Rostami, C.H. Jones, B.A. Pfeifer. 2017. Pressing diseases that represent promising targets for gene therapy. *Discov Med* 24(134):313–322.

Bentley, G., R. Higuchi, B. Hoglund, D. Goodridge, D. Sayer, E.A. Trachtenberg, and H.A. Erlich. 2009. High-resolution, high-throughput HLA genotyping by next-generation sequencing. *Tissue Antigens* 74:393–403.

Binitha Raj, R.V., M.A. Shajahan, M.V. Sudhakaran, G.N. Sreedeepthi, K.S. Rajesh and A.R. Resny 2018. Pharmacognostic studies of Lobelia alsinoides Lam. *J Pharmacogn Phytochem* 7(2):1264–1268.

Bird, A. 2007. Perceptions of epigenetics. *Nature* 447:396–398.

Brennan, C., Y. Zhang, C. Leo, B. Feng, C. Cauwels, A.J. Aguirre, M. Kim, A. Protopopov, and L. Chin. 2004. High-resolution global profiling of genomic alterations with long oligonucleotide microarray. *Cancer Res* 64:4744–4748.

Buehler, B., H.H. Hogrefe, G. Scott, H. Ravi, C. Pabon-Pena, S. O'Brien, R. Fermosa, and S. Happe. 2010. Rapid quantification of DNA libraries for next-generation sequencing. *Methods* 50:S15–S18.

Bumgarner, R. 2013. DNA microarrays: Types, applications and their future. *Curr Protoc Mol Biol* 0 22: Unit–22.1. doi:10.1002/0471142727.mb2201s101.

Carey, M.F., C.L. Peterson, and S.T. Smale. 2009. Chromatin immunoprecipitation (ChIP). *Cold Spring Harb Protoc* 2009:pdb. prot5279.

Chan, E.Y. 2009. Next-generation sequencing methods: Impact of sequencing accuracy on SNP discovery. *Methods Mol Biol* 578:95–111.

Cho, H.Y., Y. Cho, Y. Shin, J. Park, S. Shim, Y. Jung, S. Shim, and D. Cha. 2018. Functional analysis of cell-free RNA using mid-trimester amniotic fluid supernatant in pregnancy with the fetal growth restriction. *Medicine* 97(2):1–9. e9572.

Choi, B.G., S. Hwang, J.E. Kwon, and Y.H. Kim. 2018. Array comparative genomic hybridization as the first-line investigation for neonates with congenital heart disease: Experience in a single tertiary center. *Korean Circ J* 48(3):1–8. https://doi.org/10.4070/kcj.2017.0166.

Chow, B.Y., C.J. Emig, and J.M. Jacobson 2009. Photoelectrochemical synthesis of DNA microarrays. *PNAS* 106(36):15219–15224. Doi.10.1073pnas.0813011106.

Cimaglia, F., M. Tristezza, A. Saccomanno, P. Rampino, C. Perrotta, V. Capozzi, G. Spano, M. Chiesa, G. Mita, F. Grieco 2018. An innovative oligonucleotide microarray to detect spoilage microorganisms in wine. *Food Control* 87: 169–179.

Cook-Deegan, R.M. 1991. The genesis of the Human Genome Project. *Mol Genet Med* 1:1–75.

Crowther, D.J. 2002. Applications of microarrays in the pharmaceutical industry. *Curr Opin Pharmacol* 2:551–554.

Debouck, C., and P.N. Goodfellow. 1999. DNA microarrays in drug discovery and development. *Nat Genet* 21:48–50.

Della, B.C., F. Cordero, and R.A. Calogero. 2008. Dissecting an alternative splicing analysis workflow for GeneChip Exon 1.0 ST Affymetrix arrays. *BMC Genomics* 9:571.

Dey, B., S. Thukral, S. Krishnan, M. Chakrobarty, S. Gupta, C. Manghani, V. Rani 2012. DNA–protein interactions: methods for detection and analysis. *Mol Cell Biochem* DOI 10.1007/s11010-012-1269-z.

Dominguez Del Angel, V., E. Hjerde, L. Sterck, et al. 2018. Ten steps to get started in genome assembly and annotation. *F1000Research* 7(ELIXIR):148, 1–19. doi:10.12688/f1000research.13598.1.

Dumbrava, E. I., F. Meric-Bernstam, and T.A. Yap 2018. Challenges with biomarkers in cancer drug discovery and development. *Expert Opin Drug Discov.* DOI:10.1080/17460441.2018.1479740.

Elo, L.L., L. Lahti, H. Skottman, M. Kylaniemi, R. Lahesmaa, and T. Aittokallio. 2005. Integrating probe-level expression changes across generations of Affymetrix arrays. *Nucleic Acids Res* 33:e193.

Evans, J.P. 2010. The human genome project at 10 years: A teachable moment. *Genet Med* 12:477.

Fakruddin, Md., R.M. Mazumdar, A. Chowdhury, N. Hossain, S. Mahajan, and S. Islam 2013. Pyrosequencing-A next generation sequencing technology. *World Appl Sci J* 24 (12): 1558–1571.

Fallahshahroudi, A., P. Løtvedt, J. Bélteky, J. Altimiras, and P. Jensen 2018. Changes in pituitary gene expression may underlie multiple domesticated traits in chickens. *Heredity*.

Feinberg, A.P. 2018. The key role of epigenetics in human disease prevention and mitigation. *N Engl J Med* 378:1323–1334. DOI:10.1056/NEJMra1402513.

Forrest, A.R., and P. Carninci. 2009. Whole genome transcriptome analysis. *RNA Biol* 6:107–112.

Fuselli, S., R. P. Baptista, A. Panziera, A. Magi, S. Guglielmi, R. Tonin, A. Benazzo, L.G. Bauzer, C.J. Mazzoni, G. Bertorelle 2018. A new hybrid approach for MHC genotyping: High-throughput NGS and long read MinION nanopore sequencing, with application to the non-model vertebrate Alpine chamois (*Rupicapra rupicapra*). *Heredity* https://doi.org/10.1038/s41437-018-0070-5.

Gebauer, M. 2004. Microarray applications: Emerging technologies and perspectives. *Drug Discov Today* 9:915–917.

Gill, S.R., M. Pop, R.T. Deboy, P.B. Eckburg, PJ. Turnbaugh, B.S. Samuel, J.I. Gordon, D.A. Relman, C.M. Fraser-Liggett, and K.E. Nelson. 2006. Metagenomic analysis of the human distal gut microbiome. *Science* 312:1355–1359.

Glyde, R., F. Ye, M. Jovanovic, I. Kotta-Loizou, M. Buck, and X. Zhang 2018. Structures of bacterial RNA polymerase complexes reveal the mechanism of DNA loading and transcription initiation. *Molecular Cell* 70:1111–1120. https://doi.org/10.1016/j.molcel.2018.05.021.

Gong, W., I. Kwak, P. Pota, N. Koyano-Nakagawa, and D.J. Garry 2018. DrImpute: Imputing dropout events in single cell RNA sequencing data. *BMC Bioinformatics* 19(220):1–10. https://doi.org/10.1186/s12859-018-2226-y.

Green, P. 1998. Human genome project: Data quality. *Science* 279:1115–1116.

Greenhalgh, T. 2005. The human genome project. *J Royal Soc Med* 98:545.

Hardiman, G. 2004. Microarray platforms—comparisons and contrasts. *Pharmacogenomics* 5:487–502.

Hashmi, U., S. Shafqat, F. Khan, M. Majid, H. Hussain, A.G. Kazi, R. John and P. Ahmad 2015. Plant exomics: Concepts, applications and methodologies in crop improvement. *Plant Signal Behav* 10:1. DOI: 10.4161/15592324.2014.976152.

Heather, J.M., M. Ismail, T. Oakes, and B. Chain 2017. High-throughput sequencing of the T-cell receptor repertoire: Pitfalls and opportunities. *Brief Bioinfor* 1–12. https://doi.org/10.1093/bib/bbw138.

Hester, S.D., L. Reid, N. Nowak, W.D. Jones, J.S. Parker, and K. Knudtson. 2009. Comparison of comparative genomic hybridization technologies across microarray platforms. *J Biomol Technol* 20:135–151.

Heyduk, E., and T. Heyduk 2018. DNA template sequence control of bacterial RNA polymerase escape from the promoter. *Nucleic Acids Research* 46(9):4469–4486. https://doi.org/10.1093/nar/gky172.

Horner, D.S., G. Pavesi, T. Castrignano, P.D. De Meo, S. Liuni, M. Sammeth, E. Picardi, and G. Pesole. 2010. Bioinformatics approaches for genomics and post genomics applications of next-generation sequencing. *Brief Bioinform* 11:181–197.

Huang, P.J., Y.C. Liu, C.C. Lee, W.C. Lin, R.R. Gan, P.C. Lyu, and P. Tang. 2010. DSAP: Deep-sequencing small RNA analysis pipeline. *Nucleic Acids Res* 38:W385–391.

Huang, X., Q. Feng, Q. Qian, Q. Zhao, L. Wang, A. Wang, J. Guan, D. Fan, Q. Weng, T. Huang, G. Dong, T. Sang, and B. Han. 2009. High-throughput genotyping by whole-genome resequencing. *Genome Res* 19:1068–1076.

Jares, P. 2006. DNA microarray applications in functional genomics. *Ultrastruct Pathol* 30:209–219.

Johnson, W.E., W. Li, C.A. Meyer, R. Gottardo, J.S. Carroll, M. Brown, and X.S. Liu 2006. Model-based analysis of tiling-arrays for ChIP-chip. *PNAS* 103(33):12457–12462. Doi 10.1073pnas.0601180103.

Johnson, V.P. 1992. Human genome project. *S D J Med* 45:161–162.

Kamath, B.M., B.D. Thiel, X. Gai, L.K. Conlin, P.S. Munoz, J. Glessner, D. Clark, D.M. Warthen, T.H. Shaikh, E. Mihci, D.A. Piccoli, S.F. Grant, H. Hakonarson, I.D. Krantz, and

N.B. Spinner. 2009. SNP array mapping of chromosome 20p deletions: Genotypes, phenotypes, and copy number variation. *Hum Mutat* 30:371–378.

Kawaji, H., and Y. Hayashizaki. 2008. Genome annotation. *Methods Mol Biol* 452:125–139.

Khan, A.R., M.T. Pervez, M.E. Babar, N. Naveed, and M. Shoaib 2018. A Comprehensive Study of de novo genome assemblers: Current challenges and future prospective. *Evolutionary Bioinformatics* 14:1–8. https://doi.org/10.1177/1176934318758650.

Ko T., C. Wong, J. Wu, Y. and Chen 2016. Pharmacogenomics for personalized pain medicine. *Acta Anaesthesiologica Taiwanica* 54:24–30. doi.org/10.1016/j.aat.2016.02.001.

Kricka, L.J., and S.R. Master 2009. Quality control and protein microarrays. *Clin Chem* 55(6):1053–1055.

Lambert, S.A., A. Jolma, L.F. Campitelli, P.K. Das, Y. Yin, M. Albu, X. Chen, and J. Taipale 2018. The human transcription factors. *Cell* 172(4):650–665. https://doi.org/10.1016/j.cell.2018.01.029.

Lee, Y., C. Lee, L. Lai, M. Tsai, T. Lu, and E.Y. Chuang 2018. CellExpress: A comprehensive microarray-based cancer cell line and clinical sample gene expression analysis online system. *Database*, 2018, bax101. https://doi.org/10.1093/database/bax101.

Leykin, I., K. Hao, J. Cheng, N. Meyer, M.R. Pollak, R.J. Smith, W.H. Wong, C. Rosenow, and C. Li. 2005. Comparative linkage analysis and visualization of high-density oligonucleotide SNP array data. *BMC Genet* 6:7.

Li, R., H. Zhu, J. Ruan, W. Qian, X. Fang, Z. Shi, Y. Li, S. Li, G. Shan, K. Kristiansen, S. Li, H. Yang, J. Wang, and J. Wang. 2010. De novo assembly of human genomes with massively parallel short read sequencing. *Genome Res* 20:265–272.

Li, R., Y. Li, X. Fang, H. Yang, J. Wang, K. Kristiansen, and J. Wang. 2009. SNP detection for massively parallel whole-genome resequencing. *Genome Res* 19:1124–1132.

Liu, X., L. Hao, D. Li, L. Zhu, S. Hu 2015. Long non-coding RNAs and their biological roles in plants. *Genomics Proteomics Bioinformatics* 13:137–147. http://dx.doi.org/10.1016/j.gpb.2015.02.003.

Lopez-Romero, P., M.A. Gonzalez, S. Callejas, A. Dopazo, and R.A. Irizarry. 2010. Processing of agilent microRNA array data. *BMC Res Notes* 3:18.

Lü, J., H. Li, H. An, G. Wang, Y. Wang, M. Li, Y. Zhang, and J. Hu 2004. Positioning isolation and biochemical analysis of single DNA molecules based on nanomanipulation and single-molecule PCR. *J Am Chem Soc* 126(36):11136–11137. DOI:10.1021/ja047124m.

Ma, S., T.W. Murphy, and C. Lu 2017. Microfluidics for genome-wide studies involving next generation sequencing. *Biomicrofluidics* 11(021501):1–25.

Mamanova, L., A.J. Coffey, C.E. Scott, I. Kozarewa, E.H. Turner, A. Kumar, E. Howard, J. Shendure, and D.J. Turner. 2010. Target-enrichment strategies for next-generation sequencing. *Nat Methods* 7:111–118.

MAQC Consortium 2006. The MicroArray Quality Control (MAQC) project shows inter- and intraplatform reproducibility of gene expression measurements MAQC Consortium. *Nat Biotechnol* 24(9):1151–1161. doi:10.1038/nbt1239.

Mardis, E.R. 2008a. Next-generation DNA sequencing methods. *Annu Rev Genomics Hum Genet* 9:387–402.

Mardis, E.R. 2008b. The impact of next-generation sequencing technology on genetics. *Trends Genet* 24:133–141.

Marguerat, S., B.T. Wilhelm, and J. Bahler. 2008. Next-generation sequencing: Applications beyond genomes. *Biochem Soc Trans* 36:1091–1096.

McGranahan, N., R. Rosenthal, C.T. Hiley, A.J. Rowan, T.B.K. Watkins, G.A. Wilson, N.J. Birkbak, S. Veeriah, P.V. Loo, J. Herrero, C. Swanton, and the TRACERx Consortium. 2017. Allele-specific HLA loss and immune escape in lung cancer evolution. *Cell* 171:1259–1271. https://doi.org/10.1016/j.cell.2017.10.001.

Metzker, M.L. 2010. Sequencing technologies—The next generation. *Nat Rev Genet* 11:31–46.

Milos, P. 2008. Helicos bioSciences. *Pharmacogenomics* 9:477–80.

Misra, B.B. 2018. New tools and resources in metabolomics: 2016–2017. *Electrophoresis* 39:909–923. DOI 10.1002/elps.201700441.

Miyake, N., and N. Matsumoto. 2005. Microarray CGH. *Nippon Rinsho* 63(Suppl 12):167–170.

Morishita, S. 2009. How can we combine next-generation DNA sequencing and bioinformatics to reveal novel findings?. *Tanpakushitsu Kakusan Koso* 54:1239–1247.

Morozova, O., and M.A. Marra. 2008a. Applications of next-generation sequencing technologies in functional genomics. *Genomics* 92:255–264.

Morozova, O., and M.A. Marra. 2008b. From cytogenetics to next-generation sequencing technologies: Advances in the detection of genome rearrangements in tumors. *Biochem Cell Biol* 86:81–91.

Morozova, O., M. Hirst, and M.A. Marra. 2009. Applications of new sequencing technologies for transcriptome analysis. *Annu Rev Genomics Hum Genet* 10:135–151.

Morrissy, A.S., R.D. Morin, A. Delaney, T. Zeng, H. McDonald, S. Jones, Y. Zhao, M. Hirst, and M.A. Marra. 2009. Next-generation tag sequencing for cancer gene expression profiling. *Genome Res* 19:1825–1835.

Mueckstein, U., G.G. Leparc, A. Posekany, I. Hofacker, and D.P. Kreil. 2010. Hybridization thermodynamics of NimbleGen microarrays. *BMC Bioinformatics* 11:35.

Mustapha, G., A. Nafi'u, J. Ukomadu, M.U. Hizbullah, I.J. Fatima, and H.M. Maiturare 2017. Review on the role of DNA microarrays in cancer therapy. *J Adv Med Life Sci* 5(3):1–4.

Mychaleckyj, J.C. 2007. Genome mapping statistics and bioinformatics. *Methods Mol Biol* 404:461–488.

Nelson, J.D., O. Denisenko, and K. Bomsztyk. 2006. Protocol for the fast chromatin immunoprecipitation (ChIP) method. *Nat Protoc* 1:179–185.

Ng, P.C., and E.F. Kirkness. 2010. Whole genome sequencing. *Methods Mol Biol* 628:215–26.

Niedzicka, M., A. Fijarczyk, K. Dudek, M. Stuglik, and W. Babik 2016. Molecular inversion probes for targeted resequencing in nonmodel organisms. *Sci Rep* 6:24051:1–9. DOI:10.1038/srep24051.

Olova, N., F. Krueger, S. Andrews, D. Oxley, R.V. Berrens, M.R. Branco, and W. Reik 2018. Comparison of whole-genome bisulfite sequencing library preparation strategies identifies sources of biases affecting DNA methylation data. *Genome Biol* 19:33. https://doi.org/10.1186/s13059-018-1408-2.

Park, P.J. 2008. Epigenetics meets next-generation sequencing. *Epigenetics* 3:318–321.

Paszkiewicz, K., and D.J. Studholme. 2010. *De novo* assembly of short sequence reads. *Brief Bioinform* 11:457–472.

Patterson, N., and S. Gabriel. 2009. Combinatorics and next-generation sequencing. *Nat Biotechnol* 27:826–827.

Pevzner, P.A., H. Tang, and G. Tesler. 2004. *De novo* repeat classification and fragment assembly. *Genome Res* 14:1786–1796.

Pinkel, D., and D.G. Albertson. 2005. Comparative genomic hybridization. *Annu Rev Genomics Hum Genet* 6:331–354.

Prabhu, S., and I. Pe'er. 2009. Overlapping pools for high-throughput targeted resequencing. *Genome Res* 19:1254–1261.

Putonti, C. 2007. The diverse and informative future of microarray applications. *Pharmacogenomics* 8:137–140.

Rao, S., T. Chiu, J.F. Kribelbauer, R.S. Mann, H.J. Bussemaker, and R. Rohs 2018. Systematic prediction of DNA shape changes due to CpG methylation explains epigenetic effects on protein–DNA binding. *Epigenet Chromatin* 11:6.

Rauch, A., F. Ruschendorf, J. Huang, U. Trautmann, C. Becker, C. Thiel, K.W. Jones, A. Reis, and P. Nurnberg. 2004. Molecular karyotyping using an SNP array for genomewide genotyping. *J Med Genet* 41:916–922.

Raza, K. 2010. Application of data mining in bioinformatics. *Indian J Comput Sci Eng* 1(2):114–118.

Reis-Filho, J.S. 2009. Next-generation sequencing. *Breast Cancer Res* 11(Suppl 3):S12.

Reuterswärd, P. 2018. *Development of Array Systems for Molecular Diagnostic Assays*. Doctoral Thesis in Biotechnology, KTH Royal Institute of Technology School of Engineering Sciences in Chemistry, Biotechnology and Health, Stockholm, Sweden.

Rogers, S., M. Girolami, C. Campbell, and R. Breitling. 2005. The latent process decomposition of cDNA microarray data sets. *IEEE/ACM Trans Comput Biol Bioinform* 2:143–156.

Salk, J.J., M.W. Schmitt, and L.A. Loeb 2018. Enhancing the accuracy of next-generation sequencing for detecting rare and subclonal mutations. *Nat Rev Genet.* 19:269–285. doi:10.1038/nrg.2017.117.

Salser, W.A. 1974. DNA sequencing techniques. *Annu Rev Biochem* 43:923–965.

Shaffer, C. 2007. Next-generation sequencing outpaces expectations. *Nat Biotechnol* 25:149.

Shen, Y., Z. Wan, C. Coarfa, R. Drabek, L. Chen, E.A. Ostrowski, Y. Liu, G.M. Weinstock, D.A. Wheeler, R.A. Gibbs, and F. Yu. 2010. A SNP discovery method to assess variant allele probability from next-generation resequencing data. *Genome Res* 20:273–280.

Shendure, J., and H. Ji. 2008. Next-generation DNA sequencing. *Nat Biotechnol* 26:1135–1145.

Shittu, U., M. Abu Naser, Z. Idris, and S.A. Maryam 2018. Microarray gene expression statistical data analysis of three different clinical forms of human tuberculosis stimulated samples in the bioconductor R package. *J Proteomics Bioinform.* 11(2): 51–56.

Shiu, S.H., and J.O. Borevitz. 2008. The next generation of microarray research: Applications in evolutionary and ecological genomics. *Heredity* 100:141–149.

Shumway, M., G. Cochrane, and H. Sugawara. 2010. Archiving next generation sequencing data. *Nucleic Acids Res* 38:D870–871.

Sigdel, T.K., M. Nguyen, D. Dobi, S. Hsieh, J.M. Liberto, F. Vincenti, M.M. Sarwal, and Z. Laszik 2018. Targeted transcriptional profiling of kidney transplant biopsies. *Kidney Int Rep* 3:722–731; https://doi.org/10.1016/j.ekir.2018.01.014.

Srinivasan, B.S., N.B. Caberoy, G. Suen, R.G. Taylor, R. Shah, F. Tengra, B.S. Goldman, A.G. Garza, and R.D. Welch. 2005. Functional genome annotation through phylogenomic mapping. *Nat Biotechnol* 23:691–698.

Steibel, J.P., and G.J. Rosa. 2005. On reference designs for microarray experiments. *Stat Appl Genet Mol Biol* 4: Article36.

Sugano, S. 2009. Introduction: Next-generation DNA sequencing and bioinformatics]. *Tanpakushitsu Kakusan Koso* 54:1233–1237.

Sundquist, A., M. Ronaghi, H. Tang, P. Pevzner, and S. Batzoglou. 2007. Whole-genome sequencing and assembly with high-throughput, short-read technologies. *PLoS One* 2:e484.

Suzuki, M.M., and A. Bird. 2008. DNA methylation landscapes: Provocative insights from epigenomics. *Nat Rev Genet* 9:465–476.

Swidan, F., E.P. Rocha, M. Shmoish, and R.Y. Pinter. 2006. An integrative method for accurate comparative genome mapping. *PLoS Comput Biol* 2:e75.

Tanaka, T. 2000. Transcriptome analysis and pharmacogenomics. *Nippon Yakurigaku Zasshi* 116:241–246.

Teng, X., and H. Xiao. 2009. Perspectives of DNA microarray and next-generation DNA sequencing technologies. *Sci China C Life Sci* 52:7–16.

Thompson, J.F., and K.E. Steinmann. 2010. Single molecule sequencing with a HeliScope genetic analysis system. *Curr Protoc Mol Biol* Chapter 7:Unit7.10.

Tringe, S.G., and E.M. Rubin. 2005. Metagenomics: DNA sequencing of environmental samples. *Nat Rev Genet* 6:805–814.

Tuteja, R. and N. Tutej 2004. Serial analysis of gene expression (SAGE): Unraveling the bioinformatics tools. *BioEssays* 26:916–922. DOI 10.1002/bies.20070.

Velculescu, V.E., L. Zhang, B. Vogelstein, and K.W. Kinzler. 1995. Serial analysis of gene expression. *Science* 270:484–487.

Voelkerding, K.V., S.A. Dames, and J.D. Durtschi 2009. Next-generation sequencing: From basic research to diagnostics. *Clinical Chemistry* 55(4):641–658. DOI:10.1373/clinchem.2008.112789.

Von, B.A. 2008. Next-generation sequencing: The race is on. *Cell* 132:721–723.

Wang, K.C., and H.Y. Chang 2011. Molecular mechanisms of long noncoding RNAs. *Mol Cell* 43(6): 904–914. doi:10.1016/j.molcel.2011.08.018.

Weickardt, I., A. Zehnsdorf, and W. Durka. 2018. Development and characterization of simple sequence repeat markers for the invasive tetraploid waterweed Elodea nuttallii (Hydrocharitaceae). *Applications in Plant Sciences* 6(4):1–6. e1146. doi:10.1002/aps3.1146.

Wommack, K.E., J. Bhavsar, and J. Ravel. 2008. Metagenomics: Read length matters. *Appl Environ Microbiol* 74:1453–1463.

Wooley, J.C., A. Godzik, and I. Friedberg. 2010. A primer on metagenomics. *PLoS Comput Biol* 6:e1000667.

Wu, L., P.M. Williams, and W. Koch. 2005. Clinical applications of microarray-based diagnostic tests. *Biotechniques* 39:S577–582.

Yang, H., R. Ratnapriya, T. Cogliati, J. Kim, and A. Swaroop 2015. Vision from next generation sequencing: Multi-dimensional genome-wide analysis for producing gene regulatory networks underlying retinal development, aging and disease. *Prog Retin Eye Res*, 1–30.

You, Q., X. Yang, Z. Peng, L. Xu and J. Wang 2018. Development and applications of a high throughput genotyping tool for polyploid crops: Single Nucleotide Polymorphism (SNP) array. *Front. Plant Sci.* 9:104. https://doi.org/10.3389/fpls.2018.00104.

Yue, T., and H. Wang 2018. *Deep Learning for Genomics: A Concise Overview*. http://arxiv.org/abs/1802.00810v2.

Zahurak, M., G. Parmigiani, W. Yu, R.B. Scharpf, D. Berman, E. Schaeffer, S. Shabbeer, L. Cope. 2007. Pre-processing agilent microarray data. *BMC Bioinformatics* 8:142.

Zhao, J., and S.F. Grant. 2010. Advances in whole genome sequencing technology. *Curr Pham Biotechnol* 12:293–305.

Zhao, Z., T. Miki, A. Van Oort-Jansen, T. Matsumoto, D.S. Loose, and C.C. Lee. 2011. Hepatic gene expression profiling of 5'-AMP induced hypometabolism in mice. *Physiol Genomics* 43:325–345.

Zhou, X., L. Ren, Q. Meng, Y. Li, Y. Yu, and J. Yu. 2010. The next-generation sequencing technology and application. *Protein Cell* 1:520–536.

15 Bioreactors
Design, Operation, and Applications

Tony Allman

CONTENTS

"In the field of observation, chance favors only the pre-pared minds."

Louis Pasteur, 1854

15.1 BIOREACTORS: AN OVERVIEW

This chapter aims to provide an understanding of the factors governing bioreactor design, operation, and applications with special emphasis on the following aspects:

- Unraveling and describing the scientific principles underpinning bioreactor design, bioreactor instrumentation, and control
- Describing bioreactor assembly and operation
- Describing how a bioreactor can be adapted to a range of specific applications
- Providing application examples, which show new uses for simpler systems.

15.2 COMPONENT PARTS OF BIOREACTORS

The main subdivisions of a standard stirred-tank reactor (STR) are

- A base unit connected to pipe work components for control of temperature, stirring, gassing, and additions of reagents
- A culture vessel with fittings and ports to aid gas transfer, liquid addition/removal, mixing, sampling, and fitting of sensors

- Peripheral equipment such as reagent containers, additional sensors, and special systems for sterilization, cooling, separation, and removal of culture constituents
- Instrumentation for measurement and/or control of key process parameters with links to supervisory software and remote control facility

The example used to illustrate this aspect is the smallest and simplest of all fermentation systems (i.e. a bench-top bioreactor). More specialized versions for the propagation of photosynthetic organisms and fermentation of solid substrates will also be described, albeit briefly.

A number of modifications are also illustrated to facilitate reconfiguration and adaptation of bioreactors for the applications discussed in the following sections.

15.3 COMPONENT PARTS OF A "TYPICAL" VESSEL

The vessel can be constructed as a single-walled cylinder of borosilicate glass or as a glass-jacketed system, which typically has a rounded or dished base. The top plate is made from "316" L stainless steel and is compressed onto the vessel flange by nuts or a quick-release clamping system. A seal separates the vessel glass from the top plate. Port fittings of various sizes are provided for insertion of probes, inlet pipes, exit gas cooler, cold fingers, sample pipes, and so on. They work by compressing the sides of the probe/pipe against an O-ring seal. A special inoculation port will have a membrane seal held in place with a collar. Culture can be withdrawn into a sampling device or a reservoir bottle via a sample pipe situated in the bulk of the bioreactor fluid.

A gas sparger is also fixed into the top plate and this terminates in a special assembly that ensures that incoming air is dispersed efficiently within the culture by flat-bladed "Rushton-type" impellors fixed to the drive shaft. This arrangement breaks large bubbles into smaller ones with a corresponding increase in surface area for gas transfer (as measured by a physical property of the vessel—its Kla or volumetric mass transfer coefficient). A drive motor provides stirring power to the drive shaft and is usually fitted directly to the drive hub on the vessel top plate. An exit gas cooler works as a condenser to remove as much moisture as possible from the gas leaving the bioreactor to prevent excessive liquid losses during the fermentation and wetting of the exit air filter.

A narrow platinum resistance *Pt-100 temperature sensor* completes the list of minimum essential fittings. Heating is achieved through direct heating using a heater pad or by circulating warm water around the vessel jacket. If direct heating is used, a cold finger is used to control temperature by cooling the vessel contents (more than one could be used, if need be).

Pt-100 temperature sensor: A platinum resistance electrode used to give an accurate indication of vessel temperature by relating changes in electrical resistance of the sensor to temperature, 100 ohms being the value for a temperature of 0°C.

The sensors are directly coupled through a thread on the body of the electrode, as is the case in the gel-filled type of pH electrode, or through a special fitting on the vessel top plate, which provides a clamping mechanism for long sensors. Another system involves the use of a simple compression fitting that holds the body of the electrode, as with the foam probe. In this case, the height is variable and the tip of the probe must be adjusted so that it is above the surface of culture. Figure 15.1 illustrates the main components of a typical stirred-tank bioreactor.

15.4 PERIPHERAL PARTS AND ACCESSORIES

15.4.1 PERISTALTIC PUMPS

Peristaltic pumps are normally part of the instrumentation system for pH and antifoam control. The flow rate, which depends upon the bore size of the tubing used, is controlled through a "shot and delay" feeding mechanism. The feed is delivered via peristaltic tubes, which link the reservoir bottles to the vessel multiway inlet and the tubing can be aseptically connected after autoclaving.

For accurate monitoring of additives into the culture vessel, the respective reservoir bottles can be placed on analytical balances, thus allowing accurate additions of reagents.

15.4.2 MEDIUM FEED PUMPS AND RESERVOIR BOTTLES

Media feed pumps are often of variable speed to cover the desired possible range of flow rate. Pump speed can be set manually or computer controlled. The reservoir bottles are usually large (e.g., 5–20L) but are prepared in the same way as normal reagent bottles. These bottles may have to be changed

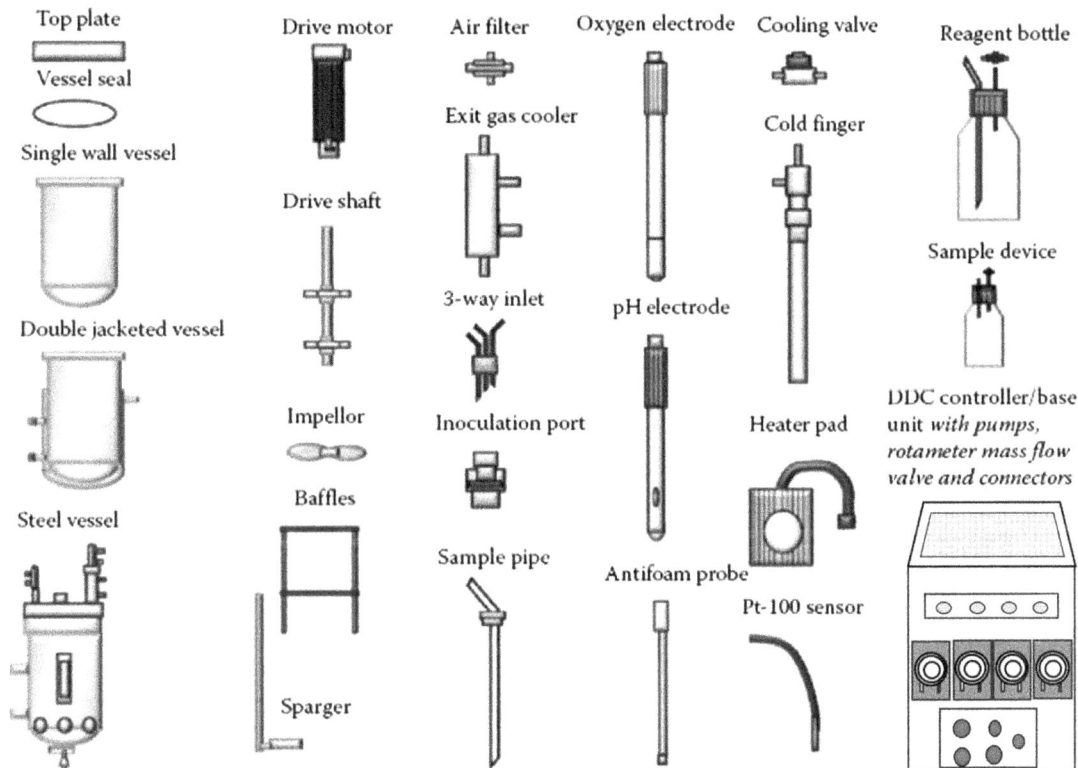

FIGURE 15.1 Major components of a bioreactor.

several times during a fermentation experiment and so tubing is often fitted with aseptic coupling devices. A harvest pump can also be added to remove culture fluid from the bioreactor vessel into a storage reservoir.

15.4.3 ROTAMETER/GAS SUPPLY

The flow of gas supply, which is oil and dust free, is controlled by a manually set flow meter (*rotameter*) fitted with a pressure regulator valve to ensure safety. In most modern systems, manual control is replaced by automatic control requiring a thermal mass flow control valve (MFV). A sterile filter (usually 0.22 µm) is fitted to ensure the removal of bacteria in the air delivered to the sparger and in turn the culture. Another filter is fitted onto the exit gas cooler (often 0.45 µm) prevents the releasing of microbes into the laboratory. In fermentation processes leading to the release of unpleasant odors or potentially dangerous gases, an extraction system should be fitted to ensure the safe disposal of the undesirable or harmful gases.

> *Rotameter:* A variable area flow meter that indicates the rate of gas flow into a bioreactor. A manual valve is adjusted until an indicator ball rises up a tube of increasing width until the required flow rate value is reached on a calibrated scale marked on the glass wall of the tube. The bottom of the ball should rest on the calibration line.

15.4.4 SAMPLING DEVICE

This allows culture fluid to be withdrawn for analysis aseptically at intervals decided by the user. Sampling may be linked to a button on the controller to automatically increment sample number and create a space for data from the sample results in a database.

15.5 ALTERNATIVE VESSEL DESIGNS

Alternatives to the conventional stirred-tank bioreactor have been designed to overcome the configuration problem, which does not allow for adequate growth of certain organisms (e.g., animal cells are often disrupted by shear forces in bioreactors with turbine impellors). Also, large-scale fermentation necessitates different designs of bioreactor for efficiency and economic reasons (e.g., production of large quantities of single-cell protein is cheaper on a large scale if air lift bioreactors are used to eliminate energy costs associated with a drive system). Several of these special designs are available at the bench/pilot scale of operation to allow for small-scale research into the suitability of a specific method.

The use of lights surrounding a standard stirred-tank bioreactor vessel is somewhat critical because it is essential in algal growth. Variable light intensity and the ability to simulate day/night cycles are common requirements. Special designs for commercial scale production of photosynthetic organisms exist, but these tend to be simple tubes or tubs with external light sources. A "flat-panel" vessel design allows for maximum light transfer with an even distribution and the opportunity to directly measure light intensity on the side of the vessel away from a freestanding lighting panel. The high yield of oil in some species of algae (up to 60% of biomass) makes them an ideal resource for biofuel generation; a domain that is currently enjoying a center stage in biotechnology.

15.5.1 AIRLIFT

Airlift fermentors are generally tall and thin with a vessel *aspect ratio* of approximately 10:1 (height-to-base diameter). It does not contain stirring blades and sometimes may have a "conical" section at the top of the vessel to facilitate effective gas exchange. Sensors are generally mounted on a steel section, typically at the top or base of the vessel. The culture fluid is aerated, and in turn mixed, by a stream of air that enters near the base of the vessel. A hollow pipe or draft tube in the center of the vessel provides a "riser" for the air (which is full of bubbles) to move upward to the top of the vessel. If a very large vessel is used, the hydrostatic head of the fluid provides a pressurizing effect to the lowest region of the culture where the air enters and so increases the dissolved oxygen concentration. The draft tube is usually double-walled to allow for heating and cooling using a thermocirculator system.

> *Aspect ratio:* The ratio of the height of a fermentation vessel to its diameter. Typically, vessels for microbial work have an aspect ratio of 2.5 up to 3:1, whereas vessels for animal cell culture tend to have an aspect ratio closer to 1:1.

When the aerated culture fluid reaches the top of the draft tube, it "spills over" and begins to fall toward the bottom of the vessel via the space between the outer wall of the draft tube and the inner wall of the vessel. A large headspace above the top of the draft tube allows for easy gas transfer from the liquid to the gas phase, which in turn causes the density/specific gravity of the liquid to increase and so it descends down to the bottom of the vessel. The descending liquid returns to the base of the vessel where it is re-aerated and begins to rise again (see Figure 15.2).

A common use of airlift bioreactors is the growth of shear-sensitive cells such as plant and animal cell cultures. Also, the design can be scaled up to produce large amounts of biomass as single-cell protein. The mixing required for a flat-panel photo-bioreactor vessel is usually provided by a modified air lift design using interior baffles rather than a full draft tube.

15.5.2 FLUIDIZED BED, IMMOBILIZED, AND SOLID-STATE SYSTEMS

This important area of bioreactor design and application will only be mentioned in outline here because Chapters 13 and 17 are devoted to these aspects.

The microbes/cells are trapped in a physical medium (e.g., alginate beads) and held in the vessel by a mesh. The medium is fed and recycled via a pump to give a continuous/semi-continuous flow to allow the entrapped cells to affect the desired biochemical reactions without the cells being washed out

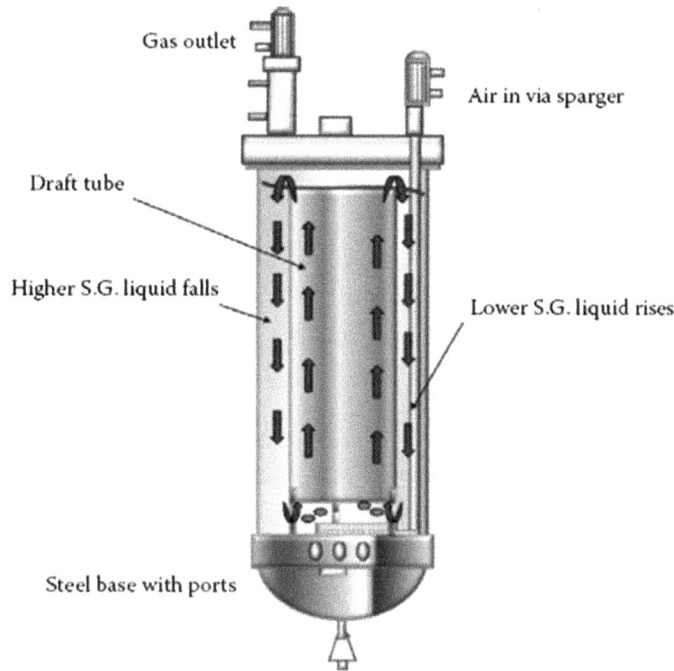

FIGURE 15.2 Airlift bioreactor.

along with spent medium. This system is well suited to growth of animal cells on the smaller scale and has large-scale application in effluent/decontamination treatment plants.

Some animal cell lines benefit from being immobilized on polymer beads in a STR system because it provides a surface for attachment of anchorage-dependent cells and affords protection to the cells against damage by bubbles and shear forces. Bacteria and fungi can also be grown in immobilized systems, using entrapment, or on solid surface.

Solid-state fermentation is a biological process in which solid or semi-solid substrates are used for growth. In some designs, the solids remain static on trays, whereas others require mixing to provide aeration and homogeneous distribution of added reagents. The substrate is rotated or mechanically mixed and temperature control is achieved by air and water circulation. It is possible to create *in situ* sterilizable versions of these systems using direct injection of steam into the vessel and rapid airflow to provide subsequent drying.

15.5.3 *In Situ* Sterilizable Bioreactors

Sterilization of large bioreactor vessels (>10 L working) is difficult and impractical. Such vessels are currently made of stainless steel (316 L) so that it can be sterilized *in situ* using an electrical heating element or steam generator, which may be built into the base unit of the bioreactor. The heating for the vessel is normally provided via a double jacket, which can be the full length of the vessel or cover just the bottom third. Internal coils within the vessel are sometimes used for heating and/or cooling where rapid changes in temperature are needed. Because the vessel body is steel, a sight window and a light must be fitted to see the culture.

Most animal cell culture vessels are top-driven to minimize shear damage to the delicate cells. The vessel top plate has port fittings, which use a membrane seal and port closure.

Options such as "push valves" and steam-sterilizable inlet lines (which create a "sterile cross" of valves and pipe work) are becoming used more commonly to remove the need for needle injection of transfer lines for medium, reagent, etc.

The medium in the vessel is heated to 121°C or above and often supplies the steam for sterilization of the exit gas filter (see Figure 15.3). The steam for vessel sterilization can be relatively impure, but sterilization of air filters and valves has to use clean steam (i.e., it must pass through a 5-μm filter).

Another increasingly common element in the configuration of larger *in situ* sterilizable (ISS) systems is the use of clean-in-place (CIP) systems. A CIP system may be integrated into the support frame or can be a freestanding unit, which can be transferred from one vessel to another as needed. Typically, a powerful centrifugal pump circulates liquid through the vessel using spray balls located near the top to clean the inner walls of the vessel. Several different cycles are used to ensure adequate level of cleanliness, including rinsing with deionized water and cleaning with a caustic solution followed by rinsing with pure water. Measuring the electrical conductivity of final effluent after cleaning ensures full cleanliness of the vessel and the water is used.

15.5.4 Containment of Pathogenic and/or Genetically Modified Organisms

An ISS bioreactor may have to be altered in certain ways if the organism to be cultured is pathogenic or genetically modified. The alterations are designed to contain any release of

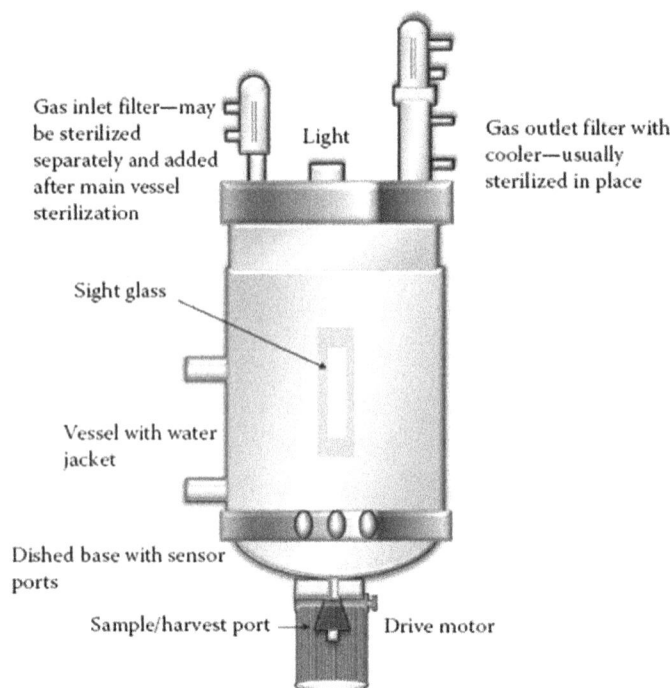

FIGURE 15.3 Major components of an ISS bioreactor vessel.

microbes into the environment by using features such as a double mechanical seal, magnetic coupling, additional air filters, extra foam control systems, and special sampling devices. More elaborate precautions include steam lines to all vessel fittings and direct discharge of any released liquid to a tank of disinfectant (a "kill tank"). Applications for this sort of technology are medical research and vaccine manufacturing.

15.6 BIOREACTOR INSTRUMENTATION

Modern instrumentation now almost exclusively uses digital controllers of one sort or another. The trend is toward color touch screens with a graphical user interface, and this can accomplish several functions previously reserved for remote software (e.g., creation of synoptic displays and real-time trend graphs).

15.6.1 DIGITAL CONTROLLERS—PROCESS CONTROLLER

A complete process controller (usually from a production control environment) is added to a housing containing all of the signal processing and control actuators for the bioreactors. It exerts control in the same way as an embedded controller but is essentially a "plug-in" component. The controller is usually programmed using simple commands to input set-point values and so on. Again, links may be provided via common industrial communications protocols to a local or remote HMI providing additional functions.

Digital controller: A digital controller uses a processor to store information about control output characteristics as mathematical algorithms. Consequently, changing the

characteristics of such a controller is achieved by reprogramming the processor.

≈

15.7 OVERVIEW OF COMMON MEASUREMENT AND CONTROL SYSTEMS

Schematic overviews and descriptions for key parameters control loops as used primarily in bench scale bioreactors are provided below.

15.7.1 SPEED CONTROL

Speed control relies on the feedback from a *tachometer* located within the drive motor. Actual speed in revolutions per minute is displayed, as determined by the tachometer signal. A power meter is sometimes included and indicates how hard the motor has to work to maintain the set speed and, thereby, indirectly, the viscosity or "density" of the culture fluid. A DC, low-voltage (24–50 V) motor is often used for safety reasons. Speed range is typically from 50 to 1,500 rpm for bacterial systems and 10–300 rpm for cell culture units. It is now possible to use "universal motors" that can cover the complete range of speeds for microbial and cell culture applications for bench-sale bioreactors.

Tachometer: An electronic device usually integrated into a drive motor to provide feedback about rotational speed in the form of an analog signal (the sensor in Figure 15.4).

Where speed is used to control the level of dissolved oxygen, an external signal from the oxygen controller can

Bioreactor Speed Control

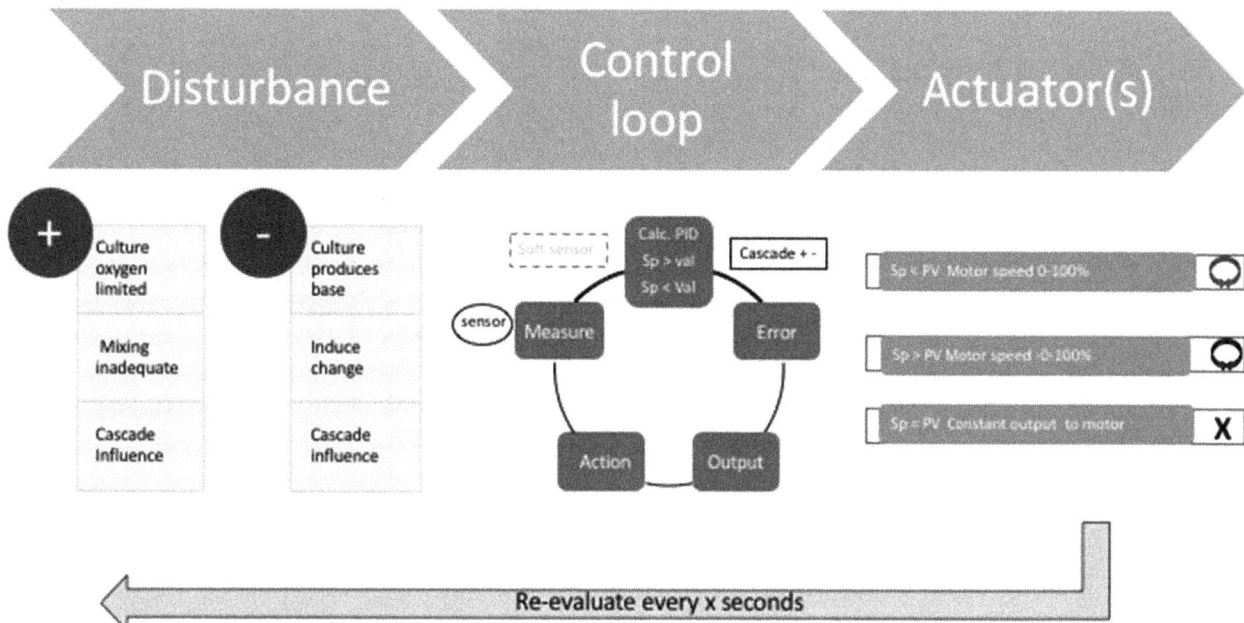

FIGURE 15.4 Speed control system.

influence the stirrer speed. In this case, an absolute maximum and minimum value for speed can be set on the speed control module to limit the effects of the oxygen controller.

15.7.2 TEMPERATURE CONTROL

A thermos-circulation system around a vessel jacket has been chosen as an example here because it is the most complex of all of the methods of temperature control. For direct heating such as via a heater pad, it is simply a matter of fitting the heater, setting the desired temperature, and switching on. Cooling is normally via a *cold finger* and flow of cooling water is controlled via the action of a solenoid valve. The Pt-100 sensor provides the feedback signal, which causes the controller to take one of the following actions:

- Heat at full power because the actual temperature is some way below the set point.
- Pulse the heater power because the actual temperature is close to set point.
- Turn on the cooling valve because the actual temperature is above set point.

Cold finger: A closed pipe or coil that passes through the bioreactor top plate and allows cooling water to circulate to act as a heat exchanger with the culture.

There is usually some indication to show which action the controller is taking at any given moment. A circulation pump and pipe work are added to the system for water circulation, and any heating is indirect (i.e., on the water circulating in the vessel jacket and not direct heating of the culture). In this case, a connection to a cold- water supply must be made (securely, using jubilee clips or cable ties). A drainpipe should also be provided from the overflow point to a sink with a clear fall to the drain (i.e., the sink must be the lowest point for the whole length of this pipe). The water should be delivered from the mains at a minimum pressure of 1.5–2 bar and a flow rate of greater than 5 L/min. The water hardness should be no more than 50 ppm suspended solids to protect the heating elements from "furring." The vessel must be connected to the circulation loop (normally by rapid coupling connectors and flexible pressure tubing). See Figure 15.5.

Use of chillers for cooling water circulation is becoming more common and this usually requires the addition of a bypass and pressure relief valve to accommodate the cooling water valve being closed at the bioreactor. Unless the cooling system has a large cooling capacity, it will be unsuitable for rapid cooling after vessel sterilization, so a supply of house water may still be needed. Larger vessels can use twin heat exchangers for heating (by steam) and cooling (by chilled water) with a closed-loop thermocirculator to transfer heat to and from the vessel contents. Water is first supplied by opening a manual valve until the jacket is filled. The heating and cooling is controlled in the same way as a directly heated system, but only the water in the jacket is affected. The jacket provides a large surface area in contact with the vessel wall for heat exchange. Good temperature control can be achieved from approximately 5–8°C above the ambient temperature or above the temperature of the cooling water. Counter cooling with water ensures stable temperature control when operating near ambient temperatures. Measured range is typically from 0 to 60°C (exceptionally up to 90°C).

Bioreactor Temperature Control

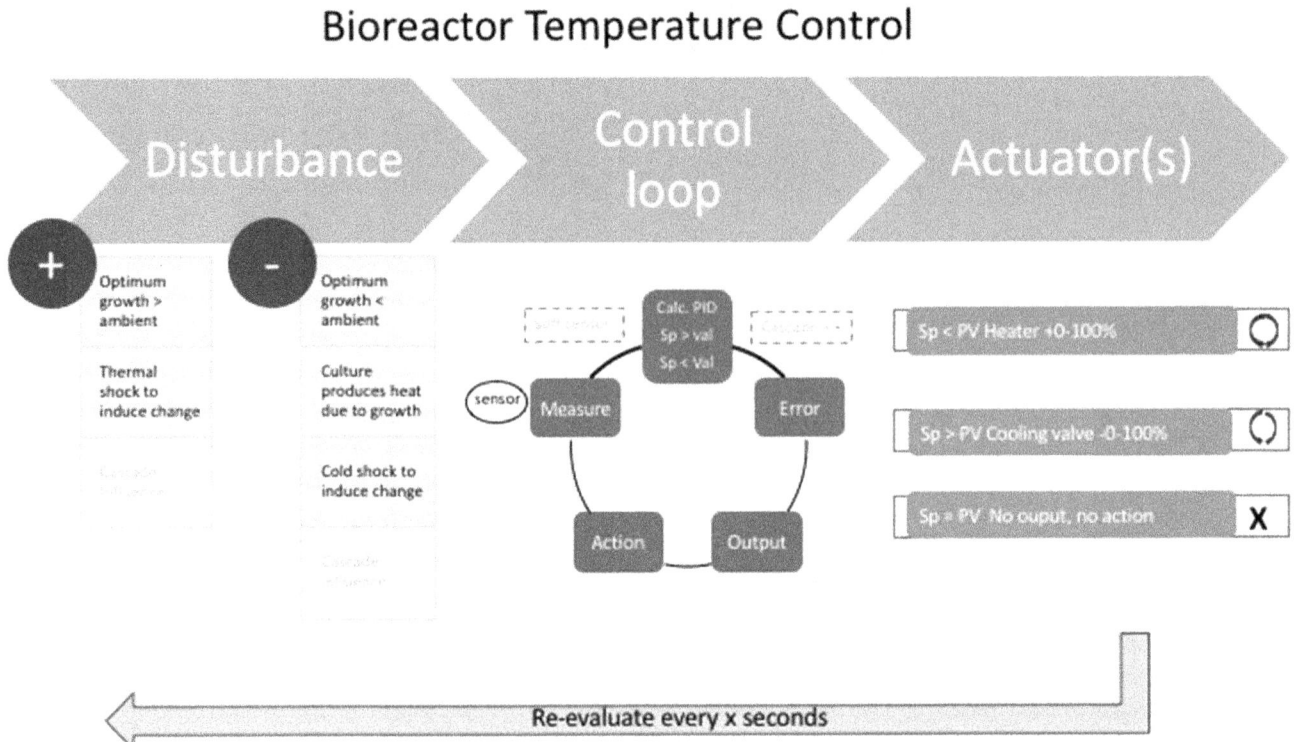

FIGURE 15.5 Temperature control system using water circulation.

A common alternative for bench scale bioreactors is to use electrical heating using a silicone mat. This is wrapped around a single-walled vessel after autoclaving and secured in place. A coil or a cold finger dipping into the vessel provides cooling. A less common alternative is a heater block with an electrical element and a cooling coil. This, in turn, allows for easy handling and can be used on multiple, parallel bioreactors.

15.7.3 Control of Gas Supply

A compressed gas (normally oil-free air) is supplied to the bioreactor at a maximum of 0.5–0.75 bar. The rotameter controls the actual flow rate of air through the bioreactor. This should not exceed 1.5 vessel volumes per minute otherwise droplets of water may be carried out with the gas leaving the bioreactor, thus wetting the exit gas filter, which in turn causing it to block. A valve at the bottom of the rotameter is turned and the indicator ball in the rotameter tube rises or falls in proportion to the valve position. A scale on the tube gives flow rates in milliliters per minute or liters per hour.

The air passes through the inlet air filter, which prevents any microbes from entering the vessel via this path. The end of the sparger is typically a ring with small holes through which the air is forced. The bubbles are immediately broken up and dispersed by the impellors on the drive shaft and the baffles, which can be fitted near the wall of the vessel. The use of several impellors ensures that all regions of the vessel receive good aeration. A "headspace" of approximately

20–30% is normally left between the culture level and the vessel top plate. Sometimes, gas can also be introduced into this region via a short pipe in the bioreactor top plate (e.g., CO_2). For certain types of fermentation (e.g., mammalian cell culture), a *gas mixing station* can be used to premix several gases before they are introduced into the bioreactor.

Gas mixing station: A device used for animal cell culture that allows a mixture of air, oxygen, nitrogen, and carbon dioxide gases to be blended into any desired combination before they are introduced into a bioreactor. This allows great flexibility in how dissolved oxygen concentration and pH are controlled within the culture.

15.7.4 Control of pH

The hydrogen ion concentration (pH) is controlled by the addition of acid or alkali as the conditions change with growth. The controller uses a pH electrode (typically a gel type) to sense these pH changes and provide a feedback signal, which activates the supply of acid or alkali to bring the pH back to the set point. The latest generation of probes can store calibration data, and so on, within the prober and can communicate via a serial protocol such as Modbus. The pH meter is calibrated before the electrode is autoclaved. Steam-sterilizable electrodes have a limited life cycle (20–50 sterilizations cycles). The pumps supplying the acid and alkali are normally built into the instrumentation or base unit housing.

The reagent bottles are connected to the bioreactor via silicone tubing; the bore size of which determine the volume of

Bioreactor Air Flow Control

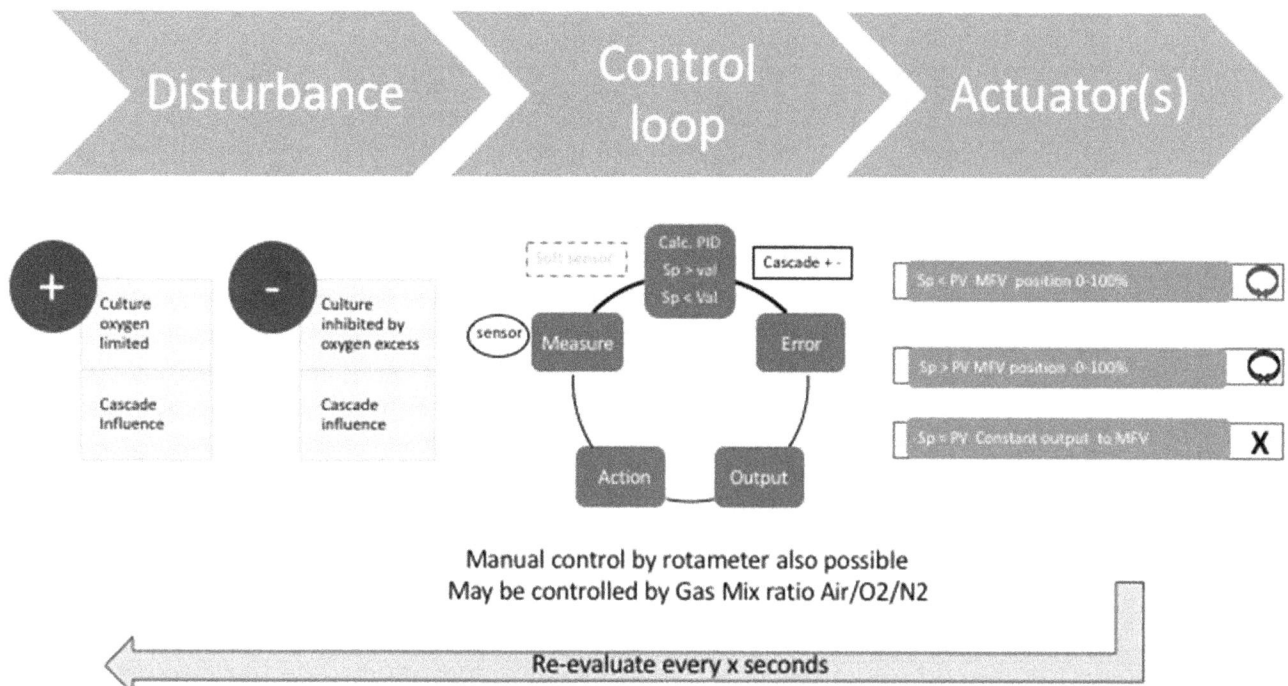

FIGURE 15.6 Gas supply system.

acid or base that can be added when the pumps are turned on. Selecting the concentration of the acid or alkali will determine how much effect each dose has on the vessel contents. Normally, the concentration of acids and alkali are in the region of 0.5–2 M. The use of ammonium salt as an alkali has the added advantage of adding extra nitrogen for the growing culture. Care should be taken when using ammonia water because the ammonia can become gaseous in the tubing.

A set-point value as well as an upper and lower limit is fed into the controller to provide a *"dead band"* range in which the controller is inactive. This band is normally ±0.5 pH units of the desired value. A proportional band adjustment may be present to widen or tighten the range of pH value over which the controller acts (see Figure 15.7).

Dead band: An area around a set-point value that can be set where no control action will take place even if the actual value deviates from the set point. This is used especially for a parameter such as pH, where small changes are often not critical and attempts to control them would lead to excessive controller action.

15.7.5 CONTROL OF DISSOLVED OXYGEN

Dissolved oxygen is one of the most difficult parameters to control. The electrodes used to measure dissolved oxygen are commonly of the polarographic type, which respond rapidly, robustly, and accurately to changes in the oxygen concentrations. The key point with this type of electrode is that it requires a voltage to polarize the anode and cathode of the

detecting cell. This polarization can take between 2 and 6 h to complete. During this time, the electrode must be connected to its relevant module, which in turn must be switched on. The latest generation of dissolved oxygen probes use a fluorescence technique which does not require any polarization time. These probes can store data regarding calibration and communicate with the bioreactor controller via a serial protocol such as Modbus.

To set the electrode to zero after autoclaving, first pass oxygen-free nitrogen through the culture vessel for a few minutes and once all oxygen is expelled, set the zero point.

The 100% value is a relative setting made after autoclaving and polarization of the electrode by turning on the airflow and stirrer speed to the maximum speed needed for a few minutes and then adjusting the controller to display 100%.

Both types of electrode have a consumable component, which needs to be replaced periodically and a special cartridge kit is available from the manufacturer to make this a simple task.

Control of dissolved oxygen can simply be achieved through influencing speed control, adjusting airflow, or by a combination of both. Increasing the speed of mixing and/or airflow may increase foaming to a level that becomes problematic. The most accurate form of flow control is to use a thermal mass flow control valve, which measures and controls airflow based on the cooling effect the gas exerts when passed over a heated element.

An additional option for high-density cultures (e.g., *Escherichia coli* or *Pichia pastoris*) is the use of a solenoid valve to supplement the airflow with pure oxygen in pulses

Bioreactor pH Control

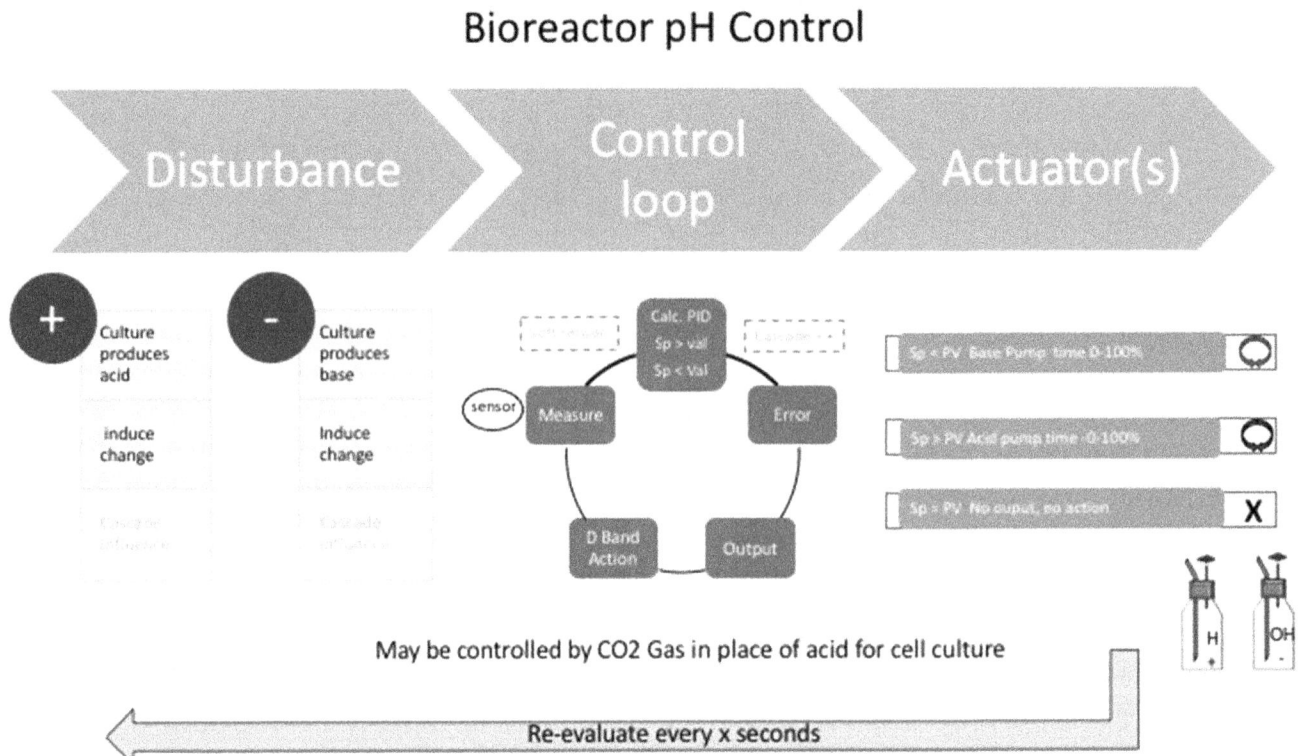

FIGURE 15.7   pH measurement and control.

(see Figure 15.8). For larger steel bioreactors, pressure control is also used to increase the amount of dissolved oxygen. Several control strategies are often used sequentially during fermentation as a cascade. This can include the option to control feeding to match the available dissolved oxygen if a limit has been reached. The range for control is typically 0–150%, but in some brewing applications, the range increases to 0–500% (addition of pure oxygen).

15.7.6   Antifoam Control

A conductance-type probe that is fitted in the vessel headspace detects the formation of foam. Once foam is detected, the probe gives the controller the signal to dispense a dose of antifoam. A delay timer ensures the antifoam reagent has adequate time to reduce the foam level before another dose of antifoam is added. The sensitivity of foam detection should be adjusted to suit the conditions prevailing in the bioreactor. A sheath of inert material around the probe prevents splashes of foam from giving "false positives" (see Figure 15.9). Normally, the metal top plate is used to provide the electrical circuit for the probe to operate so a flying lead is provided that fits into a socket somewhere on the top plate.

Antifoam reagents can be mineral oils, vegetable oils, or certain alcohols. Commercial preparations are available for use in pharmaceutical fermentations. The key thing with using oils is that they can form a skin on the surface of the culture and interfere with gas transfer at the liquid/air interface. If foam builds up unchecked, then it can reach the exit gas filter, blocking it and providing a path for contamination.

15.7.7   Feed Control

The addition of fresh nutrients and mineral salts to a culture vessel can be achieved in one of three ways:

1. *Fed-batch:* Fresh medium is added at a key point in the fermentation (usually when the initial supply of carbon source has been exhausted) and continues to the end of the fermentation without withdrawal of culture (except samples). See Figure 15.10.
2. *Perfusion:* Fresh medium is added discontinuously several times across the spectrum of fermentation. The cells are retained and the culture supernatant is drawn-off to be replaced with an equivalent quantity of fresh medium. This technique is sometimes used for animal cell culture, especially when the cells have been immobilized.
3. *Continuous:* In this case, the medium removed from the culture is continuously replaced by a fresh nutrient feed at an identical flow rate. In a chemostat, the medium usually contains a growth-limiting substrate and the rate of growth of the whole culture is subsequently determined by the flow rate with which this limiting nutrient is added. However, in a Turbidostat, the nutrient supply is plentiful and the organism grows at its maximum specific growth rate, thus maintaining a constant concentration of biomass throughout fermentation. The spent medium is removed via an overflow weir in the side of the vessel or using a level sensor and dip tube to draw culture from the bulk

Bioreactor Dissolved Oxygen Control

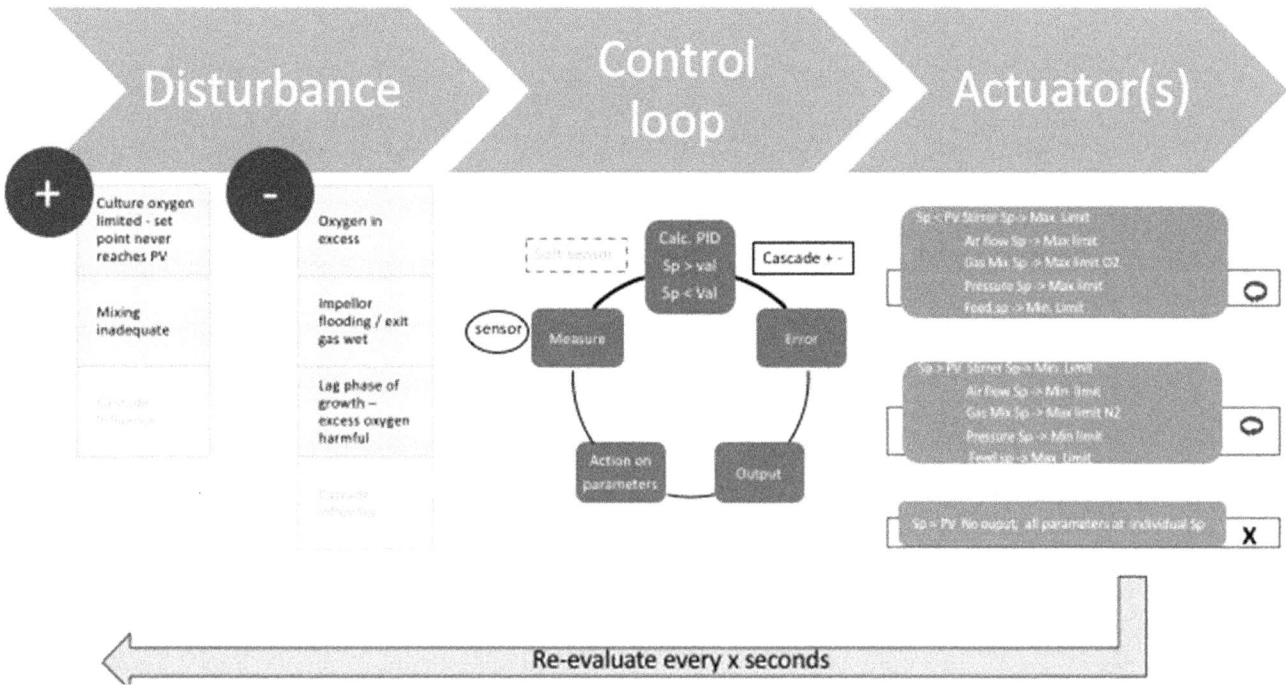

FIGURE 15.8 Dissolved oxygen measurement and control.

Bioreactor Antifoam Control

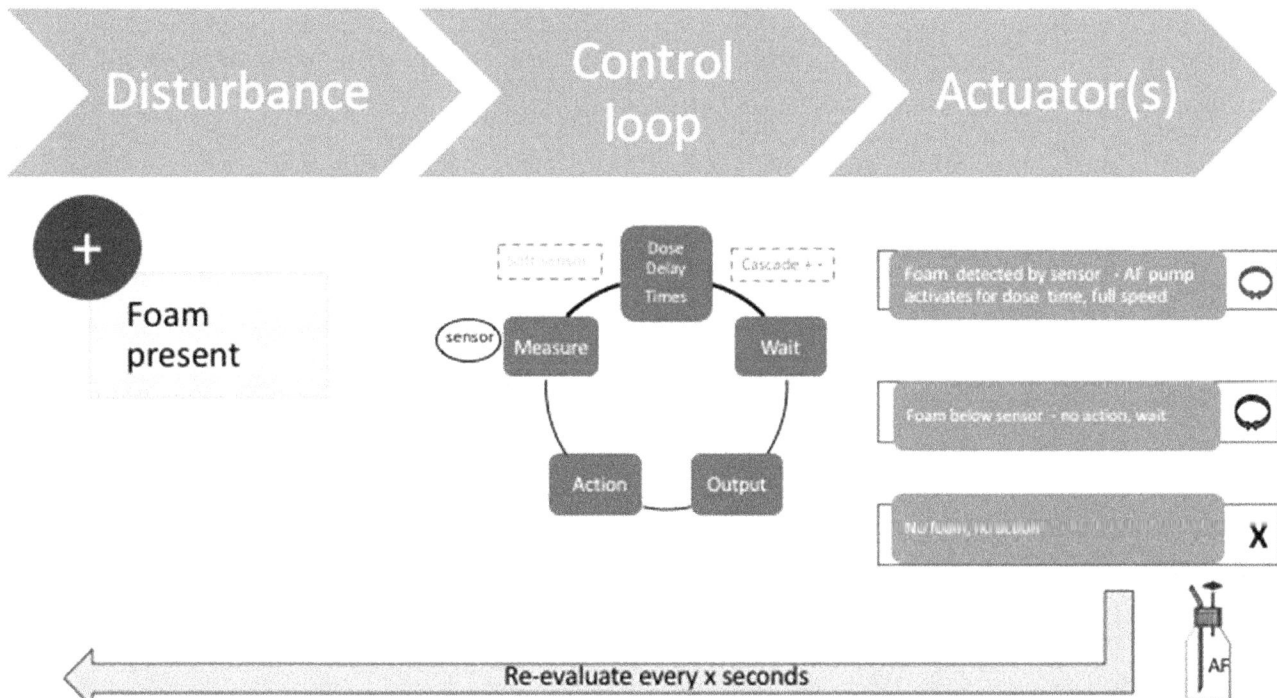

FIGURE 15.9 Foam control system.

Bioreactor Feed Control

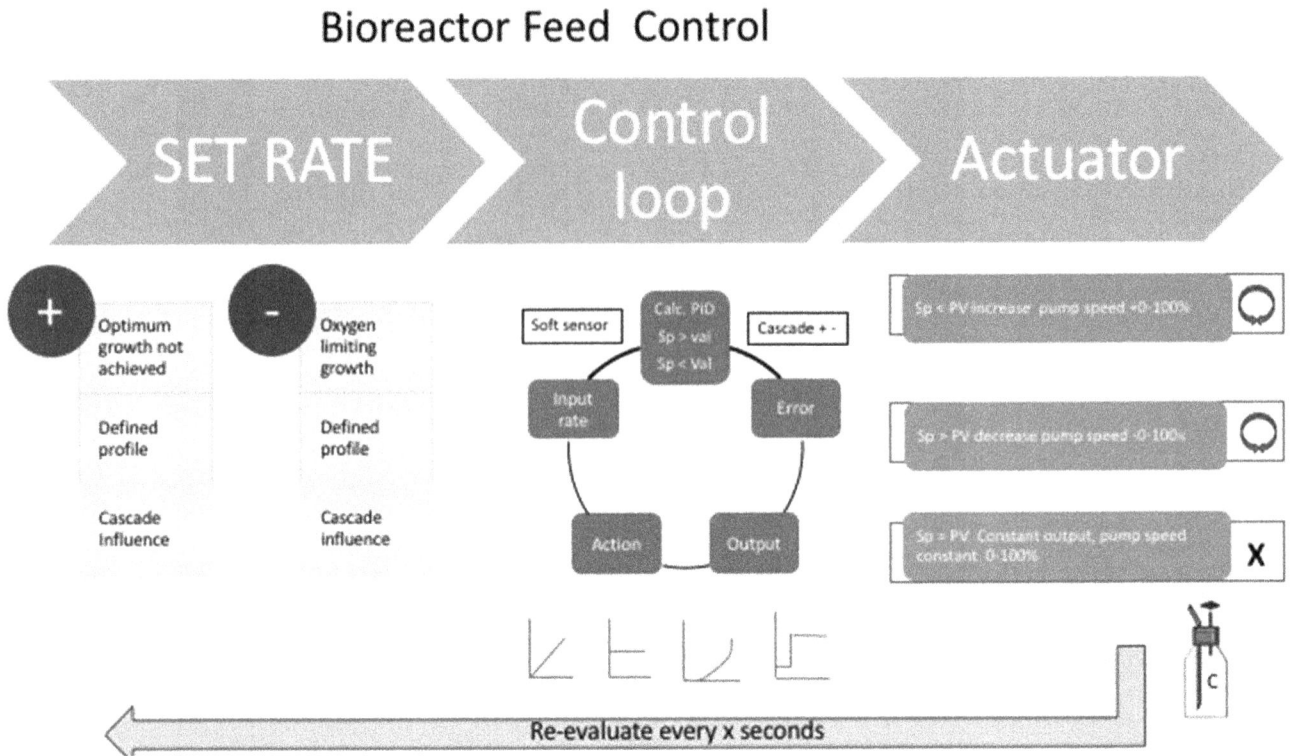

FIGURE 15.10 A feed control system.

of the liquid using a peristaltic pump. Meiera et al. (2014) describe a semi-continuous system.

15.7.8 Factors Influencing Chemostat Operation

- *Dilution rate:* The ratio of flow rate to volume (rate of change of concentration).
- *Residence time:* Volume/flow rate (i.e., how long a molecule of substrate stays in the vessel).
- *Steady state:* When the number of cells produced balances the number removed. It can typically take several residence times before a culture reaches a steady state.
- *Specific growth rate:* Defined by the Monod equation $\mu = \mu m \, S/(K_s + S)$, where μm is the maximum specific growth rate, S is the concentration of growth-limiting substrate, and K_s is the half saturation coefficient (i.e., growth will increase as substrate concentration increases until μm is reached).

15.7.8.1 Advantages and Disadvantages of Continuous Culture

The advantages of *continuous culture* are

- Cells are in a constant physiological state (steady state).
- It can be used with immobilization for high cell concentration and low feed.
- Most downstream processing works best in a continuous mode.

- Cleaning and shutdown times are much less frequent.
- It is well suited to water treatment and environmental processes.

Continuous culture: A method of allowing culture to be grown in a bioreactor at a specific growth rate. The growth rate is determined by the flow of medium through the vessel (dilution rate).

The disadvantages of continuous culture are

- Not as efficient as fed-batch cultures in the production of secondary metabolites (antibiotics) and recombinant proteins, the production of which is generally triggered at the end of the growth phase.
- Contamination can be detrimental to the process.
- Continuous processes have yet to be licensed by regulatory authorities for the production of pharmaceuticals because no there is no batch-lot traceability.
- Weir systems are more susceptible to contaminations, especially if the culture level rises above the desired level.

15.7.9 Fed-Batch Fermentation

Fed-batch fermentation is most commonly used for industrial production of secondary metabolites and recombinant proteins. Fed-batch fermentation increases biomass and yield of desired product far beyond that obtained in even the best optimized batch or continuous cultures. For example, during growth in batch culture, *E. coli* can attain a biomass concentration in the region of 16–20 OD_{600} units, compared with fed-batch cultures, which can exceed 350 OD_{600} units. Also, in the case of the yeast *P. pastoris*, the organism is initially grown on glycerol for high biomass concentration, and once this is achieved, the organism is challenged with methanol, after starvation, to trigger the induction of the AOX gene and, in turn, recombinant protein production.

The rate of feed can be controlled in several different ways:

- *Time:* A flow rate is set and maintained until a fixed time has elapsed and then it is increased or decreased accordingly.
- *Using the pO_2 value based on the available dissolved oxygen:* When the nutrient is supplied too quickly, the culture may not be able to metabolize it aerobically and switch to anaerobic pathways. To prevent this, additional oxygen must be added or the feed rate limited to match the dissolved oxygen concentration.
- *By pH:* If the microbe produces acid in, for example, an excess of glucose, then the fall in pH can be used as trigger to adjust the supply of feed.
- *Exponential feeding:* This is designed to allow the organism to grow at a chosen, specific growth rate, up to its maximum.

All of these methods can be easily achieved using process control software, which is an essential component of *modern bioreactor* process control.

Modern bioreactor: A system consisting of a few pieces of equipment that provide controlled environmental conditions for the growth of microbes (and/or production of specific metabolites) in liquid culture while preventing entry and growth of contaminating microbes from the outside environment. Note this definition includes single-use systems.

15.8 ADDITIONAL SENSORS

There are several additional parameters that can be measured and, in turn, controlled during the fermentation. The most important of these are discussed in the following synopsis.

15.8.1 REDOX

This refers to the reduction/oxidation (redox) potential of a system or a chemical, usually expressed relative to a standard hydrogen half-cell with a redox potential of 0.00V. It follows that a reducing half-cell will have a negative value whereas an oxidizing agent will have a positive one. Oxidation/reduction reactions are generally reversible.

Although the redox value obtained during aerobic fermentation is not generally informative because of the complex nature and multiplicity of biochemical reactions, the opposite is true in anaerobic fermentations because the redox electrode, and, in turn, the value obtained, is very sensitive to oxygen, thus providing a safety indicator for anaerobiosis; a redox value below –200 mV is a good indicator of anaerobic conditions. The redox electrode closely resembles a pH electrode and the conditions for handling and care are almost identical. An electrical zero can usually be set and shorting out the electrode connections should give a reading close to 0 mV. However, pH buffer 4 should never be used with redox electrodes because it will cause the electrode to malfunction and soaking for 24 h in electrolyte may be needed to restore the probe's function.

15.8.2 AIRFLOW

This has already been discussed earlier in connection with dissolved oxygen control. Most bioreactors use a thermal mass flow controller in which airflow is measured by detecting the change in temperature of a heating element before and after the air passes over it. The temperature difference is proportional to flow rate, and this can be processed to give a signal for control. Of course, the flow control systems can measure other gas flows, not just air.

A simpler system uses solenoid valves to introduce each gas into a premixing chamber at manually set flow rates and then controls the final gas flow into the bioreactor with a single mass flow control valve. Anaerobic cultures, requiring no oxygen, can also be catered for.

15.8.3 WEIGHT

This parameter is useful for continuous and/or fed-batch cultures in which the rate of addition of feed must be known and controlled very accurately. It is possible to mount a whole bioreactor on a balance and "tare" out its weight. However, a more precise approach is to use a system with a small *load cell* mounted in such a way that only the vessel and its contents are measured. The deformation of the load cell provides an output signal, which then transduced, and in turn transmitted to the controller module in the form of electrical signal. Control is normally by setting a maximum and/or minimum weight in a manner that is analogous to feed pump control.

Load cell: A method for measuring changes in weight in a bioreactor vessel using deformation of a crystal as an indicator of changes in the load. Load cells are used when it would be impractical to use a conventional balance.

A commonly used variant of this system is to measure very accurately the amount of reagent (e.g., alkali) added to a bioreactor. In this case, the reagent bottle is placed on an analytical balance, which has a computer output (normally RS232), and the decrease in weight is used as a measure of the amount of reagent added.

15.8.4 PRESSURE

The use of pressure as a control parameter is more common in larger ISS bioreactors; an increased pressure increases the concentration of dissolved oxygen in the fermentation broth.

Although a small glass vessel can often be operated under conditions of overpressure (e.g., up to 0.5–0.7 bar), it is advisable to seek advice from the manufacturer.

A piezoelectric-electric sensor normally provides an electrical signal, as a consequence of deformation of the crystal by the internal pressure, typically in the range of 0–2 bar. The control element is a *proportional valve*, which restricts the flow of gas out of the bioreactor and thereby creates a back-pressure. A mechanical overpressure valve or burst-disc is an essential safety requirement if overpressure is to be used.

> *Proportional valve:* A valve, which can be adjusted electrically or pneumatically from 0 to 100% open or closed. For example, the action of the valve is in proportion to the degree of change required by the controller to maintain a certain level of dissolved oxygen by adjusting the airflow rate.

15.8.5   Online Measurement of Biomass

A direct measurement of the number of organisms in a culture is clearly desirable for the control of feed-rate, oxygenation, and general process optimization. Normally, measurements of total cell numbers, wet weight, dry weight, and viable cell counts are made by taking samples intermittently and carrying out the relevant laboratory analysis. Fluorescence microscopy has also been used to measure the number of viable cells internally with a probe or externally through the vessel glass window. The probe-based system is usually suited to bench-scale vessels, similar in design to the optical density probe described below.

15.8.5.1   Optical Density/Turbidity Systems

The increase in turbidity and the subsequent scattering of light by the microbial populations provide a good mean to measure microbial growth. An OD sensor, by comparison, measures light scattering and this can be related to cell density (see Chapter 16 for more details). Either can be directly mounted inside of the vessel or attached to the outside where a clear optical path is available. A two-point calibration procedure is used and results can be expressed in various units. A range up to 200 g/L dry weight is measurable with a good deal of accuracy. This method is cost-effective and works well for cultures in the logarithmic phase of growth. However, it should be remembered that this method does not distinguish between live or dead cells.

15.8.5.2   Capacitance/Conductance-Based Biomass Monitor

This uses a totally novel approach to determine the numbers of viable cells of all types of organisms (bacteria, yeast, filamentous fungi, and animal cells). A probe in the vessel uses a radiofrequency electrical field to measure the natural capacitance of living cells with an intact plasma membrane, thus building up a charge the intensity that is proportional to the number of viable cells, with each type of cell having its own specific range. The signal is processed and can be expressed as dry weight or concentration in cells per milliliter. During the fermentation, an antifouling system is used to maintain the ability of the probe to give accurate readings over long time periods. This type of instrument makes possible the control of feed pumps by biomass concentration and could form the basis for an automated transfer system for inoculating larger vessels.

15.8.6   Exit Gas Analysis

Measurement of the amounts of different gases leaving a bioreactor vessel can provide valuable information about the metabolic processes taking place under the conditions in which the culture is growing. For example, the ratio of oxygen and carbon dioxide entering and leaving the vessel, or *respiratory quotient* (RQ), can be used to determine whether yeasts are producing biomass or alcohol. The RQ can be calculated by most fermentation software packages and can then be used as part of a control algorithm so that the flow rate of a feed pump can be altered accordingly. The entry gas does not need to be analyzed providing it is air because the amounts of oxygen and carbon dioxide will be those of the atmosphere. The flow rate of the air into the vessel will need to be accurately measured, normally by using a thermal mass flow controller. The exit gas may need to be conditioned (e.g., moisture removed) before going into the analyzer depending on the type of instrument used. Recent interest in biofuels has led to a requirement for methane, and in some cases hydrogen, to be measured in the exit gas; several commercial systems can meet this requirement.

> *Respiratory quotient:* RQ is a mathematically derived value related to the use of oxygen by a microbial culture as compared with the evolution of carbon dioxide. This value can be used to adjust feed rates of sugars to manipulate microbial physiology.

15.8.6.1   Infrared Carbon Dioxide Analyzer

A hot wire is used to generate a source of infrared radiation that passes through the gas that leaves the bioreactor into the sample chamber. An infrared detector measures the amount of radiation reaching it after the gas absorbs some. An optical filter is used to make sure the detector only responds to the gas of interest. A "chopper" or rotating shutter is used to allow the detector to see a reference source at regular intervals. This type of detector allows for continuous measurement with good sensitivity, accuracy, and selectivity. Output signals are normally provided as analogue signals (e.g., 0–10 V) or a string of ASCII characters for printing or transmission to computer software.

15.8.6.2   Paramagnetic Oxygen Analyzer

Oxygen has a physical property that this type of analyzer utilizes: it is more susceptible to a magnetic field than other gases. This property is measured using a finely balanced test apparatus suspended in the test chamber of the analyzer. A dumbbell of gas-filled spheres is linked to a support mechanism suspended in a magnetic field created by permanent magnets. If oxygen is present in the test gas, its attraction to the magnetic field will cause the dumbbell to be displaced. The movement is detected using a mirror in the center of the balance system, which displaces a beam of light shone via the mirror onto a photocell. The signal from the photocell will be proportional to the concentration of oxygen present in the sample. A low flow rate is needed for this system to work properly, so inlet gas is typically pumped into the detector cell with most being discarded through bypass pipeline. Once again, suitable analogue and computer outputs are often provided. Infrared detectors and paramagnetic oxygen analyzers need to be calibrated before use.

15.8.6.3 Mass Spectrometer

This represents a step upward in versatility, capability, and cost. A wide range of gases can be analyzed in the gaseous and dissolved form. These include oxygen, carbon dioxide, argon, nitrogen, ammonia, hydrogen, methanol, ethanol, and several other organic volatiles. Isotope ratios (e.g., C^{12}/C^{13}) can also be detected.

This method of analysis is rapid, thus allowing or a multi-inlet system to be used, which "shares" the analyzer among several separate vessels within a bank of bioreactors. The physical principle involved is the ionization of gas molecules by an electron source, usually a hot filament, followed by their separation in a magnetic field (the quadropole analyzer uses a combination of RF and DC electrical fields) by their mass/charge ratio. This separation takes place in a near vacuum to minimize collisions before sorting. The magnetic field is tuned so that only the ions of interest will be focused onto the detector system, a Faraday cup; a metal plate, which generates a tiny electric current whenever gaseous-ions strike it. This signal can be amplified and displayed. By retuning the magnetic field very quickly (milliseconds), different ions can be detected and measured sequentially.

15.9 "SUBSTRATE SENSORS"

Sensors for the detection of substrates usually depend on chemical or enzymic reactions within the fermentation vessel or by sensing the release of volatile components into the exit gas (see Chapter 16 in this edition for extensive coverage of this aspect). Two of the most common examples are given in Sections 15.9.1 and 15.9.2.

15.9.1 Glucose Measurement and Control

The more sophisticated versions incorporate a small dialysis unit in a sterilizable probe to take minute quantities of liquid from the medium for analysis. An enzymic reaction linked to a transducer provides a measurable current that can be amplified to give an output signal. This can provide a measurement of glucose from less than 1g/L of glucose to over 100 g/L.

15.9.2 Methanol Measurement and Control

In this case, a chemical sensor is used to measure the concentration of methanol vapor in the exit gas stream (the sensors are usually variants of the type used in "breathalyzer" systems). The concentration of methanol in the vapor can be directly related to the concentration in the culture medium. A probe-type alternative uses a carrier gas to remove the alcohol vapor from the culture. A signal transducer and amplifier system provides the output-measured value. This type of sensor is typically used to monitor growth and protein overexpression by the yeast *P. pastoris*.

15.10 BIOREACTOR PREPARATION AND USE

15.10.1 Disassembly of the Vessel

The fermentation is shut down from the control unit and transfer lines plus cable connections removed. Reagents lines are emptied and may be refilled with water. After fermentation, the vessel and reagent tubing should be re-autoclaved, ensuring that inlets and outlets are properly prepared. The culture should be disposed of with regard to local health and safety regulations. The clip/clamps/bolts, which retain the vessel top plate, are undone until the whole assembly springs free. The top plate can now be lifted upward away from the glass vessel, taking care that the air sparger, drive shaft/impellors, and temperature probe completely clear the vessel safely.

15.10.2 Cleaning

The pH and dissolved oxygen electrodes should be removed and stored in suitable reagents, as described in the manufacturer's instructions. Periodic cleaning and regeneration of the electrodes are also covered by these instructions. Doing this maintenance is very cost-effective. The vessel should be rinsed several times in distilled water to remove any loose culture residues. Cleaning of growths of culture on the vessel walls may require disassembly and light brushing of the glass. At this point, an examination of any chips or cracks in the vessel glass can be carried out and a replacement made if necessary. Vessels must be stored clean and dry. In use, any spillages of reagents or medium should be wiped up immediately with a damp cloth and not be allowed to dry out. Contact between the top plate and liquids with high chloride ion content (e.g., common salt solutions, hydrochloric acid) should be avoided to prevent corrosion. The pump heads and covers must be thoroughly cleaned if a tube breaks and reagent leaks out. Peristaltic tubing should be sterilized using water in the line and not strong acid or base.

15.10.3 Preparations for Autoclaving

The vessel seal should be removed checked for damage and can be dipped in water to aid relocation. On replacing the seal, it must be correctly located so that there is no chance of any part lifting or kinking. At this point, the vessel can be filled with medium to a maximum of 70–80% full (if active aeration is to be used, this space is vital for gas exchange). The minimum medium volume is the amount needed to adequately cover the electrodes. The vessel top plate can now be replaced and any clamping ring or bolts tightened firmly. The ports for electrodes have O-ring seals, which should be checked for damage and may be wetted with a little water to aid relocation. Electrodes normally push directly into the port fitting, and the collar is tightened down to compress the O-ring seal. All other fittings such as pipes are fitted in the same way. Ports not in use have "stoppers" fitted, and their O-ring seals should be checked also. The pH electrode should be calibrated in appropriate buffers (i.e., pH 7, then pH 4 or pH 9) for the usual two-point calibration. The pH and dissolved oxygen electrodes should be fitted, taking care not to damage them by careless insertion into the port. Both must be tightly capped to prevent moisture getting into the electrical contacts. For the dissolved oxygen electrode, a cap may have to be improvised from aluminum foil. The Pt-100 temperature sensor must be fitted and capped unless it fits into a pocket and so can be removed totally. If used, the foam probe is fitted so that it is above the liquid level. A foam probe can be pulled out of a vessel after autoclaving with little risk of contamination but cannot be pushed down. If the vessel has a conventional rotating mechanical seal, the lubricant reservoir must be checked and topped up with a suitable reagent if necessary (usually glycerin).

Reagent bottles are prepared in a similar way to the bioreactor vessel. A cap or head plate (including a seal) is fitted with a short tube and longer dip tube. A disposable filter is then connected to the short tube with silicone tubing. The shorter pipe must not dip into the liquid, and nothing must block the free passage of air through the filter. The long pipe dips into the liquid as far as possible, usually with a plastic/ silicone tubing extension. This pipe should be fitted with a length of silicone tubing that is long enough to reach the peristaltic pump. The tubing is clamped so that no liquid can escape during autoclaving. A similar procedure is used for sampling and/or harvest bottles except that two short pipes are used so neither dips into the collected culture.

The exit gas cooler should be fitted to one of the larger available ports. A short length of silicone tubing should be attached to the top of the air outlet, and a small 0.22- or 0.45-μm filter fitted. The air outlet line must be kept open during autoclaving. A short length of silicone tubing must be fitted to the air sparger inlet pipe with a 0.2-μm disposable filter mounted on top. The tubing between the sparger pipe and the filter must be clamped shut during autoclaving. If a port is to be used for inoculation or piercing with a needle, a silicone membrane must be fitted into the empty port and a clamping collar/cap used to hold it in place.

15.10.4 AUTOCLAVING

The vessel and any reagent/sampling bottles already connected by silicone tubing are assembled together on a steel tray or in an autoclave basket. A final check should be made that at least one route is available for air to enter and leave the vessel(s) and that all lines dipping into liquid are clamped closed. If the vessel has top drive and a mechanical seal, the seal must be lubricated (normally with glycerin).

If the medium cannot be autoclaved, a suitable volume of distilled water should be used (e.g., 10–20 mL/L of working volume to keep the electrodes wet). If a larger volume of, for

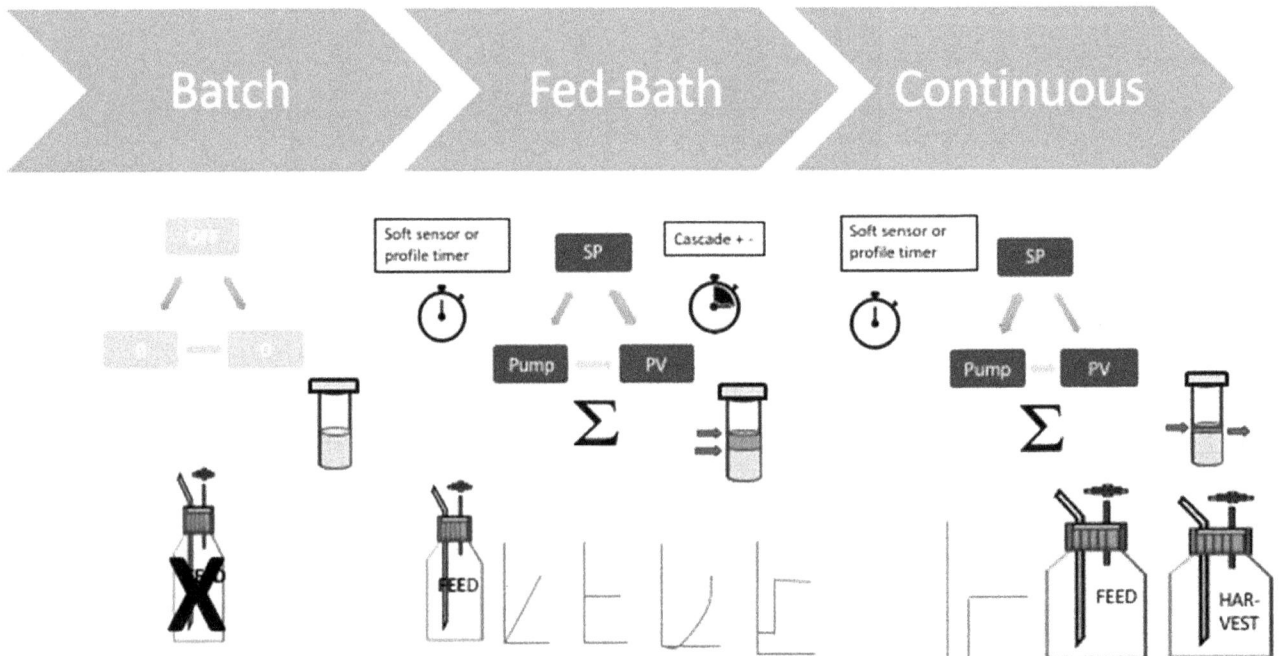

FIGURE 15.11 Options for feed control strategies.

example, phosphate-buffered saline (PBS), is used, this must be removed via a sample line before the actual medium and inoculum are aseptically transferred. A quantity of liquid is certain to be lost during autoclaving (~10%) so the medium is *over-diluted* to compensate for this or sterile distilled water is added afterward to restore the volume. Some form of indicator such as autoclave tape should be included to provide a warning if the correct sterilization procedure has not been carried out. Autoclaving at 121°C for a minimum of 30 min up to 1 h is normally considered adequate for vessel sterilization but consider potential damage to the constituent chemicals of the medium. However, some work may require temperatures of 134°C for several hours to ensure sterility. If in doubt, a safety committee should be consulted. Also, the autoclave used must have good pressure equalization during the cooling-down phase of operation to prevent medium being boiled off. The vessel and any accessory bottles must be allowed to cool completely before handling.

15.10.5　Setup After Autoclaving

The air sparger is connected to the rotameter by a piece of silicone tubing from the top of the filter to the air outlet of the rotameter. The air sparger line is unclipped between the metal pipe and the air filter. The exit gas cooler is connected to the water supply directly or via the bioreactor base unit. The tubing for water in, water out, and drain is connected to the vessel jacket for a water system, and the water is turned on so that the vessel jacket is filled. Alternatively, any pads or heater cartridges are connected to the base unit or temperature control module; the cold finger is connected to the water supply.

The tubing from the reagent bottles is connected to the multiway inlet (if necessary), and the silicone tubing from the reagent bottles is located in the relevant peristaltic pump. Any aseptic connections must be made first if the reagent bottles were autoclaved separately from the vessel. The clamps are removed so liquid can flow freely. A manual switch is often fitted, which allows the pumps to be primed with liquid before use. The drive motor is located onto the top plate (if appropriate), ensuring a good connection is made to the drive shaft. The Pt-100 temperature sensor is connected to the control module and removing the shorting cap and screwing in the cable connect the pH electrode. The dissolved oxygen electrode is connected to the appropriate cable (this requires some care, but the connector should lock firmly when it is correctly positioned by aligning the marks on the connector collar and the probe top). Connections to the foam probe are made, usually one wire into the electrode and one on the vessel top plate to make a circuit. If necessary, the dissolved oxygen electrode should be polarized. Setting the temperature control at this stage will ensure the bioreactor is ready to inoculate after calibration of the dissolved oxygen electrode. The dissolved oxygen electrode is calibrated for the zero point using nitrogen gas and then the air supply is turned on. The maximum stirrer speed to be used and the maximum airflow required on the rotameter are set. After leaving for approximately 15 min, the 100% level is set. The bioreactor is now ready to inoculate.

15.10.6　Inoculation of a Bioreactor Vessel

This section assumes that all the set-up procedures listed above have been carried out. The simplest way to inoculate a fermentor is to have a dedicated port fitted with a membrane, which is capped off before autoclaving. The inoculum (which should normally be no more than 5–10% of the total culture volume) is aseptically transferred to a sterile, disposable syringe of a suitable size. The port fitting is removed and held vertically to prevent contamination of the bottom end. The syringe needle is quickly pushed through the membrane, and the inoculum is transferred into the vessel. The vessel may be actively aerated during this procedure to minimize the risk of a contaminant getting into the vessel (safety considerations permitting). The syringe needle is quickly withdrawn and the silicone membrane reseals. The port fitting is now replaced. For added security, a couple of drops of 70% ethanol can be placed on the membrane surface before piercing and ignited to provide a thermal barrier. Alternatively, an "aseptic connection" can be made to an inlet pipe. If the line connected has a "Y" coupling in it, then the same aseptic connection could be used to introduce medium. This technique is useful if a dedicated inoculation port cannot be provided. Figure 15.12 shows a typical top plate and Figure 15.13 depicts alternative inoculation systems.

15.10.7　Sampling from a Bioreactor Vessel

All sampling starts with a sample pipe, which should dip into the bulk of the culture liquid. At its simplest, a sampling device consisting of a bottle connected to two metal needles/pipes permanently fixed through the metal and rubber seals of its cap. One pipe is connected to a 0.22-μm-air filter, whereas the other, which is linked by silicone tubing (clamped off until a sample is needed), is connected to the sampling pipe. A syringe is fitted to the air filter after autoclaving. The sample device is usually attached to the vessel top plate so that the glass bottle hangs down vertically beneath the cap and a supply of bottles of the same size are autoclaved ready for use. The use of disposable sterile syringes and one-way valves manufactured for medical applications can now make an inexpensive, reusable, and reliable sampling system for bench scale bioreactors. Resterilizable sampling systems for ISS vessels provide similar functions in large-scale industrial bioreactors (Figure 15.14 shows a simple bench-scale bioreactor sampling system).

15.11　EXAMPLES OF COMMON BIOREACTOR APPLICATIONS

The following examples provide some insight into how to configure, optimize, and refine the use of bioreactors for a range of applications.

15.11.1　Multiple, Parallel Bioreactors for Process Analytics

This category of bioreactors has some distinctive features that make it particularly suited to rapid process development and

Key

1. Gas inlet clamped shut

2. Inoculation port capped

3. Reagent inlets clamped

4. Pt-100 removed or capped

5. Sample line clamped shut

6. Exit gas outlet open

7. pO$_2$ electrode covered

8. Foam probe positioned

9. pH electrode capped

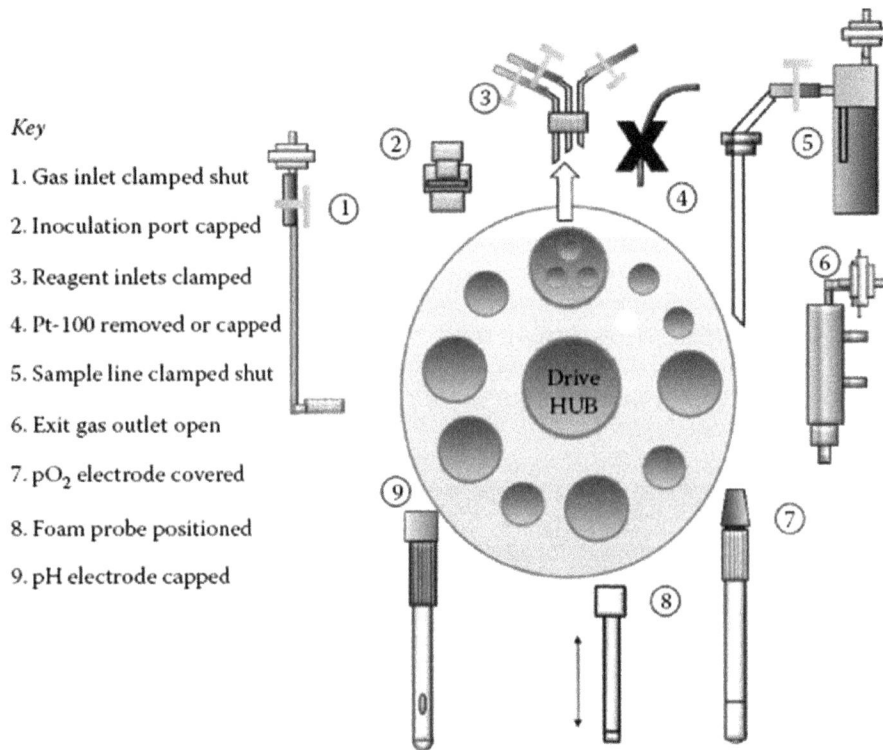

FIGURE 15.12   Vessel top plate prepared for autoclaving.

obtaining experimental results with statistical validity. This is particularly relevant with the promotion of the Process Analytic Technology (PAT) initiative advocated by the US Food and Drug Administration agency in the US (FDA, 2004). The objective of the initiative is to champion the need to generate more data on biological production processes at an earlier stage of development. The use of small, multiple bioreactors is appropriate for this work.

These systems are usually easier to set up than a conventional bioreactor, provide measurement/control of the major process parameters and can be adapted for the growth of virtually all types of microbes and cells. Key features to consider are

- Size: The vessel has similar construction, configuration, and fittings to a conventional bench-scale unit, but they are significantly smaller (typical working volumes of 100–1000 mL) and housed in a base unit capable of accepting four to eight vessels (possibly in groups of two to four).

- *Flexibility:* Multiple peristaltic pumps independently provide reagent and medium feeds to each vessel. Separate heaters/cooling actuators and stirrer systems provide individual control of temperature and stirrer speed. A single controller unit provides for different process parameters, profiles, and links

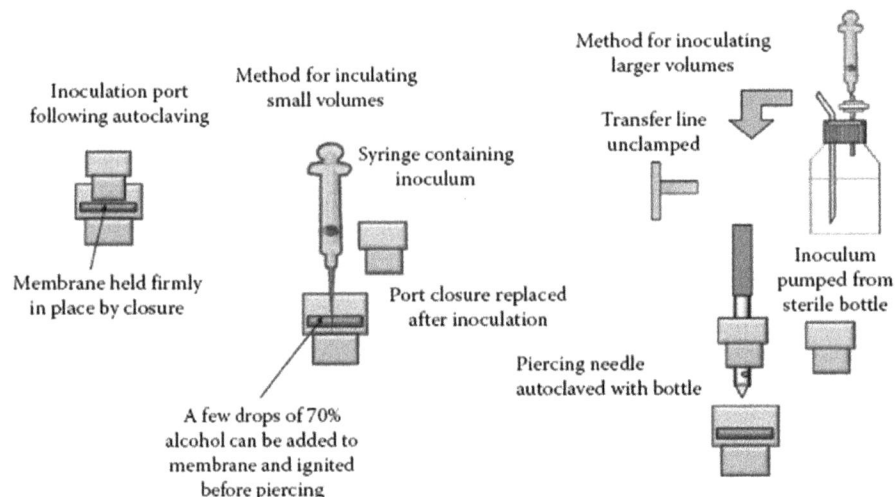

FIGURE 15.13   Inoculation of a bioreactor vessel.

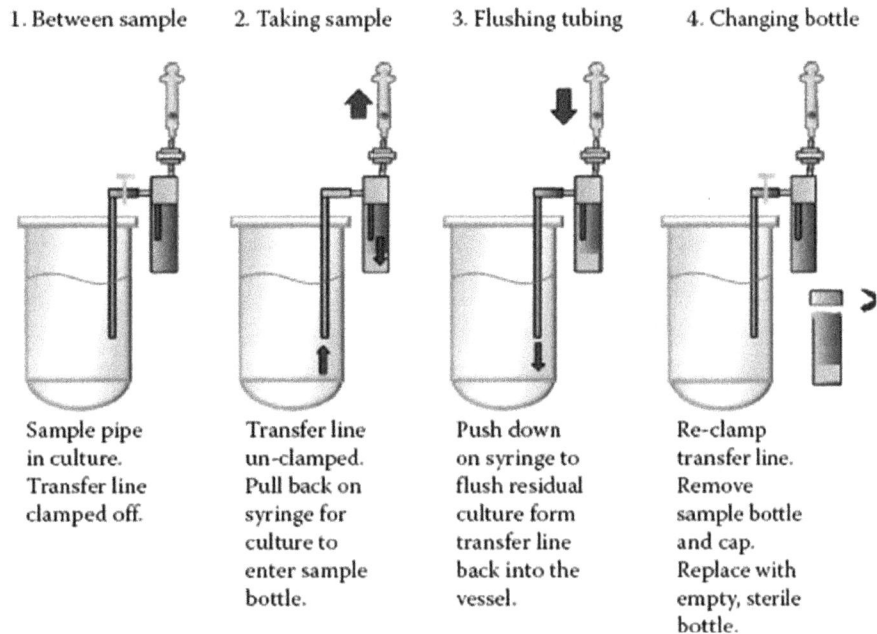

1. Between sample   2. Taking sample   3. Flushing tubing   4. Changing bottle

| Sample pipe in culture. Transfer line clamped off. | Transfer line un-clamped. Pull back on syringe for culture to enter sample bottle. | Push down on syringe to flush residual culture form transfer line back into the vessel. | Re-clamp transfer line. Remove sample bottle and cap. Replace with empty, sterile bottle. |

FIGURE 15.14 Sampling.

to external supervisory software. Several standard probes can be accommodated, including optical density for biomass monitoring or redox for anaerobic applications. The instrumentation must also be flexible and expandable enough to accommodate additional process control elements.

- *Easy handling:* This is certainly the most crucial point. A lot of vessels mean a lot of flexible tubing, a lot of sensors to calibrate, and a lot of items to take to and from the autoclave. An advantage with autoclavable systems is that they can be operated using two complete sets of vessels, so there is minimal downtime for cleaning and preparation. Small-scale ISS systems are an alternative, but these are necessarily large and complex. Small, single-use vessels are the latest development, which eliminates cleaning and re-validation steps between experiments. Chemical cleaning and sterilization in place systems (CIP-SIP) for overnight vessel and transfer line preparation provide a novel "third way" approach.
- *Validation:* The equipment should be suitable for process validation especially for pilot and production-scale bioreactors.

15.11.1.1 How Do Multiple Bioreactors Fit into the Philosophy of PAT?

Cell culture for production of therapeutic proteins has advanced dramatically in recent years. Bioreactors of several thousand liters can be used, so process optimization is of key importance. A cell culture bioreactor must account for vessel aspect ratio, internal fittings, and considerable flexibility in gas mixing and flow control. A magnetically coupled drive system provides a means to help minimize the risk of contamination during prolonged culture times. The drive system

must not damage the cells or support substrates and allow for good mixing even at speeds well below 100 rpm. For monoclonal antibody and therapeutic protein production, separation systems such as a *spin filter* are usually required, even at this small scale.

> *Spin filter:* This device is normally attached to the drive shaft of an animal cell bioreactor and allows culture liquid to be removed while leaving the cells inside of the bioreactor. For example, it allows for slow-growing cells to produce antibodies over a long period by regular harvesting of culture supernatant and replenishment by a controlled medium feed.

Rapid selection and screening of clones for productivity can be achieved using microtiter plates or shake flasks rather than bioreactors (Betts and Baganz, 2006). Statistical studies using specialist Design of Experiment (DoE) software can determine the likely key parameters to be measured and controlled. Temperature, pH, and rate of feed addition are often the most critical.

The desired clones can then be tested in multiple, parallel bioreactor system, thus facilitating sufficient replicates for statistical validity. Excursion testing of key parameters can also be performed at this stage to find the limits of productivity for the process. Additional process optimization and fine-tuning can be made in terms of medium composition, control strategies, and downstream processing.

The accumulated data will be presented along with information from production-scale fermentation as part of a submission to obtain approval for a license to manufacture.

15.11.1.2 Advantages of Using Multiple, Parallel Fermentation Systems for PAT

- Allows for the use of replicates for producing statistically valid results.

- Because many conditions may remain to be tested after computer simulations, time can be saved compared with repeated experiments using a single vessel in series.
- Several different parameter changes can be performed at once for excursion testing (e.g., different temperatures).
- Handling and preparation will be optimized for practical details such as fitting tubing post-sterilization and priming pumps, etc.
- A small-scale system can be accommodated in a research laboratory and the work can be done without the need for special facilities.

Small-scale, live fermentations form a key part the successful implementation of the PAT process by

- Confirming predicted results from statistical analysis and computer simulations
- Discovering other factors not included in the simulations (e.g., media composition or shear-sensitivity of clones)
- Providing data directly applicable to scale-up
- Use with larger-scale process trials to provide real-time optimization using scale-down techniques
- Allowing refinement of process strategies related to practical aspects of the process such as feeding rates, etc.

Irrespective of PAT requirements, testing at this scale can optimize many of the process parameters for applications such as high-density culture (see Section 15.11.2 below).

15.11.2 High-Density Cultures for Biomass and Proteins

15.11.2.1 What Is Special About High-Density Culture?

High-density culture is a strategy that allows maximal product formation (usually a recombinant protein) in the minimum time and/or smallest volume. This has several elements and emphasizes the importance of biological steps along with optimization of the process within the bioreactor. The essential steps are

1. Creation of clones that produce high yields of product and/or biomass. However, the same clone may not always meet these two criteria. One strategy is to split the process so that biomass formation is an initial step under one set of conditions, followed by a change to induce product formation. The clones will have been subject to various manipulation and analyses.
2. Multiple gene copies leading to overexpression (increases productivity but no other changes).
3. Deletion mutations so cells produce mainly product (redirecting cell machinery).

4. Use of "helper" genes to improve productivity.
5. Use of fusion proteins to aid movement of target proteins across the cell membrane (see Chapter 7).
6. Use of metabolic flux analysis to find out key pathways for nutrient use with defined media.

15.11.2.1.1 Process Optimization

Process optimization in the bioreactor, especially with regard to medium composition and the strategy used. Fed-batch culture is often to be preferred because it allows for rapid production of biomass in one medium in an initial batch phase, followed by induction and product formation, maybe using a different medium composition as a feed. Factors to consider in the optimization process are

- *Excursion testing:* As mentioned above for PAT, excursion testing is also applicable here to find the limits of productivity. Temperature, pH, and feed are usual choices for optimization, if dissolved oxygen is in excess or feed rates are balanced to match available pO_2.
- *Choice of medium:* A defined chemical medium with an additional carbon source is usually preferred because it can enhance growth by supplying essential trace elements in the correct quantities for optimal growth, is easier to remove from the product, and does not promote growth of contaminants.
- *Control of process parameters specifically associated with high-density culture:* These include high temperatures during the logarithmic phase of growth that may require enhanced cooling (use of a chiller, cooling coils, additional cold fingers, etc.) and addition of oxygen to supplement the quantity that can be attained by aeration alone.
- *High-density culture:* This depends on efficient control of key process parameters, especially during the logarithmic phase of growth, and involves the use of cascades of process parameters such as stirrer speed, gas flow, and oxygen supplementation to maximize the availability of dissolved oxygen for cell metabolism. An alternative approach is to limit the availability of the carbon source to match the quantity of dissolved oxygen available to the culture. This strategy means culture times are increased, but it has a definite advantage at the large scale where oxygen supplementation would be prohibitively expensive. An exponential feeding strategy would allow feed rates to increase synchronously with increasing cell numbers.
- *Monitoring of biomass formation:* This is done using an OD measurement (typically OD_{600}) or indirectly by exit gas analysis to calculate oxygen uptake rates.
- *Choice of inducer:* For example, isopropyl thiogalactopyranoside (IPGT) for *E. coli* and methanol for *P. pastoris*.

• *Change of physical conditions (starvation, oxygen spike, gas supplementation, temperature rise or fall):* These strategies were used successfully for decades before genetic modification was possible and usually stress the organism so that its metabolism switches on new genes as a survival response to harsh growth conditions.

15.11.2.1.2 Product Removal

This depends on where the desired substance has been produced. If it is in the cell, harvest of the biomass followed by a cell disruption step is usually necessary. If the product has been released into the culture supernatant, then it is possible to remove the product during the fermentation by a filtration or perfusion step. Subsequent downstream processing would optimally involve

• The fewest number of steps means fewest losses.
• Concentration of product in the cell or in the culture for harvest (continuous removal or isolation in inclusion bodies or outer membranes).
• Use of markers/fusion proteins to aid purification; for example, an affinity column to isolate the fusion protein or green fluorescent protein (GFP) to show activity of selected genes.

Additional factors influencing high-density microbial cultures include

• Adaptation of productive clones to bioreactor conditions (e.g., may be shear sensitive).
• Choice of inorganic supplements in the feed; for example, phosphates and/or reagents and use of ammonia water as the base reagent for *E. coli* high-density fermentations.
• Fed-batch fermentation with growth rate controlled below maximum specific growth rate to help prevent oxygen limitation/substrate excess. Use of exponential feeding with OD measurement can provide for direct control of cell growth.
• Use of indirect feedback control for feed rate (e.g., by operating the bioreactor as a pH-stat or on the basis of an "oxygen spike," which requires care because a dying culture will also have a lower oxygen requirement).
• Direct, accurate measurement and tight control of feed concentration in the bulk culture (e.g., if methanol feed reaches a concentration >4–6% in the culture medium, it can kill a culture of *P. pastoris*).

15.11.2.2 Typical Examples

One of the most common microbes for recombinant culture is *E. coli*. This is because the genome is well understood and the addition of new genes using plasmids is relatively easy. The disadvantage is that *E. coli* does not have all the necessary glycosylation pathways necessary to ensure a specific protein will be folded correctly to produce a biologically active molecule (reviewed by Choi et al., 2006).

Some aspects of *E. coli* metabolism limit growth and product formation. In addition, some genetically modified strains differ significantly from wild-type strains in their physical and physiological requirements. Examples include an enhanced sensitivity to shear and the requirement for an initial period of growth in an environment relatively rich in carbon dioxide, requiring a lowering of the flow rate of air supply and decreasing the stirring speed in the first few hours of fermentation to achieve this.

The yeast most commonly used for recombinant work is *P. pastoris*. This also has a well-characterized genome and a range of commercial applications for recombinant protein production. The method of induction for product formation is well documented and the mechanism of induction is achieved by switching from glucose/glycerol to methanol as a carbon source. One major advantage of *P. pastoris* is the ability of the AOX promoter mechanism to be used for production of a wide range of different recombinant proteins; a suitable protocol has been reported (Minning et al., 2001).

P. pastoris high-density fermentations require methanol measurement, which can be carried out directly or indirectly. If measurement of methanol concentration in the culture is made indirectly (exit gas), the methanol feed has to be introduced via a dip tube under the surface of the culture to avoid false reading as a result of methanol splashing.

15.11.3 MAMMALIAN CELL CULTURE FOR PRODUCTION OF THERAPEUTIC PROTEIN

Commercial production of recombinant proteins in mammalian cell culture has been a modern success story, with productivity rising from milligrams per day to better than 1 g/day in only a few years. This has been achieved by careful selection of cell lines suited to growth in bioreactors, intensive process optimization for a small panel of these cell lines (with fed batch being the preferred method), and increase in volume of production vessels from hundreds of liters into the thousands. The use of mammalian cells ensures that glycosylation and other posttranslational modifications are fully functional and appropriate for the desired therapeutic proteins. Current trends show more new bioreactors being acquired for cell culture than for microbial fermentations.

The CHO (Chinese Hamster Ovary) cell line is one of the most commonly used cell lines for research and production of therapeutic proteins. *Hybridoma cell lines* have similar growth requirements but are used for production of specific proteins such as monoclonal antibodies. A range of culture volumes, vessels, and techniques have been successfully used (Chu et al., 2005), some of which are described in the following subsections.

Hybridoma cell line: A cell derived from the artificial fusion of a normal (e.g., an anti-body-producing cell line from the spleen) with a transformed (immortal) cell line (e.g., a myeloma). The resulting hybrid is immortal and can produce a specific antibody if the correct spleen cell was selected after challenge of an animal with a specific antigen.

15.11.3.1 Small-Scale Cultures: Less Than 1-L Volumes

15.11.3.1.1 Deep-Well Plates

Although the standard 96-well plates can be used (Warr et al., 2011), they are not usually ideal for cell culture. The larger size of the deep well variety allows for better mixing at slow speeds, which are necessary for cell culture work because of cell sensitivity to shearing.

15.11.3.1.2 Shake Flasks

Because growth is typically in suspension, moving to an incubator shaker will provide an easy path to better mixing and gas transfer than would be available in a static cultivation system. It is also a good first step if scale-up to disposable bags or a bioreactor is contemplated. Any small-scale system will usually require an incubation chamber with some (or all) of the following features:

- *Temperature control:* The sensitivity of mammalian cells to changes in temperature makes accurate temperature control in the incubation chamber imperative. An external Pt-100 temperature sensor can be used in a test flask to ensure that control is based on liquid temperature.
- *Humidity control:* Humidity of cell culture affects osmolarity, which plays a significant role on growth and productivity of cell lines. Deep-well plate cultures can decrease in volume significantly over time because of evaporation and high surface area.
- *High speed, short throw shaking for use with micro-well plates:* Good mixing in a well containing 1–2 ml of liquid is not possible with the speed and throws used for conventional shake flasks, e.g., 50-400 rpm and 25 or 50 mm. A dedicated, high-speed shaking platform and short throw (1,000 rpm and 3 mm), provides sufficient energy to move the culture up the sides of the well provides a lager surface area for gas transfer and through mixing even for small volumes.
- *CO_2 measurement and control:* This is to be expected for mammalian cell culture with measurement and control of this key parameter. Gas usage is kept to the minimum possible, and the atmosphere is quickly regenerated on closing the incubation chamber after manipulation of culture.
- *Direct biomass monitoring:* Noninvasive systems are available for direct measurement of biomass by light scattering without the need for invasive probes or proprietary sensor spots in the flask.
- *Feeding or reagent addition:* Latest developments in this area rely on feeding caps with directly mounted controllers for addition of sugars, inducing agents, base, and so on, relying on simple, time-based dosing strategies (linear, ramp. step, bolus).
- *Validation of cleaning regimes:* The risk of unwanted microbial contamination is a major issue for cell culture work. Normal methods of cleaning an incubation chamber can be ineffective or can leave chemical residues. A suitable decontamination system would use a process that could be validated with electronic sensors and by using a biological test system. Decontamination systems based on the use of hydrogen peroxide offer this capability for incubators and other equipment.
- *Qualification for GMP validation:* This is not a physical feature but has an impact on the validation of a bioprocess. The quality of control for key parameters, such as an even temperature through the incubation chamber (in three dimensions) can be key for reproducible results. The ability to provide quality protocols for specification, operation, and regular checks when in use is common for fermenters but not usual for incubator shakers.

These features can be combined with another element of modern incubator shakers, stacking, to provide a unique tool for both industrial and academic research for the right applications. A multi-deck system can be specified to provide different elements for high-throughput bioprocess development, for example:

- One incubator shaker deck dedicated to screening of large number of clones in 96-well micro-well plates. With a dedicated tray to allow stacking in three dimensions, humidification and high-speed/short-throw shaking, several thousand individual experiments can be set up at one time.
- A second deck to use small shake flasks for scale-up studies. This can use flasks as small as 100 ml with 15–20 ml of culture and utilizes the best clones selected from the screening experiments. Process parameters can be adjusted to further increase yields and the data used to provide optimization criteria for larger-scale cultivation.
- A third deck can, where applicable, become a "production unit," with large flasks (up to 5 L total volume 1 L working volume) generating sufficient material for stage one clinical trials. This capability requires two key elements
  a. High-yield clones with at least 1–5 gm/liter of product.
  b. A qualified system for a fully validated process with reproducibility and batch traceability for quality control.

This is achieved in a relatively small space available to almost any laboratory. A large degree of automation is required externally to the incubator shaker elements for clone generation and subsequent selection from the screening process. The advantages of using a micro-well/flask system lie in its familiarity and ease of setup/use. Combined with bioprocess software able to cope with large projects, parallel experiments, and data collection/control from multiple incubator shakers, the whole "package" provides a relatively low-cost way to make high throughput techniques available to a wide range of potential users (A. Magno, 2017, personal communication).

15.11.3.2 Large-Scale Culturing in Bioreactors

Mammalian and insect cell culture can be carried out in STRs with relatively few modifications (Chu et al., 2005). The key modifications required are

- Low shear impellors, typically 0.35–0.5 vessel diameters
- A round-bottom vessel (hemispherical or, at least, dished)
- Removal of any baffles
- Special gas blending and possibly a modified sparger with a sinter or small holes for small bubbles
- Use of nonstick coating on the glass of the vessel to prevent cells from sticking
- Use of gassing and stirring regimes specifically intended to reduce foaming (e.g., pulsing of gas rather than a continuous flow)

Beyond these basic modifications, a range of other adaptations can be made to specific requirements.

15.11.3.2.1 Links to an External Analyzer

Some of the key parameters for cell culture cannot be measured directly in the vessel (e.g., glucose, glutamate, lactate, and ammonia concentrations). Cell-free culture supernatant can be removed via a simple static filter and fed to an autoanalyzer for these chemicals to be measured online but outside of the vessel.

15.11.3.2.2 Conductivity Measurement for CIP

Large *in situ* vessels are often cleaned using CIP; traces of chemicals are then removed by rinsing with water. Because cell cultures are extremely sensitive to chemical contamination, a conductivity probe can be used to check that all traces of ionic materials have been removed before starting.

15.11.3.2.3 Use of a Gas Basket

A gas basket is usually only suitable for bench scale fermentations because a pressure drop in the tubing is problematic in larger systems. The gas basket uses porous silicone tubing to pass a blended gas into the bioreactor vessel without causing foam. The tubing is wrapped around a support frame to make a tightly wrapped coil, which presents a large surface area to the medium. This replaces a more conventional sparger.

15.11.3.2.4 Headspace Gassing

This is the traditional method for cell culture because foaming is much reduced and it accounts for the typical squat aspect ratio of a cell culture vessel (1:1). This provides a large surface area relative to the depth of the culture. Headspace gassing is still commonly used for addition of carbon dioxide for pH control, so leaving the sparger as the outlet for the blended gas is required for control of dissolved oxygen.

However, any combination of gas or gases can be directed into the headspace.

15.11.3.2.5 Custom Gas Mixing

The standard "3 + 1" [air/N_2/O_2 for dissolved oxygen control and carbon dioxide (CO_2) for the acid side of pH control] gas-mixing unit with mass flow control for the blended gas is applicable to most applications. If a different type of blending is required, mass flow valves can be fitted for each inlet gas and the composition of the blend controlled via a local controller or remote software. Of course, this can be linked to the values for dissolved oxygen and pH. Gassing of cell cultures is often started with a high concentration of nitrogen and only a little air because too much oxygen can be harmful to the cells.

15.11.3.2.6 Immobilization Systems

If specific cell lines do not grow well in suspension, they can be immobilized on discs or beads of a supporting matrix and grow on the surface and inside of these structures. Support discs are often placed in a special holder within the vessel and angled impellors used to direct a flow of medium through this central chamber. The advantage with the use of beads for cell immobilization is that it is scale independent and does not rely on a solution proprietary to a single vessel manufacturer.

15.11.3.2.7 Spin Filter/Perfusion

This is used for immobilized cell cultures and those with cells in suspension. A rotating filter (typically 10–20 μm mesh size) keeps cells from entering and creates a pool of cell-free medium inside of the filter cup that can be removed continuously or occasionally for replenishment with fresh media.

15.11.3.2.8 Secure Sampling

A requirement is to have a sampling device that can be used repeatedly without risk of contamination. Several systems are available that meet this criterion and disposable options use components more usually found in medical applications.

15.11.3.3 What Are the Advantages of Performing Cell Culture in Conventional Bioreactors?

- Scale-up issues can be directly addressed because large-scale cultures (e.g., 20,000 L) will typically be performed using modified STRs.
- The potential to add sensors, inlets, outlets, perfusion systems, and other peripherals is far greater than for disposable systems.
- A large body of application data already exists for many common cell lines going back more than two decades.
- Process control capabilities are far better than for disposable systems, which often have limitations in terms of the number and type of control loops possible.
- The hardware is reusable and this has cost advantages in the long-term
- Sterilization of contaminated batches before disposal is easier to deal with in standard bioreactors.

15.11.3.4 What Are the Key Operational Factors for Mammalian Cell Culture in Bioreactors?

- Cell numbers not as high as microbial culture
- Temperature for growth in a narrow range
- Cells are more sensitive to shear and osmotic shock
- CO_2 and oxygen supplementation may be needed
- Cells can grow immobilized on beads/carriers or suspended
- Thick foam destroys cells (i.e., low gas flow rates are required)
- Longer time scale (e.g., 10–14 d)

15.11.3.5 Which Choices Can Improve Productivity in Recombinant Cell Culture?

- Selections of clones that adaptable to bioreactors—not necessarily those, which perform well in shake flasks.
- Careful selection of clones and adapting recombinant techniques to work with a strictly limited number of choices. This can help to reduce time for scale-up and allow some issues to be bypassed by working with well-characterized cell lines.
- Suspension culture is preferred in large-scale systems rather than immobilizing cells because this reduces the number of steps, costs, and the chance of poor results due to lack of attachment, etc.
- Fed-batch culture provides similar benefits in mammalian cell culture as those found in microbial cultures (Altamirano et al., 2004) (i.e., productivity is best if substrate concentration is limited) toxin dilution, and the opportunity to use perfusion techniques for material expressed into the culture supernatant.
- Choice of defined, serum-free growth media on grounds of cost-saving and eliminating steps in downstream processing to remove these substances.
- A low gas flow rate and the use of a conventional ring sparger can limit the damage caused to the cells by the formation of thick foam. Chemicals such as Pluronic (Sigma Chemicals) can help protect cells. Many different gassing strategies are used, including only adding oxygen in pulses when required and gas blends starting mainly with nitrogen.
- Use of real-time analyzers to simultaneously measure glutamate and glucose consumption along with levels of lactate and ammonia production by the culture (the latter can attain toxic levels).

15.11.4 Solid-State, Algal, and Other Bioreactors for Biofuels

Biofuel production, biorefineries, and solid-state fermentations (SSFs) have been dealt with extensively are dealt with in other chapters. Good review articles can provide additional details such as Kruger and Muller-Langer (2014) and Prabuddha et al. (2015).

15.11.4.1 Biofuels

Biofuel technologies for production of bioethanol, biogas (methanol), biodiesel, and hydrogen have recently attracted significant public interest. Biodiesel is typically produced by chemical treatment of plant oils or algal biomass as feedstock.

15.11.4.1.1 Biogas

Biogas is typically methane generated by anaerobic digestion of manure or sewage waste material. These fermentations can be carried out in any standard bioreactor at the laboratory scale, with options for commercial sensors to measure key gas production. Because gases such as methane and hydrogen are explosive in air at concentrations beginning at approximately 5%, care is needed to ensure a buildup of exhaust gases is not possible. Commercial-scale processes are invariably in digesters, which are much simpler than pharmaceutical bioreactors.

Most forms of digestion for methane production fall into this category. Because control of dissolved oxygen is not usually required, the status of the fermentation can be measured using redox potential as an alternative. Mixtures of gas can be introduced to keep a reducing atmosphere, and even a limited control of redox is possible by liquid or gas in some circumstances.

15.11.4.1.2 Bioethanol

Bioethanol production is principally a fermentation process with substrates such as sugar and starch crops, lignocellulose materials (Mansour et al., 2017) such as wood chips, and agricultural materials such as straw or grass fiber. Woody feedstock often require pretreatment with enzymes or physicochemical agents to break down long-chain molecules such as celluloses or lignin to smaller sugars. This pretreatment allows for the maximal yield from the subsequent fermentation process of converting the resulting sugars to ethanol.

15.11.4.2 Use of Photosynthetic Organisms for Biofuel Production

Algae have an important role to play because they are not food crops and do not even require land-based cultivation. They can provide a rich source of energy in the form of biomass, oils, and even direct production of hydrogen. Dependent on light for growth, algal cultures require bioreactors with lighting that can be flexibly controlled. Again, the STR is a valuable research tool for small-scale studies, but large-scale commercialization typically involves a lower level of technology and use of natural resources such as sunlight in desert regions. Factors important for photosynthetic culture include:

- *Good temperature control:* Because lighting invariably adds heat directly in the vessel or around it, good temperature control is required. A water jacket is a good choice for efficient removal of heat, but it needs modification if it is not to interfere with external lighting.

- *The quantity and quality of light:* A warm, white light will provide the widest range of available wavelengths. Alternatives may provide a spectrum optimized for photosynthesis. Conventional fluorescent tubes may be used or recent developments based on LED lighting can provide a flexible alternative.
- *Control of light intensity:* A stepped or continuously variable control of light intensity is desirable so that the light intensity can match the growth of the culture. Linking light intensity to OD is a good strategy for automated control.
- *Scale-up:* Some form of scale-up strategy may be needed. Glass vessels are clearly a good choice for bench-scale research, but structural lighting installations may also be needed above, for example, a 10-L working volume. Production-scale processes are often based on relatively simple tanks or long lengths of tubing and so do not require matching vessel geometry or mixing strategy at the laboratory scale.
- *Mixing:* Mixing may be at low speed, and a marine impellor could give better results than flat-bladed impellors in this case.

15.11.4.3 Bioremediation

The construction of a standardized environmental chamber with control of key physical and chemical parameters has been lacking for some time. Environmentally controlled chambers for solid substrates have recently been constructed and should prove invaluable in the following domains:

- Wastewater treatment
- Breakdown of liquid pollutants in soils
- Bioremediation of soils in landfill
- Decontamination/recycling of solid/semi-solid waste products (e.g., slurries, fats)
- Composting and anaerobic digestion
- Solid-state enzyme production
- Production of enhanced animal feeds

15.12 CURRENT TRENDS AND FUTURE PROSPECTS IN FERMENTOR DESIGN AND APPLICATIONS

Fermentors have come a long way from their earliest days in the 1940s. The objective was to raise productivity per unit area by using submerged culture for penicillin production to replace surface culture. The conventional STR has held a dominant role in the pharmaceutical industry, academic research, the food industry, and many other areas ever since that time. However, STRs have never been the only choice, and specialized reactors for wastewater treatment, enzyme production, and bioremediation have been available in some form for decades. However, recent interest in microbial processes involving more specialized and extreme environments has caused the STR to be adapted in various ways. They include

- Greater temperature range, especially for the culture of thermophiles at temperatures of 80°C and beyond (review by Delve and Sandroman, 2017)
- Production of structured tissues in the fermentor, produced in areas contained by a mesh or within three-dimensional scaffolds (Haycock, 2011)
- Use of more aggressive chemical environments, necessitating vessels designed without any metal parts that cannot corrode or influence culture conditions
- A rise in the need for unusual environments, with the use of inert gas overlays in the vessel and feed bottles, control based on redox, and use of totally nonporous tubing for reagent lines

These examples are in addition to the SSFs, photo-bioreactors, biofuel, and cell culture systems mentioned above.

One important point is that research-scale work is often performed using laboratory-scale fermentors, but the final production scale will not usually be based on STRs. Taking biofuel as an example, commercial systems use tubing that can be attached to fences in sunny geographical regions for algal culture. In medicine, *in vitro* cell culture systems now use disposable cartridges tagged with patent details that provide the same result as a culture in a fermentor but with a totally different design. A simple biogas digester for domestic use needs to be no more sophisticated than a simple plastic container and an electrical heating strip. These systems are cost-effective on the large scale or mass market but are poorly monitored and controlled in comparison with fully equipped fermentors.

Conversely, simpler cultivation systems such as incubator shakers can be used with a variety of containers to provide ideal conditions for high throughput, parallel applications where rapid development needs militate for simple set up and operation. High-quality control systems and novel sensors for noninvasive measurements add sophistication and increase the potential of such systems for rapid development and even production.

Regarding instrumentation and control, the trend is toward controlling processes based on microbial metabolism rather than simple feedback control of a few key process parameters such as temperature, pH, and feed rates. Software exists for metabolic monitoring and control, with multifunctional analyzers providing biochemical data such as lactate and glucose concentrations in real time. Existing sensors are also changing with the rise of optical probes for dissolved oxygen and pH. They require less in the way of calibration and maintenance, improving reliability and making validation easier. Another advantage of these sensors is their small size, allowing well-instrumented vessels of just a few milliliters to be used for many of the tasks that would have previously required a full-sized bench fermentor.

A recent revival in continuous culture for production environments is leading to the development of new ways of thinking about "whole bioprocess control" in addition to simple control loops for individual parameters. It also means providing answers to questions related to quality assurance and traceability when individual batches are no longer being produced.

This trend combines perfectly with the rise of single-use culture systems that have a clear role in rapid development of validated processes without the need for expensive infrastructure. An interesting hybrid technology is developing, with single-use vessels being combined with conventional STR services and control equipment to provide a "reversible" solution for disposable and reusable applications. To date, these disposable technologies have concentrated on mammalian cell culture, but interest in adapting the technology for bacterial cultures is growing. The fact that so many applications exist, with still new modifications of the STR being made, is validation of its basic concept and inherent flexibility.

SUMMARY

Bioreactors are composed of several different components that can be grouped by function (i.e., temperature control, pH measurement). A wide range of peripheral devices can enhance the basic facilities of the bioreactor.

Bioreactor instrumentation is based on DDC systems that offer advantages in terms of the flexibility of control and the ability to store operational protocols.

Many different types of vessels exist and include airlift, fluidized bed, and hollow fiber and specially modified STRs. Special categories of STRs include those for *in situ* sterilization and containment of pathogenic or genetically manipulated organisms.

Practical guidance has been provided for autoclaving, inoculation, and sampling to ensure safe and reliable operation for any make or type of bioreactor.

Examples have been provided for the use of bioreactors for a wide range of common applications in research and industry, including cultivation of microbial cells, algal fermentations, mammalian cell culture, and bacterial and yeast culture at high cell densities.

Current topics regarding new culture methodologies such as process analytics, exponential feeding, solid-state substrates, and the place of single-use systems have been considered.

Use of systems based on enhanced incubator shakers allows for highly parallelized rapid development studies from screening, through scale-up to basic production.

REFERENCES

Altamirano, C., C. Paredes, A. Illanes, J.J. Cairo, and F. Godia. (2004). Strategies for fed-batch cultivation of t-PA producing CHO cells: Substitution of glucose and glutamine and rational design of culture medium. *J Biotechnol* 110:171–179.

Betts, J.I. and F. Baganz (2006) Miniature bioreactors: Current practices and future opportunities *Microbial Cell Factories* 5:21.

Choi, J.H., K.C. Keum, and S.Y. Lee. (2006). Production of recombinant proteins by high cell density culture of *Escherichia coli*. *Chem Eng Sci* 61:876–885.

Chu, L., I. Blumentals, and G. Maheshwari. (2005) Production of recombinant therapeutic proteins by mammalian cells in suspension culture. In: Smales C.M., James D.C. (eds) *Therapeutic Proteins. Methods in Molecular Biology*, vol 308. Humana Press.

Deive, F.J. and M. Sanroman. (2017). *Bioreactor Development for the Cultivation of Extremophilic Microorganisms*. In: Larroche, C., Sanroman, M.A., Du, G., Pandey, A. (eds) *Current Development for the Cultivation and Extremophilic Microorganisms* 403–432 Elsevier.

FDA. (2004). *Guidance for Industry: PAT—A Framework for Innovative Pharmaceutical Development, Manufacturing, assnd Quality Assurance*. Washington, DC: US Food and Drug Administration.

Haycock, J.W. (2011) 3D cell culture: A review of current approaches and techniques. In: Haycock J. (ed) *3D Cell Culture. Methods in Molecular Biology (Methods and Protocols)*, vol 695. 1–15. Humana Press.

Kruger, M. and F.M. Muller-Langer. (2014) Review on possible algal-biofuel production processes *Biofuels* 3:333–349.

Mansour, A.A., T. Arnaud, T.A. Lu-Chau, M. Fdz-Polanco, M.T. Moreira, J. Andres, and C. Rivero. (2016). Review of solid state fermentation for lignocellulolytic enzyme production: Challenges for environmental applications. *Rev Environ Sci Biotechnol* 15: 31–46.

Meier, K., F. Carstensen, M. Wessling, L. Regestein, and J. Büchs (2014). Quasi continuous fermentation in a reverse-flow diafiltration bioreactor *Biochem Eng J* 91, 265–275.

Minning, S., A. Serrano, P. Ferrer. C. Sola, R.D. Schmid, and F. Valero. (2001). Optimization of the high-level production of *Rhizopus oryzae* lipase in *Pichia pastoris*. *J Biotechnol* 86:59–70.

Gupta, P.L.,S.M. Lee, and H.J. Choi. (2015) A mini review: Photobioreactors for large scale algal cultivation *World J Microbiol Biotechnology*. 31:1409–1417.

Warr, S.R.C., J. Patel, R. Ho, and K.V. Newell. (2011) Use of micro bioreactor systems to streamline cell line evaluation and upstream process development for monoclonal antibody production *BMC Proc* 5 (Suppl 8):14–16.

16 Biosensors in Bioprocess Monitoring and Control
Current Trends and Future Prospects

Chris E. French

CONTENTS

"Measure what can be measured, and make measurable what cannot be measured."

Galileo Galilei

16.1 INTRODUCTION

In almost all fermentations, parameters such as pH, temperature, and dissolved oxygen, are monitored as standard. Other parameters, such as substrate utilization and product formation, can be monitored offline using enzymic assays and analytical techniques such as HPLC, GCMS, or ELISA, but this involves inevitable delays, with implications for process control and management. Biosensors offer the potential for rapid or even real-time monitoring of such parameters. In this chapter, we will look at the nature of biosensors and the potential benefits of their use in these applications.

16.2 GENERAL CONSIDERATIONS

The term "biosensor" means slightly different things to different people. Generally speaking, a biosensor is an analytical device that couples a biological recognition element, which specifically detects the target analyte, to a transducer, which generates a detectable and preferably quantitative signal. This is converted into an electrical signal which is used in monitoring and control of the process in question (Figure 16.1). The application of biosensor technology to fermentation processes potentially allows rapid, highly specific, quantitative response, with minimal or no requirements for sample pre-processing.

The biological recognition element may be an enzyme, an antibody, another binding molecule, or a living cell. The transducer may detect a redox reaction, light emission, altered levels of an ion, a luminescent or fluorescent signal, or a change in mass associated with a surface (Figure 16.2). Biosensors

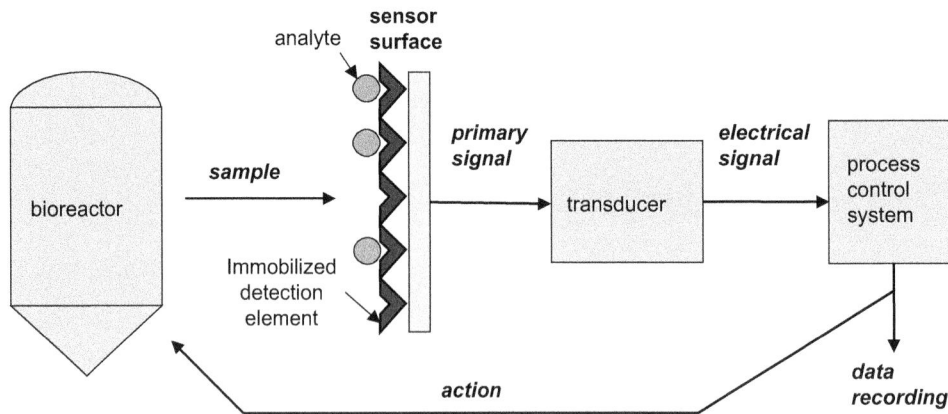

FIGURE 16.1 General scheme for a biosensor in bioprocess monitoring and control.

are potentially superior to other analytical methods such as HPLC or ELISA, because they can provide a much higher specificity within a short time span. For example, HPLC utilizes relatively nonspecific detector systems such as ultraviolet absorption or refractive index, and identification of specific substances is based on retention time within a column. This leads to an unavoidable delay before the results are obtained. By contrast, a biosensor uses a specific biological detection element to detect a single analyte, and ideally supplies a steady-state response in less than 60 seconds or so. Techniques such as ELISA also use a biological detection element, but require considerable sample processing over a number of steps, again leading to an unavoidable delay. The ideal of a biosensor (not always achieved) is to apply such a biological detection element in a configuration that gives a rapid automatic output without such sample processing.

The biological element of the biosensor is critical in that it provides the specificity of response. This relies on the evolved ability of biological molecules (especially proteins) to recognize their interaction partners with extremely high specificity. The main types of recognition element used are enzymes and antibodies. In some cases, nucleic acids or whole living cells, usually bacteria, can be used. The use of a biological sensing element is both the strength and the weakness of biosensors. Biological sensing elements are relatively labile and do not tolerate high temperatures or other sterilizing agents. The efficiency of biosensors may also diminish with time, and as such, their performance must be monitored and they must be replaced on a regular basis. Expected lifetime of the biological element is an important parameter to bear in mind when considering the use of a biosensor. Some types of biosensor, especially enzyme electrodes, may be inserted directly into the bioreactor, with the sensing element separated from the broth by a semi-permeable membrane. Other types can be used for *in situ* measurements if coupled with a suitable device for separation from the cells, such as a tangential flow microfiltration unit. Some types of biosensor, such as amperometric enzyme electrodes, are well suited to continuous monitoring; others,

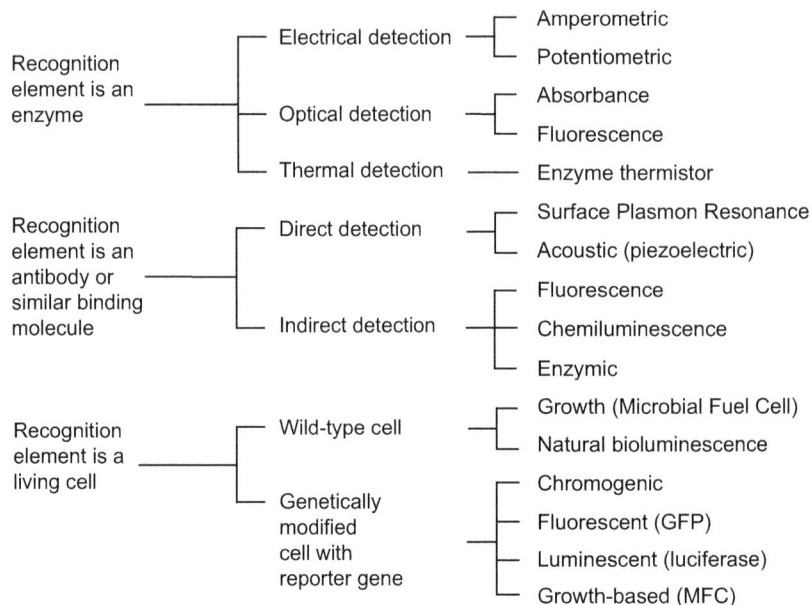

FIGURE 16.2 Overview of common types of biosensor by recognition element and transduction modality.

especially those based on antibodies, may require regeneration of the detection surface by removal of bound ligand, or even replacement of the detection surface, between measurements. Also, some types of biosensor, particularly those based on redox enzymes, may require the addition of small soluble cofactors or co-substrates along with the sample, and are therefore not suitable for *in situ* use. One of the most popular ways of using biosensors in bioprocess monitoring is flow injection analysis (FIA). In this system, the samples are withdrawn automatically and injected into a flow of a carrier liquid which is passed through the detection element, thus ensuring continuous monitoring (Schugerl, 2001).

As in any other type of sensor, it is important to consider the sensitivity of a biosensor (the lowest concentration of analyte which can be detected), as well as the dynamic range of the response (range of responses over which the response can be used to estimate the analyte concentration, i.e., between the lowest concentration that can be detected and the level at which the response saturates). These are related to the affinity of binding of the substrate and the recognition element; for example, K_m (Michaelis constant) for enzymes, and K_d (dissociation constant) for antibodies. However, the recognition element is usually immobilized on a detection surface, and the dynamic response of the system may be limited by the rate of diffusion of the analyte from bulk fluid to the surface, in which case the response may be proportional to the difference in analyte concentration between the bulk fluid and the detection surface.

16.3 BIOSENSORS USING ENZYMES AS THE RECOGNITION ELEMENT

16.3.1 GENERAL CONSIDERATIONS

Enzymes are the most popular choice of recognition element for biosensors, and the easiest to transduce, since they recognize a specific substrate (the analyte) and catalyze a reaction which causes a well-defined change in the system. Of course, this relies on the availability of a suitable enzyme which acts specifically on the target analyte. The enzyme should also be easy to manufacture, and reasonably stable when immobilized on a sensor surface. Generally, enzyme-based biosensors are only suitable for detection of small molecules, rather than proteins or cells. This makes them widely applicable to monitoring levels of substrates (such as glucose and glutamate) as well as small-molecule products such as lactate, ethanol, and penicillin, but not for monitoring production of therapeutic proteins, for which antibody-based systems are more suitable.

When compared with antibodies and other recognition elements, enzymes offer a second layer of recognition specificity; a non-target compound with roughly similar shape to the target analyte may be able to bind to the enzyme's active site, but unless it can also undergo the same reaction, it will not be detected, so will not give a false positive. The downside of this is that enzyme reactions are subject to various types of inhibition. The possible presence of inhibitors in the sample should always be considered when using an enzyme in a biosensor or any other type of enzyme-based assay, and some sample pretreatment may be necessary to remove any inhibitors that may be present.

Sensitivity and dynamic range of enzyme-based biosensors are related to the binding and reaction kinetics of the enzyme-substrate couple. Figure 16.3 shows a simplified enzyme reaction scheme, as well as standard equations used to describe enzyme activity. One important parameter is the K_m value, otherwise known as the Michaelis constant, which is a measure of the affinity of the substrate for the enzyme. In the simplest case, where substrate binding and unbinding is much more rapid than reaction, it is equal to K_d, the dissociation constant; in more realistic cases, it also takes account of the rate of reaction. In practical terms, K_m is the substrate concentration at which the rate of reaction, and therefore signal formation, is half the maximal rate (V_{max}). At substrate concentrations below K_m, the reaction rate and, in turn, signal formation, follow first-order kinetics, i.e., the signal produced is proportional to substrate concentration. On the other hand, however, at substrate concentrations much higher than the K_m (in practice, three times the K_m or higher), the reaction rate, and in turn the signal, show zero order kinetics and as such are no longer dependent on substrate concentration. This assumes that diffusion of substrates or products is not rate-limiting. In a biosensor context, where the enzyme and perhaps cofactors are immobilized, kinetics may be controlled by the rate of diffusion of substrate to the sensor surface, which is proportional to the difference between the substrate concentration at the surface of the sensor and that of the medium in the bioreactor, in which case the dynamic range of the system may be much greater than would be expected based on K_m alone. It is therefore important to characterize the sensor response in its final configuration.

The use of an enzyme as a recognition element in a biosensor depends on its ability to generate a signal which can easily be detected and converted into an electrical response. In this regard, not all enzymes are equally convenient. For example, isomerases, which simply alter the configuration of their substrates, are unlikely to generate a useful signal. Generally, the most useful class of enzymes for biosensors are oxidoreductases, which oxidize or reduce substrates; by their very nature they are involved in the movement of electrons and lend themselves to amperometric detection methods. Hydrolytic and lyase enzymes split substrate molecules into smaller parts; they may be useful if they generate or consume a substance which can be detected potentiometrically. Thermal detection is based on the release of energy during an enzyme-catalyzed reaction, as detected by an extremely sensitive thermistor. In principle, this type of device can be used for almost any type of enzyme-catalyzed reaction, though clearly it is likely to work better for highly exothermic reactions.

In some cases, enzyme-based systems require additional small soluble molecules, such as cofactors or co-substrates, to be added for enzyme activity. For example, many dehydrogenases require stoichiometric amounts of cofactors such as NAD$^+$ (nicotinamide adenine dinucleotide) or PQQ (pyrroloquinoline quinone), and these must be provided to the system.

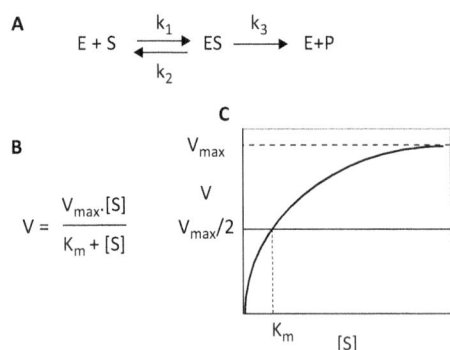

FIGURE 16.3 Enzyme kinetics. (a) Scheme for single-substrate enzyme reactions. The enzyme and substrate interact to form an enzyme-substrate complex, which reacts to form an enzyme-product complex, from which the product is released. The enzyme-substrate complex and enzyme-product complex are not distinguishable by steady-state kinetics. (b) The Michaelis-Menten equation for simple enzyme-catalyzed reactions. V = rate of reaction; V_{max} = maximum rate of reaction at saturating substrate concentrations; $[S]$ = substrate concentration; K_m = Michaelis constant, the substrate concentration at which V is half of V_{max}. This equation makes several assumptions, one of which is that the enzyme concentration is much smaller than the substrate concentration. (c) Graph showing the shape of the rate dependence on substrate concentration.

16.3.2 AMPEROMETRIC SENSORS BASED ON ENZYMES

Amperometric detection is the best established and most widely used type of transduction in biosensors. The first biosensors were amperometric sensors based on glucose oxidase (Yoo and Lee, 2010), and these are still commercially

by far the most important type of biosensor due to their application in blood glucose monitoring in diabetic patients. Of course, they are also highly applicable to monitoring glucose levels in fermentations, and a number of systems are commercially available for this purpose, as described below.

In an amperometric transducer, an electrode is held poised at a constant voltage relative to a reference electrode (such as Ag/AgCl) by the use of a circuit known as a potentiostat. Redox-active molecules can be oxidized or reduced at the electrode surface, generating a current that is measured. The potential at which the electrode is held is chosen so as to allow oxidation or reduction of the target analyte whilst minimizing the oxidation or reduction of other substances which may be present. Potential interference by redox-active molecules such as ascorbic acid, which may be present in fermentation broths, must always be considered. Amperometric transduction is ideal where an enzyme reaction generates or consumes a redox-active substance such as O_2, H_2O_2, NAD(P)H or PQQ. For recent reviews, see Grieshaber et al. (2008) and Ronkainen et al. (2008).

As noted above, the most important biosensors are those used to detect glucose, which have wide medical applications in the control of diabetes. These are usually based on the enzyme glucose oxidase (GOX) from the fungus *Aspergillus niger*, though glucose dehydrogenases linked to NAD or PQQ can also be used. GOX is a flavoprotein with a FAD prosthetic group, which is reduced by glucose, yielding gluconolactone (which is rapidly hydrolyzed to gluconic acid) and $FADH_2$, which is then reoxidized by O_2, yielding H_2O_2 (Figure 16.4a). Early glucose biosensors were based on detecting decreased

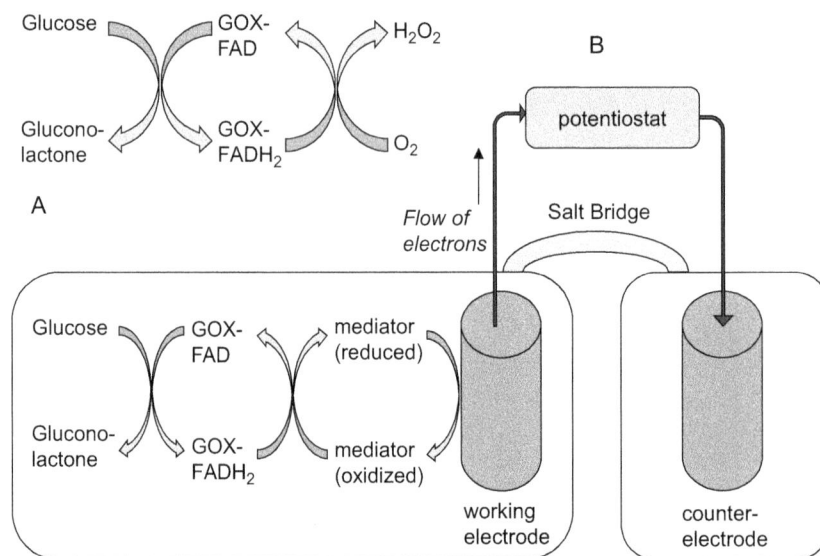

FIGURE 16.4 Oxidase-based amperometric biosensors. (a) Reaction catalyzed by glucose oxidase. Other oxidases catalyze similar reactions, with their own native substrate replacing glucose. (b) Schematic diagram of a mediated amperometric biosensor based on glucose oxidase. Glucose reduces the flavin FAD to $FADH_2$, which is then re-oxidized by a soluble mediator. The reduced form of the mediator is then re-oxidized at the sensor electrode surface, giving up its electrons. The potentiostat passes the electrons to a counter-electrode in order to maintain the set-point potential relative to the reference electrode (not shown) and measures the flow of current thus created. In modern systems, the mediator is a conductive polymer in which GOX is embedded, rather than a freely diffusible small molecule. GOX may be replaced by other oxidases such as lactate oxidase, alcohol oxidase or glutamate oxidase, to detect other analytes.

oxygen levels due to GOX activity in the presence of glucose, using a standard Clark oxygen electrode. However, this led to issues with maintaining a constant level of O_2, which is poorly soluble in water. Current devices are based on GOX immobilized at an electrode surface together with a soluble artificial mediator, which is capable of oxidizing the $FADH_2$ generated by glucose oxidation, and carrying electrons to the electrode, which is held poised at a suitable potential relative to a reference electrode by the potentiostat (Figure 16.4b).

As electrons are transferred from $FADH_2$ to the working electrode by the mediator, the potentiostat will maintain its set point potential by transferring electrons to a second electrode, creating a flow of electrons, which is to say, an electrical current. This is related to glucose concentration in the vicinity of the working electrode.

Glucose oxidase belongs to a large class of oxidase enzymes (EC1.1.3), which share a common reaction mechanism, with a flavin prosthetic group being initially reduced by the substrate, and then re-oxidized by O_2 to generate H_2O_2. Thus, similar biosensors can easily be generated to detect other small molecules by replacing glucose oxidase with a different oxidase enzyme. For example, ethanol, lactate, and glutamate may be detected using alcohol oxidase, lactate oxidase and glutamate oxidase respectively. It is also possible to further expand the substrate range using accessory enzymes; for example, Kamanin et al. (2015) reported use of screen-printed electrodes for simultaneous detection of glucose, ethanol, lactate and starch (using a mixture of α-amylase and glucose oxidase) in fermentation broth, and Moser et al. (2002) described a micro-flow sensor using glucose oxidase, lactate oxidase, glutamate oxidase, and glutamate oxidase in conjunction with glutaminase, for the simultaneous determination of glucose, lactate, glutamate, and glutamine. NAD-dependent dehydrogenases can also be used as the basis for such devices, with use of a mediator which can transfer electrons from NADH to the electrode surface.

Amperometric enzyme systems are the best developed commercial biosensor devices, and a number of systems specifically aimed at bioprocess monitoring are commercially available. For example, the GlucCell glucose detection system is basically similar to a standard home blood glucose monitoring system, except that it is optimized for offline measurement of glucose in cell culture media, using sample volumes as small as a few microliters, while the BioProfile system of Nova Biomedical, also aimed at cell culture monitoring, detects glucose, glutamate, glutamine, and lactate. Other systems are also capable of online monitoring, either with *in situ* electrodes or with automated abstraction of samples from the reactor. For example, Yellow Springs Instruments, manufacturer of the original glucose biosensor, offers the 2900M online monitor and control system, which includes an autosampler and 2900D biochemistry analyzer capable of measuring glucose, lactate, glutamate, glutamine, and ethanol, among other substances. The Trace C2 Control system of TRACE Analytics Gmbh similarly includes an autosampler and analyzer for glucose and lactate, while the CITSens Bio offers online measurement of glucose and lactate in bioreactors

using an in-dwelling electrode with wireless transmission of output to an external monitor.

16.3.3 POTENTIOMETRIC SENSORS

Potentiometric sensors detect changes in voltage caused by build-up of particular ions at an electrode surface. The simplest type of potentiometric transducer utilizes an ion-selective electrode, and the build-up of the appropriate ion causes a change in potential (voltage) with very little current flow. The standard pH electrode is an example of an ion-selective electrode. Another type of device is the ion-selective field effect transistor (ISFET), a transistor with an ion-selective membrane over the gate electrode, which provides similar functionality (Miao et al., 2003). When the ion-selective membrane is coated with a layer of immobilized enzyme to give a potentiometric biosensor, the device is sometimes called an EnFET (Enzyme Field Effect Transistor).

The simplest case is where the enzyme reaction alters the local pH sufficiently that it can be detected directly using a pH electrode or ISFET (allowing for the buffering capacity of the fermentation medium). For example, Park et al. (2004) described a potentiometric biosensor using penicillinase (β-lactamase) to quantify penicillin G and derivatives. As little as 1 μM penicillin could be detected. Likewise, Poghossian et al. (2001) reported a penicillin-sensitive EnFET (dubbed a PenFET) based on immobilized β-lactamase. Glucose can also be detected potentiometrically using glucose oxidase immobilized on a pH electrode or ISFET, since the glucose oxidase reaction generates gluconic acid, which decreases the local pH (Tinkilic et al., 2002, Luo et al., 2004). NAD-dependent dehydrogenases can also form the basis of potentiometric sensors; such devices have been used to quantify lactate and ethanol using lactate dehydrogenase and alcohol dehydrogenase (Pogoroleva et al., 2003).

A variety of other ion or gas-selective electrodes are available, and these can also be used as the basis for a biosensor if an enzyme is available which acts on the target analyte to generate or consume the appropriate ion or gas, for example, when the reaction generates or consumes NH_4^+ (NH_3), or HCO_3^- (CO_2). For example, Matsuszewki et al. (1991) reported such a device based on immobilized glutaminase for detection of L-glutamine in growth medium.

16.3.4 THERMAL BIOSENSORS

Thermal biosensors use an extremely sensitive temperature-sensing element (thermistor) to detect the increase in temperature due to heat released by an enzyme-catalyzed reaction. This is potentially more widely applicable than amperometric or potentiometric transduction, in that it can potentially be used with any enzyme-catalyzed reaction, though obviously it is likely to be most sensitive when the reaction is highly exothermic (Ramanathan and Danielsson, 2001). Usually, the enzyme is immobilized on resin beads packed in a column which is attached to the thermistor. A control column lacking enzyme can be used to control for background temperature changes. The combination of thermistor and immobilized enzyme is sometimes referred to as an enzyme thermistor.

In principle, thermal detection of enzyme reactions is possible with any class of enzyme, provided that the heat released by the reaction can be measured with sufficient sensitivity. For example, Rank et al. (1992) reported the use of a thermal biosensor for monitoring penicillin V production in a Novo-Nordisk production-scale bioreactor (160m³). Energy release due to penicillin hydrolysis was measured in a column packed with immobilized β-lactamase or penicillin acylase. Lawung et al. (2001) described flow-injection analysis for the quantitation of both penicillin and cephalosporin-based antibiotics using two different β-lactamases immobilized on an affinity column. In another example, Navratil et al. (2001) used an online thermal biosensor with immobilized glycerokinase and galactose oxidase to monitor the bioconversion of glycerol to dihydroxyacetone by cells of *Gluconobacter oxydans* in an airlift bioreactor. Finally, thermal biosensors provide yet another option for the quantitation of glucose with glucose oxidase. Ramanathan et al. (2001) described the use of immobilized glucose oxidase in an enzyme thermistor. To increase the amount of heat released per molecule of glucose oxidized, a second enzyme, catalase, was co-immobilized with the glucose oxidase to catalyze degradation of the hydrogen peroxide formed in the oxidase reaction.

16.3.5 Optical Biosensors

Optical biosensors are based on the detection of colorimetric, fluorescent or luminescent signals at the tip of a bundle of optical fibers. The fibers relay the light signal to a detector, which generates an appropriate electrical signal. A sensor based on optical fibers is sometimes referred to as an optode or optrode (by analogy with "electrode"). For some enzymes, a choice between optical and electrical transduction is possible. Optical transduction has the advantage that it is not affected by electrically noisy environments, and an optical fiber can be inserted, for example, directly into a bioreactor. However, optical transduction systems may be affected by turbidity (Bosch et al., 2007; Borisov and Wolfbeis, 2008).

While oxidases are ideally suited to amperometric detection, as described above, they are also suitable for use in optical fiber biosensors. Hydrogen peroxide may be detected with great sensitivity by chemi-luminescence, based on its peroxidase-catalyzed reaction with luminol, enhanced by *p*-iodophenol. A variety of optical biosensors based on this reaction have been reported. For example, Blankenstein et al. (1994) reported simultaneous online analysis of multiple analytes including glucose, lactate, and glutamate in animal cell culture using a multi-channel oxidase-based flow injection analysis system.

The activity of NAD(P)-dependent dehydrogenases may also be detected by optical fiber biosensors. The signal in this case is based on the fact that the reduced form of the cofactor, NAD(P)H, absorbs light at around 340 nm and fluoresces at around 450 nm. It follows that the presence of a dehydrogenase substrate leads to an increase in the NAD(P)H concentration,

and increased levels of light absorption and fluorescence. To cite a single example, Dominguez et al. (2010) described a sequential injection (SIA) sensor for measuring glycerol in the *Saccharomyces cerevisiae* fermentation process, based on the enzyme glycerol dehydrogenase, which reduces NAD⁺ to NADH. Enzymes which cause a change in the local pH, such as β-lactamase, can also be transduced optically, using indicator dyes which alter their fluorescence characteristics according to the local pH, thus providing an alternative to potentiometric transduction.

16.4 BIOSENSORS BASED ON ANTIBODIES AND OTHER BINDING MOLECULES

16.4.1 General Considerations

Compared to enzymes, antibodies and similar binding molecules have the advantage of much greater versatility. In principle, an antibody or other binding molecule can be generated for almost any target analyte. The disadvantage is that, unlike enzymic reactions, a binding event does not intrinsically generate a strong specific signal, thus making it harder to detect than an enzymic reaction. In the context of biosensors, binding events can be detected directly, as in surface plasmon resonance (SPR) or piezoelectric systems, or indirectly, using competitive assays with labeled competing ligands, or sandwich assays with labeled second antibodies. Biosensors based on antibodies are sometimes called immunosensors. For a recent review, see Byrne et al. (2009).

While antibodies are the most commonly used type of affinity element in biosensors, other possibilities exist. For example, aptamers are short single-stranded nuclei acids which can be generated to bind any target molecule through a process of systematic evolution. Once developed, aptamers can easily be manufactured in large quantities by chemical methods, and are more robust than antibodies, so can have advantages. Another option is the use of molecular imprinting. In this approach, a monomer is polymerized in the presence of the target ligand. Removal of the ligand leaves cavities which may be capable of specifically binding the same ligand. A third option is to use a protein, such as a receptor, which is known to specifically bind the target ligand. For example, lectins (carbohydrate-binding proteins) can be used to detect specific carbohydrates. In some cases, it may even be possible to use a small molecule which specifically binds the ligand of interest. For example, Yang and Chen (2002) reported the use of immobilized polymyxin B in an acoustic biosensor (Section 18.4.4) for the quantitation of endotoxin (toxic lipo-polysaccharide shed from the cell walls of Gram negative bacteria, a great concern in the manufacture of injectable biological products).

In all cases, the objective is to detect binding of the analyte to the immobilized binding molecule. Two types of method are available—label-based (indirect) and label-free (direct). Indirect detection of antibody-binding events is based on the use of labeled molecules which are detectable optically or electrically. The two major types of label-based assay are sandwich assays and competitive assays (Figure 16.5).

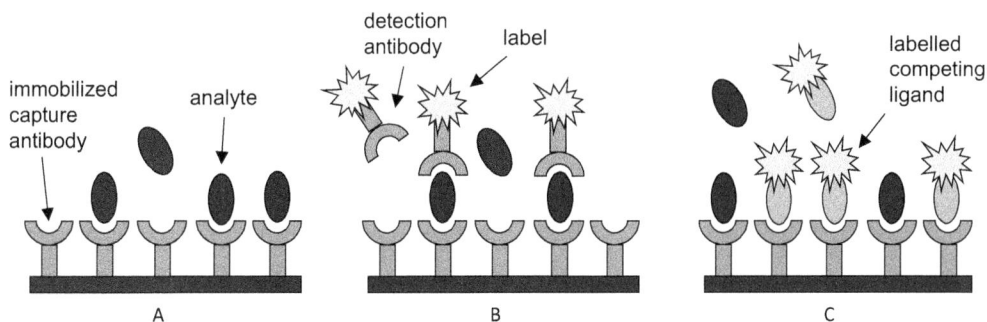

FIGURE 16.5 Affinity-based assays. (a) Direct assay. Binding of the analyte ligand (A) to the immobilized capture molecule is detected directly through some change in the properties of the immobilized film due to ligand binding. This is suited to surface plasmon resonance and piezoelectric transduction and is best suited to relatively large analytes, which cause a large relative mass change on binding. (b) Sandwich assay. A second, labeled affinity molecule (e.g., antibody) binds to the captured analyte ligand. Detection is based on the quantity of label at the capture surface. The label may be a fluorophore or an enzyme; detection may be optical or electrical. This requires that the analyte be large enough to bind two antibodies (or other binding molecules) simultaneously. (c) Competitive assay. There are a number of variants. In the version depicted here, the analyte ligand and a labeled competing ligand compete for binding sites on the film of immobilized capture molecules. Increased quantities of analyte in the sample mean that less label is detected at the capture surface. This method can be adapted for detection of relatively small molecules.

The label may be a fluorescent molecule, in which case detection is by fluorescence, or an enzyme, in which case a variety of transduction methods are possible, as described in Section 16.3, including chemiluminescence (using peroxidase as the label and adding luminol as substrate) and amperometric detection (Section 16.3.2). In sandwich assays, one antibody (the capture antibody) is bound to the surface, and a second, labeled antibody, which binds to another epitope on the same antigen, is present in solution. Ligand (analyte) binds to the immobilized antibody, and the labeled antibody then binds to this. Detection is based on the concentration of label bound at the film surface. Sandwich assays only work where the analyte ligand is large enough to bind two antibody molecules simultaneously. Competitive assays can be used for smaller analytes. A variety of configurations are possible, but all rely on competition between the analyte and a labeled ligand for binding sites. In one common version, the capture antibody is immobilized, and a mixture of analyte ligand and a known concentration of competing labeled ligand are applied to the system. These compete for binding sites. As the amount of analyte ligand increases, the amount of labeled ligand which becomes bound decreases. Detection may be based either on the amount of label which ends up bound at the surface (in which case a larger signal corresponds to a smaller amount of analyte ligand present in the assay) or on the amount of label which remains unbound (in which case a larger signal corresponds to a higher concentration of analyte). In an alternative configuration, the competing ligand is immobilized, and a labeled antibody is present in solution. The presence of the analyte (ligand) in solution reduces the amount of labeled antibody which binds at the surface.

Direct detection methods, such as surface plasmon resonance and acoustic biosensors, are sometimes referred to as "label-free" since they detect the binding event directly rather than via a labeled additive. These have the advantage of not requiring addition of a soluble component, but are more demanding than detecting a label, since the mass change will usually only be a small fractional change compared to the mass of antibody bound to the surface. It follows that these techniques are best suited to the quantitation of relatively large analytes such as proteins. Another problem with such techniques is nonspecific binding. In contrast to label-based methods, which detect only binding at the proper binding site, mass-based techniques will detect any binding to the surface whatsoever. Fermentation broths are often complex and contain many large molecules, such as proteins, which are capable of binding nonspecifically to surfaces or to other proteins. Such potential interference must always be considered.

Binding of a ligand to a molecule, assuming a homogeneous preparation and that sufficient time is allowed for the system to reach equilibrium, is governed by the dissociation constant (K_d). Under conditions where steady state is not achieved, the parameters which determines the binding of a ligand to a substrate are the association rate k_1, and the dissociation rate k_{-1}, which are related to K_d by the simple formula $K_d = k_{-1} / k_1$. This also assumes that diffusion of the ligand to the binding molecule is not rate limiting. If the interactions between the ligand and binding molecule are relatively weak, then continuous monitoring of the ligand concentration may be possible by following the instantaneous level of bound ligand (Ohlson et al., 2000). Where binding interactions are stronger, it is necessary to take a series of measurements, replacing or regenerating the detection film between measurements.

16.4.2 Optical Biosensors Based on Fluorescence

As noted above (Section 16.3.5), transduction in optical fiber biosensors is based on the transmission of light along an optical fiber to a detector. Light may be generated by fluorescent or luminescent reactions associated with enzymes immobilized at the tip of the fiber optic bundle. Optical fibers can also be applied to indirect methods of antibody-based detection using fluorescently labeled competing ligands or second antibodies. This is essentially similar to ELISA, but should give rapid responses without requiring numerous binding and washing steps before readout. While there are many laboratory reports of the use of such systems for detecting pathogens in food, for example, they do not seem to have been widely applied to process monitoring; however, they should be borne in mind as a potential alternative to techniques such as SPR and acoustic sensors, which require more complex instrumentation.

16.4.3 Surface Plasmon Resonance

SPR is an altogether different class of optical technique that is applied to the detection of small changes in mass due to ligand binding in a film of immobilized antibodies (or other binding molecules). Essentially, a ligand (the analyte) binds to a film of immobilized antibody, and the change in mass of the film due to ligand binding is detected based (usually) on a minute change in the incidence angle at which reflectance dips as photons are absorbed to generate surface plasmons. SPR is most suitable for direct detection of relatively large analytes such as proteins; it may be useful, for example, in monitoring the production of recombinant proteins in a bioreactor. Small molecules can also be detected using indirect (competitive) assays.

To describe the process in a little more detail, a beam of laser light is shone through a transparent medium coated with a thin film of gold, with a fluid layer on the other side of the gold film. Above a certain angle of incidence, total reflectance from the surface will occur, because of the difference in refractive index between the transparent medium and the fluid film. The reflected radiation intensity can be easily measured. During the process of reflectance, an evanescent wave, essentially an electrical field, is generated on the fluid side of the interface. At one particular narrow band of angles of incidence in this range, the phenomenon of surface plasmon resonance occurs, as photons are absorbed and converted to surface plasmons, leading to a sharp reduction in the intensity of the reflected light. The angle of incidence at which this effect peaks is strongly affected by the mass of material, such as antibodies plus any bound ligands, which is bound at the surface. Thus, as more ligand binds to a layer of immobilized antibody in such a system, the angle at which resonance occurs changes fractionally, and this can be detected with very high sensitivity. The measurement is given in Resonance Units (RU), where 1 RU corresponds to a change of 0.0001 degree in the angle of incidence at which the reflected radiation intensity minimum occurs.

Applications of SPR in the biopharmaceutical industry have been reviewed by Thillaivinayagalingam et al. (2010). As discussed above, the sensitivity of SPR detection depends on the fractional change in mass caused by binding of the ligand to the detection surface. Thus, SPR is better suited to the detection of large molecules, such as proteins, than to small molecules. SPR may be suited to monitoring the production of recombinant proteins. This is especially true for proteins which are secreted into the medium; for example, Chavane et al. (2008) used SPR to measure secretion of bioactive antibodies from hybridoma cell culture in a 3.5 liter bioreactor, using immobilized epitope, and Hsieh et al. (1998) reported offline monitoring of the production of *Clostridium perfringens* β-toxin by applying fermentation broth directly to an SPR sensor chip. However, intracellular proteins may also be monitored if a cell disruption step is included. For example, Vostiar et al. (2005) reported monitoring of intracellular production of recombinant human superoxide dismutase in *Escherichia coli* using SPR with immobilized monoclonal antibody, following in-line disruption using a non-ionic surfactant.

SPR can also be applied to the detection of small molecules using sandwich or competition assays similar to those described above. Sandwich assays are essentially the same as those described above except that the second antibody need not be labeled; the increase in surface mass due to the mass of the second antibody is sufficient for detection. For competitive assays, instead of an enzyme or fluorescent label, a large protein is used, to provide a sufficient mass change for detection. It has also been reported that recent SPR instruments are sufficiently sensitive for direct detection of the binding of small molecules to immobilized proteins (Rich and Myszka, 2000).

Some reports have also described integration of multiple sensors of different types in a single chip. For example, Suzuki et al. (1999) reported the construction of an SPR immunosensors chip for human IgG, also bearing integrated amperometric sensors for glucose and lactate; this was used for simultaneous on-line monitoring of glucose, lactate, and IgG in a cell culture bioreactor.

16.4.4 Acoustic (piezoelectric) Biosensors

Acoustic wave (piezoelectric) systems can also be used to measure mass changes due to ligand binding in a film of immobilized antibody. These devices are generally based on detection of changes in the nature of vibration of a piezoelectric material such as quartz when an electrical potential is applied. In this case, the antibody film is immobilized on the surface of a piezoelectric crystal. An applied voltage causes vibration of the crystal, and various characteristics of this vibration can be detected and used to infer the quantity of bound ligand. A variety of configurations are available. The most widely discussed type for biosensor use is the bulk acoustic wave (BAW) thickness shear mode (TSM) device, also known as a quartz crystal microbalance (QCM). Biosensors of this type have been recently reviewed by Ferriera et al. (2009). An alternative configuration is known as the surface acoustic wave (SAW) biosensor (Länge et al., 2008). SAW devices are in principle more sensitive but generally seem to be less suited to detection in aqueous environments than BAW devices, due to greater damping of oscillations by the liquid medium, but a number of recent publications have reported their use for detecting antibody-protein, antibody-bacteria, and DNA-DNA interactions. To give one relevant example, Berger et al. (2010) reported development of a SAW-based acoustic biosensor to detect hepatocyte growth factor in cell culture medium, based on an immobilized layer of sulphated polysaccharides, to which this protein binds with high affinity.

16.5 BIOSENSORS BASED ON WHOLE CELLS

16.5.1 GENERAL CONSIDERATIONS

It is also possible to use whole living cells as the recognition component in a biosensor. In principle this allows for more complex analysis, since the cells can respond to any of the wide variety of inputs which may alter growth or gene expression. It also raises certain practical difficulties in terms of noise (variability), and the requirement to keep cells alive throughout the process. At the moment, such systems are only used for specialized purposes, as described below, but with our increasing ability to engineer cells for specific applications, this may be an area which will expand in the future (French et al., 2011).

Cells can detect and respond to a wide variety of inputs, including concentrations of specific chemicals, as well as changes in environmental conditions such as temperature, pH, ionic strength, and so on. Outputs from cell-based systems can be in various forms. When wild-type (non-genetically modified) cells are used, the transduction system must detect some response which is naturally produced by the cells; for example, consumption of oxygen, passage of electrons through the respiratory system, change in pH caused by production of acids and so on. The use of genetically modified cells as biosensors opens a range of new possibilities. In particular, such systems often detect increased activity of a particular promoter, by activation of a reporter gene, which produces a protein which can be easily detected. In practice this is most commonly green fluorescent protein (GFP) or one of its many variants, which can easily be detected by fluorescence, or luciferase, a general term for enzymes which generate light by acting on a specific substrate (luciferin). Finally, a recent development has been the inclusion of biosensor-like genetic circuits within microorganisms which have been genetically engineered to produce a particular product. This ensures that the product is produced only at the right time, when conditions are correct, and can lead to substantial increases in yield. Such systems are sometimes referred to as dynamic sensor-regulator systems.

16.5.2 BIOSENSORS BASED ON METABOLIC ACTIVITY OF WILD-TYPE CELLS

Biosensor systems based on wild-type cells are the simplest to implement, and do not come with the regulatory issues related to the use of genetically modified cells. However, they are relatively limited in the types of stimulus which can generate an easily detectable response. The most common application is in BOD electrodes, which provide a rapid measurement of biological oxygen demand, the amount of oxygen consumed by microorganisms when a wastewater is disposed to the environment. This is a critical parameter in wastewater

treatment, which is normally measured by a five-day incubation of diluted samples followed by chemical measurement of residual O_2 levels. BOD electrodes, if properly calibrated, can in principle provide a much more rapid response, which correlates well with standard BOD measurements. BOD electrodes may consist of a standard Clark oxygen electrode, an amperometric system which measures O_2 levels at the electrode surface, with a coating of immobilized microorganisms. This is dipped into a wastewater sample, and the rate of O_2 consumption by the immobilized microorganisms is measured, giving an indication of the level of oxidizable organic material, which is to say, the BOD. Alternatively, they may be amperometric systems in which electrons are transferred directly from the respiratory chain of microorganisms to an electrode via a soluble mediator, as in a microbial fuel cell (MFC), giving a more direct measurement of microbial metabolic activity. As in amperometric glucose sensors (Section 16.3.2), this has the advantage of removing the dependence on oxygen concentration. Numerous reports describe such systems, using bacteria (for example, Hu et al., 2017) or yeast (for example, Zaitseva et al., 2017) as the immobilized sensor organism. Since disposal of spent culture media, which have a very high BOD due to their high organic content, may represent a significant cost element for biological processes, these devices may be a useful adjunct to biological processing.

The other main class of commercially available devices based on wild-type microorganisms are general-purpose toxicity sensors based on naturally luminescent bacteria. In such organisms, light is produced by an enzyme system known as bacterial luciferase. This system requires NADPH for light emission, and ATP for regeneration of the luciferin substrate (a long-chain fatty aldehyde), so that any toxic substance, which interferes with generation of ATP or NADPH will rapidly reduce the light output. Such systems, such as MicroTox, are used for rapid screening of environmental samples for toxicity, for example prior to or during remediation programs.

The metabolic activity of particular types of microorganism may also be used to monitor important parameters in certain types of biological process. The most important of these appears to be in wastewater treatment, which is invariably operated on a continuous basis and on a very large scale. For example, anaerobic digestion (AD) is a wastewater treatment process in which organic components of wastewater are degraded by microorganisms under anaerobic conditions to yield methane. For proper control of AD processes, it is valuable to monitor levels of volatile fatty acids (VFA), especially acetate, propionate and butyrate. A recent report (Kretzschmar et al., 2018) describes the use of biofilms of *Geobacter* spp. on an electrode surface to monitor VFA levels. *Geobacter*, like a number of other bacteria capable of anaerobic respiration, has the ability to pass electrons from its respiratory chain directly to an electrode surface held at a suitable potential. In this case the electrode acts as a sink for electrons derived from respiration, replacing use of a respiratory electron acceptor such as O_2 or nitrate. This generates an amperometric signal, just as in amperometric enzyme biosensors (section 18.3.2), which is related to the metabolic activity of the cells. This is essentially a miniature microbial fuel cell (MFC) (Figure 16.6). Since VFA are preferred substrates for *Geobacter* growth, the signal is related to the VFA concentration. Similar systems can be used whenever metabolic activity of a particular type of microorganism is directly related to a property of interest. In cases where an organism is not naturally capable of transferring electrons from its respiratory chain to an electrode surface, a small-molecule mediator can be used, just as in amperometric enzyme biosensors (Section 16.3.2). For example. Favre et al. (2009) reported use of an MFC-based biosensor to monitor metabolic activity during ethanol production by *Saccharomyces cerevisiae*.

16.5.3 BIOSENSORS BASED ON DETECTION OF SPECIFIC ANALYTES BY GENETICALLY MODIFIED CELLS

More versatile and specific biosensors are possible with the use of genetically modified cells. All cells respond to particular stimuli in their environment, including the presence or absence of certain chemical substances, by altering the expression of particular genes. These genes are controlled by DNA sequences known as promoters, which are switched on and off by binding of proteins known as transcription factors. Transcription factors respond to the presence of particular substances, which they may bind directly or detect by indirect mechanisms, altering their affinity for their target DNA sequences (Figure 16.7a).

For example, the genes controlling degradation of lactose in the model bacterium *Escherichia coli* are controlled by the *lac* promoter, which is regulated by binding of the transcription factor LacI. In the absence of lactose, LacI protein binds to the *lac* promoter-operator region sequence and in so doing it prevents RNA polymerase from initiating transcription, thus ensuring that these genes are not expressed; in the presence of lactose, LacI releases the promoter allowing gene expression. Such a system is easily converted to a biosensor by attaching an analyte-responsive promoter to a reporter gene, which encodes a gene product which can be easily detected. Detection and quantitation of the reporter gene product indicates activity of the promoter, giving an indication of the concentration of the analyte to which the promoter responds.

Such systems are widely used to study gene activity in molecular biology. The first practical application proposed seems to have been in environmental toxicology, detecting specific pollutants such as arsenic and mercury, for which many bacteria possess highly specific promoters controlling detoxification mechanisms. One commercial application is in the measurement of mutagenicity of new chemicals. Systems such as SOS-Chromotest couple a reporter gene to a promoter responsive to DNA damage, which normally controls a DNA repair pathway. Expression of the reporter gene on exposure to a particular chemical indicates that DNA damage has occurred, suggesting that the chemical is a potential mutagen or carcinogen.

For laboratory systems, the reporter genes most commonly used are fluorescent proteins such as green fluorescent protein (GFP) and its many homologues and derivatives, each with its own distinctive absorption and emission spectrum. These are very stable and easily detected in culture media. A popular alternative is the use of luciferases, a general term for enzymes which generate light through their activity on a target substrate (generically referred to as "luciferin," though different luciferases use luciferins of completely different structures). Luminescence generated by luciferases can be detected with very high sensitivity, since the background is much lower than for fluorescence. Popular choices include bacterial luciferase, LuxAB, which uses a long-chain fatty aldehyde as substrate, and firefly luciferase, LucFF, which uses a more complex and

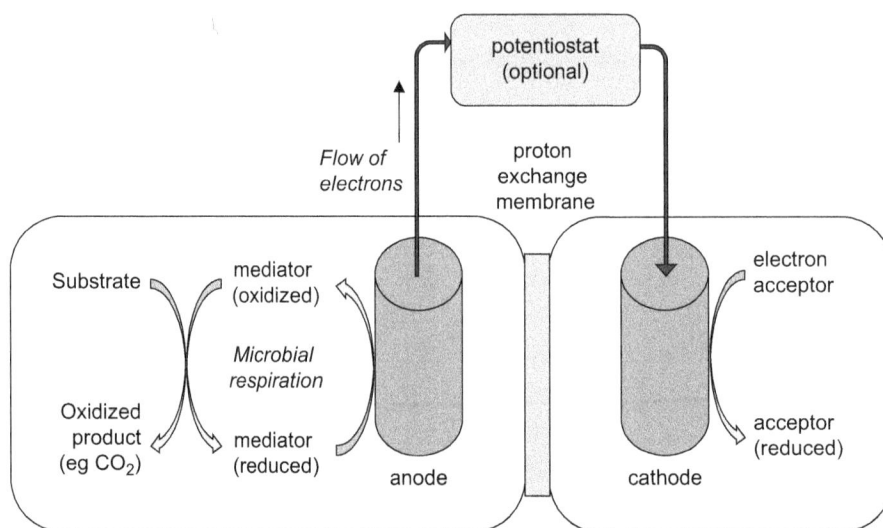

FIGURE 16.6 General scheme for a biosensor based on a microbial fuel cell (MFC). A potentiostat may be used to maintain the anode at a suitable potential, and to measure current, giving an amperometric system; alternatively, the potentiostat may be omitted and the anode connected directly to the cathode, in which case the system will settle down to a steady-state potential difference, measured as a voltage. The scheme shown here includes a soluble mediator in the anode chamber, but this may be omitted if the microorganism used has the capability to transfer electrons directly to the electrode, as seen in *Geobacter* and *Shewanella* spp.

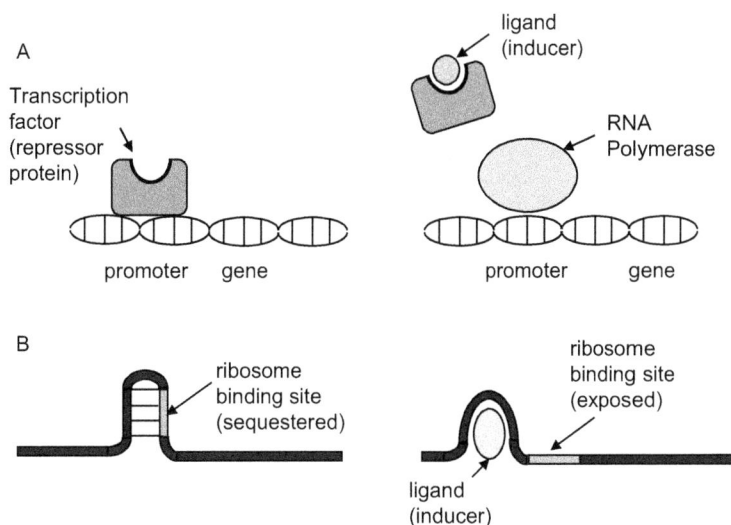

FIGURE 16.7 General schemes for genetically modified whole-cell biosensors based on expression of a reporter gene. (a) Transcriptional sensor in which expression of the reporter gene is regulated by controlling mRNA production using a promoter element controlled by a transcription factor (activator or repressor protein) which responds, directly or indirectly, to the presence of the target analyte. The system here shows negative regulation by a repressor protein which binds DNA, inhibiting transcription, in the absence of the ligand, but releases DNA, allowing transcription, when the ligand (inducer) is present. This is the most common situation in bacteria, but it is also possible to have an activator protein that enhances transcription when bound, and/or for binding of DNA to be favored rather than prevented by binding of the ligand. More complex situations are also possible, for example, "two-component" systems, in which the ligand interacts with a receptor protein, which may be extracellular, and this then covalently modifies a second "response regulator" protein, altering its DNA-binding properties. (b) Translational sensor in which mRNA is always produced but its translation to protein is controlled by a riboswitch element in the mRNA, which binds the target analyte altering the ability of the mRNA to interact with ribosomes. Again, a variety of configurations are possible. In the variant shown here, in the absence of the ligand, the mRNA folds in such a way as to prevent interaction with the ribosome; when the riboswitch element of the RNA binds ligand, the new state exposes the sites necessary for interaction with the ribosome and production of the reporter protein.

expensive substrate but has a much higher quantum yield, giving increased brightness. However, a number of other luciferases from marine creatures have become available in recent years and may have advantages in some applications.

For systems to be used in long-term process monitoring, for example, in wastewater treatment plants, other types of reporter are available. For example, if the analyte-responsive promoter is coupled to a gene essential for growth under particular conditions (for example, an enzyme required for assimilation of a particular sugar, or a component of the respiratory chain) then expression of the reporter gene is detected as increased growth, which can be measured potentiometrically by pH change due to acid production (de Mora et al., 2011), or amperometrically using a miniature MFC as described in Section 18.5.2. (Figure 16.6).

To cite one example, Goers et al. (2017) reported use of a whole-cell *E. coli*-based lactate biosensor to monitor lactate levels in a mammalian cell bioreactor. The reporter gene was GFP, and measurements were made off-line in a plate-reader. It was reported that the sensitivity was considerably superior to that of industry-standard amperometric enzyme-based lactate sensors.

Application of such biosensors to bioprocess control appears to be in its infancy and will require some serious attention as to the most suitable formats, but such devices offer potential advantages in cases where a promoter exists which is known to respond specifically to the target analyte.

Where no such promoter is known to exist, it may be possible to engineer an equivalent by generation of an analyte-specific riboswitch (Hallberg et al., 2017). A riboswitch is an element located at the beginning of a messenger RNA molecule, which, when folded in a particular way, prevents the RNA from being translated into protein (in this case, the reporter protein). The riboswitch is essentially an aptamer which can also fold in such a way as to bind a particular small molecule. Binding of the target molecule to the mRNA causes a change in overall folding so as to allow or prevent expression of the encoded protein (Figure 16.7b). In principle, riboswitches can be generated in the same way as aptamers (see Section 16.4.1) to bind to almost any given target molecule; in practice the process appears to be somewhat complex, and relatively few well-behaved examples have been described. However, as methods improve, this will be an area to watch in the future. To cite some recent examples, Jang et al. (2017) described generation of artificial riboswitches which could be used in *E. coli* to monitor levels of the flavonoid naringenin, and Xiu et al. (2017) reported use of such sensors to assay naringenin production levels in co-culture with production strains.

In addition to their potential role in bioprocess monitoring, such biosensors can also be very useful in bioprocess development. The techniques of synthetic biology allow rapid production of many variants of a genetically engineered production pathway, and screening for production levels by HPLC or similar methods can be laborious. Use of genetically encoded biosensors, on the other hand, allows many variants to be screened simultaneously, for example in 96 or 384 well plates, or even by flow cytometry if the biosensor module is included in the same cells as the production pathway, so that individual cells producing the highest level of the target product are most strongly fluorescent (Zhang et al, 2015).

16.5.4   Internal Biosensors in Engineered Metabolic Pathways: Dynamic Regulation Systems

Perhaps the most cutting-edge application of biosensors for process control is the use of internal biosensors to control gene expression within a genetically engineered pathway for manufacture of a particular product. Essentially, this is just an engineered form of the regulation which is native to naturally occurring biosynthetic pathways, but frequently absent from engineered pathways, to the detriment of production. The principle is simple – expression of one part of the pathway is controlled by a promoter or riboswitch regulated by the concentration of a molecule produced by another part of the pathway. This ensures that different parts of the product formation pathway are coordinated, reducing metabolic burden on the producer cells and potentially leading to considerable increases in yield (Liu et al., 2018).

One of the earliest examples reported was the dynamic sensor-regulator system (DSRS) described by Zhang et al. (2012) to regulate production of "microdiesel," a product consisting of ethyl esters of long-chain fatty acids, which can be used as a renewable replacement for diesel, not requiring the use of food-grade vegetable oils as in standard biodiesel. Microdiesel is produced by engineered strains of bacteria such as *E. coli* in a three-part process: biosynthesis of free fatty acids, biosynthesis of ethanol, and condensation of the two to generate the ester. Zhang et al. (2012) reported that placing the ethanol production and condensation steps under the control of a promoter which responds to the concentration of free fatty acids, so that the relevant enzymes were only produced when fatty acid levels were high enough to make them useful, led to greatly improved growth of the organism and considerably higher yields of the product. Since this report, numerous other examples have been described (Liu et al., 2018). In one interesting variant, Dahl et al. (2013) avoided the necessity to find or create an analyte-specific promoter by using a stress promoter to detect levels of a toxic intermediate in the manufacture of isoprenoids in *E. coli* and used this signal to control later steps in the pathway, increasing levels of the desired final product. As our ability to engineer complex metabolic pathways improves, introduction of such regulatory elements will be increasingly important in leading to economically viable processes.

SUMMARY

- Biosensors offer a method for automated monitoring of metabolites and reaction products during fermentation and can be used in control systems. The analyte of interest is recognized specifically by a biological molecule such as an enzyme or antibody, and an electrical signal is generated by a transducer. A very wide range of transduction methods have been reported in the literature, though relatively few seem to have been commercially adopted for process control yet.

- For small molecules such as sugars, alcohols and hydroxy-acids for which oxidase or dehydrogenase enzymes are available, the most obvious choice is an amperometric enzyme electrode. This is a well-established technology. Indeed, amperometric biosensors for the detection of common substrates and metabolites such as glucose and other sugars, ethanol, lactate, and glutamate, are commercially available in systems specifically designed for process monitoring and control.

- For small molecules where no redox enzyme is available, but which can be acted upon by an enzyme (such as a hydrolase) yielding or consuming H^+, NH_4^+ (NH_3) or HCO_3^- (CO_2), a potentiometric enzyme electrode or EnFET would be a good choice.

- For small molecules for which no enzyme is available, but against which an antibody could be raised, a competitive or sandwich immunoassay, using either optical or electrical transduction, might be suitable. For larger molecules, such as recombinant proteins, a direct affinity sensor based on SPR, piezoelectric, or mechanical transduction might offer a suitable detection system.

- Biosensors based on whole cells can be used in specific applications, where a cell's activity is related to a parameter of interest. Whole-cell biosensors can be genetically modified so as to specifically generate a response to a target analyte. A variation on this is the use of an internal genetic biosensor circuit within a genetically modified production strain to control the level of production.

ABBREVIATIONS

AD: Anaerobic Digestion
BAW: Bulk Acoustic Wave
BOD: Biological Oxygen Demand
DSRS: Dynamic Sensor-Regulator System
ELISA: Enzyme-Linked Immunosorbent Assay
EnFET: Enzyme Field Effect Transistor
FAD: Flavin Adenine Dinucleotide
FADH₂: Flavin Adenine Dinucleotide (reduced form)
FIA: Flow Injection Analysis
GC/MS: Gas Chromatography-Mass Spectrometry
GFP: Green Fluorescent Protein
GOX: Glucose Oxidase
HPLC: High-Performance Liquid Chromatography
ISFET: Ion-Selective Field Effect Transistor
MFC: Microbial Fuel Cell
NAD: Nicotinamide Adenine Dinucleotide

NADH: Nicotinamide Adenine Dinucleotide (reduced form)
PQQ: Pyrroloquinoline Quinone
QCM: Quartz Crystal Microbalance
SAW: Surface Acoustic Wave
SIA: Sequential Injection Analysis
SPR: Surface Plasmon Resonance
TSM: Thickness Shear Mode
VFA: Volatile Fatty Acids

DEFINITIONS

Analyte: The target substance to be detected.
Aptamer: A nucleic acid molecule which has been designed or evolved in such a way as to fold into a three-dimensional shape capable of binding a specific molecule.
Dynamic range: The range of sensor responses over which the analyte concentration can be determined from the sensor output, i.e., between the sensitivity and the saturation level.
Immunosensor: A biosensor using an antibody as the recognition element.
Luciferase: An enzyme which produces light by acting on a substrate known as a luciferin.
Luciferin: The substrate for a luciferase. Different luciferases use completely different and unrelated molecules as luciferins.
Mediator: A redox-active small molecule which can transfer electrons between an enzyme and an electrode.
Optode, optrode: An optical element which acts as a sensor (by analogy with 'electrode').
Potentiostat: A circuit which maintains an electrode at a constant potential relative to a reference electrode; used in amperometric sensors.
Promoter: A DNA sequence element which controls expression of one or more genes.
Reporter gene: A gene producing an easily detectable product, used to measure the activity of a promoter.
Riboswitch: An element at one end of a messenger RNA molecule which, like an aptamer, folds into a shape capable of binding a particular molecule. Binding of the riboswitch to the target molecule allows or prevents expression of the protein encoded by the remaining part of the mRNA.
Sensitivity: The lowest level of analyte which can be detected reliably.
Thermistor: A sensitive solid-state temperature measuring device, in which resistance is strongly dependent on temperature.
Transcription factor: A protein which controls the activity of a promoter by binding a specific DNA element within or near the promoter sequence.

ACKNOWLEDGMENTS

The author acknowledges the contributions made by co-authors of previous editions of this chapter, Dr. Marco Cardosi and Dr. Chris Gwenin, who provided valuable insights into the workings of biosensors and their uses in process monitoring.

REFERENCES

Berger, M., Welle, A., Gottwald, E., Rapp, M., and Lange, K. (2010) Biosensors coated with sulfated polysaccharides for the detection of hepatocyte growth factor/scatter factor in cell culture medium. *Biosensors and Bioelectronics*, **26**, 1706–1709.

Blankenstein, G., Spohn, U., Preuschoff, F., Thommes, J., and Kula, M.R. (1994) Multi-channel flow injection analysis biosensor system for on-line monitoring of glucose, lactate, glutamine, glutamate and ammonia in animal cell culture. *Biotechnology and Applied Biochemistry*, **20**, 291–307.

Borisov, S.M., and Wolfbeis, O.S. (2008) Optical Biosensors. *Chemical Reviews* **108**, 423–461.

Bosch, M.E., Sánchez, A.J.R., Rojas, F.S., and Ojeda, C.B. (2007) Recent development in optical fibre biosensors. *Sensors*, **7**, 797–859.

Byrne, B., Stack, E., Gilmartin, N., and O'Kennedy, R. (2009) Antibody-based sensors: Principles, problems and potential for detection of pathogens and associated toxins. *Sensors* **9**, 4407–4445.

Chavane, N., Jacquemart, R., Hoemann, C.D., Jolicoeur, M., and De Crescenzo, G. (2008) At-line quantification of bioactive antibody in bioreactor by surface plasmon resonance using epitope detection. *Analytical Biochemistry*, **378**, 158–165.

Dahl, R.H., Zhang, F., Alonso-Gutierrez, J., Baidoo, E., Batth, T.S., Redding-Johanson, A.M., Petzold, C.J., Mukhopadyay, A., Lee, T.S., Adams, P.D., and Keasling, J.D. (2013) Engineering dynamic pathway regulation using stress-response promoters. *Nature Biotechnology*, **31**, 1039–1046.

de Mora, K., Joshi, N., Balint, B.L., Ward, F.B., Elfick, A., and French, C.E. (2011) A pH-based sensor for detection of arsenic in drinking water. *Analytical and Bioanalytical Chemistry*, **400**, 1031–1039.

Dominguez, K., Toth, I.V., Souto, M.R.S., Mendes, F., Maria, C.G., Vasconcelos, I., and Rangel, A.O.S.S. (2010) Sequential injection kinetic flow assay for monitoring glycerol in a sugar fermentation process by *Saccharomyces cerevisiae*. *Applied Microbiology and Biotechnology*, **160**, 1664–1673.

Favre, M.F., Carrard, D., Ducommun, R., and Fischer, F. (2009) Online monitoring of yeast cultivation using a fuel-cell-type activity sensor. *Journal of Industrial Microbiology and Biotechnology*, **36**, 1307–1314.

Ferreira, G.N.M., Da Silva, A.C., and Tome, B. (2009) Acoustic wave biosensors: physical models and biological applications of quartz crystal microbalance. *Trends in Biotechnology*, **27**, 689–697.

French, C.E., de Mora, K., Joshi, N., Elfick, A., Haseloff, J., and Ajioka, J. (2011) Synthetic biology and the art of biosensor design. In *The Science and Applications of Synthetic and Systems Biology*, Institute of Medicine, National Academies Press, Washington DC.

Goers, L., Ainsworth, C., Goey, C.H., Kontoravdi, C., Freemont, P.S., and Polizzi, K.M. (2017) Whole cell *Escherichia coli* lactate biosensor for monitoring mammalian cell cultures during biopharmaceutical production. *Biotechnology & Bioengineering*, **114**, 1290–1300.

Grieshaber, D., MacKenzie, R., Vörös, J., and Reimhut, E. (2008) Electrochemical biosensors – sensor principles and architectures. *Sensors*, **8**, 1400–1458.

Gronewold, T.M.A. (2007) Surface acoustic wave sensors in the bioanalytical field: recent trends and challenges. *Analytica Chimica Acta*, **603**, 119–128.

Hallberg, Z.F., Su, Y.C., Kitto, R.Z., and Hammond, M.C. (2017) Engineering and in vivo applications of ribnoswitches. *Annual Review of Biochemistry*, **86**, 515–539.

Hsieh, H.V., Stewart, B., Hauer, P., Haaland, P., and Campbell, R. (1998) Measurement of *Clostridium perfringens* β- toxin production by surface plasmon resonance immunoassay. *Vaccine*, **16**, 997–1003.

Hu, J.F., Li, Y.Q., Gao, G.W., and Xia, S.H. (2017) A mediated BOD biosensor based on immobilized Bacillus subtilis on three-dimensional porous graphene-polypyrrole composite. *Sensors*, **17**, 2594.

Jang, S., Jang, S., Xiu, Y., Kang, T.J., Lee, S.H., Koffas, M.A.G., and Jung, G.Y. (2017) Development of artificial riboswitches for monitoring of naringenin in vivo. *ACS Synthetic Biology*, **6**, 2077–2085.

Kamanin, S.S., Arlyapov, V.A., Machulin, A.V., Alferov, V.A., and Reshetilov, A.N. (2015) Biosensors based on modified screen-printed enzyme electrodes for monitoring of fermentation processes. *Russian Journal of Applied Chemistry*, **88**, 463–472.

Kretzschmar, J., Bohme, P., Liebetrau, J., Mertig, M., and Harnisch, F. (2018) Microbial electrochemical sensors for anaerobic digestion process control – performance of electroactive biofilms under real conditions. *Chemical Engineering & Technology*, **41**, 687–695.

Länge, K., Rapp, B.E., and Rapp, M. (2008) Surface Acoustic Wave biosensors: a review. *Analytical and Bioanalytical Chemistry*, **391**, 1509–1519.

Lawung, R., Danielsson, B., Prachayasittikul, V., and Bulow, L. (2001) Calorimetric analysis of cephalosporins using an immobilised TEM-1 β--lactamase on Ni^{2+} chelating sepharose fastflow. *Analytical Biochemistry*, **296**, 57–62.

Liu, D., Mannan, A.A., Han, Y., Oyarzun. D.A., and Zhang, F.Z. (2018) Dynamic metabolic control: towards precision engineering of metabolism. *Journal of Industrial Microbiology and Biotechnology*. **45**, 535–543. doi: 0.1007/s10295-018-2013-9.

Luo, X.L., Xu, J.J., Zhao, W., and Chen, H.Y. (2004) Glucose biosensor based on ENFET doped with SiO$_2$ nanoparticles. *Sensors and Actuators B – Chemical*, **97**, 249–255.

Matuszewski, W., Rosario, S.A., and Meyerhoff, M.E. (1991). Operation of ion-selective electrode detectors in the sub-Nernstian linear response range – application to flow-injection enzymatic determination of L-glutamine in bioreactor media. *Analytical Chemistry*, **63**, 1906–1090.

Moser, I., Jobst, G., and Urban, G.A. (2002) Biosensor arrays for simultaneous measurement of glucose, lactate, glutamate, and glutamine. *Biosensors and Bioelectronics*, **17**, 297–302.

Miao, Y.Q., Guan, J.G., and Chen, J.R. (2003) Ion sensitive field effect based biosensors. *Biotechnology Advances*, **21**, 527–534.

Navratil, M., Tkac, J., Svitel, J., Danielsson, B., and Sturdik, E. (2001) Monitoring of the bioconversion of glycerol to dihydroxyacetone with immobilized *Gluconobacter oxydans* cell using thermometric flow injection analysis. *Process Biochemistry*, **36**, 1045–1052.

Ohlson, S., Jungar, C., Strandh, M., and Mandenius, C.-F. (2000) Continuous weak affinity immunosensing. *Trends in Biotechnology*, **18**, 49–52.

Park, I.S., Kim, D.K., and Kim, N. (2004) Characterization and food application of a potentiometric biosensor measuring β—lactam antibiotics. *Journal of Microbiology and Biotechnology*, **14**, 698–706.

Poghossian, A., Schoning, M.J., Schroth, P., Simonis, A., and Lüth, H. (2001) An ISFET-based penicillin sensor with high sensitivity, low detection limit and long lifetime. *Sensors and Actuators B – Chemical*, **76**, 519–526.

Pogoroleva, S.P., Zayats, M., Kharitnov, A.B., Katz, E., and Willner, I. (2003) Analysis of NAD(P)$^+$ cofactors by redox-functionalized ISFET devices. *Sensors and Actuators B – Chemical*, **89**, 40–47.

Ramanathan, K., and Danielsson, B. (2001) Principles and applications of thermal biosensors. *Biosensors and Bioelectronics*, **16**, 417–423.

Ramanathan, K., Jonsson, B.R., and Danielsson, B. (2001) Sol-gel based thermal biosensor for glucose. *Analytica Chimica Acta*, **427**, 1–10.

Rank, M., Danielsson, B., and Gram, J. (1992) Implementation of a thermal biosensor in a process environment: On-line monitoring of penicillin V in production scale fermentations. *Biosensors and Bioelectronics*, **7**, 631–635.

Rich, R.L. and Myszka, D.G. (2000) Advances in Surface Plasmon Resonance Biosensor analysis. Current Opinion in Biotechnology, 11, 54–61.

Ronkainen, N.J., Halsall, H.B., and Heineman, W.R. (2008) Electrochemical biosensors. *Chemical Society Reviews*, **39**, 1747–1763.

Schugerl, K. (2001) Progress in monitoring, modelling and control of bioprocesses during the last 20 years. *Journal of Biotechnology*, **85**, 149–173.

Suzuki, M., Nakashima, Y., and Mori, Y. (1999) SPR immunosensor integrated two miniature enzyme sensors. *Sensors and Actuators B – Chemical*, **54**, 176–181.

Thillaivinayagalingam, P., Gommeaux, J., McLoughlin, M., Collins, D., and Newcombe, A.R. (2010) Biopharmaceutical production: Applications of surface plasmon resonance biosensors. *Journal of Chromatography B – Analytical Technologies in the Biomedical and Life Sciences*, **878**, 149–153.

Tinkilic, N., Cubuk, O., and Isildak, I. (2002) Glucose and urea biosensors based on all solid-state PVC-NH$_2$ membrane electrodes. *Analytica Chimica Acta*, **452**, 29–34.

Vostiar, I., Tkac, J., and Mandenius, C.F. (2005) Intracellular monitoring of superoxide dismutase expression in an *Escherichia coli* fed-batch cultivation using on-line disruption with at-line surface plasmon resonance detection. *Analytical Biochemistry*, **342**, 152–159.

Xiu, Y., Jang, S., Jones, J.A., Zill, N.A., Linhardt, R.J., Yuan, Q.P., Jung, G.Y., and Koffas, M.A.G. (2017) Naringenin-responsive riboswitch-based fluorescent biosensor module for *Escherichia coli* co-cultures. *Biotechnology and Bioengineering*, **114**, 2235–2244.

Yang, M.X., and Chen, J.R. (2002) Self assembled monolayer based quartz crystal biosensors for the detection of endotoxin. *Analytical Letters*, **35**, 1775–1784.

Yoo, E.H., and Lee, S.Y. (2010) Glucose biosensors: an overview of use in clinical practice. *Sensors*, **10**, 4558–4576.

Zaitseva, A.S., Arlyapov, V.A., Yudina, N.Y., Alferov, S.V., and Retshilov, A.N. (2017) Use of one- and two-mediator systems for developing a BOD biosensor based on the yeast Debaromyces hansenii. *Enzyme and Microbial Technology*, **98**, 43–51.

Zhang, F.Z., Carothers, J.M., and Keasling, J.D. (2012) Design of a dynamic sensor-regulator system for production of chemicals and fuels derived from fatty acids. *Nature Biotechnology*, **30**, 354–359.

Zhang, J., Jensen, F.K., and Keasling, J.D. (2015) Development of biosensors and their application in metabolic engineering. *Current Opinion in Chemical Biology*, **28**, 1–8.

17 Cell Immobilization and Its Applications in Biotechnology
Current Trends and Future Prospects

Ronnie G. Willaert

CONTENTS

"The great men of science are supreme artists."

Martin H. Fischer

17.1 INTRODUCTION

The immobilization of whole cells can be defined as "the physical confinement or localization of intact cells to a certain region of space; without loss of desired biological activity." When cells are encapsulated in an immobilized cell system, the term "bioencapsulation" or "microencapsulation" is used; the latter is used when cells are immobilized in microcapsules (i.e., micrometer-sized systems surrounded by a barrier membrane). Recently, nanoencapsulation of biological molecules such as proteins have also been successful.

Immobilizing individual enzymes for simple reactions such as hydrolysis and isomerization, has been used to create biocatalysts for the production of various chemicals through simple and conjugated reactions. Many applications have also been developed for single and multicellular organisms (Eş et al., 2015).

The type of application, the physical and biochemical characteristics of the immobilizing matrix/agent dictate the suitability of a given system of immobilization. Accordingly, requirements will differ from one case to another; however, in any system, the following characteristics are generally desirable:

- High cell mass-loading capacity
- Affords easy access to nutrient media
- Is a simple and "nontoxic" immobilization procedure
- Affords high surface-to-volume ratio
- Facilitates optimum mass transfer
- Is sterilizable and reusable

- Facilitates easy separation of cells and carrier from media
- Is suitable for conventional reactor systems as well as cell suspension and anchorage-dependent cells
- Should be biocompatible for animal cells
- Contains immunoprotection barrier
- Should be economically viable.

17.2 IMMOBILIZED CELL SYSTEMS

On the bases of physical localization and the nature of micro-environment, immobilized cell systems can be classified into four categories (Figure 17.1).

17.2.1 SURFACE ATTACHMENT OF CELLS

Although this is not suitable where cell-free effluent is desired, immobilization of cells by adsorption to a support material can be achieved naturally or induced artificially by using linking agents (metal oxides or covalent bonding agents such as glutaraldehyde or aminosilane) (Figure 17.2). A suitable adsorbent for spontaneous attachment should possess a high affinity toward the biocatalyst and cause minimal denaturation. The adsorption of cells to an organic or inorganic support material is achieved *via* the Van der Waals forces and ionic interactions.

The adhesion behavior of viable cells is influenced by

- The physical and chemical properties of the adsorption matrix
- The identity and the biochemical characteristics of the immobilized organism (especially the outer surface of the cell wall)
- The composition and chemical and physical properties of the surrounding mobile phase

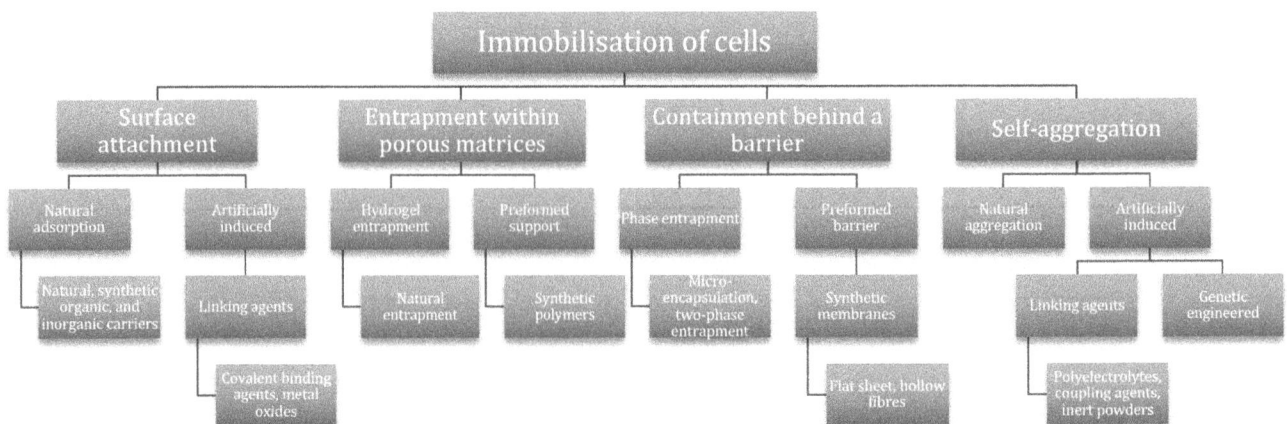

FIGURE 17.1 Classification of immobilized cell systems according to the physical localization and the nature of the microenvironment.

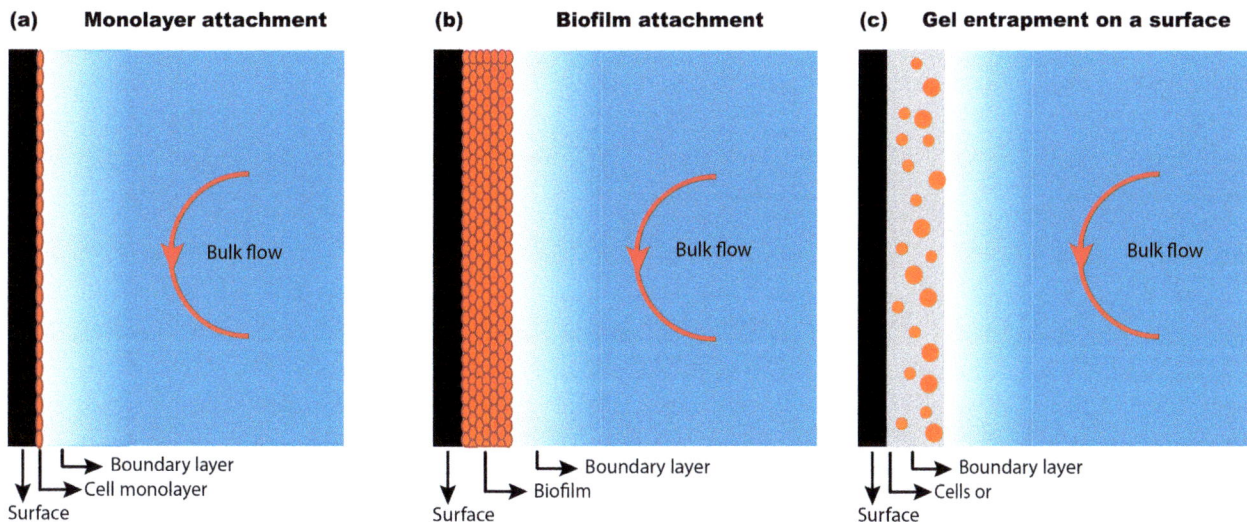

FIGURE 17.2 Cell-immobilization by adsorption/attachment to a surface: (a) adsorption of a monolayer; (b) adsorption of a biofilm; and (c) adsorption of an "artificial" biofilm (i.e., adsorption of a gel layer containing immobilized cells or cell aggregates).

The effects of different environmental and/or physiological conditions on the adhesion mechanisms of different bacteria have been intensively studied. Recent studies revealed that cells interact with their surrounding through specialized structures [e.g., pilus-associated adhesions, exopolymers (glycocalyx in the case of bacteria), and complex ligand interactions involving signaling molecules and quorum-sensing mechanisms]. Therefore, it follows that sensing a biotic or an abiotic surface may turn on genetic switches, and this in turn may lead to changes in the organism's phenotype (Loo et al., 2000).

Biofilms are communities of microorganisms in which cells stick to each other and often adhere to a surface. These adherent cells are usually embedded within a self-produced matrix of extracellular polymeric substance (EPS). For example, the Gram-negative bacterium *Pseudomonas aeruginosa* biofilm formation causes widespread diseases in humans. Pel, psl, and alg operons present in *P. aeruginosa* are responsible for the biosynthesis of extracellular polysaccharide which plays an important role in cell surface interactions during biofilm formation. Recent studies suggested that cAMP signaling pathway, quorum-sensing pathway, Gac/Rsm pathway, and c-di-GMP signaling pathway are the main mechanism that leads to the biofilm formation (Skariyachan et al., 2018). Biofilm structures can be flat or mushroom-shaped depending on the nutrient source, which seems to influence the interactions between localized clonal growth and the subsequent rearrangement of cells through type IV pilus-mediated gliding motility (Klausen et al., 2003).

Most animal cells from solid tissue grow as adherent monolayers, unless transformed into anchorage-independent cells. However, anchorage-dependent cells are often diploid and exhibit **contact inhibition**. After tissue disaggregation or subculturing, they will need to attach and spread out on the substrate before proliferating.

Contact inhibition: Animal cells that exhibit contact inhibition stop growing when cell-cell contact takes place as the culture reaches confluence.

Whereas cell adhesion is mediated by specific cell-surface receptors, cell-substrate interactions are mediated primarily by integrins, receptors for extracellular matrix (ECM) molecules such as fibronectin, entactin, laminin, and collagen, which bind them *via* a specific motif usually containing the arginine-glycine-aspartic acid (RDG) sequence (Maheshwari et al., 2000). Each integrin comprises one α and one β subunit, both of which are highly polymorphic, thus generating considerable diversity among integrins.

The first step of the cell-surface interaction of anchorage-dependent cells is attachment, in which the cells retain the round shape they possessed in suspension. The cells undergo conformational change, known as spreading, in which the cells increase their surface area before attachment to the surface. The kinetics of attachment and spreading have been determined by measuring the effective refractive index of the waveguide, the number of cells per unit area, and a parameter uniquely characterizing their shape, such as the area in contact with the surface. Examples of cell immobilization by attachment to the surface are given in Table 17.1.

17.2.2 ENTRAPMENT WITHIN POROUS MATRICES

Cell entrapment can be achieved through *in situ* immobilization in the presence of the porous matrix (i.e., gel entrapment) or by allowing the cells to move into a preformed porous matrix (Figure 17.3). Entrapped cells can reach high densities in the matrix and, compared with surface immobilization, cells are well protected from fluid shear. However, these dense cell packings may lead to mass transport limitations.

17.2.2.1 Hydrogel Entrapment

Most research in the domain of immobilized cells has used gel entrapment because of its simplicity and excellent cell containment. A wide variety of natural (polysaccharides and proteins) and synthetic polymers can be gelled into hydrophilic matrices under mild conditions to allow cell entrapment

TABLE 17.1

Examples of Cell-Immobilization by Attachment to a Surface

Material	Cell Type	Application
	Bacteria	
Ion exchange resin	*Bacillus stearothermophilus*	Amylase production
Coke	*Zymomonas mobilis*	Ethanol production
Seashell pieces	*Bacillus* sp. + *Aeromonas* sp. + *Alcaligenes* sp.	Decolorization and degradation of triphenyl methane dyes
	Fungi	
Celite	*Penicillium chrysogenum*	Penicillin production
Stainless steel fiber cloth	*Saccharomyces cerevisiae*	Beer production
Sugarcane bagasse	*Candida guilliermondii*	Xylitol production
Straw	*Agaricus* sp.	Laccacase production
Structural fibrous network of papaya wood	*Aspergillus terreus*	Itaconic acid production
	Animal Cells	
Surface modified polyethylene film	Midbrain cells	Neural differentiation
Poly(lactide-co-glycolide), poly(d,l-lactide)	Chondrocyte	Growth on biodegradable scaffolds (tissue engineering)
Pyrex glass, polysterene, glass beads	*Trichoplusia ni*	Recombinant protein production

Source: Adapted from Willaert, R.G. and Baron, G.V., *Rev Chem Eng*, 12:1–205, 1996; Willaert, R. *Fermentation Microbiology and Biotechnology*, 3rd ed., pp. 313–367, 2012. Boca Raton, FL: CRC Press.

with minimal loss of viability. Gel formation mechanisms of frequently used gels are shown in Table 17.2. The polymer-cell mixture can be formed in different shapes and sizes. The most common forms are small beads approximately 1–5 mm in diameter. Although natural polymers dominate, synthetic polymers have recently been developed and applied for the immobilization of living cells. The synthetic polymers can be easily and artificially designed for adequate properties. The porosity of the gel as well as the ionic and hydrophobic or hydrophilic properties can be adjusted. Additionally, the mechanical strength and longevity of the gels formed from synthetic polymers are generally superior to those from natural polymers.

Gel entrapment has the disadvantage of limited mechanical stability. It has been frequently observed that the gel structure is easily destroyed by cell growth in the gel matrix and carbon dioxide production. However, the gel structure can usually be reinforced (e.g., alginate gel was made stronger by the reaction of alginate with other molecules such as polyethylenei-mine, glutaraldehyde crosslinking, silica, genipin, poly(vinyl alcohol), or by partial drying of the gel). Another disadvantage of cell entrapment, compared with other immobilized

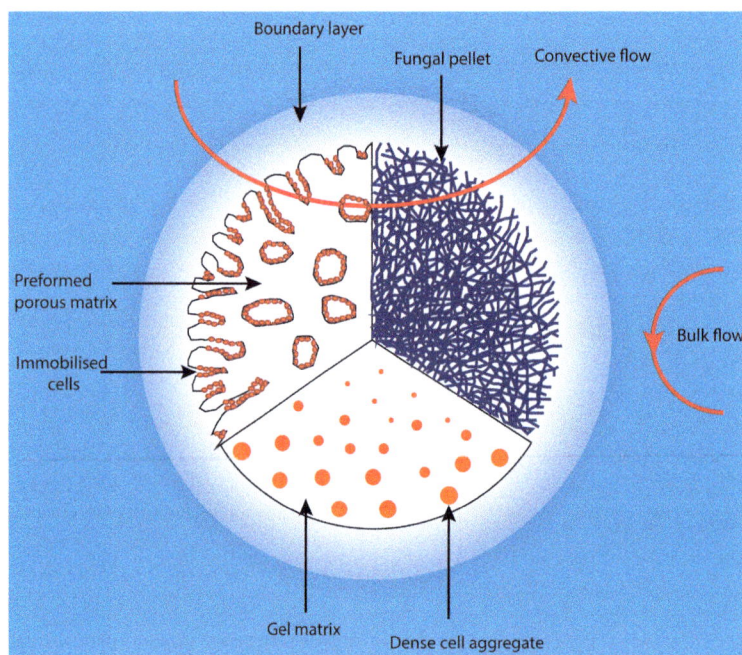

FIGURE 17.3 Spherical particle showing three immobilization methods: (1) cell immobilization in a preformed porous particle (upper left); (2) immobilization by self-aggregation (i.e., a fungal pellet); and (3) immobilization in a gel matrix (lower part).

TABLE 17.2
Gel Formation Mechanisms for Cell Entrapment

Principle of Gelation	Material
Ionotropic gelation	Alginate, chitosan
Thermal gelation	Agar, agarose, κ-carrageenan, collagen, gelatin, gellan gum, curdlan
Precipitation	Cellulose, cellulose triacetate
Polymerization with crosslinking reagent	Polyacrylamide, polymethacrylate, polyacrylamide-hydrazide
Polycondensation	Polyurethane, epoxy resin
Radical-mediated polymerization by irradiation with near-ultraviolet light	Photocrosslinkable resin prepolymers
Crosslinking through photodimerization by irradiation with visible or ultraviolet light	Photosensitive resin prepolymers
Radiation polymerization	Poly(2-hydroxyethyl methacrylate/acrylate), Poly(vinyl alcohol), poly(ethylene glycol) diacrylate/dimethacrylate
Gelation by iterative freezing and thawing or crosslinking with boric acid (and calcium alginate)	Polyvinyl alcohol

cell systems, is oxygen limitations in the matrix. Examples of cell entrapment applications are given in Table 17.3.

17.2.2.1.1 Hydrogels From Natural Polymers

Various natural hydrogels have been used to immobilize living cells. The most important gel materials will be discussed in more detail.

17.2.2.1.1.1 Alginates

Alginates constitute a family of unbranched copolymers of 1,4-linked β-d-mannuronic (M) and α-l-guluronic acid (G) of widely varying composition. The monomers are arranged in a pattern of blocks along the chain, with homopolymeric regions interspersed with regions of alternating structure (MG blocks). Divalent cation-induced gelling of alginates in solution reflects their specific ion binding capacity and the conformational change associated with it.

Entrapment of cells within spherical beads of calcium alginate has become one of the most widely used methods for immobilizing living cells. The success of this method is mainly due to the very mild conditions under which immobilization is performed, and it is a fast, simple, and cost-effective technique. The cell suspension is mixed with a sodium alginate solution, and the mixture is dripped into a solution containing calcium ions. The droplets instantaneously form

TABLE 17.3
Some Applications of Cell Entrapment in Hydrogels

Hydrogel Material	Organism	Application
Bacterial Cells		
Alginate	*Bifidobacterium longum*	Lactic acid production from whey
	Lactococcus lactis	Cell growth and release
	Streptococcus thermophilus	Inoculation of milk
Alginate-whey protein	*Lactococcus lactis*	Nisin production
Cellulose acetate phtalate	*Bifidobacterium pseudolongum*	Probiotics production
NPAE alginate	*Lactobacillus rhamnosus*	Probiotics production
PVA cryogel	*Bacillus agaradhaerens*	β-Cyclodextrin production
Microalgae		
Calcium alginate	*Chlorella vulgaris*	Removal of the pollutants: nitrogen and phosphorus, heavy metals (Cd, Cr, Cu, Fe, Pb, Ni), biocides
κ-Carrageenan	*Scenedesmus acutus*	Removal of heavy metals: cadmium, chromium, zinc
Chitosan	*Scenedesmus bicellaris*	Removal of nitrogen and phosphorus
Fungi		
Calcium alginate	*Saccharomyces cerevisiae*	Ethanol production
Polyethylene oxide	*Candida versatilis* and *Zygosaccharomyces rouxii*	Bioflavor of soy sauce
Mammalian Cells		
Barium alginate	Engineered NIH/3T3 cells	Continuous release of interleukin 12 for cancer therapy
Calcium alginate	Engineered HEK 293 EBNA	Angiostatin release for cancer therapy
	Rat bone marrow cells	Cell proliferation in a 3D scaffold
Gelatin-HPA	MDCK and NIH/3T3 cells	Scaffold and carrier for tissue engineering
RGDS chitosan	Rat osteosarcoma cells	Regeneration of bone-like tissue

Source: Adapted from Willaert, R.G. and Baron, G.V., *Rev Chem Eng*, 12:1–205, 1996; Willaert, R. *Fermentation Microbiology and Biotechnology*, 3rd ed., pp. 313–367, 2012. Boca Raton, FL: CRC Press.

Note: PVA, polyvinyl alcohol; RGDS, Arg-Gly-Asp-Ser; NPAE, N-palmitoylaminoethyl; HPA, hydroxyphenylpropionic acid.

gel spheres, entrapping the cells in a three-dimensional lattice of ionically crosslinked alginate. A major disadvantage of the use of calcium alginate beads is its sensitivity toward chelating agents such as phosphate, citrate, EDTA, and lactate, or antigelling cations such as Na+ or Mg^{2+}. Various ways to overcome this limitation are to keep the beads in a medium containing a few millimolar free calcium ions and keep the Na^+/Ca^{2+} ratio low. Replacing Ca^{2+} with other divalent ions (e.g., $Ba^{2+} > Sr^{2+} > Ca^{2+} >> Mg^{2+}$) or multivalent cations (e.g., Al^{3+} or Ti^{3+}), stabilizes alginate beads. Other stabilization methods involve the use of **polyelectrolytes** [e.g., polyethyleneimine (PEI) and polypropyleneimine (PPI), glutaraldehyde, colloidal silica, propylene glycol ester of alginic acid with PEI, and potassium poly(vinyl alcohol) sulfate] and trimethylammonium glycol chitosan iodide.

> **Polyelectrolyte**: A macromolecular substance that, on dissolving in water or another ionizing solvent, dissociates to give polyions (polycations or polyanions, i.e., multiply charged ions) and an equivalent number of ions of small charge and opposite sign. Polyelectrolytes dissociating into polycations and polyanions with no ions of small charge are also conceivable. A polyelectrolyte can be a polyacid, a polybase, a polysalt, or a polyampholyte.

17.2.2.1.1.2 Carrageenans Carrageenans possess a backbone of alternating 1,3-linked β-d-galactose and 1,4-linked α-d-galactose. Differences in structure arise from the number and location of ester sulfate groups on these sugars and the extent to which the 1,4-linked residues exist as the 3,6-anhydro derivative. Gelation can be achieved by cooling or by contact with a solution containing gel-inducing reagents such as K^+, NH_4^+, Ca^{2+}, Cu^{2+}, Mg^{2+}, Fe^{3+}, amines, and water-miscible organic solvents. Both procedures are easy to perform and facilitate high viable cell content.

There are three main types of carrageenan [i.e., lambda (λ)-, kappa (κ)-, and iota (ι)], all of which are extracted from red seaweeds, with the κ form considered to be the most suitable for cell immobilization. κ-Carrageenan gel can easily be produced in different shapes (bead, cube, membrane) according to the particular application. Bead formation can be accomplished by the dripping technique. The resonance nozzle technique has also been used to produce beads of a more spherical and uniform shape and to scale up the process (Buitelaar et al., 1990).

17.2.2.1.1.3 Agar and Agarose Agar and agarose are polysaccharides isolated from marine red algae. Two kinds of agar can be distinguished by their gelling temperatures and their methoxyl content. Whereas the first type gels in the 40 s (°C) and contains methyl ether groups, the second type gels in the low to mid-30 s (°C) and is essentially devoid of methyl ether groups.

Agarose represents the basic gel-forming component of agar. Agarose is a linear polysaccharide composed of repeating agarobiose units consisting of alternating 1,3-linked β-d-galactopyranose and 1,4-linked

3,6-anhydro-α-l-galactopyranose (i.e., l-galactose anhydride). The other species present in agar include compounds derived from the β-d-galactose by substitution of anhydride in position 6 (e.g., the 6-methyl ether and 6-sulfate) and the compounds obtained by substitution of anhydride in position 2 (the 2-sulfate derivative). Pentagonal pores are the essential feature of agarose supports. The pores are large enough to be readily penetrated by a protein with a molecular mass of several millions. The stability of the pores is dependent on the hydrogen bond formation between the strands of the triple helix of agarose chains. Disruption of these bonds results in dissolving the soluble monomeric agarose. Urea, guanidine hydrochloride, chaotropic agents, and certain detergents may disrupt the hydrogen bonds. The strength of the bonds, and therefore the porosity and size of the beads, are altered by a change in ionic strength.

Agar or agarose (2–5% w/w) is dissolved in a suitable buffer/medium by heating followed by cooling to a temperature 5–10°C above the gelling temperature before mixing or adding the cells. The physical form can be cast as a sheet, bead, or cylinder. A mold can be used or a "gel-block" can be produced with subsequent mechanical disintegration into smaller particles. Dropwise addition of molten preparation to ice-cold buffer lead to the formation of spherical beads (Nilsson et al., 1987) or using the resonance nozzle immobilization technique (Buitelaar et al., 1990).

17.2.2.1.1.4 Chitin and Chitosan Chitin is a fibrous glucan derivative, 1,4-linked 2-deoxy-β-d-glucan, that is partially acetylated, is water insoluble, and contains amino groups. Chitosan is artificially deacetylated chitin that is water soluble with nitrogen contents higher than 7% (w/w). Chitosan can be formed ionotropically in a manner similar to that described earlier for alginate: gel formation will occur using a chitosan solution with a pH value lower than 6 to protonate the amino (NH_2 groups and multivalent anion **counterions**.

> **Counterions**: Ions of low relative molecular mass, with a charge opposite to that of the colloidal ion, are called counterions; if their charge has the same sign as that of the colloidal ion, they are called co-ions.

Crosslinking of chitosan with high-molecular-weight counterions results in capsules, whereas crosslinking with low-molecular-weight counterions results in globules in which the cells are entrapped. Low-molecular-weight ions (e.g., ferricyanide, ferrocyanide, and polyphosphates) and high-molecular-weight ions [e.g., poly(aldehydocarbonic acid), poly(1-hydroxy-1-sulfonate-2-propene), and alginate] can be used. With more hydrophobic counterions (e.g., octyl sulfate, lauryl sulfate, hexadecyl sulfate, and cetylstearyl sulfate) it is also possible to produce hydrophobic gels. Bead formation in a crosslinking solution can also occur at a pH above 7.5, but in this case, it is merely a precipitation of chitosan. At pH values greater than 7.5, chitosan is totally deprotonated and becomes water insoluble.

Combining chitosan with other polymers led to the development of a wide range of tissues such as bone, liver, neural tissue, vascular grafts, cartilage, and skin. Chitosan's ability to stabilize and deliver proteins has allowed for the incorporation of growth factors into many of these structures, thus promoting angiogenesis in tissue-engineered structures. Chitosan ionotropic gel beads are, unlike calcium alginate and potassium carrageenan, stable in phosphate-buffered media, and their mechanical stability is comparable to that of calcium alginate beads.

17.2.2.1.1.5 Pectins, Pectates, and Pectinates Pectins are acidic, structural polysaccharides that are present in the cell wall of plant cells. Although pectins are branched in their native form, when extracted they are predominantly linear polymers that are based on a 1,4-linked α-d-galacturonate backbone that is randomly interrupted by 1,2-linked l-rhamnose. Like alginate, pectin is based on a diaxially linked backbone and forms gels with calcium ions. Gel formation studies indicate a very strong binding of these counterions. Because of the variability of the chemical structure of the commercially available pectins, gels can be formed in several ways. Polygalacturonic acid as the principal constituent is partly esterified with methoxyl groups. The free acid groups may be partly or fully neutralized with monovalent ions (i.e., Na^+, K^+, NH_4^+). The degree of methoxylation (DM) has an essential influence on the properties of pectin, especially on its solubility and its requirements for gelation, which are directly derived from the solubility. Pectins with a DM lower than 5% are called "pectic acids," whereas those with higher DM value are called "pectinic acids." Pectinic acids with a DM greater than 50% are called "high-methoxyl (HM) pectins," and those with lower DM value are called "low-methoxyl (LM) pectins."

The ionotropic gelation of pectate is simple, mild, and inexpensive. Dripping the polymercell solution into a crosslinking solution can easily produce beads with calcium or aluminum as counterions. Calcium pectate and aluminum pectate beads are much less sensitive to small mono- and multivalent anions [i.e., citrate, phosphate, lactate, gluconate, and chloride, and Ca^{2+} (Al^{3+}) complexing agents], which diminish the stability of the beads. Stabilization and hardening of pectate beads can be performed using a treatment with polyethyleneimine followed by glutaraldehyde (Gemeiner et al., 1994).

17.2.2.1.1.6 Gellan Gum Gellan gum is a gel-forming polysaccharide secreted by the bacterium *Pseudomonas elodea* and is produced via aerobic fermentation. Gellan gum is a linear, anionic heteropolysaccharide with tetrasaccharide repeating units consisting of two β-d-glucose, one β-d-guluronic acid, and one α-l-rhamnose residue. The polymer contains approximately 1.5 acyl substituents per tetrasaccharide repeating unit. These substituents have been identified as an l-glyceric ester on C2 of the 3-linked d-glucose and an acetic ester on C6 of the same glucose residue. The presence of these substituents, in particular the bulky glycerate groups, hinders chain association and accounts for the change in gel texture brought about by de-esterification. The substituted form produces soft, elastic, and cohesive gels, whereas hard, firm, and brittle gels are obtained from the unsubstituted

form. The rheological properties of unsubstituted gels are superior to those of other common polysaccharides such as agar, κ-carrageenan, and alginate at equimolar concentrations (Sanderson et al., 1989).

Gelation initially occurs by the formation of double helices followed by ion-induced association of these helices. Gel formation occurs when the fibrils associate in the presence of gel-promoting cations. Gelation depends upon the gum concentration, ionic strength, and the type of stabilizing cation (divalent are more effective than monovalent cations). A dispersion of gellan gum with a minimum amount of sequestrant and divalent cations will form a coherent, demoldable gel on cooling to ambient temperature. The gelation temperature increases from 35 to 55°C with increasing cation concentration. Gellan gum is suitable for the immobilization of thermophilic bacteria because of its high setting temperature (>50°C), which can be decreased by the addition of sequestrants such as citrate, metaphosphate, and EDTA.

17.2.2.1.1.7 Hyaluronic Acid Hyaluronic acid is the largest glycosaminoglycan (GAG) found in nature. It is well suited for tissue engineering because it shows a minimal inflammatory or foreign body reaction upon implantation (Gutowska et al., 2001).

17.2.2.1.1.8 Collagen Collagens are a family of highly characteristic fibrous proteins found in all multicellular animals. The central feature of all collagen molecules is their stiff, triple-stranded helical structure. Collagen chains are extremely rich in glycine and proline, both of which are important in the formation of the stable triple helix. Collagen is hydrophilic; it swells in the presence of water, is soluble at low pH, and is insoluble at high pH values.

The mechanism of cell immobilization in collagen involves the formation of multiple ionic interactions, hydrogen bonds, and van der Waals forces between the cells and collagen. Preparation of the collagen solution and mixing with cells has to be performed at low temperatures (4°C). Gelification is accomplished by raising the pH and ionic strength of the collagen solution and exposure to 37°C. Its natural ability to bind cells makes it a promising material for controlling cellular distribution within immunoisolated devices, and its enzymatic degradation can provide appropriate degradation kinetics for tissue regeneration in micro- and macroporous scaffolds (Riddle and Mooney, 2004). Although collagen is widely used in cell immobilization, it is expensive to purify for use in tissue engineering.

17.2.2.1.1.9 Gelatine Gelatine is a hydrolytic derivative of collagen. Gelification is accomplished by cooling the gelatine solution below a temperature of 30–35°C. The sol-gel transformation is reversible. Stability of the gel structure can be achieved by adding organic (e.g., glutaraldehyde or formaldehyde) or inorganic (e.g., acetate or sulfate chromium salts) compounds (Sungur and Akbulut, 1994). The three-dimensional structure of gelatin is formed by secondary interactions between the polypeptide chains. These interactions are broken upon heating. The stabilization by aldehydes is based on covalent bond formations between the gelatine strands.

In a typical procedure, the cell suspension is mixed with an aqueous gelatin solution at 40°C, the suspension is cooled, and the gel is lyophilized. Subsequently, the dry preparation is disintegrated into small particles. The glutaraldehyde treatment can be performed before the cooling of the suspension or after the lyophylization process. Dripping the hot suspension into hydrophobic liquid such as butyl acetate leads to the formation of uniform beads.

17.2.2.1.2 Hydrogels from Synthetic Polymers

The application of synthetic polymers for the immobilization of living cells has some interesting benefits compared with the use of natural polymers because synthetic polymers of adequate properties can be easily and artificially designed. The porosity of the gel as well as the ionic and hydrophobic or hydrophilic properties can be easily adjusted. Additionally, the mechanical strength and longevity of the gels formed from synthetic polymers are generally superior to those from natural polymers.

17.2.2.1.2.1 Polyacrylamide

The first synthetic gel used to entrap living microbial cells was polyacrylamide. Entrapment is performed by the polymerization of an aqueous solution of acrylamide monomers in which the cells are suspended. Polymerization of polyacrylamide is a free radical process in which linear chains of polyacrylamide are crosslinked by inclusion of a bifunctional reagent (e.g., N,N'-methylene-bisacrylamide). The crosslinking degree is a function of the relative amounts of acrylamide and bifunctional reagent and determines the porosity and fragility of the gel. Either chemical or a photochemical reaction can be used to initiate of the free radical polymerization. Persulfate and β-dimethylaminopropionitrile or N,N,N'N'-tetramethylenediamine (TEMED) are chemical catalysts, whereas sodium hydrosulfite, riboflavin, and TEMED act as photochemical polymerization initiators.

It is an easy immobilization technique, but the polymerization of the acrylamide monomers in the presence of viable cells usually results in a reduction of the viability of the entrapped cells because of the toxicity of the monomers (e.g., acrylamide and bisacrylamide) and the heat evolved during polymerization. The level of the "immobilization shock" of polyacrylamide-entrapped cells depends to a large extent on the initial physiological state of the population.

17.2.2.1.2.2 Polyacrylamide Hydrazide

To eliminate the unfavorable influences caused by the acrylamide monomer, techniques have been developed that use prepolymerized linear polyacrylamides, partially substituted with acylhydrazide groups, for the entrapment of living cells with good retention of viability. The prepolymerized material is crosslinked in the presence of viable cells by the addition of controlled amounts of dialdehydes [glyoxal, glutaraldehyde, periodate-oxidized poly(vinyl alcohol)]. The crosslinking agent affects the porosity of these gels. The best results were obtained using glyoxal.

The concentration of polymeric backbone also affects gel porosity. The mechanical stability of this gel is superior to gels made with similar concentrations of polymeric backbone from acrylamide-bisacrylamide copolymerization. This polyacrylamide hydrazide (PAAH) gel is less brittle, is chemically stable, and does not undergo deformation as a result of changes in salinity or pH.

17.2.2.1.2.3 Methacrylates

The preparation of methacrylate gels is analogous to that of polyacrylamide gels. Methacrylate monomers such as methylacrylamide, hydroxyethylmethacrylate, or methylmethacrylate are polymerized in the presence of a crosslinking agent (e.g., tetraethyleneglycol dimethacrylate) to form a porous gel.

Poly(ethylene glycol) dimethacrylate has also been used as a crosslinking agent for the entrapment of biocatalysts by radical polymerization of acrylic acid and N,N-dimethylaminoethyl methacrylate. Living cells can also be immobilized in methacrylate by γ-ray irradiation at low temperatures (Carenza and Veronese, 1994).

17.2.2.1.2.4 Photocrosslinkable Resin Prepolymers

Illumination with near-ultraviolet light of photocrosslinkable resin prepolymers initiates radical polymerization of the prepolymers and completes gel formation within 3–5 min. Various types of prepolymers possessing photosensitive functional groups have been developed. Poly(ethylene glycol) dimethacrylate (PEGM) was synthesized from poly(ethylene glycol) (PEG) and methacrylate. Hydrophylic (ENT) and hydrophobic (ENTP) photocrosslinkable prepolymer resins were prepared from hydroxyethylacrylate, isophrone diisocyanate, and PEG or poly(propylene glycol), respectively. Each prepolymer has a linear skeleton of optional length at both terminals, to which are attached the photosensitive functional groups such as acryloyl or methacroyl. PEGM and ENT containing PEG as the main skeleton are water soluble and give hydrophilic gels, whereas ENTP with poly(propylene glycol) as the main skeleton is water insoluble and forms hydrophobic gels. Using PEG or poly(propylene glycol) of different molecular weight, prepolymers of different chain lengths can be prepared: from PEGM-1000 to PEGM-4000 (molecular weight of main chain from ~1000 to 4000, respectively), ENT-1000 to ENT-6000, and ENTP-1000 to ENTP-4000. The chain length of the prepolymers corresponds to the size of the network of gels formed from these prepolymers. Anionic and cationic prepolymers can also be prepared by introducing anionic and cationic functional group(s) to the main skeleton of the prepolymers.

The entrapment of cells can be achieved by illumination of a mixture consisting of a prepolymer, a photosensitizer (e.g., benzoin ethyl ether or benzoin isobutyl ether), and the cell suspension. A suitable buffer is used for the hydrophilic prepolymer and an adequate organic solvent for the hydrophobic prepolymer. Suitable mixtures of these two types of prepolymers can also be used. In some cases, a detergent is used to mix a hydrophobic prepolymer with the suspension of biocatalysts.

Laser-induced photopolymerization of PEG diacrylates and multiacrylates has been used to entrap living mammalian cells. These molecules of various molecular weights were synthesized by reaction of PEG with acryloyl chloride using triethylamine as a proton acceptor. Tetrahydroxy-PEG was used for the multiacrylate synthesis. The advantages of this entrapment method are that the laser light is not absorbed by the cells in the absence of an exogenous, cell-binding chromophore; there is no significant heat of polymerization because of the nature, size, and dilution of the macromers used; the polymerization can proceed extremely rapidly in oxygen-containing aqueous environments at physiological pH; and gels with the proper formulation are capable of being immunoprotective and can be used for cell therapy purposes.

Synthetic resin prepolymers have the following advantages as gel-forming starting materials:

- Entrapment procedures are very simple under very mild conditions.
- Prepolymers do not contain monomers that have adverse effects on the biocatalysts to be entrapped.
- The network structure of gels can be controlled by using prepolymers of optional chain length.
- Optional physicochemical properties of gels (e.g., hydrophobicity-hydrophilicity balance and ionic nature) can be changed by selecting suitable prepolymers that were synthesized in advance in the absence of biocatalysts.

17.2.2.1.2.5 Polyurethane Urethane prepolymers have isocyanate groups at both terminals of the linear chain and are synthesized by heating for 1–2 h at 80°C from toluene diisocyanate and polyether diols composed of PEG and poly(propylene glycol) or PEG alone. Prepolymers with a different hydrophilic or hydrophobic character can be obtained by changing the ratio of PEG and poly(propylene glycol) in the polyether diol moiety of the prepolymers. The chain length and the content of isocyanate group can also be changed.

The "self-crosslinking" gel can entrap cells. The prepolymers are water miscible and when a liquid prepolymer is mixed with an aqueous cell suspension, the isocyanate functional groups at both terminals of the molecule react with each other only in the presence of water, forming urea linkages with liberation of carbon dioxide. Cells have also been entrapped in conventional polyurethanes, which were obtained by polycondensation of polyisocyanates. The polyurethane can be made in foam or in a gel structure depending on the type and concentration of the polycyanate used.

17.2.2.1.2.6 Poly (vinyl alcohol) Poly(vinyl alcohol) (PVA) is a raw material of vinylon and is a low-cost material. PVA is nontoxic to microorganisms and, consequently, can be used to entrap living cells. A PVA solution becomes gelatinous by freezing, and the gel strength increases during iterations of freezing and thawing. Using this technique, a rubber-like, elastic hydrogel can be obtained without using any chemical reagent. The gel strength increases with iteration number of freezing-thawing until seven iterations. Adding cryoprotectants such as glycerol and skim milk to the PVA-cell solution can prevent decrease of activity due to the freezing and thawing. PVA cryogels can be used up to a temperature of 65°C and have also been used to immobilize thermophilic microorganisms like *Clostridium thermocellum*, *Clostridium thermosaccharolyticum*, and *Clostridium thermoautotrophicum* to perform fermentations at 60°C (Varfolomeyev et al., 1990).

Elastic PVA gels with a high strength and durability can also be formed by crosslinking PVA with a boric acid solution to produce a monodiol-type PVA-boric acid gel lattice. This technique has two potential problems. First, the saturated boric acid solution used to crosslink the PVA is highly acidic and could cause difficulty in maintaining cell viability. Second, PVA is an extremely sticky material. As a result, PVA beads have a tendency to agglomerate, which can cause problems in fluidized-bed reactors. Such a problem can be avoided by adding a combination of PVA-boric acid and a small amount of calcium alginate (0.02%) (Wu and Wisecarver, 1992).

Chen and Lin (1994) developed a method based on the usage of phosphorylated PVA, in which PVA was first crosslinked with boric acid for a short time to form a spherical structure, which was followed by solidification of the gel beads by esterification of PVA with phosphate. The short contact time with boric acid prevented severe damage to the entrapped microorganisms.

PVA has been used to immobilize cells in Lentikats®, which are lens-shaped hydrogel particles (Krasňan et al., 2016). Because of their lenticular shape (diameter of 3–4 mm and a thickness of 200–400 μm), Lentikats have the advantage of improved mass transport compared with beads with the same diameter. They are prepared by mixing a sol solution with the biocatalyst solution, and small droplets are floored on a suitable surface where the gelation takes place. These particles have also been produced on an industrial scale (production output between 2 and 50 kg/h) using a conveyer-belt system.

17.2.2.1.2.7 Photosensitive Resin Prepolymers Photosensitive resin prepolymers are derivatives of PVA introduced by styrylpyridinium (SbQ) groups as photosensitive sites and are polymerized by photodimerization with irradiation of visible or ultraviolet light (PVA-SbQ gel). Interestingly, changing the saponification degree of PVA can control hydrophilicity of the prepolymers.

17.2.2.1.2.8 Radiation Polymers Living cells can be entrapped by γ-irradiation of a wide variety of functional monomers or prepolymers at low temperature. Utilization of this method is limited because radiation equipment is required. Irradiation of cells at low temperatures is necessary to avoid radiation damage. The rigidity of the polymer matrix can be increased and the porosity decreased by increasing the monomer concentration.

17.2.2.1.2.9 Stimuli-sensitive Hydrogels Stimuli-sensitive polymer hydrogels, which swell or shrink in response to changes in environmental conditions, have been extensively investigated and used as "smart" biomaterials and drug-delivery systems [e.g., copoly(N-isopropylacrylamide/acrylamide) (NIAAm/ AAm)], which is a "lower critical solution temperature" (LCST) hydrogel (Hoffman, 2004). This gel gradually shrinks as temperature is raised and then collapses when it is warmed through the LCST region. It expands and reswells as it is cooled below LCST. Cells containing gel beads have been prepared by inverse suspension polymerization. The conversion and activity of the immobilized cells may be enhanced by thermal cycling of the gel below its LCST because of the reduced mass transfer resistance as the gel bead "squeezes out" and "draws in" substrate when it shrinks and swells during the thermal cycling. Pore sizes and their interconnections will change significantly as the gel is shrunk or swelled. This can significantly affect the diffusion rates of substrate in and product out of the gel.

17.2.2.2 Preformed Support Materials

Cell immobilization in preformed carriers involves passive/ natural immobilization usually *in situ* in the bioreactor or the culture environment. Most of the carriers are porous with a wide range of pore sizes to suit immobilization of various organisms or tissues. For passive immobilization, cells, flocs, mycelia, cell aggregates, or spores are inoculated into the sterilized medium containing empty preformed carriers. Depending on the cell and the carrier type, immobilization then takes place in a combination of filtration, adsorption, growth, and colonization processes. Furthermore, surfaces of the carriers can be modified by various pretreatments to enhance immobilization efficiency.

The cells are entrapped in a matrix that protects them from the shear field outside of the particles. This is of particular importance for fragile cells such as mammalian cells. Unlike gel systems, porous supports can be inoculated directly from the bulk medium. As with the adsorption method, cells are not completely separated from the effluent in these systems. Mass transport of substrates and products can be achieved by molecular diffusion and convection by proper particle design and organization of external flow. Consequently, mass transport limitations are less severe under optimal conditions. When the colonized porous matrix ideally retains some free space for flow, immobilization occurs partly by attachment to the internal surface, self-aggregation, and retention in dead-end pockets within the material. This is only possible when cell adhesion is not very strong and the application of high external flow rates reversibly removes cells from the matrix. When high cell densities are obtained, convection is no longer possible and the cell system behaves as dense cell agglomerates with strong diffusion limitations. Cell immobilization methods are simple, and a high degree of cell viability is retained upon entrapment. The preformed matrix is chemically inert, resistant to microbial attack, and is incompressible. Steam sterilization is often possible and the matrix can be reused. Usually, the matrix takes up a significant volume fraction resulting in a lower immobilized cell density compared with other immobilization methods.

Various porous matrices have been described for living cell immobilization as shown in Table 17.4. The choice will usually depend on the cell type used and the kind of application. For example, the immobilization of microbial cells for the production of biochemicals in large packed-bed bioreactors requires a matrix with excellent mechanical characteristics to withstand the high-pressure drop in the reactor; tissue engineering porous matrices need to be an excellent scaffold for cell attachment, and growth and must be characterized by excellent biocompatibility characteristics. Usually, guided by the following parameters, a choice has to be made from several suitable matrices.

- What matrix material is the cheapest?
- Which matrix is the most suitable for our purpose and is it not patented?
- Is it reusable?

In tissue engineering, some applications require porous biodegradable scaffolds. The repair of large cartilage defects requires the use of a three-dimensional scaffold to provide a structure for cell proliferation and control the shape of the regenerated tissue. For example, cartilage implants based on chondrocytes and three-dimensional fibrous polyglycolic acid scaffolds closely resembled normal cartilage histologically as well as with respect to cell density and tissue composition (Freed et al., 1994).

17.2.3 CONTAINMENT BEHIND A BARRIER

When cell separation from the effluent is required or when some high-molecular-weight or specific product (permaselectivity) needs to be separated from the effluent, these systems are highly useful. The barrier can be preformed (hollow fiber systems and flat membrane reactors) or formed around the cells to be immobilized (microcapsules and two-phase entrapment). The synthetic membranes are usually polymeric microfiltration or ultrafiltration membranes, although other types of membranes have been used, such as ceramic, silicone rubber, or ion exchange membranes. Mass transfer through the membrane is not only dependent on the pore size and structure but also on the hydrophobicity/hydrophilicity and charge. Transport can be by diffusion and/or by flow induced by application of a pressure difference over the membrane. Various mild micro-encapsulation methods have been developed to entrap living cells, and some are combined with gel immobilization within the microcapsule. The barrier can be as simple as the liquid/liquid phase interface between two immiscible fluids.

Entrapment behind preformed membranes represents a gentle immobilization method because no chemical agents or harsh conditions are used. Cells are often immobilized by filtration of a cell suspension followed by some growth in the seeded reactor. Two-phase systems can be used in applications where substrates or products are partitioned separately (product inhibition, water-insoluble substrates). These systems also allow for the recycling of the cell-containing phase, which is difficult with other immobilization methods. The small spheres involved in phase entrapment have

TABLE 17.4

Some Examples of Cell Immobilization in Preformed Porous Support Materials

Material	Microorganism Natural Organic Polymers	Application/Product
Bacteria		
Cotton towel	*Propionibacterium acidipropionici*	Propionic acid production
Microalgae		
Luffa sponge	Chlorella sorokiniana	Removal of heavy metals (Cd, Cr, Ni)
Fungi		
Cellulose carrier	*Rhizopus niveus*	Wax ester production
Luffa sponge	*Saccharomyces cerevisiae*	Ethanol production
Synthetic Organic Polymers		
Bacteria		
Silicone carrier (ImmobSil®)	*Lactobacillus rhamnosus*	Exopolysaccharide production
Fungi		
Polystyrene foam	*Phanerochaete chrysosporium*	Peroxidase production
Polyurethane foam	*Agaricus* sp.	Laccacase production
	Aspergillus niger	Citric acid production
	Yarrowia lipolytica	Oil degradation
Mammalian cell		
Polyester fibrous matrix	Hybridoma cells	Monoclonal antibody production
Polyethylene terephthalate	Human trophoblast	Tissue engineering
Anorganic Materials		
Bacteria		
Kiezelguhr (Celite)	*Xanthomonas campestris*	Xanthan gum production
Fungi		
Ceramics	*Saccharomyces cerevisiae*	Ethanol production
Kiezelguhr (Celite)	*Penicillium chrysogenum*	Penicillin production
Porous glass	*Saccharomyces cerevisiae*	Beer maturation
Metallics		
Bacteria		
Alumina pellets	*Zymomonas mobilis*	Ethanol production
Stainless steel knitted mesh	*Zymomonas mobilis*	Levan and ethanol production
Fungi		
Stainless steel fiber cloth	*Saccharomyces cerevisiae*	Beer production
Stainless steel knitted mesh	*Phanerochaete chrysosporium*	Peroxidase production

Source: Adapted from Willaert, R.G. and Baron, G.V., *Rev Chem Eng*, 12:1–205, 1996; Baron, G.V. and Willaert, R.G., *Fundamentals of Cell Immobilization Biotechnology*, pp. 229–44, 2004. Dordrecht, The Netherlands: Kluwer Academic Publishers; and Willaert, R. *Fermentation Microbiology and Biotechnology*, 3rd ed., pp. 313–367, 2012. Boca Raton, FL: CRC Press.

a superior surface-to-volume ratio compared with flat sheets or hollow fibers and can be used in conventional bioreactors. Also, membranes have been used to contact aqueous and organic streams in bioprocesses. The high membrane surface area in hollow fiber reactors is especially advantageous for the culture of anchorage-dependent mammalian cells, without the drawbacks occurring with immobilized growing microorganisms. The maintenance of high-density cultures of animal and plant cells in membrane reactors resulted in long-term production with reduced cell growth. Membrane entrapment may be particularly helpful if aggregation is an advantage, as suggested for plant cells. In the biomedical engineering field, microencapsulation or macroencapsulation lead to immunoprotection for artificial organs. In most cases, nutrients are supplied to, and products are removed from, the cell mass by diffusion. Consequently, mass transfer limitations can reduce the efficiency of these systems. However, in the case of hollow fiber systems, mass transfer can be governed by convective mass transport, otherwise known as "**Starling flow**."

Starling flow: In a hollow fiber reactor, some fluid will flow from the lumen into the extracapillary space (ECS) in the entrance half of the reactor, along the fibers in the ECS, and return to the lumen in the exit half of the reactor because of the pressure gradient in the lumen (highest pressure at the entrance). This type of flow is called Starling (or toroidal) flow, in honor of the discoverer of this same fluid behavior in tissues surrounding blood capillaries.

In these systems, the cells are usually situated in the membrane porous support layer or the shell space and medium flows in the lumen. Alternatively, the cells can be contained in the lumen space in the case of macroencapsulation (also known as "diffusion chambers") for encapsulated cell therapy such as for diabetes, alleviation of chronic pain, treatment of neurodegenerative disorders, and delivery of neurotrophic factors. Here, the hollow fiber membrane allows for the diffusion of low-molecular-weight molecules (lower than the molecular weight cutoff value)—such as oxygen, glucose, nutrients, and waste product—but prevents the passage of larger molecules such as antibodies.

Molecular weight cutoff: The molecular weight cutoff (MWCO) of a membrane is defined as the molecular weight at which the membrane rejects 90% of solute.

17.2.3.1 Microencapsulation

A typical microencapsulation process involves the formation of a spherical gel mold containing cells on which is deposited a polymeric membrane. The internal gel matrix can also be liquefied and allowed to diffuse out of the capsule, leaving behind the membrane and the contained cells. The type and porosity of the membrane and size of the microcapsules can be varied to accommodate many reactant-product systems. Capsule diameters from 20 μm to 2 mm are possible. The porosity of the membrane can be varied over a range of several orders of magnitude; from glucose (180 Da) to IgG (155,000 Da) can be made to be freely permeable. The polymer membrane should offer minimal resistance to the mass transfer of essential molecules as well as the toxic end products of cell metabolism in order for the encapsulated mass to maintain normal physiological activity. For **immunoisolation** purposes, the capsule must be impermeable to host cells and soluble components of the immune system (Song and Roy, 2016).

Immunoisolation: By using membranes, the free passage of immunoglobulins and complement proteins can be prevented from interacting with implanted "foreign" biological material.

The demand for specific properties of the capsules vary; for example, proliferating cells need a stronger capsule membrane than nonproliferating cells, but it is noteworthy that the former may tolerate a tougher encapsulation procedure because viable cells will grow and replace the dead cells in the capsules. This immobilization technique is of particular interest for the immobilization of animal cells. Large process intensification over conventional cell suspension culture with low density and low productivity can be achieved. Cell encapsulation and long-term continuous culture lead to significantly higher cell densities, which results in higher productivities.

The high-culture densities provide a high degree of cell-cell contact and interaction, resulting in more favorable possible microenvironmental conditions. In addition, this technique can provide protection from shear for the sensitive animal cells and other sudden changes in the culture medium. It also permits direct aeration by air bubbles without risk of damaging the cells. The produced toxic metabolites, such as lactic acid and ammonium, will diffuse out of the capsule because of the concentration gradient, resulting in higher growth and product formation rates. Microencapsulation can provide simultaneous product separation and cell cultivation, resulting in concentration of high-molecular-weight metabolic products (e.g., monoclonal antibodies) within the capsule.

Microencapsulation of human cells or tissues is a recent technology to overcome biomedical problems because the membrane may create an immunological barrier between the host and the transplanted cells. The immunoisolation of the encapsulated cells or tissue from the elements of the immune system prevents the rejection of the transplanted cells/tissue. Consequently, the necessity of immunosuppressive drugs in allo- and xenotransplantations can be avoided. The idea of using an ultrathin polymer membrane for the immunoprotection of transplanted cells, despite its apparent promise in preclinical trials, failed to live up to expectations and continues to elude the scientific community.

17.2.3.1.1 Microencapsulation Techniques

The techniques that are used to produce the semipermeable microcapsule membranes are classified as phase inversion, polyelectroyte coacervation, and interfacial precipitation. Phase inversion involves the induction of phase separation in a previously homogeneous polymer solution by a temperature change or by exposing the solution to a nonsolvent component in a bath (wet process) or in a saturated atmosphere (dry process). Polymer precipitation time, polymer-diluent compatibility, and diluent concentration all influence phase separation and membrane porosity. Examples of polymer materials are polyacrylate and poly(hydroxyethylmet-hacrylate-co-methyl methacrylate).

In the polyelectrolyte coacervation process, a hydrogel membrane is formed by the complexation of oppositely charged polymers to yield an interpenetrating network. Mass transport characteristics can be modulated by osmotic conditions, diluents, and the molecular-weight distribution of the polyionic species. To reduce membrane permeability and to improve biocompatibility as well as mechanical properties, one or more additional coating layers with oppositely charged polymer can be added.

The interfacial precipitation technique involves the coating of a hydrogel bead with a semipermeable membrane. A lot of research has been focused on the alginate-poly-l-lysine (PLL)-PEI microcapsule, originally developed by Lim and coworkers in the early 1980s (Lim and Sun, 1980). Immunoisolation of the islets of Langerhans (bioartificial pancreas) has been studied extensively and evaluated thoroughly. Examples of applications of cell immobilization in microcapsules are listed in Table 17.5.

TABLE 17.5

Cell Immobilization in Microcapsules: Some Examples

Material	Cell Type Bacterial Cells	Application
Alginate-PLL-alginate	*Escherichia coli*	Urea and ammonia removal
Alginate-PLL-alginate	*Erwinia herbicola*	Tyrosine production
Alginate-PLL-alginate	Various bacteria	Therapeutic delivery of live bacteria
Alginate-alginate	*Lactococcus lactis*	Bacteriocin production
Cellulose acetate phthalate	*Lactobacillus acidophilus* + *Bifidobacterium lactis*	Probiotics
	Fungal Cells	
1,6-hexanediamine-poly-(allylamone) -dodecane-dioyl dichloride	*Saccharomyces cerevisiae*	Bioconversion in organic solvents
CS^c-PDMDAAC	*Yarrowia lipolytica*	Citrate production
	Animal Cells	
Alginate/agarose-PLL-alginate	BHK fibroblast, C_2Q_{12} myoblast	Viability assessment
Alginate-PLL-alginate	Engineered mouse myoblast	Tumor suppression
Alginate-PLL-alginate	Engineered C_2C_{12} myoblast to secrete erythropoietin	*In vivo* erythropoietin delivery
Alginate-PLL-alginate	Engineered 293-EBNA JN3 myeloma	Endostatin production Hepatocyte growth factor production
Alginate-PLL-alginate	Islets of Langerhans	Diabetes treatment
Alginate-PLL-alginate	Murine fibroblast	High throughput GMP encapsulation using JetCutter technology
Alginate/agarose/cellulose sulfate/ pectin-PLL-alginate	GDNF secreting 3T3 fibroblast	Treatment central nervous system diseases
Alginate-poly-L-ornithine-alginate	HEK 293, HCT 116 and HEP G2 Cell Spheroids	Cell implantation
Alginate-CS-pDADMAC	Engineered CHO	Erytropoietine production
Alginate-PMCG	Islets of Langerhans	Diabetes treatment
Collagen-HEMA-MMA-MAA- MeOCPMA	Rat hepatocytes	Bioartificial liver

Source: Adapted from Willaert, R.G. and Baron, G.V., *Rev Chem Eng*, 12:1–205, 1996; Willaert, R. *Fermentation Microbiology and Biotechnology*, 3rd ed., pp. 313–367, 2012. Boca Raton, FL: CRC Press.

Note: GDNF, glial cell line–derived neurotrophic factor; CS, cellulose sulfate; pDADMAC, poly-diallyl-dimethyl-ammoniumchloride; HEMA-MMA-MAA-MeOCPMA, hydroxyethylmethacrylate-methacrylate-methacy lic acid-4-(4-methoxycinnamoyl)phenyl methacrylate; PDMDAAC, poly(dimethyldiallylammonium chloride); PMCG, polymethyl-co-guanidine.

17.2.3.2 Cell Immobilization Using Membranes

The main advantage of membrane bioreactors is that they provide simultaneous bioconversion and product separation, which is especially attractive for the production of high-value biological molecules. As compared with conventional reactor types, the design of membrane reactors is relatively more complex and more expensive (mainly because of the high cost of the membrane material). However, because the attainment of highly concentrated products could eliminate the need for some steps of costly product purification, the utilization of these reactors can be favorable. For low-value biological products, conventional bioreactors are usually more appropriate. An exception is the case in which cofactors need to be co-entrapped to perform coenzyme-dependent bioconversion reactions.

Membrane reactors can be configured as flat-sheet or hollow-fiber modules. Hollow-fiber modules provide a higher surface-to-volume ratio without the need for membrane support. However, the geometry of flat-sheet modules is simpler, providing an accurate regulation of the distances between the membranes. Additionally, these modules can be easily disassembled, providing an easy access to module compartments and options for membrane cleaning and replacement.

Membranes can be used for three types of cell immobilization as illustrated in panels A, B, and C of Figure 17.4

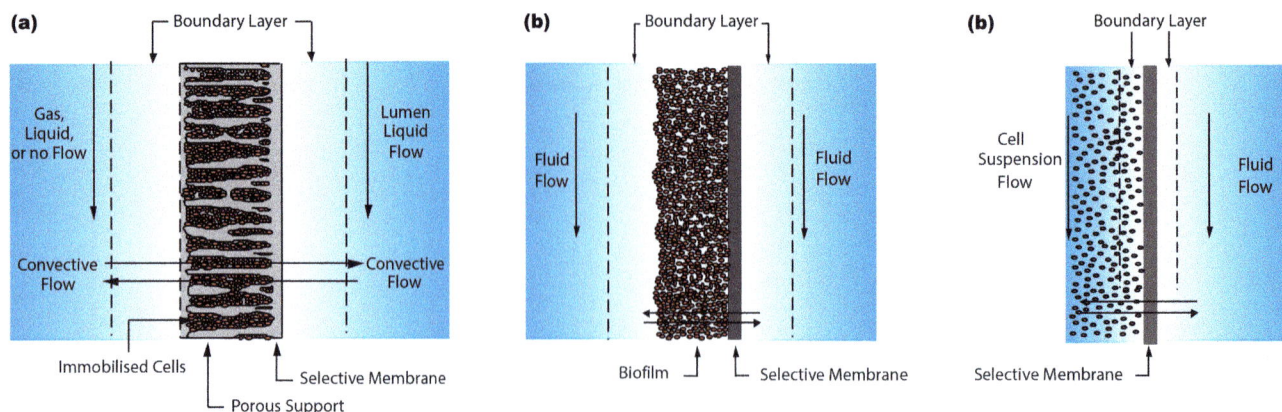

FIGURE 17.4 Synthetic membranes for cell immobilization: (a) immobilization in the porous structure of an asymmetric membrane; (b) attachment of a biofilm on a permeable membrane; and (c) a membrane to separate the cell compartment from a cell-free liquid compartment.

(i.e., immobilization within the membrane, immobilization on the membrane where a biofilm is formed, and immobilization in a cell compartment, which is separated by a membrane).

17.2.3.2.1 Immobilization within the Membrane

Cell immobilization in a membrane can give very high cell densities. The permaselective membrane permits the transport of nutrients from the bulk medium to the cells and the removal of products, whereas the release of cells into the bulk liquid is prevented. Various setups can be used, and depending on the membrane type, even convective transport through the membrane is possible. Preformed asymmetric polymeric membranes are usually used. A hollow fiber module consists of a bundle of porous hollow-fiber membranes potted at both ends in a cylindrical module. They were originally developed for separation processes.

17.2.3.2.2 Immobilization on the Membrane

Membranes can act as support for biofilm development with direct oxygen transfer through the membrane wall in one direction and nutrient diffusion from the bulk liquid phase into the biofilm in the other direction. This type of immobilization has successfully been used in aerobic cultivation of mammalian and microbial cells.

17.2.3.2.3 Immobilization in a Cell Compartment

In this case, a high suspended cell concentration is obtained by preventing the cells from escaping from the bioreactor. Membrane filters, which can be positioned internally (e.g., a spin filter for the continuous cultivation of mammalian cells) or externally, are used to keep the cells in the reactor. Alternatives for an external membrane filter are a centrifuge or a settling tank (Table 17.6).

TABLE 17.6
Cell Immobilization Using a Membrane as a Barrier: Some Examples

Material	Cell Type	Application
Immobilization within the Membrane		
Polypropylene	Bovine aortic endothelial cells	Artificial lung
Polysulfone	*Saccharomyces cerevisiae*	Ethanol production
Polyvinyl chloride/polyvinylidine chloride/polyacrylonitrile	*Escherichia coli*	Production of β-lactamase
Silicon carbide	*Saccharomyces cerevisiae*	Beer production
Immobilization on the Membrane		
Polypropylene	*Aspergillus niger*	Citric acid production
Polylysine-coated polysulfone	H1 fibroblast	Aerobic cell growth
Immobilization in a Cell Compartment (Cell Recycle)		
Ceramic	*Bacillus stearothermophilus*	Lactic acid production
Polydimethylsiloxane layer on polysulfone	*Saccharomyces cerevisiae*	Ethanol production
Polyethersulfone	*Leuconostoc mesenteroides*	d-mannitol production
Polysulfone	*Halobacterium halobium*	Bacteriorhodopsin production
	Lactobacillus rhamnosus	Lactic acid production

Source: Adapted from Willaert, R.G. and Baron, G.V., *Rev Chem Eng*, 12:1–205, 1996; Willaert, R. *Fermentation Microbiology and Biotechnology*, 3rd ed., pp. 313–367, 2012. Boca Raton, FL: CRC Press.

Very high cell densities (and high productivities) have been reported for membrane-based cell recycle systems. These systems have also the advantage of being homogeneous. Although none of the intrinsic benefits of cell immobilization are obtained (e.g., shear protection), most other advantages such as high productivity, avoiding wash-out, and simpler product recovery are retained. Some of the drawbacks of immobilization, and especially the substrate transport limitation or product inhibition, can be avoided or reduced. Oxygen transport can be a major problem at high cell densities and limits the attainable cell densities.

17.2.4 Self-Aggregation of Cells

Cells that naturally aggregate, clump, form pellets, or flocculate can also be considered as immobilized (Figure 17.3). Many industrially important products (e.g., antibiotics) are produced during secondary metabolism by fungal pellets. Microbial aggregates can be encountered in wine-making and brewing, where yeast cells flocculate at the end of fermentation. The culturing of algae, plant cells, and animal cells can also result in aggregation phenomena. Simple serum-free medium appeared to be adequate to support the growth of anchorage-dependent animal aggregates of several commercially important cell types such as African Green Monkey Cells (Vero), Baby Hamster Kidney (BHK) cells, and Chinese Hamster Ovary (CHO) cells. During callus culture—in the absence of shear fields—aggregates may reach several centimeters across. Consequently, plant cell aggregates are very susceptible to the hydrodynamic conditions in the bioreactor.

The cell-wall region is directly and indirectly influenced by biological and environmental factors through metabolism. Biological factors that affect microbial aggregation are the cell wall, extracellular secretions, genetics, growth rate, nutrition, and physiological age. Environmental factors can be subdivided into physical (hydrodynamic properties, interfacial phenomena, ionic properties, and temperature), chemical (presence of chelating agents, carbon-to-nitrogen ratio, enzymes, ferrocyanide, nitrogenous substances, oils, sugars, and trace metals), and biological factors (inoculum size, presence of other organisms or strains). Artificial flocculating agents or crosslinkers may be added to enhance the aggregation process for cells that do not naturally flocculate. Polyelectrolytes (i.e., coupling agents by covalent bond formation) or inert powders can be used as linking agents.

An example of self-aggregation is the flocculation of yeast cells. Many fungi contain a family of cell wall glycoproteins (called "adhesins") that confer unique adhesion properties. These molecules are required for the interactions of fungal cells with each other (flocculation and filamentation), inert surfaces such as agar and plastic, and mammalian tissues; they are also crucial for the formation of fungal biofilms. In hazardous environmental conditions, *Saccharomyces cerevisiae* cells possess the remarkable property to adhere to other cells or to substrates such as agar or plastics. Adhesion to surfaces is a mechanism that may lead to biofilm formation. It is often used as a model to study biofilm formation of pathogenic

yeasts. The flocculation phenomenon is exploited in the brewery industry as an easy, convenient, and cost-effective way to separate the aggregated yeast cells from the beer at the end of the primary fermentation. The timing of flocculation is crucial for brewers because the quality of beer highly depends on it. When cells start to flocculate too early, the fermentation will be incomplete with undesirable aromas and too many residual sugars. On the other hand, when the flocculation is delayed, problems can arise during beer filtration (Willaert, 2007a).

The *S. cerevisiae FLO* genes encode a family of adhesion proteins (flocculins) responsible for its flocculation under unfavorable conditions (Goossens and Willaert, 2010). Flocculin proteins promote cell-cell adhesion to form multicellular clumps, thus protecting the cells from harsh environmental stresses. The floc formation is based on the self-interaction of Flo proteins via a N-terminal PA14 lectin domain. The carbohydrate specificity and affinity are determined by the accessibility of the binding site of the Flo proteins where the external loops in the ligand-binding domains are involved in glycan recognition specificity (Goossens et al., 2015). Besides the Flo lectin—glycan interaction, also glycan—glycan interactions contribute significantly to cell-cell recognition and interaction. Table 17.7 lists some examples of cell immobilization by self-aggregation.

17.3 DESIGN OF IMMOBILIZED CELL REACTORS

17.3.1 Mass Transport Phenomena in Immobilized Cell Systems

The analysis of the influence of mass transfer on the reactor performance in immobilized-cell reactors can be quite useful because the performance of these reactors is often limited by the rate of transport of reactants to and products from (external mass transfer limitation), and by the rate of transport inside of the immobilized cell system (internal mass transfer limitation) (Willaert, 2009). External mass transfer limitations can be reduced or eliminated by a proper choice or design of the reactor and immobilized cell system. Internal mass transfer limitations are often more difficult to eliminate. An estimation of the significance of mass transport limitations is a prerequisite to optimize the performance of an immobilized-cell bioreactor.

17.3.1.1 Diffusion Coefficient

Fick's law defines mass transport by molecular diffusion; that is the rate of transfer of the diffusing substance through a unit area is proportional to the concentration gradient measured normal to the section:

$$J = -D \frac{\partial C}{\partial x} \tag{17.1}$$

where J is the mass transfer rate per unit area of a section, C the concentration of diffusing substance (amount per total volume of the system), x the space coordinate, and D the diffusion coefficient. It is general practice to use an effective diffusion coefficient (D_e), which can be readily used in the

TABLE 17.7

Cell Immobilization by Self-Aggregation of Cells

Microorganism	Application Bacteria	Bioreactor Type
Zymomonas mobilis	Ethanol production	Fluidized-bed reactor
	Fungi	
Aspergillus awamori	Enzyme production	Stirred-tank reactor
Aspergillus oryzae	α-Amylase production	Stirred-tank reactor
Penicillium chrysogenum	Penicillin production	Stirred-tank reactor
Saccharomyces cerevisiae	Beer production	Cylindroconical tank (batch)
Trichoderma reesei	Cellulolytic enzymes	Stirred-tank reactor
	Mammalian Cells	
BHK cells	Recombinant protein	Stirred-tank reactor
Neural stem cells	High-cell-density expansion	Stirred-tank reactor
Vero cells	Recombinant protein	Stirred-tank reactor
293 cells	Recombinant protein	Stirred-tank reactor

Source: Adapted from Willaert, R.G. and Baron, G.V., *Rev Chem Eng*, 12:1–205, 1996; Willaert, R. *Fermentation Microbiology and Biotechnology*, 3rd ed., pp. 313–367, 2012. Boca Raton, FL: CRC Press.

expression for the Thiele modulus and for the determination of the efficiency factor of a porous biocatalyst. When D is replaced by D_e in Equation 17.1, the corresponding concentration (Cl) is expressed as the amount of solute per unit volume of the liquid void phase. Concentration C may be correlated with C_L by using the void fraction (ε), which is the accessible fraction of a porous particle to the diffusion solute as $C = \varepsilon C_L$. Hence, the relationship between the effective diffusion coefficient and the diffusion coefficient is

$$D_e = \varepsilon D \qquad (17.2)$$

17.3.1.2 Diffusion in Immobilized Cell Systems

The effective diffusion coefficient through a porous support material (matrix) is lower than the corresponding diffusion coefficient in the aqueous phase (D_a) because of the exclusion and obstruction effect. By the presence of the support, a fraction of the total volume ($1 - \varepsilon$) is excluded for the diffusing solute. The impermeable support material obstructs the movement of the solute and results in a longer diffusional path length, which can be represented by a tortuosity factor (τ), which equals the square of the tortuosity (Epstein, 1989). The influence of both effects on the effective diffusion coefficient can be represented by

$$D_e = \frac{\varepsilon}{\tau} D_a \qquad (17.3)$$

This equation holds as long as there is no specific interaction of the diffusion species with the porous carrier. In the case of gel matrices, predictions using the polymer volume fractions are recommended by the Equation 17.4 because neither e nor t can be measured for a gel in a simple way:

$$D = \frac{\left(1-\phi_p\right)^2}{\left(1+\phi_p\right)^2} D_a \qquad (17.4)$$

where ϕ_p is the polymer volume fraction. For low-molecular-weight solutes in cell-free gels, an approximate measure of ε can be given as

$$\varepsilon = 1 - \phi_p \qquad (17.5)$$

D_e can also be expressed as a function ϕ_p by combining Equations 17.2 through 17.5 to give

$$D_e = \frac{\left(1-\phi_p\right)^3}{\left(1+\phi_p\right)^3} D_a \qquad (17.6)$$

17.3.1.3 External Mass Transfer

In the case of permeable spheres, the effect of the external mass transfer resistance on the overall uptake and/or release rate by the beads may be quantitatively evaluated by calculating the time constant for the external film (τ_e) and comparing it with the time constant for diffusion in the sphere (τ_j). The internal time constant can be calculated using Equation 17.7:

$$\tau_i = \frac{R^2}{15 D_e} \qquad (17.7)$$

where R is the radius of the bead. The Biot number (*Bi*) for beads is defined by Equation 17.8 as the ratio of the characteristic film transport rate to the characteristic intraparticle diffusion rate:

$$Bi = \frac{k_s R}{D_e} \qquad (17.8)$$

An estimation of the external mass transfer coefficient (k_s) is required to calculate τ_e, *Bi*, or the film thickness. The value of k_s can be calculated by a procedure recommended by Harriot and coworkers in the mid-1970s (Sherwood et al., 1975). Merchant and coworkers (1987) determined *Bi* for a rotating sphere. Using the empirical correlation of Noordsij and Rotte (1967), k_s could be estimated using Equation 17.9:

$$Sh = \sqrt{4 + 1.21\left(Re_p Sc\right)^{0.67}} \qquad (17.9)$$

where Sh is the Sherwood number, Re_r is the rotational Reynolds number, and Sc is the Schmidt number. In the case of diffusion through a membrane or thin disc, Bi can also be calculated. k_s can be determined for free-moving particles using the following correlations (van't Riet and Tramper, 1991):

$$Sh = \sqrt{4 + 1.21\left(Re_p Sc\right)^{0.67}} \text{ for } Re_p Sc > 10^4 \qquad (17.10)$$

$$Sh = 2 + 0.6 Re_p^{0.5} Sc^{0.33} \text{ for } Re_p < 10^3 \qquad (17.11)$$

where Re_p is the (particle) Reynolds number, which can be estimated using the following correlations:

$$Re_p = {Gr}\!/\!{18} \text{ for } Gr < 36 \qquad (17.12)$$

$$Re_p = 0.153 Gr^{0.71} \text{ for } 36 < Gr < 8 \times 10^4 \qquad (17.13)$$

$$Re_p = 1.74 Gr^{0.5} \text{ for } 8 \times 10^4 < Gr < 3 \times 10^9 \qquad (17.14)$$

where Gr is the Grashof number. Another correlation used to estimate k_s for gel beads in agitated reactors is (Kikuchi et al., 1988)

$$Sh = 2 + 0.52\left(e_s^{1/3} d_p^{4/3} / v\right)^{0.59} Sc^{1/3} \qquad (17.15)$$

where d_p is the average diameter of the particle, n is the kinematic viscosity, e_s is the energy dissipation given as $e_s = N_p n_i^3 D_i^5/V$ for a stirred tank (where N_p is the power number, n_i is the impeller speed, D_i is the impeller diameter, and V is the volume of the reactor). The ranges of validity for this correlation are

$$10 < \left(e_s^{1/3} d_p^{4/3} / v\right) < 1500 \text{ and } 120 < Sc < 1450 \qquad (17.16)$$

Also, a correlation has been recommended for agitated dispersions of small, low-density solids (Øyaas et al., 1995):

$$k_s = \frac{2D_{e0}}{d_p} + 0.31\left(Sc\right)^{-2/3}\left(\frac{\Delta\rho\upsilon g}{\rho_l}\right)^{1/3} \qquad (17.17)$$

where $\Delta\rho$ is the particle/liquid density difference and ρ_l is the density of the bulk liquid. For spherical particles in a packed bed, k_s depends on the liquid velocity around the particles. For the range $10 < Re_p < 10^4$, the Sherwood number has been correlated by Equation 17.18 (Moo-Young and Blanch, 1981):

$$Sh = 0.95 Re_p^{0.5} Sc^{0.33} \qquad (17.18)$$

An estimation of k_s can be calculated if the stirred chambers have the shape of flat cylinders using the correlation in Equation 17.19 (Sherwood et al., 1975):

$$k_s = 0.62 D_a^{2/3} v^{-1/6} \omega^{1/2} \qquad (17.19)$$

where v is the kinematic viscosity and ω is the rotational speed of the stirrer (in rad/s). The external mass transfer limitation can be experimentally investigated by observing the concentration-time profile at different mixing regimes in the bioreactor (e.g., rotation speeds of the stirrer).

17.3.2 Reaction and Diffusion in Immobilized Cell Systems

In immobilized cell systems, cellular reactions can take place in the presence of significant concentration gradients. These reactions are also called "heterogeneous reactions." Reactions can only take place when the substrate molecules are transported to the reaction place (i.e., mass transport phenomena can have a profound effect on the overall conversion rate). The concentration in each internal position must be known to determine the local rates. In most cases, these internal concentrations cannot be measured but can be estimated using a reaction-diffusion model.

17.3.2.1 Reaction-Diffusion Models

The major issues involved in modeling immobilized cell reactors are very similar to those in heterogeneous chemical reactors. This analogy has encouraged a rapid development in the model building for immobilized cell systems, even if the level of understanding of biocatalysts is lower than for chemical catalysis. The ability to predict the behavior of immobilized cell systems is required for the understanding, design, and optimization of an appropriate bioreactor. It is necessary to consider the bioreactor performance and microbial kinetics. Description of the bioreactor performance involves modeling of mass transfer effects and the flow pattern in gas and liquid phases whereas microbial kinetics deals with the kinetics on the individual cell level and on the level on the whole cell population. Single-cell kinetics can be described with an unstructured model (no intracellular components considered) or with a structured model (intracellular components considered). The population model may be either unstructured (all cells in the whole population assumed to be identical, i.e., only one morphological form) or morphologically structured (with an infinite number of morphological forms, the term "segregated population model" is often used). The models describing immobilized cell behavior are usually of the unstructured type.

Models that described immobilized cell kinetics were initially based on the steady-state models for immobilized enzymes. Steady-state models can give valuable information for design purposes, but they fail to describe transient phenomena (such as the start-up dynamics and response to changing conditions in the reactor) encountered in growing immobilized cell systems. Therefore, dynamic models have been developed to simulate the transient behavior of growing immobilized cells.

In general, gel-immobilized cell systems are considered as effective continua. However, it has been observed that when gel beads are inoculated with a low cell concentration, each growing cell will be the origin of a microcolony and growth results in the formation of expanding microcolonies (e.g., Willaert and Baron, 1993). A rigorous modeling approach of this microcolony system requires consideration of the microstructure of the immobilized cell system: diffusion in the gel phase and reaction and diffusion in the microcolony ("two-phase" system).

17.3.2.1.1 Intrinsic Kinetics

Intrinsic kinetics describes the growth and product formation rates of cells in the immobilized (or free) state as a function of the local concentrations. A typically simple unstructured model of microbial kinetics for growth on a single substrate can be described by Equations 17.20 through 17.22:

$$\text{Biomass growth}: \mu = f\left(C_\text{S}\right) \qquad (17.20)$$

$$\text{Substrate consumption}: q_\text{s} = \frac{1}{Y_{\text{X/S}}}\mu + m_\text{s} \qquad (17.21)$$

$$\text{Product formation}: q_\text{p} = \frac{1}{Y_{\text{X/P}}}\mu + m_\text{p} \qquad (17.22)$$

where μ is the specific growth rate of the cells (gDW/gDW per hour), C_S is the substrate concentration, q_s is the specific substrate utilization rate (g substrate used/gDW per hour), q_p is the specific product formation rate (g product formed/gDW per hour), $Y_{\text{X/S}}$ (gDW/g substrate) and $F_{\text{X/P}}$ (gDW/g product) are the yield coefficients, and m_s (g substrate/gDW per hour) and m_p (g product/gDW per hour) are the specific maintenance rates for substrate and product, respectively. The specific growth rate is a function of the substrate concentration and is usually of the Monod kinetics form. The model can also be extended to include growth inhibition by the product (and biomass), and $f(C_\text{S})$ becomes some function of substrate and product (and biomass). The Monod equation is bound by zero-order (at high substrate concentrations relative to the Monod constant K_S) and first-order (at vanishingly small substrate concentrations) kinetics. The solutions of reaction-diffusion problems with these two simple rate equations are valuable in that they can be applied as lower or upper bounds to the general problem without requiring detailed knowledge of the rate expressions and thus considerably facilitating the calculations.

In the interpretation of kinetic data for immobilized cells, it is important to assess the significance of mass transfer limitations. If negligible mass transfer limitation is present, the externally observed kinetics is the intrinsic cell kinetics. Any external or internal mass transfer limitation will lead to externally observed lower conversion rates. Mass transfer limitations may appear in the external film around the support matrix, or within the gel matrix, or in both.

Various claims have been made regarding changes in the intrinsic growth rate of immobilized cells, primarily regarding cells adhering to a surface. It has been asserted that the growth rate for immobilized cells is much higher than that for free-living cells. For gel-immobilized cell systems, it has been observed that the metabolic rates of gel-immobilized cells depend only on the local solution concentrations and are identical to those for free cells if diffusional limitations are absent, although some reports showed a decreased growth rate upon entrapment. Some researchers found no significant difference between the maximum specific growth rates for immobilized yeast and bacteria and those for free cells (Table 17.8). On the contrary, a significant decrease has been noted by other researchers for the same microorganisms. This change of metabolic activity upon immobilization may be due to diffusional limitations or to a change in cellular physiology.

17.3.2.1.2 Modeling

By the entrapment of cells in gel matrices, an additional barrier to mass transfer relative to free cells is introduced. This tends to lower the overall reaction rate and creating a specific microenvironment around the cells. Immobilized cells can grow in the gel matrix and the mass transfer limitations on substrate delivery and product removal lead to time-dependent spatial variations in growth rates and biomass densities, which may be accompanied by alterations in cellular physiology and biocatalytic activity. Because the local effective diffusion coefficient depends on the local biomass density, this nonhomogeneous growth will influence the local diffusive rates. The existence of chemical environmental gradients in immobilized cell systems has been verified experimentally with various microprobe techniques.

17.3.2.1.2.1 Dynamic Modeling A general dynamic model that describes the growth of the immobilized cells and the resulting time-dependent spatial variation of substrate and product in the system can be constructed by writing the mass balances over the immobilization matrix (it is usually assumed that diffusion in the system is governed by Fick's law, the cells are initially distributed homogeneously over the carrier, and there is no deformation of the matrix due to cell growth or gas production):

$$\frac{\partial}{\partial t}\left(\varepsilon\beta C_i\right) = z^{-n}\frac{\partial}{\partial z}\left(z^n D_{e,i}\frac{\partial C_i}{\partial z}\right) \pm \varepsilon\beta r_i \qquad (17.23)$$

where ε is the ratio of the volume of the pores of the matrix to the total volume; β is the ratio of the volume of the pores minus the volume of the cells to the volume of the pores in the matrix; C_i is the substrate ($i = S$) or product ($i = P$) concentration (expressed per volume available for substrate; n is a shape factor of value 0 for planar, 1 for cylindrical, or 2 for spherical geometry; and $D_{e,i}$ is the effective diffusion coefficient for species i. The substrate consumption rate (r_s) and the product formation rate (r_p) are linked to the growth rate (r_x) by Equations 17.24 and 17.25:

TABLE 17.8

Comparison of the Specific Growth Rates for Gel-Immobilized (μ_i) and Free (μ_f) Cells

Microorganism	Gel System	μ_i (h^{-1})	μ_f (h^{-1})
Bacteria			
Escherichia coli	Carrageenan (2%)	2.04[a]	2.08
		1.69[b]	1.63
E. coli B/pTG201	Carrageenan (2%)	0.24[c]	0.30
		0.18[d]	
E. coli B/pTG201	Carrageenan (2%)	0.18	0.36
Fungi			
Candida guilliermondii	Ba-alginate	0.021	0.029
Saccharomyces cerevisiae	Calcium alginate (2%)	0.30[e]	0.31
		0.27[f]	
S. cerevisiae	Gelatin (25–30%)	0.28	0.51
S. cerevisiae	Calcium alginate (1.5%)	0.115	0.126
	Carrageenan (2.5%)	0.100	
S. cerevisiae	Calcium alginate (2%)	0.25	0.41
S. cerevisiae	Calcium alginate (2%)	0.46	0.50
Thiosphaera pantotropha	Agarose (5%)	0.45[g]	0.45
		0.58[h]	

Source: Willaert, R., Nedovic, V., and Baron G.V., *Fundamentals of Cell Immobilization Biotechnology*, 2004. Dordrecht, The Netherlands: Kluwer Academic Publishers; Willaert, R. *Fermentation Microbiology and Biotechnology*, 3rd ed., pp. 313–367, 2012. Boca Raton, FL: CRC Press.

[a] Supply of 21% oxygen; [b] supply of 100% oxygen; [c] growth in gel slabs; [d] growth in gel beads; [e] single immobilized cells; [f] cells in a microcolony; [g] Growth in stirred-tank reactor; [h] Growth in Kluyver flask.

$$\frac{\partial}{\partial t}\left(\varepsilon C_X\right) = \varepsilon r_x = \varepsilon \mu C_X \qquad (17.24)$$

$$r_s = \frac{1}{Y_{X/S}} r_x \quad \text{and} \quad r_p = \frac{1}{Y_{X/P}} r_x \qquad (17.25)$$

where C_X is the biomass concentration expressed per volume available for the cells. Because ε is constant with time, the left-hand side of Equation 17.23 can be written as

$$\frac{\partial}{\partial t}\left(\varepsilon\beta C_i\right) = \varepsilon\beta\frac{\partial C_i}{\partial t} + \varepsilon C_i\frac{\partial\beta}{\partial t} \qquad (17.26)$$

If a dry weight cell density ρ_c is defined as the ratio of the cell dry mass per cell volume, β can be expressed as function of C_X:

$$\beta = 1 - \frac{C_X}{\rho_c} \qquad (17.27)$$

Using the relationship $Y_{X/S} = dC_X/dC_S$, the second term on the right side of Equation 17.25 is negligible when C_S is much smaller than $\beta\rho_c/Y_{X/S}$, and under those conditions substitution into Equation 17.23 gives

$$\varepsilon\beta\frac{\partial C_i}{\partial t} = z^{-n}\frac{\partial}{\partial z}\left(z^n D_{e,i}\frac{\partial C_i}{\partial z}\right) \pm \varepsilon\beta r_i \qquad (17.28)$$

These equations can be integrated to yield the substrate and biomass profiles as a function of time (usually together with the reactor model) using the correct initial and boundary conditions. These equations are valid for a wide range of immobilized cell systems.

17.3.2.1.2.2 Pseudo-steady-state Modeling Under certain conditions, the full dynamic modeling to describe transient behavior can be simplified to "pseudo-steady-state" modeling. Therefore, the biomass growth and the substrate consumption rate and/or product formation rate are treated separately. This approach is valid as long as the time scale for growth is much larger than the time scale for consumption and product formation. Hence, a pseudo-steady-state substrate/product distribution is assumed at each instant. As a result, the system of partial differential equations is reduced to a system of ordinary differential equations that facilitates the numerical solution.

17.3.2.1.2.3 Steady-state Modeling If the cell mass does not vary rapidly, or is fairly uniform, the concentration profiles in a gel matrix with entrapped cells can be simulated using a steady-state model at any point in time. These models can give valuable information for design purposes or can be combined with experimental *in situ* measurements (e.g., microelectrodes). In this case, mathematical calculations can be very simple, and straightforward analytical solutions can be obtained for simple reaction kinetics (e.g., zero- or first-order kinetics).

17.3.2.1.2.4 Effectiveness Factor The effectiveness factor (η) can be calculated to obtain a numerical measure of the influence of mass transfer on the reaction rate. The effectiveness factor is defined as

$$\eta = \frac{\text{observed reaction rate}}{\begin{array}{c}\text{rate which would be obtained}\\\text{without mass transfer resistance}\end{array}} \qquad (17.29)$$

Effectiveness factor calculations can be based on steady-state or dynamic reaction-diffusion models with the assumption of a homogeneous distribution of cells over the carrier. The effectiveness factor for substrate consumption can mathematically be expressed as the volume-averaged reaction rate relative to the rate at bulk-phase concentration:

$$\eta = (n+1)\frac{\left(D_e \frac{dC'_S}{dz'}\right)_{z'=1}}{r_s(1)} \qquad (17.30)$$

or

$$\eta = (n+1)\frac{\int_0^1 z'^{n'} r_s(C'_S) dz'}{r_s(1)} \qquad (17.31)$$

where n is a shape factor (0 for planar, 1 for cylindrical, or 2 for spherical geometry) and Z is the dimensionless position coordinate; and r_s is the substrate reaction rate, which is a function of the dimensionless substrate concentration (C_S'; and also of the position in the case of transient effectiveness factor).

17.3.3 BIOREACTOR DESIGN

A classification of immobilized cell reactors with their advantages and disadvantages is given in Table 17.9. Three categories can be distinguished: bioreactors filled with (1) mixed, suspended particles; (2) fixed particles or large surfaces; or (3) moving surfaces.

Most bioreactors (Figure 17.5) contain three phases: solid (the carrier or cell aggregate), bulk liquid, and gas (air/oxygen or gas feed, gaseous products).

Criteria for the selection of an appropriate bioreactor for immobilized cells and reactor types satisfying certain criteria are summarized in Table 17.10. The cell aggregate can only be fully active if the external supply or removal rates match the internal transport, utilization, and production rates. The high cell densities in the reactors put higher demands on nutrient supply and transport rates.

In aerobic fermentations, oxygen is often the limiting substrate because the liquid film around the gas bubbles is a major resistance to oxygen transfer. Gas-liquid mass transport is characterized by the liquid-phase mass transfer coefficient k_L (usually expressed as $k_L a$, with a the area of the bubble per volume). Additionally, intraparticle resistance of oxygen in the cell aggregate can also become significant depending on the Thiele modulus. Consequently, the Thiele modulus and $k_L a$ are important parameters for the design and scale-up of immobilized cell bioreactors. The parameters, which are grouped in the Thiele modulus, are the size of the aggregate, cell kinetics (kinetic parameters), and the diffusion coefficient. Therefore, the particle size or biofilm thickness are major design variables. Because substrate concentration is often not a free parameter, only aggregate size and biomass loading are engineering parameters. Unless the substrate concentration is

very low (e.g., degradation of pollutants), diffusion limitation for substrate only occurs for particles of more than several millimeters in diameter. In contrast, the penetration depth of oxygen in particles for aerobic processes is only 50–100 μm.

17.4 PHYSIOLOGY OF IMMOBILIZED MICROBIAL CELLS

The microenvironment of microbial cells is altered upon immobilization depending on the method of immobilization (Polakovič et al., 2017). The changed chemical composition and/or the physical interaction between the matrix material and the cell can have a profound effect on the physiology of the cells (Żur et al., 2016). Some examples of the effect of immobilization upon the cell's physiology are discussed for bacteria and fungi. An overview of observed effects (discussed in Section 17.4.1) upon immobilization of bacterial and fungal cells is summarized in Tables 17.11 and 17.12, respectively.

17.4.1 BACTERIAL CELLS

17.4.1.1 Plasmid Stability

Recombinant plasmid stability in host cells can be increased upon immobilization (Table 17.10), presumably because of the significant drop in the rate of cell division.

17.4.1.2 Protective Microenvironment

Immobilization can confer protection to cells exposed to toxic or inhibitory substrates or environments (Table 17.10). Gel immobilized *Trichosporium* sp. and *Pseudomonas putida* showed higher rates of phenol degradation and phenol tolerance. Alginate-entrapped *Escherichia coli* cells grown entrapped in calcium alginate showed low lipid-to-protein ratios even without phenol in the growth medium. Immobilization of cells also markedly changed the protein pattern of the outer membrane. Calcium-alginate-immobilized cultures of *Streptococcus* cells were protected from bacteriophage attack because of the exclusion of phage particles from the gel matrix. These cultures were also functionally proteinase deficient when immobilized and grown in milk, which resulted in a lower acid production due in part to the inability of the immobilized cells to hydrolyze milk proteins and to diffusional limitations of substrate into the beads. A different protein profile was found after submitting agar-entrapped *E. coli* to a cold shock (Perrot et al., 2001). It was suggested that such induction of specific molecular mechanisms in immobilized bacteria might explain the high resistance of sessile-like organisms to stresses. The degradation rate of alkyl benzene sulfonate by calcium-alginate-entrapped *P. aeruginosa* was considerably increased compared with free cells and could be further increased using low-intensity ultrasonic irradiation (Lijun et al., 2005) because ultrasound can improve the osmosis of the cell membrane, cell growth, and enzyme activities. A new ultrasound-based cell immobilization technique was described that allows manipulation and positioning of cells/particles within various gel matrices

TABLE 17.9
Classification of Immobilized Cell Reactors

Reactor Type	Advantages	Disadvantages
	Mixed, Suspended Particles	
Stirred-tank reactor	Flexible	High power consumption
	Variable mixing intensity	Shear damage to sensitive matrices
	Suitable for high viscosity	High cost
Gas/air lift reactor	No moving parts	Low local mixing intensity
Bubble column reactor	Simple	Only for low viscosities
	High solids fraction possible High gas transfer Good heat transfer	Excessive foaming possible
Fluidized-bed reactor	No moving parts in reactor	Difficult matching of feed and fluidization rates
	Simple, low cost	Requirements on particle density (dense support)
	Very high solids contents	Good local mixing intensity
	Good heat transfer	Only for low viscosity
	Variable mixing characteristics for liquid and solid	
	Fixed Particles or Large Surfaces	
Packed-bed reactor	Simple, low cost	Plugging by solids at low flow rates
Monolith reactor	Plug flow characteristics possible	High pressure drop
	Large surface to volume ratio	Aeration only externally possible Channeling and maldistribution Gas build-up and formation of "dry" spots Low external mass transfer rate at low flow rates
Trickle-bed reactor	Simple, very low cost	Plugging by excessive growth (cleaning possible)
	Plug flow approached	Only large supports possible
	Large surface-to-volume ratio High oxygen transfer rate Suitable for gas cleaning (biofilter)	Only for low viscosities
Solid (surface) culture	Simple	High cost (often manual)
	Flexible	Only batch
	Low humidity (mold fermentation)	Difficult control over operation
	Low contamination risk (by bacteria or yeasts)	Usually limited to solid substrates
Membrane reactor	Very high cell densities	Sterilization problems
	Very high productivities	Microbial damage (membrane perforation)
	Perfusion operation possible	Low capacities only
	Simultaneous product separation possible	High cost
	Separate feed of gas and liquid possible	
	Low shear	
	Moving Surfaces	
Rotating surfaces (disc, cylinder, or packing)	Low shear on biofilm	Power consumption
	Batch or continuous	Maintenance
	Excellent aeration	
	High productivity	
	Suitable for high viscosity	

Source: Baron, G.V., Willaert, R.G., and De Backer, L., *Immobilized Living Cell Systems: Modeling and Experimental Methods*, pp. 67–95, 1996. Chichester, UK: John Wiley & Sons. With permission.

before polymerization (Gherardini et al., 2005). Proteomic analysis of agar-entrapped *P. aeruginosa* showed that the immobilized bacteria were physiologically different from free cells (Vilain et al., 2004).

17.4.1.3 Effect of Mass Transport Limitation

Mass transport limitations can have an influence on the morphology of immobilized *Lactobacillus helveticus* because of the long response time of entrapped cells to random pH changes. The citrate metabolism (Cachon and Divies, 1993) and lactate production (Klinkenberg et al., 2001) of

Lactococcus lactis were altered because of the concentration and pH gradients in the gel beads. The germination time of *Bacillus subtilis* cells was significantly longer than for free cells; after a time lag due to encapsulation, the growth of the cells was uninhibited and no differences between entrapped and free cells were found.

17.4.1.4 Enhanced Productivity of Enzymes and Other Products

It has been observed that the production of several enzymes is increased upon immobilization (Table 17.11). In one case

(a)

(b)

(c)

(d)

(e)

(f)

(g)

(h)

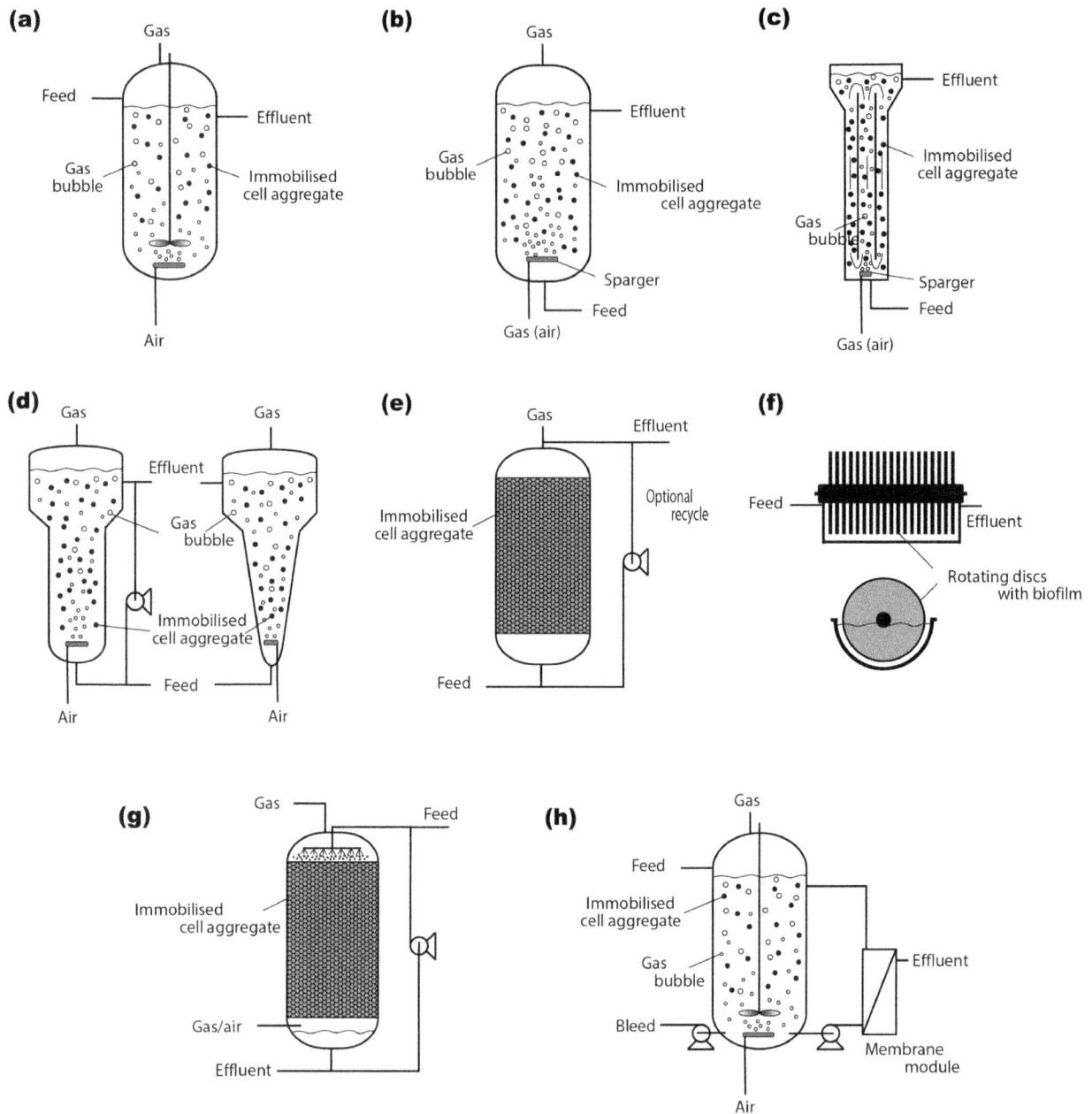

FIGURE 17.5 Immobilized cell bioreactors: (a) stirred-tank reactor; (b) bubble column reactor; (c) gas (air) lift reactor; (d) fluidized-bed reactor (left), tapered fluidized reactor (right); (e) packed-bed reactor with optional recycle loop; (f) rotating drum reactor; (g) trickle bed reactor; and (h) membrane cell-recycle reactor.

(α-amylase by alginate-entrapped *Bacillus amyloliquefaciens*), enzyme production was decreased (Argirakos et al., 1992). Protease production by alginate-entrapped *Myxococcus xanthus* cells was increased because of a reduced inhibition by peptone and gelatin as result of mass transfer limitation in the gel (Fortin and Vuillemard, 1990).

17.4.2 FUNGAL CELLS

17.4.2.1 Plasmid Stability

As in the case of recombinant bacteria, improved plasmid stability in immobilized *S. cerevisiae* has also been demonstrated.

17.4.2.2 Protective Microenvironment

Immobilization can create a protective microenvironment for the cells (Table 17.11). In the case of co-immobilization of *S. cerevisiae* cells with vegetable oils, cells could be protected against inhibitory substances because of a better solubility of the inhibitory compounds in the oil phase. An increased tolerance for ethanol by immobilized brewer's yeast (Norton et al., 1995) and a decreased inhibition of ethanol productivity in entrapped *Kluyveromyces marxianus* at high osmolality (Dale et al., 1994) have been observed. Immobilized *S. cerevisiae* contained significantly higher percentages of saturated fatty acids because of altered osmotic conditions in the microenvironment of the cells; other examples are shown in Table 17.11.

TABLE 17.10

Criteria for the Selection of the Reactor Type

Criterion		Stirred Tank	Air Lift Bubble Column	Fluidized Bed	Packed Bed	Trickle Bed	Membrane Reactor	Rotating Biological Contactor
Matrix	Strong	x	x	x	x	x		
	Weak		x	x				
Biocatalyst	Biofilm				x	x		X
	Particles	x	x	x	x			
	Membrane						x	
Sterilizability	Good	x	x	x			x	
	Poor				x	x		X
High capacity		x	X	x	x	x		
High feed viscosity		x						X
Solids in feed		x	X	x				X
Flexibility required		x	X					
Equipment cost	High	x		x			x	X
	Low		X		x	x		
High oxygen (gas) requirement		x	X		x	x		
High cell growth rate		x	X	x				
High gas production		x	X	x		x		X
Cell free effluent								
Low shear rate	Low		X		X	x	x	

Source: Modified from Baron, G.V., Willaert, R.G., and De Backer, L., *Immobilized Living Cell Systems: Modeling and Experimental Methods*, pp. 67–95, 1996. Chichester, UK: John Wiley & Sons.

17.4.2.3 Influence of Mass Transport Limitation

A high invertase activity, which was exhibited by immobilized cells, was due to a maintained expression of the *SUC2* gene and a reduced susceptibility of the enzyme to endogenous proteolytic attack (de Alteriis et al., 1999). These results have been interpreted in terms of diffusional limitations and changes in the pattern of invertase glycosylation due to growth of yeast in an immobilized state.

17.4.2.4 Enhanced Productivity

The increase in ethanol productivity by gel-entrapped and co-entrapped cells of *S. cerevisiae* was attributed to stimulation in cell permeability by Si^{4+} whereas higher ethanol production by κ-carrageenan-entrapped *Saccharomyces bayanus* was attributed to a favorable media supplement in the aqueous phase of the matrix (Brito et al., 1990). However, the reduced yield of ethanol in calcium-alginate-entrapped *S. cerevisiae* was due to lower substrate concentrations toward the center of the bead because of mass transport limitations (Gilson and Thomas, 1995).

17.4.2.5 Enhanced Enzyme Stability

A hydroxylase from entrapped *Mortierella isabellina* was found to retain its activity over a longer period compared with free mycelia. The effect of immobilization on the physiology of yeast cells and fungi is given in Table 17.12.

17.4.3 STEM CELLS

A stem cell is an unspecialized cell that can self-renew (reproduce itself) and differentiate into functional phenotypes. Stem cells originate from embryonic, fetal, or adult tissue and are broadly categorized accordingly. The integration of stem cells and tissue-engineered scaffolds has the potential to revolutionize the field of regenerative medicine. It promises great things, including the ability to grow organs composed of multiple cell types and complex structures, therapies for the correction of congenital diseases, and the promise of readily obtainable immuno-compatible tissues. Creating reserves of undifferentiated stem cells and subsequently driving their differentiation to a lineage of choice in an efficient and scalable manner is critical for the ultimate clinical success of cellular therapeutics. In recent years, various biomaterials have been incorporated in stem cell cultures, primarily to provide a conducive microenvironment for their growth and differentiation and to ultimately mimic the stem cell niche.

With the inherent plasticity and multilineage potential provided by stem cells comes an increased need for regulating

TABLE 17.11

Some Observed Effects of Bacterial Cell Immobilization on Its Physiology

Immobilization Effect	Microorganism Increased	Immobilization System
	Plasmid Stability	
	Escherichia coli	Calcium alginate, agarose, κ-carrageenan, cotton cloth, hollow fiber, polyacrylamide/hydrazide, PVA, silicone beads
	Lactococcus lactis	κ-Carrageenan/locust bean gum
	Myxococcus xanthus	κ-Carrageenan
	Protective Microenvironment	
Increased phenol degradation	*Pseudomonas putida*	Calcium alginate, polyacryl amidehydrazide
	Trichosporium sp.	Calcium alginate
Increased antibiotics tolerance	*Pseudomonas aeruginosa*	Calcium alginate
Increased phenol tolerance	*Escherichia coli Stapylococcus aureus*	Calcium alginate
Increased benzene degradation	*Pseudomonas putida* and *Pseudomonas fluorescens*	Cotton terry cloth
Increased degradation of linear alkyl benzene sulfonate	*Pseudomonas aeruginosa*	Calcium alginate
Increased stress tolerance	*Bifidobacterium longum* and *Lactococcus lactis*	κ-Carrageenan/locust bean gum
Metabolic differences	*Marinobacter* sp.	Porous glass
Degradation of DMP at higher concentration	*Bacillus* sp.	Calcium alginate, polyurethane foam
Protection from bacteriophage attack	*Streptococcus lactis Streptococcus cremoris*	Calcium alginate
Different protein profile after cold shock	*Escherichia coli*	Agar
Increased survival rate at low pH	*Bifidobacterium longum*	Gellan-xanthan, Ca-alginate
	Influence of Mass Transport Limitation	
Decreased pH-dependent morphology change	*Lactobacillus helveticus*	κ-Carrageenan/locust bean gum
Changed citrate metabolism and lactate production	*Lactococcus lactis*	Ca-alginate
Loss of SDS metabolization	*Pseudomonas fluorescens*	Polyacrylamide
	Changed Productivity of Enzymes and Other Products	
Increased α-amylase production	*Bacillus* sp.	κ-Carrageenan
	Bacillus subtilis	Polyacrylamide
Reduced α-amylase production	*B. amyloliquefaciens*	Calcium alginate
Increased dextransucrase production	*Leuconostoc mesenteroides*	Calcium alginate
Increased α-galactosidase	*Streptomyces griseoloalbus*	Calcium alginate
Increased β-galactosidase	*Escherichia coli*	Calcium alginate
Increased proteinase production	*Humicola lutea*	Polyhydroxyethylmethacrylate
Increased protease production	*Myxococcus xanthus*	Calcium alginate
Increased alginate production	*Pseudomonas aeruginosa*	κ-Carrageenan
Increased gellan gum production	*Pseudomonas elodea*	κ-Carrageenan
	Changed Morphology	
Escherichia coli	Agar	
Lactobacillus helveticus	κ-Carrageenan/locust bean gum	
Pseudomonas putida	Cotton terry cloth	
P. fluorescens	Cotton terry cloth	

Source: Adapted from Willaert, R., Nedovic, V., and Baron G.V., *Fundamentals of Cell Immobilization Biotechnology*, 2004. Dordrecht, The Netherlands: Kluwer Academic Publishers; Willaert, R. *Fermentation Microbiology and Biotechnology*, 3rd ed., pp. 313–367, 2012. Boca Raton, FL: CRC Press.

Note: DMP, dimethylphtalate; SDS, sodium dodecyl sulfate.

cell differentiation, growth, and phenotypic expression. Stem cells, like all cells, are influenced by their microenvironment, including chemical and physical cues. Until recently, chemical cues have been the primary means by which stem cells self-renew and their differentiation has been influenced. Soluble factors and substrate coating have been used in maintaining undifferentiated stem cells, as well as in promoting a particular differentiation pathway. Recent efforts have begun focusing on controlling the cellular microenvironment by engineering three-dimensional biomaterials and/or applying physical forces. Biomaterials are rapidly being developed to display and deliver stem-cell-regulatory signals in a precise

TABLE 17.12

Some Observed Effects of Fungal Cell Immobilization on Its Physiology

Immobilization Effect	Microorganism	Immobilization System
Increased Plasmid Stability		
	Saccharomyces cerevisiae	Calcium alginate bead, cotton cloth sheet, gelatin beads, porous glass beads
Protective Microenvironment		
Altered osmotic conditions in microenvironment	*Saccharomyces cerevisiae*	Calcium alginate
Decreased ethanol tolerance	*Pichia stipitis*	κ-Carrageenan
Increased ethanol tolerance	*Saccharomyces cerevisiae*	κ-Carrageenan
Increased organic solvent tolerance	*Saccharomyces cerevisiae*	Polyhydroxylated silane
Decreased inhibition of cell growth and metabolism by better carbon dioxide removal	*Saccharomyces cerevisiae*	Stainless steel fiber cloth
Increased tolerance for nitriles and amides	*Candida guilliermondii*	Barium alginate
More efficient treatment of dairy effluents	*Candida pseudotropicalis*	Calcium-alginate
More sensitive for sucrose	*Aspergillus niger*	Calcium alginate
Protection against inhibitory substances	*Saccharomyces cerevisiae*	Calcium alginate
Retention of high metabolic activity during long-term fermentation	*Saccharomyces cerevisiae*	Calcium alginate
Influence of Mass Transport Limitation		
Less susceptible of enzyme to endogenous proteolysis	*Saccharomyces cerevisiae*	Gelatin
Lower specific productivity	*Kluyveromyces lactis*	Calcium alginate
Changed Productivity		
Increased ethanol production	*Saccharomyces formosensis*	Calcium alginate and co-entrapped sand
	Saccharomyces bayanus	κ-Carrageenan
Increased laccacase production	*Agaricus* sp.	Polyurethane foam, straw, textile strips
Increased penicillin production	*Penicillium chrysogenum*	κ-Carrageenan
Shifted phosphate concentration optimum for alkaloid production	*Claviceps purpurea*	Calcium alginate
Changed secondary metabolite production	*Fusarium moniliforme*	Alginate, carrageenan, polyurethane
Changed Morphology/Composition		
	Aspergillus niger	Calcium alginate
	Claviceps fusiformis	Calcium alginate
	Gibberella fujikuroi	Calcium alginate
	Saccharomyces cerevisiae	Calcium alginate, agar, gelatine
	Saccharomyces cerevisiae	Calcium alginate
Enhanced Enzyme Stability		
Hydroxylase	*Mortierella isabellina*	Calcium alginate, polyurethane foam

Source: Adapted from Willaert, R., Nedovic, V., and Baron G.V., *Fundamentals of Cell Immobilization Biotechnology*, 2004. Dordrecht, The Netherlands: Kluwer Academic Publishers; Willaert, R. *Fermentation Microbiology and Biotechnology*, 3rd ed., pp. 313–367, 2012. Boca Raton, FL: CRC Press.

and near-physiological fashion and serve as powerful artificial microenvironments in which to study and instruct stem-cell fate in culture and *in vivo*. Table 17.13 shows some examples of observed effects of immobilization materials on stem cell culture.

17.5 BEER PRODUCTION USING IMMOBILIZED CELL TECHNOLOGY: A CASE STUDY

17.5.1 FLAVOR MATURATION OF GREEN BEER

One of the objectives of the maturation (or secondary fermentation) of green beer is the removal of unwanted aroma compounds (Stewart, 2017). The removal of the vicinal diketones diacetyl and 2,3-pentanedi-one is especially important

because these compounds have very low flavor thresholds. Diacetyl is quantitatively more important than 2,3-pentanedione and is therefore used as a marker compound. It has a taste threshold around 0.10–0.15 mg/ml in lager beer, approximately 10 times lower than that of pentanedione. These compounds impart a "buttery," "butterscotch" aroma to the beer. During the primary fermentation, these flavor-active compounds are produced as by-products of the synthesis pathway of isoleucine, leucine, and valine (ILV pathway) and are linked to the amino acid metabolism and the synthesis of higher alcohols. The excreted α-acetohydroxy acids are overflow products of the ILV pathway and are nonenzymatically converted to the corresponding vicinal diketones (Figure 17.6). This nonenzymatic oxidative decarboxylation step is the rate-limiting step and proceeds faster at a high temperature and lower pH. The

TABLE 17.13

Some Examples of Observed Effects of Immobilization Biomaterials on Stem Cell Culture

Biomaterial Effect	Stem Cell Type	Immobilization System
Natural Materials		
Maintaining pluripotency and undifferentiated state	hESC	Hyaluronic acid hydrogel
Differentiation into neural lineage cells	mESC	Fibrin gels
More efficient collagen II and proteoglycan synthesis of differentiated cells	MSC	Fibrin gels
Encouraging osteogenic differentiation	mESC	Hydroxyapatite
Promoted differentiation toward a hepatic lineage without the need for embryoid body formation	mESC	Alginate poly-l-lysine
Differentiation into endothelial cells	hBMC	Crosslinked pullulan/dextran/fucoidan
Assemble and reorganize of the pellet structure	hMSC	Type II collagen
Improve overall viability, longer maintenance of hema-topoitic functions, ability to engraft into bone marrow	hUCBC	Collagen microbeads
Synthetic Materials		
Polymers		
Pore size and polymer composition influence differentiation into the hematopoietic	mESC	Poly(lactic-co-glycolic acid)/poly(L-Lactic acid) (50/50 blend)
Encouraging cell proliferation and cell-cell interaction	MSC, mESC	Nanofibrous poly(s-caprolactone)
Upregulation of chondrogenic markers	EB from mESC	PEG hydrogel
Ceramics		
Continued osteoblastic phenotypic properties	MSC	Biphasic calcium phosphate
Metals		
Attachment, proliferation, enhanced osteogenic differentiation in bioreactor (shear stress)	MSC	Titanium scaffold (fiber meshes)
Induction of EB formation, significant differences in gene expression	mESC	Cytomatrix™ (tatalum-based scaffold)

Source: Adapted from Willaert, R. *Fermentation Microbiology and Biotechnology*, 3rd ed., pp. 313–367, 2012. Boca Raton, FL: CRC Press.

Note: hESC, human embryonic stem cell; mESC, mouse embryonic stem cell; MSC, mesenchymal stem cell; hBMC, human bone marrow cell; hMSC, human mesenchymal stem cell; hUCB, human umbilical cord blood cells; EB, embryonic body.

produced amount of α-acetolactate ((S)-2-acetolactate) is very dependent on the used strain. The production also increases with increasing yeast growth. For classical lager fermentation, 0.6 ppm α-acetolactate is typically formed. At high aeration, this value can raise to 0.9 ppm and in cylindro-conical fermentation tanks even to 1.2–1.5 ppm. Yeast cells possess the necessary enzymes (reductases) to reduce diacetyl to acetoin and further to 2,3-butanediol and to reduce 2,3-pentanedione to 2,3-pentanediol. These reduced compounds have much higher taste thresholds and have no effect on the beer flavor. The reduction reactions are dependent on the yeast strain and occur during the course of maturation after fermentation. Sufficient yeast cells in suspension are necessary to obtain an efficient reduction. Yeast strains, which flocculate early during the main fermentation, need a long maturation time to reduce the vicinal diketones.

> **Green beer**: During the beer fermentation, wort is fermented to beer. Wort is the carbohydrate extract from grinded barley malt. The fermentation proceeds in two stages. The first stage is called the "primary fermentation" and the second one is the "secondary fermentation" or "maturation." The obtained beer after the primary fermentation is called "green" beer.

The traditional maturation process is characterized by a near-zero temperature, low pH, and low yeast concentration, resulting in a very long maturation period of 3–4 weeks. Nowadays, the maturation phase is considerably reduced to approximately one week because several strategies are used to accelerate the vicinal diketone removal. Using immobilized cell technology, this maturation period can be further reduced to a time range of a few hours. Examples of immobilized cell technology (ICT) maturation processes are illustrated in Table 17.14. An example of a process scheme is illustrated in Figure 17.7.

It should also be mentioned that accelerated beer maturation can also be performed by increasing the maturation temperature, which can be realized by integrating the secondary fermentation in the primary fermentation. In this way, the production of beer can be accomplished in less than two weeks in one cylindroconical vessel.

17.5.2 PRODUCTION OF ALCOHOL-FREE OR LOW-ALCOHOL BEER

The technology to produce alcohol-free or low-alcohol beer is based on the suppression of alcohol formation by arrested

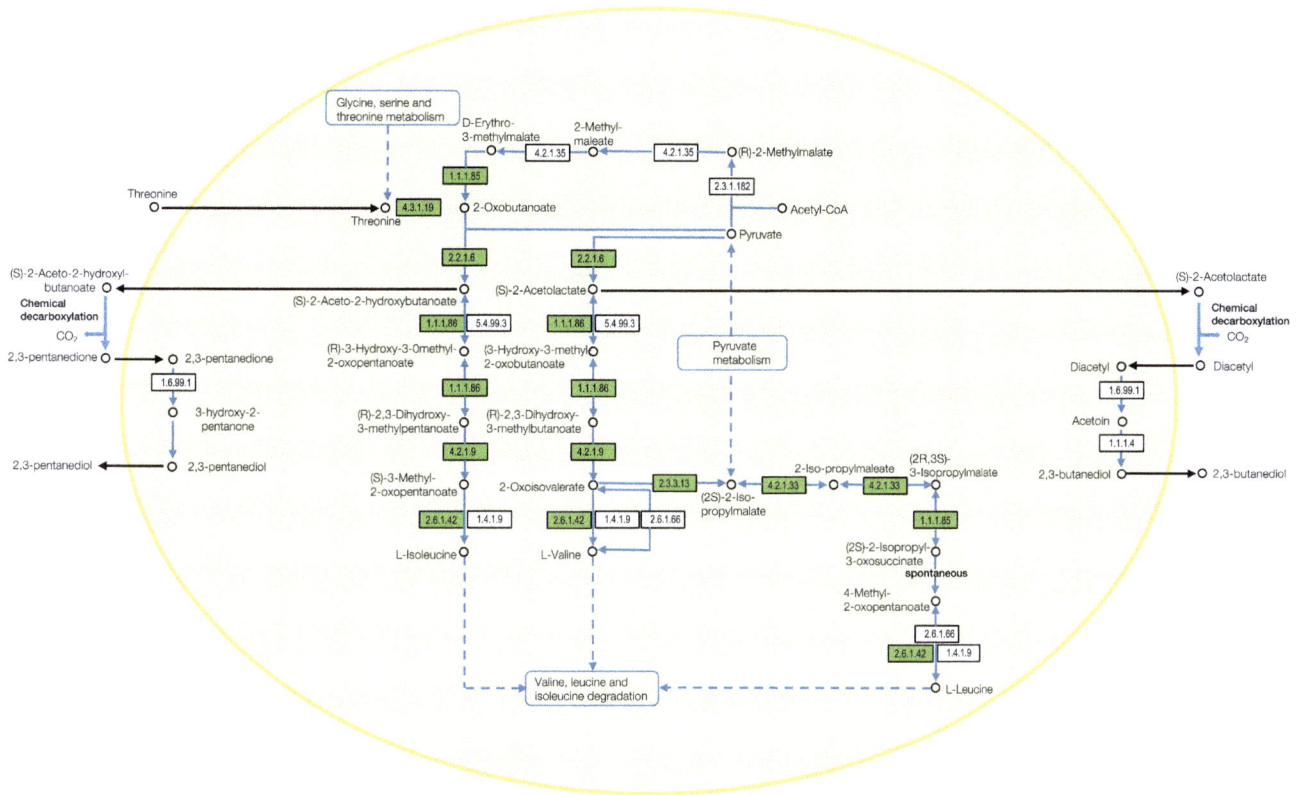

FIGURE 17.6 The synthesis and reduction of the vicinal diketones diacetyl and 2,3-pentanedione in *S. cerevisiae*. Adapted KEGG pathway sce00290 (kegg.jp).

(restricted) free-cell batch fermentation. However, the resulting beers are characterized by an undesirable wort aroma because the wort aldehydes have only been reduced to a limited degree. An alternative method of producing these beers is based on the removal of ethanol from stronger beers by using membrane, distillation, or vacuum evaporation processes. These methods have the disadvantage that the production cost is increased.

Controlled ethanol production for low-alcohol and alcohol-free beers have been successfully achieved by partial fermentation using immobilized yeast (Table 17.15). The reduction of the wort aldehydes can be quickly achieved by a short contact with the immobilized yeast cells. The process is performed at a low temperature to avoid undesirable cell growth

and ethanol production. A disadvantage of this short-contact process is the production of only a small amount of desirable esters. Nuclear mutants of *S. cerevisiae*, which are defective in the synthesis of tricarboxylic acid cycle enzymes (i.e., fumarase or 2-oxoglutarate dehydrogenase) can be used for the production of nonalcohol beer because they produce minimal amounts of ethanol, but much lactic acid. These mutants have been used in a continuous immobilized-cell process (Navratil et al., 2000).

17.5.3 Continuous Main Fermentation

Traditional beer fermentation technology uses freely suspended yeast cells to ferment wort in a non-stirred batch

TABLE 17.14

ICT for Beer Maturation

Immobilization Method	Immobilization Matrix	Reactor Type	Scale
Surface attachment	DEAE-cellulose beads	Packed bed	Industrial
Entrapment	Porous glass beads	Packed-bed	Industrial
Entrapment	Polyvinyl alcohol beads	Gas lift	Pilot
Entrapment	Calcium alginate beads	Fluidised-bed	Laboratory

Source: Adapted from Willaert, R., *Handbook of Food Products Manufacturing*, pp. 443–506, 2007. Hoboken, NJ: John Wiley & Sons; Willaert, R. *Fermentation Microbiology and Biotechnology*, 3rd ed., pp. 313–367, 2012. Boca Raton, FL: CRC Press.

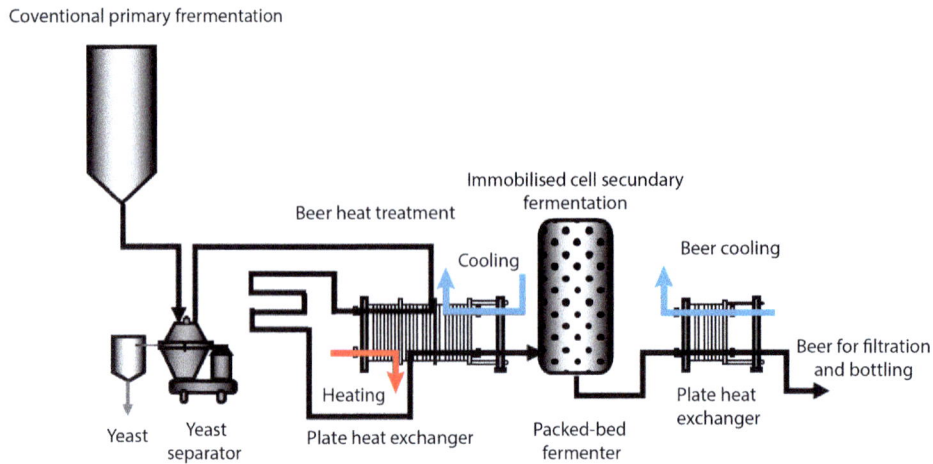

FIGURE 17.7 Process scheme for the maturation of beer using immobilized yeast cells.(Modified after Willaert, R., *Minerva Biotechnot*, 12:319–330, 2000).

TABLE 17.15
ICT for the Production of Alcohol-Free or Low-Alcohol Beer

Immobilization Method	Immobilization Matrix	Reactor Type	Scale
Surface attachment	DEAE-cellulose	Packed bed	Laboratory
Surface attachment	DEAE-cellulose	Packed bed	Industrial
Entrapment	Porous glass	Fluidized bed	Pilot
Entrapment	Silicon carbide rods	Cartridge loop reactor	Pilot
Entrapment	Calcium alginate	Gas lift	Laboratory
Entrapment	Calcium pectate	Packed bed	Laboratory

Source: Adapted from Willaert, R., *Handbook of Food Products Manufacturing*, pp. 443–506, 2007. Hoboken, NJ: John Wiley & Sons; Willaert, R. *Fermentation Microbiology and Biotechnology*, 3rd ed., pp. 313–367, 2012. Boca Raton, FL: CRC Press.

reactor. These fermentations are very time-consuming. The traditional primary fermentation for lager beer takes approximately seven days with a subsequent secondary fermentation (maturation) of several weeks. However, the resulting beer has a well-balanced flavor profile that is very well accepted and appreciated by the consumer. Nowadays, large breweries use an accelerated fermentation scheme that is based on using a higher fermentation temperature and specific yeast strains, thus facilitating the speedy production of beer—approximately 15 days.

Narziss and Hellich (1971) developed one of the first well-described ICT processes for beer production. Yeast cells were immobilized in kieselguhr (which is widely used in the brewing industry as a filter aid), and a kieselguhr filter was used as bioreactor (called the "bio-brew bioreactor"). This process was characterized by a very low residence time of 2.5 h, but required the addition of viable yeast and a 7-d maturation period to reduce the high concentration of vicinal diketones in the green beer. Although this result looked very good, the bio-brew bioreactor gave no satisfying result overall because it contained a high amount of α-acetolactate; this was later optimized (Dembowski et al., 1993). An aerobic reactor was installed in front of the bio-brew reactor, the beer flow though

the filter was optimized, and a cooling plate was installed in the filter reactor to control the temperature, thus increasing cell viability and improving the organolyptic qualities of beer. However, the concentration of the low-molecular-weight nitrogenous substances in the beer remained too high.

Baker and Kirsop (1973) were the first to use heat treatment of green beer to considerably accelerate the chemical conversion of α-acetolactate to diacetyl. They designed a two-step continuous process. The first reactor was a packed-bed reactor, also containing kieselguhr as immobilization matrix, to perform the primary fermentation. The green beer was heated using a heating coil to accelerate the α-acetolactate conversion. It was next cooled before it entered a smaller packed-bed reactor to perform the secondary fermentation. Problems associated with this process were a gradual blocking of the packed bed and a changed beer flavor.

The design and optimization of an ICT process for primary and secondary fermentations remain a challenging task, although encouraging results have been obtained recently albeit only on laboratory and pilot scales (Table 17.16). The reasons why these ICT processes have not yet been adopted in the brewing industry include complexity of operations compared with batch processes, flavor problems (because of a lack

TABLE 17.16
ICT for the Main Beer Fermentation

Immobilization Method	Immobilization Matrix	Reactor Type
Entrapment	Calcium alginate beads	Stirred tank[a] + 2 packed bed[b]
Entrapment	Ceramic beads	Stirred tank[a] + 2 packed bed[b]
Entrapment	Calcium alginate microbeads	Gas lift
Entrapment	Silicon carbide rods	Cartridge loop reactor[b] + stirred tank[a]
Entrapment	Porous organic or glass beads	Packed bed
Entrapment	κ-Carrageenan beads	Gas lift
Entrapment	Chitosan beads	Fluidized bed
Entrapment	Calcium pectate beads	Gas lift
Entrapment	PVA beads	Gas lift
Adsorption	Woodchips (aspen, beech)	Packed bed
Adsorption	DEAE-cellulose	Packed bed
Adsorption	Gluten pellets	Fluidized bed
Adsorption	Spent grains	Gas lift

Source: Adapted from Willaert, R., *Handbook of Food Products Manufacturing*, pp. 443–506, 2007. Hoboken, NJ: John Wiley & Sons; Willaert, R. *Fermentation Microbiology and Biotechnology*, 3rd ed., pp. 313–367, 2012. Boca Raton, FL: CRC Press. [a] Free cells. [b] Immobilized cells.

of understanding and controllability of the changed metabolism), yeast viability, and carrier price. The altered metabolism and the knowledge to tune the metabolism to the desired flavor especially await further investigation.

17.6 CELL IMMOBILIZATION NANOBIOTECHNOLOGY

17.6.1 NANOBIOTECHNOLOGY

Nanotechnology is the ability to work at the atomic, molecular, and supramolecular levels (on a scale of ~1–100 nm) to understand, create, and use material structures, devices, and systems with fundamentally new properties and functions resulting from their small structure. Nanobiotechnology is defined as a field that applies the nanoscale principles and techniques to understand and transform biosystems (living or nonliving) and that uses biological principles and materials to create new devices and systems integrated from the nanoscale. The biological and physical sciences share a common interest in small structures (the definition of "small" depends on the application but can range from 1 nm to 1 mm). Biological structures are relatively large compared with structures in electronics and in physical nanosciences. A microbial cell is approximately 1 μm, a fungal cell is 5 μm, and a mammalian cell is 10 μm when rounded and 50 μm when fully spread in attached culture. A vigorous trade across the borders of these areas of science is developing around new materials and tools (largely from the physical sciences) and new phenomena (largely from the biological sciences). The physical sciences offer tools for synthesis and fabrication of devices for measuring the characteristics of cells and subcellular components and of materials useful in cell and molecular biology. Biology

offers a window into the most sophisticated collection of functional nanostructures that exist.

Cell immobilization applications are directed to the control of the interaction of cells with its microenvironment (cell-substrate and cell-cell interaction) by using nanofabrication technologies. Some examples of applications of nanotechnology that are linked to cell immobilization are listed in Table 17.17. These technologies are emerging as powerful tools for tissue engineering and regenerative medicine. Precise control of the cellular environment provides new opportunities for understanding biochemical and mechanical processes responsible for changes in cell behavior; for example, the effects of cell shape on the anchorage dependence of cell growth (Chen et al., 1997), engineering stem cell microenvironment to control stem-cell fate *in vitro* and *in vivo* (see Section 19.4.3), and the effect of surface patterning for spatially controlling cell attachment (Table 17.17).

Several techniques are currently being used to create nanoscale topographies for cell scaffolding. These techniques fall into two main categories: techniques that create ordered topographies and those that create unordered topographies. Electron beam lithography and photolithography are two standard techniques for creating ordered features. Polymer demixing, phase separation, colloidal lithography, and chemical etching are most typically used for creating unordered surface patterns.

17.6.2 MICROSCALE TECHNOLOGIES IN CELL IMMOBILIZATION

Because of the micromolar scale of cells, microscale technologies are potential tools for addressing some of the challenges in cell immobilization biotechnology. In particular,

TABLE 17.17

Some Examples of Cell Immobilization Applications of Nanotechnology

Application

Surfaces patterned with self-assembled monolayers (SAMs) to guide cell attachment and growth. SAMs terminating in methyl groups are hydrophobic and adsorb proteins from solution. SAMs terminating in oligo(ethylene glycol) moieties resist the adsorption of proteins from solution. When cells attach to a surface, they do not, in general, attach to the surface directly: they attach to proteins adsorbed on the surface. The combination of SAMs with "printing" using elastomeric stamps ("soft lithography") or dip-pen nano-lithography, allows the surface to be patterned into regions to which cells attach and regions to which they do not.

Self-assembling peptide-based biomaterials for use as three-dimensional scaffolds for cell cultures, tissue engineering, and regenerative medicine.

Nanostructured polymer scaffolds for tissue engineering and regenerative medicine.

The structural features of scaffolds are engineered to support cell adhesion.

Engineering substrate mechanics to change the behavior of cells. The mechanical environment surrounding the cell and the intracellular cytoskeletal mechanics play an important role in determining the magnitude of these forces and the resulting changes in cellular behavior.

Dynamically changing the adhesive environment during cell culture to locally regulate cell adhesion:
- using thermally responsive materials
- using electroactive substrates

Engineering stem cell microenvironment to control stem cell differentiation. Two- and three-dimensional materials can be designed to control stem-cell fate *in vitro* and *in vivo* (see also Section 19.4.3).

Source: Adapted from Willaert, R. *Fermentation Microbiology and Biotechnology*, 3rd ed., pp. 313–367, 2012. Boca Raton, FL: CRC Press.

they are potentially powerful tools for addressing some of the challenges in tissue engineering. MEMS (microelectro-mechanical systems), which are an extension of the semiconductor ad microelectronics industries, can be used to control features at length scales from less than 1 µm to more than 1 cm. These techniques are compatible with cells and are now being integrated with biomaterials to facilitate fabrication of cell-material composites that can be used for tissue engineering. In addition, microsystems create new opportunities for the spatial and temporal control of cell growth and stimuli by combining surfaces that mimic complex biochemistries and geometries of the extracellular matrix with microfluidic channels that regulate transport of fluids and soluble factors.

Microfluidics involves the manipulation of very small fluid volumes, enabling the creation and control of micro- to nano-liter volume reactors. Microfluidic devices can be fabricated using soft lithography and rapid prototyping techniques. Soft lithography refers to a collection of techniques for creating microstructures and nanostructures that are based on printing, molding, and embossing. The advantages of soft lithography to fabricate microfluidic devices include low cost of production, the ease and speed of fabrication, the reduction in the amount of reagents consumed, and compatibility with cells. In rapid prototyping, a computer-aided design program is used to create a design for channels, which are printed at high resolution onto transparency film. The transparency film then serves as the photomask. The master molds are generated by using the photomask in contact lithography to produce a positive relief of photoresist. In replica molding, poly(dimethylsiloxane) (PDMS) is poured over the master and heat-cured to generate a negative replica of the master. The PDMS is then removed from the mold and sealed against a glass coverslip to form the device features and channels.

Microsystems can be integrated with bioanalytic microsystems resulting in multifunctional platforms for basic biological insights into cells and tissues, as well as for cell-based sensors with biochemical, biomedical, and environmental functions. Highly integrated microdevices show great promise for basic biomedical and pharmaceutical research. Much cell-based microsystem research takes place under "lab-on-a-chip" or "micro-total-analysis-system" (µTAS) framework that seeks to create microsystems incorporating several steps of an assay into a single system. Integrated microfluidic devices perform rapid and reproducible measurements on small sample volumes while eliminating the need for labor-intensive and potentially error-prone laboratory manipulations. Microscale technologies can miniaturize assays and facilitate high-throughput experimentation and therefore provide a tool for screening libraries. Robotic spotters capable of dispensing and immobilizing nanoliters of material have been used to fabricate cellular microarrays in which cells can be screened in a high-throughput manner (Willaert and Goossens, 2015). Recently, dip-pen nanolithography (DPN) has been used to immobilize motile bacterial cells in a cellular microarray (Nyamjav et al., 2010). DPN is a scanning probe microscopy-based nanofabrication technique that uniquely combines direct-write soft-matter compatibility with the high resolution and registry of atomic force microscopy (AFM) (see Section 17.6.3), which makes it a powerful tool for depositing soft and hard materials in the form of stable and functional architectures on various surfaces (Kumar et al., 2016).

17.6.3 Microscopic Techniques for Nanoscale Imaging and Manipulation

We use microscopy to see objects in more detail. The best distance that one can resolve with optical instruments, disregarding all aberrations, is approximately 0.5 times the wavelength of light, or on the order of 250 nm with visible radiation. High-resolution microscopy techniques that are used for nanoimaging and nanoscale characterization have been developed. They can be divided into three categories: optical microscopes, scanning probe microscopes (SPMs), and electron microscopes. Recently developed microscopy-based technologies can also be used to control and manipulate objects at the nanoscale. The unique imaging and manipulation properties of AFMs have prompted the emergence of several probe-based nanolithographies. Scanning-probe-based patterning techniques, such as dip-pen lithography, local force-induced patterning, and local-probe oxidation-based techniques are highly promising because of their relative ease and widespread availability. The latter of these is especially interesting because of the possibility of producing nanopatterns for a broad range of chemical and physical modification and functionalization processes.

17.6.3.1 Electron Microscopy

Microscopes consist of an illumination source, a condenser lens to converge the beam on the sample, an objective lens to magnify the image, and a projector lens to project the image onto an image plane, which can be photographed or stored. In electron microscopes, the wave nature of the electron is used to obtain an image. There are two important forms of electron microscopy: scanning electron microscopy (SEM) and transmission electron microscopy (TEM). Both use electrons as the source for illuminating the sample. The lenses used in electron microscopes are electromagnetic lenses.

For high-resolution surface investigations, two commonly used techniques are atomic force microscopy and scanning electron microscopy SEM. The operation of the SEM consists of applying a voltage between a conductive sample and filament, resulting in electron emission from the filament to the sample. This occurs in a vacuum environment. The electrons are guided to the sample by a series of electromagnetic lenses in the electron column. The resolution and depth of field of the image are determined by the beam current and the final spot size. The electrons interact with the sample within a few nanometers to several microns of the surface, depending on the beam parameters and sample type.

Along with the secondary electron emission, which is used to form a morphological image of the surface in the SEM, several other signals are emitted as a result of the electron beam impinging on the surface. Each of these signals carries information about the sample that provides clues to its composition. Two of the most commonly used signals for investigating composition are x-rays and backscattered electrons. X-ray signals are commonly used to provide elemental analysis. The percentage of beam electrons that become backscattered electrons has been found to be dependent on the atomic number of the material, which makes it a useful signal for analyzing the material composition.

Electron microscopy is conducted in a vacuum environment. This is a disadvantage to study hydrated samples, like attached/immobilized cells and some immobilization materials (i.e., hydrogels). To image poorly conductive surfaces without sample charging may require conductive coatings or staining, which may alter or obscure the features of interest;

or it may require low-voltage operation or an environmental chamber, which may sacrifice resolution. Recently, electron microscopy techniques have been developed for the imaging of whole cells in liquid that offers nanometer spatial resolution and a high imaging speed using a scanning transmission electron microscope (STEM)

In TEM, the transmitted electrons are used to create an image of the sample. Scattering occurs when the electron beam interacts with matter. Scattering can be elastic (no energy change) or inelastic (energy change). Elastic scattering can be coherent and incoherent (with and without phase relationship). TEMs with resolving powers in the vicinity of 1 Å are now common. A relative recent electron microscopy technique that can be used to study immobilized cells at the nanoscale is electron tomography.

Electron tomography (ET) is the most widely applicable method for obtaining three-dimensional information by electron microscopy. A tomogram is a three-dimensional volume computed from a series of projection images that are recorded as the object in question is tilted at different orientations. ET has the potential to fill the gap between global cellular localization and the detailed three-dimensional molecular structure because it can reveal the localization within the cellular context at true molecular resolution and the shapes and three-dimensional architecture of large molecular machines. It can also reveal the interaction of individual proteins and protein complexes with other cellular components, such as DNA and membranes. A recent development is cryo-electron tomography (cryo-ET), which allows the visualization of cellular structures under close-to-life conditions (Lucić et al., 2008). Rapid freezing followed by the investigation of the frozen-hydrated samples avoids artifacts notorious to chemical fixation and dehydration procedures. Furthermore, the biological material is observed directly, without heavy metal staining, avoiding problems in interpretation caused by unpredictable accumulation of staining material. Consequently, cryo-ET of whole cells has the advantage that the supramolecular architecture can be studied in unperturbed cellular environments.

17.6.3.2 Atomic Force Microscopy

Scanning probe microscopies are a family of instruments that are used to measure properties of surfaces, including AFMs and STMs. The main feature that all scanning probe microscopies have in common is that the measurements are performed with a sharp probe operating in the near field (i.e., scanning over the surface while maintaining a very close spacing to the surface). The STM, invented in the early 1980s by Binnig and Rohrer (1982), was the first to produce real space images of atomic arrangements on flat surfaces. The development of the STM arose from an interest in the study of the electrical properties of thin insulating layers. This led to an apparatus in which the probe-surface separation was monitored by measuring electron tunneling between a conducting surface and a conducting probe. A few years later, Binnig and colleagues (1986) announced the birth of the second member of the SPM family, the AFM (also known as the scanning probe microscopy, SPM). Numerous variations of these techniques have been developed later on.

Atomic force microscopy is extensively used for imaging surfaces ranging from micro- to nanometer scales, with the objective of visualizing and characterizing surface textures and shapes. It has evolved into an imaging method that yields structural details of biological samples such as proteins, nucleic acids, membranes, and cells in their native environment. AFM is a unique technique for providing subnanometer resolution at a reasonable signal-to-noise ratio under physiological conditions. It complements EM by allowing visualization of biological samples in buffers that preserve their native structure over extended time periods. Unlike EM, atomic force microscopy yields three-dimensional maps with an exceptionally good vertical resolution (<1 nm). Additionally, the measurement of mechanical forces at the molecular level provides detailed insights into the function and structure of biomolecular systems. Inter- and intramolecular interactions can be studied directly at the molecular level.

17.6.3.3 Light Microscopy

Since the earliest examination of cellular structures, biologists have been fascinated by observing cells using a light microscope. Being able to observe processes as they happen by light microscopy adds a vital extra dimension to our understanding of cell behavior and function. Microscopy has evolved to provide not only quantitative images but also a significant capability to perturb structure-function relationships in cells. These advances have been especially useful in the study of cell adhesion and migration: molecular interactions and dynamics, local perturbation of actin-based structures, and the traction forces exerted by motile cells on substrates can be measured.

Recent advances in fluorescence microscopy allowed imaging of structures at extremely high resolutions (Dersch and Graumann, 2017). The past decade witnessed an explosion of fluorescence-microscopy-based approaches to image protein dynamics and interactions, including fluorescence recovery after photobleaching (FRAP) or photoactivation using photoconvertible fluorescent proteins to assay protein mobility and maturation in cells and Forster resonance energy transfer (FRET) to monitor physical intra- or intermolecular associations in space and time.

Despite the advantages of standard fluorescence microscopy, ultrastructural imaging is not possible because of a resolution limit set by the diffraction of light. Therefore, the maximal spatial resolution of standard optical microscopy is approximately 200 nm, which is more than an order of magnitude higher than the length of cellular molecules. Several approaches have been used to break this diffraction limit (Table 17.18); for example, by exploiting the distribution of fluorescence intensity from a single molecule. When imaged, a fluorophore behaves as a point source with an Airy disc point spread function. The center of mass of the function, and therefore the position of the molecule, can be obtained by performing a least-squares fit of an appropriate function to the measured fluorescence intensity profile of the spot

(Thompson et al., 2002). With a sufficient number of photons, these methods can provide a localization of 1–2 nm, allowing the measurement of distances on the scale of individual proteins. Single-molecule detection offers new possibilities for obtaining subdiffraction-limit spatial resolution.

17.6.3.4 Force Microscopy

The ability to apply force to or measure forces generated by multiple (cell adhesion) up to a single biopolymer opens up new avenues for single cell interaction and manipulating biomolecules and interrogating cellular processes. Three forms of force microscopy are commonly used to study single molecules: optical tweezers, magnetic tweezers, and atomic force microscopy. An important advantage of single-molecule techniques is that they do not suffer from problems associated with population averaging inherent in ensemble measurements. Rare or transient phenomena that would otherwise be obscured by averaging can be resolved provided that the measurement technique has the required resolution and that the events can be captured often enough to ensure that they are not artifactual.

Optical tweezers use light to levitate a transparent bead of distinct refractive index. The trapped bead is suspended at the waist of the focused (typically infrared) laser beam. The displacement of the bead from the focal center results in a proportional restoring force and can be measured by interferometry or back-focal plane detection. A single biopolymer can be suspended between two beads or a bead and a motorized platform. Magnetic tweezers use an external, controllable magnetic field to exert force and/or torque on a superparamagnetic bead that is tethered to a surface via a single molecule. Atomic force microscopy-based force spectroscopy exerts pulling forces on a single attached molecule by retraction of the tip in the z direction (perpendicular to the x-y scanning plane). Cantilever bending is detected by the deflection of a laser beam onto a position-sensitive detector such as a quadrant photodiode. A piezoelectric actuator stage is used to control the positioning of the sample relative to the tip. Atomic force microscopy-based force spectroscopy is also used to study single-cell interactions (cell-cell and cell-substrate adhesion).

17.6.4 Case Study: Cell-Cell Adhesion Nanobiomechanics

Various immobilization technologies require cell-substrate and/or cell-cell adhesion. Cell adhesion in general is commonly defined as the binding of a cell to a substrate, which can be another cell, a surface, or an organic matrix. The adhesion of a biological cell involves complex couplings at the level of cellular biochemistry, structural mechanics, and/or surface bonding. The interactions are dynamic and act through association and dissociation of bonds between very large molecules at rates that change considerably under stress. Combining molecular cell biology with single-molecule force spectroscopy provides a powerful tool for exploring the complexity of cell adhesion; that is, how cell signaling processing

TABLE 17.18

Super-Resolution Optical Microscopy Techniques

Technique	Description	Spatial Resolution	Time Scale
Fluorescence imaging with one-nanometer accuracy (FIONA)	Localizes and tracks single-molecule emitters by finding the center of their diffraction-limited point-spread function (PSF).	~1.5 nm	~0.3 ms
Single-molecule high-resolution colocalization (SHREC)	Two-color version of FIONA. Two fluorescent probes with different spectra are imaged separately and then localized and mapped onto the plane of the microscope.	< 10 nm	~1 s per frame
Single-molecule high-resolution imaging with photobleaching (SHRImP)	Uses the strategy that upon photobleaching of two or more closely spaced identical fluorophores their position is sequentially determined by FIONA starting from the last bleached fluorophore.	~5 nm	~0.5 s per frame
Nanometer-localized multiple single molecules (NALMS)	Uses a similar principle as SHRImP to measure distances between identical fluorescent probes that overlap within a diffraction-limited spot.	~8 nm	~1 s per frame
Photoactivatable localization microscopy (PALM)	Serially photoactivates and photodeactivates many sparse subsets of photoactivatable fluoro-phores to produce a sequence of images that are combined into a super-resolution composite	~2 nm	~1 min
PALM with independently running acquisition (PALMIRA)	Records nontriggered spontaneous off-on-off cycles of photoswitchable fluorophores without synchronizing the detector to reach faster acquisition.	~50 nm	~2.5 min
Single particle tracking PALM	Combines PALM with live-cell single fluorescent particle tracking.		
Stimulated emission depletion (STED)	Reduces the excitation volume below that dictated by the diffraction limit, by coaligning one beam of light capable of fluorophore excitation with another that induces de-excitation by stimulated emission	~16 nm	~10 min
Stochastic optical reconstruction microscopy (STORM)	Small subpopulations of photoswitchable fluorophores are turned on and off using light of different colors, permitting the localization of single molecules. Repeated activation cycles produce a composite image of the entire sample.	<0 nm	~ min

Source: Adapted from Willaert, R. *Fermentation Microbiology and Biotechnology*, 3rd ed., pp. 313–367, 2012. Boca Raton, FL: CRC Press

strengthens adhesion bonds and how forces applied to cell-surface bonds act on intracellular sites to catalyze chemical processes or switch molecular interactions on and off.

Molecular and genetic approaches have identified various cell adhesion molecules (CAMs) with their ligand specificities and have determined the processes in which they are involved. To understand cell adhesion, the vast amount of qualitative data that is available must be augmented with quantitative data of the biophysics of adhesion. Historically, the strength of cell adhesion to a substrate has been studied using simple washing assays (Klebe, 1974). Washing assays have proven to be versatile and useful in identifying CAMs, important extra-cellular matrix components, and other proteins that are involved in various forms of cell adhesion. To estimate the force to which cells are subjected, various assays that are based on the regulated flow of media have been implemented, including flow chamber methods (Kaplanski et al., 1993). However, these assays only give estimates of adhesion forces because the shear force that is exerted on the cells depends on parameters such as cell size, cell shape, and how the cell is attached to the substrate. Recently, single-cell and single-molecule techniques have been developed to obtain more controlled and quantitative measurements of adhesion strength.

Single-cell force spectroscopy assays on living cells have been applied to measure the strength of cell adhesion down to single-molecule levels. A living cell can be attached to a tipless cantilever of an AFM and the interacting partner (molecule or cell) on a substrate-coated surface. Alternatively, the tip is functionalized with the interacting molecule and the interaction with the attached living cell on a surface is monitored. Atomic force microscopy-based force spectroscopy with a cantilever-bound cell can be used to investigate cell-cell and cell-matrix interactions. The approach and withdrawal of this cell to and from its surface can be precisely controlled by parameters such as applied force, contact time, and pulling speed by benefiting from the AFM's high-force sensitivity and spatial resolution. The data collected in these experiments include information on repulsive forces before contact, cell deformability, maximum unbinding forces, individual unbinding events, and the total work required to remove a cell from the surface. Force spectroscopy can identify cell subpopulations and characterize the regulation of cell adhesion events with single-molecule resolution.

Examples of using AFM-based force spectroscopy to study cell adhesion include single *S. cerevisiae* cells that have been attached to a tipless AFM cantilever and used as a living single-cell probe to perform single-cell force spectroscopy. For example, the contributions of several galectin family members in cell-substratum adhesion of Madin-Darby canine kidney cells have been studied using quantitative single cell atomic force microscopy-based force spectroscopy (Friedrichs et al., 2007). Optical tweezers have been used to orient uropathogenic *E. coli* (which present a FimH lectin at the tip of their type 1 pilus) relative to a mannose-presenting surface, and thus limit the number of points of attachment (Liang et al., 2000). It was possible to quantify the forces required to break a single interaction between pilus and mannose groups. Yeast cell flocculation is an example of self-aggregation of cells. The Flo proteins Flo1p, Lg-Flo1p, Flo5p, Flo9p, and Flo10p are lectins because they have an affinity toward specific sugar moieties. Recently, AFM-based force spectroscopy was used to unravel the molecular mechanism of flocculation (Goossens et al., 2015). It was demonstrated that the self-interaction of Flo proteins is established in two stages, involving both glycan-glycan and protein-glycan interactions. There is a crucial role of calcium in both types of interaction: Ca^{2+} takes part in the binding of the carbohydrate to the protein and the glycans aggregate only in the presence of Ca^{2+}.

SUMMARY

- Selections of immobilization matrix and method as well as bioreactor type and design are usually interrelated.
- For optimization, knowledge of fluid dynamics and external/internal mass transfer is a good starting point.
- The use of mathematical models to quantitatively describe and analyze the behavior of immobilized cells is an important component; steady-state, pseudo-steady state, and dynamic models are discussed.
- The physiology of living cells upon immobilization can be changed because they are present in a different microenvironment compared with free living cells.
- Different physiological effects due to the changed chemical composition and/or the physical interaction between the matrix material and immobilized bacterial, fungal, and stem cells are discussed.
- The production of beer using immobilized cell technology has been described by way of a case study.
- Cell immobilization has been implemented for the production of alcohol-free lager, flavor maturation of green beer, and continuous main stream fermentation.
- The use of cell immobilization technology has been successful at the industrial level in the production of biological reagents from animal cell cultures and tissue engineering. On the other hand, the full potential of cell immobilization for the production of biopharmaceuticals using microbial cells has yet to be fully realized and exploited. A few reasons are that the biopharmaceutical (fermentation) industry is very conservative and reluctant to replace the existing robust systems using free cells with this new technology; an immobilized cell system is more complex and means a higher risk; and validation, regulatory, and health and safety issues are costly and time-consuming.
- Successful industrial/commercial fermentation processes/products that are based on immobilized living cells have recently been introduced. Examples include the production of alcohol-free beer, stabilized probiotic products, protective bacterial cultures for meat products, and bioethanol as well as wastewater treatment.
- For the production of low-added-value fermentation products, the added cost of immobilization techniques and in some cases higher complexity *versus* free cell culture will not be justified unless the volumetric productivity is strongly increased.
- Future research will increase our understanding of these "complex" systems and will lower the barrier for exploitation on an industrial/commercial and medical scale. Nanobiotechnology techniques will be applied more and more to characterize these complex systems and used to create nanoscale topographies for cell immobilization.

REFERENCES

Argirakos, G., K. Thayanithy, and D.A. John Wase. 1992. Effect of immobilization on the production of α-amylase by an industrial strain of *Bacillus*. *J Chem Technol Biotechnol* 53:33–38.

Baker, D.A., and B.H. Kirsop. 1973. Rapid beer production and conditioning using a plug fermentor. *J Inst Brew* 79:487–494.

Baron, G.V., R.G. Willaert, and L. De Backer. 1996. Immobilized cell reactors. In Willaert, R.G., Baron, G.V., and De Backer, L. eds., *Immobilized Living Cell Systems: Modeling and Experimental Methods*, pp. 67–95. Chichester, UK: John Wiley & Sons.

Binnig, G., and H. Rohrer. 1982. Scanning tunneling microscopy. *Helv Phys Acta* 55:726–735.

Binnig, G., C.F. Quate, and C. Gerber. 1986. Atomic force microscope. *Phys Rev Lett* 56:930–933.

Brito, L.C., A.M. Vieira, J.G. Leitão, I. Sá-Correia, J.M. Novais, and J.M.S. Cabral. 1990. Effect of the aqueous soluble components of the immobilization matrix on ethanol and microbial exopolysaccharides production. In de Bont J.A.M., Visser J., Mattiasson B., Tramper J. eds., *Physiology of Immobilized Cells*, pp. 399–404. Amsterdam, The Netherlands: Elsevier.

Buitelaar, R.M., A.C. Hulst, and J. Tramper. 1990. Cell immobilization in thermogels: Activity retention after gelling in various organic solvents. In de Bont, J.A.M., Visser, J., Mattiasson, B., and Tramper, J. eds., *Physiology of Immobilized Cells*, pp. 205–208. Dordrecht, The Netherlands: Elsevier.

Cachon, R., and C. Divies. 1993. Localization of *Lactococcus lactis* ssp. *lactis* bv. *diacetylactis* in alginate gel beads affects biomass density and synthesis of several enzymes involved in lactose and citrate metabolism. *Biotechnol Technol* 7:453–456.

Carenza, M., and F.M. Veronese. 1994. Entrapment of biomolecules into hydrogels obtained by radiation-induced polymerization. *J Control Rel* 29:187–193.

Chen, K.C., and Y.F. Lin. 1994. Immobilization of microorganisms with phosphorylated polyvinyl alcohol (PVA) gel. *Enzyme Microb Technol* 16:79.

Chen, C.S., M. Mrksich, S. Huang, G.M. Whitesides, and D.E. Ingber. 1997. Geometric control of cell life and death. *Science* 276:1425–1428.

Dale, M.C., A. Eagger, M.R. Okos. 1994. Osmotic inhibition of free and immobilized *K. marxianus* anaerobic growth and ethanol productivity in whey permeate concentrate. *Process Biochem* 29:535–544.

de Alteriis, E., P.M. Alepuz, F. Estruch, and P. Parascandola. 1999. Clues to the origin of high external invertase activity in immobilized growing yeast: Prolonged *SUC2* transcription and less susceptibility of the enzyme to endogenous proteolysis. *Can J Microbiol* 45:413–417.

Dembowski, K., L. Narziss, and H. Miedaner. 1993. Technologisch optimierte Bierherstellung im Festbettfermentor bei sehr kurzer Produktionszeit. *Proceedings 24th European Brewery Convention Congress*, Oslo, Norway; pp. 299–306.

Dersch S., and P.L. Graumann. 2017. The ultimate picture-the combination of live cell superresolution microscopy and single molecule tracking yields highest spatio-temporal resolution. *Curr Opin Microbiol* 43:55–61.

Eş, I., J.D. Vieira, and A.C Amaral. 2015. Principles, techniques, and applications of biocatalyst immobilization for industrial application. *Appl Microbiol Biotechnol* 99:2065–2082.

Epstein, N. 1989. On tortuosity and the tortuosity factor in flow and diffusion through porous media. *Chem Eng Sci* 44:777–779.

Fortin, C., and J.C. Vuillemard. 1990. Elucidation of the mechanism involved in the regulation of protease production by immobilized *Myxococcus xanthus* cells. *Biotechnol Lett* 12:913.

Freed, L.E., J.C. Marquis, G. Vunjak-Novakovic, J. Emmanual, and R. Langer. 1994. Composite of cell-polymer cartilage implants. *Biotechnol Bioeng* 43:605–614.

Friedrichs, J., J.M. Torkko, J. Helenius, T.P. Teräväinen, J. Füllekrug, D.J. Muller, K. Simons, and A. Manninen. 2007. Contributions of galectin-3 and -9 to epithelial cell adhesion analyzed by single cell force spectroscopy. *J Biol Chem* 282:29375–29383.

Gemeiner, P., L. Texova-Benkova, F. Svec, and O. Norrlöw. 1994. Natural and synthetic carriers suitable for immobilization of viable cells, active organelles, and molecules. In Veliky, I.A., and McLean, R.J.C., eds., *Immobilized Biosystems: Theory and Practical Applications*, pp. 1–128. Glasgow, UK: Blackie Academic & Professional.

Gherardini, L., C.M. Cousins, J.J. Hawkes, J. Spengler, S. Radel, H. Lawler, B. Devcic-Kuhar, M. Groschl, W.T. Coakley, and A.J. McLoughlin, 2005. A new immobilization method to arrange particles in a gel matrix by ultrasound standing waves. *Ultrasound Med Biol* 31:261–272.

Gilson, C., and A. Thomas. 1995. Ethanol production by alginate immobilized yeast in a fluidised bed bioreactor. *J Chem Technol Biotechnol* 62:38–45.

Goossens, K.V.Y., and R.G. Willaert. 2010. Flocculation protein structure and cell-cell adhesion mechanism in *Saccharomyces cerevisiae*. *Biotechnol Lett* 32:1571–1585.

Goossens, K.V., F.S. Ielasi, I. Nookaew, I. Stals, L. Alonso-Sarduy, L. Daenen, S.E. Van Mulders, C. Stassen, R.G. van Eijsden, V. Siewers, F.R. Delvaux, S. Kasas, J. Nielsen, B. Devreese, and R.G. Willaert. 2015. Molecular mechanism of flocculation self-recognition in yeast and its role in mating and survival. *MBio* 6:e00427–15.

Gutowska, A., B. Jeong, and M. Jasionowski. 2001. Injectable gels for tissue engineering. *Anat Rec* 263:342–349.

Hart, T., A. Zhao, A. Garg, S. Bolusani, and E.M. Marcotte. 2009. Human cell chips: Adapting DNA microarray spotting technology to cell-based imaging assays. *PLoS One* 4:e7088.

Hoffman, A.S. 2004. Applications of "smart polymers" as biomaterials. In Ratner, B.D., Hoffman, A.S., Schoen, F.J., and Lemons, J.E. eds., *Biomaterials Science—An Introduction to Materials in Medicine*, pp. 107–115. London: Elsevier Academic Press.

Kaplanski, G., C. Farnarier, O. Tissot, A. Pierres, A.M. Benoliel, M.C. Alessi, S. Kaplanski, and P. Bongrand. 1993. Granulocyte-endothelium initial adhesion. Analysis of transient binding events mediated by E-selectin in a laminar shear flow. *Biophys J* 64:1922–1933.

Kikuchi, K.I., T. Sugarawa, and H. Ohashi. 1988. Correlation of liquid-side mass transfer coefficient based on the new concept of specific power group. *Chem Eng Sci* 43:2533–2540.

Klausen, M., A. Heydorn, P. Ragas, L. Lambertsen, A. Aaes-Jørgensen, S. Molin, and T. Tolker-Nielsen. 2003. Biofilm formation by *Pseudomonas aeruginosa* wild type, flagella and type IV pili mutants. *Mol Microbiol* 48:1511–1524.

Klebe, R.J. 1974. Isolation of a collagen-dependent cell attachment factor. *Nature* 250:248–251.

Klinkenberg, G., K.Q. Lystad, D.W. Levine, and N. Dyrset. 2001. pH-controlled cell release and biomass distribution of alginate-immobilized *Lactococcus lactis* subsp. *lactis. J Appl Microbiol* 91:705–714.

Krasňan, V., R., Stloukal, M., Rosenberg, and M. Rebroš. 2016. Immobilization of cells and enzymes to LentiKats®. *Appl Microbiol Biotechnol* 100:2535–2553.

Kumar, R., S. Weigel, R. Meyer, C.M. Niemeyer, H. Fuchs, Hirtz M. 2016. Multi-color polymer pen lithography for oligonucleotide arrays. *Chem Commun (Camb)* 52:12310–12313.

Liang, M.N., S.P. Smith, S.J. Metallo, I.S. Choi, M. Prentiss, and G.M. Whitesides. 2000. Measuring the forces involved in polyvalent adhesion of uropathogenic *Escherichia coli* to mannose-presenting surfaces. *Proc Natl Acad Sci USA* 97:13092–13096.

Lijun, X., W. Bochu, L. Zhimin, D. Chuanren, W. Qinghong, and L. Liu. 2005. Linear alkyl benzene sulphonate (LAS) degradation by immobilized *Pseudomonas aeruginosa* under low intensity ultrasound. *Colloids Surf B Biointerfaces* 40:25–29.

Lim, F., and A.M. Sun, 1980. Microencapsulated islets as a bioartificial endocrine pancreas. *Science* 210:908–910.

Loo, C.Y., D.A. Corliss, and N. Ganeshkumar. 2000. *S. gordonii* biofilm formation: Identification of genes that code for biofilm phenotypes. *J Bacteriol* 182:1374–1382.

Lucić, V., A. Leis, and W. Baumeister. 2008. Cryo-electron tomography of cells: Connecting structure and function. *Histochem Cell Biol* 130:185–196.

Maheshwari, G., G. Brown, D.A. Lauffenburger, A. Wells, L.G. Griffith. 2000. Cell 2018adhesion and motility depend on nanoscale RGD clustering. *J Cell Sci* 113:1677–1686.

Merchant, F.J.A., A. Margaritis, J.B. Wallace. 1987. A novel technique for measuring solute diffusivities in entrapment matrices used in immobilization. *Biotechnol Bioeng* 30:936–945.

Moo-Young, M., and H.W. Blanch. 1981. Design of biochemical reactors: Mass transfer criteria for simple and complex systems. *Adv Biochem Eng* 19.1–69.

Narziss, L., and P. Hellich. 1971. Ein Beitrag zur wesentlichen Beschleunigung der Gärung und Reifung des Bieres. *Brauwelt* 111:1491–1500.

Navrátil, M., P. Gemeiner, E. Sturd'k, Z. Dömény, D. Smogrovicová, and Z. Antalova. 2000. Fermented beverages produced by yeast cells entrapped in ionotropic hydrogels of polysaccharide nature. *Minerva Biotec* 12:337–344.

Nilsson, K., P. Brodelius, and K. Mosbach. 1987. Entrapment of microbial and plant cells in beaded polymers. *Methods Enzymol* 135:222–230.

Noordsij, P., and J.W. Rotte. 1967. Mass transfer coefficients to a rotating and a vibrating sphere. *Chem Eng Sci* 22:1475–1481.

Norton, S., K. Watson, and T. D'Amore. 1995. Ethanol tolerance of immobilized brewers' yeast cells. *Appl Microbiol Biotechnol* 43:18–24.

Nyamjav, D., S. Rozhok, and R.C. Holz. 2010. Immobilization of motile bacterial cells via dip-pen nanolithography. *Nanotechnology* 21:235105.

Øyaas, J., I. Storrø, M. Lysberg, H. Svendsen, and D.W. Levine. 1995. Determination of effective diffusion coefficients and distribution constants in polysaccharide gels with non-steady-state measurements. *Biotechnol Bioeng* 47:501–507.

Polakovič, M., J. Švitel, M. Bučko, J. Filip, V. Neděla, M.B. Ansorge-Schumacher, P. Gemeiner. 2017. Progress in biocatalysis with immobilized viable whole cells: Systems development, reaction engineering and applications. *Biotechnol Lett* 39:667–683.

Perrot, F., M. Hebraud, R. Charlionet, G.A. Junter, and T. Jouenne. 2001. Cell immobilization induces changes in the protein response of *Escherichia coli* K-12 to a cold shock. *Electrophoresis* 22:2110–2119.

Riddle, K.W., and D. Mooney. 2004. Biomaterials for cell immobilization: A look at carrier design. In Nedovic, V., and Willaert, R. eds., *Fundamentals of Cell Immobilization Biotechnology*, pp. 15–32. Dordrecht, The Netherlands: Kluwer Academic Publishers.

Sanderson, G.R., V.L. Bell, and D.A. Ortega. 1989. A comparison of gellan gum, agar, κ-carrageenan and algin. *Cereal Foods World* 34:991–998.

Sherwood, T.K.A., R.L. Pigford, and C.R. Wilke. 1975. *Mass Transfer*, London, UK: McGraw-Hill.

Skariyachan, S., V.S. Sridhar, S. Packirisamy, S.T. Kumargowda, and S.B. Challapilli. 2018. Recent perspectives on the molecular basis of biofilm formation by *Pseudomonas aeruginosa* and approaches for treatment and biofilm dispersal. *Folia Microbiol* (Praha). Jan 19. doi:10.1007/s12223-018-0585-4.

Song, S., and S. Roy. 2016. Progress and challenges in macroencapsulation approaches for type 1 diabetes (T1D) treatment: Cells, biomaterials, and devices. *Biotechnol Bioeng* 113:1381–1402.

Stewart, G.G. 2017. The production of secondary metabolites with flavour potential during brewing and distilling wort fermentations. *Fermentation* 3:63.

Sungur, S., and U. Akbulut. 1994. Immobilization of β-galactosidase onto gelatin by glutaraldehyde and chromium(III) acetate. *J Chem Technol Biotechnol* 59:303–306.

Thompson, R.E., D.R. Larson, and W.W. Webb. 2002. Precise nanometer localization analysis for individual fluorescent probes. *Biophys J* 82:2775–2783.

van't Riet, K., and J. Tramper. 1991. *Basic Bioreactor Design*. New York: Marcel Dekker.

Varfolomeyev, SD., E.I. Rainina, V.I. Lozinsky, S.V. Kalyuzhny, A.P. Sinitsyn, T.A. Makhlis, G.P. Bachurina, I.G. Bokova, O.A. Sklyankina, and E.B. Agafonov. 1990. Application of polyvinyl alcohol cryogels for immobilization of mesophilic and thermophilic micro-organisms. In de Bont, J.A.M., Visser, J., Mattiasson, B., Tramper, J. eds., *Physiology of Immobilized Cells*, pp. 325–330. Dordrecht, The Netherlands: Elsevier.

Vilain, S., P. Cosette, M. Hubert, C. Lange, G.A. Junter, and T. Jouenne. 2004. Proteomic analysis of agar gel entrapped *Pseudomonas aeruginosa*. *Proteomics* 4:1996–2004.

Willaert, R. 2000. Beer production using immobilised cell technology. *Minerva Biotechnol* 12:319–330.

Willaert, R. 2007. The beer brewing process: Wort production and beer fermentation. In Hui, Y.N., ed., *Handbook of Food Products Manufacturing*, pp. 443–506. Hoboken, NJ: John Wiley & Sons.

Willaert, R. 2012. Cell immobilization and its applications in biotechnology: Current trends and future prospects. In El-Mansi, E.M.T., Bryce, C.F.A., Demain, A.L., and Allman, A.R. eds., *Fermentation Microbiology and Biotechnology*, 3rd ed., pp. 313–367, Boca Raton, FL: CRC Press.

Willaert, R. 2009. Engineering aspects of cell immobilization. In *Encyclopedia of Industrial Biotechnology*, pp. 1385–1413. Hoboken, NJ: John Wiley & Sons.

Willaert, R.G., and G.V. Baron. 1996. Gel entrapment and microencapsulation: Methods, applications and engineering principles. *Rev Chem Eng* 12:1–205.

Willaert, R., and G.V. Baron. 1993. Growth kinetics of gel-immobilized yeast cells studied by on-line microscopy. *Appl Microbiol Biotechnol* 39:347–352.

Willaert, R.G., and K. Goossens. 2015. Microfluidic bioreactors for cellular microarrays. *Fermentation* 1(1):38–78.

Wu, K.Y.A., and K.D. Wisecarver. 1992. Cell immobilization using PVA crosslinked with boric acid. *Biotechnol Bioeng* 39:447–449.

Żur, J., D. Wojcieszyńska, and U. Guzik. 2016. Metabolic responses of bacterial cells to immobilization. *Molecules* 21(7).

18 Recent Advances and Impacts of Microtiter Plate-Based Fermentations in Synthetic Biology and Bioprocess Development

Patrick Wilk, Murni Halim, and Leonardo Rios-Solis

CONTENTS

"Wine is sunlight, held together by water."

Galileo Galilei

18.1 INTRODUCTION

Microscale fermentation has totally revolutionized the way microbial cell cultures are studied and developed, especially in the fields of industrial microbiology, synthetic biology, and bioprocess design among others, where fermentation usually needs to be optimized and higher throughput is thus beneficial (Long et al., 2014; Hemmerich et al., 2018). Currently, it is increasingly common for laboratories and industrial facilities in those fields to have replaced shake flasks and culture tubes, to varying degrees, with traditional microtiter plates (MTPs) fermenters (Pollard, Mcdonald, and Hesslein, 2016). This has greatly contributed to unlocking the bioprocess potential by significantly minimizing the resources and time required for research and development (Bareither and Pollard, 2011; Lattermann and Büchs, 2015).

The development of MTPs has run parallel to the development of synthetic and systems biology, which makes possible a rational design of biological systems (Cameron, Bashor, and Collins, 2014). MTP fermentations have played a key role in the automation of synthetic biology, exemplified by multi-million genome foundries that have been setup worldwide with the capacity to rationally synthetize and assemble complex DNA arrangements, creating strain libraries with multiple phenotypes for the synthesis of new drugs in a matter of months (Chao et al., 2017; Casini et al., 2018). The key advantage of MTPs for automation is that standard geometries and designs have been widely adopted, allowing for easy integration with automated platforms (Ladner et al., 2016).

Nevertheless, MTP fermentation speed and throughput has fallen behind in comparison with other components of the gene foundries, specifically due to complexities in microscale sensing technologies and high oxygen supply (Shih and Moraes, 2016). In addition, even the most advanced MTP fermentation systems are still not able to fully reproduce lab-bench scale fermenters for full quantitative characterizations of strain performance under process-relevant conditions (Kensy, Engelbrecht, and Büchs, 2009; Kirk and Szita, 2013; T. Ladner et al., 2016). Therefore, the speed of bioprocess development is severely lagging behind the capacity to generate biological systems. This review focuses on the state of the art of MTP technology for microbial fermentation. Current uses of MTPs in the design, build, test, and learn (DBTL) cycle of synthetic biology are analyzed as well as their impact in different stages of bioprocess development, and the most pressing challenges and future trends in the field are discussed.

18.2 ENGINEERING MTPS FOR IMPROVED FERMENTATION PERFORMANCE

The potential savings in time and resources that MTP fermentations could deliver with the provided increase in throughput has led to several studies focusing on the characterization of their engineering principles (Micheletti and Lye, 2006). Consequently, the knowledge on several key engineering principles of MTPs has drastically increased in the last 15 years, allowing for improved applications, specifically in the area of fermentations with high oxygen demand (Duetz and Witholt, 2004; Kensy et al., 2005; Kirk and Szita, 2013).

MTPs are commercially available in different formats (4 to 9,600 wells), geometries (deep or shallow plates, circular or rectangular), materials (glass, polystyrene, polypropylene) and in volumes ranging from 0.2 μL to 5.0 mL (Micheletti and Lye, 2006). Among these, the 24-, 48, and specially the 96-well plates, with a working volume of 200 μL, are the most commonly used format for MTP fermentations (Duetz, 2007; Wong et al., 2018).

Fluid mixing is a crucial property for MTP fermentations as the diffusion and convection phenomena behave differently from larger-scale vessels (Micheletti et al., 2006; Kirk and Szita, 2013). Mixing in MTPs can be achieved by several ways, including magnetics, micro-stirrers and microbubbles, but shaking is still the most commonly used method (Doig, Diep, and Baganz, 2005; Dürauer et al., 2016). Shaking can be operated by using either orbital or linear reciprocal movement; nevertheless, in the past decade, several studies have favored orbital shaking (Duetz and Witholt, 2004; Klöckner and Büchs, 2012). The understanding of the "out-of-phase" phenomenon was an important advancement to promote optimum mixing in MTPs (Büchs, Lotter, and Milbradt, 2001). When using "normal" operating and shaking conditions in a MTP, the liquid moves at the same time, or in-phase, with the rotation of the shaker device, however, if the viscosity increases, the movement of the fluid gradually stops until a complete breakdown of the liquid is achieved, resulting in an undefined swashing at the bottom of the flask or MTP (Peter et al., 2004). This observed phenomenon was coined the "out-of-phase" condition, which leads to low power input and lower mass transfer (Büchs, Lotter, and Milbradt, 2001; Zhang et al., 2008). Therefore, in order to avoid poorer mixing conditions, the determination of the critical orbital shaking speeds to assure uniform mixing in different geometries of MTPs has been largely achieved (Micheletti et al., 2006; Betts et al., 2014).

Numerical simulations of the hydrodynamic features of MTPs subjected to orbital motion have been done using various computational fluid dynamics tools, and have concluded that the mass transfer coefficient (k_La) and oxygen transfer rate (OTR) are strongly influenced by the fill volume, viscosity of the sample, the shape geometry of the well (including diameter and height), as well as by the shaking diameter and frequency (Doig et al., 2005; H. F. Zimmermann et al., 2006) Figure 18.1 displays the working principle of an orbital shaker, highlighting the importance of shaking diameter,

usually recommended to be in the range of 2 to 4 mm for MTPs, which is an order of magnitude smaller compared to most shaking flasks incubators (Wewetzer et al., 2015).

Hence, although shake flask incubators are easily adapted to provide mixing for MTP fermentations in temperature-controlled chambers, special care should be taken to adjust the shaking diameter to lower values if possible, and many dedicated MTP shakers are commonly used nowadays.

Hydrodynamic stress differs between different formats and geometries of microtiter plates; k_La values in square microtiter wells have been observed to be 30–200% higher than those in round wells (Duetz and Witholt, 2001; Hermann, Lehmann, and Büchs, 2003; Micheletti et al., 2006). It has been proposed that the square side walls provide a similar function to baffles in shake flasks, increasing mass transfer in the MTPs (Funke et al., 2009). In general, it has been found that the higher the throughput capacity of the plate, the lower the mixing performance using shaking methods (Micheletti et al., 2006). For example, based on the estimation of critical shaking frequencies for flock destruction, mixing in 6- and 24-well MTPs was found to be almost twice as efficient as that in the 96-well MTPs (Dürauer et al., 2016), but other parameters can alter this general observation (Latermann et al., 2014).

When dealing with small media volumes for microbial cell cultures, an appropriate plate sealing method is required to avoid excessive evaporation and insufficient OTR. Several covers made of different layers of materials (microfiber, ePTFE, and silicone) have shown a great capacity for reducing evaporation (Zimmermann et al., 2003; Rios-Solis et al., 2011; Halim et al., 2013). In the meantime, the use of gas permeable seals has been shown to allow sufficient OTRs in MTP fermentations while maintaining low evaporation rates (Bos et al., 2015). Novel MTP seals have been designed to allow sampling during experiments, enabling the integration of MTPs with liquid handling systems or directly with auto-samplers from analytical tools, including high-performance liquid chromatography for accurate determination of kinetic parameters (Chen et al., 2008; Matosevic, Szita, and Baganz, 2011; Rios-Solis et al., 2013). Newer shaking devices specifically designed for MTP fermentations include humidity control systems, which, combined with the use of the previously mentioned 'breathing' seals, have largely reduced the problem of water evaporation (Huber et al. 2009).

Overall, the engineering characterization of MTPs has led to improved understanding and has helped determine the optimal experimental conditions for MTP fermentations, providing an excellent guidance toward improving scale translation.

18.3 IMPACT OF MTP FERMENTATION IN SYNTHETIC BIOLOGY AND BIOPROCESS DEVELOPMENT

Synthetic biology is a recent multidisciplinary field focusing on developing foundational principles and technologies to ultimately enable a systematic forward-engineering of

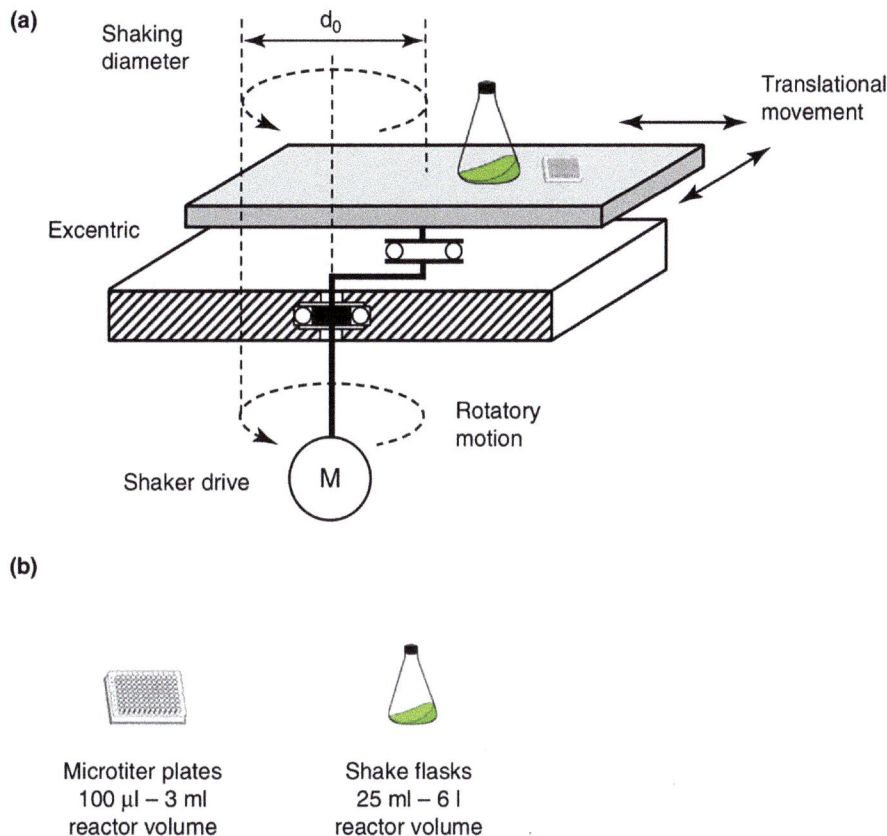

FIGURE 18.1 (a) Working principle of an orbital shaker and (b) shake flasks and MTPs bioreactor systems. Figure modified with permission from Klöckner and Büchs (2012).

biological parts and systems for improved and novel applications (Church et al., 2014). It follows a common flow process called the design, build, test, and learn cycle (DBTL), which is nowadays possible to iterate in an automated way in a genome foundry to promote the understanding and development of novel bio-systems as shown in Figure 18.2.

Previously, the bottleneck to build intricate DNA designs for biological engineering lay in the capacity of DNA synthesis, in addition to the considerable manual throughput required to perform all the traditional molecular and cell biology experiments involved in the "Design" and "Build" phases of the DBTL cycle. Due to the advancements in using technology for gene synthesis and the great improvements in automated molecular biology platforms in the forms of fully fledged genome foundries (Chao et al., 2017), the bottleneck has now arguably shifted to the "Test" phase in the cycle (Shih and Moraes, 2016). Following the same way in which the design and build steps benefited from high-throughput and automation, MTP fermentation devices are playing a critical role in "unburdening" the "Test" step by allowing researchers to accurately mimic the conditions and information obtained from shake flask fermentations (Neubauer et al., 2013). Nowadays, MTP fermentations are even capable of detailed multi-omics analysis at single well level, allowing researchers to gather superior data to feed the "Learn" step of the DBTL cycle. (Gorochowski et al., 2014; Marcellin and Nielsen, 2018). Nevertheless, as the field of synthetic biology developed, the

research focus has gradually expanded from fundamental studies to applying the gathered knowledge to develop novel bio-systems with real world applications (Campbell, Xia, and Nielsen, 2017; Flores Bueso and Tangney, 2017). Therefore, a current challenge of MTP fermentations is that they still do not mimic larger-scale operations, and the screening performed using MTPs is still categorized as primary or early stage characterization only as shown in Figure 18.3 (Long et al., 2014).

This has a tremendous impact for synthetic biology, as the "Learn" step in the DBTL cycle is only iterated with early stage data not completely relevant for larger-scale conditions. This implies that many of the apparent top-performing strains in the primary and secondary characterizations selected for further scale-up optimization may not be ideal, causing expensive surprises and a heavy burden in the bioprocess development (Long et al., 2014; Hemmerich et al., 2018).

The third round of screening is usually performed in lab scale bioreactors in the range of 500 mL to 5 L (Kensy, Engelbrecht, and Büchs, 2009). These types of systems allow for high levels of process control and closely mimic larger-scale bioreactors (Schmidt, 2005). They also allow key variables such as pH, DO, gas contents, and biomass to be measured continuously. Nevertheless, lab-bench scale bioreactors are a bottleneck in bioprocess development as they demand an important reduction of throughput (Micheletti and Lye, 2006). It is in this bottleneck of bioprocess development

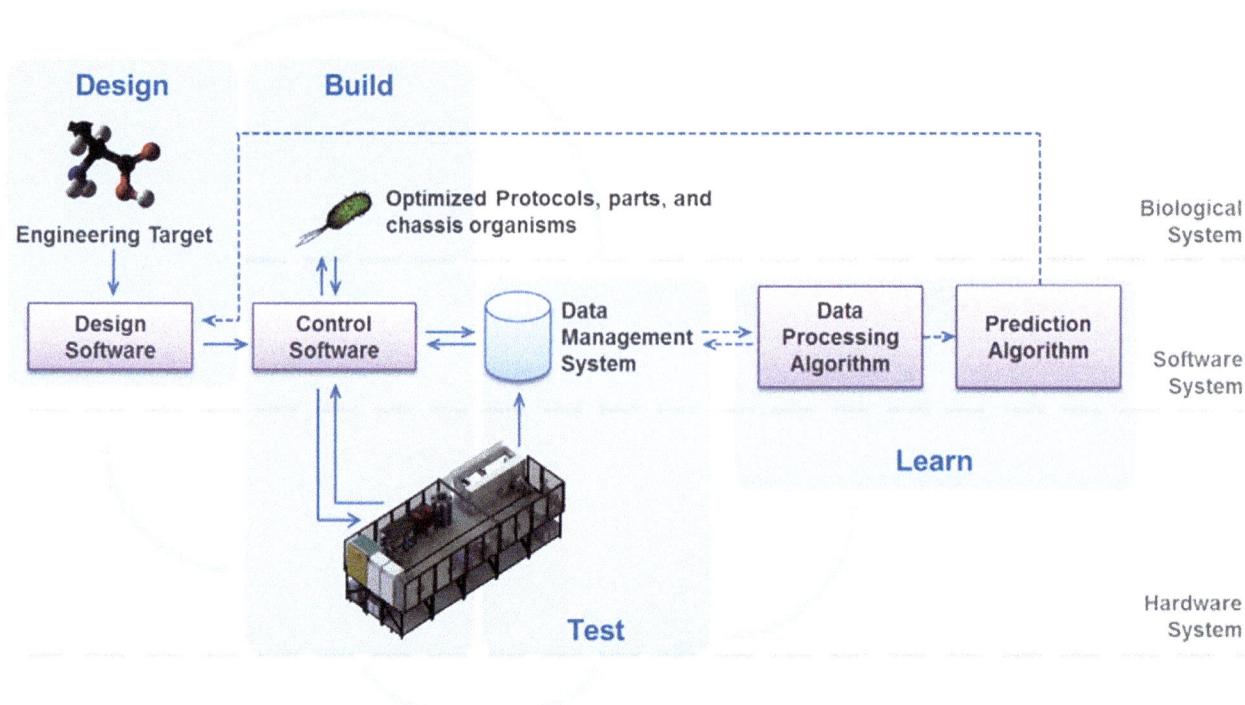

FIGURE 18.2 The impact of MTP fermentations in the architecture of a typical genome biofoundry. A genome biofoundry platform involves the interaction of biological components with software and automated hardware systems (Chao et al., 2017). MTP fermentation devices are a key hardware element to conduct several tasks in the in the build and test phases. The software system orchestrates the automation processes and potentially can assist in the experimental design as well as data processing for the MTP fermentations. Figure modified with permission from Chao et al. (2017).

that the novel MTP based devices are increasingly having an important impact (Neubauer et al., 2013). The development of micro-sensors for MTPs now allows the continuous monitoring of variables such as pH, biomass, and DO with great accuracy, and more importantly, biomass concentrations of up to 70 g/L can be reached due to the incorporation of fed-batch capacities to MTPs, allowing the high cell densities of industrial-scale bioreactors to be more closely mimicked (Buchenauer et al., 2009; Faust et al., 2014; Wilming et al., 2014).

18.4 ROUTINE MTP FERMENTATION DEVICES

Among the most commonly used devices to perform primary screening of microbial cell cultures using MTPs are the commercial microplate readers. The versatility of these devices has made them very popular for a diverse set of experimental procedures, (i.e., *in vitro* protein characterization), and they have proven to be a reliable technology for MTP fermentations due to their high-quality optical sensors for continuous biomass density monitoring, as well as delivering temperature control and limited shaking capacities (Duetz, 2007; Morris et al., 2016). Primary screening for growth rate is usually easily determined within these devices, which is an important indication of microbial fitness to estimate the burden or toxicity of newly inserted metabolic pathways or genetic modifications (Blomberg et al., 2011).

The easy automation of these systems is also an important advantage. They allow the continuous, high-throughput, parallel measurement of kinetic growth data and have the potential to be easily integrated into automated platforms for further sample processing (Baboo et al., 2012). There are two common sensors used to estimate the biomass concentration, which are based on optical density or light scattering. The optical density measurement allows an easy comparison with optical density data from shake flasks or bioreactors by using the correct path length of the measuring chamber. Nonetheless, the relationship of optical density and biomass in MTPs becomes nonlinear at relatively low cell densities (1 > OD_{600}) (Begot et al., 1996), which can become a problem for online measurements of higher cell densities without sample dilution. In comparison, light scattering signal depends on cell characteristics including diameter, shape, or granularity among others (Kunze et al., 2014). Usually, it is assumed that these parameters do not change significantly during fermentation and in this case the data presents an almost linear relationship between biomass and signal scatter intensity, therefore this method has been favored for online biomass monitoring in MTPs, where methods have shown a linear relationship of up to 50 g L^{-1} of dry cell weight (Kensy et al., 2009). Nevertheless, it has been reported that the scattering signal is sensitive to morphological changes of the cells, and proper measures should be taken in case this phenomenon may be present during the MTP cultures (Kunze et al., 2014)

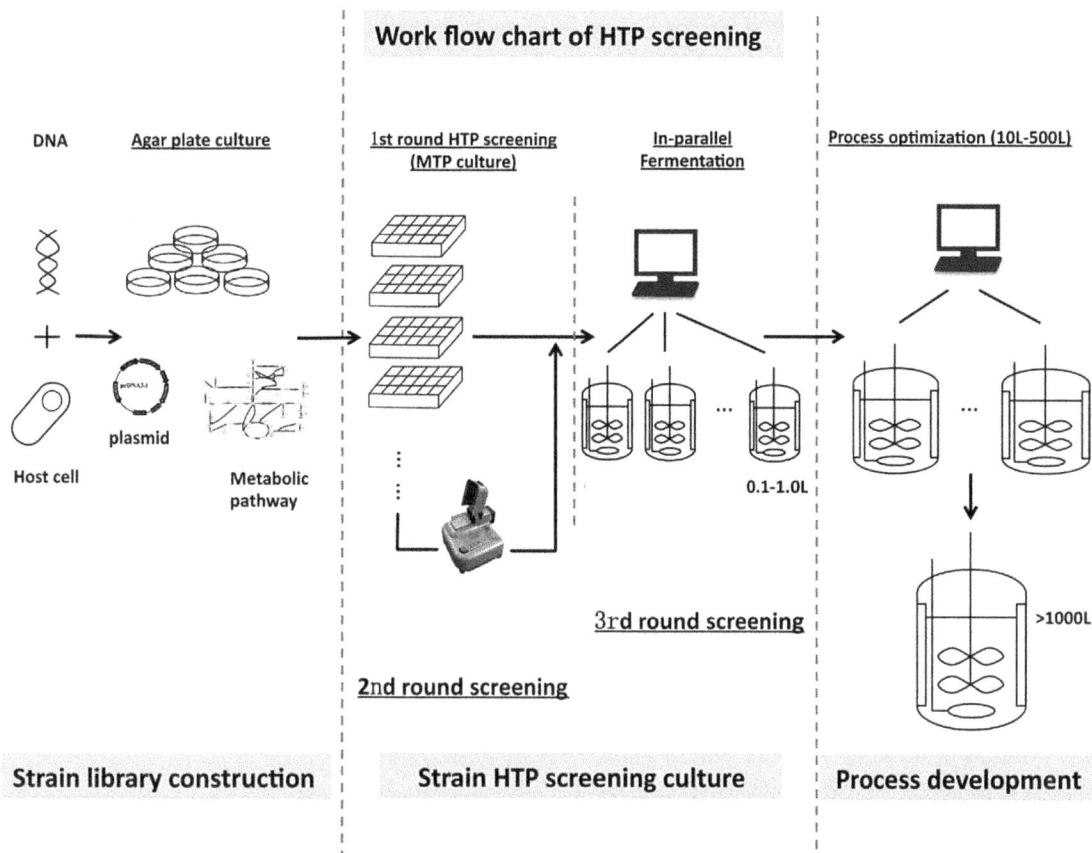

FIGURE 18.3 A representative flow chart of bioprocess development. The process starts with the creation of a strain library. MTP fermentations are currently capable of performing the primary and secondary screenings. Further rounds of screening to mimic larger-scale conditions are performed using in parallel lab-bench scale fermenters to determine the final clone. The final process optimization is performed in larger-scale pilot plant facilities. Figure modified with permission from Long et al. (2014).

MTP devices equipped with additional optical fluorescence capabilities have become very popular nowadays. This has been proven very useful in synthetic biology, allowing researchers to harness the properties of fluorescent reporter proteins to create whole cell biosensors for the assessment of several biological performance parameters (Brognaux et al., 2012). This has been key to enable the characterization of several types of synthetic biology DNA parts, such as promoters, terminators, and ribosome binding sites among others, which have been linked to a reporter protein so that expression can be estimated and normalized against the biomass measured during the MTP fermentation (Gorochowski et al., 2014). Following this technique, Lee et al. (2015) established and characterized a full synthetic biology toolkit for *S. cerevisiae* parts, including promoters, terminators, and plasmid copy number, using MTP fermentations. Rei Apel et al. (2017) went even further, characterizing promoters at different physiological stages (exponential and stationary phase) in addition to characterizing the expression capacity of different loci using CRISPR Cas9.

The drawback of such popular and versatile MTP devices like the microplate readers is that they are usually equipped with poor shaking capacities and no humidity control. This causes the oxygen transfer rate to be poor while sample water evaporation can critically affect results in longer experiments (Zimmermann et al., 2003). Due to the low shaking speeds, sedimentation of cells can often occur, affecting the online measurements. $k_L a$ values in the range of 22–25 h^{-1} are usually expected, thus limiting their application to low-density cultures that do not mimic larger-scale performance (John et al., 2003).

18.5 NEXT GENERATION OF MTP FERMENTERS

In order to reduce the previously mentioned problems of poor oxygenation of MTPs, in the last few years several groups have focused on optimizing mixing for better mass and energy transfer (Kirk and Szita, 2013; Dürauer et al., 2016). A microscale bioreactor designed to work with 24 well disposable cassettes (up to 10 ml volume) called μ-24 System was commercialized by Pall Corporation (New York, USA) achieving $k_L a$ values of up to 80 h^{-1}, which were in the lower range of those observed in larger bioreactors (Ramirez-Vargas et al., 2014). Oxygen control was achieved through addition of air or blended gas into the wells, which allowed the desired oxygen level to be maintained throughout the cultivation run (Tang et al., 2006). Although the μ-24 system showed important improvements for high-throughput microbial fermentations under industrially relevant conditions, reproducibility among individual wells was an issue, as well as difficulties for automation due to its complex setup (Holmes et al., 2009).

A system called "Growth Profiler" capable of supporting microbial cultures in 12 parallel 96-well MTPs was commercialized by Enzyscreen (Haarlem, The Netherlands), where the manufacturer claims values of k_La up to 100-200 h^{-1}could be achieved. For biomass measurements, the microscale shaker inconveniently needed to slow down to 30 rpm for a few seconds in order to take photos of the bottom of the transparent wells to estimate the cell densities. (Lennen et al., 2015; M. Klein et al., 2016).

The degree of baffling in the wells can have an important impact on oxygen supply in MTP fermentations (Lattermann et al., 2014). More than 30 well geometries with different degrees of baffling were studied in a 48-well MTP to determine the maximum oxygen transfer capacity (Funke et al., 2009). The best performing geometry was a six-petal flower shaped well as shown in Figure 18.4, reaching an OTR of up to 100 mmol L^{-1}h^{-1} ($k_La > 600$ h^{-1}). Nevertheless, as mentioned for the Growth Profiler system, one of the disadvantages of those setups was the necessity to stop the shaking while the measurements were performed, leading to undesired changes in the culture conditions (Huber et al., 2009). Novel methods to allow continuous monitoring without disturbing the shaking frequencies were therefore explored, specifically by coupling the measuring electronic optic sensors directly to the shaker platform (Samorski, Müller-Newen, and Büchs, 2005). This concept was further applied and commercialized by m2p-labs GmbH in the 0.8–2.5 mL working volume MTP shaker called BioLector (Baesweiler, Germany). This MTP

device uses light scattering and fluorescent optical sensors to accurately determine biomass, DO, and pH. The optical sensor was attached to the bottom of the MTP, acting simultaneously as a light emitting and collection unit to enable non-invasive measurements (Funke et al., 2009; Kensy et al., 2009). The spectrometer producing the scattering light beam was fixed at an angle of 30° to one arm of the linear motion module to avoid interference by light reflection, as shown in Figure 18.4.

The optical sensors for pH and DO were set up vertically, allowing quasi-continuous measurement without the need to stop the shaking (Buchenauer et al., 2009). In addition, the device allowed humidified air to be continuously introduced to the shaker and combined with the use of the previously discussed flower shape 48-well MTPs, allowed OTRs of up to 100 mmol L^{-1} h^{-1} (k_La values of up to 600 L^{-1}) while minimizing media evaporation (Huber et al., 2009). The system was validated for to perform quantitative MTP fermentations to optimize the secretory production of a cutinase from *Fusarium solani pisi* with *Corynebacterium glutamicum* (Rohe et al., 2012), as well as to perform a 7000-fold scale up from MTP to stirred-tank fermenter for both *Escherichia coli* and the yeast *Hansenula polymorpha* (Kensy, Engelbrecht, and Büchs, 2009).

18.6 MTP FERMENTATIONS WITH FED-BATCH CAPACITIES

An important aspect of the lab-bench scale bioreactor is the fed-batch operation capacity, which allows microbial substrate-limited growth to be performed and minimizes carbon

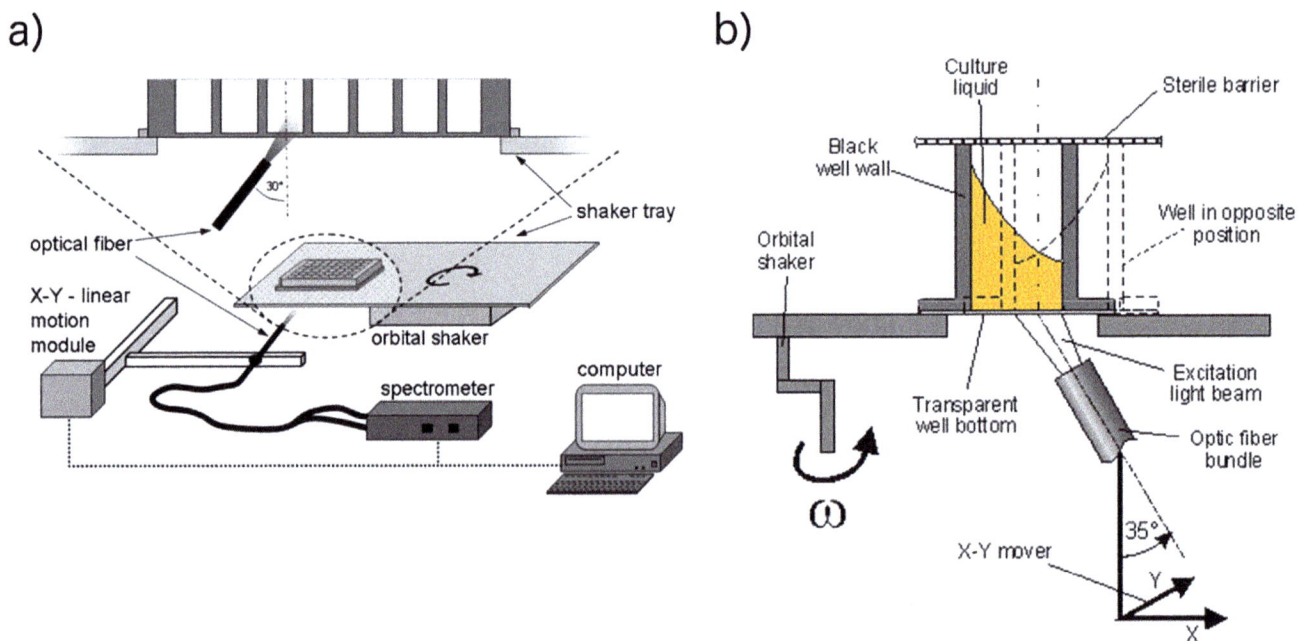

FIGURE 18.4 (a) BioLector measurement system; the optical fiber is fixed on the x–y linear motion module at an angle of approximately 30–35°, to avoid the interference by light directly reflected from the well bottom. (b) Zoom of a BioLector MTP well to showcase the measurement principle via back scattering of light from cells and fluorescence emission of molecules. Figure modified with permission from Funke et al. (2009) and Kensy et al. (2009).

overflow while greatly improving production yields (de Maré et al., 2005). Until recently, fed-batch capacities were only accessible in the lab-bench scale bioreactors, meaning that screening experiments using MTPs were traditionally done using batch operating systems, which could lead to wrong strain selection hampering bioprocess development (Krause, Neubauer, and Neubauer, 2016).

To alleviate this discrepancy of information delivered by the MTPs, novel methods allowing fed-batch operating capabilities for MTP fermentations have been recently developed (Faust et al., 2014). Some of the recent technologies to enable microscale fed-batch operation are based on the use of enzymatic release systems (Panula-Perälä et al., 2008), hydrogel substrate release systems (Wilming et al., 2014), and microfluidic systems (Buchenauer et al., 2009). Wilming et al. (2014) described a 44-well MTP with continuous optical measurements and fed-batch capabilities using a hydrogel channel that connected each well with the diffused substrate. The rate of substrate feeding could be tuned by changing the substrate concentrations, channel geometry, and gel properties and was successfully shown to support the growth of *E. coli* and *H. polymorpha*. A microfluidic chip inserted at the bottom of a 48-well MTP has been reported, allowing volumes on the nanolitre scale to be dispensed to single wells for pH control in *E. coli* cultures (Buchenauer et al., 2009). This was further improved by Funke et al. (2010) who introduced a micro-pump to allow substrate fed-batch feeding intro the microfluidic chip. A key advantage of this system is the user-friendly setup based on using disposable microfluidic chips, which avoids the cumbersome necessity to connect the chip to pneumatic connections, as these are robustly directly integrated into the shaker. This technology has been integrated with the previously described BioLector system, and as shown in Figure 18.5, the external valves lead pressurized air to the connections on top of each of the MTP wells.

The "microfluidic BioLector," nowadays commercialized a "RoboLector" by m2p-labs GmbH (Baesweiler, Germany) has been validated for various strains and conditions, including

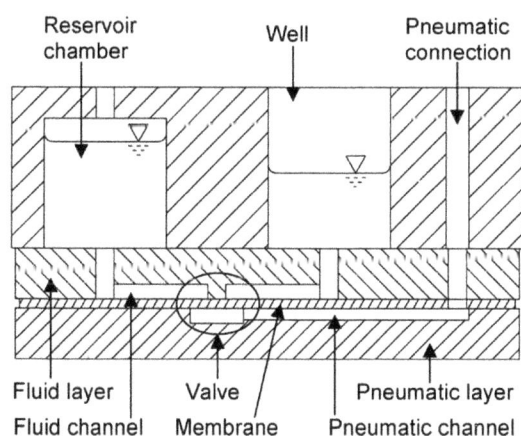

FIGURE 18.5 Schematic cross-section of the microfluidic cassette inserted in the microfluidic BioLector device. Figure modified with permission from Buchenauer et al. (2009).

a successful comparison with data from 2L lab-bench scale *E. coli* fed-batch fermentations (Kensy et al., 2009; Rohe et al., 2012; Hemmerich et al., 2014; Wewetzer et al., 2015; Mühlmann, Forsten, et al. 2017).

18.7 MTP FERMENTATION INTEGRATED INTO AUTOMATED BIOPROCESS PLATFORMS

The integration of MTP fermentation systems into robotic platforms vastly increases their capacity to gather quantitative data for bioprocess development (Hemmerich et al., 2018). Standardization of MTP formats and devices was promoted by the Society for Laboratory Automation and Screening (SLAS, www.slas.org), allowing the MTP formats to be easily handled by multiple commercially available robotic platforms (Wu and Zhou, 2014).

Automation not only allows the sample throughput to be increased, but also increases fermentation capabilities and complexity, e.g., by allowing sample collection, induction of genes, or the addition of nutrients at different intervals (Rohe et al., 2012). All these tasks could be programmed to happen in response to online measurements (e.g. biomass, pH, protein fluorescence) (Unthan et al., 2015). This offers new opportunities in bioprocess control, for example, by optimizing the induction of protein expression as a function of the individual biomass density in each well (Mühlmann, Kunze, et al. 2017). Integrating the MTPs into fully fledged automated platforms has allowed MTP fermentations to be linked with multi-omics analytical tools to improve the characterization of the engineered strains (Marcellin and Nielsen, 2018). It is becoming more common to include parallel proteomics, fluxomics, transcriptomics and metabolomics analysis for characterization of engineered strains to detect bottlenecks and improve bio-product formation (Brunk et al., 2016). Heux et al. (2014) successfully linked the experimental gathering and data flow between microscale cell cultivation and flux quantification, achieving considerable advances in throughput and metabolic characterization to facilitate the development of microbial cell factories. Nevertheless, the processing of the samples post-fermentation for multi-omics analysis is laborious, and research is still focusing on the difficult automation of those analytical techniques, which will undoubtedly facilitate strain characterization (de Raad, Fischer, and Northen, 2016; Heux et al., 2017; Marcellin and Nielsen, 2018).

A major development of the automated microscale platforms has been the integration of the MTP fermentations with other unit operations (Baumann, Hahn, and Hubbuch, 2015). This has led to the setup of novel automated platforms that allow the study of process relationships, which are usually ignored or poorly explored; for instance, between upstream and downstream processing (Baumann et al., 2015; Unthan et al., 2015). The benefits of several automated platforms linking MTP fermentations with bio-catalysis optimization steps using whole cells, lysates, and purified enzymes have been clearly described (Rios-Solis, Morris, et al. 2015b). Several successful reports have highlighted the benefits of automated platforms linking fermentation with automated induction and bioconversion steps to evaluate trade-offs in activities between enzyme production and specific catalytic activities including transketolase, transaminase, and oxygenases (Du et al., 2014; Rios-Solis, Mothia, et al. 2015a). Recently, an automated platform for high-throughput enzyme library screening was described integrating automated colony picking, MTP fermentation with continuous monitoring of biomass density, induction, subsequent enzyme purification steps and biochemical assay (Fibinger et al., 2016).

The previously described "RoboLector" device was successfully extended with a vacuum filtration module allowing researchers to perform a cellulase screening experiment consisting of upstream, downstream, and analysis steps as shown in the scheme of Figure 18.6 (Mühlmann, Kunze, et al. 2017).

Microscale normal flow filtration devices have been successfully applied in MTPs, allowing researchers to quantitatively evaluate the influence of upstream processing conditions on the microfiltration behavior of *E. coli* fermentation broths (Jackson, Liddell, and Lye, 2006). The integration of MTP fermentation with other units of operation is constantly increasing, for example, the integration with microscale automated chromatographic steps (Chhatre and

Titchener-Hooker, 2009; Konstantinidis et al., 2018), protein precipitation (Baumgartner et al., 2015) or cell separation in two phases systems (Zimmermann et al., 2017) offer great potential to discover novel bioprocess relationships.

18.8   CHALLENGES AND FUTURE TRENDS

18.8.1   HIGH THROUGHPUT EXPERIMENTAL DESIGN AND DATA PROCESSING

MTP fermentations have enabled a considerable gain in throughput turnover using minimal resources; nevertheless, the design of experiments to take advantage of the higher throughput capacity is not always obvious, and in many cases the design is still done manually by the users using heuristic criteria, which hampers the quantity of information that could have potentially been gained (Galvanin, Macchietto, and Bezzo, 2007). Therefore, the use of user-friendly algorithms and tools for designing the experiments is required to maximize the information gained (Weuster-botz, 2000). One of the most common tools used for this is "Design-of-Experiments" (DoE), which was developed to optimize information gained

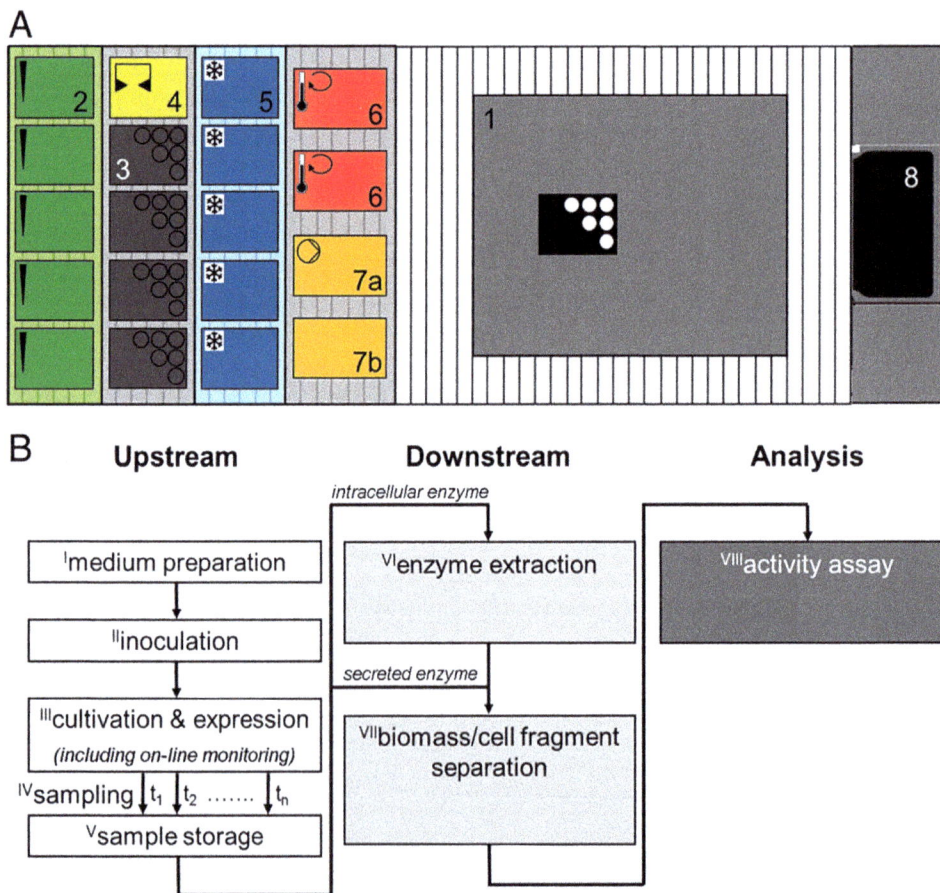

FIGURE 18.6   Automated high throughput platform of recombinant cellulase. (a) Layout of the automated high-through put platform: 1, MTP fermentor device; 2, pipetting tip carrier; 3, carrier with 4 MTP storage positions; 4, gripping tool parking position; 5, cooling carrier for storage of up to 5 MTPs containing samples and/or chemicals; 6, heater shakers for 2X MTPs; 7a, MTP Vacuum filtration module, 7b, lid parking position; 8, waste bag. (b) Schematic protocol for the high throughput–automated platform procedure for recombinant cellulase expression with processing steps I–VIII. Figure modified with permission from Mühlmann et al. (2017).

from a limited number of experiments and infer functional relationships, for example between the fermentation input factors (e.g., selected strain, media composition, induction time) and output factors (e.g., specific growth rate) (Harms et al., 2008; Mandenius and Brundin, 2008). A key advantage of DoE is enabling the systematic analysis and identification of non-intuitive input factors, which are usually overlooked due to preconceived user bias (Olsson et al., 2006). There is a clear necessity to integrate such DoE algorithms into the native software of the MTP devices in a user-friendly way, to reduce the barrier for optimum experimental design of the MTP fermentations.

In terms of data processing, larger datasets of useful information can nowadays be gathered easily in shorter periods of time, with minimal resources. The processing of this data, however, is still mainly done manually, which is a laborious and daunting task for the users (Hall et al., 2014; Sprouffske and Wagner, 2016). The current challenge is to develop and incorporate novel software and algorithms into the MTP based devices in a way that allows the information to be processed automatically and efficiently for rapid clear visualization, which would allow the experimental hypothesis to be validated or rejected instantly (Vasser et al., 2004; Dörr and Bornscheuer, 2018). Usually, the first step of processing the raw data involves using a calibration curve to translate the measured data intensity into a meaningful value for the user (Sprouffske and Wagner, 2016). While light scattering values usually still require a manual calibration to determine the biomass density, microscale probes for pH, and DO can nowadays be supplied with automatic internal calibration (Kensy et al., 2009). The pH and DO calibration for the BioLector are MTP dependent, meaning each MTP comes with a unique barcode that needs to be entered into the system for automatic calibration of the fluorescent pH and DO sensors (Mühlmann, Kunze, et al. 2017). Secondly, the automatic software-enabled calculation of critical fermentation parameters such as specific growth rate, biomass and product titer, and specific yield would really enhance the decision-making process (Hemmerich, Wiechert, and Oldiges, 2017). It is important to highlight that in order for the MTP fermentations to be readily adopted by industry, the metadata from the high-throughput experiments should be safely stored and organized along the experimental data in order to be available to the quality and control departments for validation procedures. An integrated robotic platform allowed automatic fitting of E. coli fed-batch experimental data into kinetic differential equations (Cruz Bournazou et al., 2017). This enabled the continuous re-design of the running cultivations using the information generated by periodic parameter estimations, promoting more efficient computer aided bioprocess development. Several of those advanced mathematical models and algorithms are likely to be adopted soon by the majority of the commercial MTP fermentation software packages (Yordanov et al., 2016; Ladner et al., 2017).

18.8.2 NOVEL SENSORS FOR MTP FERMENTATIONS

The development of novel microscale sensing technologies and their robust incorporation into MTPs will revolutionize the way fermentations are characterized using high-throughput techniques (Gruber et al., 2017). Developments include improved sensors for determining complete and accurate respiration activities in individual wells, for example, an MTP based device called "µRamos" has been recently developed allowing OTR determination with similar accuracy as the standard RAMOS (Flitsch et al., 2016). Novel integrated sensors for impedance measurements in MTPs have been recently developed, which with further optimization could give the opportunity for online continuous monitoring of viable cells and even infer their cell size (Hofmann et al., 2010; Luchterhand et al., 2015). Other exciting developments include obtaining highly detailed information for direct metabolite quantification using multi-wavelength optical techniques (Marcellin and Nielsen, 2018). A method to perform 2D fluorescence spectroscopy in each well of a 48-well MTP has been reported, requiring 0.5 h per measurement cycle (Ladner et al., 2016; Paquet-Durand et al., 2017). The 2D fluorescent sensor allowed the parallel determination of glycerol, glucose, and acetate in H. polymorpha and E. coli cultures. A proof of concept has also been recently been published using Fourier transform infrared spectroscopy (FTIR) in MTPs for the parallel detection of glucose, citrate, and triglyceride lipid in cultures of an oleaginous filamentous fungus (Kosa et al., 2017). Furthermore, the latest Raman devices have shown the capability to provide accurate multi-analyte data in real time within complex biological mixtures, which could provide another layer of information if incorporated into MTPs (Esmonde-White et al., 2017; Voss and Mittelheuser, 2017). Raman spectroscopy has been recently applied to allow label-free, continuous monitoring of cell growth and physiological state of lactic acid bacteria, potentially facilitating more accurate estimations of the growth states of microorganisms and their heterogeneities during fermentations (Ren et al., 2017). The information gathered from these multi-analyte sensors usually requires comparing the gathered data with previously curated models and validated datasets, again highlighting the need for novel data processing methodologies for easy data evaluation (Rocke, 2004).

18.8.3 INDIVIDUAL WELL CONTROL IN MPT FERMENTATIONS

Another important challenge in MTP fermentations is the lack of control of some parameters at the individual well level, for example temperature, maximal oxygen transfer rate, shaking, and external gas supply. Individual temperature control in a single well is usually not possible, therefore incubation chambers controlled by a thermostat are generally used instead (Hemmerich et al., 2018). Novel thermostat technologies have been recently developed for MTPs, for example, a 24-well MTP linked to a thermoregulator that can be used on any type of shaking platform without any external temperature-controlled environment (Nunes, Fernandes, and Ribeiro, 2013). In the µ-24 system (10 ml working volume) previously described from Pall Biosystems (New York, USA), each individual well within the single use-cassette included a temperature sensor and thermal heat conductor, allowing control of temperature at the individual well level, in addition to DO, pH and gas blending control (Ramirez-Vargas et al., 2014). Nevertheless, in general, it is not possible to evaluate the full impact of these parameters in single MTP wells, and therefore multiple MTP fermentation experiments may be required, critically reducing the throughput advantages (Mühlmann, Forsten, et al. 2017).

18.8.4 Microbioreactors

A new trend is aiming at solving part of the previously described drawbacks by developing a novel type of microscale bioreactor, commonly denominated "microbioreactors," which aim to scale down industrial bioreactors into micro-stirred vessels in the volume range of 1–15 mL (Hemmerich et al., 2018). For example, the "micro-Matrix" system, which is commercialized by Applikon (Delft, Netherlands), comprises 24 parallel micro-stirred bioreactors (2–5 mL) with individual temperature, Ph, and DO control (Wiegmann, Martinez, and Baganz, 2018). A novel 48 parallel micro-stirred bioreactor system called "bioREACTOR 48" was commercialized by 2mag AG (Munich, Germany), which integrates novel micro-impellers to optimize the hydrolysis of biomass, which is a feature not yet available in MTPs (Riedlberger and Weuster-Botz, 2012). The k_La of such system has been reported to reach up to 1400 h^{-1}, and they have been successfully applied in aerobic and anaerobic bioprocess development studies (Schmideder et al., 2016). Despite the impressive OTR achieved by the bioreactor 48 micro-stirrers, the system requires a robotic arm for the intermittent control of pH and fed-batch operation, which can lead to short-term oxygen limitations and discrepancies in bioprocess control (Faust et al., 2014; Schmideder et al., 2016). A system of eight parallel micro-stirred bioreactors that allow individual mixing and off-gas analysis via mass spectrometer has also been described, providing accurate individual respiratory activity data (Klein, Schneider, and Heinzle, 2013). Satorius (Göttingen, Germany) commercializes a system previously developed by Tap Biosystems (Royston, UK) which comprises a 24 × 15 mL single use fed-batch microbioreactor system called "ambr 15 fermentation," designed especially for microbial culture, allowing to expose the cells to individual levels of shear similar to the expected in an industrial-scale bioreactors (Velez-Suberbie et al., 2018). A similar version of this system has been developed for mammalian cell cultures too (ambr 15 cell culture).

Although microbioreactors can offer several advantages in comparison to their MTP based counterparts, some of their drawbacks are a usually lower throughout requiring higher volumes, higher consumable costs, and difficult integration into automated commercial platforms (Lattermann and Büchs, 2015; Hemmerich et al., 2018). In the coming years, it will be interesting to see which technology (MTP based bioreactors vs microbioreactors) will be more widely adopted by academia and industry, or whether a combination of both types of microscale fermenters will lead to better solutions for bioprocess development.

18.9 CONCLUSIONS

To summarize, the development of the synthetic biology field and its automation in genome foundries has created a capacity to generate a great diversity of rationally designed strain libraries with minimum resources and time. The experimental capacity to characterize the performance of such strains is still a bottleneck for bioprocess development. The information gathered during the primary and secondary characterization of those libraries, which used to be done in shake flasks, can now easily be obtained with the same quality of information using MTP fermentations, allowing for great savings in time and resources. One of the biggest challenges of MTP fermentations is that they still do not mimic larger-scale conditions, and the screening performed using MTPs is still categorized as a primary screen. This critically impacts synthetic biology, as the "Learn" step in the DBTL cycle is only iterated with early stage data not completely relevant for larger-scale conditions.

Nevertheless, novel MTP technologies are closing the gap to mimic lab-bench scale bioreactors, especially in terms of the achievable biomass densities, due to improvements in mass transfer technologies and in the addition of fed-batch capacities. Improvements in MTPs' online monitoring technologies, combined with the incorporation of multi-wavelength optical sensors and the additional integration with high throughput multi-omics analytical tools, will have a profound effect on synthetic biology due to the unprecedented levels of superior information that could potentially be fed into the DBTL cycle.

Recent advances in algorithms for experimental design and data processing, combined with easy automation and successful standardization of the MTPs, is allowing genome foundries to integrate the use of these novel MTP fermentation devices seamlessly into their workflow, radically increasing the impact of synthetic biology on bioprocess development.

In addition, the integration of MTP fermentations with other unit operations is revolutionizing the way bioprocess development is performed. It allows the parallel optimization of fully integrated bio-manufacturing processes by linking pre-processing treatments to fermentation and downstream culture operations previously optimized independently. This encourages the discovery of unexpected bioprocess relationships, thus leading to the development of faster global optimization methodologies.

SUMMARY

The field of synthetic biology has experienced a rapid development in terms of its capacity to design and build complex DNA constructs and strain libraries with minimum resources and time. The experimental capacity to characterize and test such strains is still a bottleneck, which is severely hampering the learning outcomes of the design, build, test, and learn cycle (DBTL). Microtiter plate (MTP) based fermentations have arisen as a solution for this bottleneck, due to their high level of standardization and high-throughput automation potential.

This review highlights the recent developments and current challenges of MTP fermentations and their impact in synthetic biology and bioprocess development. A special focus is given to the recent advances in MTPs that allow them to mimic larger-scale bioreactors, due to improvements in mass transfer technologies and the incorporation of microscale fed-batch capacities. Novel developments in MTPs' online monitoring technologies are discussed, including the recent incorporation of multi-wavelength optical sensors and the additional integration with high-throughput multi-omics analytical tools.

Finally, the integration of MTP fermentations with other units of operation is reviewed, which allows the integrated parallel optimization of both upstream and downstream operations. This is encouraging the discovery of unexpected bioprocess relationships, thus leading to faster global optimization methodologies.

Those recent advancements in MTP fermentations will allow an unprecedented level of superior information to be fed into the DBTL cycle, which will unlock the potential of synthetic biology and revolutionize its impact in bioprocess development.

REFERENCES

Apel, A. R. et al. A Cas9-based toolkit to program gene expression in Saccharomyces cerevisiae. **45**, 496–508 (2017).

Baboo, J. Z. et al. An automated microscale platform for evaluation and optimization of oxidative bioconversion processes. *Biotechnol. Prog.* **28**, 392–405 (2012).

Bareither, R. & Pollard, D. A review of advanced small-scale parallel bioreactor technology for accelerated process development: Current state and future need. *Biotechnol. Prog.* **27**, 2–14 (2011).

Baumann, P. et al. Integrated development of up- and downstream processes supported by the Cherry-Tag™ for real-time tracking of stability and solubility of proteins. *J. Biotechnol.* **200**, 27–37 (2015).

Baumann, P., Hahn, T. & Hubbuch, J. High-throughput micro-scale cultivations and chromatography modeling: Powerful tools for integrated process development. *Biotechnol. Bioeng.* **112**, 2123–2133 (2015).

Baumgartner, K. et al. Determination of protein phase diagrams by microbatch experiments: Exploring the influence of precipitants and pH. *Int. J. Pharm.* **479**, 28–40 (2015).

Betts, J. P. J., Warr, S. R. C., Finka, G. B., Udenh, M., Towna, M., Jandaa, J. M., Baganz, F., & Lye, G. J. Impact of aeration strategies on fed-batch cell culture kinetics in a single-use 24-well miniature bioreactor. *Biochem. Eng. J.* **82**, 105–116 (2013).

Blomberg, A. Measuring growth rate in high-throughput growth phenotyping. *Curr. Opin. Biotechnol.* **22**, 94–102 (2013).

Brognaux, A., Neubauer, P., Gorret, N., Thonart, P. & Delvigne, F. Direct and indirect use of GFP whole cell biosensors for the assessment of bioprocess performances : Design of milliliter scale-down bioreactors. *Biotechnol. Prog.* **29**, 48–59. (2012).

Bos, A. B. et al. Optimization and automation of an end-to-end high throughput microscale transient protein production process. *Biotechnol. Bioeng.* **112**, 1832–1842 (2015).

Brunk, E. et al. Characterizing strain variation in engineered *E. coli* using a multi-omics-based workflow. *Cell Syst.* **2**, 335–346 (2016).

Buchenauer, A. et al. Micro-bioreactors for fed-batch fermentations with integrated online monitoring and microfluidic devices. *Biosens. Bioelectron.* **24**, 1411–1416 (2009).

Büchs, J., Lotter, S. & Milbradt, C. Out-of-phase operating conditions, a hitherto unknown phenomenon in shaking bioreactors. *Biochem. Eng. J.* **7**, 135–141 (2001).

Cameron, D. E., Bashor, C. J. & Collins, J. J. A brief history of synthetic biology. *Nat. Rev. Microbiol.* **12**, 381–390 (2014).

Campbell, K., Xia, J. & Nielsen, J. The Impact of Systems Biology on Bioprocessing. *Trends Biotechnol.* **35**, 1156–1168 (2017).

Casini, A. et al. A pressure test to make 10 molecules in 90 days: external evaluation of methods to engineer biology. *J. Am. Chem. Soc.* **140**, 4302–4316 (2018).

Chao, R., Mishra, S., Si, T. & Zhao, H. Engineering biological systems using automated biofoundries. *Metab. Eng.* **42**, 98–108 (2017).

Chhatre, S. & Titchener-Hooker, N. J. Review: Microscale methods for high-throughput chromatography development in the pharmaceutical industry. *J. Chem. Technol. Biotechnol.* **84**, 927–940 (2009).

Chen, B. H., Hibbert, E. G., Dalby, P. A. & Woodley, J. M. A new approach to bioconversion reaction kinetic parameter identification. *AIChE J.* **54**, 2155–2163 (2008).

Church, G. M., Elowitz, M. B., Smolke, C. D., Voigt, C. A. & Weiss, R. Realizing the potential of synthetic biology. *Nat. Rev. Mol. Cell Biol.* **15**, 289–294 (2014).

Cruz Bournazou, M. N. et al. Online optimal experimental re-design in robotic parallel fed-batch cultivation facilities. *Biotechnol. Bioeng.* **114**, 610–619 (2017).

de Maré, L. et al. A cultivation technique for E. coli fed-batch cultivations operating close to the maximum oxygen transfer capacity of the reactor. *Biotechnol. Lett.* **27**, 983–90 (2005).

de Raad, M., Fischer, C. R. & Northen, T. R. High-throughput platforms for metabolomics. *Curr. Opin. Chem. Biol.* **30**, 7–13 (2016).

Doig, S. D., Diep, A. & Baganz, F. Characterisation of a novel miniaturised bubble column bioreactor for high throughput cell cultivation. *Biochem. Eng. J.* **23**, 97–105 (2005).

Doig, S. D., Pickering, S. C. R., Lye, G. J. & Baganz, F. Modelling surface aeration rates in shaken microtitre plates using dimensionless groups. *Chem. Eng. Sci.* **60**, 2741–2750 (2005).

Dörr, M. & Bornscheuer, U. T. Program-guided design of high-throughput enzyme screening experiments and automated data analysis/evaluation. In *Protein Engineering: Methods and Protocols* (eds. Bornscheuer, U. T. & Höhne, M.), pp. 269–282. Springer, New York (2018). doi:10.1007/978-1-4939-7366-8_16

Du, C. J., Rios-Solis, L., Ward, J. M., Dalby, P. A. & Lye, G. J. Evaluation of CV2025 ω-transaminase for the bioconversion of lignin breakdown products into value-added chemicals: Synthesis of vanillylamine from vanillin. *Biocatal. Biotransformation* **32**, 302–313 (2014).

Duetz, W. A. Microtiter plates as mini-bioreactors: miniaturization of fermentation methods. *Trends Microbiol.* **15**, 469–75 (2007).

Duetz, W. A. & Witholt, B. Effectiveness of orbital shaking for the aeration of suspended bacterial cultures in square-deepwell microtiter plates. *Biochem. Eng. J.* **7**, 113–115 (2001).

Duetz, W. A. & Witholt, B. Oxygen transfer by orbital shaking of square vessels and deepwell microtiter plates of various dimensions. *Biochem. Eng. J.* **17**, 181–185 (2004).

Dürauer, A., Hobiger, S., Walther, C. & Jungbauer, A. Mixing at the microscale: Power input in shaken microtiter plates. *Biotechnol. J.* **11**, 1539–1549 (2016).

Esmonde-White, K. A., Cuellar, M., Uerpmann, C., Lenain, B. & Lewis, I. R. Raman spectroscopy as a process analytical technology for pharmaceutical manufacturing and bioprocessing. *Anal. Bioanal. Chem.* **409**, 637–649 (2017).

Faust, G. et al. Feeding strategies enhance high cell density cultivation and protein expression in milliliter scale bioreactors. *Biotechnol. J.* **9**, 1293–1303 (2014).

Fibinger, M. P. C. et al. Fully automatized high-throughput enzyme library screening using a robotic platform. **9999**, 1–12 (2016).

Flitsch, D. et al. Respiration activity monitoring system for any individual well of a 48-well microtiter plate. *J. Biol. Eng.* **10**, 1–14 (2016).

Flores Bueso, Y. & Tangney, M. Synthetic biology in the driving seat of the bioeconomy. *Trends Biotechnol.* **35**, 373–378 (2017).

Funke, M. et al. Bioprocess control in microscale: Scalable fermentations in disposable and user-friendly microfluidic systems. *Microb. Cell Fact.* **9**, 86 (2010).

Funke, M., Diederichs, S., Kensy, F., Müller, C. & Büchs, J. The baffled microtiter plate: Increased oxygen transfer and improved online monitoring in small scale fermentations. *Biotechnol. Bioeng.* **103**, 1118–1128 (2009).

Galvanin, F., Macchietto, S. & Bezzo, F. Model-Based Design of Parallel Experiments. *Ind. Eng. Chem. Res.* **46**, 871–882 (2007).

Gorochowski, T. E., Van Den Berg, E., Kerkman, R., Roubos, J. A. & Bovenberg, R. A. L. Using synthetic biological parts and microbioreactors to explore the protein expression characteristics of escherichia coli. *ACS Synth. Biol.* **3**, 129–139 (2014).

Gruber, P., Marques, M. P. C., Szita, N. & Mayr, T. Integration and application of optical chemical sensors in microbioreactors. *Lab Chip* **17**, 2693–2712 (2017).

Halim, M., Rios-Solis, L., Micheletti, M., Ward, J. M. & Lye, G. J. Microscale methods to rapidly evaluate bioprocess options for increasing bioconversion yields: application to the ω-transaminase synthesis of chiral amines. *Bioprocess Biosyst. Eng.* **37**, 931–941 (2013).

Hall, B. G., Acar, H., Nandipati, A. & Barlow, M. Growth rates made easy. *Mol. Biol. Evol.* **31**, 232–238 (2014).

Harms, J., Wang, X., Kim, T., Yang, X. & Rathore, A. S. Defining process design space for biotech products: Case study of Pichia pastoris fermentation. *Biotechnol. Prog.* **24**, 655–662 (2008).

Hemmerich, J. et al. Comprehensive clone screening and evaluation of fed-batch strategies in a microbioreactor and lab scale stirred tank bioreactor system: Application on Pichia pastoris producing Rhizopus oryzae lipase. *Microb. Cell Fact.* **13**, 1–16 (2014).

Hemmerich, J., Noack, S., Wiechert, W. & Oldiges, M. Microbioreactor systems for accelerated bioprocess development. *Biotechnol. J.* **13**, 1–9 (2018). doi:10.1002/biot.201700141

Hemmerich, J., Wiechert, W. & Oldiges, M. Automated growth rate determination in high-throughput microbioreactor systems. *BMC Res. Notes* **10**, 1–7 (2017).

Hermann, R., Lehmann, M. & Büchs, J. Characterization of gas-liquid mass transfer phenomena in microtiter plates. *Biotechnol. Bioeng.* **81**, 178–186 (2003).

Heux, S., Bergès, C., Millard, P., Portais, J. C. & Létisse, F. Recent advances in high-throughput 13C-fluxomics. *Curr. Opin. Biotechnol.* **43**, 104–109 (2017).

Heux, S., Juliette, P., Stéphane, M., Serguei, S. & Jean-Charles, P. A novel platform for automated high-throughput fluxome profiling of metabolic variants. *Metab. Eng.* **25**, 8–19 (2014).

Hofmann, M. C., Funke, M., Büchs, J., Mokwa, W. & Schnakenberg, U. Development of a four electrode sensor array for impedance spectroscopy in high content screenings of fermentation processes. *Sensors Actuators, B Chem.* **147**, 93–99 (2010).

Holmes, W. J., Darby, R. A. J., Wilks, M. D. B., Smith, R. & Bill, R. M. Developing a scalable model of recombinant protein yield from Pichia pastoris: The influence of culture conditions, biomass and induction regime. *Microb. Cell Fact.* **8**, 1–14 (2009).

Huber, R. et al. Robo-Lector - A novel platform for automated high-throughput cultivations in microtiter plates with high information content. *Microb. Cell Fact.* **8**, 1–15 (2009).

Jackson, N. B., Liddell, J. M. & Lye, G. J. An automated microscale technique for the quantitative and parallel analysis of microfiltration operations. *J. Memb. Sci.* **276**, 31–41 (2006).

John, G. T., Klimant, I., Wittmann, C. & Heinzle, E. Integrated optical sensing of dissolved oxygen in microtiter plates: A novel tool for microbial cultivation. *Biotechnol. Bioeng.* **81**, 829–836 (2003).

Kensy, F. et al. Oxygen transfer phenomena in 48-well microtiter plates: Determination by optical monitoring of sulfite oxidation and verification by real-time measurement during microbial growth. *Biotechnol. Bioeng.* **89**, 698–708 (2005).

Kensy, F., Engelbrecht, C. & Büchs, J. Scale-up from microtiter plate to laboratory fermenter: Evaluation by online monitoring techniques of growth and protein expression in Escherichia coli and *Hansenula polymorpha* fermentations. *Microb. Cell Fact.* **8**, 1–15 (2009).

Kensy, F., Zang, E., Faulhammer, C., Tan, R. K. & Büchs, J. Validation of a high-throughput fermentation system based on online monitoring of biomass and fluorescence in continuously shaken microtiter plates. *Microb. Cell Fact.* **8**, 1–17 (2009).

Kirk, T. V & Szita, N. Oxygen transfer characteristics of miniaturized bioreactor systems. *Biotechnol. Bioeng.* **110**, 1005–1019 (2013).

Klein, M. et al. The expression of glycerol facilitators from various yeast species improves growth on glycerol of Saccharomyces cerevisiae. *Metab. Eng. Commun.* **3**, 252–257 (2016).

Klein, T., Schneider, K. & Heinzle, E. A system of miniaturized stirred bioreactors for parallel continuous cultivation of yeast with online measurement of dissolved oxygen and off-gas. *Biotechnol. Bioeng.* **110**, 535–542 (2013).

Klöckner, W. & Büchs, J. Advances in shaking technologies. *Trends Biotechnol.* **30**, 307–314 (2012).

Konstantinidis, S. et al. Flexible and accessible automated operation of miniature chromatography columns on a liquid handling station. *Biotechnol. J.* **13**, 1–10 (2018).

Kosa, G., Shapaval, V., Kohler, A. & Zimmermann, B. FTIR spectroscopy as a unified method for simultaneous analysis of intra- and extracellular metabolites in high-throughput screening of microbial bioprocesses. *Microb. Cell Fact.* **16**, 1–11 (2017).

Krause, M., Neubauer, A. & Neubauer, P. The fed-batch principle for the molecular biology lab: controlled nutrient diets in ready-made media improve production of recombinant proteins in Escherichia coli. *Microb. Cell Fact.* **15**, 1–13 (2016).

Kunze, M., Roth, S., Gartz, E. & Büchs, J. Pitfalls in optical on-line monitoring for high-throughput screening of microbial systems. *Microb. Cell Fact.* **13**, 53 (2014).

Ladner, T. et al. Application of mini- and micro-bioreactors for microbial bioprocesses. In *Current Developments in Biotechnology and Bioengineering: Bioprocesses, Bioreactors and Controls*, (eds. Larroche C, Sanromán MÁ, Du G, Pandey A.), pp. 433–461. Amsterdam: Elsevier B.V. (2016). doi:10.1016/B978-0-444-63663-8.00015-X

Ladner, T., Beckers, M., Hitzmann, B. & Büchs, J. Parallel online multi-wavelength (2D) fluorescence spectroscopy in each well of a continuously shaken microtiter plate. *Biotechnol. J.* **11**, 1605–1616 (2016).

Lattermann, C. & Büchs, J. Microscale and miniscale fermentation and screening. *Curr. Opin. Biotechnol.* **35**, 1–6 (2015).

Lattermann, C., Funke, M., Hansen, S., Diederichs, S. & Büchs, J. Cross-section perimeter is a suitable parameter to describe the effects of different baffle geometries in shaken microtiter plates. *J. Biol. Eng.* **8**, 1–10 (2014).

Lee, M. E., DeLoache, W. C., Cervantes, B. & Dueber, J. E. A highly characterized yeast toolkit for modular, multipart assembly. *ACS Synth. Biol.* **4**, 975–986 (2015).

Lennen, R. M. et al. Transient overexpression of DNA adenine methylase enables efficient and mobile genome engineering with reduced off-target effects. *Nucleic Acids Res.* **44**, e36 (2015).

Long, Q. et al. The development and application of high throughput cultivation technology in bioprocess development. *J. Biotechnol.* **192**, 323–338 (2014).

Luchterhand, B. et al. Newly designed and validated impedance spectroscopy setup in microtiter plates successfully monitors viable biomass online. *Biotechnol. J.* **10**, 1259–1268 (2015).

Mandenius, C. & Brundin, A. Review : Biocatalysis and bioreactor design optimization, bioprocess methodology, using design-of-experiments. *Biotechnol Progr* **24**, 1191–1203 (2008).

Marcellin, E. & Nielsen, L. K. Advances in analytical tools for high throughput strain engineering. *Curr. Opin. Biotechnol.* **54**, 33–40 (2018).

Matosevic, S., Szita, N. & Baganz, F. Fundamentals and applications of immobilized microfluidic enzymatic reactors. *J. Chem. Technol. Biotechnol.* **86**, 325–334 (2011).

Micheletti, M. et al. Fluid mixing in shaken bioreactors: Implications for scale-up predictions from microlitre-scale microbial and mammalian cell cultures. *Chem. Eng. Sci.* **61**, 2939–2949 (2006).

Micheletti, M. & Lye, G. J. Microscale bioprocess optimisation. *Curr. Opin. Biotechnol.* **17**, 611–8 (2006).

Morris, P., Rios-Solis, L., García-Arrazola, R., Lye, G. J. & Dalby, P. A. Impact of cofactor-binding loop mutations on thermotolerance and activity of E. coli transketolase. *Enzyme Microb. Technol.* **89**, 85–91 (2016).

Mühlmann, M. et al. Cellulolytic roboLector - towards an automated high-throughput screening platform for recombinant cellulase expression. *J. Biol. Eng.* **11**, 1–18 (2017).

Mühlmann, M., Forsten, E., Noack, S. & Büchs, J. Optimizing recombinant protein expression via automated induction profiling in microtiter plates at different temperatures. *Microb. Cell Fact.* **16**, 1–12 (2017).

Neubauer, P. et al. Consistent development of bioprocesses from microliter cultures to the industrial scale. *Eng. Life Sci.* **13**, 224–238 (2013).

Nunes, M. A. P., Fernandes, P. C. B. & Ribeiro, M. H. L. Microtiter plates versus stirred mini-bioreactors in biocatalysis: A scalable approach. *Bioresource Technology* **136**, 30–40 (2013).

Olsson, I. M. et al. Rational DOE protocols for 96-well plates. *Chemom. Intell. Lab. Syst.* **83**, 66–74 (2006).

Panula-Perälä, J. et al. Enzyme controlled glucose auto-delivery for high cell density cultivations in microplates and shake flasks. *Microb. Cell Fact.* **7**, 1–12 (2008).

Paquet-Durand, O., Ladner, T., Büchs, J. & Hitzmann, B. Calibration of a chemometric model by using a mathematical process model instead of offline measurements in case of a H. polymorpha cultivation. *Chemom. Intell. Lab. Syst.* **171**, 74–79 (2017).

Peter, C. P., Lotter, S., Maier, U. & Büchs, J. Impact of out-of-phase conditions on screening results in shaking flask experiments. *Biochem. Eng. J.* **17**, 205–215 (2004).

Pollard, J., Mcdonald, P. & Hesslein, A. Lessons learned in building high-throughput process development capabilities. *Eng. Life Sci.* **16**, 93–98 (2016).

Ramirez-Vargas, R., Vital-Jacome, M., Camacho-Perez, E., Hubbard, L. & Thalasso, F. Characterization of oxygen transfer in a 24-well microbioreactor system and potential respirometric applications. *J. Biotechnol.* **186**, 58–65 (2014).

Ren, Y., Ji, Y., Teng, L. & Zhang, H. Using Raman spectroscopy and chemometrics to identify the growth phase of Lactobacillus casei Zhang during batch culture at the single-cell level. *Microb. Cell Fact.* **16**, 1–10 (2017).

Riedlberger, P. & Weuster-Botz, D. New miniature stirred-tank bioreactors for parallel study of enzymatic biomass hydrolysis. *Bioresour. Technol.* **106**, 138–146 (2012).

Rios-Solis, L. et al. A toolbox approach for the rapid evaluation of multi-step enzymatic syntheses comprising a 'mix and match' E. coli expression system with microscale experimentation. *Biocatal. Biotransformation.* **9**, 192–203 (2011).

Rios-Solis, L. et al. High throughput screening of monoamine oxidase (MAO-N-D5) substrate selectivity and rapid kinetic model generation. *J. Mol. Catal. B Enzym.* **120**, 100–110 (2015a).

Rios-Solis, L. et al. Modelling and optimisation of the one-pot, multi-enzymatic synthesis of chiral amino-alcohols based on microscale kinetic parameter determination. *Chem. Eng. Sci.* **122**, 360–372 (2015b).

Rios-Solis, L. et al. Non-linear kinetic modelling of reversible bioconversions: Application to the transaminase catalyzed synthesis of chiral amino-alcohols. *Biochem. Eng. J.* **73**, 38–48 (2013).

Rocke, D. M. Design and analysis of experiments with high throughput biological assay data. *Semin. Cell Dev. Biol.* **15**, 703–713 (2004).

Rohe, P., Venkanna, D., Kleine, B., Freudl, R. & Oldiges, M. An automated workflow for enhancing microbial bioprocess optimization on a novel microbioreactor platform. *Microb. Cell Fact.* **11**, 1–13 (2012).

Samorski, M., Müller-Newen, G. & Büchs, J. Quasi-continuous combined scattered light and fluorescence measurements: A novel measurement technique for shaken microtiter plates. *Biotechnol. Bioeng.* **92**, 61–68 (2005).

Schmieder, A. et al. High-cell-density cultivation and recombinant protein production with Komagataella pastoris in stirred-tank bioreactors from milliliter to cubic meter scale. *Process Biochemistry* **51**, 177–184 (2016).

Schmidt, F. R. Optimization and scale up of industrial fermentation processes. *Appl. Microbiol. Biotechnol.* **68**, 425–35 (2005).

Shih, S. C. C. & Moraes, C. Next generation tools to accelerate the synthetic biology process. *Integr. Biol.* **8**, 585–588 (2016).

Sprouffske, K. & Wagner, A. Growthcurver: An R package for obtaining interpretable metrics from microbial growth curves. *BMC Bioinformatics* **17**, 17–20 (2016).

Tang, Y., Laidlaw, D., Gani, K., & Keasling, J. D. Evaluation of the effects of various culture conditions on Cr(VI) reduction by shewanella oneidensis MR-1 in a novel high-throughput mini-bioreactor. *Biotechnol. Bioeng.* **95**, 176-184 (2006).

Unthan, S., Radek, A., Wiechert, W., Oldiges, M. & Noack, S. Bioprocess automation on a mini pilot plant enables fast quantitative microbial phenotyping. *Microb. Cell Fact.* **14**, 1–11 (2015).

Vasser, M., Vasser, M., Polakis, P. & Polakis, P. Letters to nature. *Nature* **429**, 2–6 (2004).

Velez-Suberbie, M. L. et al. High throughput automated microbial bioreactor system used for clone selection and rapid scale-down process optimization. *Biotechnol. Prog.* **34**, 58–68 (2018).

Voss, J. & Mittelheuser, N. E. Advanced monitoring and control of pharmaceutical production processes with Pichia pastoris by using Raman spectroscopy and multivariate calibration methods. *Journal of Bioscience and Bioengineering.* 1281–1294 (2017). doi:10.1002/elsc.201600229

Weuster-botz, D. Experimental design for fermentation media development: Statistical design or global random search ? **90**, 473–483 (2000).

Wewetzer, S. J. et al. Parallel use of shake flask and microtiter plate online measuring devices (RAMOS and BioLector) reduces the number of experiments in laboratory-scale stirred tank bioreactors. *J. Biol. Eng.* **9**, 1–18 (2015).

Wiegmann, V., Martinez, C. B. & Baganz, F. A simple method to determine evaporation and compensate for liquid losses in small-scale cell culture systems. *Biotechnol. Lett.* 1–8 (2018). **40**:1029–1036. doi:10.1007/s10529-018-2556-x

Wilming, A., Bähr, C., Kamerke, C. & Büchs, J. Fed-batch operation in special microtiter plates: A new method for screening under production conditions. *J. Ind. Microbiol. Biotechnol.* **41**, 513–525 (2014).

Wong, J. et al. High-titer production of lathyrane diterpenoids from sugar by engineered Saccharomyces cerevisiae. *Metab. Eng.* **45**, 142–148 (2018).

Yordanov, B., Dunn, S-J., Kugler, H., Smith, A., Martello, G., & Emmott, S. A method to identify and analyze biological programs through automated reasoning. *Systems Biology and Applications* **2**, 1–16 (2016).

Zhang, H., Lamping, S. R., Pickering, S. C. R., Lye, G. J. & Shamlou, P. A. Engineering characterisation of a single well from 24-well and 96-well microtitre plates. *Biochem. Eng. J.* **40**, 138–149 (2008).

Zimmermann, S. et al. Cell separation in aqueous two-phase systems - influence of polymer molecular weight and tie-line length on the resolution of five model cell lines. *Biotechnol. J.* **1700250**, 1–12 (2017).

Zimmermann, H. F., Anderlei, T., Büchs, J. & Binder, M. Oxygen limitation is a pitfall during screening for industrial strains. *Appl. Microbiol. Biotechnol.* **72**, 1157–1160 (2006).

Zimmermann, H. F., John, G. T., Trauthwein, H., Dingerdissen, U. & Huthmacher, K. Rapid evaluation of oxygen and water permeation through microplate sealing tapes. *Biotechnol. Prog.* **19**, 1061–1063 (2003).

19 Control and Monitoring of Industrial Fermentations
An Industrial Perspective and Recent Trends

Gilles Roux, Cesar Arturo Aceves-Lara, Zetao Li,
Boutaib Dahhou, and Craig J.L. Gershater

CONTENTS

"A person, who never made a mistake, never tried any-thing new."

Albert Einstein

19.1 REQUIREMENT FOR CONTROL

In the early 1970s, fermentation processes were perceived to be as much art as science. The best indicator for a fermentation progress was the experience operator looking through a site glass and observing the color and texture of the foam layer on top and sometimes by smelling near to the bioreactor. Much progress in control bioprocess has occurred during the past 40 years.

Control is generally defined as the power to direct or influence. In the case of fermentation control, the requirement to control a given biotechnological process is generally dictated by the need to bring about a desired outcome such as maximizing output of product formation. Control is applied to provide a near optimal environment for microorganisms. Microorganisms used in industrial processes have been isolated by virtue of their ability to overproduce certain commercially significant attributes. These attributes can range from the ability to produce carbon dioxide for leavening activity in the case of *Saccharomyces cerevisiae* (baker's yeast) to the production of useful microbial metabolites such as alcohol from yeast or antibiotics from filamentous bacteria.

Whatever the reason for commercial exploitation, the microorganism will have been obtained directly or indirectly from an ecosystem very remote from that of the fermentation laboratory and process plant. The metabolic attribute to be exploited will have evolved as a result of environmental factors unknown to the fermentation scientist, and therefore, process optimization must seek to replicate those environmental factors responsible for expression of the desired attribute in the totally "artificial" environment of the industrial bioreactor or fermentor.

19.1.1 MICROBIAL GROWTH

The principal fermentation control requirement is population growth (see Chapters 2 and 3 for more details) of the microorganism of interest. In its natural habitat, the microbe will respond to environmental stimuli such as excess nutrients by synthesizing enzymes and biomass capable of exploiting the resource as effectively as possible. In the fermentor system, the microbe will be inoculated into the fermentation medium "feast" and will thus attempt to colonize this environment through rapid growth.

The role of the fermentation control strategy is to provide—by control of environmental effectors such as temperature, aeration, pH, and dissolved oxygen—the optimum conditions for growth and colonization. The fermentation scientist uses an environment that is capable of being controlled to a limited degree (i.e., the fermentor) to develop a control strategy that will modify inputs to the fermentor system to achieve the desired outputs. This system of modifying inputs to obtain a desired output from a control system is often described as a control loop.

19.1.2 NATURE OF CONTROL

A fermentation development program seeks to establish what control set points are needed for the control loops in the control system. It is easy to forget that the control system can be human.

19.1.3 CONTROL LOOP STRATEGY

The basic element of a control system is the control loop. Figure 19.1 summarizes the various components that make up the control loop. At the center of the control loop is the system requiring control, the fermentor. The system can be affected by several influences, and in this case, it is the temperature of the system that is to be controlled. A very simple control loop is shown in which only cooling may be applied to the

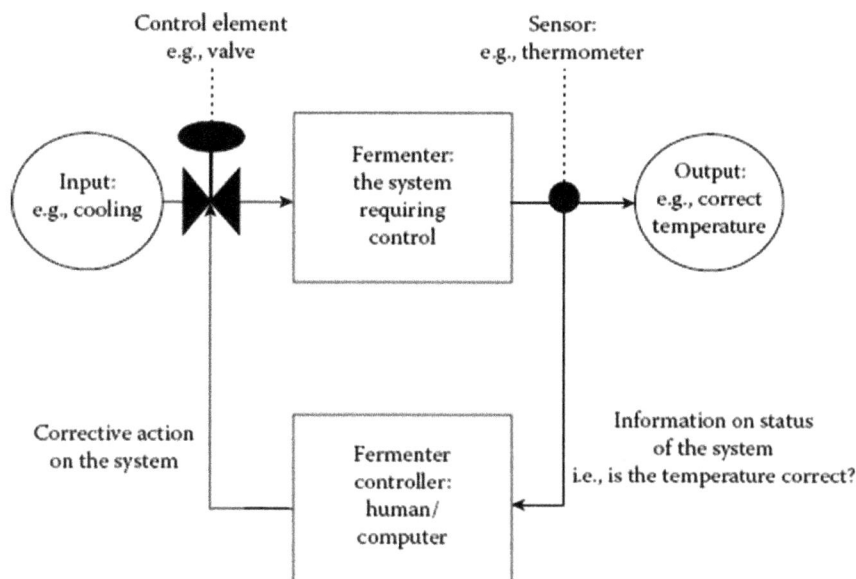

FIGURE 19.1 Simple control loop.

fermentor principally to remove metabolic heat during incubation. The temperature of the fermentor is measured using a thermometer, and a human operator or a control system can monitor this measurement and make adjustments if the desired temperature (the set point) and the actual temperature are not equal (error in the system). If the temperature of the fermentor is above the set point, the flow of cooling water is turned on (input to the system) and once the temperature equals the set point, the control valve is switched off. In reality, dynamic fermentor systems are never fully stable and constant monitoring and control are required, hence the advantage of automated systems, which maintain constant vigil on the process and adjust inputs to the system as and when required automatically.

19.2 SENSORS

Development of fermentation processes thus far has been accomplished primarily in large-scale reactors (>10 L) with the aid of relatively few sensors. The problem with monitoring the fermentation system is maintaining the sterile integrity of the fermentor, hence any sensor has to be capable of sterilization *in situ* during steam sterilization or remotely and then aseptically introduced to the fermentor.

19.2.1 Historical Perspective

In the 1940s, the sensors were largely based on manual sampling and offline analysis; control was improved through the 1950s and 1960s by the use of limited electrical signals controlling pneumatic outputs (valves, etc.). In the 1970s, new sensors capable of being sterilized were introduced, including pH and dissolved oxygen probes. In addition, more accurate methods of measuring flow rates (both liquid and gas) became available and other engineering parameters such as motor power to the agitator and enhanced nutrient feed addition

systems were developed. In addition, improvements to sensors and other measuring devices mini-computers were introduced to provide simple control and data logging.

Recently, a wide range of new sensors (see Chapter 18 for more details) has emerged for online analysis of fermentation parameters. Recent techniques that have become available include biomass probes, online liquid chromatography systems, near infrared spectroscopy, and so on. For the most part, the reliability and relevance of some of the measurements together with prohibitive cost tend to preclude these sensors from everyday fermentations.

19.2.2 Typical Fermentation Sensors

The control elements (sensors) that should be considered routine for most (aerobic) fermentation systems are

1. *Temperature:* Temperature is measured using a platinum resistance thermometer (PRT probe), where the increase in temperature is proportional to the increase in electrical resistance in the probe. Temperature will be controlled by the addition of cooling water to a jacket or cooling finger of a fermentor; heat will be added by direct heating of the vessel or its contents (electrical heating mantle or "hot finger") or by the injection of hot water or steam to the circulating water in a jacket or heat exchanger.

2. *Airflow rate:* Airflow rate is measured using a standard pressure drop device such as variable area flow meters or more often a mass flow sensor. Airflow will generally be controlled using a proportional (0 to 100% open) valve upstream of the sterile inlet filter on a fermentor. Airflow is frequently expressed as VVM, or the volume of gas per volume of liquid per minute; fermentor design generally permits up to 2 VVM.

3. *Vessel pressure:* Vessel pressure is measured using diaphragm-protected Bourdon gauges or strain-gauge pressure transducers. Pressure in the vessel is induced during *in situ* steam sterilization and during normal incubation with the introduction of air. Control of vessel pressure is by regulation of the vent gas from a fermentor. Pressure is generally a negatively acting loop in that a fully driven output valve (100% open) results in minimal pressure in the fermentor. The units of pressure are generally bar gauge (i.e., pressure within the reactor above atmospheric).

4. *Vessel agitation rate:* Vessel agitation rate is measured using proximity detectors to detect the speed of the shaft. Impellors or mixers are controlled by standard motor controllers, and the units are revolutions per minute. Agitation power is sometimes measured using current transformers measuring the electrical power consumption of the motor; the units of power are Watts. pH is measured using steam-sterilizable combined glass electrodes. pH is controlled by the use of buffers or by the addition of acids or base titrants.

5. *Dissolved oxygen:* Dissolved oxygen is measured using polargraphic-type probes; here galvanic voltages on a membrane-covered oxygen-reducing cathode induce a current (amperometric) proportional to the amount of oxygen diffusing through the membrane. Altering the status of specific control loops controls dissolved oxygen. An increase in dissolved oxygen is induced in fermentation medium by (1) increasing the airflow rate (volumetric increase in oxygen), (2) increasing agitator speed (smaller air bubbles increasing the surface area available for diffusion), and (3) increasing overpressure in the fermentor (generally increasing the residence time for air bubbles). Dissolved oxygen is measured as the partial pressure of oxygen at the electrode surface and hence is expressed as percentage saturation (often referred to as dissolved oxygen tension, DOT). To obtain a mass, a fully saturated solution of oxygen in water is approximately equivalent to 1.2 mM/L.

6. *Foam:* Foam is detected by conductance, or capacitance probes completing an electric circuit when foam is contacted. Foam may then be controlled by reducing high aeration and/or agitation rates or by the addition of antifoaming agents.

Figure 19.2 shows the main control elements of a fermentor. There are many different configurations that may be specified, but the one shown would be capable of providing data and control options for an aerobic fermentation.

19.2.3 CONTROL ACTION

Temperature control would be achieved by the regulated supply of temperature-controlled water to the jacket of the fermentor. In the case illustrated, there are two thermometers in the system—one indicating the temperature of the medium/broth and the other indicating the temperature of the return water flow. In this configuration, the two signals may be

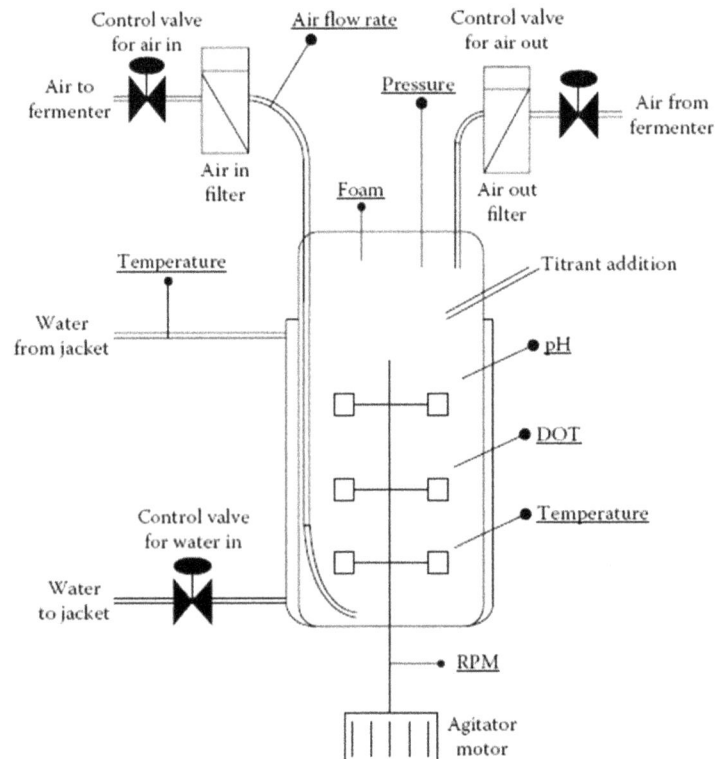

FIGURE 19.2 Basic elements for control of a fermentor.

compared and control options could include linking their function to provide a regulated flow of temperature-controlled water to the jacket via the control valve shown.

Airflow control would be achieved by regulating the linear gas flow rate to the system by adjusting the airflow control valve shown in response to signals coming from the airflow sensor. Pressure control is achieved by regulating the flow of off gas from the fermentor. It is probably beneficial to have the airflow and pressure control functions independent of each other. For the most part these control loops will act to the limit of their engineering configuration, and if a particular set point (i.e., for airflow) cannot be achieved because of excess back-pressure, then this may be noted as a process constraint and the operating protocol adjusted accordingly. Agitator speed is controlled by direct feedback of the revolutions per minute using a suitable tachometer.

pH is generally controlled by the addition of acid or base titrant in response to changes in pH value during the fermentation. One important aspect of the pH loop is the calibration of the probe before and after sterilization. These are achieved before sterilization by the adjustment of slope and intercept values on the pH controller to be used during the fermentation. Subsequent to sterilization it is possible to check the calibration of the pH probe by comparing the pH of a sample of medium withdrawn from the fermentor using a calibrated external pH meter. The function of the pH-control loop will be discussed in more detail later.

Dissolved oxygen control is achieved in several different ways. It is possible to rely totally on increasing agitator speed to shear air bubbles rising through the broth to increase surface area and thus mass transfer of oxygen from the gaseous to liquid phase. However, it is also possible to cascade the control output from the dissolved oxygen controller to agitator speed, airflow, and pressure or any combination of these. In a particularly high-oxygen-demanding fermentation, all three outputs may be specified, and care has to be exercised in the use of independent control loops such as agitator speed that this loop will be the "slave" to the DOT "master" control loop function. The function of the DOT control loop will be discussed in more detail later. Foam is controlled by the addition of antifoam agents via the feed addition system in response to a contact probe detecting rising foam.

19.3 CONTROLLERS

Control instrumentation has developed rapidly in recent years, and the range of options available for fermentation control is extensive. The development of integrated circuits in the last 30 years or so has meant that complex control functions can be devolved to cheaper instruments, or conversely more sophisticated control options have become increasingly available to the fermentation scientist.

19.3.1 Types of Control

Two fundamental types of control may be incorporated into a fermentation system: sequence control and loop control. Sequence control is that part of control that permits automation of the fermentor operation such as sterilization and other valve automation sequences; that is, for the most part providing an ordered array of digital (on/off) signal control. Loop control is generally associated with that part of control dealing with combinations of digital and analogue control signals, and although used in sequence control (particularly automated sterilization sequences), is most often associated with control of incubation. The industrial fermentation scientist will frequently wish to adapt the control strategy for optimal process performance and to ensure versatility; modified control programs may be specified, although the costs and potential delays associated with this approach are likely to be significant.

Control of fermentation systems can be achieved by the use of discrete single-loop controllers, by programmable logic controllers controlling sequence and loop functions, and by the use of specific software packages to control all aspects of the fermentation. This type of control is sometimes called *distributed digital control* (DDC) and defines boundaries of operation in which whole plant control can be achieved using computer-based systems.

19.3.2 Control Algorithms

Whatever type of controller is selected for fermentation control, effective action will depend on the response of the controller. This response is determined by the nature of the control algorithms programmed into the system. A control algorithm is a mathematical representation of the steps required to achieve effective control, most often programmed as equations in which the controller output is a function of signal deviation away from a set point. As indicated in Figure 19.1, most controllers will function as feedback control systems; that is, the deviation of measured variable compared with the desired set point will determine how large the control effect should be on the fermentor system via the appropriate control algorithm.

19.3.3 PID

The types of control algorithm most frequently encountered are three-term or PID controllers. The PID controller is made up of three elements—P, proportional; I, integral; and D, derivative/differential—the purpose of these functions is to provide a fast-acting response to process deviation and scale the response to the output to achieve smooth control action. The characteristics of PID control are

- Proportional control provides an output, the magnitude of which is proportional to the deviation between the measured variable and the set point.
- Integral control tends to reduce the effect of proportional control alone, helping to bring the measured variable back to the set point faster by minimizing the integral of control error.
- Derivative action also tends to reduce the effect of proportional control alone, this time by estimating

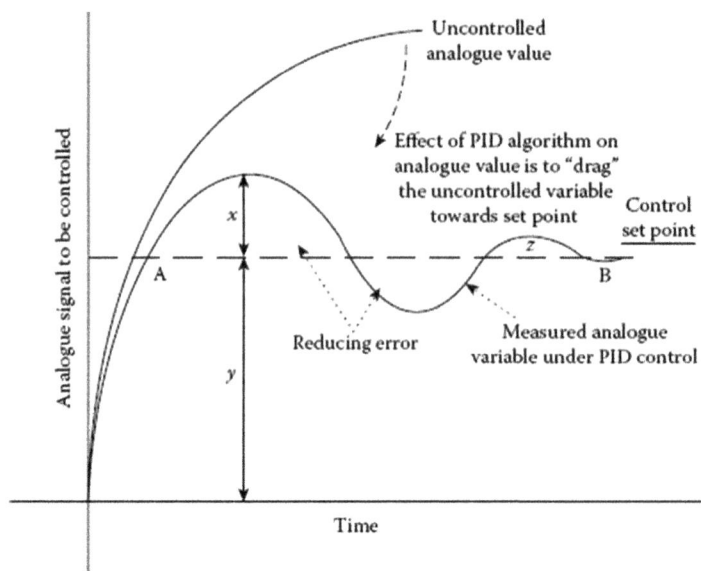

FIGURE 19.3   PID control.

the slope of measured variable with time and maximizing the slope of the measured variable compared with the set point.

The effect of PID control on an uncontrolled variable is shown in Figure 19.3. As indicated, the algorithm tends to "drag" the analogue value toward the set point, the speed and effectiveness of this control response are a function of the parameters specified by the PID algorithm, and the process of setting these parameters is termed *tuning the loop*. The method is generally carried out by trial and error with the loop online. There are three parameters associated with PID loop tuning: the proportional band (usually set as a percentage of full-scale deflection of the analogue signal), the integral time, and the derivative time. The method chosen for loop tuning will be recommended in control instrument vendors' instructions, and the sequence is generally as follows:

1. Set the proportional band first with no integral (maximum value) or derivative function (minimum value).
2. When oscillations occur with small process perturbations, decrease the integral time (from its maximum) until oscillations reoccur, and then reset the integral time to minimize oscillations.
3. Establish whether control is adequate with PID control alone, If not satisfactory following process perturbations, increase the derivative time constant (from its minimum) to minimize oscillations.

PID loop performance can be assessed by measuring the initial overshoot as a ratio of x/y. A measure of PID loop effectiveness can be assessed by examining the control error decay ratio of z/x. In addition, the "gain" of the controller (i.e., how fast the measured variable attains the set point when full control is applied) is described by point A. Point B describes the

"settling time" for the loop (i.e., when the measured variable is within a fixed percentage of the set-point value, e.g., ±5%).

19.4   DESIGN OF A FERMENTATION CONTROL SYSTEM

The first stage in designing a fermentation control system is to clearly establish what the process objectives are for the automation project.

19.4.1   CONTROL SYSTEM OBJECTIVES

Objectives may be typically divided into various categories of control.

1. Control of basic incubation functions only, typically airflow, agitator speed, and temperature;
2. Automation of the control of incubation only relying on manual operation of valves associated with sterilization;
3. Full automation of the fermentor system including all sterilization and auxiliary vessel control;
4. Advanced control options including event-based control; and
5. Advanced computing methods for inferential control.

Whatever type of controller is considered necessary, the process scientist will have to specify the type of control and then make this specification available to the manufacturer of various control systems. Although the more complex the fermentor system is to be controlled, the more detailed the specification will need to be.

1. Basic control will typically be achieved using single-loop controllers and is frequently associated with autoclavable fermentors. These controllers tend to

combine amplification of process signals and control function in one "box"; this functionality may also include a local readout function in engineering units (i.e., values recognizable to the operator). There is a wide range of commercially available single-loop controllers, some of which have been developed with control of fermentation processes in mind.

2. Incubation control may include other features, including operation of feed delivery systems and control of more complex loops such as dissolved oxygen by multiple outputs that may be specified by the operator. The specification in this case may be more complex and subject to negotiation with the instrument vendor. Sterilization of this type of vessel may be by autoclaving or by *in situ* steam sterilization, but this will be done manually, probably to minimize cost.

3. Full incubation and sterilization control of the fermentor will require a control function for valve sequencing (digital control) and control of analogue parameters for incubation (process variables and control of proportional valves). The requirement for valve sequencing and more complex patterns of analogue control is met using programmable logic controllers (PLCs) or often fermentor manufacturer's own control software running on a personal computer. This type of control is generally required for larger fermentors (>20 L) or when several identical vessels are to be purchased and manual sterilization will be too manpower-intensive. The need for a detailed specification is determined by whether the vendor's own software meets the needs of the fermentation scientist. The benefit of using the vendor's

own software is speed of implementation and probably lower costs; the disadvantages are that you get what the vendor thinks you want, not what you might actually need. The opposite is generally the case with PLCs; costs and time scales will be higher, but if you get the specification right, you get exactly what you want and need.

4. Advanced incubation control regimes may be required where complex fermentation patterns are required and changes to set points may be needed online, particularly in response to specified "events." This will be discussed in more detail later. Large fermentor control tasks may be related to complexity of control and numbers of fermentors within a system. If the task involves many fermentors to be controlled, then a system of distributed control will probably be required. The architecture for distributed control is shown in Figure 19.4.

Figure 19.4 illustrates one possible control option for the control of four fermentors. In this configuration, each fermentor is independent of the next and control is affected by each fermentor being equipped with its own controller (in this case a PLC). This type of distributed control has the advantage that failure of a control system will only affect one fermentor; the experimental design may be ruined by such an occurrence, but without a distributed system, failure of a single controller will result in failure of the entire run. The disadvantage of distributed control is that the cost may be prohibitive to implement one controller per vessel. In the author's laboratory, a compromise has been reached with vessels larger

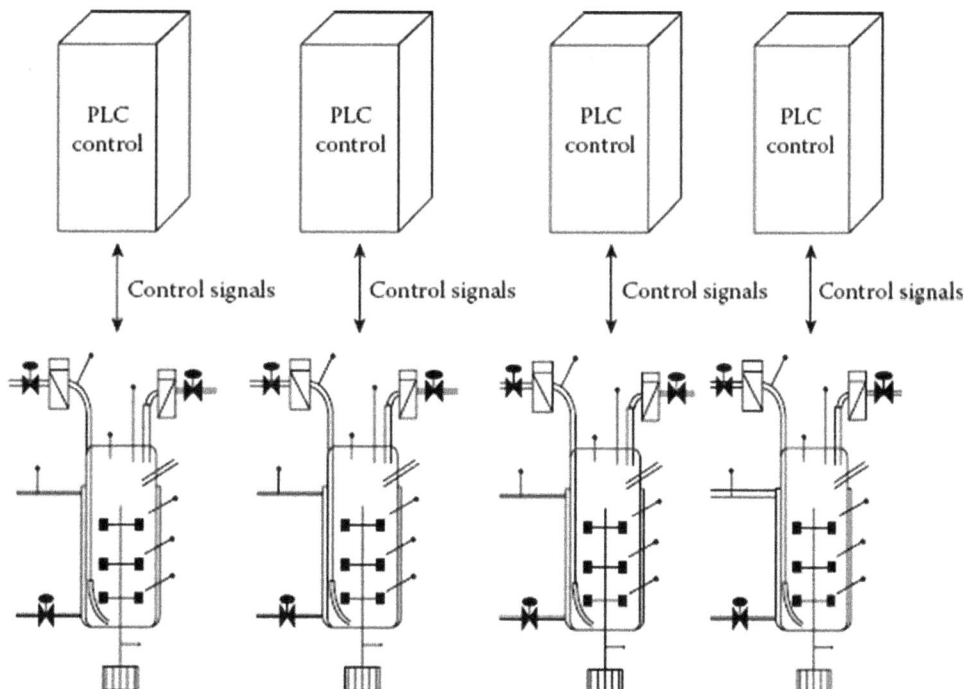

FIGURE 19.4 Distributed fermentor control.

than 100 L in working volume where a single PLC controls a pair of identical fermentors.

5. Advanced computing methods may be required where analytical systems are inadequate to optimize the fermentation process and cost of goods or value of product warrant investment in "next-generation" computing methods. Inference methods are available that algorithmic sensors can function to estimate for analytes and processes for which no other sensor or probe exists. This will be discussed in more detail later. The sort of computing methods being developed include expert systems, metaheuristic algorithms, artificial neural networks, and model-based systems. These self-learning or decision-making systems rely on process pattern recognition to identify regions of the fermentation process more or less susceptible to perturbations. Once identified remedial action beyond the scope of experienced operators or less sophisticated controllers can be initiated to reduce cost, avoid catastrophes, or simply improve the process. These systems can often only be justified in production environments where the value of the product warrants large investment in time and money to maximize productivity. As these methods become more widely available, the costs of implementation of advanced computing methods will decrease and wider applications will be sought and introduced.

19.4.2 Fermentation Computer Control System Architecture

In defining the overall strategy for computer control of fermentations, system architecture will emerge to indicate how the system will be operated; how data will flow around the system; and how that data can be captured, stored, and interrogated when required. The control system when configured is an information exchange system.

Information about the progress of the fermentation is detected by sensors and transmitted to amplification/signal conditioning units that forward this information to the controller. Figure 19.5 shows an example of a fermentor control system architecture in which the information exchange is indicated.

In Figure 19.5, there is flow of analogue and digital sensor data from the fermentor vessel to the control cabinet. The PLC in Figure 19.5 is connected to the operator terminal, which may be a PC with a video display unit (VDU) or a standalone plant terminal. If located on the process plant floor, then the whole cabinet has to be splash- and dust-proof, often referred to as an IP55 rating.

The PLC is also connected to a data highway onto which other fermentor systems are connected. In Figure 19.5, all of the fermentors are in communication with a supervisory and data acquisition system (SCADA), the function of which is to collect all of the plant data and store it for later retrieval and possible interrogation. It should be understood that the SCADA is not a database; it generally cannot give answers to database-type queries, but using advanced graphics and trending graphs, can present the operator with comprehensive information on plant status.

19.4.3 Fermentation Plant Safety

One of the overriding functions of a fermentation control system is monitoring and action relevant to safety. A fermentor is a potentially dangerous piece of equipment because it uses live steam (to 135°C), overpressure (to ≥ 2 bar gauge),

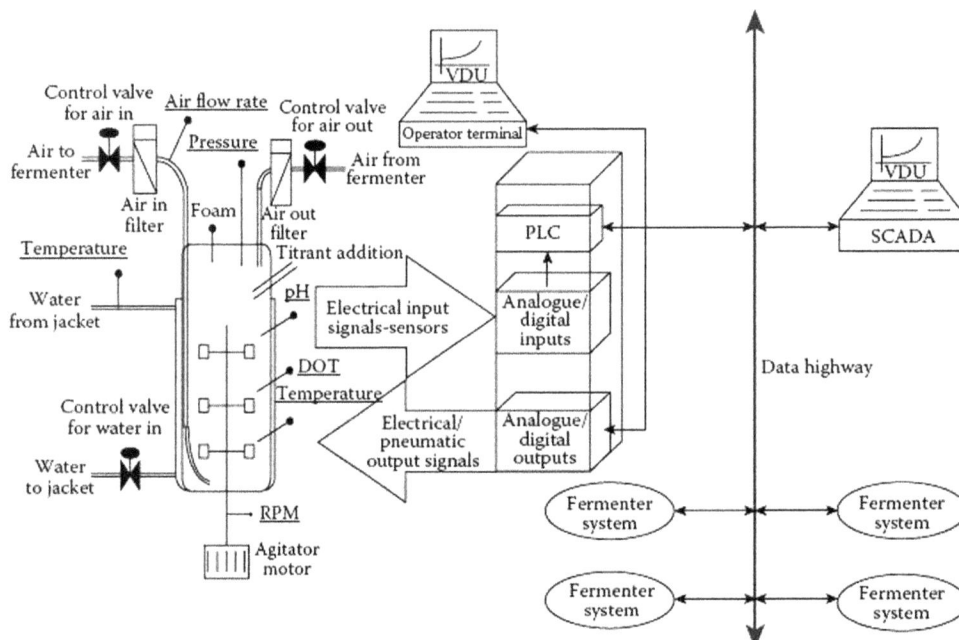

FIGURE 19.5 Fermentor control system architecture indicating information exchange.

acids, bases, and the microorganism and its products. The automated control system can ensure by program interlocks, and so on, that the plant is operated with the maximum safety margins on every function, and specifying this is perhaps the most important part of the acquisition of a new control system. This will be discussed in more detail later.

19.5 FERMENTOR CONTROL SPECIFICATION

When planning for a new fermentor control system, the specification for the system is crucial to the success of the project. The specification will be used to judge vendor's response and to subsequently quote and oversee the project to completion. Time and effort spent generating a comprehensive system specification are rewarded many times over during implementation and commissioning.

19.5.1 SPECIFYING SEQUENCE CONTROL

One of the most important tasks when planning to automate a fermentor system, particularly for sterilization and other automated sequences, is to accurately specify what the sequences are to be automated and how this will be achieved. There are fixed unit operations associated with fermentor control and the specification will define these and indicate how this control will be achieved.

19.5.2 FERMENTATION UNIT OPERATIONS

Taking batch fermentation as a typical example, the unit operations to be selected might include

- *Blank sterilization:* This is generally used as a pre-batching cleansing sequence when all steam and drain valves are opened.
- *Medium batching:* The fermentor is usually in a safe state for opening and preparation. In the author's laboratory, the safe state is defined by "standby." In standby, all valves are shut with the exception of valves venting the vessel interior to atmosphere. During batching, probe calibration and insertion into the vessel must be complete before water is added to the vessel!
- *Medium sterilization:* The sterilization sequence will be partly dictated by the fermentor vessel configuration and geometry. Steam sterilization of the medium can be achieved by direct steam injection or by indirect heating via a heat exchanger or vessel jacket. Whatever the mode of sterilization chosen, it is crucial that the correct valves are operated in the correct sequence, not just for process integrity but also for safety.
- *Medium hold:* It is useful to maintain the fermentation medium in a safe state after sterilization. It is at this time that final adjustments to the fermentor setup can be made (e.g., pH adjustment via a tartan feed system or temperature alteration before inoculation).

- *Fermentor inoculation:* To introduce inoculum into the sterilized fermentor requires opening the vessel in a controlled manner and introducing inoculum using strict aseptic techniques throughout. If introducing the inoculum via peristaltic pumps, then a reduction in vessel pressure will be required.
- *Incubation:* Although not strictly a sequence, there are clearly certain valves that must be opened to permit incubation of the culture to proceed under specified conditions. The principal function of the incubation sequence is to start the clock counting the hours elapsed since inoculation. During incubation, the full functionality of the control system may be used to program the correct fermentation "trajectory" for the run. This will involve specifying set points for all of the main controllers, including agitator speed, airflow, temperature, and feeds if available, as well as controls responding to more metabolic influences such as pH and dissolved oxygen.
- *Harvest:* It may be necessary to specify a sequence dealing with harvest operations. This could entail prechilling or suitable pH adjustment of the broth to permit easier product recovery operations. Harvest could also include "killing" the vessel; that is, sterilizing the vessel interior (with or without cells) before safe opening and cleaning.
- *Cleaning:* Several options may be available here, including full sterilization or heating in the presence of caustic detergents to fully automated clean-in-place (CIP) systems with complex valve operations of their own.

19.5.3 VESSEL STATES

Fermentation unit operations may be termed *vessel states,* and the transition between them must be strictly regulated in an automated plant; for example, initiating automatic sterilization of the batch during the incubation should be avoided at all costs. State changes can be summarized in a state diagram (Figure 19.6).

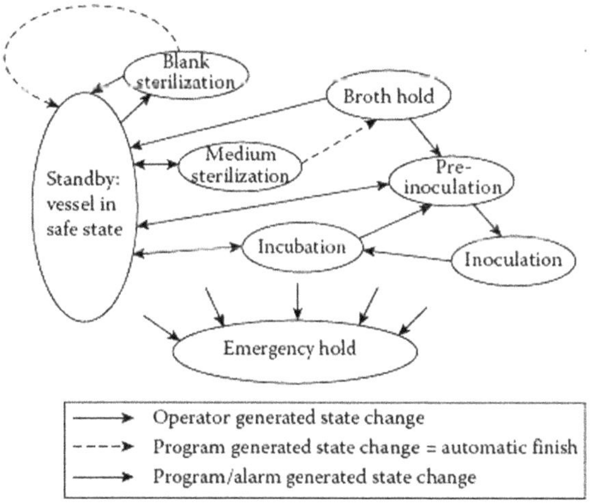

FIGURE 19.6 Control state diagram.

Each one of the states in the diagram defines a part of the control program specifying sequences to be executed for effective control of the fermentation system. Navigating around the state diagram defines the safe operation of the vessel under automatic control. Certain transitions are under operator control such as from standby to medium sterilization. Other states are attained automatically as a result of a particular sequence finishing, such as the transition from sterilization of the medium sterilization to broth hold. Within an automated control system there is the opportunity to initiate emergency action that could come about as a result of alarm settings or operator intervention. This is indicated in Figure 19.6, where a state called emergency hold may be attained from anywhere in the program. Such a state can be programmed to do whatever is deemed appropriate for safe containment of the process; in the author's laboratory *emergency hold* constitutes a safe universal shutdown of the fermentation process sealing the vessel but retaining set points for possible restart of incubation.

19.5.4 SEQUENCE LOGIC

Control of multiple valves on an automated fermentor requires the program to control the opening and closing of valves such that the state operation is effectively and safely completed. Each of the vessel states indicated previously and possibly many others have to be programmed taking best human operator practice and engineering constraints into account. Defining the sequence follows a pattern of working that ensures ambiguities are minimized and objectives are clearly stated. A comprehensive description will come from accurate drawings of the plant; these drawings are often referred to as piping and instrumentation drawings (P&ID, not to be confused with PID for control). The drawing will identify the units of control to be defined in the program. Modern operating systems will tend to function with structured code, and it is possible to consider individual blocks of code controlling individual units of control.

This can be illustrated with reference to Figure 19.7, a typical fermentor configuration (as a notional P&ID) is shown for sterilization, and only those valves that are relevant to this highly simplified representation are shown.

The P&ID identifies the key elements for the following groups of valves:

- Air in group V1.X
- Air out group V2.X
- Jacket group V3.X

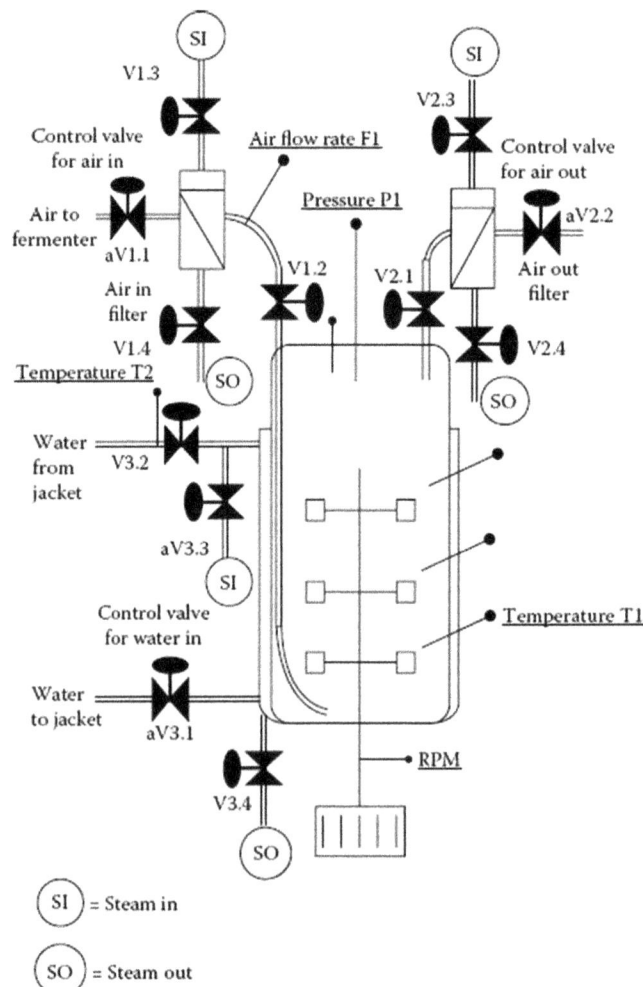

FIGURE 19.7 Control of sterilization example.

SEQUENCE LOGIC for sterilization operations:

1. START: all valves closed
2. Drain jacket
3. Heat-up phase
4. Direct steam injection
5. Air filter sterilization
6. Sterilization temperature
7. Air filter pressurization
8. Crash cool
9. Ballast air
10. Broth hold

19.5.5 Flow Charting

P&ID drawings, valve descriptions (Table 19.1), and status charts (Table 19.2) are essential in defining the operation to be automated. However, before code can be written, the operation is translated into a flow chart of valve operation and decision gates that must be followed. The flow of code to accomplish the task will follow the chart (Figure 19.8).

In this simple example, many different factors have been taken into account, but the flow chart is far from complete; for example, there is no mechanism in this for operators to enter set points. Again, with this example, there are only 12 valves; on a 4500-L pilot-scale fermentor used for research and production there may be 70 or more valves organized in many functional groups, hence programming such a system

will require many days/weeks of specifying, programming, and testing before the fermentor control system may be commissioned.

19.6 CONTROL OF INCUBATION

Specifying the sequence logic of a fermentor control system is only part of the control task, arguably the most important function of the fermentor control system is to control the incubation of the microorganism of interest. Returning to the case made for the control of fermentations at the start of the chapter, the purpose of the control system is to provide an optimal environment for growth and expression of an attribute associated with that growth. This is very difficult; one can pose the rhetorical question, what is the natural habitat for *Escherichia coli*? The answer given by many might be "the intestinal tract of most vertebrate animals," but enteric bacteria can be isolated from many mesophillic environments and evolved probably billions of years before the advent of colons.

Within the manufacturing sector fermentation development will be investigating the factors associated with a well-established fermentation process where the objectives of a development fermentor will be to obtain relatively modest increases in productivity (leading to substantial financial savings on large-scale production) or to achieve reduction in "cost of goods." In a research and development environment, the constraints usually come from working with a wide range of culture types, limited/no knowledge of those cultures, limited/no knowledge of the fermentation systems, and with very short development times required.

19.6.1 Specification for Incubation Control

Specifying the control options for a fermentor system must then take into account not only the type of organisms to be grown but also what sort of control options will actually be required. The control options for a typical research and development fermentor system are described in Sections 19.6.1.1–19.6.1.7.

19.6.1.1 Temperature

Control of incubation temperature will be achieved by the use of system of addition of heat and cooling. This control loop is fundamental to fermentor systems. The degree of accuracy of control during incubation will probably be on the order of a set point in the range 20–50°C (± 0.1°C). Response times may also be important, and controlled ramp rates of better than 1°C/min may be necessary.

19.6.1.2 Aeration

For most industrially significant processes the microorganisms will be aerobic. Supply of air to the fermentor will need to be controlled typically in the range 0–2 VVM. The accuracy of the control system will need to be on the order of $\pm 1\%$ of full-scale deflection (FSD). This is generally a fast-acting loop and obtaining adequate ramp rates is not a problem.

TABLE 19.1

Valve Descriptions for Example Fermentor (Figure 19.7)

Air in Group	Air Out Group	Jacket Group
V2.1	Digital air out valve	
aV2.2	Analogue control valve for pressure	
V2.3	Digital steam to air out filter	
V2.4	Digital steam out from air out filter	
aV3.1	Analogue water to jacked valve	
V3.2	Digital water from jacket valve	
aV3.3	Analogue control valve steam to jacket	
V3.4	Digital jacket drain/condensate out valve	
aV1.1	Analogue control valve: air to vessel	
V1.2	Digital air ex: filter block valve	
V1.3	Digital steam to air in filter	
V1.4	Digital steam out from air in filter	
	BOTH: Analogue and time i.e., pH > 6.5 AND < 24 h true	
	EITHER or BOTH: i.e., pH > 6.5 true OR time < 24 h OR both true	
	EITHER or TRUE (but not both): either pH > 6.5 true OR time < 24 h true	
	EITHER or NEITHER (but not both): i.e., neither pH > 6.5 OR time < 24 h, OR pH, OR time, but NOT pH AND ...time true	
	NEITHER: neither pH > 6.5 OR time < 24 h true	

TABLE 19.2
Valve Status Chart for Sterilization Sequence Logic

					Valve Status					
Valves	START	Drain	Heat	Filter	Direct	STER	Press	Cool	Ballast	Hold
					Air in					
aV1.1	0	0	0	0	0	0	$a = 1$	$a = 1$	a	a
V1.2	0	0	0	0	1	1	0	0	1	1
V1.3	0	0	0	1	1	1	0	0	0	0
V1.4	0	0	0	1	0	0	$1 \Rightarrow 0$	0	0	0
					Air out					
V2.1	0	0	1	1	1	1	1	1	1	1
aV2.2	0	0	$a = 1$	a	a	a	A	a	a	a
V2.3	0	0	0	0	0	1	0	0	0	0
V2.4	0	0	0	0	0	1	0	0	0	0
					Jacket					
aV3.1	0	0	0	0	0	0	0	$a = 1$	a	a
V3.2	0	0	0	0	0	0	0	1	1	1
aV3.3	0	0	$a = 1$	a	a	a	A	0	0	0
V3.4	0	1	1	1	1	1	1	0	0	0

19.6.1.3 Agitation

The stirrer for a traditional fermentor is used to minimize gradients within the bulk broth, the larger the vessel the greater the potential for gradient (mass transfer) problems. Sufficient motor power has to be available to obtain uniform mixing times to be on the order of 1–2 min or less (determined from the addition of a marker into the bulk liquid and the time taken to homogeneity recorded). These fast mixing times may not be achievable in pilot-scale vessels (3000 L + working volumes) and these system constraints must be identified and resolved, for example, in scaled-down experimental designs (modeling large-scale parameters in small-scale vessels). Agitation ramp

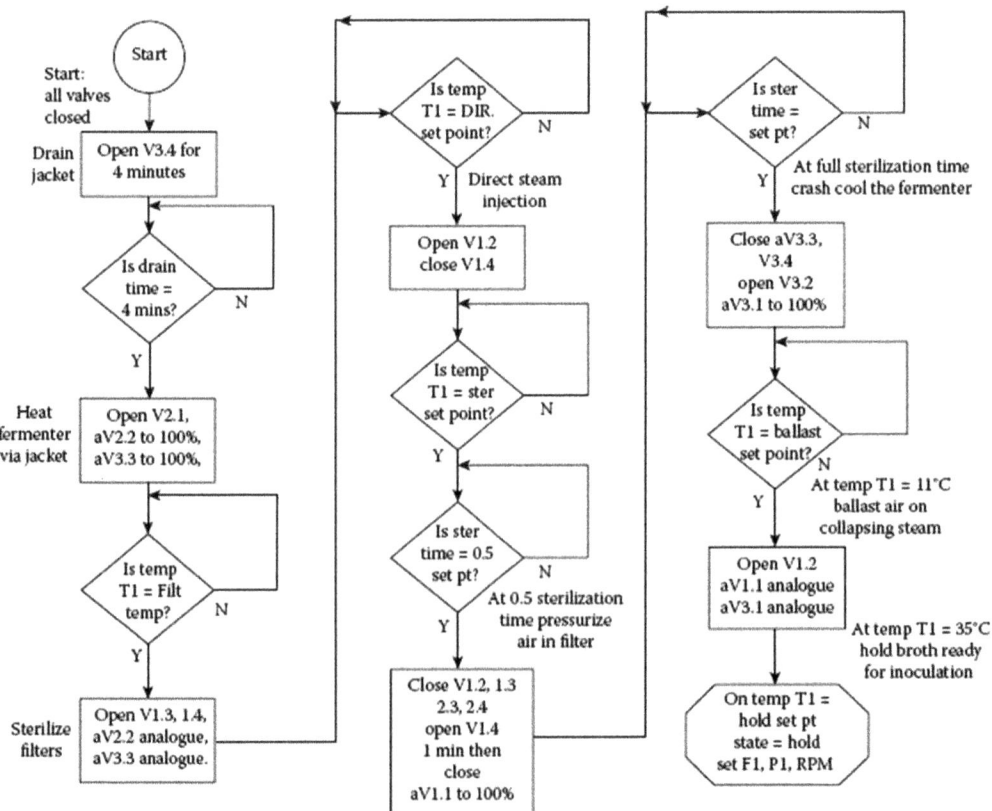

FIGURE 19.8 Example flow chart for sterilization of a fermentor.

rates may be high, and the control system should be capable of ramp rates equivalent to 10% of FSD per minute.

19.6.1.4 Pressure

This control loop only applies to pressure-rated vessels. The control of pressure is required for accurate control of *in situ* steam sterilization as well as for some incubation regimes. The control requirement during incubation may come from maintaining adequate pressure for high oxygen transfer or to simulate hydrostatic head pressure in large-scale vessels in scale-down experiments in smaller scale vessels. Control of pressure is critically linked to safety, and safeguards within the control system must ensure that the vessel cannot be overpressurized. Pressure can be a very fast-acting loop and obtaining adequate ramp rates is not a problem. The function of this loop may be in conflict with the aeration loop, and the fermentation scientist has to bear in mind that control of these two loops may be incompatible. For example, a set point for pressure near the safe operating limit of the vessel may be impossible to achieve with extremely low airflow rates (0.1 VVM).

19.6.1.5 Hydrogen Ion Concentrations, The pH

Most industrial fermentations need to be run within a certain range of pH values for maximal productivity. Although media are formulated in such a way that it ensures that a certain level of buffering capacity is "built in," there is often a requirement to control the pH away from this "natural" value. This is achieved most often by the addition of acid or base titrants. Addition of acid or base titrants requires a feed system of some description. Under these circumstances, the control of the feed system is subordinated to the pH controller; in other circumstances the feed system may act under the direction of another controller or independently, which will be discussed in more detail below. When tuning a pH control loop it is very difficult to define the "correct" PID settings because the strength of the titrants will vary. Therefore, the "gain" of the controller (i.e., how quickly a response will be achieved) will also vary greatly. User definable dead bands around the set point can bring about adequate pH control. The purpose of the dead band is to allow for the natural buffering capacity of the medium to have sufficient time to act so that titrants are not added close to the set point on a detuned loop, which will be the case with variable titrant concentrations.

In Figure 19.9, the response of a pH control loop after acid or base addition is shown. It will be noted that there is a proportional action on the additions of titrant made (i.e., the larger the "error" the more frequent the control action). The closer the measured pH gets to the set point, the longer the delay between each addition of titrant; however, when the dead band threshold is crossed, all titrant addition stops and the natural buffering capacity of the medium results in an approach to the set point within the dead band limits. The setting of the dead band limits may be asymmetrical about the set point depending on the criticality or tendency of the fermentation to be acid or base tolerant. Mineral acid titrants may be substituted for organic acids or the principal carbohydrate energy source such as glucose. In this case, the buffering capacity or delay equates to the metabolic rate for the consumption of the sugar, and protons generated by catabolism will require a delay of some minutes before a noticeable effect on the pH will be detected. This type of control of pH by metabolic action is really a means of permitting the microorganism to autoregulate the supply of carbon.

19.6.1.6 Dissolved Oxygen

The control of dissolved oxygen in aerobic fermentations may be critical to the successful outcome of that fermentation. The requirement for oxygen may be very high during the rapid growth phase of a batch culture, and oxygen limitation may result in inadequate growth and incomplete oxidation of the primary energy source. The control of dissolved oxygen is generally achieved by increasing the "driving force" for the mass transfer of oxygen from the gaseous phase to the liquid phase. The methods or outputs available to the control system to achieve this include increasing the agitator speed, which

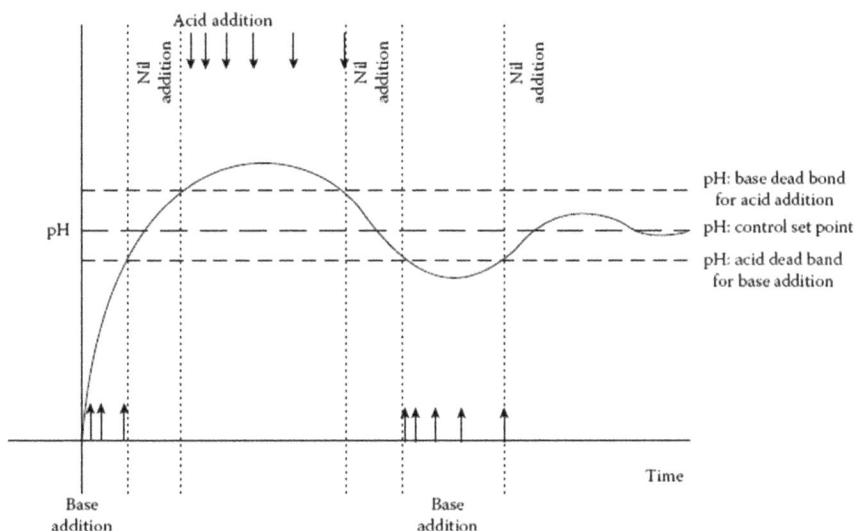

FIGURE 19.9 Hydrogen ion (pH) control by titrant addition.

generally increases the shear on the air bubbles, making them smaller and increasing their surface area available for gaseous exchange.

The second output that may be programmed is to simply increase the airflow rate, which ensures a greater volumetric airflow through the liquid. The last output that may be incorporated into a dissolved oxygen controller is vessel overpressure, which will have the effect of increasing the hydrostatic head pressure and preventing rapid flushing of "precious" air from the fermentor.

All of these outputs can be specified for a dissolved oxygen controller individually or in any combination of all three. The output of the dissolved oxygen controller (master) must be cascaded onto the closed-loop controllers for agitator, airflow, and pressure (slaves) as required. The consequence of this is that the function of the closed-loop controller will be subordinated to that of the dissolved oxygen controller to achieve the desired dissolved oxygen tension in the fermentation broth.

In Figure 19.10, the dissolved oxygen controller (master) is linked to all three closed-loop controllers (slaves). As the DOT falls, the oxygen controller output will rise, and this will cause an increase in agitator speed from a default closed-loop set point to a maximum revolutions per minute specified for the DOT output. In the example, the DOT continues to fall as the output of the controller passes a threshold equivalent to one-third of the total output range of the DOT controller. At this point, the second specified closed-loop output is subordinated to the DOT controller and takes over from agitator speed, leaving the set point for the first output at its specified maximum. The second output, in this case airflow, increases to try and compensate for the fall in dissolved oxygen, and

when this output limit is reached (two-third of the DOT controller output), then the third closed loop (in the example, vessel pressure) takes over, leaving airflow at a maximum value.

In addition to control by oxygen replenishment, it is also possible to initiate dissolved oxygen control by depletion. With the slave outputs set to relatively high output values, medium batched with little or no energy source present can have the DOT controlled by the regulated feed of, for example, carbohydrate. Under these circumstances, the feed system becomes the cascaded slave output and only delivers glucose at a rate equivalent to the metabolic rate of the organism under fixed oxygen mass transfer limitations set by the closed-loop outputs (speed, air, and pressure).

19.6.1.7 Feed Systems and Antifoam

Most fermentor systems will be equipped with a mechanism for introducing various feed solutions under aseptic conditions. The method for achieving this will depend largely on the size and configuration of the fermentor itself. On smaller laboratory-scale vessels, feeds will be controlled by the use of peristaltic pumps working on flexible-walled tubing. On larger vessels such as those found in the pilot plant (>10–20 L), it is possible that purpose designed and constructed addition vessels independently sterilizable will deliver feeds using shot-wise additions of feed solutions.

Whatever the method of addition chosen, it is very likely that feeds will be pulsed in some way and therefore the variables available for control are pulse width (i.e., size of liquid 'shot') and interval between pulses. The set point for a feed controller can be several shots in a specified interval of control or the feed controller can be cascaded onto a master controller

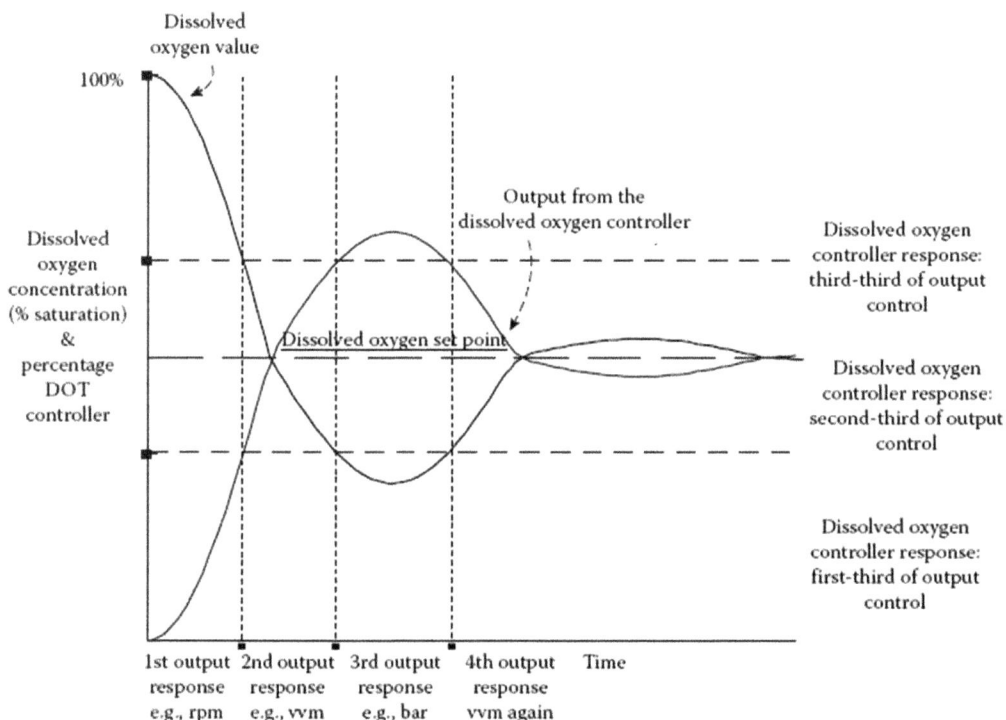

FIGURE 19.10 Dissolved oxygen control by master/slave cascade control.

such as pH or dissolved oxygen. The output of the feed controller can be linked to the interval such that the larger the output of the controller (further away from the desired number of shots) the shorter the interval and hence the more rapid the feed rate. Calibration of a feed system is difficult and will be affected by tubing age and vessel back-pressure. It is possible to introduce another sensor to take over the control of feed additions (i.e., a balance for the feed reservoir), and this type of control loop for feeds can provide very accurate and nonpulsed control of feed flows, but costs may be high, and it is only really applicable to smaller-scale vessels (<20 L).

Antifoam addition is variation of the feed control loop where the sensor is a contact probe detecting rising foam. The control variables here may be time related. To prevent overaddition of antifoam agent (which may destroy dissolved oxygen control by bursting air bubbles) a splash time interval can be specified in which the contact probe would need to be covered for more than a few seconds to initiate antifoam addition. Similarly, if the probe remains covered after antifoam pumps being switched on, then secondary control action can be specified in which the airflow rate and/or agitation speed may be reduced to minimize the generation of foam. Clearly, if this happens then the dissolved oxygen controller will be affected, but the output of the antifoam controller may be the ultimate master controller of closed loops to prevent loss of broth by foam-out.

19.7 ADVANCED INCUBATION CONTROL

The control options specified above represent the basic control elements for most fermentation development purposes. However, many fermentation development programs will require more than just a system of fixed set-point control, we present a description of an advanced PLC in the author's laboratory that is capable of more advanced fermentation control options.

The wild-type microorganism in its natural habitat (whatever that is), as previously discussed, is subject to transient environments. When a fermentation scientist attempts to grow the microbe and encourages it to express the desired attribute,

a fermentation regime will be imposed that will only represent a tiny fraction of all of the influences that the organism will encounter. Given that most fermentations will be run under fixed set-point regimes, the environment that is thus imposed will effectively be a "snapshot" of the full range of conditions that have brought about the expression, through evolutionary pressure, of the desired phenotype. Another option for control is therefore to provide a mechanism in which events beyond the fixed set-point control regime can be specified.

19.7.1 Fermentation Profiles

A typical fermentation control profile is shown below (Figure 19.11). The main features of the profile are captured in Table 19.3. When analyzing the progress of fermentation, it is sometimes useful to identify the key features and in which phase of the growth of the organism they occur. For this purpose, it is useful to use standard definitions for the growth phases, although their strict interpretation is obviously open to debate. In the example given, the following features become apparent:

- Harvest time is a long time after product accretion has ceased (this is frequently the case with old processes that have been inherited).
- Carbohydrate uptake is largely linear over the growth phase of the organism and is depleted by 60 h.
- Biomass accretion occurs between 0 and 60 h and then after substrate depletion goes into the decline phase.
- Dissolved oxygen profile is the mirror image of the growth profile.
- Product accretion has its onset at 24 h and is complete by 72 h.
- Harvest time using this protocol can be at 72–84 h.
- Dissolved oxygen is not limiting; therefore, under this fixed set-point regime would higher biomass yield more product (increasing fermentor vessel volumetric productivity)?

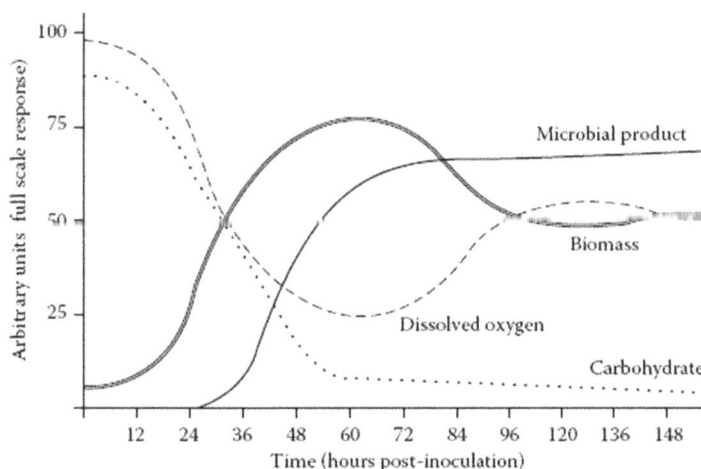

FIGURE 19.11 A typical fermentation profile for a filamentous microorganism producing a secondary metabolite.

TABLE 19.3

Fermentation Profile Data for Key Analytes at Various Growth Phases ("Snapshots")

Analyte	Time (h)	Valve (Arbitrary)	Grow Phase
Biomass	12	10	Lag
	24	25	Acceleration
	36	50	Exponential
	48	70	Deceleration
	60	75	Stationary
	84	60	Decline
	148	50	Harvest
Carbohydrate	0	85	Inoculation
	24	70	Acceleration
	36	35	Exponential
	48	20	Deceleration
	60	10	Stationary
	148	5	Harvest
Dissolved oxygen	0	100%	Inoculation
	24	80%	Acceleration
	36	40%	Exponential
	48	30%	Deceleration
	60	25%	Stationary
Microbial product	24	0	Acceleration
	36	10	Exponential
	48	35	Deceleration
	60	50	Stationary
	72	60	Decline
	148	65	Harvest

- Carbohydrate is depleted by 60 h, a further carbohydrate feed at either this point or when the DOT was less than 40% of saturation may promote a further product accretion phase.
- pH is not shown here, but the level is likely to fall between 12 and 60 h and this may require pH control by titrant addition.

This type of profile is typical of the kind that would be obtained with fixed set-point control for the principal closed-loop controllers (not including DOT control).

19.7.2 Event-Tracking Control

As indicated above, with the carbohydrate feed option it is possible to identify an event in the fermentation that serves to trigger another control action. Hence an event is a specific change in state or time or any combination of changes that can initiate a new event. In the case cited, a carbohydrate feed could be initiated at 60 hours postinoculation (time-based event) or if the dissolved oxygen measurement fell below 40% of saturation (analogue value event). Computer control systems, particularly those specified by the customer, can easily accommodate program decision gates that will test the status of the fermentation and apply another control action on the fermentation if the decision gate criteria are met. To define what the decision gates should be and what values of analyte or combinations of analyte values should be used requires the fermentation scientist to establish set points for control loops that he or she can control and then observe the effect on analytes that do not have online sensors (biomass, carbohydrate, and product in this example).

As soon as the fermentation scientist wishes to use event-based control, then further options may be sought to extend this capability. In the author's laboratory the PLC fermentor control system has been programmed with four types of user-definable events:

1. *Time-based events:* These become true at a specified number of hours postinoculation.
2. *Analogue value events:* These become true when a process value or combination of two process values exceed a threshold limit.
3. *Elapsed time events:* These become true at a defined time after another event has occurred.
4. *Boolean events:* These are logical combinations of any two other events using standard Boolean operators.

The trigger events for the most part can be almost any process signal or event "flag" indicating the status of that event. In addition, system events or alarm levels can be specified (this apart from warning alarms is a very powerful use of alarms signals). The events can be organized in any combination giving virtually limitless fermentation control strategies. The events themselves can initiate new set points, ramp rates (rates of change between set points), or new events. It is possible to "latch" initiating events such that dependent events will remain true even if the initiating event is no longer true.

Figure 19.12 shows three graphs for event-based control. Graph 3 shows the status (true/false) of three fermentation process events. Graph 2 shows set points for two output controllers (these could be any type of controller as discussed previously) to be driven by the specified events (1, 2, and 3). Graph 1 is the fermentation profile against which the new set-point control regime is to be imposed. A summary of the main changes is given below.

- Event 2 becomes true first (e.g., time value = 12 h) and the set point for controller 1 is changed from the default set point (arbitrary value 25) to the event 2 set-point (50) control action initiated at the end of the lag phase.
- The rate of change from one set point to the other can also be specified, and in the graph, it is comparatively slow—ramp-rate control action over acceleration phase.
- At approximately 40 h event 1 becomes true. The set point of controller 1 changes from event set point 2 to event set-point 1 (arbitrary value 75) with a faster ramp rate—control action at end of the exponential phase, start of deceleration phase.

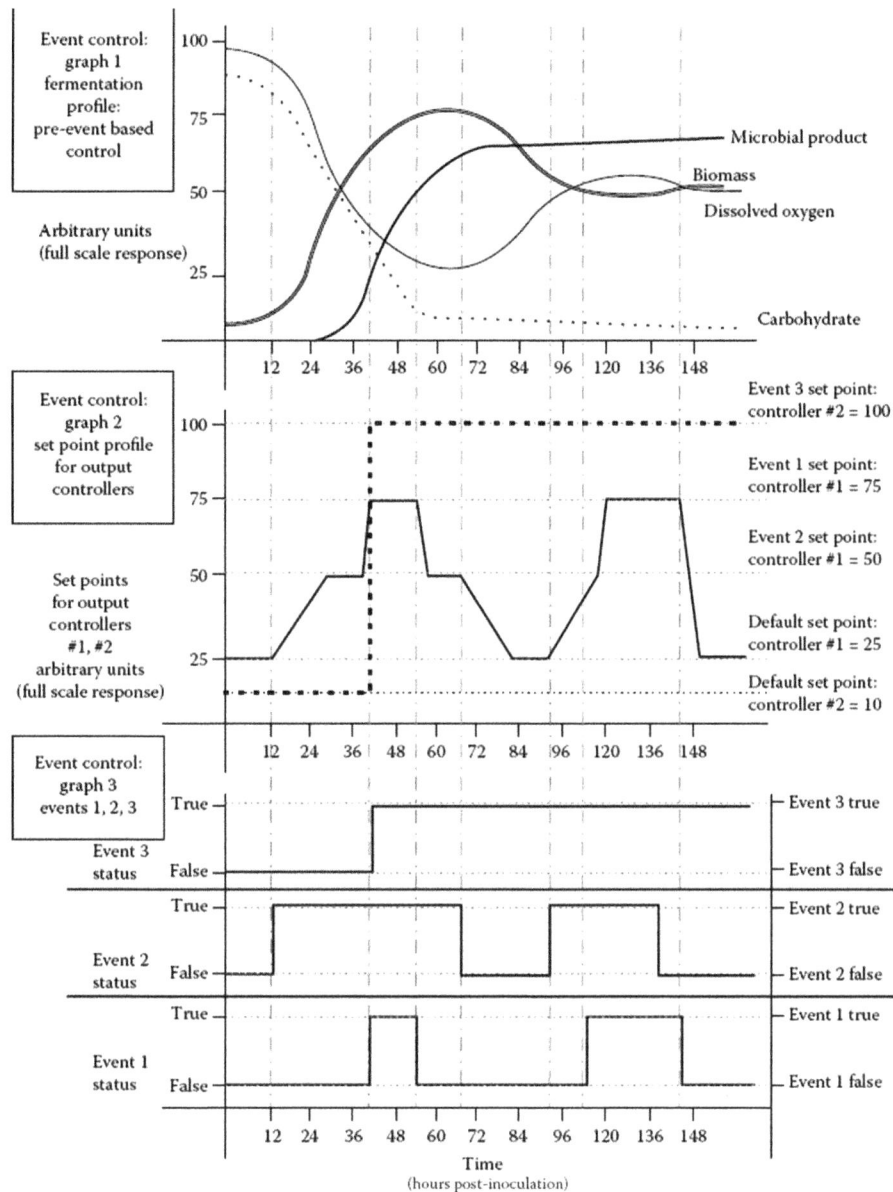

FIGURE 19.12 Example of event-based control for two controllers.

- Events 1 and 2 are true at this time. This initiates event 3 also becoming true, the set point for controller 2 changes from arbitrary value 10 to 100 (for example, could be a feed rate)—example feed rate response could be in response to deceleration phase.
- At approximately 50 h, event 1 then no longer is true and set-point controller 1 returns to event set point 2 level (50). Note that event 3 set point is latched and remains true for the remainder of the batch—start of stationary phase.
- At approximately 70 h, events 1 and 2 are no longer true and the default set point for controller 1 returns to the default level of 25 (~84 h)—70 h end of product accretion phase.
- At approximately 90 h, event 2 becomes true again, and the event 2 set point (50) for controller 1 is used. Before the event 2, set point is reached, event 1

becomes true and the event 1 set point (75) is used again (faster ramp rate). Event 1 set point is in use for controller 1 by 120 h—recovery from decline/death phase by 90 h (scavenged substrates?).
- At approximately 140 h, event 2 is no longer true, and a short time later event 1 is no longer true; at this point the controller 1 set point is reduced from 100 to 25 (default set point) — harvest time.

The results of this event-based control will be judged by comparing the fermentation profile (as shown in Figure 19.12: graph 1) with the one generated for the event control regime.

19.7.3 BOOLEAN CONTROL AND RULE GENERATION

It may be seen from the example in Figure 19.12 that complex patterns of control can be imposed on the fermentation and more or less significant changes to the growth environment and expression of phenotype will follow. However, it is difficult for the fermentation scientist to know which fermentation control regime to impose. This lack of knowledge comes from not knowing how the organism will respond to changes in individual control loops but also what the interactions with other control loops will be. To address this, Boolean logic can be used to introduce an element of control by "choice." With Boolean logic, it is possible to present options or choices from which the control system can select a preprogrammed path for control. This type of experiment is then generating a new kind of response variable—one based on the control path selected by the metabolism or response of the total fermentations system in which

Fermentation system = stainless steel vessel + valves + sensors + services (air, electricity, etc.) + medium + microorganism + microbial metabolism + expressed phenotype

If Boolean options or choices are presented to the system, then the path selected by the total system can represent a "rule" by which the response of the microorganism to an imposed environment can be described. To illustrate the principle of control by rules, Figure 19.13 shows a simple XOR rule statement.

The rule defined in Figure 19.13 is remarkably simple and helps to define glucose feed regime by pH and dissolved oxygen control. Having established this rule a fermentation experiment can then test which route is chosen under defined conditions of medium or mass transfer for oxygen, and so on. Another powerful use for control by rules comes with culture or mutant evaluation in which the chosen route may indicate a propensity toward one pattern of metabolism or another, which may have been induced within a putative mutant culture. Clearly, this is just one of many rules that could be introduced, and libraries of rules or elements that make up rules can be constructed to provide an array of control options to explore microbial metabolism under controlled conditions (Table 19.4).

19.7.4 SUMMARY OF EVENT AND NONSTABLE SET-POINT CONTROL

A summation of the possible interactions for event-based control is shown in Figure 19.14; here control options are defined as possible inputs and outputs to an experimental system in which events become control loops themselves evaluating whether conditions defined by rules are satisfied. This type of control system is an extension of the standard media and control recipes normally prepared and fermentation response data will produce a data set of online, offline, and derived data that generate rules for high productivity.

The fermentation scientist is being asked to establish the correct conditions for a microorganism to express a desirable attribute or phenotype in a totally artificial environment; event and rule-based control together with transient and variable set-point control (unstable) may help establish what unique set of factors and their interactions support the expression of a rare and valuable phenotype.

19.8 OTHER ADVANCED FERMENTATION CONTROL OPTIONS

The reader should be aware that other control philosophies are in use, and this technology is certain to continue to change and develop. Some of the control philosophies currently in use or in advanced development are described in the following subsections.

19.8.1 KNOWLEDGE-BASED SYSTEMS

Knowledge-based systems (KBSs) can be considered an extension of the sort of control capability described previously in which previously held "expert knowledge" is captured in a control system often referred to as an *expert system*. With this type of control already established, facts and data associated with a process can be held in a database of knowledge against which future decisions can be made. The principal difference with expert systems and the event-based control described here is that KBSs will only work on existing knowledge and will not necessarily generate new knowledge. The power of expert systems lies in their ability to execute potentially hundreds of rules per second controlling many functions associated with plant operation simultaneously. Early warning patterns can be recognized from this previous knowledge to

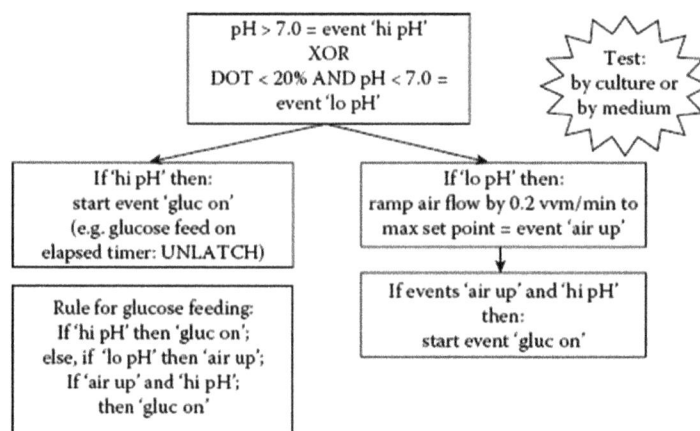

FIGURE 19.13 Example of control by rule: Specifying a rule for glucose addition.

TABLE 19.4

Boolean Truth Table for Fermentation Control

Boolean Operator	Event a (pH > 6.5) Status	Event b (Time < 24 h) Status	Comments (Example Event Combinations)
AND	Yes	Yes	
OR	Yes	No	
	No	Yes	
	Yes	Yes	
XOR	Yes	No	
	No	Yes	
NAND	No	No	
	No	Yes	
	Yes	No	
NOR	No	No	

alert either human or computer control operations of an excursion from normal operating parameters.

19.8.2 Artificial Neural Networks

Much discussion has been made in recent years of "intelligent" computer systems. By intelligent systems we mean systems capable of learning. The mechanism by which this occurs is beyond the scope of this chapter but for the most part is very similar to the control loop principle in which an input to a system is modified by attempting to minimize the error between desired and actual outcome. The use of neural networks in fermentation will come from the ability of these systems to recognize patterns and take actions on unknown data on the basis of those learned patterns. It is possible that the event-based system described here will be augmented by neural net ideas to help with obtaining data or process inferences not available by direct measurement. Artificial neural networks (ANNs) do not in themselves assist in providing data on the exact nature of process input interactions because the training and optimization of an ANN is a pure black box model of the process.

19.8.3 Metaheuristic Algorithms

Optimization based on natural systems, such as genetic algorithms (GAs), ants colony optimization (ACO) algorithms, and particle swarm optimization (PSO), dates from the beginning of the 1990s. These algorithms are important when the objective function has several optima and one is interested in a global minimum. Evolutionary algorithms, also of stochastic nature, can be used for nonlinear optimization and may indeed find global optima.

19.8.3.1 GAs

In this case, control philosophy has been influenced by modern biological genetics. Again, it is not possible to describe completely how these systems work here, but a brief description will be given. GA are algorithms that operate on a finite set of points, called a *population*. The different populations

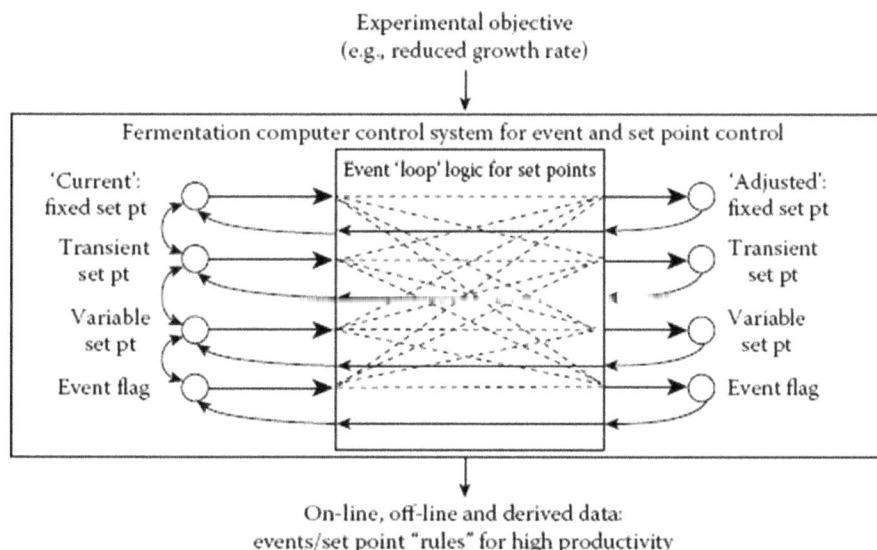

FIGURE 19.14   Control of fermentation by event loops and set-point rules.

are interpreted as generations. They are derived on the principles of natural selection and incorporate operators for fitness assignment, selection of points for recombination, recombination of points, and mutation of a point.

GAs are capable of searching vast areas of experimental space and by a process of natural selection of rules or algorithms that come closest to minimizing some process cost function tend to optimize on a process goal. This type of system may assist in rule induction by eliminating rules not fit for the process environment.

19.8.3.2 PSO

The PSO is a new search method based on the simulation of animal social behavior. This method has been applied in several studies to solve constrained and nonconstrained optimization problems (e.g., Broth nutrient optimization). The PSO algorithm work as follows: A position and velocity vector in the search space are associated with each individual from a population, with a dimension equal to the total parameters, resulting in a swarm size equal to the particle number. The particle position vector behaves under the restriction of three conditions (actual particle position vector, better previously assigned particle position vector, and better previously assigned swarm position vector) that allow delimiting a geographically searching space according to a linear vectorial combination, bringing out the particle to assume a new particle position vector. Starting at the swarm lower limit, each particle pathway is calculated in iterative way, obtaining new particle position vector.

19.8.4 Modeling

For many years, there has been a disparity between what modeling could apparently offer (see Chapter 6 for more details) and the acceptance and use of modeling by the process community for everyday control and optimization problems. This may now be changing with the advent of advanced user-friendly software to assist in forming and testing model predictions. There are many kinds of models that may be constructed in attempting to describe a complex system the key to approximation is simple models (i.e., relating carbon dioxide evolution to biomass or growth estimations). These types of relationships can be defined relatively easily by curve-fitting raw data and then predicting based on alteration of parameters defined in function generated from curve fits. When several functions describing microbial growth and productivity have been created, then they can be combined in a mathematical relationship potentially useful in predicting fermentation outcomes. Biological models are usually divided into two categories: nonstructured and structured models. Nonstructural models do not consider the biological structure and the state of the cells. They are usually represented with empirical equations. On the other hand, structured models taken into account the cell physiology. Most of the models used in bioprocess are nonstructural, but an increasing number of structured models have appeared recently.

Optimization, monitoring, and control can be applied combining successful models to online measurements. Then thousands of fermentations can be correctly run in software before committing to expensive stainless steel vessels.

The unstructured model (the most widely used unstructured models to describe cell growth are the Monod kinetic model) is based on the observation of the macroscopic kinetics within the reactor. We consider that the biomass activity can be sufficiently specified using only a single variable. Generally, the microbial concentration in the medium and changes in the biomass composition are completely ignored. These models are generally based on original Monod equation (Monod, 1942) or a similar equation, including the various enzymatic reactions. The constants appearing in these mathematical equations are empirical and often determined by optimization based on experimental data. Other equations describing consumption of the substrate and the biosynthesis of products are necessary to write the model.

In our case, a model developed using a fuzzy classification method to describe the process behavior (Piera et al., 1989). A fuzzy classification method is an algorithm (based on fuzzy logic) that operates in homogeneous classes of combinations of a set of abstract objects or signals. This type of model does not require any mathematical knowledge of the process and builds a representation of the process in the form of physiological states. The physiological states of a biological process can be defined as being a defined interval in which the principal physiological variables remain constant. This model does not try to describe the evolution of the process variables but rather to determine the current situation on the basis of these variables. The goal of this methodology is to design a process model in which supervision and diagnosis can be exercised effectively.

19.8.4.1 Unstructured Models

We used our own software, which was developed under MATLAB® (Bâati et al., 2004). To solve the system of nonlinear algebraic equations representing the culture, the Gauss-Newton method (which is a method used to solve nonlinear least-squares problems) with a mixed quadratic and cubic line search procedure was applied. For numerical integration, Runge-Kutta algorithms are a family of implicit and explicit iterative methods for the approximation of solutions of ordinary differential equations. Several algorithms with different orders, according to the difficulty of the modeling equations, were used (which check for integrity and thus prevent frequent numerical problems). The optimization runs were carried out using a multitasking Pentium computer.

A learning process in this case meaning a heuristic search strategy that allows selection of the most appropriate model from a range of models stored in a database. A heuristic search is a method that might not always find the best solution, but it guarantees to find a good solution in reasonable time. A fermentation process, which is nonlinear, can be modeled by the following dynamic equations:

$$\dot{X}(t) = \Omega\big(X(t), u(t), \eta(t)\big)$$
$$Y(t) = HX(t)$$

(19.1)

where $X(t)$ is the state vector (set of concentrations), generally including biomass, substrate, and product concentration); $Y(t)$ is the observation vector (set of measured concentrations), which can be measured; $u(t)$ is the input vector (set of control variables), which can be used to take into account the effect of environmental variables; and $\eta(t)$ is the kinetic vector (set of kinetic parameters), which contains the main biological parameters of the fermentation reaction. The vector $\eta(t)$ is composed of complex functions of the state variables and of several biological constants and its expression is varied for different fermentation processes. Therefore, the primary task of modeling is to identify which model of $\eta(t)$ is suited to the specifics of a process and then to determine the corresponding biological constants. A set $\equiv$ of models of $\eta(t)$ is stored in the database: $\Xi = \left\{ f_1(\theta_1, X(t)), f_2(\theta_2, X(t)), \ldots \right\}$ where, θ_i is the unknown parameter vector (set of unknown kinetic parameters). Minimization of the criteria between the output of the model $Y^m(t)$ and the output of the process $Y(t)$ allows the best match parameter vector θ_i^* for i selection of the model of $\eta(t)$ to be obtained:

$$J_i = \min_{\theta_i \to \theta_i^*} \int_0^{t_F} \left(Y^m(t) - Y(t)\right)^T Q\left(Y^m(t) - Y(t)\right) \quad (19.2)$$

A parameter adjustment rule facilitates this task, and an optimization algorithm can be used. The basic idea of parameter adjustment is to start with some estimate of the correct weight settings, modify the weight in the algorithm on the basis of accumulated experiences, and promote the features that appear to be good predictors by increasing their weights and by decreasing bad ones.

The minimization of the criterion $J = \min_{k \to k^*}\{J_k\}$ allows suitable model k^* and the corresponding parameter vector θ^* for the real fermentation process to be obtained [i.e., $\eta(t) = f_{k^*}(\theta^*, X(t))$]. A heuristic search strategy from the database containing the different models of $\eta(t)$ allows the "model match" task to be realized from any k to k^*.

Fermentation processes are characterized by biological degradation of substrate $S(t)$ (glucose) by a population of microorganisms $C(t)$ (biomass) into metabolites, such as alcohol $P(t)$ (ethanol). The physical model of the process is usually described by a set of nonlinear differential equations derived from the material mass-balances and involves modeling of the growth rate. These equations are

$$\begin{cases} \dfrac{dC(t)}{dt} = \mu(t)C(t) - D(t)C(t) \\[2mm] \dfrac{dS(t)}{dt} = \dfrac{1}{Y_{C/S}}\mu(t)C(t) + D(t)S_{in}(t) \\[2mm] \qquad\qquad - D(t)S(t) \\[2mm] \dfrac{dP(t)}{dt} = \dfrac{Y_{P/S}}{Y_{C/S}}\mu(t)C(t) - D(t)P(t) \end{cases} \quad (19.3)$$

In a batch fermentation in which all nutrients are initially introduced into the bioreactor the dilution rate, $D(t) = 0$.

In fed-batch fermentation, fresh nutrient in the culture medium is added as and when required. Here, the dilution rate $D(t)$ is given by

$$\begin{cases} D(t) = \dfrac{F(t)}{V(t)} \\[3mm] \dfrac{dV(t)}{dt} = F(t) \end{cases} \quad (19.4)$$

The specific growth rate $\mu(t)$ is a function of the process state and several biological parameters θ ($\theta \in \mathbb{R}^n$). These parameters are time dependent and moreover, they depend on the environmental conditions, which are then held fixed.

The usual approach in bioprocess modeling is to adopt particular analytical structures for specific growth rate and calibrate the kinetic coefficients from experimental data. However, this modeling is often hazardous because the reproducibility of experiments is often uncertain because the same environmental conditions may be difficult to obtain and so prevent changes in the internal state of the organism.

Many analytical laws have been suggested in the literature for specific growth rate modeling that take into consideration the limitation and/or inhibition of the growth by certain process variables.

The aim of model validation was to seek a model of the specific speed of growth which is able to take into account the specifics of alcoholic fermentation, while remaining mathematically simple. This model will be used in the design of estimation, control, fault-detection, and isolation algorithms for the automatic control of the fermentation process. The selected model is

$$\mu(t) = \mu(\theta, S) = \mu_m \frac{S(t)}{K_S + S(t)} \quad (19.5)$$

The specific growth rate is of the Monod type. The influence of the substrate concentration on the growth is defined by a time limitation of an absence of substrate. Specific rates of degradation and production are coupled with the growth by yield coefficients. Using this choice, a simple model in which the parameters are easy to determine is provided. Experimental data were used to calculate the numerical values of the parameters μ_m, K_S and the yield coefficients $Y_{C/S}$ and $Y_{P/S}$. The results obtained are given in Table 19.5.

The determination of the model is carried out by looking for the best minimization of the criterion needed (Equation 19.2) by using experimental data available for fermentation in discontinuous mode. The discontinuous mode is much richer in kinetic information than the continuous one, which provides only stationary states. The results obtained by applying the selected model in an experiment are presented in Figure 19.15. This figure represents the comparison of the measured values and those given by the model. These results show that the selected model correctly simulates the process

TABLE 19.5
The Process Model Parameters

μ_m	$0.38\ h^{-1}$
K_S	$5\ g.dm^{-3}.h^{-1}$
$Y_{C/S}$	0.07
$Y_{P/S}$	0.16

as defined by the experimental data. The validation of the model parameters was carried out by using other experiments. This model [cf. Equations 19.3 and 19.5] will be used in other sections for the development of estimation, control, and fault detection, and isolation algorithms.

19.8.4.2 Behavioral Models

The complexity of this task requires a combination of classification techniques and expert knowledge. The process states, their causal relations, and the transition conditions are identified using classification. However, expert knowledge is necessary to validate the results in agreement with the nature of the process. It is possible, starting from expert knowledge, to give validation to the supervisory model. Suggested methodology uses classification under the supervision of the expert system to obtain a process model. This is based on the iterative application of fuzzy techniques. The objective of the method is to identify a set of significant states for the expert system. It can be summarized by the stages shown in Figure 19.16 (Waissman-Vilanova et al., 1999; Waissman-Vilanova, 2000).

At the beginning, given a set of measured data, no knowledge on the process states is available (i.e., all data correspond to the same general state). Unsupervised learning of data is applied and a set of classes is obtained. The expert must then map the set of classes to a set of physiological states. Three situations are possible:

1. One class is equivalent to a physiological state.
2. A set of classes is equivalent to a physiological state.
3. Any class is equivalent to any known state.

If a class exists of this final type, a new unsupervised learning process is applied, considering just the data assigned to that class. This procedure is called *data refinement*. When all data are classified in known states, all possible sequencing paths of the state are identified. This is accomplished by looking on every possible temporal correlation observed on all data available.

As application we consider the database considered here is constructed by real data extracted from batch-mode biotechnological process. Several signals are available by online measurement. Among them, the expert system chooses a subset

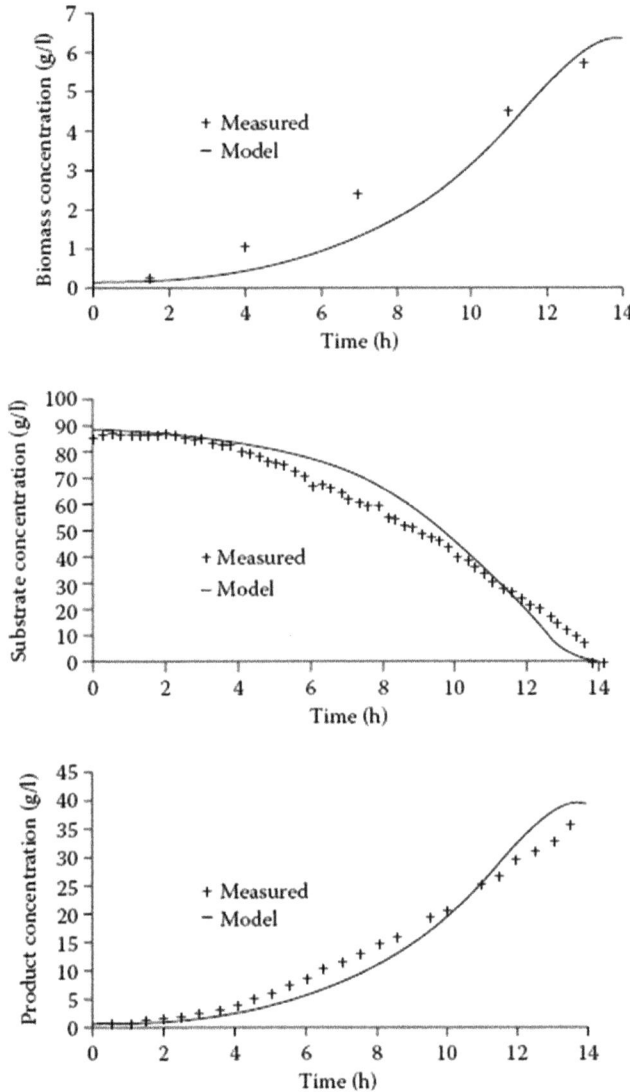

FIGURE 19.15 Model validation.

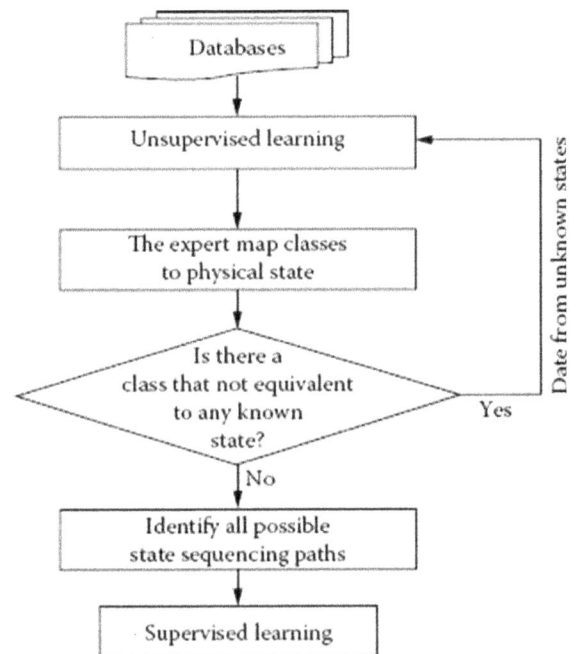

FIGURE 19.16 Methodology application.

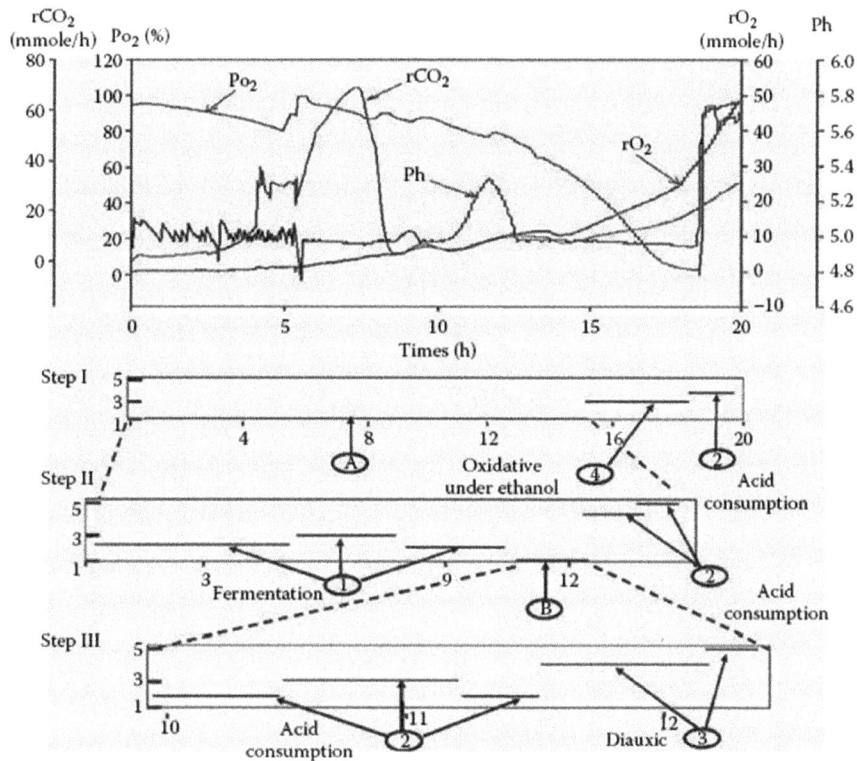

FIGURE 19.17 Behavioral model: Construction steps.

of four signals that contain the most relevant information to determinate the physiological state in the process, including

- Percentage of dissolved oxygen pO_2
- pH
- Percentage of rO_2 in output gas
- Percentage of rCO_2 in output gas

For the methodology applied to the application, four data set records are considered. Three data sets are considered for learning and one data set for testing the results obtained by the methodology. All of the physiological states are determined by the expert system with the help of offline measurement analysis (i.e., intra- and extracellular analysis). Figure 19.17 presents the different steps for extracting a process behavioral model. In step I, with unsupervised learning of all of the data available, we obtain a set of three classes, and two physiological states (2 and 4) are identified (consumption of acid and oxidative metabolism under ethanol). Class A does not present a physiological state but a set of several states. To refine the data in class A, unsupervised learning is used (step II). Amongst the six extracted classes, the expert system identified two states. Union of three classes identifies state 1, i.e., fermentation, while state 2 (acid consumption) is identified by union of two classes. In step III, the set of data in B is used. By means of unsupervised learning, states 3 (diauxic) and 2 are recognized. The direct observation of all state sequences leads to building an automatic control structure representing the process model (Nakkabi et al., 2002) as shown in Figure 19.18. The transitions conditions are presented in Table 19.6.

19.9 RECENT TRENDS IN FERMENTATION CONTROL

Advances in several fields have allowed for new possibilities in the areas of measurement and control. New sensor technologies (for more details see Chapter 18) such as the use of fluorescence and the micro sensor arrays allowed for shrinking of sensor systems for use in a wide range of applications (e.g., in microtiter plates).

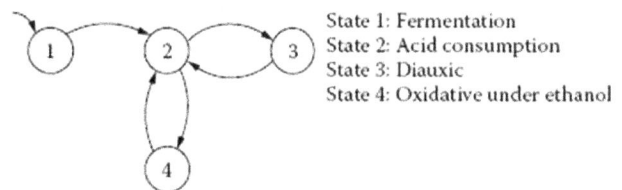

State 1: Fermentation
State 2: Acid consumption
State 3: Diauxic
State 4: Oxidative under ethanol

FIGURE 19.18 Behavioral model: automaton.

TABLE 19.6
Behavioral Model: Transitions

Transitions	Conditions
Transition 1 to 2	$-1.037*pH + 0.118*pO_2 - 0.008*rO_2 > 0$
Transition 2 to 3	$1.082*pH - 0.104*pO_2 + 0.04*rO_2 + 0.001*rCO_2 > 0$
Transition 3 to 2	$-0.88*pH + 0.018*pO_2 + 0.001*rO_2 - 0.05*rCO_2 > 0$
Transition 2 to 4	$1.33*pH - 0.21*pO_2 + 0.06*rO_2 + 0.067*rCO_2 > 0$
Transition 4 to 2	$60.13*pH + 17.59*pO_2 - 0.027*rO_2 - 0.027*rCO_2 > 0$

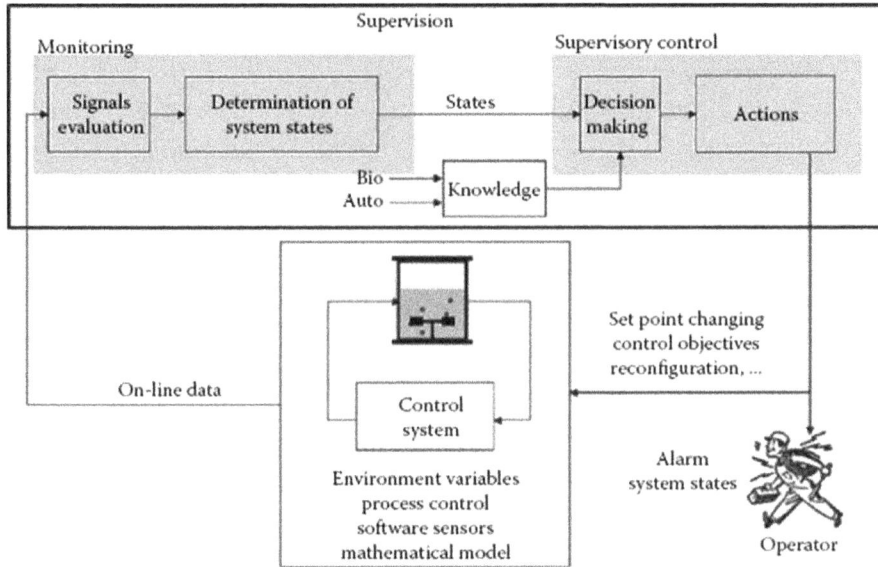

FIGURE 19.19   Supervision scheme.

All available information on the process [e.g., environmental variables (temperature, pH, stirring, etc.)] and control loop signals must be exploited to ensure that the process is operating reliably under the given conditions. For fermentation processes, supervision becomes an issue of primary importance for increasing the reliability, availability, and safety of these systems. Unfortunately, the lack of process knowledge, the absence of reliable sensors, and the unpredictable behavior of microorganisms make this task very difficult, sometimes impossible, for the human operator. Several schemes of supervision applied to various domains (chemical process, petroleum process, wastewater treatment process, etc.) have been proposed (Antsaklis and Passino, 1993; Aguilar-Martin, 1996; Dojat et al., 1998). However, all of these can be described in general by the scheme shown in the Figure 19.19.

According to Kotch (1993), two main tasks are to be considered in a supervision system: process monitoring and supervisory control. The process monitoring part uses data collection and signal processing to decide if the process is in an abnormal state and if a corrective action must be taken. Such intervention, along with diagnosis, is regarded as a supervisory control task.

Within this framework, the development of an integrated methodology for control of the fermentation processes requires approaching modeling, adaptive techniques, and supervision for process control. In this section, the integrated methodology of the process control is formulated by adding a supervisory block (see Diagram 19.1). This block is dedicated to the acquisition of data (measurements, alarms, back to working condition) and process control parameters. In other words, this block collects and exploits all information resulting from the modeling, estimation, observation, control loop, environment, and the expert knowledge of the processes.

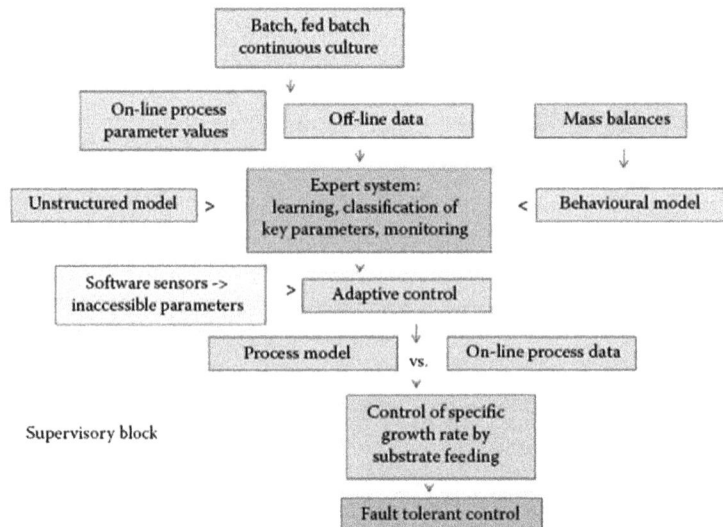

DIAGRAM 19.1   Integrated process control methodology.

19.9.1 New Sensor Technology

The ability to measure yield of product directly online has long been a fondly held objective of fermentation technologists and researchers alike. Use of new sensors (see Chapter 18 for more details) with simple optical fiber connections that fit in the fermentor vessels and detect changing amounts of fluorescence in a dye make this possible. Light of a specific wavelength is used to excite the fluorescence, which can then be measured fluorimetrically.

Linking the expression of the desired protein with a fluorescent detection system can provide direct, online measurements related to yield. A simple analogue output signal from the detector system allows for integration of the new data with process control software. This, in turn, can modify other measured parameters to maximize production.

This system can also be used as a new way to detect standard parameters (e.g., pH, dissolved oxygen, and dissolved carbon dioxide concentration). The fluorescent detection systems can be incorporated in the sensor probes or separately as tiny "patches." These can be placed in simple disposables such as microtiter plate wells or added to incubation flasks. This enables rapid screening based on small volumes with some indication of the prevailing environmental conditions in the culture. These data can influence judgments about which isolates to use for a particular process and generate data of potential use in scale-up.

Another factor aiding real-time measurement of novel process parameters is the development of simple microfiltration systems to allow for separation of cells from the supernatant culture medium that can then be taken to an external analyzer. In this case, the separation system is key, rather than the analytical processes. However, the common feature is the acquisition of data in real time to inform control profiles designed to optimize yield and/or extend productivity in terms of metabolites or biomass.

Finally, many sensors are becoming much smaller, leading to micro sensors arrays. These micro devices are particularly useful for bioprocess to capture information on transient changes related to stress production situations. They are based on methods that allow Nano volume handling of nucleic acids, proteins, or other cellular materials. These Nano volumes are analyzed using miniature electronic devices.

19.9.2 Software Sensors

The impossibility of measurement online of the reactant and product concentrations (e.g., biomass) stops the optimization and control applications. An interesting alternative is the use of software sensors currently named "state observers." Software sensors use the available online information and a model to predict the state variables online.

There are two classes of software sensors used commonly in bioprocess. The first ones are based on a perfect knowledge of a model structure (e.g., Luenberger and Kalman). These software sensors have as advantage a short convergences time but their weakness is in the parameters uncertainty. The second classes are the asymptotic observers. Asymptotic observers are based only on the mass and energy balance, which is the main advantage. On the other hand, asymptotic observers have long time of convergences.

Generally, software sensors are quite accurate for some parts of the process (e.g., growth stage). Unfortunately, they are often less accurate during certain product synthesis portions of the process because not all cellular metabolisms are well understood. Then, it is necessary to develop more robust models that take into account these variations.

In the context of this problem, the techniques of adaptive filtering and estimation, commonly called *software sensors*, seem to be an inevitable alternative (Ben Youssef, 1996; Ben Youssef et al., 1996; Nejjari et al., 1999a, b). The objective is to recreate the unavailable variables by using online measurements, offline experimental data, and the physicochemical model of the process.

The nonlinear and nonstationary characteristics of these processes, bringing into play the living microorganisms, constitute a considerable limitation of the performance obtained by the estimation procedure based on a linear approach. Given the strongly nonlinear character of these processes, we chose the use of nonlinear, adaptive techniques based on the structure of the model for this type of process. The model used for developing estimation methods and software sensor algorithms is that obtained from experimental data, as explained in Section 19.8.4. As mentioned earlier, the two principal problems that appear in the control of the fermentation processes are the inaccessibility of certain biological variables and the critical temporal variability of the kinetic parameters. The motivation is to know how to use measurements and the model's equations to minimize the errors introduced by measurement noise. For that, we use the techniques of filtering, which take into account the measurements available (substrate concentration) and the information resulting from unstructured models. Thus, we make a joint estimate of states and parameters. Figure 19.20 represents the diagram of the developed algorithm.

The software sensor proposed uses the available data to rebuild the state variables and/or the estimated kinetic parameters. The estimator (an algorithm for calculating an estimate of a given quantity by using the observed data) described in

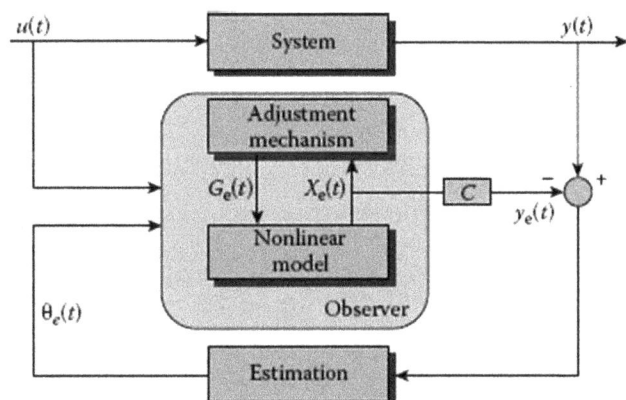

FIGURE 19.20 Estimation: Adaptive scheme.

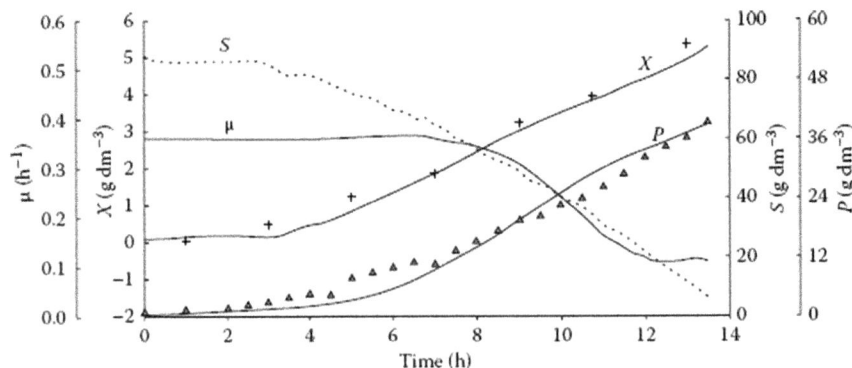

FIGURE 19.21 Estimation: Batch mode.

Zeng et al. (1993b) was applied to three types of fermentation (batch, fed-batch, and continuous modes). Figure 19.21 shows the estimated biomass concentration, the estimated product concentration, and the estimated specific growth rate obtained by the adaptive estimation algorithm. The estimation results are satisfactory compared with the offline analysis (symbols) of biomass and product concentrations.

Controlled fed-batch fermentation is shown in Figure 19.22. The substrate feed flow rate controlled to regulate the substrate concentration in the reactor at 44 g/dm³. The input substrate concentration $S_{in}(t)$ was set at 160 g/dm³. The estimation results of biomass concentration, product concentration, and specific growth rate are given in Figure 19.22. Good agreement was found between the online estimations and measurements obtained by offline analysis (symbols).

The results obtained from a controlled continuous fermentation process are given in Figure 19.23. In this application, the substrate concentration $S(t)$. and the biomass concentration $C(t)$. were controlled by manipulating the dilution rate $D(t)$ and the input substrate concentration $S_{in}(t)$. The evolution of estimated biomass concentration, product concentration, and specific growth rate are illustrated in Figure 19.23. These estimated results are again satisfactory, as shown by the offline biomass analysis (symbols).

19.9.3 ADAPTIVE CONTROL

The achievement of a good control and an effective monitoring of fermentation processes require the availability of

real-time information regarding the physiological states of the process. However, the bioprocesses' dynamics are still badly understood, and many methodological problems of modeling are yet to be solved. In the laboratory, certain variables can be evaluated using offline analysis. In most cases, they require too long a time to be of direct use in real-time control of biological reactions. Moreover, commercial bioprocesses instrumentation suffers from a crucial lack of reliable and inexpensive direct sensors.

The nonstationary character of these processes encourages the use of methods such as adaptive control (linear or nonlinear), making it possible to represent the nonlinearity of the model and take into account the variations of their parameters. It is the nature of these processes that leads to the development of algorithms that exploit the inherent nonlinear structure of their models.

This section presents results obtained from the linear approach in which we used the nonlinear model of the fermentation process operating in continuous mode (see Section 19.8.4) to determine the synthesis parameters of the linear controller.

We used the approach of the indirect adaptive control scheme (Dahhou et al., 1991a, b, c); the system parameters are estimated directly, as shown in Figure 19.24. These estimates are used for the readjustment of the regulatory parameters. In this type of algorithm, the plant parameters are estimated online and used to calculate the controller parameters. The controller considers these parameters as if they were true parameters. This approach is founded on the certainty

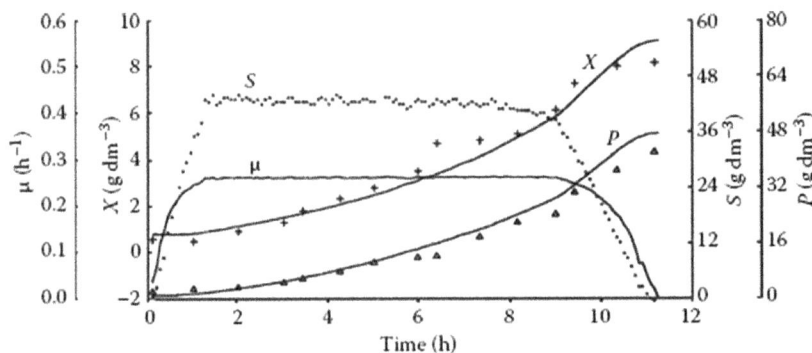

FIGURE 19.22 Estimation: Fed-batch mode.

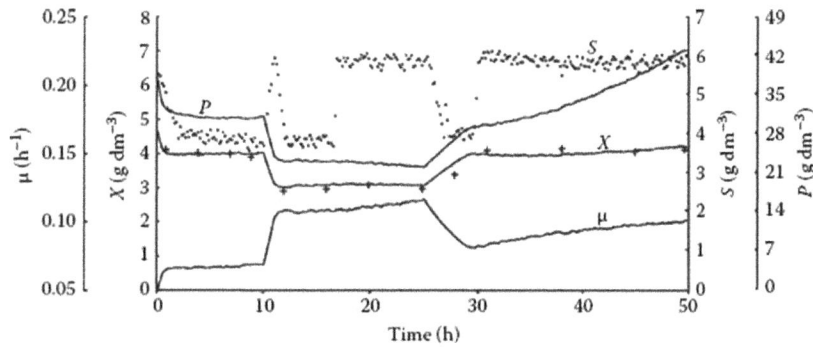

FIGURE 19.23 Estimation: Continuous mode.

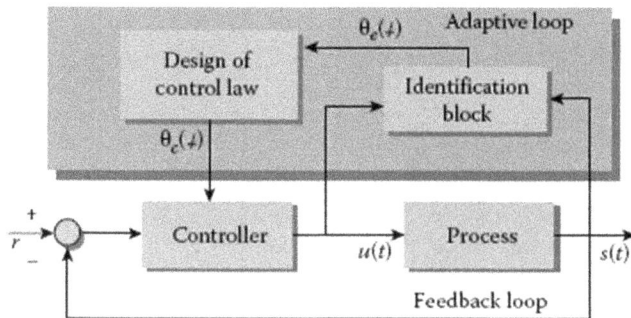

FIGURE 19.24 Control: Adaptive scheme.

equivalence principle [the problem can be separated into two stages (1) obtain the values of the estimated parameter, and (2) at time t, solve the controller optimization problem using the estimates parameters in place of the true parameters].

We have chosen to present the experimental results obtained from the control of alcoholic fermentation in continuous mode. The principal objective is the regulation and the tracking of the substrate concentration inside of the bioreactor as the dilution rate changes (Dahhou et al., 1993). The control objective is the minimization of a specific quadratic criterion (M'Saad et al., 1990). The control law is coupled with a robust estimator. Input and output data are first filtered and normalized, which reduces the effects of un-modeled dynamics and noise. The parameter adaptation algorithm takes the form of standard recursive least-squares with forgetting factor, data normalization factor, and adaptation freezing associated with the definition of an available information signal (M'Saad et al., 1989). The forgetting factor approach enables the recursive algorithm to reduce the effect of the older error data by multiplying the error data by a discounting factor. The adaptation freezing allows for stopping the estimation algorithm in the absence of persistent excitation. The results obtained are represented by the graphs in Figure 19.25. The substrate concentration profile follows the requested model reference output. The inlet substrate concentration change generates a perturbation on the output, which has been rejected after 2 h. The second graph of Figure 19.25 shows the corresponding manipulated input profile, which does not saturate. The estimated model is represented by parameters, the evolutions of which are given by the graphs in Figure 19.26. We note that

the freezing parameter is sometimes activated (zones 1, 2, and 3) and this depends on the available information signal. At this time, we believe that the activation of the freezing parameter is justified according to the measurement of the information signal (i.e., a required condition of persistent excitation is not satisfied, but we can suppose that the system dynamic is changing also). For us, the fact that this condition is not checked can cause a drift of the parameters and thus it can have a risk of instability. The freezing of the parameters was a means to mitigate this problem. However, this drift of the parameters can be justified by a change of system dynamics. This change can be due to the nature of the process (e.g., by the aging of the microorganisms and contamination). In these circumstances, freezing the parameters can be a serious error of appreciation and comprehension. This reasoning leads us along to think that consideration should be given to the possibility that the process can be subjected to operational faults of operation that result in abnormal situations.

19.9.4 Expansion of the Capability of DDC Instrumentation

Virtually all commercial fermentors now use control instrumentation (see Chapter 17 for more details) based on DDC. A few manufacturers have developed systems with all control functions directly handled by the computer. This has some advantages or additional of new capabilities and library routines if a standard process control software package is used. Also, modern computers have more than adequate processing power to cope with real-time process control. The downside of this approach is, of course, the potential for a computer failure leaving the fermentor with no control whatsoever.

A "halfway" position is to allow external software to interact directly with supervisory process control software. Calculation of some control characteristics is made externally, and the result is passed to the fermentor software as, for example, a revised set point for one or more parameters. Windows™-based operating systems allow for just this process through Object Linking and Embedding (OLE). For example, a short script in Visual Basic™ could link supervisory software with an Excel™ spreadsheet making calculations and generating set points in real time based on fermentation data.

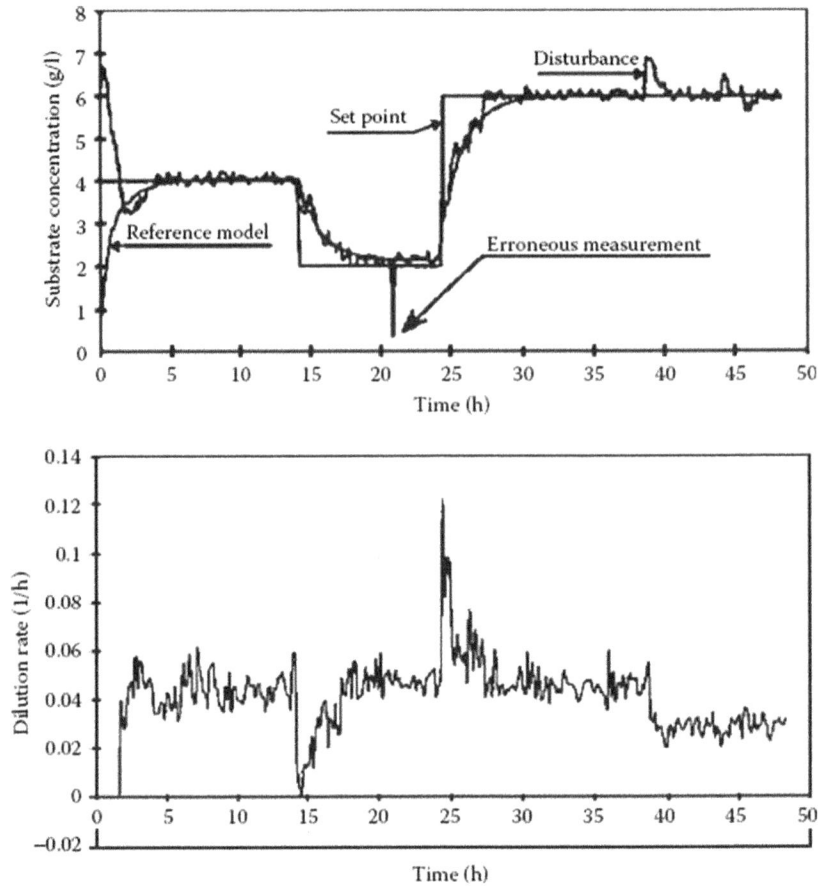

FIGURE 19.25 Evolution of substrate concentration and dilution rate.

This introduces another layer in the control process but has the advantage of one or more independent microprocessors providing control at the fermentor.

An increasing trend regarding usability is the use of color touch screens, even for bench scale fermentors. They typically use a graphical user interface and incorporate some of the features previously only found in PC-based software, such as trend graphing. Alternative approaches involve enhancements to navigation between menus and the use of tabs for

rapid access to options. Often, the more advanced functions of the controller, such as configuration of PID loops, are now placed in restricted sections where the casual user cannot accidentally make fundamental changes.

19.9.5 USE OF COMMON COMMUNICATION PROTOCOLS

A general trend in fermentation control is to add more peripheral items such as balances, pumps, and additional gases almost

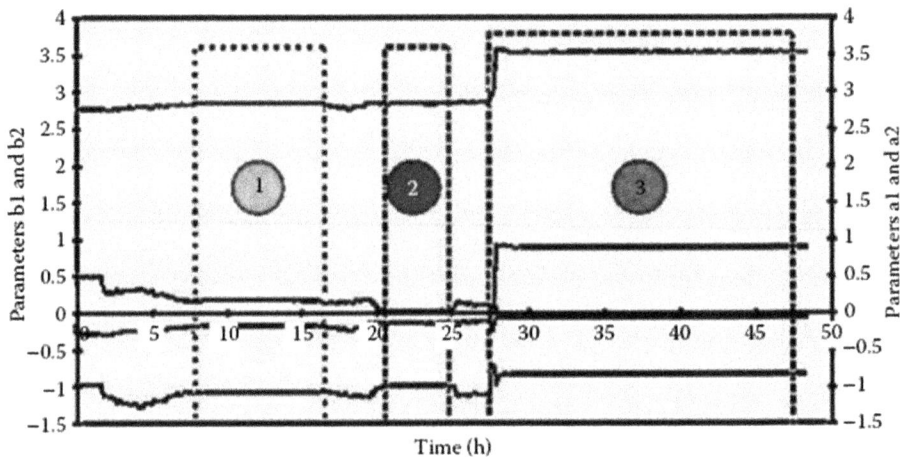

FIGURE 19.26 Estimation of system parameter.

irrespective of the scale of the process. Several manufacturers of components such as mass flow control valves and peristaltic pumps are offering them for use with the Modbus serial protocol. This allows for great flexibility in adding these items to a fermentor as the physical connection and firmware changes are very easy to implement. The Modbus protocol is a genuine common standard allowing for the potential range of equipment that can be added to a standard fermentor to be greatly increased.

The trend for interactivity is also shown at a higher level of process control. Fermentor instrumentation and supervisory software can be linked to other devices or programs using the OLE for Process Control (OPC) protocol. This requires each device has a driver written in its firmware or software and this allows two-way communication with any other enabled systems. Client and server systems are catered for within the specifications of the OPC standards.

The potential use for this system in industrial process control of fermentation is to realize another long-held desire—linking of upstream and downstream processes together with the actual incubation phase in the fermentor for optimization which is truly process-wide and fully integrated. Of course, Microsoft Office™ applications are OPC enabled so; for example, a controller could transfer data directly to an Excel spreadsheet via OPC.

19.9.6 Use of Databases for Storage Bioprocess Data

Actually, computer progress enables the storage of large volumes of bioprocess data. These data bioprocess can be visualized with historian utilities. Furthermore, data can be studied with mathematical tools, as principal component analysis (PCA), partial least-squares regression (PLS), statistical analysis, and so on. These mathematical tools can be used for

several purposes, including to find abnormal operational conditions of a bioprocess, to establish a set point, and to optimize a bioprocess operating condition.

Unfortunately, there are two bottlenecks that stop the correct use of databases. First, the current commercial software with mathematical tools is not very much in use in the bioprocess industry. Secondly, there is a need for tools to facilitate data set preparation.

19.9.7 Supervision for Process Control

Modeling, estimation, filtering, and control aspects mentioned previously constitute a lower layer level of control that does not account for all of the information resulting from the process. We adopted a strategy that consists of adding a layer of higher level supervisory block in which all of this information can be exploited.

If a hierarchical order is established, the feedback loop and the adaptation loop will provide the lower level. We considered the layer immediately above the adaptive loop, which is the supervision layer. In this level, the evolution of the signals coming from the adaptation and feedback loops are used to recognize specific situations and to act on the parameters of the different algorithms of control and estimation. The general idea of the supervision is the evaluation of the significant signals of the system to test its performances on the basis of certain predefined criteria, which can be inherent to the controller or within the process. The violation of these criteria starts a second task for the supervisor: it must act on the system in a way not envisaged by the controller to improve its performances or to help the operator to make a decision at the time when an indication of some dysfunction appears. These anomalies can

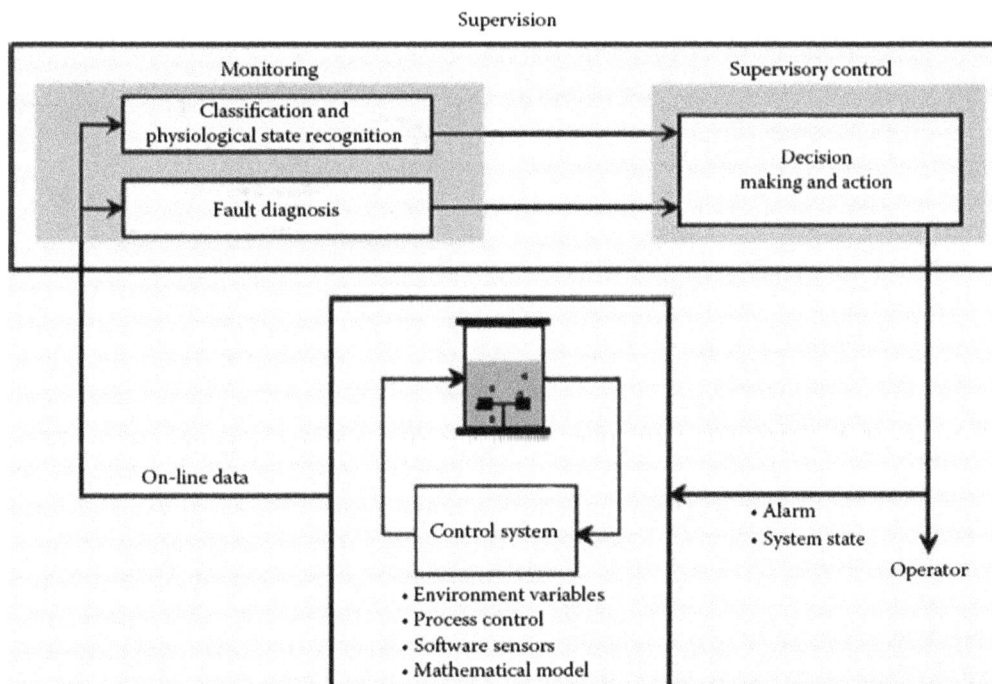

FIGURE 19.27 Physiological states and fault detection.

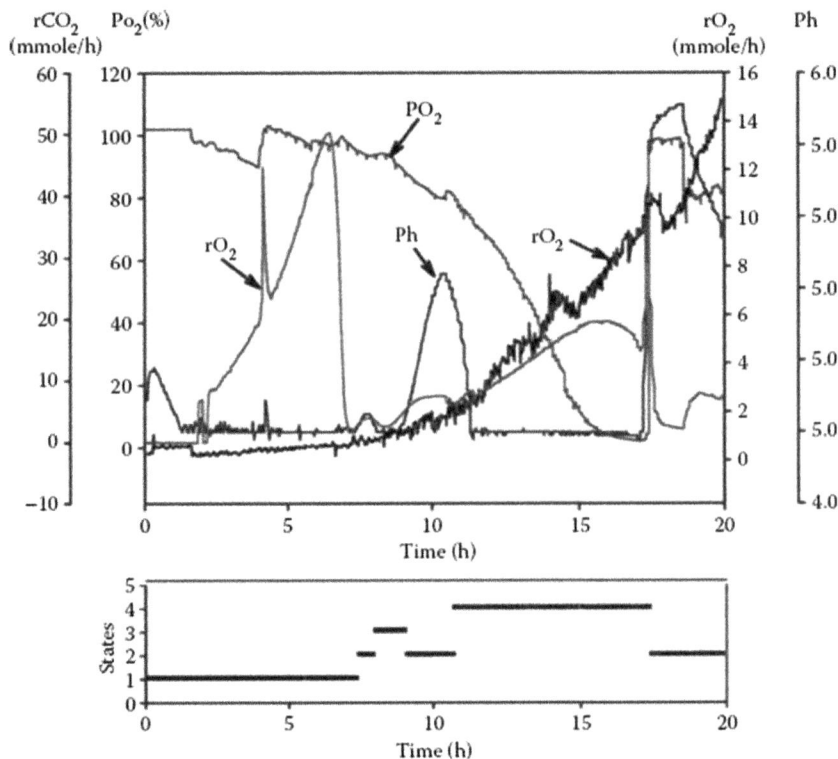

FIGURE 19.28 Good learning.

be related to the biological reaction or ascribed to the operation of the hardware (actuators, sensors, etc.). Figure 19.27 illustrates an example of such a supervision block.

19.9.7.1 Classification

We used the model derived in the modeling section for supervisory purposes (Nakkabi et al., 2002). This model is validated by experiment to confirm the expected optimal production. Thereafter, we use this model like a reference in the supervisory system for real-time monitoring of an alcoholic fermentation process. The results of the first experiment of the supervised process are presented in Figure 19.28.

Table 19.7 presents the different functions concerning the acknowledged four physiological states. Note that the difference between $t + 1$ and t corresponds to one acquisition. By using the automatic control obtained in Figure 19.29 and Table 19.7, it is possible to know the state of the system and

its tendency. Actually, the membership's function (for a fuzzy set, this is a generalization of the indicator function in classical sets) of state 1 (fermentation), during the instant t, is the highest followed by the one of state 2 (consumption of acid). This order indicates that the processes are in functional state 1 with tendency to switch to state 2; this transition is authorized by the global model of the process (Figure 19.29); therefore, the situation is considered as normal. The same analysis keeps its validity for the instants $t + 1$ and $t + 2$. At $t + 3$ the membership's function is the greatest; the system is, then in state 2 with an inclination to shift to state 3. The situation is always considered normal because this transition is authorized by the process model (Figure 19.29). Furthermore, the supervision system did not detect any unusual physiological behavior compared with the reference model, suggesting that the system is operationally satisfactory.

Table 19.8 shows the different membership function. At the instant t, the process is at state 3 (diauxic state), which represents the highest value of the membership function. The membership function of state 2 (consumption of acid), comes after, indicating the process tendency. The situation is considered as normal because the process model allows the transition. At the instant $t + 1$, the process remains at state 3 but changes its tendency to state 4 (oxidative under ethanol state). According to the process model (Figure 19.29), this transition is not authorized. In this case, an alarm is set off. This analysis remains true for the instant $t + 2$ At the instant $t + 3$, the tendency of the process is confirmed and the system switches from state 3 to state 4. The situation is abnormal, pointing to an abnormality in the process. Then, the supervision system can readily detect

TABLE 19.7
Membership's Function for Good Learning

States	Time			
	t	$t + 1$	$t + 2$	$t + 3$
Fermentation	0.29	0.35	0.32	0.24
Acid consumption	0.26	0.32	0.31	0.29
Diauxic	0.22	0.11	0.15	0.28
Oxidative under ethanol	0.23	0.22	0.22	0.19
	Normal	Normal	Normal	Normal
Situation				

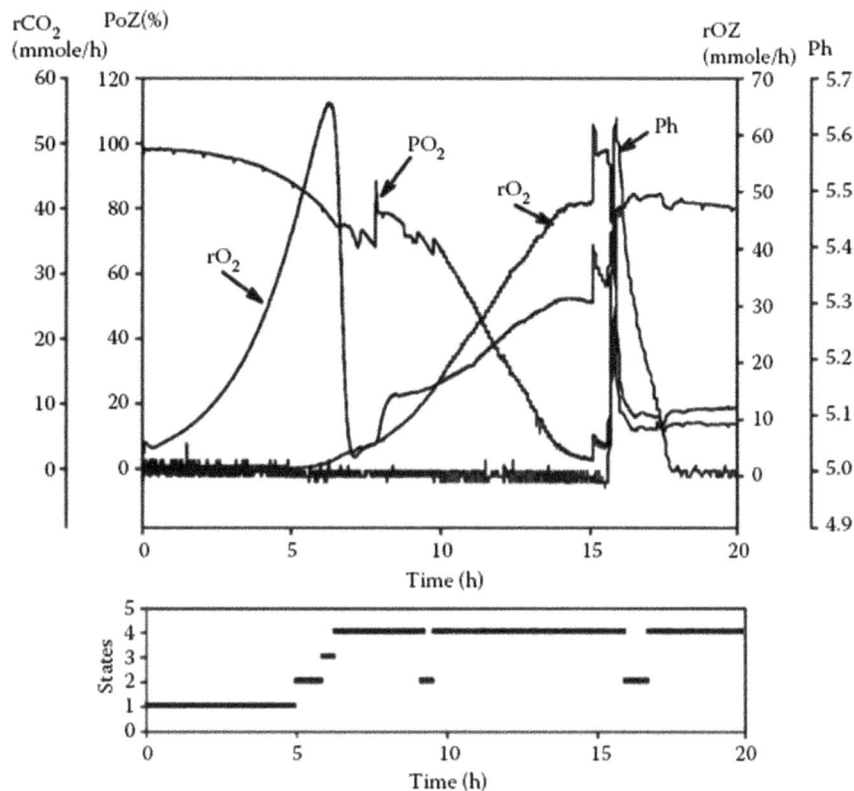

FIGURE 19.29 Bad learning.

a different physiological behavior from those usual ones; this is explained by the fact that the phase of diauxic is shorter. Indeed, one notices in the figure that there is no intermediate consumption of acid (there is no increase in pH) between the ethanol consumption and the glucose consumption. This may indicate a better capacity for oxidation (Table 19.8).

19.9.7.2 FAULT DETECTION AND ISOLATION

Fault detection and isolation (FDI) is one of the most important tasks assigned to intelligent supervisory control systems. A fault is understood as any kind of dysfunction in the actual dynamic system that leads to an unacceptable anomaly in the overall system performance. Such malfunctions may occur in the sensors, in the actuators, or in the components of the process (Kabbaj, 2004).

TABLE 19.8
Membership's Function for Bad Learning

States	Time			
	t	$t+1$	$t+2$	$t+3$
Fermentation	0.01	0.03	0.01	0.01
Acid consumption	0.32	0.12	0.12	0.21
Diauxic	0.41	0.46	0.46	0.37
Oxidative under ethanol	0.26	0.41	0.41	0.41
	Normal	Alarm	Alarm	Abnormal
	Situation			

In this section, we are interested with faults in process dynamics. In model-based fault detection methods, a fault is considered as a variation of one or several parameters compared with a reference value. The problem is then to detect these parameters' variations, to distinguish between those resulting from faults and those resulting from normal behavior, and to decide if these variations are indeed significant compared with uncertainties in the model and the noise in the measured data. It is well known that the FDI procedure is explicitly divided into two stages: residual generation and residual evaluation. The principle is to use the measurements of output and input signals and the process mathematical model to generate residuals; we used these residuals as indicators of dysfunction. These residuals are defined as the difference between the estimated and real values of the various outputs. After evaluation, the fault is detected and isolated. In the model-based methods, the residuals can be generated using observers, parameter estimation, and parity relations.

We develop here a method based on adaptive observers for FDI (Zhang, 2000) in an alcoholic fermentation process (Kabbaj et al., 2001). The faults are modeled as changes in the system parameters $\theta = \left[\theta_1, \theta_2, \ldots, \theta_n\right]^T$. The fault-free operating mode is characterized by the nominal vector θ^0, which is supposed to be known. The residuals are generated using a state observer on the basis of nominal parameters θ^0, and a set of adaptive observers as shown by Figure 19.30. Each adaptive observer estimates only one parameter of the supervised system, in addition to the state variables. The residuals $\gamma_1, \gamma_2, \ldots, \gamma_n$ are defined as being the prediction error of each observer.

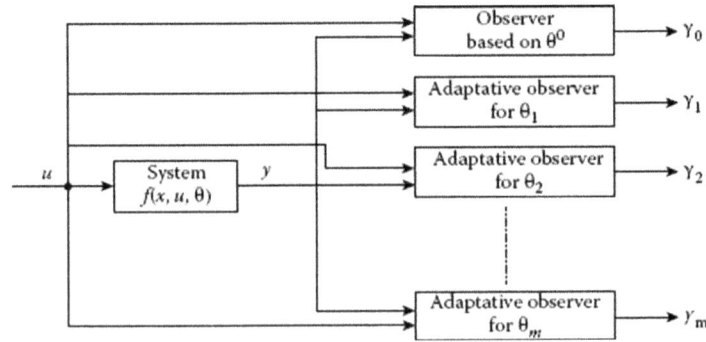

FIGURE 19.30 Bank of observers.

Faults in the process system (no fault in actuators or sensors) will be considered. As is well known in biotechnology, the growth rate is one of most important parameters to describe the evolution of the microbial population. The variation of this parameter is sensitive to the operating conditions (pH, temperature, agitation, oxygen, etc.). The modeling of these various influences within a single mathematical expression is complex. To supervise the variables of the environment, we can rather supervise the structural parameters of the growth rate (Equation 19.5). Faults are so modeled as changes in the system parameters $\theta = \left[\mu_m, K_S \right]$ (the maximum growth rate and the saturation constant). The fault-free operating mode is characterized by the nominal vector θ^0, which is supposed to be known.

19.9.7.3 Adaptive Observers for FDI

As explained earlier, the scheme consists in developing two adaptive observers denoted *observer^j* (j = 1, 2) and a state observer, based on nominal parameters θ^0, called *observer^0*. Each adaptive observer estimates only one parameter of the supervised system in addition to the state variables. We have opted for an estimator using the nonlinearity of the process and the approach of the model reference to reconstitute the state variables and the parameters of the model (Zeng et al.,

1993a). Instead of estimating the growth rate like a time-varying parameter, we instead estimate the structural parameters. This makes the estimation algorithm better adapted to the fault detection approach used.

Let $\hat{S}^0(t)$, $\hat{S}^1(t)$, and $\hat{S}^2(t)$ be the estimated outputs given by the *observer^0*, *observer^1*, and *observer^2*, respectively. The residuals may be defined as being the corresponding estimation errors

$$
\begin{aligned}
\gamma_0 &= \hat{S}^0(t) - S(t) \\
\gamma_1 &= \hat{S}^1(t) - S(t) \\
\gamma_2 &= \hat{S}^2(t) - S(t)
\end{aligned}
\tag{19.6}
$$

The residual γ_1 is associated with the maximum growth rate $\theta_1 \equiv \mu_m$ and the residual γ_2 with the saturation constant $\theta_2 \equiv K_S$.

19.9.7.4 Residual Behaviors

In the fault-free operating mode, all of the residuals γ_0, γ_1, and γ_2 are practically zero, and as such any departures from zero will be detected as a fault. If the fault corresponds to a change in a single parameter, all of the residuals except one will persistently differ from zero. If, after a transient, γ_j (j = 1,2) converges back to zero, the fault corresponds to a change

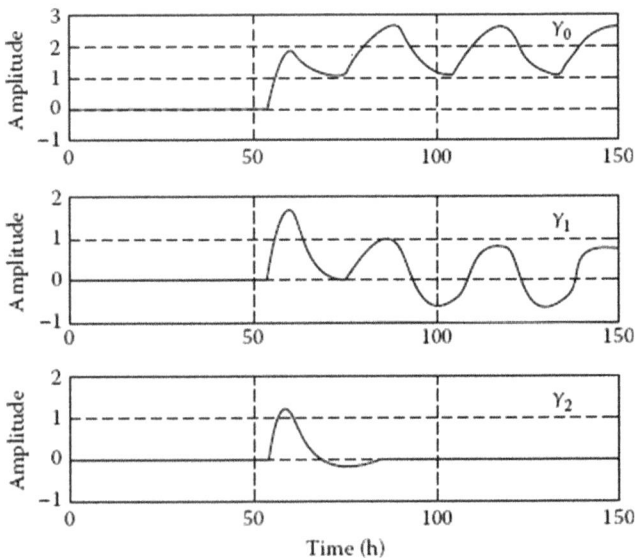

FIGURE 19.31 Residuals evaluation with fault in μ_m (without noise).

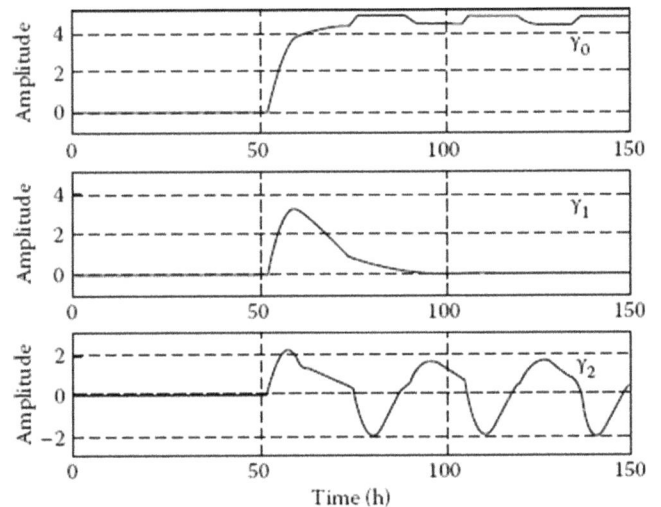

FIGURE 19.32 Residuals evaluation with fault in K_S (without noise).

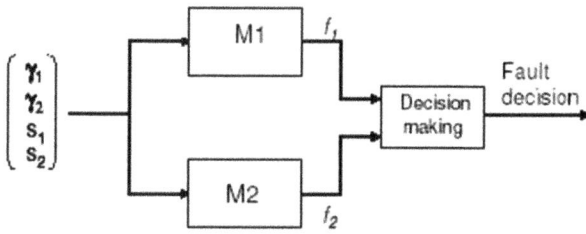

FIGURE 19.33 Decision making using behavioral models.

in θ_j. Figures 19.31 and 19.32 show the residuals' behavior when the parameter μ_m changes from 0.38 to 0.40 h^{-1} and K_S from 5 to 4.7 g/L, respectively.

Microbiologically speaking, any changes in growth conditions in the bioreactor (e.g., temperature, pH, or aeration) will have an adverse effect on substrate consumption rate (K_S) which, in turn, diminishes growth rate (μ).

This method of detection and isolation of faults on the basis of the adaptive observers gives satisfactory results without the presence of noise. However, following Kabbaj et al. (2002), we noted that it very is difficult to detect and isolate faults in the presence of noise. To mitigate this problem, we propose a new methodology that combines analytical and knowledge for FDI. This combines the knowledge about process and the information hidden in data to extract a behavioral model in the form of a decision tree. This model is then used for an automatic residual evaluation. This approach proves to be remarkably robust. Thus, faults of the type described are detected and isolated even in presence of measurement noise (Nakkabi et al., 2003).

The generated residuals are used in classification to recognize the process state (state 1: no faults, state 2: faults in μ_m,

or state 3: faults in K_S). In addition, we propose these residuals (γ_1 and γ_2) using some composite residuals defined by

$$s_1 = \gamma_0 - \gamma_1$$
$$s_2 = \gamma_0 - \gamma_2 \qquad (19.7)$$

These kinds of residuals are very significant and helpful in the classification procedure because they are robust to measurement noise. The goal is to diagnose the kind of fault by evaluating the given residuals. Therefore, two behavior models ($M1$, $M2$) are implemented. Each of them is designed to be sensitive to one fault, as shown in Figure 19.33. These models are generated as explained in Figure 19.16 (Section 19.8.4.2 Unstructured models). The models' output (f_1, f_2) denotes the presence or not of fault. For each fault the training data are collected in the presence of measurement Gaussian noise with zero mean and variance equal to 0.5. The effect of this noise on residuals is very important compared with faults. Thus, residual evaluation is delicate. The validation has been done on multiple data with different noise of variance 0.2, 0.3, 0.5, and 1. The results obtained with variance equal to 1 are presented.

As can be seen from Figure 19.34, residuals are generated as growth rate (μ_m) changes from 0.38 to 0.40 h^{-1}. As dictated by the design, a fault is detected should any residuals (e.g., γ_1) depart from a zero value. Unfortunately, this behavior is not clearly observed in the residuals γ_1 and γ_2 plotted in Figure 19.34 because of a high level of noise in the measurements. In this case the composite residuals can be used, but just for fault detection. The residual evaluation using behavioral models allows us to detect and isolate faults as shown in the bottom of Figure 19.34 by the transient from state 1 (no fault) to state 2 (fault in μ_m).

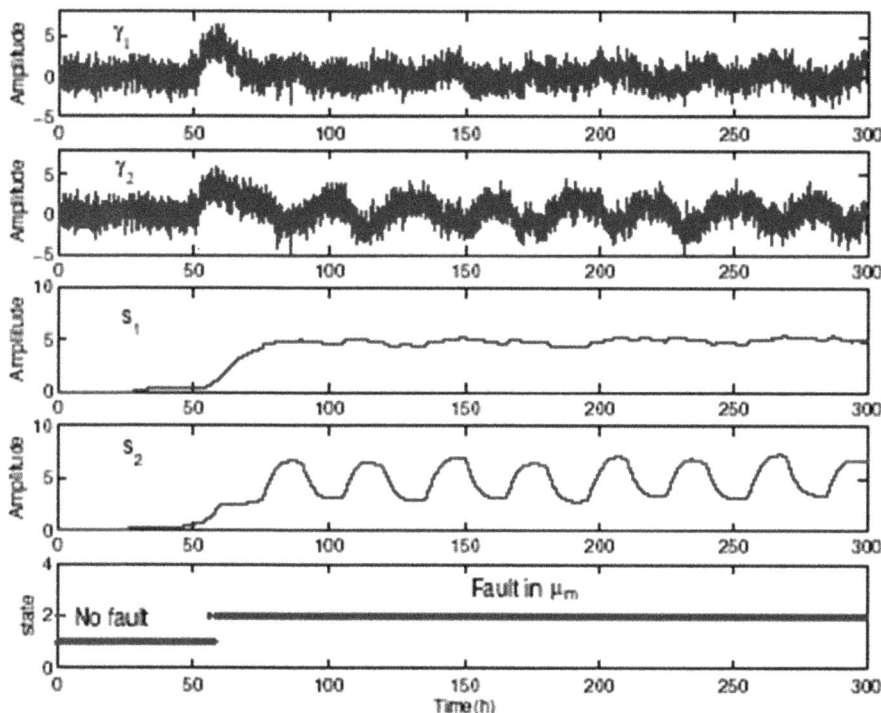

FIGURE 19.34 Residuals valuation with fault in μ_m (with noise).

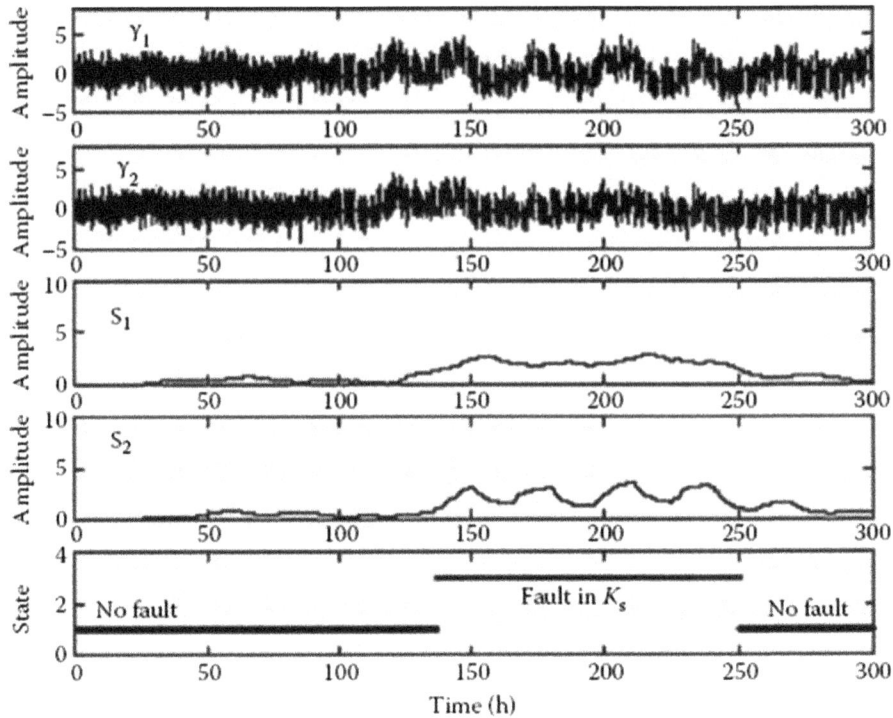

FIGURE 19.35   Residuals evaluation with fault in K_S (with noise).

The fault is now simulated by changing the parameter K_S from 5 to 4.7 g/L at $t = 130\ h$ and from 4.7 to 5 g/L at $t = 250$ h. The corresponding residuals are illustrated in Figure 19.35. Similarly, the behavioral models based on residual evaluation show clearly the transient from state 1 (no fault) to state 3 (fault in K_S) at $t = 130$ h and from state 3 to state 1 at $t = 250$ h.

19.9.7.5   Principle of Fault-Tolerant Control

The control law is of the form $u = K(x,\varphi)$, where the controller parameter vector is $\varphi \in R^q$. Using this controller, the closed-loop system is $\dfrac{dx}{dt} = f\left(x,\theta,\varphi\right)$ and $y = Cx$. The choice of the controller parameter vector φ is called *controller configuration*. One assumes that this choice is related with a cost function $J(\theta,\varphi)$. Let $p_1(\theta,\varphi)$, $p_2(\theta,\varphi)$ ... $p_i(\theta,\varphi)$, and $p_m(\theta,\varphi)$ are m parameters depending on the closed-loop system with respective to the constraint condition. These m parameters can take values of eigenvalues (these are a special set of scalars associated with a linear system of equations that are sometimes also known as characteristic roots, characteristic values, or proper values; these values describe the dynamic of the process) of the closed-loop system or other values following the application context. Then the objective of controller parameter configuration is that the closed-loop system satisfies

$$\min J\left(\theta,\varphi\right) \tag{19.8}$$

$$p_i\left(\theta,\varphi\right) \in \Omega_i, \quad \forall i = 1,\cdots,m \tag{19.9}$$

Equation 19.8 can be analytic or nonanalytic. Equation 19.9 represents the constraint condition of the controller parameters

choice. Ω_i represents a certain domain in the complex plane. For example, if p_i is an eigenvalue, then Ω_i can be chosen as the left-half s plane. The constraint condition (Equation 19.9) implies a set of crucial indexes that should be satisfied. The closed-loop system is called a system with good stability if Equation 19.9 is satisfied.

After the occurrence of the fault, we accommodate the fault by controller reconfiguration. We change the controller parameter vector according to the system information. We use fault detection, isolation, and identification procedures to get this information. Then, in different periods the controller parameter vector is chosen as follows:

1. *Before the fault occurrence*: The system parameter vector is known and equal to θ^0. Δ is only the point θ^0. The controller parameter vector should be optimized in the domain $\Psi_M(\theta^0)$. The optimum controller parameter vector in this case as φ^0. $\Psi_M(\theta^0)$ contains all possible maximum available domain $\Psi_M(\Delta^i)$, where Δ^i is any domain containing the point θ^0. The optimum result in this case is the best one of the results corresponding to all possible $\Psi_M(\Delta^i)$.

2. *After the fault occurrence and before the fault detection*: The system parameter vector in this case is θ^f, and the corresponding maximum available domain is $\Psi_M(\theta^f)$. Because we do not know the fault occurrence, the controller parameter vector φ^0 is still used, which may be not in the domain $\Psi_M(\theta^f)$. So, Equation 19.9 may be not satisfied, therefore the system may be unstable.

3. *After the fault is detected but it is not isolated*: We have known the fault occurrence; we will change

the controller parameter vector to accommodate the fault. Because the fault has not been isolated, we do not know the location of the system faulty parameter vector, so we considered that θ may be any point in its maximum possible domain Δ^M. The maximum available domain corresponds to Δ^M is $\Psi_M(\Delta^M)$. Because Δ^M contains all possible Δ, $\Psi_M(\Delta M)$ will be subset of any $\Psi_M(\Delta)$; that is to say $\Psi M(\Delta M)$ is the smallest of all possible maximum available domains $\Psi_M(\Delta)$. If the condition in Equation 19.9 is satisfied by the open-loop system, then $\Psi_M(\Delta^M) \neq \phi$ because it at least has an element φ^{off}, where φ^{off} is the controller parameter vector value, which places the system in a closed-loop state. The controller parameter vector should be optimized in the domain $\Psi_M(\Delta^M)$, which is only a small subset of $\Psi_M(\theta^f)$, but not in the whole domain $\Psi_M(\theta^f)$ because we do not know $\Psi_M(\theta^f)$ in this case. We note that the optimum controller parameter vector in this case is φ^d. The result in this case is worse than the one optimized using $\Psi_M(\theta^f)$, but it is the best choice in this case. The condition in Equation 19.9 is satisfied because $\Psi_M(\Delta^M)$ is a subset of $\Psi_M(\theta^f)$.

4. *After the fault is isolated and is identified*: We assume that the estimation of the system parameter vector is $\tilde{\theta}^f$. Because we know that the best optimum result is based on $\Psi_M(\theta^f)$, we want to get the estimated value of θ^f as quickly as possible. Usually a precise estimation cannot be obtained immediately; therefore, we assume that in the early period (when the fault is identified) the estimation is given with the error limits. In other words, the estimation is not a point, but a possible domain defined by a line segment. Because the fault has been isolated, this domain will be in a parallel of the coordinate axes that passes the point θ^0. We note this line segment as the domain Δ^{ii}, then Δ^{ii} will be a subset of Δ^M and it contains the point θ^f. Along with the operation of the fault identification, the domain Δ^{ii} will become smaller and smaller, and at the end it converges to the point θ^f. Accordingly, the correspondent maximum available domain $\Psi_M(\Delta^{ii})$ is a subset of $\Psi_M(\theta^f)$ and it contains the domain $\Psi_M(\Delta^M)$. The controller parameter vector should be optimized in $\Psi_M(\Delta^{ii})$. We note the optimum controller parameter vector in this case as φ^{ii}. The result in this case will be worse than the one optimized in $\Psi_M(\theta^f)$ but better than the ones optimized in $\Psi_M(\Delta^M)$. Along with the operation of the fault identification, $\Psi_M(\Delta^{ii})$ will become bigger and bigger and converges to $\Psi_M(\theta^f)$. So, the optimum result will become increasingly better and better and converge to the one that best corresponds to $\Psi_M(\theta^f)$.

The objective of the control is to make the specific growth rate $\mu(t)$ of the system in Equation 19.3 follow the rate $\mu_r(t)$ of a given reference model. This is done by manipulating the dilution rate $D(t)$. The reference model is chosen as (Li and Dahhou, 2006)

$$\begin{cases} \dfrac{dC_r(t)}{dt} = \mu_r(t)C_r(t) - r(t)C_r(t) \\ \dfrac{dS(t)}{dt} = \dfrac{1}{Y_{C/S}}\mu_r(t)C_r(t) + r(t)S_{in}(t) \\ \qquad\qquad - r(t)S_r(t) \\ \dfrac{dP(t)}{dt} = \dfrac{Y_{P/S}}{Y_{C/S}}\mu_r(t)C_r(t) - r(t)P_r(t) \end{cases} \quad (19.10)$$

It has a same structure as the process system model. In the reference model in Equation 19.10, $r(t)$ is input of the reference model and $\mu_r(t)$ is the same structure of $\mu(t)$.

The control variable is chosen as (Li and Dahhou, 2006)

$$D(t) = T(t)r(t) - L(t)\hat{\mu}(t) \quad (19.11)$$

with

$$T(t) = \frac{a_r(t)}{\hat{a}(t)}$$
$$L(t) = \frac{a_r(t)}{\hat{a}(t)} - 1 \quad (19.12)$$

and

$$\hat{\mu}(t) = \hat{\mu}_m \frac{S(t)}{\hat{K}_S + S(t)} \quad (19.13)$$

The terms $\hat{a}(t)$ and $a_r(t)$ are given by (Zeng et al., 1993b)

$$\hat{a}(t) = \frac{1}{Y_{P/S}}\hat{\mu}(t)\left(1 - \frac{\hat{\mu}(t)}{\mu_m}\right)\frac{C(t)}{S(t)} \quad (19.14)$$

$$a_r(t) = \frac{1}{Y_{P/S}}\mu_r(t)\left(1 - \frac{\mu_r(t)}{\mu_{mr}}\right)\frac{C_r(t)}{S_r(t)} \quad (19.15)$$

The constraint condition of Equation 19.9 of the controller parameter choice is chosen as:

- Ensuring that the eigenvalue of the closed-loop system is in the left-half s plane or, in other words, the closed-loop system is stable;
- Ensuring the gain k_c can be limited in the area between the two values k_c^0 and k_c^1:

$$k_c = \frac{T(t)}{\delta(t)L(t) + 1} \quad (19.16)$$

When the fault is associated with a parameter μ_m, we know that $K_S = K_S^0$. For the parameter μ_m, we know from its

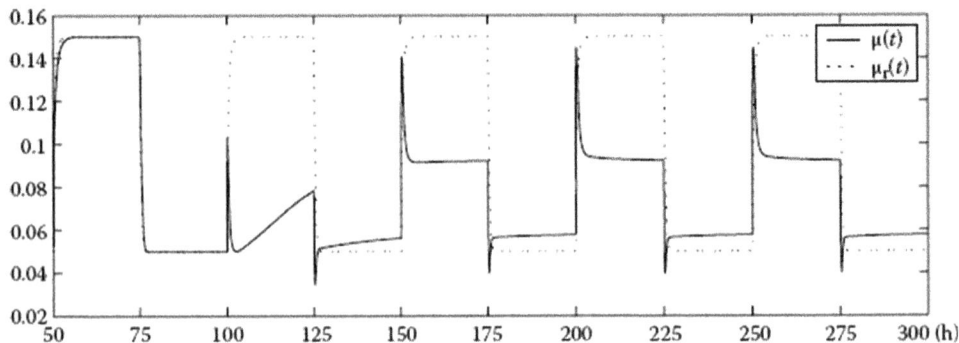

FIGURE 19.36 The control result (fixed parameter control).

estimation $\hat{\mu}_m$ and the corresponding bounds $\mu_m^b \leq \mu_m \leq \mu_m^a$ that we can get

$$\delta(t) = \frac{\hat{\mu}_m}{\mu_m} \tag{19.17}$$

According to the limits it will be

$$\delta^{\min}(t) = \frac{\hat{\mu}_m}{\mu_m^a} \text{ and } \delta^{\max}(t) = \frac{\hat{\mu}_m}{\mu_m^b} \tag{19.18}$$

When the fault is associated with a parameter K_S, we know that $\mu_m = \mu_m^0$. For the parameter K_S, we know from its estimation $\hat{K}_S$, and the corresponding limits $K_S^b \leq K_S \leq K_S^a$, we can get

$$\delta(t) = \frac{K_S + S(t)}{\hat{K}_S + S(t)} \tag{19.19}$$

According to the limits it will be

$$\delta^{\min}(t) = \frac{K_S^b + S(t)}{\hat{K}_S + S(t)} \text{ and } \delta^{\max}(t) = \frac{K_S^a + S(t)}{\hat{K}_S + S(t)} \tag{19.20}$$

For different faults, using Equation 19.14 or Equation 19.16, we can calculate the controller parameters $L(t)$ and $T(t)$.

To show the validity of the method, simulation experiments have been done on the alcoholic fermentation process. The nominal value of the system parameter vector is $[\mu_m, K_s] = [0.38, 5.0]$. The parameter vector value of the reference model

is $[\mu_{mr}, K_{sr}] = [0.30, 5.0]$. The fault takes place at $t = 100$ h. The faulty parameter vector is $[\mu_m, K_s] = [0.38, 2.0]$.

1. *Non-FTC control*: To make the comparison, in the first we give two results of non-FTC control. One is fixed parameter control and another is adaptive control. In the non-FTC control situation, the result of controller parameter calculation is not modified.
   - *Fixed parameter control*: In the controller parameter calculation, the process parameter vector $[\mu_m, K_s]$ is substituted by its nominal value $[\mu_m^0, K_S^0]$. Figure 19.36 presents the control result. It shows that after the fault occurrence at $t_f = 100$ *h* the control effect becomes very bad. There is a large follow error between $\mu(t)$ and $\mu_r(t)$. Figure 19.37 shows the system gain $k_c(t)$. It shows that after the fault occurrence the gain $k_c(t)$ has great variation and its deviation from 1 is large. This is caused by the parameter vector difference between θ^f of the postfault system and θ^0, which is used to calculate the controller parameters and the variable $\mu(t)$.
   - *Adaptive control*: In the controller parameter calculation, the process parameter vector $[\mu_m, K_s]$ is substituted by its estimation value $[\hat{\mu}_m, \hat{K}_S]$. This estimation is provided by a parameter-state joint estimate procedure. For details of the parameter-state joint estimate, the reader is refereed to Zeng et al. (1993b). Figure 19.38

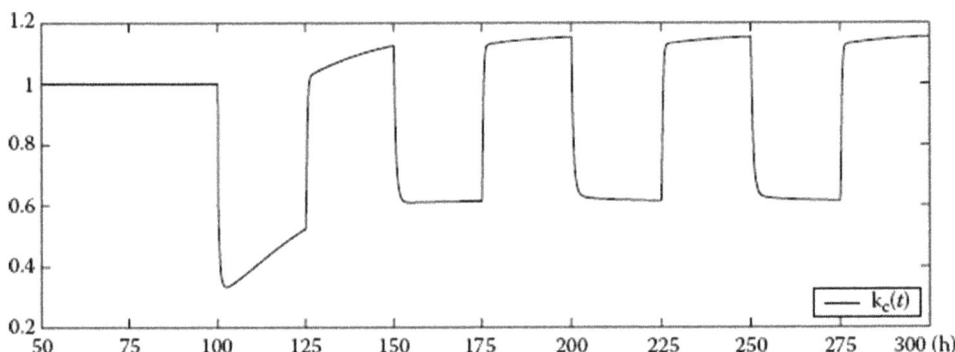

FIGURE 19.37 The system gain (fixed parameter control).

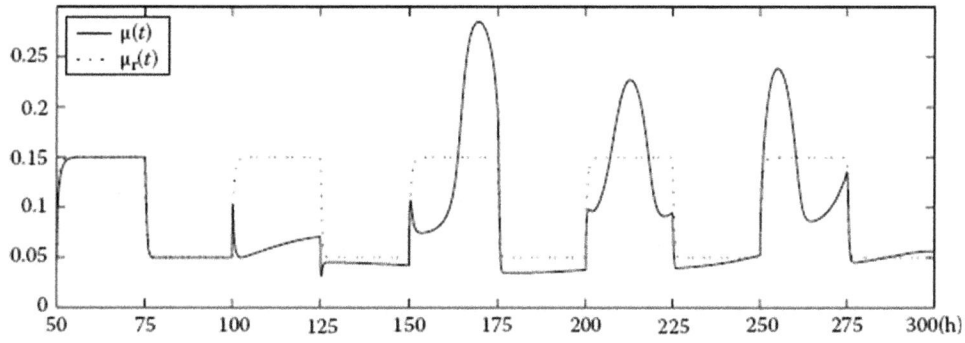

FIGURE 19.38   The control result (adaptive control).

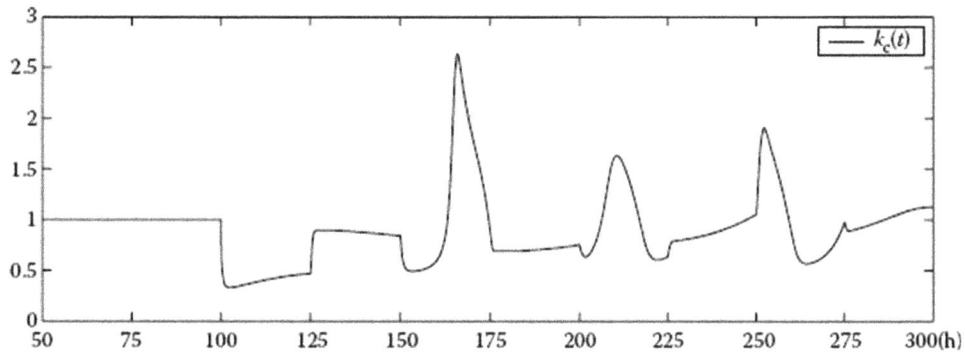

FIGURE 19.39   The system gain (adaptive control).

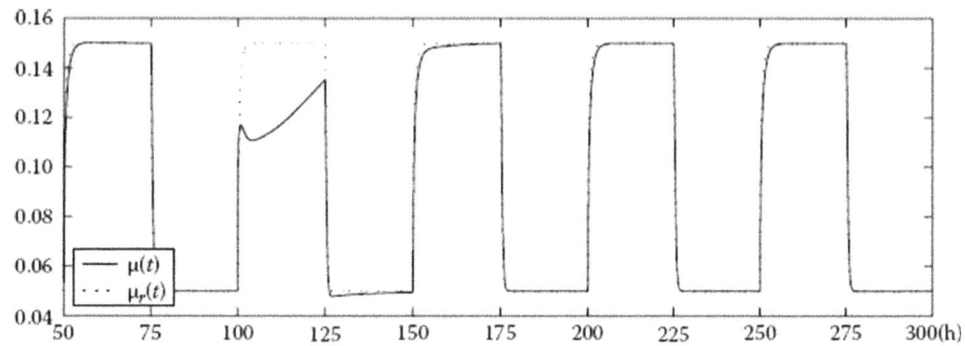

FIGURE 19.40   The control result (without constraint of the controller parameter $L(t)$).

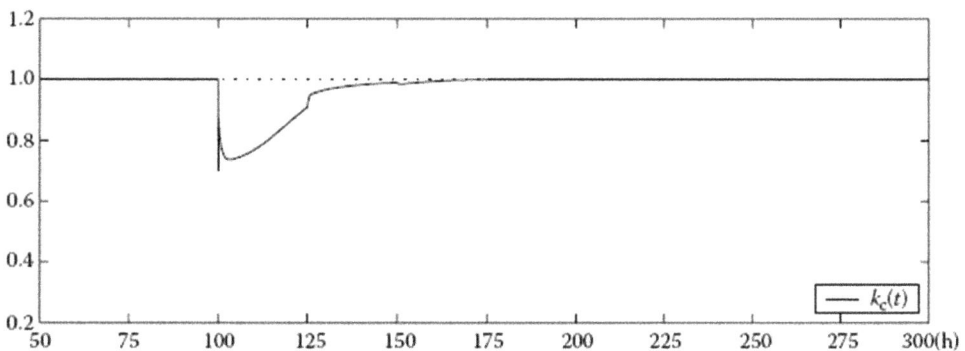

FIGURE 19.41   The system gain (without constraint of the controller parameter $L(t)$).

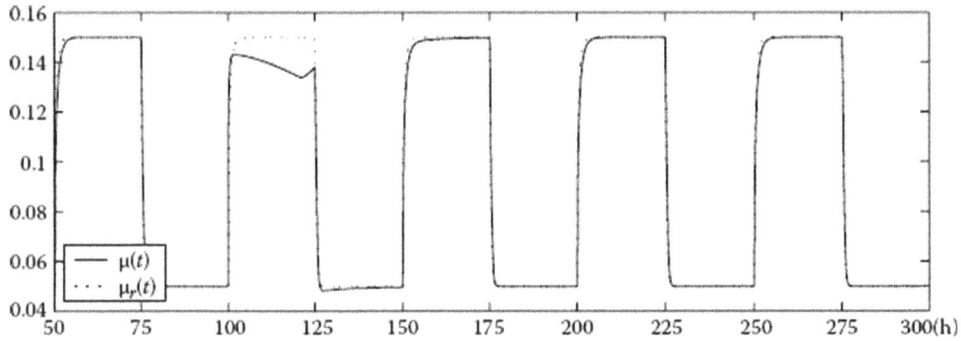

FIGURE 19.42   The control result (with constraint of the controller parameter $L(t)$).

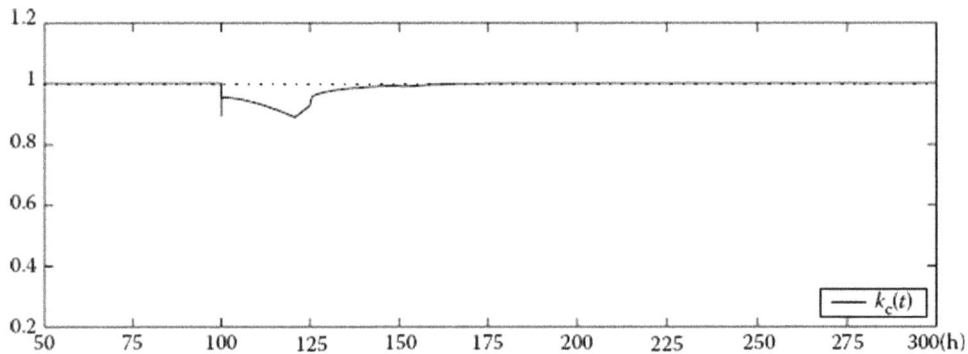

FIGURE 19.43   The system gain (with constraint of the controller parameter $L(t)$).

shows the control result. It shows that after the fault occurrence the control effect is very bad. Figure 19.39 shows that the system gain $k_c(t)$ shows great variation and the deviation from 1 is large.

2. *FTC control*: In the FTC control manner, k_c^0 is chosen as 0.85.

   • *FTC control with fault isolation and identification but without constraint to the controller parameter $L(t)$*: In the controller parameter calculation, the process parameter vector $[\mu_m, K_s]$ is substituted by its estimation value $[\hat{\mu}_m, \hat{K}_S]$. This estimation is provided by the intervals that are based on the fault isolation and identification method. Figure 19.40 shows the control result. It shows that the control effect is much better than the ones of Figures 19.36 and 19.38. However, in the beginning period after the fault occurrence there is also an evident control error. Figure 19.41 shows the gain $k_c(t)$. It shows that $k_c(t)$ has not been limited by the condition $k_c(t) \geq 0.85$.

   • *FTC control with fault isolation, identification, and with constraint to the controller parameter $L(t)$*: In the first, we calculate the controller parameters by the same way as in the previous case (without constraint), then the calculated result is modified according to the fault-tolerant control procedure (Li and Dahhou, 2006). Figure 19.42 presents the control result. It shows

that the effect is much better than the ones of preceding examples. In the begin period after the fault occurrence, the control error is much smaller than the one of Figure 19.40. Figure 19.43 presents the gain $k_c(t)$. It shows that, in the beginning period after the fault occurrence, the deviation of $k_c(t)$ from 1 is much smaller than the case of Figure 19.41, and it accords with the condition $k_c(t) \geq 0.85$. And in the following time, $k_c(t)$ always equals 1, and the control arrives at its optimum.

SUMMARY

This chapter has attempted to elucidate the complexity of fermentation systems and the difficult task facing the fermentation scientist who has to elucidate the key features of a fermentation to significantly enhance microbial productivity with a minimum expenditure of resources.

Fermentation development has traditionally used a heuristic approach to refine empirical knowledge.

The requirement for empiricism is still very much present, but modern fermentation control systems and techniques should allow for a more systematic approach to fermentation optimization (see Chapters 6 and 14 for more details).

It is likely that modern computing techniques such as event tracking and rule-based controls will augment the application of microbial physiology to solving problems of applied microbiology.

Microbial physiology knowledge itself in turn may benefit from advanced control ideas by increasing understanding how a microorganism can successfully respond to its environment.

The development of an integrated process control methodology requires the design of a supervisory block containing all available information and procedures.

This supervisory block recognizes specific indicators/parameters and acts, if necessary, on the process or by informing the operator.

Two approaches to modeling were implemented: the behavioral model and the nonstructured model. Whereas the former is based on the use of a fuzzy classification technique, the latter is based on mass balance considerations. The behavioral model is obtained by using the online measurement of environmental variables (temperature, pH, stirrer speed, etc.) and describes the physiological states of the process. The nonstructural model is obtained from mass balance considerations and is used for the development of software sensors, control scheme, and FDI algorithms.

The whole of the results obtained, and the algorithms developed using these modeling approaches, are used in a supervisory block to ensure an effective monitoring of the process.

REFERENCES

Aguilar-Martin, J. 1996. *Knowledge-Based Real Time Supervision. Tempus-Modify Workshop.* Budapest, Hungary.

Alford, J.S. (2006). Bioprocess control: advances and challenges. *Comp Chem Eng* 30(10–12):1464–1475.

Alford, J.S. (2009). Principles of bioprocess control. *CEPR* 105 (11):44–51.

Antsaklis, P.J., and K.M. Passino, eds. 1993. *An Introduction to Intelligent and Autonomous Control.* Norwell, MA: Kluwer Academic Publishers.

Bâati, L., G. Roux, B. Dahhou, and J.L. Uribelarrea. 2004. Unstructured modeling growth of *Lactobacilius acidophilus* as a function of the temperature. *Math Comput Simul* 65:137–145.

Bailey, J.E., D.F. Ollis (1986). *Instrumentation and Control, in Biochemical Engineering Fundamentals*, 2nd edn, Maidenhead: McGraw-Hill, 658–722.

Beluhan, D., D. Gosak, N. Pavlovic, and M. Vampola (1995). Biomass estimation and optimal control of the fermentation process. *Comp Chem Eng* 19(Suppl.).387–392.

Ben-Youssef, C. 1996. Filtrage, Estimation Et Commande Adaptative D'un Procédé De Traitement Des Eaux Usées. PhD thesis. Institut National Polytechnique de Toulouse.

Ben-Youssef, C., B. Dahhou, F.Y. Zeng, and J.L. Rols. 1996. Estimation and filtering of nonlinear systems: Application to a wastewater treatment process. *Int J Syst Sci* 27:497–505.

Blackmore, R.S., J.S. Blome, and J.O. Neway (1996). A complete computer monitoring and control system using commercially available, configurable software for laboratory and pilot plant *Escherichia coli* fermentations. *J Ind Microbiol* 16:383–389.

Chen, W., C. Graham, and R.B. Ciccarelli. (1997). Automated fed-batch fermentation with feed-back controls based on dissolved oxygen (DO) and pH for production of DNA vaccines. *J Ind Microbiol Biotechnol* 18:43–48.

Dahhou B., G. Chamilothoris, and G. Roux. (1991a). Adaptive predictive control of a continuous fermentation process. *Int J Adapt Cont Signal Process* 5:351–362.

Dahhou, B., G. Roux, and A. Cheruy. 1993. Linear and nonlinear adaptive control of alcoholic fermentation process. *Int J Adapt Cont Signal Process* 7:213–233.

Dahhou, B., G. Roux, and I. Queinnec. 1991b. Adaptive control of a continuous fermentation process. *Presented at the Symposium on Modeling and Control of Technological Systems*, Lille, France, pp. 738–743.

Dahhou, B., G. Roux, I. Queinnec, and J.B. Pourciel. 1991c. Adaptive pole placement control of a continuous fermentation process. *Int J Syst Sci* 22:2625–2638.

Diaz, C., P. Dieu, C. Feuillerat, P. Lelong, and M. Salome (1995). Adaptive control of dissolved oxygen concentration in a laboratory-scale bioreactor. *J Biotechnol* 43:21–32.

Dochain, D. (2003). State and parameter estimation in chemical and biochemical processes: A tutorial. *J Process Control* 13(8):801–818.

Dojat, M., N. Ramaux, and D. Fontaine. 1998. Scenario recognition for temporal reasoning in medical domains. *Artif Intell Med* 14:139–155.

Gregory, M. E., P. J. Keay, P. Dean, M. Bulmer, and N.F. Thornhill. (1994). A visual programming environment for bioprocess control. *J Biotechnol* 33:233–241.

Kabbaj, N. 2004. Developpement D'algorithmes De Detection Et D'isolation De Defauts Pour La Supervision Des Bioprocedes. PhD thesis. Université de Perpignan.

Kabbaj, N., A. Doncescu, B. Dahhou, and G. Roux. 2002. Wavelet based residual evaluation for fault detection and isolation. *Presented at the 17th IEEE International Symposium on Intelligent Control*, Vancouver, British Columbia, Canada.

Kabbaj, N., M. Polit, B. Dahhou, and G. Roux. 2001. Adaptive observers based fault detection and isolation for an alcoholic fermentation process. *Presented the 8th IEEE International Conference on Emerging Technologies and Factory Automation.* Antibes-Juan les Pins, France.

Kotch, G.-G. 1993. Modular Reasoning. A New Approach Towards Intelligent Control. PhD thesis. Swiss Federal Institute of Technology. Zurich, Switzerland.

Kong, D.Y., R. Gentz, and J.L. Zhang. (1998). Development of a versatile computer integrated control system for bioprocess controls. *Cytotechnology* 26:227–236.

Kurtanjek, Z. (1994). Modelling and control by artificial neural networks in biotechnology. *Comp Chem Eng* 18(Suppl.): 627–631.

Li, Z., and B. Dahhou. 2006. An observers based fault isolation approach for nonlinear dynamic systems. *Presented at the Second IEEE-EURASIP International Symposium on Control, Communications and Signal Processing 2006.* (ISCCSP'06), Marrakech, Morocco.

Monod, J. 1942. *Recherche sur la Croissance des Cultures Bactériennes.* Edition Hermes: Paris.

M'Saad, M., I.D. Landau, and M. Duque. 1989. Example applications of the partial state reference model adaptive control design technique. *Int J Adapt Cont Signal Process* 3: 155–165.

M'Saad, M., I.D. Landau, and M. Samaan. 1990. Further evaluation of the partial state reference model adaptive control design. *Int J Adapt Cont Signal Process* 4:133–146.

Nakkabi, Y., N. Kabbaj, B. Dahhou, G. Roux, and J. Aguilar. 2003. A combined analytical and knowledge based method for fault detection and isolation. *Presented at the 9th IEEE International Conference on Emerging Technologies and Factory Automation.* Vol. 2, pp. 161–166, Lisbon, Portugal.

Nakkabi,Y., A. Doncescu, G. Roux, M. Polit, and V. Guillou. 2002. Application of data mining in biotechnological process. Presented at the Second IEEE International Conference on Systems, Man, and Cybernetics, Hammamet, Tunisia.

Nejjari, F., B. Dahhou, A. Benhammou, and G. Roux. 1999a. Nonlinear multivariable adaptive control of an activated sludge wastewater treatment process. *Int J Adapt Cont Signal Process* 13:347–365.

Nejjari, F., G. Roux, B. Dahhou, and A. Benhammou. 1999b. Estimation and optimal control design of a biological and wastewater treatment process. *Int J Math Comp Simul* 48:269–80.

Omstead, D.R. (ed.) (1990). *Computer Control of Fermentation Processes.* Boca Raton, FL: CRC Press.

Onken, U., and P. Weiland (1985). Control and optimisation, in Rehm, H.J. and Reed, G. (eds.) *Biotechnology*, Vol. 2, Weinheim: VCR Verlag, 787–806.

Piera, N., P. Desroches, and J. Aguilar-Martin. 1989. Lamda: An incremental conceptual clustering system. Report 89420. LAAS/CNRS.

Romeu, F.J. (1995). Development of biotechnology control systems. *ISA Trans* 34:3–19.

Sys, J., A. Prell, and I. Havlik (1993). Application of the distributed control system in fermentation experiments. *Folia Microbiol.* 38:235–241.

Waissman-Vilanova, J. 2000. Construction D'un Modele Compartemental Pour La Supervision De Procedes: Application A Une Station De Traitement Des Eaux. PhD thesis. Institut National Polytechnique de Toulouse.

Waissman-Vilanova, J., R. Sarrate-Estruch, B. Dahhou, and J. Aguilar-Martin. 1999. Building an automaton for condition monitoring in a biotechnological process. *Presented at the 5th European Control Conference*, Karlsruhe, Germany.

Wang, H.Y. (1986) Bioinstrumentation and computer control of fermentation processes, in Demain, A.L. and Solomon, N.A. (eds.) *Manual of Industrial Microbiology and Biotechnology*, Washington, DC: ASM, pp. 308–320.

Zeng, F.Y., B. Dahhou, and M.T. Nihtila. (1993a). Adaptive control of nonlinear fermentation process via MRAC technique. *J Appl Math Model* 17:58–69.

Zeng, F.Y., B. Dahhou, M.T. Nihtila, and G. Goma. 1993b. Microbial specific growth rate control via MRAC method. *Int J Syst Sci* 24:1973–1985.

Zhang, Q. 2000. A new residual generation and evaluation method for detection and isolation of faults in nonlinear systems. *Int J Adapt Control Signal Process* 14:759–773.

Index